Lecture Notes in Computer Science   9587

Commenced Publication in 1973
Founding and Former Series Editors:
Gerhard Goos, Juris Hartmanis, and Jan van Leeuwen

Advanced Research in Computing and Software Science

Subline of Lecture Notes in Computer Science

Rūsiņš Mārtiņš Freivalds · Gregor Engels
Barbara Catania (Eds.)

SOFSEM 2016:
Theory and Practice
of Computer Science

42nd International Conference on Current Trends
in Theory and Practice of Computer Science
Harrachov, Czech Republic, January 23–28, 2016
Proceedings

 Springer

Editors
Rūsiņš Mārtiņš Freivalds
University of Latvia
Riga
Latvia

Barbara Catania
University of Genoa
Genoa
Italy

Gregor Engels
University of Paderborn
Paderborn
Germany

ISSN 0302-9743   ISSN 1611-3349   (electronic)
Lecture Notes in Computer Science
ISBN 978-3-662-49191-1   ISBN 978-3-662-49192-8   (eBook)
DOI 10.1007/978-3-662-49192-8

Library of Congress Control Number: 2015958904

LNCS Sublibrary: SL1 – Theoretical Computer Science and General Issues

Printed on acid-free paper

This Springer imprint is published by SpringerNature
The registered company is Springer-Verlag GmbH Berlin Heidelberg

Preface

This volume contains the invited and contributed papers selected for presentation at the 42nd Conference on Current Trends in Theory and Practice of Computer Science (SOFSEM 2016), which was held January 23–28, 2016, in Harrachov, Czech Republic.

SOFSEM (originally SOFtware SEMinar) is devoted to leading research and fosters cooperation among researchers and professionals from academia and industry in all areas of computer science. SOFSEM started in 1974 in the former Czechoslovakia as a local conference and winter school combination. The renowned invited speakers and the growing interest of the authors from abroad gradually turned SOFSEM in the mid-1990s into an international conference with proceedings published in the Springer LNCS series, in the last two years in their prestigious subline ARCOSS: *Advanced Research in Computing and Software Science*. SOFSEM became a well-established and fully international conference maintaining the best of its original winter school aspects, such as a higher number of invited talks and an in-depth coverage of novel research results in selected areas of computer science. SOFSEM 2016 was organized around the following three tracks:

- Foundations of Computer Science
  (chaired by Rūsiņš Mārtiņš Freivalds)
- Software Engineering: Methods, Tools, Applications
  (chaired by Gregor Engels)
- Data, Information, and Knowledge Engineering
  (chaired by Barbara Catania)

With its three tracks, SOFSEM 2016 covered the latest advances in research, both theoretical and applied, in selected areas of computer science. The SOFSEM 2016 Program Committee consisted of 61 international experts from 22 different countries, representing the track areas with outstanding expertise.

An integral part of SOFSEM 2016 was the traditional SOFSEM Student Research Forum (chaired by Roman Špánek), organized with the aim of presenting student projects in both the theory and practice of computer science, and to give the students feedback on the originality of their results. The papers presented at the Student Research Forum were published in separate local proceedings (together with the accepted posters). The copy of these local proceedings is available via CEUR-WS.

In response to the call for papers, SOFSEM 2016 received 150 abstracts and after withdrawals and removal of double and fake submissions, the final number of submitted papers totaled 116 from 38 different countries. The submissions were distributed in the conference three tracks as follows: 62 in the Foundations of Computer Science, 21 in the Software Engineering, and 33 in the Data, Information, and Knowledge Engineering. From these, 34 submissions fell in the student category.

After a detailed reviewing process (using the EasyChair Conference System for reviewing and discussions), a careful selection procedure was carried out within each track. Following strict criteria of quality and originality, 43 papers were selected for presentation, namely: 27 in the Foundations of Computer Science, six in the Software Engineering, and 10 in the Data, Information, and Knowledge Engineering.

Based on the recommendation of the chair of the Student Research Forum, 14 student papers were chosen for the SOFSEM 2016 Student Research Forum. Moreover, five posters were accepted for poster presentation.

As editors of these proceedings, we are grateful to everyone who contributed to the scientific program of the conference, especially the invited speakers and all the authors of contributed papers. We also thank the authors for their prompt responses to our editorial requests.

SOFSEM 2016 was the result of a considerable effort by many people. We would like to express our special thanks to:

- The members of the SOFSEM 2016 Program Committee and all external reviewers for their careful reviewing of the submissions
- Roman Špánek for his preparation and handling of the Student Research Forum
- The SOFSEM Steering Committee, chaired by Július Štuller, for guidance and support throughout the preparation of the conference
- The Organizing Committee, consisting of Martin Řimnáč (Chair), Pavel Tyl, Dana Kuželová, Július Štuller, and Milena Zeithamlová for the generous support and preparation of all aspects of the conference
- Springer for its continued support of the SOFSEM conferences
- CEUR-WS for publishing the copy of the second volume of the proceedings

We are greatly indebted to the Action M Agency, in particular Milena Zeithamlová, for the local arrangements of SOFSEM 2016. We thank the Institute of Computer Science of the Czech Academy of Sciences for its invaluable support of all aspects of SOFSEM 2016. Finally, we are very grateful for the financial support of the Czech Society for Cybernetics and Informatics.

November 2015

Barbara Catania
Gregor Engels
Rūsiņš Mārtiņš Freivalds

Organization

Steering Committee

Barbara Catania	University of Genoa, Italy
Ivana Černá	Masaryk University, Brno, Czech Republic
Miroslaw Kutylowski	Wroclaw University of Technology, Poland
Jan van Leeuwen	Utrecht University, The Netherlands
Tiziana Margaria-Steffen	University of Limerick, Ireland
Brian Matthews	STFC Rutherford Appleton Laboratory, UK
Branislav Rovan	Comenius University, Bratislava, Slovakia
Petr Šaloun	Technical University of Ostrava, Czech Republic
Július Štuller, *Chair*	Institute of Computer Science, Academy of Sciences, Czech Republic

Program Committee

Track Chairs

Barbara Catania	University of Genoa, Italy
Gregor Engels	University of Paderborn, Germany
Rūsiņš Mārtiņš Freivalds	University of Latvia, Latvia

Student Research Forum Chair

Roman Špánek	Technical University of Liberec, Czech Republic

Program Committee Members

Farid Ablayev	Kazan, Russia
Marie-Pierre Béal	Paris, France
Steffen Becker	Chemnitz, Germany
Zohra Bellahsène	Montpellier, France
Petr Berka	Prague, Czech Republic
Mária Bieliková	Bratislava, Slovakia
Jan Bouda	Brno, Czech Republic
Stephane Bressan	Singapore, Republic of Singapore
Ruth Breu	Innsbruck, Austria
Tomáš Bureš	Prague, Czech Republic
Davide Buscaldi	Paris, France
Johann Eder	Klagenfurt, Austria
Uwe Egly	Vienna, Austria
Gregor Engels	Paderborn, Germany

Additional Reviewers

Mikhail Abramskiy
Ahmad Salim Al-Sibahi
Jesús Alonso
Marcella Anselmo
Pablo Arrighi
Hauke Baller
Kaspars Balodis
Annalisa Barla
Carl Barton
Luca Bernardinello
Krists Boitmanis
Vincenzo Bonifaci
Boban Celebic
Kārlis Čerāns
Aleksandar S. Dimovski
Michael J. Dinneen
Mike Domaratzki
Mayte Giménez Fayos
Peter Floderus
Markus Frank
Bulat Gabbasov
Mohsen Ghaffari
Massimiliano Goldwurm
Alexander Golovnev
Stefan Göller
Martin Haeusler
Florian Häser
Marcus Hilbrich

Hendrik Jan Hoogeboom
Jesper Jansson
Stacey Jeffery
Zbynek Jiracek
Charles Jordan
Kamil Khadiev
Alfred Khayroullin
Dennis Komm
Christian Koncilia
Filip Krijt
Petr Kurka
Giovanna Lavado
Dimitrios Letsios
Christos Levcopoulos
Jiamou Liu
Bruno Loff
Anton Marchenko
Arnaud Malapert
Ladislav Maršík
Lukas Märtin
Vladimir Matena
Abel Molina
Debajyoti Mondal
Viviane Moreira
Michael Nieke
Bengt J. Nilsson
Francesca Odone
Maris Ozols

Peteris Paikens
Mia Persson
Ved Prakash
Julien Provillard
Renato Renner
David Roberson
Lorenzo Rosasco
Stefano Rovetta
Clemens Sauerwein
Shinnosuke Seki
Alexander Shen
Christian Sillaber
Dzmitry Sledneu
Stefan Stanciulescu
Patrick Totzke
Farouk Toumani
Leo Truksans
Bianca Truthe
Robin Kothari
Matthias Kowal
Miroslaw Kowaluk
Sergejs Kozlovics
Maksims Kravcevs
Alexander Vasiliev
Marcos Villagra
Shenggen Zheng
Mansur Ziatdinov
Wieslaw Zielonka

Organization

SOFSEM 2016 was organized by the Institute of Computer Science of the Czech Academy of Sciences and Action M Agency, Prague.

Organizing Committee

Martin Řimnáč, *Chair*   Institute of Computer Science, Prague, Czech Republic
Pavel Tyl   Technical University Liberec, Czech Republic
Dana Kuželová   Institute of Computer Science, Prague, Czech Republic
Július Štuller   Institute of Computer Science, Prague, Czech Republic
Milena Zeithamlová   Action M Agency, Prague, Czech Republic

Supported by

ČSKI – Czech Society for Cybernetics and Informatics **ČSKI**

SSCS – Slovak Society for Computer Science

Contents

Software Engineering: Methods, Tools, Applications (Regular Papers)

Data, Information, and Knowledge Engineering (Regular Papers)

Foundations of Computer Science
(Invited Talks)

Cryptography in a Quantum World

Gilles Brassard[1,2]([⊠])

[1] Département d'informatique et de recherche opérationnelle,
Université de Montréal, C.P. 6128, Succursale Centre-ville,
Montréal, QC H3C 3J7, Canada
[2] Canadian Institute for Advanced Research, Toronto, Canada
brassard@iro.umontreal.ca
http://www.iro.umontreal.ca/~brassard/en/

Abstract. Although practised as an art and science for ages, cryptography had to wait until the mid-twentieth century before Claude Shannon gave it a strong mathematical foundation. However, Shannon's approach was rooted is his own information theory, itself inspired by the classical physics of Newton and Einstein. But our world is ruled by the laws of quantum mechanics. When quantum-mechanical phenomena are taken into account, new vistas open up both for codemakers and codebreakers. Is quantum mechanics a blessing or a curse for the protection of privacy? As we shall see, the jury is still out!

Keywords: Cryptography · Quantum mechanics · Quantum computation · Post-quantum cryptography · Quantum communication · Quantum key distribution · Edgar Allan Poe

1 Introduction

For thousands of years, cryptography has been an ongoing battle between codemakers and codebreakers [1,2], who are more formally called cryptographers and cryptanalysts. Naturally, *good* and *evil* are subjective terms to designate codemakers and codebreakers. As a passionate advocate for the right to privacy, my allegiance is clearly on the side of codemakers. I admit that I laughed hysterically when I saw the *Zona Vigilada* warning that awaits visitors of the *Plaça de George Orwell* near City Hall in Barcelona [3]. Nevertheless, I recognize that codebreakers at Bletchley Park during the Second World War were definitely on the side of good. We all know about the prowess of Alan Turing, who played a key role at the routine (this word is too strong) decryption of the German Enigma cipher [4]. But who remembers Marian Rejewski, who actually used pure (and beautiful) mathematics to break Enigma with two colleagues *before* the War even started? [5] Indeed, who remembers except yours truly and nationalistic Poles such as my friend Artur Ekert? Certainly not filmmakers! [6] And who remembers William Tutte, who broke the much more difficult Lorenz cipher (codenamed Tunny by the Allies), which allowed us to probe the mind of Hitler? [7] Tutte moved on to found the Computer Science department at the

© Springer-Verlag Berlin Heidelberg 2016
R.M. Freivalds et al. (Eds.): SOFSEM 2016, LNCS 9587, pp. 3–16, 2016.
DOI: 10.1007/978-3-662-49192-8_1

University of Waterloo, Canada, now home of IQC, the Institute for Quantum Computation, but never said a word until the 1990 s about how he won the War for us [8]. The Canadian Communications Security Establishment pays homage with its Tutte Institute for Mathematics and Computing. But who else remembers those silent heroes on the codebreaking side? I am getting carried away by emotions as I type these words while flying from Tōkyō to Calgary, on my way home after the amazingly successful 5th Annual Conference on Quantum Cryptography, QCrypt 2015 [9].

Regardless of the side to which good belongs, the obvious question is: *Who will win the battle* between codemakers and codebreakers? More specifically, how do the recent advances in Quantum Information Science (QIS) change this age-old issue? Until the mid-twentieth century, History has taught us that codemakers, no matter how smart, have been systematically outsmarted by codebreakers, but it ain't always been easy. For instance, *le chiffre indéchiffrable*, usually attributed to Blaise de Vigenère in 1585, but actually invented by Giovan Batista Belaso 32 years earlier, remained invulnerable until broken by Charles Babbage in 1854, more than three centuries after its invention. (Baggage is best known for his invention of the *Analytical Engine*, which would have been the first programmable computer had the technology of his days been able to rise up to the challenge of building it.) The apparent upper hand of codebreakers, despite the still enduring invulnerability of the *chiffre indéchiffrable*, prompted American novelist and high-level amateur cryptanalyst Edgar Allan Poe to confidently declare in 1841 that "It may be roundly asserted that human ingenuity cannot concoct a cipher which human ingenuity cannot resolve" [10]. Poe, do I need to mention, was among other things the author of *The Gold-Bug* [11], published in June 1843. This extraordinary short story centring on the decryption of a secret message was instrumental on kindling the career of prominent cryptographers, such as William Friedman's, America's foremost cryptanalyst of a bygone era, who read it as a child [12, p. 146].

Cryptography was set on a firm scientific basis by Claude Shannon, the father of information theory [13], as the first half of the twentieth century was coming to a close [14]. Actually, it's likely that his groundbreaking work was achieved several years earlier but kept classified due to the War effort. In any case, Shannon's theory was resolutely set in the context of *classical physics*. In retrospect, this is odd since it was clearly established at that time that Nature is ruled not by the Laws envisioned centuries earlier by Sir Isaac Newton, and not even by those more modern of Albert Einstein, but by the counterintuitive features of the emerging quantum mechanics. Shannon was well aware of this revolution in physics, but he probably did not think it relevant to the foundations of information theory, which he developed as a purely abstract theory.

In particular, Shannon did not question the "fact" that encrypted information transmitted from a sender (codenamed Alice) to a receiver (codenamed Bob) could be copied by an eavesdropper (codenamed Eve) without causing any disturbance noticeable by Alice and Bob. From this unfounded assumption, Shannon proved a famous theorem according to which perfect secrecy requires the availability of a shared secret key as long as the message that Alice wishes to transmit securely to Bob, or more precisely as long as the *entropy* of that message, and that this key cannot be reused [14]. This theorem is mathematically impeccable, but it is nevertheless irrelevant in our quantum-mechanical world since the assumption on which its proof is based does not hold.

My purpose is to investigate the issue of whether or not Poe was right in his sweeping mid-nineteenth century statement. Could it be indeed that codebreakers will continue to have the upper hand over codemakers for the rest of eternity?

2   The Case of Classical Codemakers Against Classical Codebreakers

The first electronic computers were designed and built to implement Tutte's beautiful mathematical theory on how to break the high-level German code during World War II. They were codenamed the *Colossus* and ten of them were built in Bletchley Park [15]. As mentioned in the Introduction, they were instrumental in allowing us to win the War. However, in order to secure secrecy of the entire Bletchley Park operation, they were smashed to bits (funny expression when it concerns computers!) once the War was over. Consequently, I "learned" as a child that the first electronic computer in history had been the American ENIAC, when in fact it was the eleventh! Little did the pioneers of the Colossus imagine that, by an ironic twist of fate, they had unleashed the computing power that was to bring (temporary?) victory to the codemakers. In a sense, codebreakers had been the midwife of the instrument of their own destruction. Perhaps. Indeed, the rise of *public-key cryptography* in the 1970 s had led us to believe that an increase in computing power could only be in favour of codemakers, hence at the detriment of codebreakers.

But well before all this took place, a cryptographic method that offers perfect secrecy, which later came to be known as the *one-time pad*, had already been invented in the nineteenth century. It is usually attributed to Gilbert Vernam, who was granted a US Patent in 1919 [16]. However, according to prime historian David Kahn, Vernam had not realized the crucial importance of never using the same key twice until Joseph Mauborgne pointed it out [1, p. 398]. But it was later discovered that the one-time pad had been invented 35 years earlier by Frank Miller, a Sacramento banker [17]. Its perfect security was demonstrated subsequently by Shannon [14]. In any case, the one-time pad requires a secret key as long as the message to be transmitted, which makes it of limited practical use. It was nevertheless used in real life, for instance on the red telephone between John Kennedy and Nikita Khrushchev during the Cold War [18], as well as between Fidel Castro and Che Guevara after the latter had left Cuba for Bolivia [19]. But in our current information-driven society, we need a process by which any two citizens can enjoy confidential communication. For this, a method to establish a shared secret key is required. Could this be achieved through an *authentic public channel*, which offers no protection against eavesdropping?

The first breakthrough in the academic world came to Ralph Merkle in 1974, who designed a scheme capable of providing a quadratic advantage to codemakers over codebreakers. Merkle's scheme is secure under the sole assumption (still unproven to this day) that some problems can only be solved by exhaustive search over their space of potential solutions. At the time, Merkle was a graduate student at the University of California in Berkeley, enrolled in a computer security class. Unable to make his ideas understood by his professor, Merkle "dropped the course, but kept working on the idea" [20]. After several years, he prevailed and his landmark paper was finally published [21]. However, Whitfield Diffie, a graduate student "next door", at Stanford University, had similar ideas independently, albeit shortly after Merkle. But Diffie was lucky enough to have an advisor, Martin Hellman, who understood the genius of his student. Together, they made the concepts of public-key cryptography and digital signature immensely popular [22], two years before Merkle's publication.

A few years later, Ronald Rivest, Adi Shamir and Leonard Adleman, inspired by the Diffie-Hellman breakthrough, proposed an implementation of public-key cryptography and digital signatures that became known to all as the RSA cryptosystem [23]. And thus, history was made. The fact that the RSA cryptosystem had in fact been invented in 1973 by Clifford Cocks [24], at the British secret services known as GCHQ, is of little relevance to the practical importance of the discovery on what was to become the Internet. As long as the factorization of large numbers remained infeasible, the codemakers had finally won the battle, proving Poe wrong. Soon, electronic safety all over the Internet revolved around this RSA cryptosystem, as well as the earlier invention known as the Diffie-Hellman key establishment protocol [22]. At about the same time, Robert McEliece invented another approach, based on error-correction codes [25], which did not come into practical use because it required much longer keys than either the RSA or the Diffie-Hellman solution. Later, the same apparent level of security was obtained with significantly shorter keys by bringing in the number-theoretic notion of elliptic curves [26,27]. And the Internet was a happy place. Or so it seemed.

End of story?

3   The Unfair but Realistic Case of Classical Codemakers Against Quantum Codebreakers

End of story? Not quite! In the early 1980s, Richard Feynman [28,29] and, independently, David Deutsch [30], invented the theoretic notion of a *quantum computer*. This hypothetical device would use the counterintuitive features of quantum mechanics for computational purposes. At first, it was not clear that quantum computers, even if they could be built, could speed up calculations.

And then, in 1994, Peter Shor [31], and independently Alexis Kitaev [32], discovered that quantum computers have the power to factor large numbers and extract discrete logarithms efficiently, bringing to their knees not only the RSA cryptosystem but also the Diffie-Hellman key establishment scheme, even if based on elliptic curves. As a society, we are extremely fortunate that Shor's and Kitaev's discoveries were made before a quantum computer had already been built for some other purposes (such as computational physics and chemistry). Quite literally, this saved civilization from catastrophic collapse. But now that we have known about the looming threat for over two decades, surely we are active at deploying solutions that have at least a fighting chance to withstand the onslaught of a quantum computer.

Well, not really. :-(

The general apathy towards the quantum threat to worldwide security on the Internet and beyond is quite simply appalling. Why react today (or more appropriately twenty years ago) when we can quietly wait for disaster? After all, no serious business model looks more than five years in the future, and it *would* be expensive to change the current cryptographic infrastructure. And indeed, a full-scale quantum computer is unlikely to materialize in the next five years. Except perhaps in an ultra-secret basement somewhere, be it governmental of industrial... But when (not "if") this happens, all *past* communications will become insecure to whomever was wise enough to have stored the Internet traffic that was until then undecipherable. The fact that current cryptographic techniques are susceptible to being broken *retroactively* is their main conceptual weakness. Any secret entrusted to them today, even if it is indeed currently secure (something that we do not know how to prove), will be exposed as soon as a sufficiently large quantum computer becomes operational.

So, was Poe right after all? Are codebreakers poised to regain their upper hand? Not necessarily! Alternative encryption methods have been designed, which are not (yet) known to be vulnerable to a quantum attack, ironically including the historical McEliece approach [25], which had been scorned upon its invention because of the length of its keys. More recent approaches based on hash functions, short vectors in lattices and multivariate polynomials are being vigorously investigated. The emerging field of *post-quantum cryptography* is devoted to the study of (hopefully) quantum-resistant encryption [33,34]. Unfortunately, we cannot prove that any of these alternatives is secure, but at least they are not already known to be compromised by the advent of a quantum computer. Well, in the case of lattice-based cryptography [35], this is not so clear anymore [36–38]. But one thing is sure: we cannot hope to be protected by these techniques if we don't use them! On the other hand, some of these more recent schemes could in fact be *less secure* than RSA against a *classical* attack, simply because they have not yet stood the test of time. Therefore, a transition to these new techniques should be carried out with the utmost care. But it *must* be carried out.

Michele Mosca likes to tell the following tale. Let x denote the length of time (in years) that you want your secrets to remain secret. Let y denote the time it will take to re-tool the current infrastructure with quantum-safe encryption (assuming that such a thing actually exists). Let z denote the time it will take before a full-scale quantum computer is operational. Mosca's "theorem" tells us that if $x + y > z$, then it is time to panic! Sadly, it may even be that $y > z$, meaning that it's already too late to avoid a complete meltdown of the Internet. So, *what are we waiting for?*

It turns out that the American National Security Agency (NSA) is taking this threat *very* seriously indeed. This last August (2015), they issued a directive called "Cryptography Today" in which they announced that they "will initiate a transition to quantum resistant algorithms in the not too distant future" [39]. Most significantly, they wrote: "For those partners and vendors that have not yet made the transition to Suite B elliptic curve algorithms, we recommend not making a significant expenditure to do so at this point but instead to prepare for the upcoming quantum resistant algorithm transition". Said plainly, even though elliptic-curve cryptography is believed to be more secure than first-generation public key solutions against classical cryptanalysis, it is no longer considered to offer sufficient long-term security under the looming threat of a quantum computer to be worth implementing at this point. It's nice to see that *someone* is paying attention. For once, I'm glad that the NSA is listening! :-)

From a theoretical perspective, despite what I wrote above, it *is* possible to have provably quantum-safe encryption under the so-called random oracle model, which is essentially the model that was used by Merkle in his original 1974 invention of public key establishment [20]. In a classical world, this model roughly corresponds to the assumption that there are problems that can only be solved by exhaustive search over their space of potential solutions. In the quantum setting, exhaustive search can be replaced by a celebrated algorithm due to Lov Grover, which offers a quadratic speedup [40], but no more [41].

Recall that Merkle's original idea brought a quadratic advantage to codemakers over codebreakers. But since Grover's algorithm offers a quadratic speedup to codebreakers, this completely offsets the codemakers' advantage. As a result, codebreakers can find the key established by codemakers in the same time it took to establish it! [42] The obvious reaction is to let the codemakers use quantum powers as well, but please remember that in this section, we consider quantum codebreakers but only classical codemakers. Nevertheless, I have discovered with Peter Høyer, Kassem Kalach, Marc Kaplan, Sophie Laplante and Louis Salvail that Merkle's idea can be modified in a way that if the codemakers are willing to expend an effort proportional to some parameter N, they can obtain a shared key that cannot be discovered by a quantum codebreaker who is not willing to expend an effort proportional to $N^{7/6}$ [43]. As I said, this is purely theoretical because it is not possible to argue that such an advantage offers practical security. Indeed, N would have to be astronomical before a key that is obtained in, say, one second would require more than one year of codebreaking work. In contrast, Merkle's quadratic advantage is significant for reasonably small values of N. Nevertheless, our work should be seen as a proof of principle. Now that we know that *some* security is possible in the unfair case of classical codemakers against quantum codebreakers, it is worth trying to do better (or prove that it is not possible).

Coming back to the question asked at the end of the Abstract, quantum mechanics appears to be a curse for the protection of privacy in this unfair context, which is hardly surprising since only codebreakers were assumed to use it!

4   Allowing Codemakers to Use Quantum Computation

The previous section considered a realistic scenario in which simple citizens want to protect their information against a much more powerful adversary. Indeed, it is likely that quantum computers will initially be available only to large governmental, industrial and criminal organizations. Furthermore, it is safe cryptographic practice to assume that your adversary is computationally more powerful (and possibly also more clever) than you are.

Nevertheless, in the more distant future, one can imagine a world in which quantum computers are as ubiquitous as classical computers are today. When this happens, codemakers will no longer be limited to classical computing. Can this restore the balance? Or even better, could the availability of quantum computers turn out to be to the advantage of codemakers, just as had been the availability of ever increasing classical computational power since the inception of public-key cryptography in the mid-1970s? Unfortunately, I am not aware of any encryption technique that would benefit from quantum computation sufficiently to offset the benefits that quantum computation would bestow on codebreakers.

For instance, it is easy to *partially* repair Merkle's approach [42] if the codemakers are also allowed to use Grover's algorithm, or more precisely a variant known as BBHT [44]. Having expended an effort proportional to N in order to obtain a shared key, they can create a puzzle on which classical codebreakers would have to expend an effort proportional to N^3, a clear improvement over the quadratic advantage of the original classical Merkle approach. However, a quantum codebreaker would simply use Grover's algorithm to obtain the key after an effort proportional to $N^{3/2}$. This is not a complete break, but this quantum scheme is not as secure as Merkle's original would have been against a classical adversary. So, we see that quantum-mechanical powers have helped the codebreakers more than the codemakers. Can codemakers use quantum powers in a more clever manner? Well, we have developed a less obvious Merkle-like quantum key establishment scheme against which a quantum codebreaker needs to spend a time proportional to $N^{7/4}$ [43]. This is still not quite the quadratic advantage that was possible in an all-classical world, but it is reasonably close and possibly secure enough to be used in practice.

Nevertheless, quantum mechanics still appears to be a curse for the protection of privacy even when codemakers are also allowed to make use of it.

5   Allowing Codemakers to Use Quantum Communication

Until now, we had restricted all communication between codemakers to be classical. It turns out that quantum communication comes with a great advantage because of the no-cloning theorem [45], which says that the state of elementary particles cannot be copied even in principle. This is *precisely* what causes the demise of the "famous" theorem by Shannon mentioned at the end of the Introduction. Quantum information transmitted between codemakers can*not* be copied by an eavesdropper without causing a detectable disturbance.

Inspired by an unpublished manuscript written by Steven Wiesner in April 1968, while he was participating in the Columbia University student protests [46], Charles Bennett and I realized in 1982 that quantum mechanics provides us with a channel on which passive eavesdropping is impossible. This led us and Seth Breidbart to a write down what would become the leitmotif of the nascent field of *quantum cryptography*.

> When elementary quantum systems, such as polarized photons, are used to transmit digital information, the uncertainty principle gives rise to novel cryptographic phenomena unachievable with traditional transmission media, e.g. a communications channel on which it is impossible in principle to eavesdrop without a high probability of being detected. [47]

Armed with this idea, we devised a cryptographic protocol in which a one-time pad could be safely reused indefinitely, as long as no eavesdropping is detected. This secure reuse of a one-time pad is precisely what Shannon had mathematically demonstrated to be impossible: all security is lost as soon a "one-time" pad is used twice. Our advantage, of course, comes from the fact that *we* could detect eavesdropping and discontinue the use of a pad as soon as it had been compromised (yet providing perfect secrecy even on the last message that was sent), whereas *he* had no fundamental way to detect eavesdropping, and therefore he was forced to play safe.

In more detail, Shannon proved that the one-time pad is unconditionally secure provided the shared key is perfectly random, completely unknown of the eavesdropper, and used once only. However, even though no information leaks concerning the message in case of interception, information *would* leak concerning the key itself. This is of no consequence as long as the key is never reused. But if it is, the key-secrecy condition is no longer fulfilled the second time, which is why the system becomes insecure. It follows that a "one-time" pad *can* be reused safely, Shannon's theorem notwithstanding, provided the previous communications have not been subject to eavesdropping, and it remains secure the first time that it is.

Expounding on these ideas, we wrote our paper on "How to re-use a one-time pad safely" in 1982 and had it published... a few months ago, 2^5 years later! [47] The reason it took so long to publish is that as soon as it was about to be rejected from the *Fifteenth Annual ACM Symposium on Theory of Computing*, Bennett and I had a much better idea: we realized that it is more practical to use the quantum channel to establish a shared secret random key, and then use this key as a *classical* one-time pad to encode the actual message, rather than use the channel to transmit the message directly. The main advantage of this indirect approach is that even if most of the quantum information is lost in the channel—indeed, optical fibres are not very transparent to single photons over several kilometres—a random subset of a random key is still a (shorter) random key. In contrast, a small random subset of a meaningful message is fairly likely to be mostly random and totally useless.

Thus was born *Quantum Key Distribution*, which is now called simply QKD. We presented QKD for the first time at the 1983 IEEE International Symposium on Information Theory [48], but each paper was allowed only a one-page abstract. Consequently, our protocol had to wait another year before it could be published in the Proceedings of a conference held in Bengalūru, India, where I had been invited to present any paper of my choice [49]. I suspected that the idea of QKD was likely to be rejected if submitted to a conference with full published proceedings, which is why I seized the opportunity provided by a blank-cheque invitation to sneak it at that conference! This is how our original QKD protocol came to be known as "BB84", where the Bs stand for the authors, despite the fact that we had invented and presented it in 1983. Thirty years later, *Natural Computing* (Springer) and *Theoretical Computer Science* (Elsevier) decided to join forces and publish special BB84 commemorative issues. This is how the earlier 1982 paper came to be published [47], whereas the original "BB84 paper" was published for the first time in a journal [50]. For more information on the early history of quantum cryptography, please read Ref. [51].

It was fairly easy to show that BB84 is secure against the most obvious attacks that an eavesdropper might attempt [52]. However, it took ten years after its invention before a complete formal proof of unconditional security, taking into account *any* attack possible according to the laws of quantum mechanics, was obtained [53]. Well, not exactly. This early proof, as well as the few that followed for the purpose of simplifying it, contained a major oversight. They proved that the key established by BB84 (and other similar QKD protocols) was perfectly secret... provided it is never used! Indeed, Renato Renner and Robert König realized ten years later that a clever adversary could keep the eavesdropped information at the quantum level (unmeasured). Later, when the key is used, say as one-time pad, the information that it leaks on the key (which would not be a problem in classical cryptography since the key would not be reused) could inform the eavesdropper about the appropriate measurement to make in order to learn more of the key and, therefore something about the message itself [54]. At first, this was only a theoretical worry, but then it was shown that the danger is real because one could purposely design a QKD scheme that could be proved secure under the old definition, but that really leaked information if the "secret" key is used [55]. Fortunately, the adequate ("composable") definition was given and BB84 was correctly proven secure a few months later [56].

Et voilà! Quantum cryptography offers an unbreakable method for codemakers to win the battle once and for all against any possible attack available to codebreakers, short of violating the widely accepted laws of physics. Despite the discouraging news brought about by the previous sections, in which quantum mechanics appeared to be a curse for codemakers, in the end it is a blessing for the protection of privacy.

As my much missed dear friend Asher Peres once said, *"The quantum taketh away and the quantum giveth back"*. Indeed, quantum mechanics can be exploited to break the cryptography that is currently deployed over the world-wide Internet, via Shor's algorithm, but quantum mechanics has also provided us with the ultimately secure solution. (To be historically exact, the quantum giveth "back" ten years *before* it taketh away!)

Poe was wrong. End of story!

Oh well... Not so fast. Poe was wrong *in theory*. Now, one has to build an apparatus that implements QKD as specified by the theoretical protocol. Exactly? Not possible! Any real implementation will be at best an approximation of the ideal protocol. The first prototype was built by Bennett and me, with the help of three students (two of whom have become highly respected researchers in the field) as early as 1989, even though the journal paper was published a few years later [52,57]. This prototype was not intended to be more than a proof of principle and some of its parts made such loud noises that we could literally hear the bits fly by... and zeroes did not make the same noise as ones. So, this first implementation was secure provided the eavesdropper is deaf!

Afterwards, serious experimental physicists entered the game and ever increasingly sophisticated devices have been built, capable of establishing secret keys over longer and longer distances. This business became so serious that companies sprung up to market QKD equipment, such as ID Quantique [58] in Switzerland. China has recently announced that it has almost completed the installation of a quantum communications network stretching two thousand kilometres from Beijing to Shanghai [59]. Several countries have plans to move the quantum highway to space, so that distances will no longer be an issue.

In the mean time, a new breed of (typically friendly) pirates has sprung up: the *Quantum Hackers*. In 2009, a team lead by Vadim Makarov completed a "full-field implementation of a complete attack on a running QKD connection; an installed eavesdropper obtained the entire 'secret' key, while none of the parameters monitored by the legitimate parties indicated a security breach" [60]. Of course, this was not an attack against BB84 or any other provably secure QKD protocol, which would have been an attack against quantum mechanics itself: this was an attack against one particular imperfect *implementation* of a perfect idea. The specific flaw was eradicated... and Makarov found another weakness!

And so, the game of cat and mouse between codemakers and codebreakers continues. Only the battlefield has shifted from the realm of mathematics and computer science to the realm of physics and engineering. Nevertheless, even an imperfect implementation of QKD has a significant advantage over classical systems: it *must* be attacked while the key establishment process is taking place. There is nothing to store for subsequent codebreaking when new technology or new algorithms become available. If the technology is available today for the implementation of some imperfect version of QKD but not yet for breaking it, everlasting security is achievable. Similarly, I have not mentioned the fact that the deployment of QKD requires the availability of an authenticated classical channel between the codemakers to avoid a person-in-the-middle attack, much as was the case for Merkle's classical approach in 1974. However, if the codemakers can establish short-lived secure authentication keys by any method, those keys can give rise to everlasting security through the use of QKD, again an advantage that has no classical counterpart [61].

Nevertheless, it is legitimate to wonder if there is any hope of one day building an implementation of QKD so close to the ideal protocol that it will effectively be secure against all possible attacks, regardless of the codebreaker's technology and computing time? It is tempting to say that this would be Mission: Impossible. Surely, an army of Makarovs will spring up with increasingly clever ideas to defeat increasingly sophisticated (yet imperfect) implementations of QKD. Said otherwise, surely Poe was right in the end.

Well... Maybe not! A new approach to QKD has sprung up, based on a brilliant idea put forward by Artur Ekert as early as 1991 [62]. Instead of basing the security of QKD on the impossibility of cloning quantum information—more fundamentally the impossibility of obtaining classical information on a quantum system without disturbing it [63]—Ekert's idea was to base the security of QKD on violations of Bell inequalities [64] in entangled nonlocal quantum systems [65]. Even though Ekert's original 1991 QKD protocol cannot give rise to an apparatus that would be more secure than one based on BB84 [63], his fundamentally revolutionary idea opened the door to other theoretical QKD protocols that have the potential to be secure *even if implemented imperfectly*. The security of those so-called "device-independent QKD protocols" would depend only on the belief that information cannot travel faster than light, that codemakers are capable of choosing their own independent randomness, and of course that they live in secure private spaces (since there is no need for codebreakers if the adversary is capable to physically eavesdrop over the codemakers' shoulders!). In the extreme case, highly theoretical device-independent QKD protocols have been designed whose security does not even depend on the validity of quantum mechanics itself! A recent survey of this approach is found in Ref. [66].

The catch is that the implementation of fully device-independent QKD protocols represents formidable technological challenges. It is not clear that we shall ever reach the required sophistication to turn this dream into reality. Nevertheless, a first essential step towards this goal has been achieved very recently by Ronald Hanson and collaborators in the Netherlands when they performed a long-awaited experiment in which they closed both the locality and the detection loopholes in experimental violations of Bell inequalities [67,68].

Shall we ever be able to build such a device? If so, the codemakers will have the final laugh. But what if not?

Was Poe right in the end? The jury is still out!

Acknowledgments. I am grateful to all those with whom I have had fruitful discussions on these issues in the past 36 years, starting with my lifelong collaborators Charles Bennett and Claude Crépeau. I thank Michele Mosca for allowing me to quote his "theorem". I am also grateful to Rūsiņš Freivalds for his invitation to present this paper to this 42nd International Conference on Current Trends in Theory and Practice of Computer Science (SOFSEM) and for his involvement in my 1998 election as Foreign Member of the Latvian Academy of Sciences. This work was supported in part by Canada's Natural Sciences and Engineering Research Council of Canada (NSERC), the Institut transdisciplinaire d'informatique quantique (INTRIQ), the Canada Research Chair program and the Canadian Institute for Advanced Research (CIFAR).

References

1. Kahn, D.: The Codebreakers: the Comprehensive History of Secret Communication from Ancient Times to the Internet, 2nd revised edn. Scribner, New York (1996)
2. Singh, S.: The Code Book: the Science of Secrecy from Ancient Egypt to Quantum Cryptography. Anchor Books, New York (2000)
3. http://tumblr.radarq.net/post/16344039232/big-brother-is-watching-you-in-the-plaza-de-george. Accessed 8 October 2015
4. Hodges, A.: Alan Turing: the Enigma. Random House, London (2012)
5. Rejewski, M.: How Polish mathematicians broke the Enigma cipher. Ann. Hist. Comput. **3**(3), 213–234 (1981)
6. Tyldum, M., Moore, G.: The imitation game (2014)
7. https://en.wikipedia.org/wiki/W._T._Tutte. Accessed 8 October 2015
8. Tutte, W.T.: FISH and I. http://www.usna.edu/Users/math/wdj/_files/documents/papers/cryptoday/tutte_fish.pdf. Transcript of a lecture given at the University of Waterloo, 19 June 1998
9. QCrypt2015. http://2015.qcrypt.net. Accessed 8 October 2015
10. Poe, E.A.: A few words on secret writing. Graham's Lady's Gentleman's Mag. **XIX**(1), 33–38 (1841)
11. Poe, E.A.: The Gold-Bug. Philadelphia Dollar Newspaper, Philadelphia (1843)
12. Rosenheim, S.J.: The Cryptographic Imagination: Secret Writing from Edgar Poe to the Internet. Johns Hopkins University Press, Baltimore (1997)
13. Shannon, C.E.: A mathematical theory of communication. Bell Syst. Tech. J. **27**(3), 379–423 (1948)
14. Shannon, C.E.: Communication theory of secrecy systems. Bell Syst. Tech. J. **28**(4), 656–715 (1949)
15. Colossus Computer. https://en.wikipedia.org/wiki/Colossus_computer. Accessed 8 October 2015
16. Vernam, G.: Secret signaling system, U.S. Patent 1,310,719 (1919)
17. Bellovin, S.M.: Frank Miller: inventor of the one-time pad. Cryptologia **35**(3), 203–222 (2011)
18. Moscow-Washington Hotline. https://en.wikipedia.org/wiki/Moscow-Washington_hotline. Accessed 8 October 2015
19. James, D.: Ché Guevara: a Biography. Rowman & Littlefield, Lanham (1970)
20. Merkle, R.C.: C.S. 244 project proposal. http://www.merkle.com/1974 (1974). Accessed 8 October 2015
21. Merkle, R.C.: Secure communications over insecure channels. Commun. ACM **21**(4), 294–299 (1978)
22. Diffie, W., Hellman, M.E.: New directions in cryptography. IEEE Trans. Inf. Theory **22**(6), 644–654 (1976)
23. Rivest, R.L., Shamir, A., Adleman, L.: A method for obtaining digital signatures and public-key cryptosystems. Commun. ACM **21**(2), 120–126 (1978)
24. Wayner, P.: British document outlines early encryption discovery. http://www.nytimes.com/library/cyber/week/122497encrypt.html (1997). Accessed 8 October 2015
25. McEliece, R.J.: A public-key cryptosystem based on algebraic coding theory. DSN Prog. Rep. **42**(44), 114–116 (1978)
26. Koblitz, N.: Elliptic curve cryptosystems. Math. Comput. **48**(177), 203–209 (1987)
27. Miller, V.S.: Use of elliptic curves in cryptography. In: Williams, H.C. (ed.) CRYPTO 1985. LNCS, vol. 218, pp. 417–426. Springer, Heidelberg (1986)

28. Feynman, R.P.: Simulating physics with computers. Int. J. Theor. Phys. **21**(6/7), 467–488 (1982)
29. Feynman, R.P.: Quantum mechanical computers. Opt. News **11**(2), 11–20 (1985)
30. Deutsch, D.: Quantum theory, the Church-Turing principle and the universal quantum computer. Proc. R. Soc. London A **400**, 97–117 (1985)
31. Shor, P.W.: Polynomial-time algorithms for prime factorization and discrete logarithms on a quantum computer. SIAM J. Comput. **26**(5), 1484–1509 (1997)
32. Kitaev, A.Y.: Quantum measurements and the Abelian stabilizer problem. arXiv preprint quant-ph/9511026 (1995)
33. Bernstein, D.J., Buchmann, J., Dahmen, E. (eds.): Post-Quantum Cryptography. Springer Science & Business Media, Berlin (2009)
34. Bernstein, D.J., Lange, T.: Post-quantum cryptography. http://pqcrypto.org/. Accessed 8 October 2015
35. Micciancio, D., Regev, O.: Lattice-based cryptography, pp. 147–191. In: [33] (2009)
36. Wolchover, N.: A tricky path to quantum-safe encryption. Quanta Magazine. https://www.quantamagazine.org/20150908-quantum-safe-encryption/. Accessed 8 October 2015
37. Campbell, P., Groves, M., Shepherd, D.: Soliloquy: a cautionary tale. https:// docbox.etsi.org/Workshop/2014/201410_CRYPTO/S07_Systems_and_Attacks/ S07_Groves_Annex.pdf. Accessed 8 October 2015
38. Biasse, J.F., Song, F.: A note on the quantum attacks against schemes relying on the hardness of finding a short generator of an ideal in $\mathbb{Q}(\zeta_{p^n})$. http://cacr. uwaterloo.ca/techreports/2015/cacr2015-12.pdf. Accessed 8 October 2015
39. National Security Agency: Cryptography Today. https://www.nsa.gov/ia/ programs/suiteb_cryptography/. Accessed 8 October 2015
40. Grover, L.K.: Quantum mechanics helps in searching for a needle in a haystack. Phys. Rev. Lett. **79**(2), 325–328 (1997)
41. Bennett, C.H., Bernstein, E., Brassard, G., Vazirani, U.: Strengths and weaknesses of quantum computing. SIAM J. Comput. **26**(5), 1510–1523 (1997)
42. Brassard, G., Salvail, L.: Quantum Merkle puzzles. In: Second International Conference on Quantum, Nano and Micro Technologies, pp. 76–79 (2008)
43. Brassard, G., Høyer, P., Kalach, K., Kaplan, M., Laplante, S., Salvail, L.: Merkle puzzles in a quantum world. In: Rogaway, P. (ed.) CRYPTO 2011. LNCS, vol. 6841, pp. 391–410. Springer, Heidelberg (2011)
44. Boyer, M., Brassard, G., Høyer, P., Tapp, A.: Tight bounds on quantum searching. Fortschr. Phys. **46**(4&5), 493–505 (1998)
45. Wootters, W.K., Żurek, W.H.: A single quantum cannot be cloned. Nature **299**(5886), 802–803 (1982)
46. Wiesner, S.: Conjugate coding. ACM Sigact News **15**(1), 78–88 (1983). Original manuscript written in 1968
47. Bennett, C.H., Brassard, G., Breidbart, S.: Quantum cryptography II: how to reuse a one-time pad safely even if P=NP. Nat. Comput. **13**(4), 453–458 (2014). Original manuscript written in 1982
48. Bennett, C.H., Brassard, G.: Quantum cryptography and its application to provably secure key expansion, public-key distribution, and coin-tossing. In: Proceedings of IEEE International Symposium on Information Theory, p. 91, September 1983
49. Bennett, C.H., Brassard, G.: Quantum cryptography: public key distribution and coin tossing. In: Proceedings of International Conference on Computers, Systems and Signal Processing, pp. 175–179, December 1984

50. Bennett, C.H., Brassard, G.: Quantum cryptography: public key distribution and coin tossing. Theor. Comput. Sci. **560**(Part 1), 7–11 (2014)

51. Brassard, G.: Brief history of quantum cryptography: a personal perspective. In: Proceedings of IEEE Information Theory Workshop on Theory and Practice in Information Theoretic Security, pp. 19–23, October 2005. arxiv.org/abs/quant-ph/0604072

52. Bennett, C.H., Bessette, F., Brassard, G., Salvail, L., Smolin, J.: Experimental quantum cryptography. J. Cryptology **5**(1), 3–28 (1992)

53. Mayers, D.: On the security of the quantum oblivious transfer and key distribution protocols. In: Coppersmith, D. (ed.) CRYPTO 1995. LNCS, vol. 963, pp. 124–135. Springer, Heidelberg (1995)

54. Renner, R., König, R.: Universally composable privacy amplification against quantum adversaries. In: Kilian, J. (ed.) TCC 2005. LNCS, vol. 3378, pp. 407–425. Springer, Heidelberg (2005)

55. König, R., Renner, R., Bariska, A., Maurer, U.: Small accessible quantum information does not imply security. Phys. Rev. Lett. **98**(14), 140502 (2007)

56. Renner, R., Gisin, N., Kraus, B.: Information-theoretic security proof for quantum-key-distribution protocols. Phys. Rev. A **72**(1), 012332 (2005)

57. Bennett, C.H., Brassard, G., Ekert, A.K.: Quantum cryptography. Sci. Am. **267**(4), 50–57 (1992)

58. ID Quantique. http://www.idquantique.com

59. Fadilpašić, S.: China's quantum communications network almost ready. http://www.itproportal.com/2015/08/31/chinas-quantum-communications-network-almost-ready/. Accessed 9 October 2015

60. Gerhardt, I., Liu, Q., Lamas-Linares, A., Skaar, J., Kurtsiefer, C., Makarov, V.: Full-field implementation of a perfect eavesdropper on a quantum cryptography system. Nat. Commun. **2**, 349 (2011)

61. Unruh, D.: Everlasting multi-party computation. In: Canetti, R., Garay, J.A. (eds.) CRYPTO 2013, Part II. LNCS, vol. 8043, pp. 380–397. Springer, Heidelberg (2013)

62. Ekert, A.K.: Quantum cryptography based on Bell's theorem. Phys. Rev. Lett. **67**(6), 661–663 (1991)

63. Bennett, C.H., Brassard, G., Mermin, N.D.: Quantum cryptography without Bell's theorem. Phys. Rev. Lett. **68**(5), 557–559 (1992)

64. Bell, J.S.: On the Einstein-Podolsky-Rosen paradox. Physics **1**(3), 195–200 (1964)

65. Einstein, A., Podolsky, B., Rosen, N.: Can quantum-mechanical description of physical reality be considered complete? Phys. Rev. **47**(10), 777–780 (1935)

66. Ekert, A., Renner, R.: The ultimate physical limits of privacy. Nature **507**(7493), 443–447 (2014)

67. Hensen, B., Bernien, H., Dréau, A.E., Reiserer, A., Kalb, N., Blok, M.S., Ruitenberg, J., Vermeulen, R.F.L., Schouten, R.N., Abellán, C., Amaya, W., Pruneri, V., Mitchell, M.W., Markham, M., Twitchen, D.J., Elkouss, D., Wehner, S., Taminiau, T.H., Hanson, R.: Loophole-free Bell inequality violation using electron spins separated by 1.3 kilometres. Nature **526**(7575), 682–686 (2015)

68. Johnston, H.: Physicists claim 'loophole-free' Bell-violation experiment. Physics World (2015). http://physicsworld.com/cws/article/news/2015/sep/02/physicists-claim-loophole-free-bell-violation-experiment

Relating Sublinear Space Computability Among Graph Connectivity and Related Problems

Tatsuya Imai[1] and Osamu Watanabe[2]([⊠])

[1] Heroz, Inc., Tokyo, Japan
[2] Department of Mathematical and Computer Science,
Tokyo Institute of Technology, Tokyo 152-8552, Japan
watanabe@is.titech.ac.jp

Abstract. We investigate sublinear-space computability relation among the directed graph vertex connectivity problem and its related problems, where by "sublinear-space computability" we mean in this paper $O(n^{1-\varepsilon})$-space and polynomial-time computability w.r.t. the number n of vertices. We demonstrate algorithmic techniques to relate the sublinear-space computability of directed graph connectivity and undirected graph length bounded connectivity.

1 Introduction and Preliminaries

Space complexity is one of the important complexity measures. In general algorithms with small complexity are important, but recently, due to the increase of data size, we face demands for sublinear-space algorithms in various applications, that is, demands for algorithms using much smaller working memory than input data size. Sublinear-space computability is also important from a theoretical view point for understanding the nature of computation. For example, the famous L = NL question is about the $O(\log n)$-space computability of the following directed graph connectivity problem. (Although we formulate in this paper connectivity problems as a problem of asking the connectivity of a given pair of vertices, the space complexity is the same even if we consider the connectivity for *all* pairs of vertices.)

stConn (Directed Graph Connectivity)
input: Directed graph $G = (V, E)$ and vertices $s, t \in V$.
task: Determine whether there exists a path from s to t.
size parameter: The number of vertices, denoted by n.

In order to understand the $O(\log n)$-space (in)computability of stConn, different versions of this connectivity problem have been investigated, and various important results have been obtained. For example, the breakthrough result of Reingold [7] shows that the connectivity is $O(\log n)$-space decidable for undirected graphs. On the other hand, not so much has been studied for a bit more relaxed $o(n)$-space computability. In this paper, we consider one of such $o(n)$-space bounds, that is, $O(n^{1-\varepsilon})$-space bound defined by "saving" parameter $\varepsilon > 0$.

© Springer-Verlag Berlin Heidelberg 2016
R.M. Freivalds et al. (Eds.): SOFSEM 2016, LNCS 9587, pp. 17–28, 2016.
DOI: 10.1007/978-3-662-49192-8_2

The stConn problem may not be solvable in $O(\log n)$-space, but it may still be solvable in $o(n)$-space and *and* polynomial-time. In fact, Barnes et al. [3] gave an $O(n/2^{\sqrt{\log n}})$-space and polynomial-time algorithm. But we aim for a stronger $o(n)$-space bound, that is, $O(n^{1-\varepsilon})$-space computability as Widgeson asked in [8]. Here we also require[1] the polynomial-time computability, which is crucial from both theoretical and practical view points. In fact, we have an $O((\log n)^2)$-space (and $O(n^{\log n})$-time) algorithm for stConn from Savitch's theorem. We do not want to go beyond the polynomial-time bound for reducing working memory. Thus, in this paper we consider both polynomial-time *and* $O(n^{1-\varepsilon})$-space computability, which we will call *sublinear-space computability* throughout this paper.

Recently, sublinear-space computability has been shown for some graph classes [1,2,4,6]. For example, for the directed *planar* graph connectivity problem, we have an $O(\sqrt{n})$-word-space and polynomial-time algorithm [2]. Unfortunately, however, an essential gap seems to exist to extend it to the general case. In this paper we would like to identify a requirement/restriction that makes the problem difficult. Certainly, directedness is a key for the hardness because the connectivity is decidable in $O(\log n)$-space for undirected graphs. As an alternative to directedness, we consider "bounded length" requirement; that is, we consider the problem that asks, for a given b, whether there is a path from s to t consisting of at most b edges, in other words, s is connected to t by a path of "length" at most b. Let us use UstConn$_{lb}$ to denote this version of *undirected* graph connectivity problem. It has been known that stConn is $O(\log n)$-reducible to UstConn$_{lb}$; that is, the difficulty of solving stConn in $O(\log n)$-space can be transformed to UstConn$_{lb}$, or more specifically, we have stConn $\notin$ L $\Rightarrow$ UstConn$_{lb}$ $\notin$ L. We ask in this paper whether this type of relation holds for their sublinear-space computability.

As a main result, we show a way to relate the sublinear-space computability of UstConn$_{lb}$ to that of stConn with almost same saving. This can be regarded as an sublinear-space approximately preserving reduction. We also explain the idea of a similar sublinear-space approximately preserving reduction. Therefore, we can conclude that directedness and length bound are computationally equivalent requirements also in the sublinear-space computability context. While we leave it open to extend this technique to other NL problems, we show similar relation holds for another graph connectivity problem and its length bounded version, which we hope to give us a hint to obtain a more general technique relating sublinear-space computability.

Preliminaries. We use standard notions and notation in graph theory and computational complexity theory. In this paper we consider both directed and undirected graphs, but we may assume that a graph is directed unless it is specified as undirected. A directed edge is denoted by an order pair of vertices, e.g., (u, v),

---

[1] Any $O(\log n)$-space algorithm is (modified to) a polynomial-time algorithm; thus, it is not necessary to require the polynomial-time computability when discussing the log-space computability.

whereas an undirected edge is denoted by a set of vertices, e.g., $\{u, v\}$. In the directed case, by a "path" we mean the sequence of directed edges having one direction from its source vertex to destination vertex. For any path, its *length* is the number of edges on the path. For any vertices u and v, a *shortest path* from u to v is a path from u to v with the smallest length, and by $\operatorname{leng}(u, v)$ we denote the length of the shortest path from u to v.

We basically follow the standard machine based framework for discussing time and space complexity. We consider that input data is given separately in a read-only memory area, and space complexity is the amount of working memory used for computation. Precisely speaking, we should measure the number of bits; but in our context we may ignore a $O(\log n)$ factor and use the number of working variables to measure space complexity. Throughout this paper we use n denote the number of vertices of a given graph, which is regarded as the main size parameter for all problems considered in this paper. We do not use the number of edges as a size parameter. This is because (i) the number of edges does not seem to be so relevant for discussing polynomial-time computability and space complexity, and (ii) we indeed have a polynomial-time and $O(n)$-size algorithms for various connectivity problems.

2 Length Bounded Undirected Graph Connectivity

As explained in Introduction, motivated by the $O(\log n)$-space computability of the undirected connectivity problem, we consider its length bounded version. That is, the following problem. (In this paper we use $[k]$ to denote $\{0, 1, \ldots, k\}$ instead of $\{1, \ldots, k\}$.)

UstConn$_{\text{lb}}$
input: Undirected graph $G = (V, E)$, $s, t \in V$, and integer $b \in [n - 1]$.
task: Determine whether there exists a path between s and t of length $\leq b$.

Nothing is known for the sublinear-space computability of this problem. Here we assume the following sublinear-space computability of this problem. That is, we assume that ε saving holds for this problem. We discuss whether a similar saving can be implied from this assumption for stConn.

Assumption 1. *There is an algorithm* Algo_UstConn$_{\text{lb}}$ *that solves* UstConn$_{\text{lb}}$ *in polynomial-time and* $O(n^{1-\varepsilon})$-space.

It has been well known that UstConn$_{\text{lb}}$ is also NL-complete problem. In particular, there is a standard log-space many-one reduction from stConn to UstConn$_{\text{lb}}$, and any $O(\log n)$-space algorithm solving UstConn$_{\text{lb}}$ can be used to give an $O(\log n)$-space algorithm for stConn. Let us recall this reduction first.

Consider any instance (G, s, t) of stConn, where $G = (V, E)$ is a directed graph and s and t are vertices in V. The reduction creates a "layered" undirected graph $nG = (nV, nE)$ that is defined by

$$nV = \big\{\, (i, v) \,|\, i \in [n-1] \text{ and } v \in V \,\big\}, \text{ and}$$
$$nE = \big\{\, \{(i, u), (i+1, v)\} \,|\, i \in [n-2] \text{ and } (u, v) \in E \,\big\}$$
$$\cup \big\{\, \{(i, u), (i+1, u)\} \,|\, i \in [n-2] \text{ and } u \in V \,\big\}.$$

Then it is easy to see the following property holds. Thus, a mapping (G, s, t) to $(nG, (0, s), (n-1, t), n-1)$ is a reduction from stConn to UstConn$_\text{lb}$.

Claim. For any $s, t \in V$, there is a path from s to t in G if and only if there is a path from $(0, s)$ to $(n-1, t)$ in nG of length $n-1$.

Let `Algo_red` denote the algorithm solving stConn by using this reduction method and `Algo_UstConn`$_\text{lb}$. Note that the instance given to `Algo_UstConn`$_\text{lb}$ has n^2 vertices. Thus, we have the following complexity bounds.

Lemma 1. `Algo_red` *solves* stConn *in polynomial-time and* $O((n^2)^{1-\varepsilon})$-*space.*

Note that $O(n^{2(1-\varepsilon)}) = O(n^{1-(2\varepsilon-1)})$; this space bound is still sublinear if $\varepsilon > 0.5$. But clearly, the saving got reduced considerably due to the increase of the graph size, i.e., the number of vertices, by the reduction.

We introduce two algorithmic ideas to suppress this graph size increase of the standard reduction. For this we consider the length bounded connectivity also in a directed graph, and we introduce the notion of "bounded length" below. Consider any directed graph $G = (V, E)$, and let $\text{leng}(u, v)$ denote the length from u to v on this graph. For any $u, v \in V$ and for any $b \in [n-1]$, we define $\text{leng\_bl}(u, v, b)$ by

$$\text{leng\_bl}(u, v, b) = \begin{cases} \text{leng}(u, v), & \text{if } \text{leng}(u, v) \le b, \text{ and} \\ \bot, & \text{otherwise.} \end{cases}$$

We call this function *bounded length*. When necessary, we write $\text{leng\_bl}(G: u, v, b)$ for explicitly expressing the length is considered on G. Note that deciding whether $\text{leng\_bl}(u, v, b) \ne \bot$ is exactly the length bounded connectivity on the directed graph G. We can generalize the reduction based algorithm `Algo_red` to determine whether $\text{leng\_bl}(u, v, b) \ne \bot$ in polynomial-time and $O((bn)^{1-\varepsilon})$-space. Let us still use `Algo_red` to denote this algorithm, and let $t_\text{red}(bn)$ denote a polynomial time bound for `Algo_red` to determine $\text{leng\_bl}(u, v, b) \ne \bot$.

Clearly, the graph size increase can be suppressed if stConn can be solved by using only $\text{leng\_bl}(\cdot, \cdot, b)$ with small b. One simple idea is to compute bounded length recursively. For example, consider the following graph $G(b) = (V, E(b))$, where

$$E(b) = \big\{\, (u, v) \,|\, \text{leng\_bl}(u, v, b) \ne \bot \,\big\}.$$

Apply `Algo_red` on $G(b)$ to determine $\text{leng\_bl}(G(b): u, v, b) \ne \bot$. Whenever `Algo_red` needs to see whether an edge (u, v) exists in $G(b)$, we run `Algo_red` on G to determine $\text{leng\_bl}(G: u, v, b) \ne \bot$. Clearly, we have $\text{leng\_bl}(G(b): u, v, b) \ne \bot$ if and only if $\text{leng\_bl}(G: u, v, b^2) \ne \bot$, and it is easy to see that this depth two recursion for deciding $\text{leng\_bl}(G(b): u, v, b) \ne \bot$ can be done in $O((t_\text{red}(bn))^2)$-time and $O(2(bn)^{1-\varepsilon})$-space.

We can extend this idea and use `Algo_red` on $G^r(b) = (V, E^r(b))$, where $E^r(b)$ is defined inductively by $E^r(b) = \{\,(u, v) \mid \text{leng\_bl}(G^{r-1}(b): u, v, b) \neq \perp\,\}$. Let `Algo_red`^{r} denote the algorithm that determines $\text{leng\_bl}(G^{r-1}(b): u, v, b) \neq \perp$ by using `Algo_red` recursively up to depth r. That is, `Algo_red`^{r} determines $\text{leng\_bl}(G^{r-1}(b): u, v, b) \neq \perp \iff \text{leng\_bl}(G: u, v, b^r) \neq \perp$. Thus, for solving an stConn instance (G, s, t), it is enough to compute $\text{leng\_bl}(G^{r-1}: s, t, b)$ with $b = n^{1/r}$ by `Algo_red`^{r}. This gives the following bounds.

Lemma 2. *For any $r \geq 1$, `Algo_red`^{r} solves stConn in $O((t_{\text{red}}(n^{1+1/r})^r)$-time and $O(rn^{(1+1/r)(1-\varepsilon)})$-space.*

Remark. *Though the parameter r need not be a constant for the algorithm, it must be a constant in order to bound the running time by polynomial. Then we can simplify the above space bound by*

$$O\big(n^{(1+1/r)(1-\varepsilon)}\big) = O\big(n^{1-((1+1/r)\varepsilon - 1/r)}\big).$$

Hence, the saving of `Algo_red`^{r} is $(1 + 1/r)\varepsilon - 1/r$.

Unfortunately, the above saving is still not so good. In order to have a non-trivial saving we need to choose $r > 1/\varepsilon$, which makes the time bound very high (while it is still polynomial). We can overcome this problem by using the idea in [3]. Barnes et al. [3] gave a weak sublinear-space algorithm. Their algorithm is a combination of two algorithms **B1** and **B2**; algorithm **B2**, which is used by **B1** as a subroutine, is in fact computes the bounded length, i.e., $\text{leng\_bl}(\cdot, \cdot, L)$ for relatively small L. Here we use the **B1** part of their algorithm.

We first show that $\text{leng\_bl}(G: u, v, b^r)$ is indeed sublinear-space computable by using decision algorithm `Algo_red`^{i} and its modification. The idea is simple. Consider any $u, v \in V$. If $\text{leng\_bl}(u, v, b^r) = \perp$, then we are done. (Here and below by $\text{leng\_bl}(\cdot, \cdot, \cdot)$ we mean $\text{leng\_bl}(G: \cdot, \cdot, \cdot)$.) Otherwise, it suffices to test whether $\text{leng\_bl}(u, v, d) \neq \perp$ holds for all $d \in [b^r - 1]$; we have $\text{leng\_bl}(u, v, b^r) = d$ with the minimum d such that $\text{leng\_bl}(u, v, d) \neq \perp$ holds. Note here that $d = a_0 + a_1 b + a_2 b^2 + \cdots + a_{r-1} b^{r-1}$ for some $a_0, \ldots, a_{r-1} \in [b - 1]$. Then we can test whether $\text{leng\_bl}(u, v, d) \neq \perp$ holds by using a layered graph (rV, rE') similar to (rV, rE). Here rE' is defined as follows: for each $i \geq 0$, rE' has an edge $\{(i, u), (i + 1, v)\}$ if and only if $\text{leng\_bl}(u, v, a_i b^i) \neq \perp$, which can be tested by using a slightly modified `Algo_red`$^{i+1}$. Hence, $\text{leng\_bl}(u, v, d) \neq \perp$ can be tested in time $O(t_{\text{red}}(rn) t_{\text{red}}(bn)^r) = O(t_{\text{red}}(bn)^{r+1})$ and in space $O((rn)^{1-\varepsilon} + (bn)^{1-\varepsilon}) = O((bn)^{1-\varepsilon})$ (since we may assume that $r < b$). Therefore, $\text{leng\_bl}(u, v, b^r)$ is computable as follows.

Lemma 3. *We have an algorithm `Algo_bl` that computes $\text{leng\_bl}(G: u, v, b^r)$ in $O(b^r t_{\text{red}}(bn)^{r+1})$-time and $O((bn)^{1-\varepsilon})$-space.*

Next we introduce a key tool, namely, a small "separator." Consider any instance for stConn, i.e., $G = (V, E)$ and $s, t \in V$, and fix them in the following explanation. Let L be an algorithm parameter that is determined later. For any $j \in [L - 1]$, let V_j be a subset of V defined by

$$V_j = \{ v \mid \text{leng}(s, v) \bmod L = j \}.$$

Family $\{V_j\}_{j \in [L-1]}$ has several important properties. First, it is a partition of the set of vertices of V reachable from s; and hence, there should be some j_0 such that $|V_{j_0}| \leq n/L$. We use such V_{j_0} to record the reachability from s, thereby reducing the space complexity for memorization. Another important property, though it is trivial, is that each V_j is a separator of all shortest paths of length $\geq j$. More specifically, for any $j \in [L - 1]$, if $\text{leng}(s, v) \geq j$, then there should be some vertex u in V_j that is on one of the shortest paths from s to v and for which $\text{leng}(u, v) < L$ holds; thus, once we have V_{j_0}, we only need to compute, for each $u \in V_{j_0}$, its bounded length to t, i.e., $\text{leng\_bl}(u, t, L)$, to determine the connectivity from s to t.

These properties justify the following algorithm outline for deciding connectivity from s to t in G: (1) Compute V_j for each $j \in [L - 1]$ from $j = 0$ to $L - 1$. If the algorithm finds that $|V_j| > n/L$, then it stops the computation of V_j and moves to the computation of V_{j+1}. On the other hand, move to the next step as soon as V_j with appropriate size can be computed. (2) Compute $\text{leng\_bl}(u, t, L)$ for each $u \in V_{j_0}$. Output "yes" if there is some $u \in V_{j_0}$ such that $\text{leng\_bl}(u, t, L) \neq \perp$. Otherwise, output "no." This is essentially the B1 part of the algorithm of Barnes et al., and we name the algorithm that solves stConn following this outline as Algo_Betal; note that we assume that the algorithm Algo_bl of Lemma 3 is used here for computing the bounded length.

Now it remains to implement the above step (1). Below we give a procedure for computing V_j for a given $j \in [L - 1]$; the actual computation of (1) is to use this procedure to find j_0 for which the procedure successfully computes V_{j_0}. For this procedure, we can again use the bounded length. In [3] B1 is stated as a breadth first algorithm, but here for the sake of later explanation, we state it as a "closest vertex first" algorithm, which can be regarded as a variation of the Dijkstra's algorithm. In the following procedure, we use a variable D to denote the set of vertices in V_j whose length from s has been determined, and for any $v \in D$, we use $d[v]$ to record this length. For any $u \in D$ and $v \in V \setminus D$, we define a function cost_via(u, v) by

$$\text{cost\_via}(u, v) = d[u] + \text{leng\_bl}(u, v, L).$$

That is, cost_via(u, v) is the length of a path from s to v that has u in V_j. It is easy to see that u is the vertex in V_j on the path closest to v.

The correctness of the procedure will be explained in the next section for a more general procedure. Here we analyze the time and space complexity bounds of Algo_Betal. Note that the most time consuming part is the computation of the bounded length $\text{leng\_bl}(\cdot, \cdot, L)$, and it is easy to see that the bounded length is computed at most $O(Ln^3)$ times; thus, by using Algo_bl of Lemma 3 for computing the bounded length, we can bound the total running time by $O(Ln^3 L(t_{\text{red}}(L^{1/r}n))^{r+1})$, which is roughly poly$(n)^r$. On the other hand, we can

```
procedure for computing V_j
    d[s] ← 0; D ← { s };
    while v_min ≠ ⊥ do {
        v_min ← ⊥;  cost_min ← +∞;
        for each v ∈ V \ D do {
            u_closest ← argmin{ cost_via(u, v) | u ∈ D };
            cost ← cost_via(u_closest, v);
            if leng_bl(u_closest, v, L) = L
                ∧ cost < cost_min then {
                v_min ← v;  cost_min ← cost;
            }
        }
        if v_min ≠ ⊥ then {
            D ← D ∪ { v_min };  d[v] ← cost_min;
            if |D| > n/L then report failure and stop;
        }
    }
    return D as V_j;
```

Fig. 1. Procedure for computing V_j in Algo_Betal

bound the space complexity of Algo_Betal by $O(n/L + (L^{1/r}n)^{1-\varepsilon})$. This space bound is (approximately) minimized by choosing L to satisfy $n/L = (L^{1/r}n)^{1-\varepsilon}$, or equivalently, $n^\varepsilon = L^{1+(1-\varepsilon)/r}$. With this choice of L, we can bound the space complexity by $O(n^\alpha)$, where

$$\alpha = 1 - \frac{\varepsilon}{1 + (1 - \varepsilon)/r} = 1 - \varepsilon\left(1 - \frac{1-\varepsilon}{r + (1 - \varepsilon)}\right) < 1 - \varepsilon\left(1 - \frac{1}{r+1}\right).$$

Theorem 1. *Using the sublinear-space algorithm assumed by Assumption 1, algorithm* Algo_Betal *solves* stConn *in* $O((\text{poly}(n))^r)$*-time and* $O(n^{1-(1-1/(r+1))\varepsilon})$*-space.*

We may regard this result as a reduction from stConn to UstConn_lb that preserve approximate sublinear-space computability, where by "approximate" we mean that one can get a saving as arbitrarily close to the original saving. Here let us call intuitively our construction of algorithm Algo_Betal as a *sublinear-space approximately preserving reduction* without giving any formal definition. Naturally we may ask whether this sublinear-space approximately preserving reduction exits also from UstConn_lb to stConn. Here again we can use a similar idea to show such a reduction.

Consider any instance (G, s, t, b) of UstConn_lb, where $G = (V, E)$ is an undirected graph, $s, t \in V$, and $b \in [n-1]$. We consider a layered directed graph $nG = (nV, nE)$ defined by

$$nV = \big\{\, (i,v) \mid i \in [b] \text{ and } v \in V \,\big\}, \text{ and}$$
$$nE = \big\{\, ((i,u),(i+1,v)) \mid i \in [b-1] \text{ and } \{u,v\} \in E \,\big\}$$
$$\cup \big\{\, \{(i,u),(i+1,u)\} \mid i \in [b-1] \text{ and } u \in V \,\big\}.$$

Then again it is easy to see that $\mathrm{leng\_bl}(s,t,b) \neq \perp$ if and only if there is a directed path from $(0,s)$ to (b,t) in nG. Therefore, using an argument similar to the above, we can also define a sublinear-space approximately preserving reduction from $\mathrm{UstConn_{lb}}$ to stConn. The detail construction as well as giving a formal definition to the notion of "sublinear-space approximately preserving reduction" is left to the interest reader.

3　Another Example: Two Vertex Distance Problem

Although we have close sublinear-space computability relation between stConn and $\mathrm{UstConn_{lb}}$, it is not so clear whether similar relation holds with the other NL-problems. While we have not been able to develop a general result, we can show some example result that would give us a hint for applying our technique to other problems.

We consider here the problem of computing the "distance" between two vertices in a directed and weighted graph. A weighted graph is a graph whose edge is given a cost. In this explanation we assume that each cost is a positive integer; furthermore, in order to avoid introducing another size parameter, we also assume that each cost can be expressed by $\mathrm{poly}(\log n)$ bits so that cost computation can be trivially done in $\mathrm{poly}(\log n)$-space. For specifying a cost at each edge, we use a *cost function*, a mapping from an edge to its cost. For example, a weighted directed/undirected graph is given by $G = (V, E, c)$, where c is a mapping from E to its cost. Consider any pair of vertices u and v of some weighted graph. For any path from u to v, its *weight* is the sum of the cost of edges on the path. A *lightest path* from u to v is a path from u to v with the smallest weight, and the *distance* from u to v (denoted by $\mathrm{dist}(s,t)$) is the weight of the lightest path from u to v. We will keep using "length" to mean the number of edges and "shortest path" to mean a path (connecting a specified pair of vertices) with the smallest number of edges. In summary we consider the following problem.

stDist (Two Vertex Distance Problem)
input: Directed and weighed graph $G = (V, E, c)$, $s, t \in V$, and $d \geq 0$.
task: Determine whether $\mathrm{dist}(s,t) \leq d$.
Remark. For simplicity we assume in this paper that weights are integers expressed in $\mathrm{poly}(\log n)$ bits.

Clearly this problem is in NL. But it is not clear that a similar sublinear-space (approximately) preserving reduction from this problem to, e.g., stConn. Yet, we can still consider similar relation to its undirected and length bounded version. More specifically, consider the following problem.

UstDist$_{lb}$
input: Undirected and weighed graph $G = (V, E, c)$, $s, t \in V$, $d \geq 0$,
and $b \in [n - 1]$.
task: Determine whether dist_bl$(s, t, b) \leq d$.

Here dist_bl(s, t, b) is the length bounded distance from s to t, that is, the weight
of the lightest path from s to t of length $\leq b$; we assume that dist_bl$(s, t, b) = \bot$
if there is no path of length $\leq b$ from s to t. We use dist_bl(s, t, b) also for
directed graphs. Note that even if there is a path from s to t of length $\leq b$ (i.e.,
dist_bl$(s, t, b) \neq \bot$), we may not have dist_bl$(s, t, b) = \text{dist}(s, t)$. (*Cf.* We have
leng_bl$(s, t, b) = \text{leng}(s, t)$ if leng_bl$(s, t, b) \neq \bot$.)

We again base an assumption that UstDist$_{lb}$ has a polynomial-time and
$O(n^{1-\varepsilon})$-space algorithm, which we denote as Algo_UstDist$_{lb}$. Using this algorithm, we show a sublinear-space algorithm for stDist with an approximately
same saving. The problems are somewhat complicated compared with stConn
and UstConn$_{lb}$, which is mainly due to the above mentioned difference between
leng_bl and dist_bl. Thus, for deriving the sublinear-space computability of stDist
(based on Algo_UstDist$_{lb}$) there are some points where the previous argument
need to be modified appropriately. We explain below such points.

Let us consider one instance for stDist; that is, a directed and weighted
graph $G = (V, E, c)$ and $s, t \in V$. As before, our first step is to develop an
algorithm based on a log-space many-one reduction from stDist to UstDist$_{lb}$.
More specifically, for any $u, v \in V$ and any bound $b \geq 0$, we define a layered and
weighted undirected graph $bG = (bV, bE, c^+)$ as before, with which we can decide
whether dist_bl$(G: u, v, b) \leq d$ for a given d by using Algo_UstDist$_{lb}$. Then we
can use a binary search to determine the value of dist_bl$(G: u, v, b)$. This algorithm is polynomial-time and $O((bn)^{1-\varepsilon})$-space. The second step is also similar
to the previous argument. We use a recursive way to compute dist_bl(u, v, b^r)
in $O(r(bn)^{1-\varepsilon})$-space, while the computation time grows to poly$(bn)^r$. Here
again we first define a decision algorithm and uses it to compute the value
of dist_bl(u, v, b^r) by a binary search. Now an interesting point is the last step
where we use the idea of the algorithm of Barnes et al.

We introduce some new notions. Consider any two vertices $u, v \in V$. Note
that a lightest path from u to v may not be unique. A shortest path among such
lightest paths is called a *best path*. Note that there may be still more than one
best paths, but we do not need to distinguish them in the following discussion.
We introduce a function bpleng that gives the length of a best path; that is,
bpleng(u, v) is the length of a best path from u to v. We also consider its length
bounded version. For any integer $b \geq 0$, consider lightest paths from u to v of
length $\leq b$; then a length b bounded best path is a shortest one among such
lightest paths, and bpleng_bl(u, v, b) is the length of this length bounded best
path. In other words, bpleng_bl(u, v, b) is the length of a shortest path with
weight dist_bl(u, v, b). In the previous argument, we explained a way to compute
leng(u, v, b), i.e., the length of a shortest path from u to v within length bound
b. Here we can use a similar technique; we only need to modify the algorithm so
that it computes the length of a shortest path within length bound b *with weight*

d, for $d = \text{dist\_bl}(u, v, b)$ computed beforehand. In this way, we can compute $\text{bpleng\_bl}(u, v, b^r)$ in $\text{poly}(bn)^r$-time and $O(r(bn)^{1-\varepsilon})$-space.

With two functions dist_bl and bpleng_bl, we now explain how to generalize the idea of Barnes et al. The key point is to use the length of a best path to partition V. For a given parameter L, we consider a family $\{V_j\}_{j\in[L-1]}$, where for each $j \in [L-1], V_j$ is defined by

$$V_j = \big\{ v \mid \text{bpleng}(s, v) \bmod L = j \big\}.$$

Again $\{V_j\}_{j\in[L-1]}$ is a partition of all vertices of V connected from s. In particular, each V_j is a separator of all best paths of length $\geq j$. Note again that there must be some V_j such that $|V_j| \leq n/L$. We use one of such V_j's, say, V_{j_0}, to record necessary information to search the lightest (in fact, best) path from s to all vertices in V. The information we need to keep for each $v \in V_{j_0}$ is the weight (which is in fact distance) and the length of a best path from s to v. It is easy to see an outline similar to the one explained for Algo_Beta1 works. Here we only explain the procedure for computing V_j for a given $j \in [L-1]$.

---

procedure for computing V_j
  d[s], l[s] $\leftarrow$ 0;   D $\leftarrow$ { s };
  while v_min $\neq \perp$ **do** {
    v_min $\leftarrow \perp$;   dist_min $\leftarrow +\infty$;
    for each v $\in V \setminus$ D **do** {
      u_closest $\leftarrow$ argmin{ cost_via(u, v) $\mid u \in$ D };
      (dist, leng) $\leftarrow$ cost_via(u_closest, v);
      if leng $= L$
        $\wedge$ dist $<$ dist_min **then** {
        v_min $\leftarrow$ v;   dist_min $\leftarrow$ dist;   leng_min $\leftarrow$ leng;
      }
    }
    if v_min $\neq \perp$ **then** {
      D $\leftarrow$ D $\cup$ { v_min };   d[v] $\leftarrow$ dist_min;   l[v] $\leftarrow$ leng_min;
      if |D| $> n/L$ **then** report failure and stop;
    }
  }
  return D as V_j;

Fig. 2. Procedure for computing V_j for the stDist problem

The outline of the procedure is the same as before. We use a variable D to denote the set of vertices in V_j whose best path from s has been determined, and for any $v \in$ D, we use d[v] and l[v] to record the weight and length of its best path. For any $u \in$ D and $v \in V \setminus$ D, we define a function cost_via(u, v) by

$$\text{cost\_via}(u, v) = (\text{d}[u] + \text{dist\_bl}(u, v, L), \text{l}[u] + \text{bpleng\_bl}(u, v, L),).$$

That is, "cost" is now a pair of the weight and length of the lightest path from s to v going through u (in D). Then for comparing a pair of such costs, we use the lexicographic order; that is, compare weights first (and if they are equal) compare lengths next. For computing u_closest in the procedure, we use this comparison.

The key point for showing the correctness of this procedure is stated in the following lemma. Once the lemma is proved, the correctness of the procedure and the whole algorithm follows easily, which we omit in this paper. The lemma can be proved by an induction on k, which is also not so difficult and omitted here.

Lemma 4. *Let $v_1, v_2, \ldots$ be the enumeration of elements of V_j under the order given by our cost comparison. Then for any $k \geq 1$, v_k is the kth vertex that is selected as* v_min *and put into* D. *Furthermore, values* d[v_min] *and* l[v_min] *at the point* v_min $= v_k$ *is put into* D *are respectively* $\text{dist}(s, v_k)$ *and* $\text{bpleng}(s, v_k)$.

Since the other part and the analysis for L is almost the same, we omit the rest of the argument and state only the result.

Theorem 2. *Suppose that there is an algorithm that solves* UstDist_lb *in polynomial-time and* $O(n^{1-\varepsilon})$-space. *Then for any integer* $r > 0$, *we have an algorithm that solves* stDist *in* $O((\text{poly}(n))^r)$-time *and* $O(n^{1-(1-1/(r+1))\varepsilon})$-space.

Clearly, the sublinear preserving reduction designed for this theorem can be used as a part of a real algorithm. In fact, as explained in [5], we can modify the algorithm given in [3] (i.e., the one for B2 explained in the previous section) to design the one computing both dist_bl$(\cdot, \cdot, b^r)$ and bpleng_bl$(\cdot, \cdot, b^r)$ in polynomial-time and $O(n/b^r + r(b + n/k))$-space (ignoring a $\log n$ factor) for relatively small (but not necessarily constant) r, where k is another algorithm parameter. By using this algorithm and choosing parameters b, k, and r appropriately as explained in [3] we can derive the following weakly sublinear-space algorithm corresponding to the one for stConn of Barnes et al.

Corollary 3. [5] *There exits a polynomial-time and* $O(n/2^{\sqrt{\log n}})$-space *algorithm for* stDist.

4 Concluding Remarks

We showed a way to relate the sublinear-space computability of UstConn_lb to that of stConn with almost same saving, which can be regarded as an sublinear-space approximately preserving reduction. We extend this reduction technique for relating the sublinear-space computability of stDist and UstDist_lb. We then naturally ask whether similar relation holds for any other NL problems; we may even need to develop a framework for discussing directedness and length bound restriction in general. Also it would be interesting if we can give a similar reduction from, say, stDist to stConn.

In the context of our sublinear-space computability, we do not have to restrict ourselves to problems in NL. For example, it would be interesting if we can extend our technique for relating problems in LogCFL, etc.

Acknowledgements. The authors would like to thank Dr. Kotaro Nakagawa for his helpful comments on earlier version of this paper. This work is supported in part by the ELC project (MEXT KAKENHI Grant No. 24106008).

References

1. Asano, T., Doerr, B.: Memory-constrained algorithms for shortest path problem. In: Proceedings of the 23rd Annual Canadian Conference on Computational Geometry (CCCG 2011) (2011)
2. Asano, T., Kirkpatrick, D., Nakagawa, K., Watanabe, O.: $\tilde{O}(\sqrt{n})$-space and polynomial-time algorithm for planar directed graph reachability. In: Csuhaj-Varjú, E., Dietzfelbinger, M., Ésik, Z. (eds.) MFCS 2014, Part II. LNCS, vol. 8635, pp. 45–56. Springer, Heidelberg (2014)
3. Barnes, G., Buss, J.F., Ruzzo, W.L., Schieber, B.: A sublinear space, polynomial time algorithm for directed s-t connectivity. In: Proceedings of Structure in Complexity Theory Conference, pp. 27–33. IEEE Computer Society Press (1992)
4. Chakraborty, D., Pavan, A., Tewari, R., Vinodchandran, V., Yang, L.: New time-space upper bounds for directed reachability in high-genus and H-minor-free graphs. In: ECCC TR14-035 (2014)
5. Imai, T.: Polynomial time memory constrained shortest path algorithms for directed graphs (in Japanese). In Proceedings of 12th Forum for Informatics (FIT 2013), IEICE Japan, RA-002 (2013)
6. Imai, T., Nakagawa, K., Pavan, A., Vinodchandran, N.V., Watanabe, O.: An $O(n^{\frac{1}{2}+\epsilon})$-space and polynomial-time algorithm for directed planar reachability. In: Proceedings of the 28th Conference on Computational Complexity (CCC 2013), pp. 277–286. IEEE (2013)
7. Reingold, O.: Undirected connectivity in log-space. J. ACM **55**(4), 1–24 (2008)
8. Wigderson, A.: The complexity of graph connectivity. In: Havel, I.M., Koubek, V. (eds.) MFCS 1992. LNCS, vol. 629, pp. 112–132. Springer, Heidelberg (1992)

Learning Automatic Families of Languages

Sanjay Jain[1]([✉]) and Frank Stephan[2]

[1] School of Computing, National University of Singapore,
Singapore 117417, Singapore
sanjay@comp.nus.edu.sg
[2] Department of Mathematics and Department of Computer Science,
National University of Singapore, Singapore 119076, Singapore
fstephan@comp.nus.edu.sg

Abstract. A class of languages is automatic if it is uniformly regular using some regular index set for the languages. In this survey we report on work about the learnability in the limit of automatic classes of languages, with some special emphasis to automatic learners.

1 Introduction

A language is a set of strings over some finite alphabet. Consider the following model of language learning. A learner receives all elements of the target language, one elment at a time, repetition allowed, in arbitrary order (this form of information provided to the learner is called a text for the language). Note that no non-elements of the language are provided to the learner. For technical reasons, we allow a special symbol # as input, which denotes "no datum" (this allows for a text of empty language as an infinite sequence of #'s). After receiving each new element the learner outputs its conjecture about what the target language might be (this is usually expressed in the form of a grammar for the language, in some hypothesis space). If the sequence of grammars output by the learner converges to a correct grammar for the target language then the learner is said to identify the language (from the corresponding text). For learning a language, the learner is expected to learn it from all texts for the language. Learning of one language is not useful, as a learner which outputs just a grammar for the language, whatever the input might be, is similar to a person who predicts earthquake everyday, and is right on the day earthquake actually occurs. So what is more interesting is whether the same learner can learn all languages from a class of languages. This is essentially the model of learning proposed by Gold [11] and called **TxtEx**-learning. In Gold's original model there is no restriction on the memory of the learner. Thus, the learner can remember all its past input data when it comes up with its new hypothesis. In some cases below we will consider restrictions on the memory of the learner. Thus, we define a learner taking this into account.

Research for this work is supported in part by NUS grants C252-000-087-001 (S. Jain) and R146-000-181-112 (S. Jain and F. Stephan).

R.M. Freivalds et al. (Eds.): SOFSEM 2016, LNCS 9587, pp. 29–40, 2016.
DOI: 10.1007/978-3-662-49192-8_3

Let $\mathbb{N}$ denote the set of natural numbers.

A *text* T is a mapping from $\mathbb{N}$ to $\Sigma^* \cup \{\#\}$. Content of a text T, denoted content$(T) = \{T(i) : i \in \mathbb{N}\} - \{\#\}$. $T[n]$ denotes the initial sequence $T(0)T(1)\ldots T(n-1)$ of the text T, of length n.

A finite sequence is an initial segment of a text. SEQ denotes the set of all finite sequences. For a finite sequence $T[n]$, content$(T[n]) = \{T(i) : i \in \mathbb{N}\} - \{\#\}$.

We use $\sigma \diamond \tau$ to denote the concatenation of two finite sequences σ and τ. Similarly, $\sigma \diamond T$ denotes the concatentation of σ and T.

Definition 1 (Based on Gold [11]; see also [5,17]). Suppose Σ is the alphabet set for the languages and $\mathcal{L} = \{L_\alpha : \alpha \in I\}$ is a class of languages to be learnt, where I is an index set. Let $\mathcal{H} = \{H_\beta : \beta \in J\}$ be the hypothesis space, used by the learner, where J is the index set for the hypotheses. We always assume that $\mathcal{H}$ is uniformly r.e.; in some cases below we will put more restrictions on the hypothesis space. Let ? be a special symbol not in J which denotes "no new conjecture at this point". Suppose Γ is a finite set of alphabet used for memory by the learner.

(a) A learner is an algorithmic mapping from $\Gamma^* \times (\Sigma^* \cup \{\#\})$ to $\Gamma^* \times (J \cup \{?\})$. A learner has an initial memory $mem_0 \in \Gamma^*$ and initial conjecture $hyp_0 \in J \cup \{?\}$.

Intuitively, for a learner M, $M(mem, x) = (mem', hyp)$, means that based on old memory mem and current datum x, the new memory of the learner is mem' and hyp is its conjecture.

(b) Suppose a learner M with initial memory mem_0 and initial hypothesis hyp_0 is given. Suppose T is a text for a language L.

(i) Let $mem_0^T = mem_0$ and $hyp_0^T = hyp_0$.

(ii) For $k > 0$, let $(mem_k^T, hyp_k^T) = M(mem_{k-1}^T, T(k-1))$

(iii) Define $M(T[k]) = (mem_k^T, hyp_k^T)$.

(iv) M on T *converges* on text T to the hypothesis hyp iff, for all but finitely many k, $hyp_k^T = hyp$.

(c) M **TxtEx**-learns a language L (using hypothesis space $\mathcal{H}$) if, for all texts T for L, M on T converges to a hypothesis β such that $H_\beta = L$.

(d) M **TxtEx**-learns the class $\mathcal{L}$ (using hypothesis space $\mathcal{H}$) iff M **TxtEx**-learns all the languages in the class $\mathcal{L}$ (using hypothesis space $\mathcal{H}$).

(e) **TxtEx** $= \{\mathcal{L} :$ some learner M learns $\mathcal{L}$ using some automatic family $\mathcal{H}$ as the hypothesis space$\}$.

Intuitively, mem_k^T and hyp_k^T in part (b) above denote the memory and conjecture of the learner M after having seen the data $T[k]$.

We now consider automatic classes of languages. Intuitively, a class of languages is automatic if the class is uniformly regular. More formally, let Σ be a finite alphabet, and let @ be a special symbol not in Σ. Given two strings $x = x_0 x_1 \ldots x_{n-1}$ and $y = y_0 y_1 \ldots y_{m-1}$ over the alphabet Σ, define convolution of x and y, conv(x, y) as follows. Let $r = \max(\{m, n\})$. For $i < n$, let $x_i' = x_i$; for $n \le i < r$, let $x_i' = @$. For $i < m$, let $y_i' = y_i$; for $m \le i < r$, let $y_i' = @$. Now, convolution of x, y is defined as conv$(x, y) = z_0 z_1 \ldots z_{r-1}$, where $z_i = (x_i', y_i')$;

note that z_i is a member of the alphabet $\Sigma \cup \{@\} \times \Sigma \cup \{@\}$. One can extend the definition of convolution to multiple strings similarly.

A class of languages $\mathcal{L}$ is said to be automatic if there is an indexing $(L_\alpha)_{\alpha \in I}$, for some regular index set I such that, $\mathcal{L} = \{L_\alpha : \alpha \in I\}$ and $\{\text{conv}(\alpha, x) : x \in L_\alpha\}$ is regular [20]. A relation $R = \{(x_1, x_2, \ldots, x_n) : x_1, x_2, \ldots, x_n \in \Sigma^*\}$ is said to be automatic if $\{\text{conv}(x_1, x_2, \ldots, x_n) : (x_1, x_2, \ldots, x_n) \in R\}$ is regular. Similarly, a function f is said to be automatic if $\{\text{conv}(x, y) : f(x) = y\}$ is regular.

The following theorem is important and it also enables to derive that the first-order theory of any given automatic structure is decidable.

Theorem 2 (Blumensath and Grädel [4], Khoussainov and Nerode [23]). *Any relation that is first-order definable from existing automatic relations is automatic and there is an algorithm to construct the corresponding automaton from automata for the relations and functions of the automatic structure and the defining formula.*

Furthermore, one can characterise the automatic functions as functions which map convoluted tuples to convoluted tuples and which can be computed by a one-tape Turing machine with the output starting at the same position as originally the input started and computation time being linear [7]; for this characterization one can either use deterministic or non-deterministic Turing machines.

When learning automatic classes, we usually require that the hypothesis space $\mathcal{H}$ is also automatic. This paper surveys some of the results in learnability of countable automatic classes of languages. For learnability of uncountable classes we refer the reader to Jain et al. [18].

2   Characterization of Learnability of Autmatic Classes

For the characterization, we first consider the notion of tell-tale sets as introduced by Angluin. Let $x <_{ll} w$ denote that x is length-lexicographically smaller than w. That is $|x| < |w|$ or $|x| = |w|$ and x is lexicographically before w (based on some fixed ordering of the alphabet). Let $x \leq_{ll} w$ denote that $x <_{ll} w$ or $x = w$.

Definition 3 (Angluin's Tell Tale condition [2]). Suppose $\mathcal{L} = \{L_\alpha : \alpha \in I\}$ is a class of languages.

(a) D is a *tell-tale* of L (with respect to $\mathcal{L}$) iff D is finite and for all $L' \in \mathcal{L}$, $D \subseteq L' \subseteq L$ implies $L = L'$.

(b) $\mathcal{L}$ satisfies Angluin's Tell-Tale condition iff every $L \in \mathcal{L}$ has a tell-tale with respect to $\mathcal{L}$.

(c) [17] For all $L \in \mathcal{L}$, we say that w is a *tell-tale cut-off word* for L (with respect to $\mathcal{L}$) iff $\{x \in L : x \leq_{ll} w\}$ is a tell-tale for L (with respect to $\mathcal{L}$).

Essentially, Angluin [2] showed that if a class $\mathcal{L}$ does not satisfy Angluin's tell-tale condition, then it cannot be **TxtEx**-learnable. This result applies even for general classes of r.e. languages, and even for non-recursive learners, though Angluin's stated theorem was only for indexed families and recursive learners.

Jain, Luo and Stephan showed that a class satisfying Angluin's tell-tale condition is enough for **TxtEx**-learnability of automatic classes. In particular they showed that such a learner can have several useful additional properties.

Definition 4. Suppose M is a learner. The notation is as in Definition 1 for memory and hypothesis of M on a text T.

(a) [3] M is said to be *consistent* on L if, for all texts T for L, for all n, $H_{hyp_k^T} \supseteq \text{content}(T[k])$. M is said to be consistent on $\mathcal{L}$ if it is consistent on each $L \in \mathcal{L}$.

(b) [2] M is said to be *conservative* on L if, for all texts T for L, for all k, if $\text{content}(T[k+1]) \subseteq H_{hyp_k^T}$, then $hyp_{k+1}^T = hyp_k^T$. M is said to be conservative on $\mathcal{L}$ if it is conservative on each $L \in \mathcal{L}$.

(c) [28,31] M is said to be *set-driven* if, for all σ and τ in SEQ, if $\text{content}(\sigma) = \text{content}(\tau)$, then $M(\sigma) = M(\tau)$.

When we say that M consistently (conservatively, set-drivenly, etc.) learns $\mathcal{L}$, we mean that M **TxtEx**-learns $\mathcal{L}$, and is consistent (resepctively conservative, set-driven) on $\mathcal{L}$.

Theorem 5 (Jain et al. [17]). *Suppose $\mathcal{L}$ is automatic and satisfies Angluin's tell-tale condition. Then there exists a learner M which is set-driven, consistent and conservative on $\mathcal{L}$ and which **TxtEx**-learns $\mathcal{L}$.*

Note that if an automatic class $\mathcal{L}$ does not satisfy Angluin's tell-tale condition, then there is no learner (even non-recursive learner) which **TxtEx**-learns $\mathcal{L}$. As the tell-tale condition is first-order definable, we get the following corollary:

Corollary 6 (Jain et al. [17]). *It is decidable whether a given family $\{L_\alpha : \alpha \in I\}$ is **TxtEx**-learnable (where the decision algorithm gets as input the alphabet Σ and DFAs for regular set I and the regular set $\{conv(\alpha, x) : x \in L_\alpha\}$).*

A similar characterization for some other learning criteria can also be obtained. Let us consider Finite learning.

Definition 7 (Gold [11]).

(a) M **TxtFin**-learns L (using hypothesis space $\mathcal{H} = (H_\beta)_{\beta \in J}$) iff for all texts T for L, there exists an n and a β such that:
  (i) For $m < n$, $M(T[n]) \in \Gamma \times \{?\}$;
  (ii) For $m \geq n$, $M(T[n]) \in \Gamma \times \{\beta\}$;
  (iii) $H_\beta = L$.

(b) M **TxtFin**-learns $\mathcal{L}$ (using hypothesis space $\mathcal{H}$) iff it **TxtFin**-learns each $L \in \mathcal{L}$ (using hypothesis space $\mathcal{H}$).

(c) **TxtFin** $= \{\mathcal{L} : (\exists M)[M$ **TxtFin**-learns $\mathcal{L}$ using some automatic hypothesis space $\mathcal{H}]\}$.

A useful concept for **TxtFin**-learnability is the concept of characteristic sample.

Definition 8 (Lange and Zeugmann [26], Mukouchi [27]).

(a) A finite set S is a *characteristic sample* for L with respect to the class $\mathcal{L}$ iff
    (i) $S \subseteq L$.
    (ii) For all $L' \in \mathcal{L}$, $S \subseteq L'$ implies $L = L'$.
(b) $\mathcal{L}$ satisfies the characteristic sample property iff every $L \in \mathcal{L}$ has a characteristic sample with respect to $\mathcal{L}$.

Theorem 9 (Jain et al. [17]). *Suppose $\mathcal{L}$ is an automatic class. Then $\mathcal{L} \in$* **TxtFin** *iff it satisfies the characteristic sample property.*

3   Automatic Learners

A learner M is automatic if the function it computes is automatic. That is, the mapping (old mem, datum) $\mapsto$ (new mem, hyp) is automatic. When requiring the learners to be automatic, we add the term **Auto** in the learning criteria (for example **AutoTxtEx** means **TxtEx**-learnability using automatic learners). Besides, we often require some restrictions on the memory. These are mentioned in the following.

Definition 10. Fix a learner M. For a text T, let mem_k^T and hyp_k^T be as in Definition 1.

(a) [33] The learner M is *iterative* if, for all texts T and k, $mem_k^T = hyp_k^T$.
(b) [17] The learner M is *word-size memory limited* if there exists a constant c such that for all T and k, $|mem_k^T| \leq \max(\{|T(m)| : m < k\}) + c$.
(c) [17] The learner M is *hypothesis-size memory limited* if there exists a constant c such that for all T and k, $|mem_k^T| \leq |hyp_k^T| + c$.

We denote the above memory restrictions on a learner by using the terms **It**, **Word** and **Index** in the criteria names. For example, **AutoWordTxtEx** denotes **TxtEx**-learning by an automatic learner with word-size memory limitation. The next theorem shows that requiring learners to be automatic can be very restrictive. This is not only because automatic learners have limited memory: automatic learners cannot even learn classes which can be iteratively learnt.

Theorem 11 (Jain et al. [17]). *The automatic class*

$$\{\{0,1\}^* - \{x\} : x \in \{0,1\}^*\}$$

is **ItTxtEx**-*learnable but not* **AutoTxtEx**-*learnable.*

Theorem 12 (Jain et al. [17]).

(a) **AutoItTxtEx** $\subseteq$ **AutoWordTxtEx** $\subseteq$ **AutoTxtEx**.
(b) **AutoItTxtEx** $\subseteq$ **AutoIndexTxtEx** $\subseteq$ **AutoTxtEx**.
(c) **AutoWordTxtEx** $\not\subseteq$ **AutoIndexTxtEx**.

At the time of writing, it is still open **AutoWordTxtEx** = **AutoTxtEx** and whether **AutoIndexTxtEx** = **AutoItTxtEx**. Furthermore, it is open whether **AutoIndexTxtEx** ⊆ **AutoWordTxtEx**.

However, if the alphabet for languages is unary, then **AutoIndexTxtEx** ⊂ **AutoWordTxtEx** = **AutoTxtEx**.

An interesting class which is automatically learnable is unions of regular patterns languages which have variables only among the last n symbols.

Angluin [1] introduced the pattern languages and Shinohara [32] investigated learnability of the class of pattern languages generated by regular patterns. Automatic classes of pattern languages are a special case of classes of pattern languages generated by regular patterns and the n-th automatic class $\mathcal{P}_n$ is defined as follows: For a finite alphabet Σ, $\mathcal{P}_n$ contains all pattern languages of the form $\alpha b_1 b_2 \ldots b_n$, where $\alpha \in \Sigma^*$ and each $b_1, b_2, \ldots, b_n$ is either a member of Σ or the set Σ^*; for example, if $\Sigma = \{0, 1, 2\}$, then $0112101012\Sigma^*21\Sigma^*$ is an automatic pattern language in $\mathcal{P}_4$.

Theorem 13 (Case et al. [9]). *Suppose $|\Sigma| \geq 3$ and $n > 0$. Then, $\mathcal{L} = \{L_1 \cup L_2 : L_1, L_2 \in \mathcal{P}_n\}$ is **AutoWordTxtEx**-learnable via a learner that, furthermore, for all texts T for a language L outside $\mathcal{L}$, converges to an index for a language L' such that $L - L'$ is finite.*

4   Automatic Learning from Fat Text

A text T is called *fat* (see [29]) if every member of content(T) appears infinitely often in the text. That is, for all $x \in$ content(T), there exist infinitely many n such that $T(n) = x$.

As the automatic learners are very much memory limited, one may expect to overcome some of these limitations using a fat text. In fact that is indeed the case as shown by the following result. For learnability of a class from fat texts, we just require the learnability when the input text is fat, and do not care what happens when the input text is not fat.

Theorem 14 (Jain et al. [17]). *Suppose $\mathcal{L}$ is an automatic class which satisfies Angluin's tell-tale condition. Then there exists a learner M which **AutoWordTxtEx**-learns $\mathcal{L}$ from fat texts.*

Osherson, Stob and Weinstein [29] considered partial learning in which the learner need not converge to a correct hypothesis but instead satisfy the following for any text T for the language L being learnt:

(a) Exactly one hypothesis is output infinitely often — that is, there exists exactly one p such that $hyp_k^T = p$ for infinitely many k;
(b) The unique p which is output infinitely often is a grammar for the input language L.

Theorem 15 (Jain et al. [17]). *Every automatic class $\mathcal{L}$ can be partially learnt by an automatic learner with word-size limited memory from fat texts.*

5   Negative Counterexamples

The model of learning in which the learners get only positive data, as considered in most of the literature on inductive inference, is based on the studies by linguists that children mainly get only positive data. However, this is not entirely true as the children are often told about the errors they make. Thus, there is some negative data, in the form of counterexamples that is given to the children. In this section we consider giving the learner negative counterexamples, if any, to their conjectures. This is given in the form of a separate text, where the new datum is either an appropriate negative counterexample (if it exists) to the previous conjecture or a # (indicating no negative counterexample).

For this, the learner is considered as a mapping from (old mem, new datum, new counter example) to (new mem, new hypothesis). We can then define $M(T[n], T'[n])$ as the pair of memory and conjecture of the learner after having seen the text $T[n]$ and corresponding counterexamples given by $T'[n]$. As this definition is a straightforward generalization of Definition 1, we omit the details of the learner but concentrate on how the counterexample text is defined.

Definition 16 (Jain and Kinber [13]). Suppose M is a learner and $\mathcal{H} = (H_\beta)_{\beta \in j}$ is the hypothesis space used by M.

(a) T' is a counterexample text for M on input text T for a language L iff for all n, where $M(T[n], T'[n]) = (mem, hyp)$,
   if $H_{hyp} \subseteq L$, then $T'(n) = \#$
   if $H_{hyp} \not\subseteq L$, then $T'(n) \in H_{hyp} - L$.

(b) T' is a least-counterexample text for M on input text T for a language L iff for all n, where $M(T[n], T'[n]) = (mem, hyp)$,
   if $H_{hyp} \subseteq L$, then $T'(n) = \#$
   if $H_{hyp} \not\subseteq L$, then $T'(n) = \min(H_{hyp} - L)$.

(c) M **NCEx**-learns a language L (using hypothesis space $\mathcal{H}$) iff for all texts T for L, and all corresponding counterexample texts T' for M on the input text T, $M(T, T')$ converges to a hypothesis hyp such that $H_{hyp} = L$.

(d) M **NCEx**-learns a class $\mathcal{L}$ of languages (using hypothesis space $\mathcal{H}$) iff it **NCEx**-learns each language from $\mathcal{L}$ (using hypothesis space $\mathcal{H}$).

(e) **NCEx** = $\{\mathcal{L} : (\exists M)[M$ **NCEx**-learns $\mathcal{L}$ using some automatic family $\mathcal{H}$ as the hypothesis space$\}$.

One can similarly define **ItNCEx**, **ItLNCEx**, **LNCEx** and other learning criteria (here **LNC** stands for least counterexample).

Theorem 17 (Jain and Kinber [13]). *Suppose $\mathcal{L} = \{L_\alpha : \alpha \in I\}$ is an automatic family.*

(a) *$\mathcal{L} \in$ **AutoItNCEx** via a learner that uses a class preserving hypothesis space, $\mathcal{H} = \{H_\beta : \beta \in J\}$, where the languages in $\mathcal{H}$ are same as that in $\mathcal{L}$, though indexing may be different (with potentially several copies of the same language).*

(b) $\mathcal{L} \in$ **AutoWordNCEx** *via a learner that uses the hypothesis space* $(H_\alpha)_{\alpha \in I}$, *where* $H_\alpha = L_\alpha$.

The learners witnessing the above result are however inconsistent. For consistency, we need least negative counterexamples.

Theorem 18 (Jain et al. [16]). *Every automatic family is* **AutoWordLNCEx**-*learnable via a consistent learner.*

The learner in the above proof however uses a general automatic hypothesis space, which may contain lanugages outside the class $\mathcal{L}$. It can be shown that some automatic class $\mathcal{L}$ cannot be **AutoLNCEx**-learnt using a class preserving automatic hypothesis space.

6   Parallel Learning of Automatic Classes

In parallel learning, the learner simultaneously gets texts for n distinct languages from the target class $\mathcal{L}$, and outputs its conjectures for the corresponding texts. This study was originated for general **TxtEx**-learning by [24] and then studied by [14, 15] for the case of learning automatic families. These learning criteria are denoted by (m, n)-**TxtEx** and (m, n)-**TxtFin**-learning. Here, note that for the above learning criteria, when the input texts are for distinct languages in the class, we require the sequence of conjectures to converge on all the texts, whether the corresponding texts are actually learnt or not.

Furthermore, one can also distinguish the cases of the learner being required to specify the m texts which it learns, and the learner only being required to learn m of the n texts, without any constraint on specifying which texts it learnt. The requirement of specifying the texts which are learnt is denoted by using **Super** in the name of the learning criteria.

Theorem 19 (Jain and Kinber [14, 15]).

(a) *Suppose* $0 < m \le n$, $\mathcal{L}$ *is an automatic family and all except at most* $n - m$ *langauges* $L \in \mathcal{L}$ *have a characteristic sample with respect to* $\mathcal{L}$. *Then* $\mathcal{L}$ *is* (m, n)-**SuperTxtFin**-*learnable.*

(b) *Suppose* $0 < m \le n$ *and* $\mathcal{L}$ *is an automatic class having at least* $2n + 1 - m$ *languages. Then* (m, n)-**SuperTxtFin**-*learnability of* $\mathcal{L}$ *implies that there are at most* $n - m$ *languages in* $\mathcal{L}$ *which do not have a characteristic sample with respect to* $\mathcal{L}$.

For the case when the learner is not required to specify the languages it learns, the characterisation is slightly different.

Theorem 20 (Jain and Kinber [14, 15]).

(a) *Suppose* $0 < m \le n$, $\mathcal{L}$ *is an automatic family, and there exists a subset* $\mathcal{S}$ *of* $\mathcal{L}$ *of cardinality at most* $n - m$ *such that every language in* $\mathcal{L} - \mathcal{S}$ *has a characteristic sample with respect to* $\mathcal{L} - \mathcal{S}$. *Then* $\mathcal{L}$ *is* (m, n)-**TxtFin**-*learnable.*

(b) *Suppose $0 < m \leq n$, $\mathcal{L}$ is an automatic class having at least $2n + 1 - m$ languages. Then (m, n)-**TxtFin**-learnability of $\mathcal{L}$ implies that there exists a subclass $\mathcal{S}$ of $\mathcal{L}$ of cardinality at most $n - m$ such that every language in $\mathcal{L} - \mathcal{S}$ has a characteristic sample with respect to $\mathcal{L} - \mathcal{S}$.*

For finite automatic classes the situation becomes a bit more complicated and full characterization depends on a combinatorial argument. We refer the reader to [15] for full details. Furthermore, one can show a hierarchy for (m, n)-learnability as follows.

Theorem 21 (Jain and Kinber [14,15]).

(a) *Suppose $0 < m < n$. Then, there exists an automatic class $\mathcal{L}$ which is (m, n)-**SuperTxtFin**-learnable but not $(m + 1, n)$-**TxtFin**-learnable.*
(b) *There exists an automatic class $\mathcal{L}$ such that for all $n > 1$, $\mathcal{L}$ is $(n - 1, n)$-**TxtFin**-learnable but not $(1, n)$-**SuperTxtFin**-learnable.*

For **TxtEx**-learnability, (m, n)-**TxtEx**-learnability and (m, n)-**SuperTxtEx**-learnability coincide [15]; thus we give the following results only for (m, n)-**TxtEx**.

Theorem 22 (Jain and Kinber [14,15]).

(a) *Suppose $0 < m \leq n$ and $\mathcal{L}$ is an automatic class. Then $\mathcal{L}$ is (m, n)-**TxtEx**-learnable iff at most $n - m$ languages in $\mathcal{L}$ do not have a tell-tale with respect to $\mathcal{L}$.*
(b) *For $n > 0$ there exists an automatic $(1, n + 1)$-**TxtEx**-learnable class that is not $(1, n)$-**TxtEx**-learnable.*
(c) *For $n > 0$, $(1, n)$-**TxtFin** $\subseteq$ **TxtEx**.*

Jain and Kinber also explore the above model of parallel learning when the learners are automtic. However, here the picture becomes more complicated and full characterization is not yet known.

7   Robust Learning of Automatic Classes

Intuitively, a class of objects is robustly learnable if every computable transformation of the class is learnable. Robust learnability seems a desirable property as it indicates that learning of the class is not due to presence of some artificial coding within the data, but is due to the structure of the class itself (Bārzdiņš, in the 1970s). Bārzdiņš reasoned that if a class is learnable only due to some self-referential property then this self-referential part can be "removed" via computable transformations, and thus the class be transformed into an unlearnable class. This line of research has been explored in various papers such as [6,8,10,12,21,25,30]. However, most of these work were on function learning and there does not seem to be a good definition for robust learning of classes of languages. Using automatic classes and related transformations, [19] explored robust learning of classes of automatic languages. The translations which were considered valid for this are defined as follows:

Definition 23 (Jain et al. [19]). Let an automatic class $(L_\alpha)_{\alpha \in I}$ be given. Let Φ be a first order formula, with a distinguished variable x as a unique free variable, where Φ is allowed to use predicates "$y \in X$" and "$y \in L_\alpha$" along with the set I. Let $\Phi(L)$ be the the language consisting of all strings s such that $\Phi[s/x]$ is true, where X is taken to be L. Φ is an automatic translator (for the automatic family $\mathcal{L}$) if the following conditions hold:

(a) for all languages L, L', if $L \subseteq L'$, then $\Phi(L) \subseteq \Phi(L')$;
(b) for all languages $L, L' \in \mathcal{L}$, if $L \not\subseteq L'$, then $\Phi(L) \not\subseteq \Phi(L')$.

Let $\Phi((L_\alpha)_{\alpha \in I}) = (\Phi(L_\alpha))_{\alpha \in I}$. It is easy to verify that any translation of an automatic family is automatic. Jain et al. [19] considered various properties such as consistency, conservativeness, strong monotonicity and confidence and obtained various charaterizations on when automatic classes are robustly learnable and when some translations of automatic classes are learnable under above constraints. We consider some of these characterizations below.

Theorem 24 (Jain et al. [19]). *Given an automatic class $\mathcal{L} = (L_\alpha)_{\alpha \in I}$, the following are equivalent:*

(a) *Every translation of $\mathcal{L}$ is **TxtEx**-learnable.*
(b) *For all $\alpha \in I$, there exists a $b_\alpha \in I$ such that for all $\beta \in I$, either $L_\beta \not\subseteq L_\alpha$ or there exists a $\gamma \leq_{ll} b_\alpha$ such that $L_\alpha \not\subseteq L_\gamma$ and $L_\beta \subseteq L_\gamma$.*

Theorem 25 (Jain et al. [19]). *Suppose $\mathcal{L}$ is an automatic class all of whose translations are **TxtEx**-learnable. Then, every translation of $\mathcal{L}$ is consistently and conservatively **TxtEx**-learnable iff $\mathcal{L}$ is well founded under inclusion.*

A learner M is said to be *strong monotonic* [22] if, for all texts T, for all $m < n$, if $hyp_m^T \neq ?$ and $hyp_n^T \neq ?$, then $H_{hyp_m^T} \subseteq H_{hyp_n^T}$, where hyp_s^T denotes the hypothesis of M after seeing input $T[s]$.

Theorem 26 (Jain et al. [19]). *Given an automatic class $\mathcal{L} = (L_\alpha)_{\alpha \in I}$, the following are equivalent:*

(a) *Every translation of $\mathcal{L}$ is strong monotonically **TxtEx**-learnable.*
(b) *For all $\alpha \in I$, there exists a $b_\alpha \in I$ such that for all $\beta \in I$, either $L_\alpha \subseteq L_\beta$ or there exists a $\gamma \leq_{ll} b_\alpha$ such that $L_\alpha \not\subseteq L_\gamma$ and $L_\beta \subseteq L_\gamma$.*

Furthermore, every automatic class has some translation which is strong monotonically **TxtEx**-learnable.

Acknowledgements. This survey consists of work done with several authors: John Case, Efim Kinber, Trong Dao Le, Qinglong Luo, Eric Martin, Yuh Shin Ong, Shi Pu, Samuel Seah and Pavel Semukhin.

References

1. Angluin, D.: Finding patterns common to a set of strings. J. Comput. Syst. Sci. **21**(1), 46–62 (1980)
2. Angluin, D.: Inductive inference of formal languages from positive data. Inf. Control **45**, 117–135 (1980)
3. Bārzdiņš, J.: Inductive inference of automata, functions and programs. In: Proceedings of the 20th International Congress of Mathematicians, Vancouver, pp. 455–460 (1974). In: Russian. English translation in American Mathematical Society Translations: Series 2, vol. 109, pp. 107–112 (1977)
4. Blumensath, A., Grädel, E.: Automatic structures. In: 15th Annual IEEE Symposium on Logic in Computer Science (LICS), pp. 51–62. IEEE Computer Society (2000)
5. Case, J., Jain, S., Ong, Y.S., Semukhin, P., Stephan, F.: Automatic learners with feedback queries. J. Comput. Syst. Sci. **80**, 806–820 (2014)
6. Case, J., Jain, S., Ott, M., Sharma, A., Stephan, F.: Robust learning aided by context. J. Comput. Syst. Sci. **60**, 234–257 (2000). (Special Issue for COLT 1998)
7. Case, J., Jain, S., Seah, S., Stephan, F.: Automatic functions, linear time and learning. In: Cooper, S.B., Dawar, A., Löwe, B. (eds.) CiE 2012. LNCS, vol. 7318, pp. 96–106. Springer, Heidelberg (2012)
8. Case, J., Jain, S., Stephan, F., Wiehagen, R.: Robust learning - rich and poor. J. Comput. Syst. Sci. **69**(2), 123–165 (2004)
9. Case, J., Jain, S., Le, T.D., Ong, Y.S., Semukhin, P., Stephan, F.: Automatic learning of subclasses of pattern languages. Inf. Comput. **218**, 17–35 (2012)
10. Fulk, M.: Robust separations in inductive inference. In: 31st Annual IEEE Symposium on Foundations of Computer Science, pp. 405–410. IEEE Computer Society Press (1990)
11. Gold, E.M.: Language identification in the limit. Inf. Control **10**(5), 447–474 (1967)
12. Jain, S.: Robust behaviorally correct learning. Inf. Comput. **153**(2), 238–248 (1999)
13. Jain, S., Kinber, E.: Automatic learning from positive data and negative counterexamples. In: Bshouty, N.H., Stoltz, G., Vayatis, N., Zeugmann, T. (eds.) ALT 2012. LNCS, vol. 7568, pp. 66–80. Springer, Heidelberg (2012)
14. Jain, S., Kinber, E.: Parallel learning of automatic classes of languages. In: Auer, P., Clark, A., Zeugmann, T., Zilles, S. (eds.) ALT 2014. LNCS, vol. 8776, pp. 70–84. Springer, Heidelberg (2014)
15. Jain, S., Kinber, E.: Parallel learning of automatic classes of languages. Accepted for Theoretical Computer Science (2016). Special Issue for ALT 2014
16. Jain, S., Kinber, E., Stephan, F.: Automatic learning from positive data and negative counterexamples (2014). (Manuscript)
17. Jain, S., Luo, Q., Stephan, F.: Learnability of automatic classes. J. Comput. Syst. Sci. **78**(6), 1910–1927 (2012)
18. Jain, S., Luo, Q., Semukhin, P., Stephan, F.: Uncountable automatic classes and learning. Theoret. Comput. Sci. **412**(19), 1805–1820 (2011). Special Issue for ALT 2009
19. Jain, S., Martin, E., Stephan, F.: Robust learning of automatic classes of langauges. J. Comput. Syst. Sci. **80**, 777–795 (2014)
20. Jain, S., Ong, Pu, Y.S., Stephan, F.: On automatic families. In: Arai, T., Feng, Q., Kim, B., Wu, G., Yang, Y. (eds.) Proceedings of the 11th Asian Logic Conference, in Honor of Professor Chong Chitat's 60th birthday, 2009, pp. 94–113. World Scientific (2011)

21. Jain, S., Smith, C., Wiehagen, R.: Robust learning is rich. J. Comput. Syst. Sci. **62**(1), 178–212 (2001)
22. Jantke, K.: Monotonic and non-monotonic inductive inference of functions and patterns. In: Dix, J., Jantke, K.P., Schmitt, P.H. (eds.) NIL 2004. LNCS, vol. 543, pp. 161–177. Springer, Heidelberg (1990)
23. Khoussainov, B., Nerode, A.: Automatic presentations of structures. In: Leivant, Daniel (ed.) LCC 1994. LNCS, vol. 960, pp. 367–392. Springer, Heidelberg (1995)
24. Kinber, E., Smith, C., Velauthapillai, M., Wiehagen, R.: On learning multiple concepts in parallel. J. Comput. Syst. Sci. **50**, 41–52 (1995)
25. Kurtz, S., Smith, C.: A refutation of Barzdins' conjecture. In: Jantke, K.P. (ed.) AII 1989. LNCS, vol. 397, pp. 171–176. Springer, Heidelberg (1989)
26. Lange, S., Zeugmann, T.: Types of monotonic language learning and their characterization. In: Proceedings of the Fifth Annual Workshop on Computational Learning Theory, pp. 377–390. ACM Press (1992)
27. Mukouchi, Y.: Characterization of finite identification. In: Jantke, K. (ed.) Analogical and Inductive Inference. Proceedings of the Third International Workshop, pp. 260–267 (1992)
28. Osherson, D., Stob, M., Weinstein, S.: Learning strategies. Inf. Control **53**, 32–51 (1982)
29. Osherson, D., Stob, S.M., Weinstein, S.: Systems that Learn: an Introduction to Learning Theory for Cognitive and Computer Scientists. MIT Press, Cambridge (1986)
30. Ott, M., Stephan, F.: Avoiding coding tricks by hyperrobust learning. Theoret. Comput. Sci. **284**(1), 161–180 (2002)
31. Schäfer-Richter, G.: Über Eingabeabhängigkeit und Komplexität von Inferenzstrategien. PhD thesis, RWTH Aachen (1984)
32. Shinohara, T.: Polynomial time inference of extended regular pattern languages. In: Goto, E., Furukawa, K., Nakajima, R., Nakata, I., Yonezawa, A. (eds.) RIMS Symposia on Software Science and Engineering. LNCS, vol. 147, pp. 115–127. Springer, Berlin (1982). Kyoto, Japan
33. Wiehagen, R.: Limes-Erkennung rekursiver Funktionen durch spezielle Strategien. J. Inf. Process. Cybern. (EIK) **12**(1–2), 93–99 (1976)

Software Engineering: Methods, Tools, Applications (Invited Talks)

From ESSENCE to Theory Oriented Software Engineering

Sebastian Holtappels[✉], Michael Striewe, and Michael Goedicke

University of Duisburg-Essen, Essen, Germany
{sebastian.holtappels,michael.striewe,michael.goedicke}@uni-due.de

Abstract. The Essence standard combines a kernel and a modelling language for software engineering. It defines dynamic semantics of Essence by a mixture of formal and informal means. This paper presents a uniform formalization of the dynamic semantics based on a graph grammar and discusses various applications of this grammar. It is shown that solid formal foundation is useful for research towards theory oriented software engineering.

1 Introduction

The Essence standard [3] is one result of the work of the "Software Engineering Method and Theory" (SEMAT) Initiative[1]. It combines a kernel and a modelling language both aiming to be a part of a general theory for software engineering.

The SEMAT Initiative designed the kernel as a universal foundation for all existing Software Development Practices thus including only these aspects of software engineering that are common to all practices. The Essence language is used to describe the elements of the kernel and the practices and methods used in a software development process [14].

Besides defining static semantics and graphical notations for a language to express kernel elements and practices, the standard also defines dynamic semantics. This definition happens in a mixture of informal text, object-based data models, and formal function definitions in VDM-SL. While this provides a sufficient view on the general aims and possibilities of the dynamic semantics, it cannot directly be enacted. Hence a uniform formalization of the dynamic semantics using a graph grammar has been created in [13]. It is fully enactable via a graph-transformation engine. Thus it can be used as a basis for several applications which involve formal reasoning about practices, methods, states of software development endeavours, and much more. It is the goal of the theory track within the SEMAT Initiative to gather and explore such applications in order to make software development an engineering discipline which is based on formally founded theories.

This paper summarizes the formalization in Sect. 2 and sketches several applications in the context of theory oriented software engineering in Sect. 3. Related Work is presented in Sect. 4. Section 5 concludes this paper and illustrates possible future directions of Essence.

---

[1] http://www.semat.org/.

© Springer-Verlag Berlin Heidelberg 2016
R.M. Freivalds et al. (Eds.): SOFSEM 2016, LNCS 9587, pp. 43–50, 2016.
DOI: 10.1007/978-3-662-49192-8_4

2   Formal Graph-Based Dynamic Semantics for Essence

The Essence language defines a dynamic semantics to support the enactment of Essence based models. More precisely the dynamic semantics of Essence describes how the Level 0 Model is populated, how the overall state is determined and how to generate guidance (guidance function). The Level 0 Model contains the instances of concrete Essence Elements e.g. an instance of the Requirements Alpha. The name Level 0 Model originates from the OMG Meta Object Facility [2]. Level 0 describes the instance level of a model [2,3].

The overall state is determined by determining the Alpha state of each Alpha. The guidance function is used to create guidance for an Essence user. Therefore the guidance function compiles a list of activities which will progress an Alpha from its current state to a targeted state of the same Alpha [3].

In [13] this dynamic semantics is formalized using a graph grammar. This formal graph-based dynamic semantics of Essence enables additional application possibilities. As foundation for the graph grammar [13] defines a type graph of the relevant subset of the Essence language. The type graph is shown in Fig. 1.

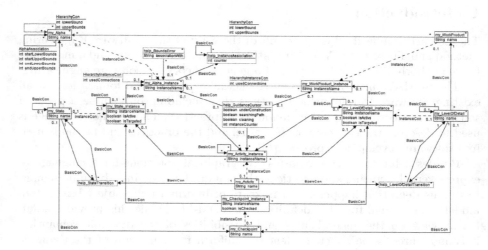

Fig. 1. Type graph of the relevant subset of the Essence language [13]

Based on the areas of the dynamic semantics of Essence [13] defines rules for populating the Level 0 Model, determining the current state of an Alpha and for generating guidance. The area for populating the Level 0 Model is extended by rules for deleting unwanted or deprecated elements. In addition to the formalization of the dynamic semantics [13] includes graph transformation rules for creating the Alphas of the Essence Kernel and the Alphas and other elements of the SCRUM [17] practice [13].

The rules defining the dynamic semantics are designed to be applicable for each practice modeled in the Essence language. Because of this generic design, most of the events and operations described by the dynamic semantics had to be divided into multiple rules. E.g. for the instantiation of one Alpha and the dependent elements, [13] defines one rule to create the instance of an Alpha, another one to create the instances of the states of one Alpha and a rule to create the instances of the checkpoints of this states. Prioritization is used to control which rule can or must be applied in a given state. The lowest priority stage 0 describes rules, that the user should apply manually e.g. the instantiation of an Alpha. Stage 1 includes all rules to be applied automatically e.g. the determination of the current Alpha state. Stages 2 and 3 include rules to ensure the consistency of the model. Stage 4 includes rules for the deletion of elements. This is necessary because multiple rules can be necessary to delete a node, e.g. an Alpha. The higher priority ensures that the deletion is finished before any other rule can be applied. This is necassary to ensure the consistency of the model [13].

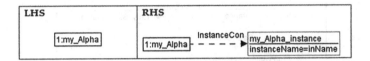

Fig. 2. Instantiation of an Alpha [13]

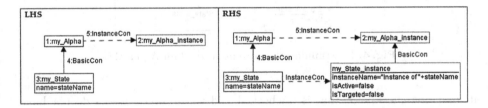

Fig. 3. Instantiation of an Alpha state [13]

Figures 2, 3 and 4 illustrate how the dynamic semantics is defined by graph transformation rules. Figure 2 defines the creation of an instance of an Alpha. To be as generic as possible this rule does not include the creation of any state or association of the instantiated Alpha. Figure 3 defines the creation of an instance of one Alpha state of the instantiated Alpha. This rule is applied until all states of the instantiated Alpha are instantiated. Figures 2 and 3 do not include the conditions under which these rules are applicable [13].

Figure 4 shows the rule for determining the current Alpha state. The state of the Alpha is determined by the checkpoints of its Alpha states. The current Alpha state is that state whose checkpoints are checked and whose successor has

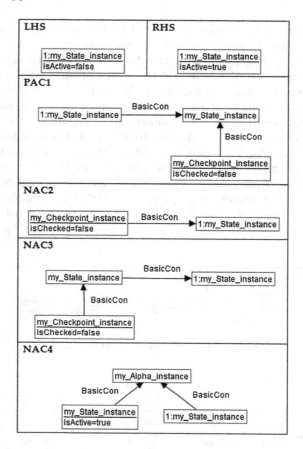

Fig. 4. Determining the current state of an Alpha [13]

at least one checkpoint that is unchecked and whose predecessors do not have any checkpoints that are unchecked. The first thing to be described by a graph transformation rule is the transformation caused when the rule is applied. In this case the transformation is the change of the value of the isActive Attribute from false to true to mark the state as the currently active state. To ensure only the right state is changed the graph transformation rule needs additional conditions under which it is applicable. The first application condition is that the successor of the watched state must have at least one checkpoint, that is not checked (PAC1). In addition neither the watched state nor its predecessor should have a state that is unchecked (isChecked = false) (NAC2 and NAC3). Finally this Alpha instance should not have any other state where isActive is true, to ensure that at each single point in time a state has not more than one state marked as the current state of the referenced Alpha instance [3,13].

In addition to the rule shown in Fig. 4 the determination needs a rule which is applicable for the last state of an Alpha and a rule to change the value of isActive from true to false [13].

3   Applications

As the formalization based on graph transformation rules is fully enactable using some graph transformation engine, it can be applied automatically or semi-automatically in various contexts [4,5,11,15]. In particular, three key areas can be named: Validation and analysis of the Essence language and kernel as such, validation and analysis of some practice or method described using the Essence language, and (retrospective) analysis of an actual software development endeavour. All three areas will be discussed in the following subsections. The applications discussed here are an extension of the applications discussed in [13].

3.1   Language and Kernel Analysis

As in many domains, formalization is valuable on its own right also for the Essence language. This way, it can be shown that the concepts included in the language are sufficient to describe both software engineering practices and methods, and any state that can occur during a software engineering endeavour. It can also be shown that the concept of Alphas and Alpha states is enactable. Moreover, it can be shown that a graph structure is able to describe any state of any software engineering endeavour in terms of the SEMAT concepts, and that thus graph transformations are sufficient to formalize any transition between these states [5,6,19].

3.2   Practice and Method Analysis

For the goal of theory oriented software engineering, the possibility to analyse practices and methods following some formal rules is probably the most valuable effect of the formalization of dynamic semantics.

As a simple application, reachability can be computed to find out whether a particular project state A can be reached using a given practice. In terms of graph grammars, this question is equal to the question whether a particular instance graph can be created from the empty start graph by applying general and practice specific transformation rules [4,7,9].

If reachability is proven for a predefined set of different project states, a practice can be considered complete with respect to the project states, indicating that all these states can be reached in the project. This is particularly interesting for defining minimal methods that should still be able to reach all project targets. In turn, reasoning about reachability can also be used to get clearness about risks in a software development endeavour. In particular, it can be analysed whether a particular state which is considered harmful is reachable or can be avoided by selecting a specific practice [4, 8, 19].

While these applications consider practices and endeavours as a whole, it can also be interesting to dig deeper into specific project slices. For example, activities defined in the Essence language can be associated with competencies. Thus it can be concluded from a static method description which competencies are needed in a team in general. However, simulation or formal reasoning can be used to get a more detailed view on which competencies are needed in which phase of the project or in which situations. In combination with the findings regarding reachability discussed above, this can be used to detect that a specific risky situation is more likely to appear if some competency is missing or that some other competency is (only) required to recover from some bad project state [5, 6, 19].

Instead of analysing just one practice or methods, comparisons between different practices or analyses of changes in one practice can also be performed. Criteria for comparison can be reachability and completeness as above, but also for example the number of reachable (risky) states or the average number of applicable transformation rules per state (where a small number indicates a closely guided process, while a big number implies more flexibility).

3.3 Endeavour Analysis

Besides formal reasoning and simulation as discussed so far, the same graph grammar can also be applied in project management tools in order to support an actually running endeavour. In this way it can help to make the tool adaptive in a way that it can suggest appropriate actions based on the situation at hand. The key point here is that simulations and formal reasoning must make some assumptions. In particular, it can either be assumed that all team members stick to the method all the time, or all possible deviations are computed. However, in an actual endeavour, both is unlikely, but just a non-empty subset of all possible deviations will occur [4, 5, 11, 15].

At the same time, a tool can also describe changes in the current situation in terms of graph transformation rules and thus provide an abstract trace of project progress for further research. In particular, these traces can be compared between different projects following the same method to help research on key factors for success and failure in software development. The knowledge gained this way can in turn be used again for the adaptive behaviour of tools sketched above in a way that predictions and suggestions for the next steps in an actual endeavour are made based on the project history [5, 6, 19].

4   Related Work

Using Graph Transformation as a means to define static and dynamic aspects of a language, especially the semantics of a language is a proven concept as [6, 11, 15, 16, 19] show.

In the particular domain of Software Process Modelling, formalization of enactment is much more rare. For example SPEM 2.0 [18] as an alternative to Essence has been criticised for its lacking enactment support [10,12]. Other standards like ISO/IEC 24744 [1] also do not include uniformly formalized dynamic semantics as we discussed them in this paper.

5 Conclusion

This paper demonstrated a formalization of the dynamic semantics of the Essence language using graph transformation rules. Several applications were sketched and discussed in which such a formalization can be used to get more insight into practices, methods, and software development endeavours. Gaining more formally founded insight is an important step towards a more theory oriented software engineering, in which decisions can be taken on the ground of solid models, previous knowledge, and proven correlations. The theory track of the SEMAT initiative will continue working in this direction.

References

1. Software Engineering Metamodel for Development Methodologies (ISO/IEC 24744)
2. Omg Meta Object Facility (April 2014)
3. Essence - Kernel and Language for Software Engineering Methods (September 2015)
4. Bardohl, R., Taentzer, G., Minas, M., Schürr, A.: Application of Graph Transformation to Visual Languages, pp. 105–180. World Scientific, London (1999)
5. Baresi, L., Heckel, R.: Tutorial introduction to graph transformation: A software engineering perspective. In: Corradini, A., Ehrig, H., Kreowski, H.-J., Rozenberg, G. (eds.) ICGT 2002. LNCS, vol. 2505, pp. 402–429. Springer, Heidelberg (2002)
6. Blostein, D., Schürr, A.: Computing with graphs and graph transformations. Softw.-Pract. Experience 29(3), 197–217 (1999)
7. Bottoni, P., Taentzer, G., Schürr, A.: Efficient parsing of visual languages based on critical pair analysis and contextual layered graph transformation. In: 2000 IEEE International Symposium on Visual Languages (VL 2000), pp. 59–60 (2000)
8. Corradini, A., Montanari, U., Rossi, F., Ehrig, H., Heckel, R., Löwe, M.: Algebraic Approaches to Graph Transformation. Part I: Basic Concepts and Double Pushout Approach, pp. 163–245. World Scientific, London (1997)
9. Ehrig, H., Löwe, M.: Compugraph. Computing by graph transformation. Final report. Technical report, ESPRIT Basic Research Working Group No. 3299, Berlin (1992)
10. Elvester, B., Benguria, G., Ilieva, S.: A comparison of the essence 1.0 and SPEM 2.0 specifications for software engineering methods. In: Proceedings of the Third Workshop on Process-Based Approaches for Model-Driven Engineering (PMDE 2013)
11. Engels, G., Hausmann, J.H., Heckel, R., Sauer, S.: Dynamic meta modeling: a graphical approach to the operational semantics of behavioral diagrams in uml. In: Evans, A., Caskurlu, B., Selic, B. (eds.) UML 2000. LNCS, vol. 1939, pp. 323–337. Springer, Heidelberg (2000)

12. Henderson-Sellers, B., Gonzalez-Perez, C.: The rationale of powertype-based meta-modelling to underpin software development methodologies. In: Proceedings of the 2nd Asia-Pacific Conference on Conceptual Modelling - vol. 43, APCCM 2005, pp. 7–16 (2005)

13. Holtappels, S.: Eine formale Beschreibung der dynamischen Semantik von ESSENCE. Master's thesis, Universität Duisburg-Essen (2014)

14. Jacobson, I., Ng, P.-W., McMahon, P.E., Spence, I., Lidman, S.: The Essence of Software Engineering: Applying the SEMAT Kernel. Addison-Wesley Professional, Reading (2013)

15. Kuske, S.: A formal semantics of UML state machines based on structured graph transformation. In: Gogolla, M., Kobryn, C. (eds.) UML 2001. LNCS, vol. 2185, pp. 241–256. Springer, Heidelberg (2001)

16. Maggiolo-Schettini, A., Peron, A.: A graph rewriting framework for Statecharts semantics. In: Cuny, J., Ehrig, H., Engels, G., Rozenberg, G. (eds.) Graph Grammars and Their Application to Computer Science. LNCS, vol. 1073, pp. 107–121. Springer, Heidelberg (1996)

17. Schwaber, K., Sutherland, J.: The scrum guide. the definitive guide to scrum: the rules of the game (2013)

18. Software and Systems Process Engineering Metamodel Specification (SPEM) Version 2.0, Document formal/2008-04-01 (April 2008). http://www.omg.org/spec/SPEM/2.0/

19. Toffetti, G., Pezzè, M.: Graph transformations and software engineering: success stories and lost chances. J. Vis. Lang. Comput. 24(3), 207–217 (2013)

Incremental Queries and Transformations: From Concepts to Industrial Applications

Dániel Varró(✉)

Department of Measurement and Information Systems,
Budapest University of Technology and Economics,
Magyar tudósok krt. 2, Budapest 1117, Hungary
varro@mit.bme.hu

Abstract. Model-driven engineering (MDE) is widely used nowadays in the design of embedded systems, especially in the automotive, avionics or telecommunication domain. Behind the scenes, design and verification tools in these domains frequently exploit advanced model query and transformation techniques to support various rich tool features. The rapid increase in the size and complexity of system models has drawn significant attention to incremental model query and transformation approaches, which enable fast and incremental reactions to model changes caused by systems engineers or automated design steps. In this paper, I overview two open source Eclipse projects, EMF-INCQUERY and VIATRA, which have been actively used as a basis for developing various academic and industrial tools for critical systems.

Keywords: Model queries · Model transformations · Incremental evaluation · Reactive programming · Software tool qualification

1 Software Tools in Model-Based Systems Engineering

Model-driven engineering plays an increasingly important role in the design of critical embedded and cyber-physical systems in various application domains including automotive, avionics or telecommunication. Advanced design and verification tools aim to *simultaneously improve quality and decrease costs by early validation* to highlight conceptual design flaws well before traditional testing phases in accordance with the correct-by-construction principle. Furthermore, they improve productivity of engineers by automatically synthesizing different design artifacts (source code, configuration tables, test cases, fault trees, etc.) necessitated by certification standards (like DO-178C or ISO 26262).

There are two main trends nowadays in the software tool market of systems engineering. On the one hand, certain market shares are dominated by very few industrial tools (e.g. Matlab Simulink, Dymola, MagicDraw, DOORS) each of which typically supports a specific development stage (requirements engineering, simulation, allocation, test generation, etc.). In order to protect the important intellectual property rights, these tools are of closed nature, which implies such

© Springer-Verlag Berlin Heidelberg 2016
R.M. Freivalds et al. (Eds.): SOFSEM 2016, LNCS 9587, pp. 51–59, 2016.
DOI: 10.1007/978-3-662-49192-8_5

huge tool integration costs for system integrators (like airframers or car manu-
facturers) that can easily exceed the total licensing costs of individual tools. On
the other hand, recent initiatives (like PolarSys) have started to promote the
development of open language standards and the systematic use of open source
software components in tools for critical systems to reduce licensing costs and
the risks of vendor lock-in.

When software tools are used for developing a critical system, the tools themselves need to be validated with the same scrutiny as the system under design
by *software tool qualification*, especially, when no further human checking is carried out on the outputs of such tools. Software tool qualification distinguishes
between *design tools* which, by definition, may introduce new errors to the system and *verification tools* which may fail to reveal existing errors of the system.

Unsurprisingly, software tool qualification is extremely costly due to the high
algorithmic complexity, tightly couple architecture and unexpected feature interaction of such tools. In fact, most companies rather opt for using tools just as aids
to highlight errors quickly and then they carry out the traditional verification&
validation process with thorough simulation and testing. Anyhow, systematic
software engineering techniques to simultaneously improve quality and reduce
the costs of software tool qualification would be highly beneficial. Existing software engineering practices may guarantee the quality of the system itself, but
they frequently *fail to ensure the quality of the software tool* used in systems
engineering. Furthermore, the rapid increase in the size and complexity of systems models introduces significant *scalability challenges* for these tools.

Language engineering aims to provide foundations, techniques and tools for
domain-specific modeling languages to capture the models. *Model transformation
engineering* aims to systematically develop queries and transformations used in
automated code generators, debuggers to process these models. Of course, a
seamless integration of these techniques is needed when developing real tools.

In this paper, I overview two open source Eclipse projects supporting model
query and transformation techniques integrated into in various industrial tools
for model-based systems engineering. EMF-INCQUERY is an incremental model
query framework while VIATRA supports reactive, event-based transformations.
Their scientifically well-founded basis enables *semantic integration of different
tool features* to (i) complement structural integration provided by the component
(plugin) architecture of Eclipse and to (ii) support tool qualification by precise
specification and execution semantics of those features.

2   Incremental Model Queries in EMF-IncQuery

EMF-INCQUERY is an open source Eclipse project[1] to define declarative graph
queries over EMF models [33] without manual coding and execute them efficiently using incremental graph pattern matching techniques over an imperative
programming language such as Java. The benefits of EMF-INCQUERY include:

---

[1] https://www.eclipse.org/incquery.

(i) a high-level and powerful declarative *graph query language* [8,39];
(ii) a highly efficient *incremental query engine* capable of evaluating queries over models with millions of elements [7,39];
(iii) an advanced *integrated development environment* [39] to construct and validate model queries supported by state-of-the-art Xtext tooling.
(iv) its modular architecture enables *easy integration with existing EMF-based modeling tools* [39].

The primary use case for model queries is to support the live validation of well-formedness constraints and design rules of a domain in order to highlight and report inconsistencies as soon as they are introduced. Efficient incremental evaluation is based on adapting Rete networks [13] to change notifications sent by EMF-based models. Additional main use cases include advanced support for incremental calculation and maintenance of base model indexers [39], derived features [26], soft traceability links [14], or incremental view maintenance [12].

Detailed scalability assessment of EMF-INCQUERY is carried out in numerous papers for validation of well-formedness constraints [7,39], detection of source code anti-patterns [40] or maintenance of soft traceability links [14] over models with 10 million elements. Ongoing development within the MONDO European project[2] aims to develop a distributed and incremental query engine [30] deployed over cloud based storages to further improve scalability.

Example. The definition of a sample well-formedness constraint (taken from [20]) for checking valid allocations of application instances to host instances (e.g. in a cloud application or a cyber-physical system) is listed in Fig. 1. The query notAllocatedButRunning captures an erroneous situation for allocation when an application *app* is running, but not allocated to a host instance (using another graph pattern allocatedApplication by negative composition). When checking this constraint on the instance model depicted in Fig. 2, app2 is the only Application-Instance which matches the pattern (thus violates the constraint) since app1 is allocated to a host instance ht1 while app3 is stopped.

```
1  //EMF-IncQuery pattern in the query definition file
2  @Constraint(
3    key = {"app"}, severity = "error"
4    message = "$app.id$ is not allocated but it is running",
5  )
6  pattern notAllocatedButRunning(app : ApplicationInstance) {
7    ApplicationInstance.state(app, ::Running);
8    neg find allocatedApplication(app);
9  }
10
11 private pattern allocatedApplication(app : ApplicationInstance) {
12   ApplicationInstance.allocatedTo(app, _host);
13 }
```

Fig. 1. Sample queries for well-formedness constraints (adapted from [20])

---

[2] http://www.mondo-project.org/.

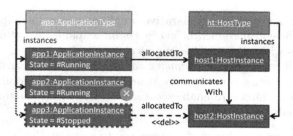

Fig. 2. Sample instance model

By using a @Constraint annotation, EMF-INCQUERY will automatically integrate the query into a model editor using the Eclipse Modeling Framework (EMF) [33] as underlying model representation. As a result, an error marker will immediately be placed on the model whenever this consistency constraint is violated, which is removed automatically once the source of the problem is corrected (e.g. the application instance is stopped or allocated to a host instance).

3   VIATRA: A Reactive Transformation Platform

VIATRA is a reactive, event-driven model transformation platform [6] where transformations are executed continuously as *reactions* to changes of the underlying model. The VIATRA project[3] provides:

(1) An *internal domain-specific language* over Xtend [38] to specify both *batch and event-driven, reactive transformations*.
(2) A *complex event-processing engine* [11] over EMF models to specify reactions upon detecting complex chains of events.
(3) A *rule-based design space exploration* framework [4,15] to explore design candidates as models satisfying multiple criteria over states and trajectories.
(4) A *model obfuscator* to remove sensitive information from a confidential model (e.g. to create bug reports).

VIATRA adopted the principles of reactive programming [5]. The core concept of reactive programming is *event-driven behavior*: components are connected to event sources and their behavior is determined by the event instances observed on *event streams*. Compared to sequential programming, the benefits of reactive programming are remarkable especially in cases when continuous interaction with the environment has to be maintained by the application based on external events without a priori knowledge on their sequence.

The *specification* of a VIATRA transformation program contains (1) *rule specifications* consisting of *model queries*, which serve as a precondition to the transformation, and *actions*, which typically prescribe model manipulations. Furthermore, (2) *execution schemas* are defined in order to orchestrate the reactive

---

[3] http://www.eclipse.org/viatra/.

behavior. VIATRA uses an internal domain-specific language for specifying transformations, i.e. an advanced API over Java and Xtend [38]. Viatra has proven
to be an efficient execution platform for incremental transformations in [20,24].

Example. A sample event-driven transformation rule (adapted from [20]) is
illustrated in Fig. 3, which removes a stopped *ApplicationInstance* from the
model if it is no longer allocated to a host instance. The execution of this rule is
triggered by a disappearance of a match of its precondition pattern stoppedAp-
plInstance. If the application instance is still stopped after the observed change,
then we remove applInst from appType.

When executing the rule over the model of Fig. 2, the rule is triggered when
the allocatedTo reference is removed between app3 and host2. Then the rule action
removes app3 together with the incoming instances reference from app (illustrated
by dotted lines in Fig. 2).

```
1  // Located query definition file .eiq
2  // Finds an application instance which is stopped and allocated to a host instance
3  pattern stoppedApplInstance(appInst: ApplicationInstance,
4              appType:ApplicationType) {
5    ApplicationType.instances(appType, appInst);
6    ApplicationInstance.state(appInst, AppState::Stopped);
7    ApplicationInstance.allocatedTo(appInst, host);
8  }
9  --------------------------------------------------------------------------
10 // Located query definition file .xtend
11 val deleteApplInstanceRule = createRule().name("deleteApplInstance")
12   .precondition(stoppedApplInstance) //a graph pattern as precondition
13   .lifeCycle(ActivationLifecycles.default)
14   //action to be executed when a pattern match gets lost
15   .action(ActivationStates.DISAPPEARED) [
16     // remove application instance from application type
17     if (appInst.state == AppState::Stopped) {
18       appType.remove(ApplicationType_Instances, appInst);
19     }
20   ].build
```

Fig. 3. A sample event-driven transformation rule

4   Selected Recent Applications

The EMF-INCQUERY and VIATRA frameworks have actively been used in different research and industrial projects carried out by various researchers and
practitioners. Below we provide a short overview of selected applications of these
frameworks within our own projects.

- A recent project aimed to define a model-driven approach and tool chain
  for the synthesis of complex, integrated Matlab Simulink models capable of
  simulating the software and hardware architecture of an airplane [14,16].
- As a bi-product of the project, the Massif (Matlab Simulink Integration
  Framework for Eclipse)[4] framework [16,19] was developed, which provides
  a bidirectional bridge between Matlab Simulink models and their EMF model
  counterpart by calling the Matlab API.

---
[4] https://github.com/FTSRG/massif/wiki.

- Formal validation of domain-specific languages is carried out in [29] by using back-end logic solvers where derived features and well-formedness constraints are captured by queries.
- Incremental queries and transformations provide foundations for incremental code generators [18] to avoid complete regeneration in case of small changes.
- Incremental recomputation of graphical views of Sirius [17] can be also be driven by reactive transformations.
- Live detection of human gestures and movements are carried out in [11] by using streaming transformations and complex event processing. Similar technology is used in ongoing work for runtime verification of cyber-physical systems and detecting critical situations in IoT applications [27].

In addition, EMF-INCQUERY and/or VIATRA is known to be *integrated into popular open source modeling tools* such as Papyrus UML [36], Capella [25], mbeddr [3], Sirius [37] or Artop [1].

5   Related Work

There are, of course, other open source technologies which support model queries or transformation used in Eclipse based tooling.

Query Technologies. EMF Model Query 2 [31] provides simple query primitives for selecting model elements that satisfy a set of conditions. The OCL development environment of the Eclipse OCL project [34] provides different ways to edit OCL constraints: an Xtext-based editor for file-based editing, an embedded editor inside Ecore model editors. The Epsilon Validation Language is dedicated to support the construction of validation rules within the Epsilon family [22], while the Acceleo Query Language (AQL) is heavily used within the Sirius project [37] to populate views from underlying models. However, relatively few academic approaches support incremental evaluation [10,28].

Transformation Technologies. The development environment of *EMF-based model transformation tools* provide support for specifying, executing and evaluation of transformations including frameworks such as ATL [32], Henshin [9], QVTo [35] or eMoflon [2]. Many industrial applications rely on Xtend [38] as a code generation and transformation language based on Java. Epsilon [22] provides the Epsilon Transformation Language and the low-level Epsilon Object Language with an advanced execution platform. Recently introduced new features of ATL include target incremental computation [21] combined into the ReactiveATL transformation engine.

6   Conclusions

While a multitude of design are verification tools is used in model-driven systems engineering of critical systems, the complexity of those tools is frequently comparable to the system under design. Certification standards of critical systems necessitate to qualify those tools, i.e. to justify that the tools themselves do not introduce new errors to the design. The complexity of the tools makes tool qualification extremely costly, and provides a strong motivation for solid foundations of integrated tool features. The paper overviews two open source projects, EMF-INCQUERY and VIATRA to serve as a precise and efficient basis by incremental model queries and reactive transformations as illustrated on various industrial applications.

Our ongoing research and development aims to develop systematic approaches to the verification and validation of tool features and language specifications. This primarily includes the automated synthesis of a large, well-formed and diverse set of instance models to serve as test cases or scalability benchmarks. Furthermore, as distinction between design-time and run-time models are being more and more blurred [23] for smart cyber-physical systems, incremental query and transformation techniques will likely be used as part of the underlying middleware, which triggers further open challenges.

Acknowledgments. The author is indebted for the continuous and deep contributions of all contributors of the EMF-INCQUERY and VIATRA project teams. In particular, I would like to highlight the 8+ year involvement of Gábor Bergmann, Ábel Hegedüs, Ákos Horváth, István Ráth and Zoltán Ujhelyi (listed in alphabetic order).

This work was partially supported by the MONDO Project (EU ICT-611125) and the MTA-BME Lendület 2015 Research Group on Cyber-Physical Systems.

References

1. Artop: The AUTOSAR tool platform (2015). https://www.artop.org/
2. eMoflon (2015). http://www.moflon.org/
3. mbeddr (2015). https://mbeddr.com/
4. Abdeen, H., Varró, D., Sahraoui, H., Nagy, A.S., Hegedüs, Á., Horváth, Á., Debreceni, C.: Multi-objective optimization in rule-based design space exploration. In: 29th IEEE/ACM International Conference on Automated Software Engineering (ASE 2014), pp. 289–300. IEEE, Vasteras (2014)
5. Bainomugisha, E., Carreton, A.L., Cutsem, T.V., Mostinckx, S., Meuter, W.D.: A survey on reactive programming. In: ACM Computing Surveys (2012)
6. Bergmann, G., Dávid, I., Hegedüs, Á., Horváth, Á., Ráth, I., Ujhelyi, Z., Varró, D.: VIATRA 3: a reactive model transformation platform. In: Kolovos, D., Wimmer, M. (eds.) ICMT 2015. LNCS, vol. 9152, pp. 101–110. Springer, Heidelberg (2015). http://dx.doi.org/10.1007/978-3-319-21155-8_8
7. Bergmann, G., Horváth, A., Ráth, I., Varró, D., Balogh, A., Balogh, Z., Ökrös, A.: Incremental evaluation of model queries over EMF models. In: Petriu, D.C., Rouquette, N., Haugen, Ø. (eds.) MODELS 2010, Part I. LNCS, vol. 6394, pp. 76–90. Springer, Heidelberg (2010). http://dx.doi.org/10.1007/978-3-642-16145-2_6
8. Bergmann, G., Ujhelyi, Z., Ráth, I., Varró, D.: A graph query language for EMF models. In: Cabot, J., Visser, E. (eds.) ICMT 2011. LNCS, vol. 6707, pp. 167–182. Springer, Heidelberg (2011)
9. Biermann, E., Ermel, C., Taentzer, G.: Precise semantics of EMF model transformations by graph transformation. In: Czarnecki, K., Ober, I., Bruel, J.-M., Uhl, A.,

Völter, M. (eds.) MODELS 2008. LNCS, vol. 5301, pp. 53–67. Springer, Heidelberg (2008)

10. Cabot, J., Teniente, E.: Incremental integrity checking of UML/OCL conceptual schemas. J. Syst. Softw. **82**(9), 1459–1478 (2009)

11. Dávid, I., Ráth, I., Varró, D.: Streaming model transformations by complex event processing. In: Dingel, J., Schulte, W., Ramos, I., Abrahão, S., Insfran, E. (eds.) MODELS 2014. LNCS, vol. 8767, pp. 68–83. Springer, Heidelberg (2014). http://dx.doi.org/10.1007/978-3-319-11653-2_5

12. Debreceni, C., Horváth, A., Hegedüs, A., Ujhelyi, Z., Ráth, I., Varró, D.: Query-driven incremental synchronization of view models. In: 2nd Workshop on View-Based, Aspect-Oriented and Orthographic Software Modelling (VAO 2014), pp. 31:31–31:38. ACM (2014). http://doi.acm.org/10.1145/2631675.2631677

13. Forgy, C.L.: RETE: a fast algorithm for the many pattern/many object pattern match problem. Artif. Intell. **19**(1), 17–37 (1982)

14. Hegedüs, Á., Horváth, Á., Ráth, I., Starr, R.R., Varró, D.: Query-driven soft traceability links for models. Softw. Syst. Model. 1–24 (2014). http://dx.doi.org/10.1007/s10270-014-0436-y

15. Hegedüs, Á., Horváth, Á., Varró, D.: A model-driven framework for guided design space exploration. Autom. Softw. Eng. **22**(3), 399–436 (2015). http://dx.doi.org/10.1007/s10515-014-0163-1

16. Horváth, Á., Hegedüs, Á., Búr, M., Varró, D., Starr, R.R., Mirachi, S.: Hardware-software allocation specification of ima systems for early simulation. In: Digital Avionics Systems Conference (DASC). IEEE, Colorado Springs (2014)

17. Horváth, A., Ráth, I.: IncQuery gets Sirius: faster and better diagrams. In: EclipseCon Europe (2015). https://www.eclipsecon.org/europe2015/session/incquery-gets-sirius-faster-and-better-diagrams

18. Horváth, A., Ráth, I., Hegedüs, A., Balogh, A.: IoT supercharged: complex event processing for MQTT with eclipse technologies. In: EclipseCon France (2015). https://www.eclipsecon.org/france2015/session/decreasing-your-coffee-consumption-incremental-code-regeneration

19. Horváth, A., Ráth, I., Starr, R.R.: Massif - the love child of Matlab Simulink and Eclipse. In: EclipseCon NA (2015). https://www.eclipsecon.org/na2015/session/massif-love-child-matlab-simulink-and-eclipse

20. IncQuery Labs Ltd.: CPS Demonstrator: a model transformation benchmark (2015). https://github.com/IncQueryLabs/incquery-examples-cps/wiki/

21. Jouault, F., Tisi, M.: Towards incremental execution of ATL transformations. In: Tratt, L., Gogolla, M. (eds.) ICMT 2010. LNCS, vol. 6142, pp. 123–137. Springer, Heidelberg (2010)

22. Kolovos, D., Rose, L., Garcia-Domnguez, A., Paige, R.: The Epsilon Book (2015). http://www.eclipse.org/epsilon/doc/book/

23. Lee, E.A., Hartmann, B., Kubiatowicz, J., Rosing, T.S., Wawrzynek, J., Wessel, D., Rabaey, J.M., Pister, K., Sangiovanni-Vincentelli, A.L., Seshia, S.A., Blaauw, D., Dutta, P., Fu, K., Guestrin, C., Taskar, B., Jafari, R., Jones, D.L., Kumar, V., Mangharam, R., Pappas, G.J., Murray, R.M., Rowe, A.: The swarm at the edge of the cloud. IEEE Des. Test **31**(3), 8–20 (2014). http://dx.doi.org/10.1109/MDAT.2014.2314600

24. van Pinxten, J., Basten, T.: Motrusca: interactive model transformation use case repository. In: 7th Doctoral Symposium on Computer Science and Electronics, p. 57 (2014)

25. Polarsys: Capella (2015). https://www.polarsys.org/capella/

26. Ráth, I., Hegedüs, A., Varró, D.: Derived features for EMF by integrating advanced model queries. In: Vallecillo, A., Tolvanen, J.-P., Kindler, E., Störrle, H., Kolovos, D. (eds.) ECMFA 2012. LNCS, vol. 7349, pp. 102–117. Springer, Heidelberg (2012)

27. Ráth, I., Horváth, A.: IoT supercharged: complex event processing for MQTT with eclipse technologies. In: EclipseCon Europe (2015). https://www. eclipsecon.org/europe2015/session/iot-supercharged-complex-event-processing-mqtt-eclipse-technologies

28. Reder, A., Egyed, A.: Incremental consistency checking for complex design rules and larger model changes. In: France, R.B., Kazmeier, J., Breu, R., Atkinson, C. (eds.) MODELS 2012. LNCS, vol. 7590, pp. 202–218. Springer, Heidelberg (2012)

29. Semeráth, O., Barta, A., Horváth, Á., Szatmári, Z., Varró, D.: Formal validation of domain-specific languages with derived features and well-formedness constraints. Softw. Syst. Model. 1–36 (2015). http://dx.doi.org/10.1007/s10270-015-0485-x

30. Szárnyas, G., Izsó, B., Ráth, I., Harmath, D., Bergmann, G., Varró, D.: IncQuery-D: a distributed incremental model query framework in the cloud. In: Dingel, J., Schulte, W., Ramos, I., Abrahão, S., Insfran, E. (eds.) MODELS 2014. LNCS, vol. 8767, pp. 653–669. Springer, Heidelberg (2014)

31. The Eclipse Foundation: EMF Model Query 2 (2012). http://wiki.eclipse.org/EMF/Query2

32. The Eclipse Foundation: ATL (2015). http://www.eclipse.org/atl/

33. The Eclipse Foundation: EMF: The eclipse modeling framework (2015). http://www.eclipse.org/emf

34. The Eclipse Foundation: MDT OCL (2015). http://www.eclipse.org/modeling/mdt/?project=ocl

35. The Eclipse Foundation: Model to model project (2015). http://www.eclipse.org/m2m/

36. The Eclipse Foundation: Papyrus (2015). https://eclipse.org/papyrus/

37. The Eclipse Foundation: Sirius (2015). http://www.eclipse.com/sirius/

38. The Eclipse Foundation: Xtend (2015). http://www.eclipse.org/xtend

39. Ujhelyi, Z., Bergmann, G., Hegedüs, Á., Horváth, Á., Izsó, B., Ráth, I., Szatmári, Z., Varró, D.: EMF-IncQuery: an integrated development environment for live model queries. Sci. Comput. Program. 98, 80–99 (2015). http://dx.doi.org/10.1016/j.scico.2014.01.004

40. Ujhelyi, Z., Szoke, G., Horváth, Á., Csiszár, N.I., Vidács, L., Varró, D., Ferenc, R.: Performance comparison of query-based techniques for anti-pattern detection. Inf. Softw. Technol. 65, 147–165 (2015). http://dx.doi.org/10.1016/j.infsof.2015.01.003

Data, Information, and Knowledge Engineering (Invited Talks)

Big Sequence Management: A glimpse of the Past, the Present, and the Future

Paris Descartes University, Paris, France
themis@mi.parisdescartes.fr

Abstract. There is an increasingly pressing need, by several applications in diverse domains, for developing techniques able to index and mine very large collections of sequences, or data series. Examples of such applications come from biology, astronomy, entomology, the web, and other domains. It is not unusual for these applications to involve numbers of data series in the order of hundreds of millions to billions, which are often times not analyzed in their full detail due to their sheer size. In this work, we describe recent efforts in designing techniques for indexing and mining truly massive collections of data series that will enable scientists to easily analyze their data. We show that the main bottleneck in mining such massive datasets is the time taken to build the index, and we thus introduce solutions to this problem. Furthermore, we discuss novel techniques that adaptively create data series indexes, allowing users to correctly answer queries before the indexing task is finished. We also show how our methods allow mining on datasets that would otherwise be completely untenable, including the first published experiments using one billion data series. Finally, we present our vision for the future in big sequence management research.

Keywords: Data management · Data indexing · Data analytics · Data series

1 Introduction

[Motivation.] Data series have gathered the attention of the data management community for almost two decades [12,35,49]. Data series are one of the most common types of data, and are present in virtually every scientific and social domain: they appear as audio sequences [26], shape and image data [54], financial [47], environmental monitoring [42] and scientific data [22], and they have many diverse applications, such as in health care, astronomy, biology, economics, and others.

Recent advances in sensing, networking, data processing and storage technologies have significantly eased the process of generating and collecting tremendous amounts of data series at extremely high rates and volumes. It is not unusual for applications to involve numbers of sequences in the order of hundreds of millions to billions [1,2].

© Springer-Verlag Berlin Heidelberg 2016
R.M. Freivalds et al. (Eds.): SOFSEM 2016, LNCS 9587, pp. 63–80, 2016.
DOI: 10.1007/978-3-662-49192-8_6

[Data Series.] A *data series*, or *data sequence*, is an ordered sequence of data points[1]. Formally, a data series $T = (p_1, ... p_n)$ is defined as a sequence of points $p_i = (v_i, t_i)$, where each point is associated with a value v_i and a time t_i in which this recording was made, and n is the size (or length) of the series. If the dimension that imposes the ordering of the sequence is time then we talk about *time series*, though, a series can also be defined over other measures (e.g., angle in radial profiles in astronomy, mass in mass spectroscopy, position in genome sequences, etc.).

A key observation is that analysts need to process and analyze a sequence (or subsequence) of values as a single object, rather than the individual points independently, which is what makes the management and analysis of data sequences a hard problem. Note that even though a sequence can be regarded as a point in n-dimensional space, traditional multi-dimensional approaches fail in this case, mainly due to the combination of the following two reasons: (a) the dimensionality is typically very high, i.e., in the order of several hundreds to several thousands, and (b) dimensions are strictly ordered (imposed by the sequence itself) and neighboring values are correlated.

[Need for Data Series Indexing.] In this context, nearest neighbor queries are of paramount importance, since they form the basis of virtually every data mining, or other complex analysis task involving data series. However, nearest neighbor queries across a large collection of data series are challenging, because data series collections grow very large in practice, with datasets including billions, or even trillions of data series [13, 40]. Thus, methods for answering nearest neighbor queries rely on two main techniques: data summarization and indexing. Data series summarization is used to reduce the dimensionality of the data series [3, 15, 27, 28, 32, 34, 39], and then indexes are built on top of these summarizations [5, 39, 46, 49, 52].

Nevertheless, as the data series collections grow in size, the operation of indexing these collections can itself become the bottleneck in the entire process. As an answer to this problem, we have developed the iSAX2.0 [12] and iSAX2+ [13], the first data series indexes that inherently support bulk loading, and thus aim to minimize the index building time. Bulk loading refers to mechanisms that allow us to insert at once a large quantity of data in an index, and as a result lead to fast index-building times. Furthermore, we describe the ADS+ index [57, 58], which is the first data series index than can start answering queries correctly before the entire index has been built. This goal is achieved by building very fast the main-memory part of the index (i.e., only the inner nodes), and deferring the materialization of the (expensive) leaf nodes to query time. This novel approach considerably shrinks the data-to-query gap, allowing users to start answering queries much faster than any previous approach, and enabling truly exploratory analysis on very large data series collections.

---

[1] For the rest of this paper, we are going to use the terms *data series* and *sequence* interchangeably.

[Need for Data Series Management Systems.] There are important reasons why data Series (or Sequence) Management Systems (SMSs) are on the cusp of becoming a focal point for research activity in data management. The solutions that are currently available require custom code and the development of ad hoc systems for various tasks, requiring huge investments in time and effort, and duplication of effort across different teams. Even existing approaches based on DBMSs [7], Column Stores [50], or Array Databases [51]) do not provide a viable solution, since they have not been designed for managing and processing sequence data. Therefore, they do not offer a suitable declarative query language, storage model, auxiliary data structures (such as indexes), and optimization mechanism that can support a variety of sequence query workloads in an efficient manner.

We argue that a SMS is necessary in order to enable big sequence analytics, since it will offer the abstractions, tools, and automations needed for achieving this goal. Just like databases abstracted the relational data management problem and offered a black box solution that is now omnipresent, the proposed system will make it feasible for analysts that are not experts in data series management, as well as common users, to tap in the goldmine of the massive and ever-growing data series collections they (already) have.

[Contributions.] The contributions of this work can be summarized as follows.

- We briefly review the work relevant to data series summarization, and data series indexing. We present in more detail the iSAX summarization method, and discuss how it can be used to construct a data series index. Furthermore, we give an overview of the first data series indexes that support bulk loading, namely, iSAX2.0 and iSAX2+, which lead to index-building times considerably faster than previous approaches, allowing us to index datasets with 1 billion data series.
- We describe the first adaptive data series index, ADS+, which reduces by an additional order of magnitude the time needed by the index before it is ready to start answering queries. The ADS+ index starts by a minimal tree structure based on summarizations of the data series. Then, the index structure is continuously enriched as more queries arrive: each query that is not covered by the current contents of the index, triggers additional data to be brought inside the index, thus adaptively and automatically expanding subtrees in the hot branches of the index. This enables ADS+ to answer several hundreds of thousands of queries by the time that state-of-the-art techniques are still in the index creation phase.
- We argue for the need to develop a general-purpose sequence management system, and discuss the features of such a system: (a) it should be able to cope with big data sequences, that is, massive collections of sequences, which can be heterogeneous (i.e., originate from disparate domains and thus exhibit very different characteristics), and which can have uncertainty in their values (e.g., due to inherent errors in the measurements); (b) it should efficiently support a wide range of sequence queries and mining operations at a scalable fashion, while exploiting the benefits of physical and logical independence; and

(c) it should support cost-based optimization, which will enable the system to automatically pick the right storage and execution strategies for answering different queries.

Paper Organization. The rest of this paper[2] is organized as follows. We structure our discussion in three main sections: we briefly review the main research directions and results in the literature in Sect. 2; we describe the current state of the art in data series indexing in Sect. 3; and we present our vision for the future in Sect. 4. Finally, we conclude in Sect. 5.

Note that the focus of this paper is on the data management problems relevant to massive sequence collections, and not on data mining and analysis, which we do not discuss here. Nevertheless, we argue that in most cases, the correct data management techniques can lead to significant time efficiency benefits for the mining and analysis algorithms.

2 The Past: Summarizations and Indexes

2.1 On Data Series Queries

There are various types of data sequence queries that analysts need to perform: (a) simple Selection-Projection-Transformation (SPT) queries, and (b) more complex Data-Mining (DM) queries. Simple SPT queries are those that select sequences and project points based on thresholds, point positions, or specific sequence properties (e.g., above, first 10 points, peaks), or queries that transform sequences using mathematical formulas (e.g., average). An example SPT query could be one that returns the first x points of all the sequences that have at least y points above a threshold. The majority of these queries could be handled (albeit not optimally) by current database management systems, which nevertheless, lack a domain specific query language that would support and facilitate such processing.

DM queries on the other hand are more complex by nature: the processing has to take into consideration the entire sequence, and treat as a single object, therefore being much more complex to process. Examples under this category are: queries by content (range and similarity queries, nearest neighbors), clustering, classification, outlier patterns, frequent sub-sequences, and others. These queries cannot be supported by current data management systems, since they require specialized data structures, algorithms and storage methods in order to be performed efficiently.

Note that the data series datasets and queries may refer to either static, or streaming data. In the case of streaming data series, we are interested in the subsequences defined by a sliding window. The same is also true for static data series of very large size (e.g., an electroencephalogram, or a genome sequence), which we

---

[2] A more detailed analysis of the topics discussed in this paper can be found in our previous studies [12, 13, 17, 18, 29, 36–38, 57–59].

divide into sub-sequences using a sliding (or shifting window). The length of these sub-sequences is chosen so that it can contain the patterns of interest.

One of the most basic data mining tasks is that of finding similar data series in a database [3]. The query comes in the form of a data series X and it says "find me the data series in the database which is most similar to X". Similarity search is an integral part of most data mining procedures, such as clustering [53], classification and deviation detection [11, 16].

2.2 On Data Series Summarizations

A common approach for answering such queries is to perform a dimensionality reduction, or summarization technique. Several such summarizations have been proposed, such as the Discrete Fourier Transform (DFT) [3], the Discrete Wavelet Transform (DWT) [15], the Piecewise Aggregate Approximation (PAA) [28, 56], the Adaptive Piecewise Constant Approximation (APCA) [14], or the Symbolic Aggregate approXimation (SAX) [34].

Note that recent studies suggest that on average, there is little to differentiate between these summarizations in terms of fidelity of approximation [19, 37] (even though it *is* the case that certain representations favor particular data types, e.g., DFT for star-light-curves, APCA for bursty data, etc.).

These summarizations are usually accompanied by distance bounding functions that relate distances in the summarized space to distances in the original space through either lower or upper-bounding. With such bounding functions, we can index data series directly in the summarized space [5, 39, 46, 49, 52], and use these indexes to efficiently answer nearest neighbor queries on large data series collections.

2.3 On Data Series Indexing

Even though recent studies have shown that in certain cases sequential scans can be performed very efficiently [40], such techniques are only applicable when the database consists of a single, long data series, and queries are looking for potential matches in small subsequences of this long data series. Such approaches, however, do not bring benefit to the general case of querying a mixed database of several data series. Therefore, indexing is required in order to efficiently support data exploration tasks, which involve ad-hoc queries, i.e., the query workload is not known in advance.

A large set of indexing methods have been proposed for the different data series summarization methods, including traditional multidimensional [9, 21, 29, 39] and specialized [5, 46, 49, 52] indexes. Moreover, various distance measures have been presented that work on top of such indexes, e.g., Discrete Time Warping (DTW) and Euclidean Distance (ED).

Indexing can significantly reduce the time to answer DM queries. Nevertheless, recent studies have observed that the mere process of building the index can be prohibitively expensive in terms of time cost [12, 13, 57]: e.g., the process of creating the index for 1 billion data series takes several days to complete. This problem can be mitigated by the bulk loading technique. Bulk-loading has been studied in the context of traditional database indexes, such as B-trees and R-trees, and other multi-dimensional index structures [4,20,23,24,30,43].

In the following section, we give an overview of iSAX 2.0 [12] and iSAX2+ [13], two data series indexes that implement a bulk loading strategy.

2.4   On the iSAX Summarization and Family of Indexes

The Piecewise Aggregate Approximation (PAA) [28,56] is a summarization technique that segments the data series in equal parts and calculates the average value for each segment. An example of a PAA representation can be seen in Fig. 1; in this case the original data series is divided into 4 equal parts. Based on PAA, Lin et al. [34] introduced the Symbolic Aggregate approXimation (SAX) representation that partitions the value space in segments of sizes that follow the normal distribution. Each PAA value can then be represented by a character (i.e., a small number of bits) that corresponds to the segment that it falls into. This leads to a representation with a very small memory footprint, an important requirement for managing very large data series collections. A segmentation of size 3 can be seen in Fig. 1, where the data series is represented with the SAX word "10 10 11".

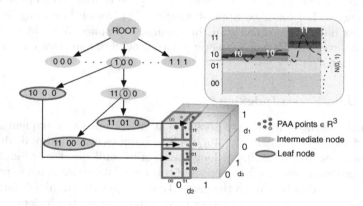

Fig. 1. An example of iSAX and SAX representations [57]

The SAX representation was later extended to indexable SAX (iSAX) [49], which allows variable cardinality for each character of a SAX representation. An iSAX representation is composed of a set of characters that form a word, and each word represents a data series. In the case of a binary alphabet, with a word size of 3 characters and a maximum cardinality of 2 bits, we could have a set of data series (two in the following example) represented with the following words: $00_2 10_2 01_2$, $00_2 11_2 01_2$, where each character has a full cardinality of 2 bits and each word corresponds to one data series. Reducing the cardinality of the second character in each word, we get for both words the same iSAX representation: $00_2 1_1 01_2$ (1_1 corresponds to both 10 and 11, since the last bit is trailed when the cardinality is reduced). By starting with a cardinality of 1 for each character in the root node and by gradually performing splits by increasing the cardinality by one character at a time, one can build a tree index [48, 49]. Such cardinality reductions can be efficiently calculated with bit mask operations.

The iSAX 2.0 and iSAX2+ Indexes. Inserting a large collection of time series into the index iteratively is a very expensive operation, involving a high number of disk I/O operations [12, 13]. This is because for each time series, we have to store the raw data series on disk, and insert into the index the corresponding iSAX representation. In order to speedup the process of building the index, we developed iSAX 2.0 [12] and iSAX2+ [13], the first data series indexes with a bulk loading strategy.

The key idea is to effectively group the data series that will end up in a particular subtree of the index, and process them all together. In order to achieve this goal, we use two main memory buffer layers, namely, the First Buffer Layer (FBL), and the Leaf Buffer Layer (LBL) [13]. The FBL corresponds to the children of the root of the index, while the LBL corresponds to the leaf nodes. The role of the buffers in FBL is to cluster together data series that will end up in the same subtree of the index, rooted in one of the direct children of the root. In contrast, the buffers in LBL are used to gather all the data series of leaf nodes, and flush them to disk.

The algorithm operates in two phases, which alternate until the entire dataset is processed, as follows (for more details, refer to [13]). During Phase 1, the algorithm reads data series and inserts them in the corresponding buffer in the FBL. This phase continues until the main memory is full. Then Phase 2 starts, where the algorithm proceeds by moving the data series contained in each FBL buffer to the appropriate LBL buffers. During this phase, the algorithm processes the buffers in FBL sequentially. For each FBL buffer, the algorithm creates all the necessary internal and leaf nodes, in order to index these data series. When all data series of a specific FBL buffer have been moved down to the corresponding LBL buffers, the algorithm flushes these LBL buffers to disk.

The difference between iSAX 2.0 [12] and iSAX2+ [13] is that the former treats the data series raw values (i.e., the detailed sequence of all the values of the data series) and their summarizations (i.e., the iSAX representations) together, while the latter uses just the summarizations in order to build the index, and only processes the raw values in order to insert them to the correct leaf node. In both cases, the goal is to minimize the random disk accesses, by making sure that the data series that end up in the same leaf node of the index are (temporarily) stored in the same (or contiguous) disk pages. Indeed, the experiments demonstrate that iSAX 2.0 and iSAX2+ significantly outperform previous approaches, reducing the time required to index 1 billion data series by 72 % and 82 %, respectively.

3   The Present: Adaptive Indexing

The target of indexing techniques is to make query processing efficient, so that analysts can repeatedly fire several exploratory queries with quick response times. However, even with a data series index that implements bulk loading, the amount of time required to build the index can be a significant bottleneck: for example, it takes more than a full day to build a state-of-the-art index over a data set of 1 billion data series in a modern server machine [57]. The main cost components of indexing are: (a) reading the data to be indexed, (b) spilling the indexed data and structures to disk, and (c) incurring the computation costs of figuring out where each new data entry belongs to (in the index structure). As the data size grows, the total indexing cost increases dramatically, to a degree where it creates a big and disruptive gap between the time when the data is available and the time when one can actually have access to the data. In fact, as the data grows, the query processing cost increasingly becomes a smaller fraction of the total cost (indexing + querying) [57].

As data sizes grow even bigger, waiting for several days before posing the first queries can be a major show-stopper for many applications both in businesses and in sciences. In addition, firing exploratory queries, i.e., queries which are not known a priori, is becoming quickly a common scenario. That is, in many cases, analysts and scientists need to explore the data before they can figure out what the next query is, or even which experiment to perform next; the output of one query inspires the formulation of the next query, and drives the experimental process.

In this section, we describe the ADS+ index, which enable fast indexing and a low data to query gap, when dealing with very large collections of data series.

3.1   The ADS+ Index

Even though iSAX 2.0 and iSAX2+ can effectively cope with very large data series collections, users still have to wait for extended periods of time before being able to start answering queries. We would instead like to allow users to answer queries much sooner.

The ADS+ index [57] answers this problem by performing only a few basic steps, mainly creating the basic skeleton of the index tree, which contains condensed information on the input data series. As queries arrive, ADS+ fetches data series from the raw data and moves only those data series needed to correctly answer the queries inside the index. Future queries may be completely covered by the contents of the index, or alternatively ADS+ adaptively and incrementally fetches any missing data series directly from the raw data set. When the workload stabilizes, ADS+ can quickly serve fully contained queries while as the workload shifts, ADS+ may temporarily need to perform some extra work to adapt before stabilizing again. In addition, ADS+ does not require a fixed leaf size; it dynamically and adaptively adjusts the leaf size in hot areas of the index; all leaves start with a reasonably big size to guarantee fast indexing times, but the more a given area is queried, the more the respective leaves are split into smaller ones to enhance query times.

Proposed Algorithm. The main intuition (for more details, refer to [57]) is that one can quickly build the index tree using a large leaf size, saving time from very expensive split operations, and rely on queries that are then going to force splits in order to reduce the leaf sizes in the hot areas of the index. ADS+ uses two different leaf sizes: a big build-time leaf size for optimal index construction, and a small query-time leaf size for optimal access costs. This allows us to make future queries benefit from every split operation performed, finding the relevant data by traversing the tree, and not by scanning larger leaves. Initially, the index tree is built as in plain ADS, with a constant leaf size, equal to build-time leaf size. In traditional indexes, this leaf size remains the same across the life-time of the index. In our case, when a query that needs to search a partial leaf arrives, ADS+ refines its index structure on-the-fly by recursively splitting the target leaf, until the target sub-leaf becomes smaller or equal to the query-time leaf size.

Adaptive and on demand leaf splitting allow ADS+ to have both fast index building and fast query processing. It does not waste time on creating fine-grained versions of each sub-tree of the index, but rather concentrates on the parts that are related to the current workload. When queries focus to a subset of the dataset, ADS+ does not need to exhaustively index and optimize all data; it rather concentrates on the most related sub-trees of the index.

Another optimization that gives ADS+ a lightweight behavior is that it delays leaf materialization even further. In particular, when traversing the tree for query processing, which leads to adaptive leaf splitting, ADS+ does not materialize the initial big leaf, nor all the leaves it creates on its way to the target small leaf. For example, when ADS+ needs to split a big leaf X and this results in X being split recursively into n new nodes until we reach the target leaf Z with a small leaf size, ADS+ fully materializes only the leaf Z. For the rest of the leaves, ADS+ uses the partial information contained in the leaves to perform the splits, i.e., the iSAX representations. This results in (a) less computation as opposed to having to split based on raw data, (b) less I/O as SAX representations are much smaller, and (c) it enhances the adaptive behavior of ADS+ as it materializes only the truly interesting data that the queries are targeting.

An example of this process is shown in Fig. 2. Figure 2(a) depicts the state of ADS+ after initialization and before any query has arrived, while Fig. 2(b) shows how a single query results in adaptive splits of the right sub-tree until the target leaf node is fully materialized; intermediate nodes remain in partial mode and with a variable leaf size.

Experimental Results. For the purposes of the experimental evaluation, we implemented from scratch an optimized version of iSAX 2.0 in C and compiled with GCC 4.6.3 under Ubuntu Linux 12.04.2. We used an Intel Xeon machine with 64 GB of RAM and 4x 2 TB, SATA, 7.2K RPM Hard Drives in RAID0. All algorithms are set such as they make maximum use of all available memory.

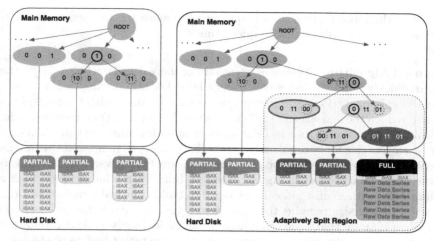

(a) ADS+ after index building.   (b) ADS+ index after a query.

Fig. 2. The ADS+ index [57]

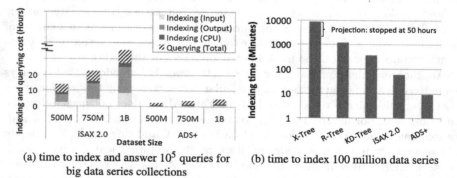

(a) time to index and answer 10^5 queries for   (b) time to index 100 million data series
big data series collections

Fig. 3. Performance comparison between ADS+ and other indexes [57]

We study the behavior up to 1 billion data series and with 10^5 random queries. Regarding leaf sizes, we use the optimal leaf size observed for each index strategy, i.e., 20K for iSAX 2.0, and for ADS+ 2K build-time and 10 query-time leaf size. Figure 3(a) shows the total time needed to build the index and answer all queries. Across all data sizes, ADS+ consistently outperforms iSAX 2.0 by a big margin. For 1 billion data series, ADS+ answers all 10^5 queries in less than 5 h, while iSAX 2.0 needs more than 35 h. By adaptively expanding the tree and adjusting leaf sizes only for the hot workload parts, ADS+ enjoys a $7x$ gain over full indexing in iSAX 2.0. Also, the rate at which the cost of ADS+ grows is significantly smaller than that of iSAX 2.0; For example, going from 500 M to 1 B data series, iSAX 2.0 needs more than twice the time, while ADS+ enjoys a sub-linear cost increase.

One interesting question is how indexes which are tailored for data series search compare against state-of-the-art spatial indexes. In this experiment, we compare ADS+ and iSAX 2.0 against KD-Tree [8], R-Tree [21], and X-Tree [9], which is a state-of-the-art adaptive version of R-Tree. Here, we use a set of 100 million data series. Figure 3(b) depicts the time needed to complete the index building phase for each index. Overall, both data series tailored indexes, iSAX 2.0 and ADS+, significantly outperform the more generic spatial indexes. For example, iSAX 2.0 is one order of magnitude faster than R-Tree while ADS+ is two orders of magnitude faster, and more than an order of magnitude faster than KD-Tree. The raw benefit comes from the fact iSAX 2.0 and ADS+ are tailored to perform efficient comparisons of SAX representations (with bitwise operations). ADS+ being adaptive enjoys further benefits as we discuss in previous experiments as well. X-Tree is significantly slower as a result of its more expensive index building phase which focuses on minimizing overlap between nodes. Naturally, this helps query processing times as less overlap allows queries to focus faster on data of interest. However, as we scale to big data, index building is the main bottleneck and thus X-Tree is prohibitively expensive.

4   The Future: Sequence Management System

Even though analysts in a variety of domains need to manage and process increasingly large data series collections, there is currently no general-purpose solution for the efficient management of sequence datasets. The techniques and tools that are available are rather fragmented, each one addressing only specific and narrow needs.

As a result, the few expert analysts need to invest heavily in the development of customized tools for processing their datasets in order to identify patterns, gain insights, detect abnormalities, and extract useful knowledge, while the many analysts that are not experts are simply not able to process their data. Consider for instance, that for several of their analysis tasks, neuroscientists are currently reducing each of their 3,000 point long sequences to a single number (the global average) in order to be able to analyze their huge datasets [1].

We note that current relational DBMSs [7], Column Stores [50], and Array Databases [51] could eventually be used to store and process sequences. Nevertheless, they cannot efficiently support complex data mining queries, (that is, queries that treat the entire sequence as a single object, such as sequence similarity queries, clustering, classification, etc.), which require fast distance computations among the sequences in the collection, since they do not natively support any mechanisms for pruning the search space.

Consequently, these systems cannot offer optimization functionality for the execution of DM queries, which is a key requirement for efficient processing and analysis of very large sequence collections. Therefore, in this section we argue for the need to design and develop a general-purpose Sequence Management System (SMS).

A key element of a SMS is the design of a cost-based optimizer for the execution of sequence queries, with a special focus on complex data mining

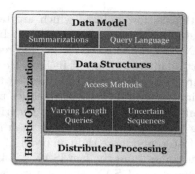

Fig. 4. The architecture of a data series management system

queries. There is currently no optimizer available for sequence queries, even though it is a necessary component for efficient and scalable processing and analytics. As we discuss next, traditional approaches fail in our setting, and therefore, major breakthroughs are needed in this direction.

The optimizer should depend on and be closely related to the storage and indexing solutions for sequences, two research areas that should also be addressed. The design of the data model should accommodate various sequence summarization techniques, including novel techniques for uncertain sequences, and innovative access methods (i.e., storage and indexing) that will be able to adapt to the user needs (i.e., the query workload). Moreover, particular attention should be paid to optimizations specific to data sequence techniques relevant to modern hardware and distributed environments.

In Fig. 4, we illustrate the general architecture of a SMS. We elaborate on the individual components of the system in the following sections. We discuss optimization last, since it touches on the rest of the components, and also include a discussion on the need for a data sequence benchmark.

4.1 Data Model

As we mentioned earlier, neither the relational model nor the array model can adequately capture the characteristics of sequences. In the case of relational data, there are various options available for translating sequences into relations and each one of them has significant limitations. On the other hand, in Array Databases we lack the expressive power to define collections of sequences, and are restricted to defining large multi-dimensional matrices that encode both sequence and meta-data on an equal basis, which hinders efficiency.

An ideal sequence model should instead be able to effectively describe collections of sequences and allow us to do operations on them. It should allow us for example to select sequences based on meta-data or based on their values, project them as complete sequences, or sub-sequences, and join them in a variety of ways for computing calculations. At the same time such a model should intuitively allow for both intra-sequence and inter-sequence aggregations, and be compatible with different sequence summarization methods. Finally, the corresponding query language could be based on previous works [31,44], suitably extended to deal with data series as single objects, as well as with DM queries.

4.2 Data Structures

A large collection of access methods has been proposed in the literature, able to evaluate different queries under various settings, including both indexes and scan-based methods. Recent work in this area is encouraging [13,57], with iSAX2+ demonstrating scalability to dataset sizes 2–3 orders of magnitude more than the current state of the art, and ADS+ exhibiting a further 7-fold improvement in the time to prepare an index on 1 billion data series and answer 100,000 *approximate* queries.

Other promising directions should also be explored, such as methods that rely on fast scans of the data [27,40]. These directions can provide viable alternatives to the indexes discussed above, and in several situations can be the access method of choice. This is especially true given the data management trend on large-scale parallelization, the usage of compression, multi-cores, SIMD architectures and the exploitation of available GPUs [41].

We also propose to extend these techniques along two orthogonal dimensions: supporting queries of varying length, and uncertain sequences. We note that existing techniques only consider collections of data series with the same length, leading to indexes that can answer queries of a fixed (predefined) length. As a result, new access methods that also consider varying length queries have to be developed. Contrary to previous approaches [25], we argue that the information already captured by certain data sequence indexes can be exploited, and is possible to develop new varying-length query answering techniques on top of this.

In several cases, data sequences can be uncertain, that is, the raw data have an inherent uncertainty in their values (e.g., because of errors introduced by the measurement devices), and integrate the solutions to the proposed system. There exist promising studies on modeling and analyzing uncertain sequences [6,45,55], but more work is needed in order to improve the quality and time performance [17]. A promising direction in this respect is the modeling of uncertain sequences with possible world semantics based on full-joint distributions, which can retain the correlation information among neighboring points [18]. Nevertheless, there are still important scalability issues to be overcome in order for such techniques to be used with large sequence collections.

4.3 Distributed Processing

During the last years there has been a lot of research on MapReduce systems, where various methods have been proposed to support the indexing of large multidimensional data [33], where an index is distributed among several compute nodes. Nevertheless, up to this point work on sequential data query processing using MapReduce has mainly concentrated on efficiently performing parallel scans of the complete dataset, while all indexing-related studies only consider read-only operations. Even though various approaches have been proposed for speeding up iterative algorithms, none of the proposed models is a suitable match for the algorithms and techniques we need, where timely communications among workers play a crucial role in reducing the amount of total work done. Therefore, there is need for more work in this area, taking into consideration new paradigms as well [10].

4.4   Cost Based Optimization

As we discussed above, there can be multiple different execution strategies for answering the same query, including the various choices of serial scans, indexes, and processing methods (e.g., parallelization, GPU, etc.). The challenge in choosing the right execution strategy is to estimate the amount of data that such a query will need to access before executing it. For example, a fast parallel SIMD-enabled scan on compressed data might be a better option than the use of a non-optimized index when SIMD instructions are available, but not a better choice when such instructions are not available. All these characteristics have to be exploited by the cost-based optimization models, and considered in a way that is transparent to the user. This problem becomes even more challenging when complex queries involving several operators need to be executed (e.g., consider an analysis task that combines a series of SPT operators as a pre-processing step, and then applies a DM operator).

While in traditional relational databases there are simple and efficient ways in order to estimate query selectivity [7], this is not the case for sequence similarity queries that lie in the heart of most sequence mining algorithms. The challenges in this context arise from the combination of the very high dimensional and sequential nature (i.e., the inherent correlations among neighboring values) of these data.

Up to this point, no efficient methods have been proposed to solve this problem, and ground-breaking work needs to be done. We believe that a promising direction is to carefully study the hardness of a query: being able to control the effort needed to answer a query can be the right step stone for solving the inverse problem, that of estimating the effort it will take to answer a query, before executing it.

4.5   Data Series Benchmarking

Despite the rich literature on methods for indexing and answering similarity queries on data sequences, we note the absence of any related benchmarks. We argue for the need of fair benchmarks that can stress-test sequence processing techniques in a controlled way and to pre-defined levels of query hardness.

Such benchmarks will be designed to capture differences in the quality of summarization methods, indexes and storage methods, when working in *combination*, which is what makes the design of such a benchmark a challenging task. Our ongoing work constitutes the first solution towards this directions: it hows that the amount of effort employed by data series indexes can be consistently captured across different indexing approaches, using implementation-invariant measures [59].

5   Conclusions

In this work, we discussed the state-of-the-art data series indexing approaches that can cope with the data deluge. We reviewed the iSAX 2.0 and iSAX2+ indexes, which are the first specifically designed for very large collections of data series, and use novel algorithms for efficient bulk loading. We also described the first adaptive indexing approach, ADS+, where the index is built incrementally and adaptively, resulting in a very fast initialization process. We experimentally validated the proposed algorithms, including the first published experiments to consider datasets of size up to one billion data series, showing that we can deliver orders of magnitude improvements in the time required to build the index, and to start answering queries.

Furthermore, we observed that even though data series are a very common data type, there is currently no system that can inherently accommodate, manage, and support complex analytics for this type of data. Therefore, in this paper we argue for the special nature of the sequences data type, and articulate the necessity for rigorous work on data series management systems. We propose a sequence management system that will employ a data model specialized to sequences. The system will be distributed by design, and consider the large volume of sequences, their heterogeneity (in terms of properties and characteristics), and possible uncertainty in their values. Finally, the system will support cost-based optimization, thus, leading to the desired scalability for big sequence analytics.

Acknowledgements. I would like to thank my collaborators (in alphabetical order): Alessandro Camerra, Johannes Gehrke, Stratos Idreos, Eamonn Keogh, Michele Linardi, and Yin Lou. Special thanks go to Kostas Zoumpatianos, who has been the driving force behind several of the ideas discussed in this paper.

References

1. Adhd-200 (2011). http://fcon_1000.projects.nitrc.org/indi/adhd200/
2. Sloan digital sky survey (2015). https://www.sdss3.org/dr10/data_access/volume.php
3. Agrawal, R., Faloutsos, C., Swami, A.N.: Efficient similarity search in sequence databases. In: Lomet, D.B. (ed.) FODO 1993. LNCS, vol. 730, pp. 69–84. Springer, Heidelberg (1993)

4. An, N., Kanth, R., Kothuri, V., Ravada, S.: Improving performance with bulk-inserts in oracle r-trees. In: VLDB, pp. 948–951. VLDB Endowment (2003)
5. Assent, L., Krieger, R., Afschari, F., Seidl, T.: The TS-tree: efficient time series search and retrieval. In EDBT (2008)
6. Aßfalg, J., Kriegel, H.-P., Kröger, P., Renz, M.: Probabilistic similarity search for uncertain time series. In: Winslett, M. (ed.) SSDBM 2009. LNCS, vol. 5566, pp. 435–443. Springer, Heidelberg (2009)
7. Astrahan, M.M., Blasgen, M.W., Chamberlin, D.D., Eswaran, K.P., Gray, J., Griffiths, P.P., King, W.F., Lorie, R.A., McJones, P.R., Mehl, J.W., Putzolu, G.R., Traiger, I.L., Wade, B.W., Watson, V.: System R: relational approach to database management. TODS **1**(2), 97–137 (1976)
8. Bentley, J.L.: Multidimensional binary search trees used for associative searching. Commun. ACM **18**(9), 509–517 (1975)
9. Berchtold, S., Keim, D.A., Kriegel, H.P.: The X-tree: an index structure for high-dimensional data. In: VLDB, pp. 28–39 (1996)
10. Bernstein, P., Bykov, S., Geller, A., Kliot, G., Thelin, J.: Orleans: distributed virtual actors for programmability and scalability. MSR-TR-2014-41 (2014)
11. Bu, Y., wing Leung, T., chee Fu, A.W., Keogh, E., Pei, J., Meshkin, S.: Wat: finding top-k discords in time series database. In: SDM, pp. 449–454 (2007)
12. Camerra, A., Palpanas, T., Shieh, J., Keogh, E.: iSAX 2.0: indexing and mining one billion time series. In: ICDM (2010)
13. Camerra, A., Shieh, J., Palpanas, T., Rakthanmanon, T., Keogh, E.J.: Beyond one billion time series: indexing and mining very large time series collections with iSAX2+. KAIS **39**(1), 123–151 (2014)
14. Chakrabarti, K., Keogh, E., Mehrotra, S., Pazzani, M.: Locally adaptive dimensionality reduction for indexing large time series databases. In: SIGMOD (2002)
15. Chan, K.-P., Fu. A.-C.: Efficient time series matching by wavelets. In: ICDE (1999)
16. Chandola, V., Banerjee, A., Kumar, V.: Anomaly detection: a survey. ACM Comput. Surv. **41**(3), 1–58 (2009)
17. Dallachiesa, M., Nushi, B., Mirylenka, K., Palpanas, T.: Uncertain time-series similarity: return to the basics. PVLDB **5**(11), 1662–1673 (2012)
18. Dallachiesa, M., Palpanas, T., Ilyas, I.F.: Top-k nearest neighbor search in uncertain data series. PVLDB **8**(1), 13–24 (2014)
19. Ding, H., Trajcevski, G., Scheuermann, P., Wang, X., Keogh, E.: Querying and mining of time series data: experimental comparison of representations and distance measures. PVLDB **1**, 1542–1552 (2008)
20. Soisalon-Soininen, E., Widmayer, P.: Single and bulk updates in stratified trees: an amortized and worst-case analysis. In: Klein, R., Six, H.-W., Wegner, L. (eds.) Computer Science in Perspective. LNCS, vol. 2598, pp. 278–292. Springer, Heidelberg (2003)
21. Guttman, A.: R-trees: a dynamic index structure for spatial searching. In: SIGMOD (1984)
22. Huijse, P., Estévez, P.A., Protopapas, P., Principe, J.C., Zegers, P.: Computational intelligence challenges and applications on large-scale astronomical time series databases. IEEE Comp. Int. Mag. **9**(3), 27–39 (2014)
23. Van den Bercken, J., Seeger, B.: An evaluation of generic bulk loading techniques. In: VLDB, pp. 461–470 (2001)
24. Van den Bercken, J., Widmayer, P., Seeger, B.: A generic approach to bulk loading multidimensional index structures. In: VLDB (1997)
25. Kadiyala, S., Shiri, N.: A compact multi-resolution index for variable length queries in time series databases. KAIS **15**(2), 131–147 (2008)

26. Kashino, K., Smith, G., Murase, H.: Time-series active search for quick retrieval of audio and video. In: ICASSP (1999)
27. Kashyap, S., Karras, P.: Scalable knn search on vertically stored time series. In: KDD (2011)
28. Keogh, E., Chakrabarti, K., Pazzani, M., Mehrotra, S.: Dimensionality reduction for fast similarity search in large time series databases. KAIS **3**(3), 263–286 (2000)
29. Keogh, E.J., Palpanas, T., Zordan, V.B., Gunopulos, D., Cardle, M.: Indexing large human-motion databases. In: VLDB, pp. 780–791 (2004)
30. Arge, L., Hinrichs, K.H., Vahrenhold, J., Vitter, J.V.: Efficient bulk operations on dynamic R-trees. Algorithmica **33**(1), 104–128 (2002)
31. Lerner, A., Shasha, D.: Aquery: query language for ordered data, optimization techniques, and experiments. In: VLDB (2003)
32. Li, C.S., Yu, P., Castelli, V.: Hierarchyscan: a hierarchical similarity search algorithm for databases of long sequences. In: ICDE (1996)
33. Liao, H., Han, J., Fang, J.: Multi-dimensional index on hadoop distributed file system. In: NAS (2010)
34. Lin, J., Keogh, E., Lonardi, S.: A symbolic representation of time series, with implications for streaming algorithms. In: DMKD (2003)
35. Lin, J., Khade, R., Li, Y.: Rotation-invariant similarity in time series using bag-of-patterns representation. J. Intell. Inf. Syst. **39**(2), 287–315 (2012)
36. Palpanas, T.: Data series management: the road to big sequence analytics. SIGMOD Rec. **44**(2), 47–52 (2015)
37. Palpanas, T., Vlachos, M., Keogh, E.J., Gunopulos, D.: Streaming time series summarization using user-defined amnesic functions. IEEE Trans. Knowl. Data Eng. **20**(7), 992–1006 (2008)
38. Palpanas, T., Vlachos, M., Keogh, E.J., Gunopulos, D., Truppel, W.: Online amnesic approximation of streaming time series. In: ICDE, pp. 339–349 (2004)
39. Rafiei, D., Mendelzon, A.: Similarity-based queries for time series data. In: SIGMOD (1997)
40. Rakthanmanon, T., Campana, B.J.L., Mueen, A., Batista, G., Westover, M.B., Zhu, Q., Zakaria, J., Keogh, E.J.: Searching and mining trillions of time series subsequences under dynamic time warping. In: KDD (2012)
41. Raman, V., Attaluri, G.K., Barber, R., Chainani, N., Kalmuk, D., KulandaiSamy, V., Leenstra, J., Lightstone, S., S. Liu, S., Lohman, G.M., Malkemus, T., Müller, R., Pandis, I., Schiefer, B., Sharpe, D., Sidle, R., Storm, A.J., Zhang, L.: DB2 with BLU acceleration: so much more than just a column store. PVLDB **6**(11), 1080–1091 (2013)
42. Raza, U., Camerra, A., Murphy, A.L., Palpanas, T., Picco, G.P.: Practical data prediction for real-world wireless sensor networks. IEEE Trans. Knowl. Data Eng. **27**(8), 2231–2244 (2015)
43. Choubey, R., Chen, L., Rundensteiner, E.A.: GBI: a generalized R-tree bulk-insertion strategy. In: Güting, R.H., Papadias, D., Lochovsky, F.H. (eds.) SSD 1999. LNCS, vol. 1651, pp. 91–108. Springer, Heidelberg (1999)
44. Sadri, R., Zaniolo, C., Zarkesh, A.M., Adibi, J.: A sequential pattern query language for supporting instant data mining for e-services. In: VLDB (2001)
45. Sarangi, S.R., Murthy, K.: DUST: a generalized notion of similarity between uncertain time series. In: KDD (2010)
46. Schäfer, P., Högqvist, M.: SFA: a symbolic fourier approximation and index for similarity search in high dimensional datasets. In: EDBT (2012)
47. Shasha, D.: Tuning time series queries in finance: case studies and recommendations. IEEE Data Eng. Bull. **22**(2), 40–46 (1999)

48. Shieh, J., Keogh, E.: iSAX: disk-aware mining and indexing of massive time series datasets. DMKD **19**(1), 24–57 (2009)
49. Shieh, J., Keogh, E.J.: *i*SAX: indexing and mining terabyte sized time series. In: KDD, pp. 623–631 (2008)
50. Stonebraker, M., Abadi, M., Batkin, D.J., Chen, J. X., Cherniack, M., Ferreira, M., Lau, E., Lin, A., Madden, S., O'Neil, E.J., O'Neil, P.E., Rasin, A., Tran, N., Zdonik, S.B.: C-store: a column-oriented DBMS. In: VLDB (2005)
51. Stonebraker, M., Brown, P., Poliakov, A., Raman, S.: The architecture of SciDB. In: Bayard Cushing, J., French, J., Bowers, S. (eds.) SSDBM 2011. LNCS, vol. 6809, pp. 1–16. Springer, Heidelberg (2011)
52. Wang, Y., Wang, P., Pei, J., Wang, W., Huang, S.: A data-adaptive and dynamic segmentation index for whole matching on time series. PVLDB **6**(10), 793–804 (2013)
53. Warren Liao, T.: Clustering of time series data - a survey. Pattern Recogn. **38**(11), 1857–1874 (2005)
54. Ye, L., Keogh, E.J.: Time series shapelets: a new primitive for data mining. In: KDD (2009)
55. Yeh, M., Wu, K., Yu, P.S., Chen, M.: PROUD: a probabilistic approach to processing similarity queries over uncertain data streams. In: EDBT (2009)
56. Yi, B., Faloutsos, C.: Fast time sequence indexing for arbitrary Lp norms. In: VLDB (2000)
57. Zoumpatianos, K., Idreos, S., Palpanas, T.: Indexing for interactive exploration of big data series. In: SIGMOD (2014)
58. Zoumpatianos, K., Idreos, S., Palpanas, T.: RINSE: interactive data series exploration with ADS+. PVLDB **8**(12), 1912–1923 (2015)
59. Zoumpatianos, K., Lou, Y., Palpanas, T., Gehrke, J.: Query workloads for data series indexes. In: KDD (2015)

Pay-as-you-go Data Integration: Experiences and Recurring Themes

Norman W. Paton[1]($\boxtimes$), Khalid Belhajjame[2], Suzanne M. Embury[1],
Alvaro A.A. Fernandes[1], and Ruhaila Maskat[1]

[1] School of Computer Science, University of Manchester,
Oxford Road, M13 9PL Manchester, UK
{npaton,suzanne.m.embury,alvaro.a.fernandes}@manchester.ac.uk
[2] Université Paris Dauphine, Place du Marchal de Lattre de Tassigny,
75775 Paris Cedex 16, France
Khalid.Belhajjame@dauphine.fr

Abstract. Data integration typically seeks to provide the illusion that data from multiple distributed sources comes from a single, well managed source. Providing this illusion in practice tends to involve the design of a global schema that captures the users data requirements, followed by manual (with tool support) construction of mappings between sources and the global schema. This overall approach can provide high quality integrations but at high cost, and tends to be unsuitable for areas with large numbers of rapidly changing sources, where users may be willing to cope with a less than perfect integration. Pay-as-you-go data integration has been proposed to overcome the need for costly manual data integration. Pay-as-you-go data integration tends to involve two steps. Initialisation: automatic creation of mappings (generally of poor quality) between sources. Improvement: the obtaining of feedback on some aspect of the integration, and the application of this feedback to revise the integration. There has been considerable research in this area over a ten year period. This paper reviews some experiences with pay-as-you-go data integration, providing a framework that can be used to compare or develop pay-as-you-go data integration techniques.

1 Introduction

Data integration brings together data from multiple sources, in ways that isolate users from inconsistent representations. Data integration has been seen as an important area for decades, as commercial organisations often find themselves with large numbers of databases, whose combined use can be important for data analysis [5]. More recently, the growing interest in big data has given rise to the realisation that data wrangling – the process of combining and cleaning the data sets that are required for analysis – is an important, and expensive, part of many big data projects [28].

In classical data integration, data integration and domain experts work together, with tool support, to capture the data requirements of an application,

© Springer-Verlag Berlin Heidelberg 2016
R.M. Freivalds et al. (Eds.): SOFSEM 2016, LNCS 9587, pp. 81–92, 2016.
DOI: 10.1007/978-3-662-49192-8_7

and to identify how data from different sources can be combined to meet these requirements. This approach, with significant expert input, is at the high-cost, high-quality end of the spectrum, and is suitable for, and targeted at, reasonably stable enterprise environments.

The classical approach is less well suited to settings in which: there are enormous numbers of sources; sources come, go or change rapidly; there are diverse or unstable requirements; or there is no budget for employing data integrators. Such settings are not uncommon. For example, in many domains there may be hundreds or thousands of potentially relevant data sets on the web, from which structured representations can be obtained using data extraction techniques [20]. In such settings, systematic manual data integration that produces a perfect solution is not a practical proposition. For example, consider an e-commerce company that is interested in price comparison with competitors; relevant data sources come and go on a daily basis, and both format and contents change regularly. A typical online retailer will struggle to manually integrate the relevant sources in order to support well-informed decisions.

Pay-as-you-go data integration, sometimes referred to as dataspaces [22], has been proposed as an alternative to the classical approach. A range of proposals have been made for pay-as-you-go approaches [23], which tend to involve: *Initialisation:* automatic creation of integrations that are generally of poor quality; followed by *Improvement:* the obtaining of feedback on some aspect of the integration, and the application of this feedback to revise the integration. Feedback may be: *explicit,* e.g. annotations on correct/incorrect result values; or *implicit,* e.g. inferring matches or rankings from query logs.

Although there have been a good many proposals for pay-as-you-go data integration techniques, we know of little work on methodologies to enable their systematic development. In this paper, we identify some themes that have recurred across multiple proposals, and describe how these themes can be used to characterise the behaviour of several representative proposals.

The paper is structured as follows. Section 2 outlines the key challenges that may need to be faced by a data integration process. Section 3 presents the main contribution of the paper, in the form of a framework that captures recurring themes that can be used for designing or comparing pay-as-you-go techniques. This framework is then illustrated in practice to describe a collection of proposals in Sect. 4. Some conclusions and areas for further investigation are provided in Sect. 5.

2   Data Integration

The overall task of data integration can be considered to consist of a series of steps, as illustrated in Fig. 1. For certain integration activities, some of these steps may not be required, there may be extra steps, or the process may be iterative. However, these are common components of an integration lifecycle:

Source Selection identifies data sources that may be relevant to a data integration task (e.g. [17]).

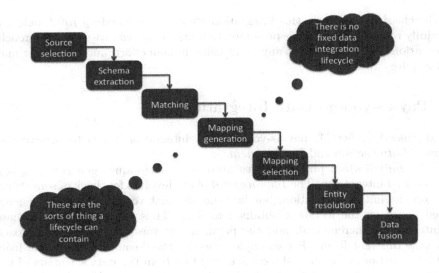

Fig. 1. Abstract data integration lifecycle

Schema Extraction identifies recurring structures (and the data that conform
   to them) in the deep web (e.g. [20]) or in sources that do not conform to
   formal schemas (e.g. in linked open data [12]).
Matching identifies correspondences between elements in different schemas, for
   example suggesting that an attribute in one represents the same notion as
   an attribute in another (e.g. [33]).
Mapping Generation produces queries that can be used to translate data
   from one schema to another (e.g. [19]).
Mapping Selection chooses between the generated mappings, to identify a
   subset that is correct and/or meets the requirements of the application
   (e.g. [7]).
Entity Resolution identifies duplicate instances within a data collection
   (e.g. [18]).
Data fusion combines information from duplicate instances to create the
   instances of a target representation (e.g. [6]).

These steps can be carried out automatically, manually or semi-automatically. In automated approaches, algorithms generate candidate solutions;
for example, for *Matching* syntactic similarity measures can be used to compare
schema elements, and for *Mapping Generation* alternative mappings can be generated that take into account the results from *Matching*. In manual approaches,
human experts create solutions by exploring the relevant information using
generic tools; this is likely to be inefficient in practice, as human decision-making
can be informed by the results of automated analyses. As a result, classical data
integration is a semi-automatic process, in which, for example, candidate matches
and mappings from automated techniques are reviewed and revised by experts.

In the classical approach, this integration effort is expended *up front*, before a carefully refined integration is presented to users. In the *pay-as-you-go* approach, integrations can be refined at any point using human effort, and that effort may not require experts.

3   Pay-as-you-go Data Integration

As discussed in Sect. 1, pay-as-you-go data integration tends to involve two phases, *Initialisation* and *Improvement*.

The *Initialisation* phase involves automated techniques generating a *best effort* initial integration. The *Improvement* phase involves feedback of some type, first on the initial integration, but later on the best version that can be generated based on the feedback obtained to date. Thus the *Improvement* phase is intrinsically incremental, and the payment for the pay-as-you-go approach can take different forms. For example, consider the e-commerce example from the introduction. One form of feedback could be from the data scientists of the e-commerce company, who annotate the different sites from which data has been retrieved using a relevance score. This form of feedback requires knowledge of the price comparison task, but not knowledge of data integration, and the payment is in the form of the time of the data scientist. Another approach to feedback could crowdsource information on entity resolution (e.g. [35]). This form of feedback requires the ability to recognise which products are the same, which might be considered to involve a common-sense comparison, and the payment is in the form of money to the crowd workers.

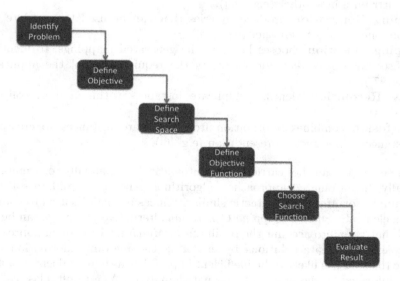

Fig. 2. Steps in the pay-as-you-go data integration process

In this paper we focus on the *Improvement* phase of pay-as-you-go data integration, and in particular discuss recurring features in the design of pay-as-you-go techniques that can be used both to characterise existing proposals and to design new ones. Recurring features of pay-as-you-go proposals are illustrated in Fig. 2, and discussed below; examples of each of these features for a series of case studies are provided in Sect. 4.

Identify problem. Individual proposals tend to relate to a single data integration step from Fig. 1, and sometimes to a specific feature within a step.

Define objective. There is a need to characterise what constitutes a good solution to the problem; this may be in the form of a generic measure, such as precision or recall, or using a metric that is specific to the problem.

Define search space. The *Improvement* phase of pay-as-you-go data integration typically refines the automatic technique used for *Initialisation*. The automatic technique uses an algorithm to generate candidate solutions. The search for candidate solutions must in some way be able to take into account the *objective*.

Define objective function in terms of feedback. The objective function is used in the search for effective solutions to assess the effectiveness of the solutions in terms of the feedback. This in turn involves pinning down the type of feedback required.

Choose search function. The search function is an algorithm that, given some feedback, explores alternative solutions to the *problem*, in a way that seeks to maximise (or minimise) the objective function in terms of the feedback.

Evaluate result. As the objective function in terms of the feedback always approximates the objective, it is important to assess empirically how much feedback is needed to allow the search to identify well behaved solutions.

4   Pay-as-you-go Case Studies

In this section, we revisit several proposals for the *improvement* stage of pay-as-you-go data integration proposals in the light of the features from Sect. 3, with a view to showing how these steps capture their key features.

4.1   Mapping Selection

In this section, we show how the framework can be applied to characterise the selection of mappings that together meet user-specified quality requirements [3].

Identify problem. Given a set of matches, it is possible to automatically generate a set of candidate mappings. For example, the following were among the mappings generated by a commercial schema mapping tool for populating a table with schema *(name, country, province)*.

M1 = SELECT name, country, province from Mondial.city
M2 = SELECT city, country, province from Mondial.located

The problem can be defined as follows: given a set of candidate mappings, and feedback on their results, identify the subset that best meets the users requirements in terms of precision and recall. Thus in this problem, we assume that the user may be willing to trade off precision (the fraction of the returned result that is correct) with recall (the fraction of the correct results that are returned).

Define objective. Following on from the problem statement, we can identify several different objectives, here cast as constrained optimization problems. For a set of candidate mappings M:

Variant 1:

maximise (for some s $\subseteq$ M) precision(s)
such that recall(s) > threshold

Variant 2:

maximise (for some s $\subseteq$ M) recall(s)
such that precision(s) > threshold

Thus we assume that the user can specify the extent to which they can tolerate a reduction in quality along one dimension, and then the objective is to do as well as possible on the other. For example, if the user thinks that they can tolerate one in five of the results being incorrect, then *Variant 2* would be used, with a threshold of 0.8.

Define search space. Here the search space is the set of all subsets of the set of the candidate mappings. A set with n elements has 2^n subsets.

Define objective function in terms of feedback. The objective is defined in terms of the precision and recall of a set of mappings. The precision and recall depend on the ground truth, but we do not know the ground truth. Thus the ground truth needs to be estimated based on the feedback. As a result, here we assume that the feedback takes the form of user annotations that tuples in mapping results are correct (*true positives*) or incorrect (*false positives*).

For example, the precision of a mapping m in the context of user feedback UF, can be estimated by counting the true positives (tp) and false positives (fp) in UF:

$$precision(m, UF) = \frac{|tp(m,UF)|}{|tp(m,UF)|+|fp(m,UF)|}$$

where the function $tp(m, UF)$ (resp. fp) returns the set of tuples from the result of m that are annotated as true positives (resp. false positives) in UF. The precision of a set of mappings can be estimated in an analogous manner.

Choose search function. Different search functions could be used to explore the sets of possible mappings; in the original paper a Mesh Adaptive Direct Search is employed [3].

Evaluate result. In the original paper [3], the results were evaluated to identify how much feedback was required to enable the search to reliably identify collections of mappings that meet the objectives. The experiments showed that results were unreliable until enough feedback has been obtained for each mapping to enable a dependable estimate of its precision and recall to be obtained, but that suitably reliable estimates could often be obtained with feedback on modest numbers of tuples per mapping (in the empirical study, this was typically around 10).

4.2 Entity Resolution

In this section, we show how the framework can be applied to characterise the pay-as-you-go configuration of entity resolution.

Identify problem. Entity resolution is the task of identifying different records that represent the same entity. It is also known as duplicate detection, instance identification and merge-purge [18]. As pairwise record comparison is $O(n^2)$ on the number of records, entity resolution tends to involve both:

1. *Blocking*: fast but approximate identification of candidate pairs; and
2. *Clustering*: more careful but costly grouping of candidate pairs into clusters, where each cluster is intended to contain all the records that represent a single entity.

Both Blocking and Clustering have control parameters such as thresholds, and clustering has a distance function. The problem can be defined as follows: given feedback on pairs of records that indicate if they represent the same entity, identify control parameter settings that lead to the most effective assignments of records to clusters.

Define objective. The objective is to maximise the correctness of the assignments of records to clusters, taking into account which records should be clustered together and which should not.

Define search space. The search space is the set of configuration parameters used by the underlying entity resolution strategy; in our work, we have built on the proposal of Costa *et al.* [13]. This particular proposal has 8 control parameters and a set of weights in a distance function, such that there is one weight per matching attribute; thus a typical search space contains at least 12 numerical dimensions.

Define objective function in terms of feedback. The objective is defined in terms of the fraction of the values that have been correctly clustered together, which depends on the ground truth, which is not available. Thus the ground truth needs to be estimated based on the feedback; we use the following measure of correctness, which requires that the feedback takes the form of *match* or *unmatch* annotations on pairs of records. The correctness of a clustering C in the context of user feedback UF, can be estimated by counting the extent to which the expectations in the feedback are met in the clustering, thus:

$$correctness(C, UF) = \frac{|mm(C,UF)| + |uu(C,UF)|}{|mm(C,UF)| + |uu(C,UF)| + |mu(C,UF)| + |um(C,UF)|}$$

where $mm(C, UF)$ returns the matched records in the feedback that appear together in clusters, $uu(C, UF)$ returns the unmatched records in the feedback that do not appear together in clusters, $mu(C, UF)$ returns the matched records in the feedback that do not appear together in clusters, and $um(C, UF)$ returns the unmatched records in the feedback that appear together in clusters.

Choose search function. Different search functions could be used to explore the space of configuration parameters; in our case we have used an evolutionary search.

Evaluate result. For entity resolution, there are a number of standard test data sets; we have evaluated our approach using several of them [29]. The results showed that even with feedback on a few percent of the records, substantial improvements in correctness can be observed.

4.3   Grouping Users

In pay-as-you-go data integration tasks, the feedback obtained from users may be subjective. For example, for one user looking for holidays, only beach holidays would be suitable participants in an answer (and thus for the user *true positives* in a mapping result), whereas a beach holiday is likely to be seen as a *false positive* if presented as an option to someone who is interested in going skiing.

In this section, we show how the framework can be applied to support a pay-as-you-go approach to grouping users, with a view to sharing feedback [4].

Identify problem. Given a collection of users with different interests, can we cluster these users in a way that allows the sharing of feedback, and thus more cost-effective pay-as-you-go integration?

Define objective. Clusters of users need to be produced that have the property that a better integration can be obtained by using the feedback of all the users in the cluster to inform the integration, than when using only the feedback of the user.

Lets assume that we are interested in sharing feedback for mapping selection, as described in Sect. 4.1. As mapping selection depends on estimates of precision and recall that use feedback, the objective is to cluster users based on their consistency in terms of precision.

Define search space. The search space is the set of possible clusters.

Define objective function in terms of feedback. Clustering depends on a distance function. In this case, the distance between users is defined as the average difference in the precision estimates obtained for mappings for which they have provided feedback:

$$distance(u_i, u_j) = \frac{\sum_{k=1}^{n} precision(m_k, UF_{u_i}) - precision(m_k, UF_{u_j})}{n}$$

where u_i, u_j are different users, each m_k is a candidate mapping, and *precision* estimates the precision of a mapping for a given user's feedback, using the definition from Sect. 4.1.

Choose search function. A hierarchical clustering algorithm was used.

Evaluate result. Experimental results showed that, when a user was within a distance of 0.1 of the centroid of a cluster, the feedback of the cluster was almost as valuable as feedback provided by the user.

5   Conclusions

Pay-as-you-go integration shows promise as a paradigm, at least in part because there seems to be no alternative in increasingly prominent cases. For data integration tasks where there are numerous sources, these sources change rapidly, or there is little budget for manual integration, the pay-as-you-go approach with its blend of automation and incremental improvement promises to provide cost-effective, best-effort solutions.

In this paper we have presented a framework for describing and designing the improvement phase of pay-as-you-go data integration, and have illustrated the framework using representative data integration tasks. There have been many proposals for pay-as-you-go data integration (e.g. [3,10,11,25,27,34]), but these have typically been developed in isolation, and without the benefit of shared methodologies or design principles. It is hoped that the proposal in this paper will prove to be helpful in leading to more systematic and efficient design of pay-as-you-go systems.

In what follows, we elaborate on related areas of ongoing investigation and future research.

Crowdsourcing. There has been significant interest in the use of crowdsourcing for obtaining information for different data management tasks (e.g. [2,8,32]), and as a source of feedback for pay-as-you-go data integration (e.g. [21,31,35]). For the most part work has focused on paid microtasks for systems such as Amazon Mechanical Turk[1] or CrowdFlower[2], but it seems entirely possible that other approaches, for example that combine domain experts with paid microtasks, could be effective (e.g. [1]). Here open issues include: identifying what feedback collection tasks are best suited to what groups of people, and the systematic design of crowdsourcing tasks.

Efficient Collection of Feedback. As explicit (as opposed to implicit) feedback involves human effort, it must be considered to be expensive to collect, and thus there is a need to obtain the most cost-effective feedback. Here there have been a range of approaches, using active learning or bespoke algorithms for identifying *which feedback to obtain next* [14,24,26,36–38], as well as investigations into *which workers should be recruited* to carry out a task [9,40]. Although there has been significant progress in both these areas, it is not always clear which forms of active learning best suit (or do not suit) different tasks, or how to decide what feedback to collect: (i) when there are several different tasks to carry out that may benefit from feedback; or (ii) how to share feedback across different parts of the data integration lifecycle.

---

[1] https://www.mturk.com.
[2] http://www.crowdflower.com.

Systematic Integration of Evidence. There is potentially a lot of evidence to inform pay-as-you-go integration, with the combination of automation that can make use of any available evidence, and the provision of feedback to refine the results of automated techniques. Evidence sources include: results of matching, mapping and quality algorithms; feedback of different sorts from different groups, of different qualities; logging information on the use of integrations; and results of analyses on integrated data sets. Thus there is also a need for an integrated approach to data integration, in which all the available evidence is used together systematically.

There are a several results on evidence accumulation for data integration (e.g. [15,39]), but most current work on pay-as-you-go data integration involves a single type of feedback for a single task. The real breakthrough may come from greater ambition, in which more sources and more techniques provide an additional opportunity rather than an additional challenge (e.g. as demonstrated in the absence of feedback in knowledge base construction [16,30]).

Acknowledgement. Research on data integration at Manchester is supported by the VADA Programme Grant of the UK Engineering and Physical Sciences Research Council, whose support we are pleased to acknowledge.

References

1. Acosta, M., Zaveri, A., Simperl, E., Kontokostas, D., Auer, S., Lehmann, J.: Crowdsourcing linked data quality assessment. In: Alani, H., Kagal, L., Fokoue, A., Groth, P., Biemann, C., Parreira, J.X., Aroyo, L., Noy, N., Welty, C., Janowicz, K. (eds.) ISWC 2013, Part II. LNCS, vol. 8219, pp. 260–276. Springer, Heidelberg (2013)
2. Amsterdamer, Y., Davidson, S.B., Milo, T., Novgorodov, S., Somech, A.: OASSIS: query driven crowd mining. In: International Conference on Management of Data, SIGMOD 2014, Snowbird, 22–27 June 2014, pp. 589–600 (2014)
3. Belhajjame, K., Paton, N.W., Embur, S.M., Fernande, A.A.A., Hedeler, C.: Incrementally improving dataspaces based on user feedback. Inf.Syst. **38**(5), 656–687 (2013)
4. Belhajjame, K., Paton, N.W., Hedeler, C., Fernandes, A.A.A.: Enabling community-driven information integration through clustering. Distrib. Parallel Databases **33**(1), 33–67 (2015)
5. Bernstein, P.A., Haas, L.M.: Information integration in the enterprise. CACM **51**(9), 72–79 (2008)
6. Bleiholder, J., Naumann, F.: Data fusion. ACM Comput. Surv., 41(1) (2008)
7. Bonifati, A., Mecca, G., Pappalardo, A., Raunich, S., Summa, G.: Schema mapping verification: the spicy way. In: Proceedings EDBT 2008, 11th International Conference on Extending Database Technology, Nantes, 25–29 March 2008, pp. 85–96 (2008)
8. Bozzon, A., Brambilla, M., Ceri, S.: Answering search queries with crowdsearcher. In: Proceeding of 21st WWW, pp. 1009–1018 (2012)
9. Cao, C.C., She, J., Tong, Y., Chen, L.: Whom to ask? jury selection for decision making tasks on micro-blog services. PVLDB **5**(11), 1495–1506 (2012)

10. Cao, H., Qi, Y., Candan, K.S., Sapino, M.L.: Feedback-driven result ranking and query refinement for exploring semi-structured data collections. In: EDBT, pp. 3–14 (2010)

11. Chai, X., Vuong, B.-Q., Doan, A., Naughton, J.F.: Efficiently incorporating user feedback into information extraction and integration programs. In: SIGMOD Conference, pp. 87–100 (2009)

12. Christodoulou, K., Paton, N.W., Fernandes, A.A.A.: Structure inference for linked data sources using clustering. Trans. Large-Scale Data- Knowl.-Centered Syst. **19**, 1–25 (2015)

13. Costa, G., Manco, G., Ortale, R.: An incremental clustering scheme for data deduplication. Data Min. Knowl. Disc. **20**(1), 152–187 (2010)

14. Crescenzi, V., Merialdo, P., Qiu, D.: Crowdsourcing large scale wrapper inference. Distributed and Parallel Databases (October 2014)

15. Demartini, G., Difallah, D.E., Cudré-Mauroux, P.: Large-scale linked data integration using probabilistic reasoning and crowdsourcing. VLDB J. **22**(5), 665–687 (2013)

16. Dong, X.L., Gabrilovich, E., Heitz, G., Horn, W., Lao, N., Murphy, K., Strohmann, T., Sun, S., Zhang, W.: Knowledge vault: a web-scale approach to probabilistic knowledge fusion. In: KDD, pp. 601–610 (2014)

17. Dong, X.L., Saha, B., Srivastava, D.: Less is more: selecting sources wisely for integration. PVLDB **6**(2), 37–48 (2012)

18. Elmagarmid, A.K., Ipeirotis, P.G., Verykios, V.S.: Duplicate record detection: a survey. IEEE TKDE **19**(1), 1–16 (2007)

19. Fagin, R., Haas, L.M., Hernández, M., Miller, R.J., Popa, L., Velegrakis, Y.: Clio: schema mapping creation and data exchange. In: Borgida, A.T., Chaudhri, V.K., Giorgini, P., Yu, E.S. (eds.) Conceptual Modeling: Foundations and Applications. LNCS, vol. 5600, pp. 198–236. Springer, Heidelberg (2009)

20. Furche, T., Gottlob, G., Grasso, G., Guo, X., Orsi, G., Schallhart, C., Wang, C.: DIADEM: thousands of websites to a single database. PVLDB **7**(14), 1845–1856 (2014)

21. Gokhale, C., Das, S., Doan, A., Naughton, J.F., Rampalli, N., Shavlik, J.W., Zhu, X.: Corleone: hands-off crowdsourcing for entity matching. In: SIGMOD Conference, pp. 601–612 (2014)

22. Halevy, A.Y., Franklin, M.J., Maie, D.: Principles of dataspace systems. In: PODS, pp. 1–9 (2006)

23. Hedeler, C., Belhajjame, K., Fernandes, A.A.A., Embury, S.M., Paton, N.W.: Dimensions of dataspaces. In: Sexton, A.P. (ed.) BNCOD 26. LNCS, vol. 5588, pp. 55–66. Springer, Heidelberg (2009)

24. Quoc, N., Hung, V., Wijaya, T.K., Miklós, Z., Aberer, K., Levy, E., Shafran, V., Gal, A., Weidlich, M.: Minimizing human effort in reconciling match networks. In: ER, pp. 212–226 (2013)

25. Isele, R., Bize, C.: Learning linkage rules using genetic programming. In: Proceeding 6th International Workshop on Ontology Matching, vol. 814 of CEUR Workshop Proceedings (2011)

26. Isele, R., Bizer, C.: Active learning of expressive linkage rules using genetic programming. J. Web Sem. **23**, 2–15 (2013)

27. Jeffery, S.R., Franklin, M.J., Halevy, A.Y.: Pay-as-you-go user feedback for dataspace systems. In: SIGMOD, pp. 847–860 (2008)

28. Kandel, S., Heer, J., Plaisant, C., Kennedy, J., van Ham, F., Riche, N.H., Weaver, C., Lee, B., Brodbeck, D., Buono, P.: Research directions in data wrangling: visuatizations and transformations for usable and credible data. Inf. Vis. **10**(4), 271–288 (2011)
29. Köpcke, H., Thor, A., Rahm, E.: Evaluation of entity resolution approaches on real-world match problems. PVLDB **3**(1), 484–493 (2010)
30. Niu, F., Zhang, C., Ré, C., Shavlik, J.W.: Elementary: large-scale knowledge-base construction via machine learning and statistical inference. Int. J. Semantic Web Inf. Syst. **8**(3), 42–73 (2012)
31. Osorno-Gutierrez, F., Paton, N.W., Fernandes, A.A.A.: Crowdsourcing feedback for pay-as-you-go data integration. In: DBCrowd, pp. 32–37 (2013)
32. Parameswaran, A.G., Park, H., Garcia-Molina, H., Polyzotis, N., Widom, J.: Deco: declarative crowdsourcing. In: Proceeding 21st CIKM, pp. 1203–1212 (2012)
33. Rahm, E., Bernstein, P.A.: A survey of approaches to automatic schema matching. VLDBJ **10**(4), 334–350 (2001)
34. Talukdar, P.P., Jacob, M., Mehmood, M.S., Crammer, K., Ives, Z.G., Pereira, F., Guha, S.: Learning to create data-integrating queries. PVLDB **1**(1), 785–796 (2008)
35. Wang, J., Kraska, T., Franklin, M.J., Feng, J.: Crowder: crowdsourcing entity resolution. Proc. VLDB Endow. **5**(11), 1483–1494 (2012)
36. Whang, S.E., Lofgren, P., Garcia-Molina, H.: Question selection for crowd entity resolution. PVLDB **6**(6), 349–360 (2013)
37. Yan, Z., Zheng, N., Ives, Z.G., Talukdar, P.P., Yu, C.: Actively soliciting feedback for query answers in keyword search-based data integration. PVLDB **6**(3), 205–216 (2013)
38. Zhang, C.J., Chen, L., Jagadish, H.V., Cao, C.C.: Reducing uncertainty of schema matching via crowdsourcing. PVLDB **6**(9), 757–768 (2013)
39. Zhao, B., Rubinstein, B.I.P., Gemmell, J., Han, J.: A bayesian approach to discovering truth from conflicting sources for data integration. PVLDB **5**(6), 550–561 (2012)
40. Zheng, Y., Cheng, R., Maniu, S., Mo, L.: On optimality of jury selection in crowdsourcing. In: Proceedings of the 18th International Conference on Extending Database Technology, EDBT 2015, Brussels, 23–27 March 2015, pp. 193–204 (2015)

Foundations of Computer Science
(Regular Papers)

Robust Recoverable Path Using Backup Nodes

Marjan van den Akker[1], Hans L. Bodlaender[1], Thomas C. van Dijk[2(✉)],
Han Hoogeveen[1], and Erik van Ommeren[1]

[1] Universiteit Utrecht, Utrecht, The Netherlands
{J.M.vandenAkker,H.L.Bodlaender,J.A.Hoogeveen}@uu.nl
[2] Lehrstuhl Für Informatik I, Universität Würzburg, Würzburg, Germany
thomas.van.dijk@uni-wuerzburg.de

Abstract. We consider routing in networks in the presence of node failures. The focus is specifically on the single-node failure model, which captures the resilience of networks in a realistic fault setting. We introduce a model of recoverable routing, where we ask for an s-t-path that can be repaired easily and locally by assigning 'backup nodes:' when a node on the path fails, it is replaced by its backup node. We resolve the basic algorithmic and complexity questions for finding paths in this model, depending on the properties we require of the backup assignment. For some cases we provide polynomial-time algorithms, and for the others we prove $\mathcal{NP}$-completeness and provide exponential-time algorithms. Lastly, we consider a problem variant where the path is given and ask for a backup assignment.

1 Introduction

We consider a network modeled by a simple graph $G = (V, E)$ with a source node $s \in V$ and a destination node $t \in V$. We write paths using square brackets, like $[p_1, p_2, \ldots, p_k]$. In order to route a packet through the network, we are looking for a path P from s to t. This problem is complicated by considering node failures, for which we want to provide a certain level of robustness. Our robustness condition—that the path must remain valid in the presence of a single node failure—would, by itself, make us overly cautious. We therefore include a recovery model which allows for the easy recovery of a valid path in case a failure makes our initial solution invalid. In contrast to earlier studies of this fault model (e.g. [1,8,10]), we focus on the crucial complexity issues of the model. Our usage of preassigned backup nodes contrasts related work on recoverable paths, which includes subgraph selection [5], network design [4] and cost uncertainty [6] (as opposed to node failure). Particularly related to the current work are the online replacement paths of Adjiashvili et al. [2]; in contrast, our strategy with backups affects the path only locally. We model the network as a graph, disregarding any spatial structure that might be present in the network [3].

In our failure model we will concern ourselves with single-node failures: any single node $v \notin \{s, t\}$ can fail. Asking for a path P that is valid in any failure scenario is then uninteresting. (Only the path $[s, t]$ would be valid, and only

© Springer-Verlag Berlin Heidelberg 2016
R.M. Freivalds et al. (Eds.): SOFSEM 2016, LNCS 9587, pp. 95–106, 2016.
DOI: 10.1007/978-3-662-49192-8_8

if $\{s, t\} \in E$). We therefore introduce a recovery procedure. In case a failure invalidates our initial path P, we want there to be an easy, local way to repair P. In this way the failure can be dealt with in an online fashion: just travel along P, and if the path turns out to be blocked by a node failure, take a local detour and then resume along on P as originally planned. We do this by preassigning backup nodes. Consider the energy usage implied by a certain path: the nodes involved expend energy to transmit the packet. Our recovery procedure changes the path only locally: the recovered path is mostly equal to the input path, even if the fault occurs early on the path. This means that the recovery procedure does not force unexpected energy expense onto many unrelated nodes.

A *path with backups* R assigns to each *main node* p_i a single *backup node* b_i. If p_i fails, then b_i will take its place. We say "p_i is backed up by b_i" and "b_i backs up p_i." We write a path with backups as $[\frac{p_1}{b_1}, \frac{p_2}{b_2}, \ldots, \frac{p_k}{b_k}]$. By $path(R)$ we denote the path formed by the main nodes of $R : path(\ [\frac{p_1}{b_1}, \frac{p_2}{b_2}, \ldots, \frac{p_k}{b_k}]\ ) = [p_1, p_2, \ldots, p_k]$.

Definition 1 (Recoverable path). *A path with backups* $R = [\frac{p_1}{b_1}, \frac{p_2}{b_2}, \ldots, \frac{p_k}{b_k}]$ *is called* recoverable *if and only if the following two properties are satisfied. First, $path(R)$ is a path in G. Secondly, the following recovery procedure succeeds for any node $v \notin \{s, t\}$. Take $path(R)$, but wherever $p_i = v$, use b_i instead: the resulting path P must be a path in $G - v$.*

Note that if a failing node v occurs in $path(R)$ more than once, then it may happen that the recovery procedure substitutes a different backup node for each occurrence: at this point, we do not require that $p_i = p_j$ implies $b_i = b_j$. A recoverable path R is called *simple* if and only if $path(R)$ is simple.

Definition 2 (Recoverable s-t-path). *A recoverable path* $R = [\frac{p_1}{b_1}, \frac{p_2}{b_2}, \ldots, \frac{p_k}{b_k}]$ *is a recoverable s-t-path if and only if $p_1 = b_1 = s$ and $p_k = b_k = t$.*

In this paper we consider the combinatorial problem of finding optimal recoverable s-t-paths. We look at several variations of this problem and give a polynomial-time algorithm or hardness result; the results are summarised at the end of this section. First we make some basic observations. See Fig. 1(a) for an illustration of the edge sets involved in the following two statements.

Lemma 1. $R = [\frac{p_1}{b_1}, \frac{p_2}{b_2}, \ldots, \frac{p_k}{b_k}]$ *is a recoverable s-t-path if and only if the following conditions all hold:*

(1) $p_1 = b_1 = s$,
(2) $p_k = b_k = t$,
(3) $p_i \notin \{s, t\} \implies p_i \neq b_i$,
(4) the   fol-
    lowing edges are in E, for all $1 < i < k$: $\{p_{i-1}, p_i\}, \{p_i, p_{i+1}\}, \{p_{i-1}, b_i\}$ and $\{b_i, p_{i+1}\}$.

Proof. The first two conditions come from the definition of recoverable s-t-path. The third condition is the observation that a node cannot backup itself (except s and t). Consider the edges mentioned in condition four. The first two edges

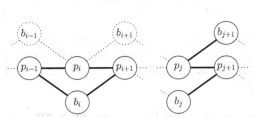
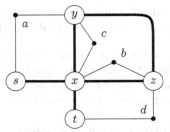

(a) Two ways to index the edges involved in a recoverable path; see Lemma 1 and Proposition 1

(b) In this graph $[s, \frac{x}{a}, \frac{y}{b}, \frac{z}{c}, \frac{x}{d}, t]$ is a recoverable s-t-path.

Fig. 1. Structures involved in recoverable paths

correspond to $path(R)$ being a path in G. The latter two edges correspond to a valid recovery path in case p_i fails. The graph G is simple, so for any i we have $p_i \neq p_{i+1}$. As only a single node can fail, no edge is required from b_i to b_{i+1}. If these edges are present in G, then R is a recoverable s-t-path. If R is a recoverable s-t-path, then these edges are present in G.   □

Proposition 1. *The edge set in condition 4 of the preceding lemma is equivalently defined as follows. For all $1 \leq i < k : \{p_i, p_{i+1}\}, \{p_i, b_{i+1}\}$ and $\{b_i, p_{i+1}\}$.*

Note that a recoverable path has $b_{i-1} \neq p_i$, since equality would imply a self-loop on p_i. We will look at four variations of the RECOVERABLE PATH problem by considering two questions.

- Is a node allowed to back up multiple main nodes, or does $b_i = b_j$ imply $p_i = p_j$? That is, is the relation from main nodes to their backup injective? If we do not enforce this, some nodes may experience high load after recovery.
- Is a node allowed to be backed up by multiple nodes if it occurs as a main node multiple times, or does $p_i = p_j$ imply $b_i = b_j$? That is, is the relation from main nodes to their backup a function? If we enforce this, the recovery procedure is truly local in the sense that it does not need to know where on the path it is in order to perform its rerouting.

Let B be the relation consisting of all pairs (p_i, b_i) occurring on a recoverable s-t-path. By *injective backup*, we mean the property that B is injective; we similarly define *functional backup*. For both or neither property we say *one-to-one* and *many-to-many* respectively. Note that one-to-one backup is both functional and injective.

Lemma 2. *A graph G has a simple recoverable s-t-path if and only if it has a recoverable s-t-path R with functional backup (that is, $p_i = p_j$ implies $b_i = b_j$).*

Proof. A simple s-t-path is trivially functional since $p_i \neq p_j$ for all i and j. In the other direction, let s-t-path R have functional backup. If R is not simple, it can be made simple by shortcuts. Let $i < j$ and $p_i = p_j$. Since the backup relation is function, we have $b_i = b_j$. Remove steps p_{i+1} through p_j from R: the result is still a recoverable path. Repeat until R is simple.   □

It may seem reasonable to restrict our attention only to simple recoverable paths. However, there exist graphs that have a recoverable s-t-path but do not have a *simple* recoverable s-t-path: see Fig. 1(b) for an example. Note that the indicated path does not have functional backup.

Results. We give a polynomial-time algorithm for RECOVERABLE PATH with many-to-many backup, including some weighted versions (Sect. 2). The three other variants are $\mathcal{NP}$-complete. We give exponential-time algorithms for these hard variants (Sect. 3). Finally, we look at a related problem: a normal path is given and we ask whether backups can be assigned to make the path recoverable (Sect. 4). We show $\mathcal{NP}$-completeness for the injective case and give an exponential time algorithm. For the other three cases we provide polynomial-time algorithms.

2   Polynomial-Time Algorithm for Many-to-Many Backup

Here we present a polynomial-time algorithm for RECOVERABLE PATH with many-to-many backups. We also solve some weighted variations. All of these problems are solved in $\mathcal{O}(nm)$, where $n = |V|$ and $m = |E|$. This improves to $\mathcal{O}(d^2 n)$ time on graphs of bounded degree d.

We will find a recoverable s-t-path in G by finding a normal path in a suitably defined auxiliary graph G^A.

Definition 3 (Auxiliary graph G^A). *The auxiliary graph of $G = (V, E)$ is the undirected graph $G^A = (V^A, E^A)$, with $V^A = \{(v, w)_V \mid v, w \in V, v \neq w \vee v \in \{s, t\}\} \cup \{(v, w)_E \mid \{v, w\} \in E\}$, and $E^A = \{\{(v, w)_V, (v, x)_E\} \mid \{w, x\} \in E\} \cup \{\{(v, w)_E, (w, x)_V\} \mid \{v, x\} \in E\}$.*

Lemma 3. *There exists a recoverable s-t-path R in G if and only if there exists a normal $(s, s)_V$-$(t, t)_V$-path P in G^A.*

Proof. Interpret P's nodes alternatingly as (main node, backup node) pairs in G and as edges in E. Lemma 1 shows that this path in G^A is equivalent to a recoverable path in G.

(1) Starting from $(s, s)_V \in V^A$ guarantees $p_1 = b_1 = s$.
(2) Going to $(t, t)_V \in V^A$ guarantees $p_k = b_k = t$.
(3) By construction, $(v, v)_V \notin V^A$ unless $v \in \{s, t\}$. (Recall that G is simple.)
(4) By construction, an edge in E^A exists if and only if the edges required by Proposition 1 exist in E.   □

Lemma 4. *The auxiliary graph G^A has $O(n^2)$ nodes, has $O(nm)$ edges, and can be constructed in $O(n^2 + nm)$ time.*

Proof. We build an adjacency matrix of G in $O(n^2)$ time. For each node v and edge $\{w, x\} \in E$, we check if $\{v, x\} \in E$, and if so, add the edge $\{(v, w)_V, (v, x)_E\}$ to G^A. For each node w and edge $\{v, x\} \in E$, we check if $\{v, w\} \in E$, and if so, add the edge $\{(v, w)_E, (w, x)_V\}$ to G^A. We can build these adjacency lists with radix sort.

Theorem 1. RECOVERABLE PATH *with many-to-many backup can be solved in* $\mathcal{O}(nm)$ *time.*

Proof. Assume $s \neq t$. First remove all isolated nodes from G, so $n \leq 2m$. Build G^A and use depth first search to check if there is a path from $(s, s)_V$ to $(t, t)_V$. Correctness follows from Lemma 3; runtime from Lemma 4. $\qquad\square$

In case G has bounded maximum degree, a linear-time algorithm exists.

Definition 4 (Type-2 auxiliary graph $G^{A'}$). *The type-2 auxiliary graph of* $G = (V, E)$ *is the directed graph* $G^{A'} = (V^{A'}, E^{A'})$, *with* $V^{A'} = \bigcup_{\{u,v\} \in E} \{(u, v), (v, u)\}$, *and* $E^{A'} = \{((v, w), (w, x)) \mid \exists y \in V : \{v, y\} \in E \wedge \{y, x\} \in E\}$.

Lemma 5. *There is a recoverable s-t-path in G if and only if $s = t$ or there are* $v, w \in V$ *such that there is a path from $\{s, v\}$ to $\{w, t\}$ in $G^{A'}$.*

Proof. Again, we check the properties required in Lemma 1; this time, the existence of an arc in $G^{A'}$ corresponds precisely to the existence of the required edges in G. $\qquad\square$

Theorem 2. RECOVERABLE PATH *with many-to-many backup can be solved in* $O(d^2n)$ *time on graphs with maximum degree d.*

Proof. The type-2 auxiliary graph $G^{A'}$ has $O(dn)$ nodes, and $O(d^2n)$ arcs. Construct it in $O(d^2n)$ time. Use depth first search to find a path from (s, v) to (w, t) in $G^{A'}$, for some $v, w \in V$. $\qquad\square$

We can extend this auxiliary-graph approach to handle several weighted versions of the problem. In a network context, this can be used to model, for example, delay times or energy costs. Consider a weight function $w : E \to \mathbb{Z}_{\geq 0}$. We first ask for a recoverable s-t-path R such that the weight of $path(R)$ is minimised and call this problem RECOVERABLE SHORTEST PATH.

Theorem 3. RECOVERABLE SHORTEST PATH *with many-to-many backup and integer weights can be solved in* $\mathcal{O}(nm)$ *time.*

Proof. We use the auxiliary graph G^A and introduce a weight function $E^A \to \mathbb{Z}_{\geq 0}$. A recoverable path R uses the edge $\{v, w\} \in E$ for $path(R)$ if and only if the corresponding path in G^A uses the node $(v, w)_E$ or $(w, v)_E$. We therefore want to weight the usage of node $(v, w)_E$ by $w(\{v, w\})$; we achieve this by assigning that weight to all of its incoming edges. With integer weights, we can use a standard linear-time algorithm [9] to find the minimum-weight (s, s)-(t, t)-path. Since G^A has $\mathcal{O}(n^2)$ nodes and $\mathcal{O}(nm)$ arcs, this gives an $\mathcal{O}(nm)$ time algorithm for RECOVERABLE SHORTEST PATH. $\qquad\square$

The preceding version of the problem only considers the weight of the path in case nothing goes wrong. If a node failure impacts the path, we are faced with the recovery procedure, which in general will give a path of different weight. To take this into account, we look at the expected length of the recovered path: EXPECTED SHORTEST RECOVERABLE PATH. We will work with a probability distribution over which node fails, if any. The case of no failure is denoted by $\varnothing$. Let $f : \{\varnothing\} \cup V - \{s, t\} \to \mathbb{Q}$ be this probability mass function.

Theorem 4. EXPECTED SHORTEST RECOVERABLE PATH *with many-to-many backup can be solved in* $\mathcal{O}(nm)$ *time.*

Proof. We introduce a weight function $E^A \to \mathbb{Q}_{\geq 0}$. An auxiliary edge in E^A corresponds directly to certain edges E (see Definition 3). We can determine the expected weight these edges contribute when included in a recoverable path. Then by linearity of expectation the shortest path in G^A is the recoverable path in G with the lowest expected length. Runtime is again $\mathcal{O}(nm)$. $\square$

3   Exponential-Time Algorithms

In contrast to the many-to-many backup case, RECOVERABLE PATH is hard when we require that the backup relation is functional, injective or one-to-one. A proof based on reduction from 3-CNF-SAT is omitted for space.

Theorem 5. RECOVERABLE PATH *with injective backup is* $\mathcal{NP}$-*complete. The same holds for functional backup and one-to-one backup.*

In this section we provide dynamic-programming algorithms that work, with minor modification, for each of the three hard variants of RECOVERABLE PATH. In the functional and injective cases, it runs in $\mathcal{O}(2^n \cdot n^3)$ time and $\mathcal{O}(2^n \cdot n^2)$ space. In the one-to-one case, it runs in $\mathcal{O}(4^n \cdot n^3)$ time and $\mathcal{O}(4^n \cdot n^2)$ space.

For notational convenience we define the concept of a *friend set*.

Definition 5 (Friend set $F(x, y, z)$). *Let x, y, z be nodes in G. The friend set of (x, y, z) is the set of nodes v that can backup node y on a recoverable path where y occurs between x and z. That is $F(x, y, z) = \{ v \in V \mid (x, v) \in E \land (v, z) \in E \land v \neq y \}$.*

Consider functional backup. This means that every occurrence of a node p as main node on a recoverable path must be backed up by the same node. By Lemma 2 we know that, in fact, there exists a recoverable s-t-path with functional backup if and only if there exists a simple recoverable s-t-path, so once we have used a node as main node we can disregard ever using it again.

We solve the problem based on a recurrence relation for a boolean function p_{fun} in the parameters $y, z \in V$ and $S \subseteq V$. The value of $p_{\text{fun}}(S, y, z)$ is the following: does there exist a simple recoverable path with $p_1 = b_1 = s$, ending with main node y followed by z, and using, besides y and z, exactly the nodes in S as main nodes.

As a base case for our recurrence relation, we can observe that $p_{\text{fun}}(\varnothing, y, z)$ is true precisely if $y = s$ and $(y, z) \in E$: the path must start at s and the edge must exist. For non-empty S, we have that $p_{\text{fun}}(S, y, z)$ is true if and only if the edge (y, z) actually exists, and there exists a predecessor x for y such that

1. a valid backup exists for y, and
2. by recursion, there exists a recoverable path ending in x and y that further uses precisely the nodes in $S - \{x\}$ as main nodes.

This gives the following equation, with the base case that $p_{\text{fun}}(\varnothing, y, z)$ is true if and only if $y = s \wedge (y, z) \in E$.

$$p_{\text{fun}}(S, y, z) = (y, z) \in E \wedge \exists x \in S : \Big(\exists b \in F(x, y, z) : p_{\text{fun}}(S - \{x\}, x, y) \Big) \quad (1)$$

Note that checking for the existence of $b \in F(x, y, z)$ corresponds to checking for the edges required in condition 4 of Lemma 1.

Theorem 6. RECOVERABLE PATH *with functional backup can be solved in* $\mathcal{O}(2^n \cdot n^3)$ *time and* $\mathcal{O}(2^n \cdot n^2)$ *space.*

Proof. Check, using dynamic programming, whether $p_{\text{fun}}(S, y, t)$ is true for any $S \subseteq V$ and $y \in V$. Because of the recurrence relation of p_{fun}, this is equivalent to the existence of a recoverable s-t-path: exactly the edges required by Lemma 1 are present. The parameter S ensures that the path is simple, which ensures functional backup by Lemma 2.

As for runtime and space, we start out by noting that the dynamic program has $\mathcal{O}(2^n \cdot n^2)$ states. These can be calculated in $\mathcal{O}(n)$ time each as follows. Checking $(y, z) \in E$ is a simple test. Then there are existential quantifiers over $x \in S$ and $b \in F(x, y, z)$, both of which might range over $\Theta(n)$ items. Note however that b is not used in the recurrence. We can therefore precompute $(\exists b \in F(x, y, z))$ for all combinations of nodes $x, y, z \in V$. Then this check runs in constant time by table lookup.   □

Theorem 7. RECOVERABLE PATH *with injective backup can be solved in* $\mathcal{O}(2^n \cdot n^3)$ *time and* $\mathcal{O}(2^n \cdot n^2)$ *space.*

Proof (sketch). The approach is similar to the functional case. The polynomial term in the runtime can be kept at n^3 using mutual recursion relations, alternately considering pairs of subsequent path nodes and pairs of a main node and its backup node (as in Definition 3).   □

Lastly we consider the one-to-one case. The established machinery unfortunately leads to a runtime of $\Theta^*(4^n)$: it seems that we need to know two things for every node, namely if it is already used as main node and, independently, if it is already used as a backup node. This is because we allow a node to be both main node and (elsewhere) backup node on the same path; this is simply something the definitions permit. (If we were to disallow nodes being both main node and backup node on the same path, an $\mathcal{O}^*(2^n)$-time algorithm like the previous ones would be possible.)

Theorem 8. RECOVERABLE PATH *with one-to-one backup can be solved in* $\mathcal{O}(4^n \cdot n^3)$ *time and* $\mathcal{O}(4^n \cdot n^2)$ *space.*

4 Backup Assignment

In this section we take a look at a problem related to finding a recoverable path. This time we are given an s-t-path P in G and the question is: does there exist a *recoverable* s-t-path R such that $path(R) = P$? We call this the BACKUP ASSIGNMENT problem.

We can again look at four variations based on what kind of backup relation we allow. We show that the injective variant of the problem is $\mathcal{NP}$-complete and give an exponential time algorithm. We give polynomial-time algorithms for the other three variants.

For the analysis of the problem, we again use *friend sets* (compare Definition 5). This time it is convenient to index them differently.

Definition 6 (Friend set $F(P, i)$). *Let $P = [p_1, \ldots, p_k]$ be a path in G and let p_{i-1}, p_i, p_{i+1} be consecutive nodes on P. The friend set of index i is the set of nodes v that can backup p_i. That is $F(P, i) = F(p_{i-1}, p_i, p_{i+1}) = \{ v \in V \mid (p_{i-1}, v) \in E \wedge (v, p_{i+1}) \in E \wedge v \neq p_i \}$.*

4.1 Polynomial Cases

We now give polynomial-time algorithms for three problem variants. The algorithm for the many-to-many variant is the simplest.

Theorem 9. BACKUP ASSIGNMENT *with many-to-many backup can be solved in polynomial time.*

Proof. According to Lemma 1, the edges required for a node to be in a friend set are exactly those that are required to be a legal backup. Because the backup relation is allowed to be many-to-many, every node can be considered separately. Therefore, in a solution to BACKUP ASSIGNMENT with many-to-many backup, any node can be backed up by any node from its friend set and only by those. If the algorithm fails—because some $F(P, i)$ is empty—no valid backup assignment exists. This greedy assignment can clearly be done in polynomial time.   □

The functional variant is not much more complicated.

Theorem 10. BACKUP ASSIGNMENT *with functional backup can be solved in polynomial time.*

Proof. Compared to the many-to-many case, we have the extra condition that every time a node p occurs on P it must be assigned the same backup. Therefore we can assign to it a certain backup node b only if that is valid for every occurrence of p. This leaves only the nodes that are in the intersection of friend sets of all occurrences of p. Among those, the choice can again be made arbitrarily. This too can clearly be done in polynomial time.   □

We will solve the one-to-one variant of the problem using bipartite matching. As the name suggests, we will use the matching to assign main nodes to backup nodes. We will now construct a bipartite graph G_m that models the right constraints.

Note that a node may occur on a recoverable path both as a main node and as a backup node. We therefore construct two sets of nodes, which together form the node set of G_m.

- A set of nodes V_P representing the main nodes of the path, with a node for every distinct node occurring on P, except s and t.
- A set of nodes V_b representing potential backup nodes, with a node for every node in $V - \{s, t\}$.

We exclude s and t because in a recoverable s-t-path these are necessarily assigned to backup themselves: by definition $p_1 = b_1 = s$ and $p_k = b_k = t$. Then with one-to-one backup the nodes s and t are fully occupied and can be disregarded.

To obtain the edge set of G_m, we insert an edge between a node $p \in V_P$ and a node $b \in V_b$ if and only if b is a legal backup for p, that is, if and only if b is in the friend set of all occurrences of p. An example of this construction can be seen in Fig. 2.

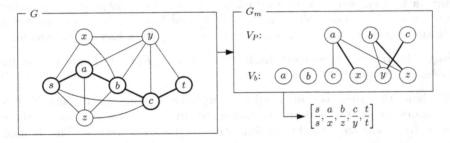

Fig. 2. Example of the matching graph G_m for one-to-one BACKUP ASSIGNMENT

Theorem 11. BACKUP ASSIGNMENT *with one-to-one backup can be solved in polynomial time.*

Proof. By construction, the graph $G_m[V_P]$ contains no edges and neither does $G_m[V_b]$. Then a matching in G_m has size at most $|V_P|$ and any edge in any matching involves exactly one node from V_P and one from V_b. We will interpret an edge in the matching as assigning a main node to a backup node.

By construction of the edge set, there exists a one-to-one backup assignment if and only if there exists a matching of size $|V_P|$ in G_m. The algorithm constructs G_m and finds a maximum-cardinality matching. If the matching has size $|V_P|$ then we have a valid backup assignment. If the maximum matching is smaller, no valid backup assignment exists. The graph G_m can clearly be constructed in polynomial time and the matching can also be found in polynomial time (see for example [7]).   □

4.2   Exponential-Time Algorithm

Now we turn to the remaining case of injective backup, which, as we mentioned already at the beginning of this section, is $\mathcal{NP}$-complete. An instance of BACKUP ASSIGNMENT consists of both a graph G and a prescribed path P. Note that in an $\mathcal{NP}$-completeness proof this path P will, in general, be nonsimple: an injective backup assignment for a simple path is necessarily one-to-one and can, by our preceding results, be found in polynomial time. The following theorem can be proved using reduction from 3-CNF-SAT.

Theorem 12. BACKUP ASSIGNMENT *with injective backup is* $\mathcal{NP}$*-complete.*

Our exponential-time algorithm for BACKUP ASSIGNMENT with injective backup is based on dynamic programming. Note that since the backup relation is not required to be functional, we need to assign backups for the *occurrences* of nodes on P, not just one backup node for every distinct node on P.

We start off with an observation about the structure of injective backup assignments and some notation.

Lemma 6. *Consider an injective backup assignment and let $p_i = p_j$ be distinct occurrences of a single node. Suppose $b_i \neq b_j$ and $b_i \in F(P, j)$. Then changing p_j's backup from b_j to b_i results in another valid injective backup assignment.*

Proof. The backup assignment itself is valid: $b_i \in F(P, j)$. There is also no problem with injectivity, since $p_i = p_j$.   □

This shows that there is some freedom in injective backup assignments: if multiple nodes are used to back up the various occurrences of a single node v, these can be freely changed within the limitation of the above lemma. This is used to argue the correctness of some arbitrary choices the algorithm makes when picking backup nodes.

Definition 7 (Index set). *The index set $\mathcal{I}(v)$ of a node v is the set of indices where the node v occurs on the path P, that is, $\mathcal{I}(v) = \{i \in \mathbb{N} \mid p_i = v\}$.*

Definition 8 (Node multiplicity). *The multiplicity $\mu(v)$ of a node v is the number of times v occurs on the path P, that is, $\mu(v) = |\mathcal{I}(v)|$. Let $\mu_{max} = \max\{\mu(v) \mid v \in V\}$.*

Definition 9 ($\prec$, v_{min}, v_{max}, $pred(v)$). *Fix an arbitrary total order $\prec$ on V. Let v_{min} be the minimum node according to $\prec$, and v_{max} be the maximum. By $pred(v)$ we denote the predecessor of v according to $\prec$.*

We will now set up a function for use in the dynamic programming algorithm. The nodes of the graph are handled one by one, in some order; for each node v, we consider the occurrences of v in the order that they occur on P.

Definition 10. *The boolean function $a(v, O, B)$ is defined for arguments $v \in V, O \subseteq \mathcal{I}(v), B \subseteq V$. It is defined to be true if and only if there is a way to assign backups that is injective, where exactly the nodes in B are used as backup, and where exactly the following occurrences have been assigned a backup: all $p_i \prec v$, and all p_j for $j \in O$. (Thus leaving all other occurrences unassigned.)*

Then we can solve BACKUP ASSIGNMENT with injective backup as follows. Check whether $a(\, v_{max},\, \mathcal{I}(v_{max}),\, B\,)$ is true for any subset B: by definition this means assigning backups to all occurrences on P, using any set of backup nodes.

Theorem 13. BACKUP ASSIGNMENT *with injective backup can be solved in* $\mathcal{O}^*(\, 2^{n + \mu_{max}}\,)$ *time.*

Proof. Calculate $a(v, O, B)$ for all combinations of $v \in V$, $O \in \mathcal{I}(v)$ and $B \subseteq V$, using the following recurrence relation.

If some occurrence of v is already backed up (that is, $O \neq \varnothing$), we can recurse on which node $b \in B$ is its backup. In view of Lemma 6, we can then immediately use b to backup as many other occurrences of v as possible: since we are already deciding to use b as backup for *some* occurrence of v, it cannot be wrong to use it for more occurrences. There is still the choice of which occurrence in O to recurse on, but this choice can be made arbitrarily. Call this node $arb(O)$. Then

$$a(\, v,\ O,\ B\,) \;=\; \exists b \in\ B \cap F(\, P, arb(O)\,)\ :\ a(\, v,\ O'(b),\ B - \{b\}\,) \qquad (2)$$

$$\textbf{where } O'(b) \;=\; \{\, i \in O \mid b \notin F(P, i)\,\}.$$

Here, O' is the set of occurrences that cannot be backed up using a particular choice of b.

The preceding case handled $O \neq \varnothing$. The case $O = \varnothing$ is quite simple, since directly from the definition of $a(\cdot)$ we have the following equality (for $v \neq v_{min}$).

$$a(v, \varnothing, B) \;=\; a(\, pred(v),\ \mathcal{I}(pred(v)),\ B\,) \qquad (3)$$

This leaves setting the base case for our recursion. This is also easily accomplished from the definition. We let $a(v_{min}, \varnothing, \varnothing) = \text{true}$: it is indeed possible to back up no occurrences using no backup nodes.

The algorithm then checks whether $a(v_{max}, \mathcal{I}(v_{max}), B)$ is true for any $B \subseteq V$. Correctness of the algorithm follows from correctness of the recurrence. Dynamic programming ensures that $a(\cdot)$ is only ever evaluated once for every value of the parameters; call these the dynamic programming states. Evaluating a single dynamic programming state can clearly be done in polynomial time. For the runtime up to polynomial factors, it then remains to bound the number of different dynamic programming states. The total number of states is

$$\sum_{v \in V} \left(2^{|\mathcal{I}(v)|} \cdot 2^{|V|} \right) \;\stackrel{def}{=}\; \sum_{v \in V} \left(2^{\mu(v)} \cdot 2^{n} \right) \;\leq\; \sum_{v \in V} \left(2^{\mu_{max}} \cdot 2^{n} \right) \;=\; n \cdot 2^{\mu_{max}} \cdot 2^{n}.$$

This gives total running time of $\mathcal{O}^*(\, 2^{n + \mu_{max}}\,)$. $\square$

5   Conclusion

We have introduced a model of recoverable routing in the single-node failure model. As with other *robust recoverability* models, the motivation is as follows. Choosing a solution that is feasible for any failure scenario is overly cautious—in our case it would only allow paths of one hop. On the other hand, unrestricted replanning in case of a failure can be too costly in terms of computational power or the information available. We therefore plan a route that, in case of failure, can be fixed easily and locally. For this model, we have resolved the basic algorithmic and complexity questions.

We have presented several algorithms. For the polynomially-solvable case of RECOVERABLE PATH we have given an $\mathcal{O}(nm)$-time algorithm. For the functional and injective cases, the runtime of $\mathcal{O}^*(2^n)$ that our algorithms achieve seems reasonable. When generalised to the one-to-one case, however, our algorithm runs in $\Theta^*(4^n)$ time. It seems to us there should be a better way to handle the one-to-one case.

References

1. Abbasi, A.A., Younis, M.F., Baroudi, U.A.: Recovering from a node failure in wireless sensor-actor networks with minimal topology changes. IEEE Trans. Veh. Technol. **62**(1), 256–271 (2013)
2. Adjiashvili, D., Oriolo, G., Senatore, M.: The online replacement path problem. In: Bodlaender, H.L., Italiano, G.F. (eds.) ESA 2013. LNCS, vol. 8125, pp. 1–12. Springer, Heidelberg (2013)
3. Álvarez-Miranda, E., Candia-Véjar, A., Carrizosa, E., Pérez-Galarce, F.: Vulnerability assessment of spatial networks: models and solutions. In: Fouilhoux, P., Gouveia, L.E.N., Mahjoub, A.R., Paschos, V.T. (eds.) ISCO 2014. LNCS, vol. 8596, pp. 433–444. Springer, Heidelberg (2014)
4. Álvarez-Miranda, E., Ljubić, I., Raghavan, S., Toth, P.: The recoverable robust two-level network design problem. INFORMS J. Comput. **27**(1), 1–19 (2015)
5. Büsing, C.: The exact subgraph recoverable robust shortest path problem. In: Ahuja, R.K., Möhring, R.H., Zaroliagis, C.D. (eds.) Robust and Online Large-Scale Optimization. LNCS, vol. 5868, pp. 231–248. Springer, Heidelberg (2009)
6. Büsing, C.: Recoverable robust shortest path problems. Networks **59**(1), 181–189 (2012)
7. Cormen, T.H., Leiserson, C.E., Rivest, R.L., Stein, C.: Introduction to Algorithms, 3rd edn. The MIT Press, Cambridge (2009)
8. Nanda, A., Rath, A.K., Rout, S.K.: Node sensing & dynamic discovering routes for wireless sensor networks. Computing Research Repository (CoRR), abs/1004.1678 (2010)
9. Thorup, M.: Undirected single-source shortest paths with positive integer weights in linear time. J. ACM **46**(3), 362–394 (1999)
10. Wang, Y.-H., Chao, C.-F.: Dynamic backup routes routing protocol for mobile ad hoc networks. Inf. Sci. **176**(2), 161–185 (2006)

On Contact Graphs with Cubes
and Proportional Boxes

M. Jawaherul Alam[1]([✉]), Michael Kaufmann[2], and Stephen G. Kobourov[1]

[1] Department of Computer Science, University of Arizona, Tucson, USA
mjalam@email.arizona.edu
[2] Wilhelm-Schickhard-Institut Für Informatik,
Universität Tübingen, Tübingen, Germany

Abstract. We study two variants of the problem of contact representation of planar graphs with axis-aligned boxes. In a *cube-contact representation* we realize each vertex with a cube, while in a *proportional box-contact representation* each vertex is an axis-aligned box with a pre-specified volume. We show how to construct such representations representation for some classes of planar graphs.

1 Introduction

We study *contact representations* of planar graphs in 3D, where vertices are represented by interior-disjoint axis-aligned boxes and edges are represented by shared boundaries between the corresponding boxes. A contact representation of a planar graph G is *proper* if for each edge (u, v) of G, the boxes for u and v have a shared boundary with non-zero area. Such a contact between two boxes is also called a *proper contact*. *Cubes* are axis-aligned boxes where all sides have the same length. A contact representation of a planar graph with boxes is called a *cube-contact* representation when all the boxes are cubes. In a weighted variant of the problem a *proportional box-contact* representation is one where each vertex v is represented with a box of volume $w(v)$, for any function $w : V \rightarrow \mathbb{R}^+$, assigning weights to the vertices V. Note that this "value-by-volume" representation is a natural generalization of the "value-by-area" cartograms in 2D.

Related Work: Koebe's 1930 theorem [9] represents planar graphs by touching disks in 2D. Proper contact representation with rectangles in 2D is the well-known *rectangular dual* problem, for which several characterizations exist [10,14]. Representations with other axis-aligned and non-axis-aligned polygons [6,15] have been studied. The weighted variant of the problem has been considered in the context of rectangular, rectilinear, and unrestricted cartograms [7,11]. Thomassen [13] shows that any planar graph has a proper contact representation with touching boxes in 3D, while Felsner and Francis [8] find a (not necessarily proper) contact representation of any planar graph with touching cubes. Recently, Bremner *et al.* [5] asked whether any planar graph can be represented by proper contacts of cubes. They answered the question positively for the case of partial planar 3-trees and

© Springer-Verlag Berlin Heidelberg 2016
R.M. Freivalds et al. (Eds.): SOFSEM 2016, LNCS 9587, pp. 107–120, 2016.
DOI: 10.1007/978-3-662-49192-8_9

some planar grids, but the problem remains open for general planar graphs. The weighted variant of the problem in 3D is less studied, although some results are known for proportional representation of special classes (e.g., outerplanar, planar bipartite, planar, complete) using 3D L-shapes [2].

Our Contribution: Here we expand the class of planar graph representable by proper contact of cubes. Specifically, we show how to compute proportional box-contact representations for plane 3-trees, and both proportional box-contact representations and a cube-contact representations for *nested maximal outerplanar graphs*, which are defined as follows. A *nested outerplanar graph* is either an outerplanar graph or a planar graph G where each component induced by the internal vertices is another nested outerplanar graph with exactly three neighbors in the outerface of G. A *nested maximal outerplanar graph* is a subclass of nested outerplanar graphs that is either a maximal outerplanar graph or a maximal planar graph in which the vertices on the outerface induce a maximal outerplanar graph and each component induced by internal vertices is another nested maximal outerplanar graph. Some proofs are sketched due to space limitations; for more details see the full paper [1].

2   Preliminaries

A *3-tree* is either a 3-cycle or a graph G with a vertex v of degree three in G such that $G - v$ is a 3-tree and the neighbors of v form a triangle. If G is planar, then it is called a *planar 3-tree*. A *plane 3-tree* is a planar 3-tree along with a fixed planar embedding. Starting with a 3-cycle, any planar 3-tree can be formed by recursively inserting a vertex inside a face and adding an edge between the newly added vertex and each of the three vertices on the face [4,12]. Using this simple construction, we can create in linear time a *representative tree* for G [12], which is an ordered rooted ternary tree T_G spanning all the internal vertices of G. The root of T_G is the first vertex we have to insert into the face of the three outer vertices. Adding a new vertex v in G will introduce three new faces belonging to v. The first vertex w we add in each of these faces will be a child of v in T_G. The correct order of T_G can be obtained by adding new vertices according to the counterclockwise order of the introduced faces.

An *outerplanar graph* is one that has a planar embedding with all vertices on one face (outerface). It is *maximal* if no edge can be added without violating outerplanarity (all faces except the outerface are triangles). For $k > 1$, a k-outerplanar graph G is an embedded graph such that deleting the outer-vertices from G yields a graph where each component is at most a $(k - 1)$-outerplanar graph; a 1-outerplanar graph is just an outerplanar graph. Note that any planar graph is a k-outerplanar graph for some $k > 0$.

Let G be a planar graph. We define the *pieces* of G as follows. If G is outerplanar, it has only one piece, the graph itself. Otherwise, let $G_1, G_2, \ldots, G_f$ be the components of the graph obtained by deleting the outer vertices (and their incident edges) from G. Then the pieces of G are all the pieces of G_i for each

$i \in \{1, 2, \ldots, f\}$, as well as the subgraph of G induced by the outer-vertices of G. Note that each piece of G is an outerplanar graph. Since G is an embedded graph, for each piece P of G, we can define the *interior* of P as the region bounded by the outer cycle of P. Then we can define a rooted tree $\mathcal{T}$ where the pieces of G are the vertices of $\mathcal{T}$ and the parent-child relationship in $\mathcal{T}$ is determined as follows: for each piece P of G, its children are all the pieces of G that are in the interior of P but not in the interior of any other pieces of G. A piece of G has *level* l if it is on the l-th level of $\mathcal{T}$. All the vertices of a piece at level l are also *l-level* vertices. A planar graph is a *nested outerplanar graph* if each of its pieces at level $l > 0$ has exactly three vertices of level $(l - 1)$ as a neighbor of some of its vertices. On the other hand a *nested maximal outerplanar graph* is a maximal planar graph where all the pieces are maximal outerplanar graphs.

3   Representations for Planar 3-Trees

Here we prove that planar 3-trees have proportional box-representations.

Theorem 1. *Let $G = (V, E)$ be a plane 3-tree with a weight function w. Then a proportional box-contact representation of G can be computed in linear time.*

Proof: Let a, b, c be the outer vertices of G. We construct a representation Γ for G where b is at the bottom side of Γ, a is at the back of $\Gamma - \{b\}$ and c is at the right side of $\Gamma - \{a, b\}$; see Fig. 1. Here for a set of vertices S, $\Gamma - S$ denotes the representation obtained from Γ by deleting the boxes representing the vertices in S. The claim is trivial when G is a triangle, so assume that G has at least one internal vertex. Let r be the root of the representation tree T_G of G. Then r is adjacent to a, b and c and thus defines three regions G_1, G_2 and G_3 inside the triangles $\Delta_1 = abr$, $\Delta_2 = bcr$ and $\Delta_3 = car$ (including the vertices of these triangles). By induction hypothesis G_i, $i = 1, 2, 3$ has a proportional box-contact representation Γ_i where the boxes for the three vertices in Δ_i occupy the

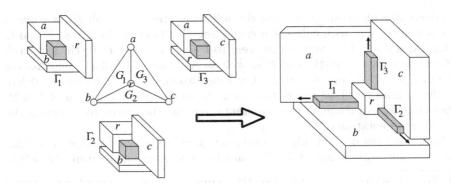

Fig. 1. Illustration for the proof of Theorem 1

bottom, back and right sides of Γ_i. Define $\Gamma'_i = \Gamma_i - \Delta_i$. We now construct the desired representation for G. Take a box for r with volume $w(r)$ and place it in a corner created by the intersection of three pairwise-touching boxes; see Fig. 1. For each Δ_i, $i = 1, 2, 3$, there is a corner p_i formed by the intersection of the three boxes for Δ_i. We now place Γ'_i (after possible stretching[1]) in the corner p_i so that it touches the boxes for the vertices in Δ_i by three planes. Note that this is always possible since we can choose the surface areas for a, b and c to be arbitrarily large and still realize their corresponding weights by appropriately changing the thickness in the third dimension. This construction requires only linear time, by keeping the stretching factor for each region in the representative tree T_G at the vertex representing that region. Then the exact coordinates can be computed with a top-down traversal of T_G.   □

4   Cube-Contacts for Nested Maximal Outerplanar Graphs

Theorem 2. *Any nested maximal outerplanar graph has a proper contact representation with cubes.*

We prove Theorem 2 by construction, starting with a representation for each piece of G, and combining the pieces to complete the representation for G. Let G be a nested maximal outerplanar graph. Augment G by adding a bounding triangular face $\{A, B, C\}$ to G and then triangulating the graph by adding dummy edges from $\{A, B, C\}$ to the outer vertices of G; see Fig. 2(a). Call this the *extended graph* of G. For consistency, let the three dummy vertices have level 0. The observation below follows from the definition of nested maximal outerplanar graphs.

Observation 1. *Let G be a nested-maximal outerplanar graph and G' be the extended graph of G. Then for each piece P of G at level l, there is a triangle of $(l-1)$-level vertices adjacent to the vertices of P and no other k-level vertices with $k < l$ are adjacent to any vertex of P.*

Given this observation, we use the following strategy to obtain a contact representation of G with cubes. For each piece P of G at level l, let A, B and C be the three $(l-1)$-level vertices adjacent to P's vertices. Let P' be the subgraph of G induced by the vertices of P as well as A, B and C; call P' the *extended piece* of G for P. We obtain a contact representation of P' with cubes and delete the three cubes for A, B and C to obtain the contact representation of P with cubes. Finally, we combine the representations for the pieces to complete the desired representation of G.

Before we describe this algorithm in detail, we present a useful lemma. This lemma is also interesting by itself, since for any outerplanar graph O, where

---

[1] In Sects. 3 and 5, the operation of *stretching* some representation involves expanding it in some axial direction and shrinking it in some other, so that the volume remains unchanged.

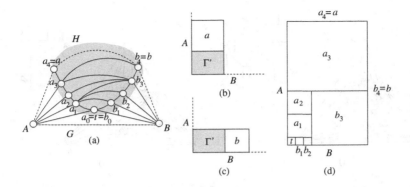

Fig. 2. Illustration for the proof of Lemma 2

each face has at least one outer edge, it provides a contact representation of O in the plane with squares and with a rectangle as the outer boundary of the representation.

Lemma 2. *Let G be planar graph with outerface $ABba$ and at least one internal vertex, such that $G - \{A, B\}$ is a maximal outerplanar graph. If there is no chord between any two neighbors of A and no chord between any two neighbors of B, then G has a contact representation Γ in 2D where each inner vertex is represented by a square, the union of these squares forms a rectangle, and the four sides of these rectangles represent A, B, b and a, respectively.*

Proof: We prove this lemma by induction on the number of vertices in G. Denote the maximal outerplanar graph $H = G - \{A, B\}$; see Fig. 2(a). If G contains only one internal vertex v, then we compute Γ by representing v by a square $R(v)$ of arbitrary size and representing A, B, b and a by the left, bottom, right and top sides of $R(v)$.

We thus assume that G has at least two internal vertices. Let u be the unique common neighbor of $\{a, b\}$ in H. If u is a neighbor of A, then $H - \{a\}$ is a maximal outerplanar graph. By induction hypothesis, $G - \{a\}$ has a contact representation Γ' where each internal vertex of $G - \{a\}$ is represented by a square and the left, bottom, right and top sides of Γ' represent A, B, b and u. Then we compute Γ from Γ' by adding a square $R(u)$ to represent u such that $R(u)$ spans the entire width of Γ' and is placed on top of Γ'; see Fig. 2(b). A similar construction can be used if u is a neighbor of B; see Fig. 2(c). We thus compute a contact representation for G; see Fig. 2(d). $\square$

4.1   Cube-Contact Representation for Extended Pieces

Lemma 3. *Let P be a piece of G at level l and P' be the extended piece for P with $(l-1)$-level vertices A, B, C. Then P' has a cube-contact representation.*

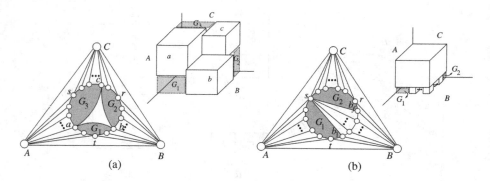

Fig. 3. Illustration for **Case A** in the proof of Lemma 3

Proof: Let r be a common neighbor of B and C; s a common neighbor of A and C; t a common neighbor of A and B. It is easy to find a contact representation of P' if r, s and t are the only vertices of P, so let P have at least four vertices. The outer cycle of P can be partitioned into three paths: P_a is the path from s to t, P_b is the path from r to t and P_c is the path from r to s. Note that all vertices on the path P_a (P_b, P_c) are adjacent to A (B, C). A chord (u, v) is a *short chord* if it is between two vertices on the same path from the set $\{P_a, P_b, P_c\}$. (Note that a chord between two vertices from the set $\{r, s, t\}$ is also a short chord.) We have the following two cases.

Case A: There is no short chord in P. In this case all the chords of P are between two different paths. We consider the following two subcases.

Case A1: There is no chord with one end-point in $\{r, s, t\}$. In this case, due to maximal-planarity there exist three vertices a, b and c, adjacent to A, B, and C, respectively such that (i) ab is the chord between vertices of P_a and P_b farthest away from t, (ii) bc is the chord between vertices of P_b and P_c farthest away from r, and (iii) ac is the chord between vertices of P_a and P_c farthest away from s; see Fig. 3(a). We can then find three interior-disjoint subgraphs of P' defined by three cycles of P': G_1 is the one induced by all vertices on or inside $ABba$; G_2 is induced by all vertices on or inside $BCcb$; and G_3 is induced by all vertices on or inside $ACca$. Each of these subgraphs has the common property that if we delete two vertices from the outerface (two vertices from the set $\{A, B, C\}$ in each subgraph), we get an outerplanar graph. From the representation with squares from the proof of Lemma 2, we find a contact representation of $G_i, i = 1, 2, 3$ where each internal vertex of G_i is represented by a cube and the union of all these cubes forms a rectangular box whose four sides realize the outer vertices. We use such a representation to obtain a contact representation of P' with cubes as follows.

We draw pairwise adjacent cubes (of arbitrary size) for A, B, C. We need to place the cubes for all the vertices of P in the a corner defined by three faces of the cubes for A, B, C. Then we place three mutually touching cubes for a, b

and c, which touch the walls for A, B and C, respectively; see Fig. 3(a). We also compute a contact representation of the internal vertices for each of the three graphs G_1, G_2 and G_3 with cubes using Lemma 2, so that the outer boundary for each of these representation forms a rectangular pipe. We adjust the sizes of the three cubes for a, b and c in such a way that the three highlighted rectangular pipes precisely fit these three representations (after some possible scaling). Note that this construction works even if one or more of the subgraphs G_1, G_2 and G_3 are empty. This completes the analysis of **Case A1**.

Case A2: There is at least one chord with one end-point in $\{r, s, t\}$. Due to planarity all such chords will have the same end-point in $\{r, s, t\}$. Suppose s is this common end point for these chords; see Fig. 3(b). Let b_1 and b_f be the first and last endpoints in the clockwise order of these chords around s. Then we can find two subgraphs G_1 and G_2 induced by the vertices on or inside two separating cycles ABb_1s and $BCsb_f$. We find contact representations for the internal vertices of these two graphs G_1 and G_2 using Lemma 2 so that the outer-boundaries of these representation form rectangular pipes. We then obtain the desired contact representation for P', starting with the three mutually touching walls for A, B and C at right angles from each other, placing the cubes for s and $b_1, \ldots, b_f$ as illustrated in Fig. 3(b), and fitting the representations for G_1 and G_2 (after some possible scaling) in the highlighted regions.

Case B: there are some shord chords in P. In this case, we find at most four subgraphs from P' as follows. At each path in $\{P_a, P_b, P_c\}$, we find the *outermost chord*, i.e., one that is not contained inside any other chords on the same path. Suppose these chords are a_1a_2, b_1b_2 and c_1c_2, on the three paths P_a, P_b, P_c, respectively. Then three of these subgraphs G_a, G_b and G_c are induced by the vertices on or inside the three triangles Aa_1a_2, Bb_1b_2 and Cc_1c_2. The fourth subgraph P^* is obtained from P' by deleting all the inner vertices of the three graphs G_a, G_b and G_c; see Fig. 4.

A cube representation of P^* can be found by the algorithm in **Case A**, as P^* fits the condition that there is no chord between any two neighbors of the same vertex in $\{A, B, C\}$. Note that by moving the cubes in the representation by an arbitrarily small amount, we can make sure that for each triangle xyz in P^*, the three cubes for x, y and z form a corner surrounded by three mutually touching walls at right angles to each other. Now observe that each of the three graphs G_a, G_b and G_c is a planar 3-tree; thus using the algorithm of either [5] or [8], we can place the internal vertices of these three graphs in their corresponding corners, thereby completing the representation. □

4.2 Cube-Contact Representation for a Nested Maximal Outerplanar Graph

Proof of Theorem 2: Let G be a nested maximal outerplanar graph. We build the contact representation of G by a top-down traversal of the rooted tree T of the pieces of G. We start by creating a corner surrounded by three mutually touching walls at right angle to each other. Then whenever we traverse any vertex

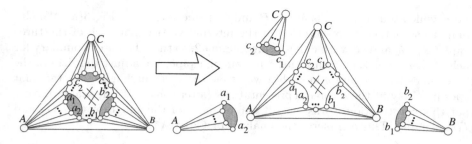

Fig. 4. Removing chords with end-vertices in the same neighborhood

of $\mathcal{T}$, we realize the corresponding piece P at level l by obtaining a representation using Lemma 3 and placing this in the corner created by the three already-placed cubes for the three $(l-1)$-level vertices adjacent to P (after possible scaling). □

5 Proportional Box-Contacts for Nested Outerplanar Graphs

In this section we prove the following main theorem.

Theorem 3. *Let $G = (V, E)$ be a nested outerplanar graph and let $w : V \to \mathbb{R}^+$ be a weight function defining weights for the vertices of G. Then G has a proportional contact representation with axis-aligned boxes with respect to w.*

We construct a proportional representation for G using a similar strategy as in the previous section: we traverse the construction tree $\mathcal{T}$ of G and deal with each piece of G separately. Each piece P of G is an outerplanar graph and hence one can easily construct a proportional box-contact representation for P as follows. Any outerplanar graph P has a contact representation with rectangles in the plane. In fact in [3], it was shown that P has a contact representation with rectangles on the plane where the rectangles realize prespecified weights by their areas. Thus by giving unit heights to all rectangles we can obtain a proportional box-contact representation of P for any given weight function. However if we construct proportional box-contact representation for each piece of G in this way, it is not clear that we can combine them all to find a proportional contact representation of the whole graph G. Instead, we use this construction idea in Lemmas 4 and 5 to build two different proportional rectangle-contact representations for outerplanar graphs and we use them in the proof of Theorem 3.

Suppose O is an outerplanar graph and Γ is a contact representation of O with rectangles in the plane. We say that a corner of a rectangle in Γ is *exposed* if it is on the outer-boundary of Γ and is not shared with any other rectangles.

Lemma 4. *Let O be a maximal outerplanar graph with a weight function w. Let $1, \ldots, n$ be the clockwise order of the vertices around the outer-cycle. Then a proportional rectangle-contact representation Γ of O for w can be computed* *so that rectangle R_1 for 1 is leftmost in Γ, rectangle R_n for n is bottommost in $\Gamma - R_1$, and the top-right corner for each rectangle is exposed in Γ.*

Proof Sketch: Constructing Γ is easy when G is a single edge $(1, n)$, so let G contain at least 3 vertices. Let x be the unique vertex adjacent to $(1, n)$. Denote by $G[1, x]$ the graph induced by all vertices between 1 and x and by $G[x, n]$ the graph induced by the vertices between x and n. Recursively draw $G[1, x]$ and $G[x, n]$ and remove the rectangles for 1, x, n to obtain the drawings Γ_1 and Γ_2. Draw rectangles R_1, R_x and R_n for 1, x, n, with required areas and place Γ_1 (possibly stretched) between R_1, R_x and Γ_2 (possibly stretched) between R_x, R_n to complete the drawing; see Fig. 5.   □

Note that in these layouts the top right corners of the rectangles for vertices $\{1, \dots, n\}$ have increasing x-coordinates and decreasing y-coordinates. Thus we refer to them as **Staircase (SC)** layouts and to the algorithm as the **SC Algorithm**.

Lemma 5. *Let O be a maximal outerplanar graph with a weight function w. Let $1, \dots, n$ be the clockwise order of the vertices around the outer-cycle. Then a proportional rectangle-contact representation Γ of O for w can be computed so that rectangle R_1 for 1 is leftmost in Γ, rectangle R_n for n is bottommost in $\Gamma - R_1$, and the top-right corners of all rectangles for vertices $\{1, \dots, i\}$ and the bottom-right corners of all rectangles for vertices $\{i, \dots, n\}$ are exposed in Γ.*

Proof Sketch: Computing Γ is easy when G is a single edge $(1, n)$, so let G have at least 3 vertices and x be the unique vertex adjacent to $(1, n)$. Define the two graphs $G[1, x]$ and $G[x, n]$ as in the proof of Lemma 4; see Fig. 6(a). If $x > i$, then recursively draw $G[1, x]$ and remove the rectangles for 1 and x from it; call the result Γ_1. Draw $G[x, n]$ using the **SC Algorithm** and remove x and n to find Γ_2. Now draw three mutually touching rectangles R_1, R_x and R_n for 1, x and n, with the necessary areas and place Γ_1 (after possible stretching) between R_1, R_x and Γ_2 (after 90° clockwise rotation and possible stretching) between

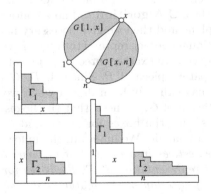

Fig. 5. Illustration for Lemma 4

R_x, R_n to complete the drawing; see Fig. 6(b). The cases when $x = i$ and $x < i$ follow similar constructions; see Fig. 6(c)–(d).   □

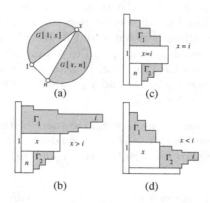

Fig. 6. Illustration for Lemma 5

Note that in the layout obtained above the top-right corners for vertices $\{1, \ldots, i\}$ and the bottom-right corners for vertices $\{i+1, \ldots, n\}$ form two staircases. Thus we refer to this as a **Double-Staircase (DSC)** layout, to the algorithm as the **DSC Algorithm**, and to vertex i as the *pivot vertex*.

Let O be a maximal outerplanar graph and let Γ be either a **SC** or a **DSC** layout. Then any triangle $\{p, q, r\}$ in O is represented by three rectangles and the shared boundaries of these rectangles define a *T-shape*. The vertex whose two shared boundaries are collinear in the T-shape is called the *pole* of the triangle $\{p, q, r\}$.

Proof of Theorem 3. Let $\mathcal{T}$ be the construction tree for G. We compute a representation for G by a top-down traversal of $\mathcal{T}$, constructing the representation for each piece as we traverse it. Let P be a piece of G at the l-th level. If P is the root of $\mathcal{T}$, then we use the **SC Algorithm** to find a contact representation of P with rectangles in the plane and then we give necessary heights to these rectangles to obtain a proportional contact representation of P with boxes. Otherwise, the vertices of P are adjacent to exactly three $(l-1)$-level vertices A, B, C that form a triangle in the parent piece of P. Since A, B, C belong to the parent piece of P, their boxes have already been drawn when we start to draw P. To find a correct representation of G, we need that the boxes for the vertices in P have correct adjacencies with the boxes for A, B, and C; hence we assume a fixed structure for such a triangle. We maintain the following invariant:

Let $\{p, q, r\}$ be three vertices in a piece P of G forming a triangle. Then in the proportional contact representation of P, the boxes for p, q, r are drawn in such a way that (i) the projection of the mutually shared boundaries for these boxes in the xy-plane forms a T-shape, (ii) the highest faces (faces with largest *z-coordinate) of the three rectangles have different z coordinates and the highest face of the pole-vertex of the triangle has the smallest z-coordinate.*

Note that by choosing the areas of the rectangles in the **SC** layout, we can maintain this invariant for the parent piece by suitably adjusting the heights of the boxes (e.g., incrementally increasing heights for the vertices in the recursive **SC Algorithm**).

We now describe the construction of a proportional box-contact representation of P with the correct adjacencies for A, B and C. By the invariant the projection of the shared boundaries for $\{A, B, C\}$ forms a T-shape in the xy-plane. Without loss of generality assume that A is the pole of the triangle and the highest faces of B, C and A are in this order according to decreasing z-coordinates. Also assume that P is a maximal outerplanar graph; we later argue that this assumption is not necessary.

Let ab be a common neighbor of A and B; bc a common neighbor of B and C; ca a common neighbor of C and A. Then the outer cycle of P can be partitioned into three paths: P_a is the path from ca to ab, P_b is the path from ab to bc and P_c is the path from bc to ca. All the vertices on the path P_a (P_b, P_c, respectively) are adjacent to A (B, C, respectively). We first assume that there is no chord in P between ca and a vertex on path P_a. We consider the following two cases.

Case 1: No vertex of P is adjacent to all of $\{A, B, C\}$. We label the vertices of P in clockwise order starting from $ca = 1$ and ending at n, where n is the number of vertices in P. Let i and j be the indices of vertices bc and ab, respectively. Let x be the index of the vertex that is the (unique) third vertex of the inner face of P containing the edge $(1, n)$. Define the two graphs $G[1, x]$ and $G[x, n]$ as in the proof of Lemma 4. We first find a proportional contact representation of P for w restricted to the vertices of P using rectangles in the plane, then we give the needed heights to these rectangles. Draw $G[x, n]$ using the **SC Algorithm** and delete the rectangles R_x and R_n, for x and n, to obtain Γ_2. Draw rectangles R_x and R_n so that the bottom side of R_x touches the top side of R_n, the left sides for both the rectangles have the same x-coordinate and the right side of R_n extends past R_x. Now place Γ_2 (possibly stretching) touching the right side of R_x and the top side of R_n; this is possible as we can make the width of R_n suitably long. Place rectangle R_1 for 1, touching the left sides of R_x and R_n, so that its bottom side is aligned with R_n and its top side is aligned with the top side of the rectangle for j. To complete the rest of the drawing, we have the following two subcases:

    Case 1a: $x \leq i$. Use the **SC Algorithm** to draw $G[1, x]$ and delete from it rectangles R_1 and R_x to obtain Γ_1. Place Γ_1 (after 90° counterclockwise rotation and possible stretching) touching the top side of R_1 and left side of R_x; this is possible by choosing a suitably large height for R_x.

    Case 1b: $x > i$. Use the **DSC Algorithm** with pivot vertex i to draw $G[1, x]$. Delete rectangles R_1 and R_x from this drawing to obtain Γ_1. Place Γ_1 (after 90° counterclockwise rotation and possible stretching) touching the top side of R_1 and left side of R_x, so that the top side of the rectangle for $(x - 1)$ extends past the top side of R_x.

So far we used the function w to assign areas for the rectangles and obtained proportional box-contact representation of P from the rectangles by assigning unit heights. However, by changing the areas for the rectangles, we can obtain different heights for the boxes. We will use this property to maintain adjacencies with $\{A, B, C\}$, as well as to maintain the invariant. Specifically, once we get the box representation of P, we stretch it by increasing the heights for the boxes, so that when we place it at the corner created by the T-shape for $\{A, B, C\}$ it will not intersect the representation for any of its sibling pieces in $\mathcal{T}$. Consider the point p which is the intersection of the lines containing the right side of the rectangle for i and the top side of the rectangle for j. We place Γ such that the point p superimposes on the corner for the T-shape in the projection on the xy-plane. Since the highest faces of B, C and A are in this order according to z-coordinate, the adjacencies of the vertices in P with $\{A, B, C\}$ are correct. By appropriately choosing the areas for the rectangles, we ensure that all the boxes for the vertices of P have their highest faces above that of B and that the invariant is maintained.

Case 2: A vertex of P is adjacent to all of $\{A, B, C\}$. In this case at least one of $\{A, B, C\}$ has only one neighbor in P. Assume first that a vertex b $(= ab = bc)$ of P is adjacent to all of $\{A, B, C\}$ and this is the only neighbor of B. Then we follow the steps for *Case 1a* with $b = j$ (and some vertex between x and b as i). But when we finally place this representation of P in the corner for the T-shape of $\{A, B, C\}$ we find the point p to superimpose on this corner as follows. The point p is on the line containing the top side of the rectangle for b and has x-coordinate between the right sides of the rectangles for b and $(b-1)$.

If a vertex c $(= bc = ca)$ is adjacent to all of $\{A, B, C\}$ and is the only neighbor of C in P, then we follow the steps of *Case 1b* with $i = 2$ and $j = ab$. We find the point p to superimpose on the corner for the T-shape of $\{A, B, C\}$ as follows. The point p is on the line containing the top side of the rectangle for $j = ab$ and has x-coordinate between the left sides of the rectangles for 1 and 2.

If a vertex a $(= ab = ca)$ is adjacent to all of $\{A, B, C\}$ and is the only neighbor of A in P, then we number the vertices of P in the clockwise order starting from the clockwise neighbor of a and ending at $a = n$. We use the **SC Algorithm** to find a representation of P with rectangles and give the needed heights to obtain a representation with boxes. In the corner for the T-shape of $\{A, B, C\}$, we superimpose the intersection point for the lines containing the top side of the rectangle of n and the right side of the rectangle for bc.

Finally, we consider the case when there is a chord between ca and another vertex on the path P_a. Take the innermost such chord and let its other end-vertex be t. Then consider the two subgraphs P_1 and P_2 induced by all the vertices outside the chord and inside the chord (along with the two vertices ca and t). P_1 does not contain any chord from ca; thus we use the algorithm above to obtain a representation of P_1; denote this by Γ'. In this representation ca and t will play the roles of 1 and n, respectively. Each vertex of P_2 is adjacent to A and we find a proportional contact representation of P_2 and attach it with Γ' as follows. We use the **SC Algorithm** to find a proportional contact representation of P_2 with rectangles in the plane and delete the rectangles for ca and t from it to obtain Γ''. In Γ', we change the height of the rectangle R_1 for $ca = 1$ to increase its area so that its bottom side extends past the bottom side of the rectangle R_n for $t = n$. Then we place Γ'' (after reflecting with respect to the x-axis and possible stretching) touching the right side of R_1 and the bottom side of R_n. Since the **SC Algorithm** can accommodate any given area for the layout, we can change the heights of the boxes for the vertices in P_2 to maintain the invariant.

Thus with the top-down traversal of $\mathcal{T}$, we obtain a proportional contact representation for O. We assumed that each piece of O is maximal outerplanar. However in the contact representation, for each edge (u, v), either a face of the box R_u for u is adjacent to the box R_v for v and no other box; or a face of R_v is adjacent to R_u and no other box. In both cases the adjacency between these two coxes can be removed without affecting any other adjacency. Thus this algorithm holds for any nested outerplanar graph O.   □

6   Conclusions and Future Work

We proved that nested maximal outerplanar graphs have cube-contact representations and nested outerplanar graphs have proportional box-contact representations. These classes are special cases of k-outerplanar graphs, and the set of k-outerplanar graphs for all $k > 0$ is equivalent to the class of all planar graphs. Thus this approach might help show that all planar graphs have proper cube-contact, or proportional box-contact representations.

Acknowledgments. We thank Therese Biedl, Steve Chaplick, Stefan Felsner, and Torsten Ueckerdt for discussions about this problem.

References

1. Alam, M.J., Kaufmann, M., Kobourov, S.G.: On contact graphs with cubes and proportional boxes. CoRR abs/1510.02484 (2015)
2. Alam, M.J., Kobourov, S.G., Liotta, G., Pupyrev, S., Veeramoni, S.: 3D proportional contact representations of graphs. In: Bourbakis, N.G., Tsihrintzis, G.A., Virvou, M. (eds.) Information, Intelligence, Systems and Applications. pp. 27–32. IEEE (2014)
3. Alam, M.J., Biedl, T.C., Felsner, S., Gerasch, A., Kaufmann, M., Kobourov, S.G.: Lineartime algorithms for hole-free rectilinear proportional contact graph representations. Algorithmica **67**(1), 3–22 (2013)
4. Biedl, T.C., Velazquez, L.E.R.: Drawing planar 3-trees with given face areas. Comput. Geom. Theory Appl. **46**(3), 276–285 (2013)
5. Bremner, D., Evans, W., Frati, F., Heyer, L., Kobourov, S.G., Lenhart, W.J., Liotta, G., Rappaport, D., Whitesides, S.H.: On representing graphs by touching cuboids. In: Didimo, W., Patrignani, M. (eds.) GD 2012. LNCS, vol. 7704, pp. 187–198. Springer, Heidelberg (2013)
6. Duncan, C.A., Gansner, E.R., Hu, Y.F., Kaufmann, M., Kobourov, S.G.: Optimal polygonal representation of planar graphs. Algorithmica **63**(3), 672–691 (2012)

7. Evans, W., Felsner, S., Kaufmann, M., Kobourov, S.G., Mondal, D., Nishat, R.I., Verbeek, K.: Table cartograms. In: Bodlaender, H.L., Italiano, G.F. (eds.) ESA 2013. LNCS, vol. 8125, pp. 421–432. Springer, Heidelberg (2013)
8. Felsner, S., Francis, M.C.: Contact representations of planar graphs with cubes. In: Hurtado, F., van Kreveld, M.J. (eds.) Symposium on Computational Geometry, pp. 315–320 (2011)
9. Koebe, P.: Kontaktprobleme der konformen Abbildung. Berichte uber die Verhandlungen der Sachsischen Akad. der Wissenschaften zu Leipzig. Math.-Phys. Klasse **88**, 141–164 (1936)
10. Kozminski, K., Kinnen, E.: Rectangular duals of planar graphs. Networks **15**(2), 145–157 (1985)
11. van Kreveld, M.J., Speckmann, B.: On rectangular cartograms. Comput. Geom. **37**(3), 175–187 (2007)
12. Mondal, D., Nishat, R.I., Rahman, M.S., Alam, M.J.: Minimum-area drawings of plane 3-trees. J. Graph Algorithms Appl. **15**(2), 177–204 (2011)
13. Thomassen, C.: Interval representations of planar graphs. J. Comb. Theory Ser. B **40**(1), 9–20 (1988)
14. Ungar, P.: On diagrams representing maps. J. Lond. Math. Soc. **28**, 336–342 (1953)
15. Yeap, K.H., Sarrafzadeh, M.: Floor-planning by graph dualization: 2-concave rectilinear modules. SIAM J. Comput. **22**, 500–526 (1993)

Orthogonal Layout
with Optimal Face Complexity

M. Jawaherul Alam[1], Stephen G. Kobourov[1], and Debajyoti Mondal[2](✉)

[1] Department of Computer Science, University of Arizona, Tucson, USA
{mjalam,kobourov}@cs.arizona.edu,
[2] Department of Computer Science, University of Manitoba, Winnipeg, Canada
jyoti@cs.umanitoba.ca

Abstract. We study a problem motivated by rectilinear schematization of geographic maps. Given a biconnected plane graph G and an integer $k \geq 0$, does G have a strict-orthogonal drawing with at most k reflex angles per face? For $k = 0$ the problem is equivalent to realizing each face as a rectangle. The problem can be reduced to a max-flow problem in some linear-size nonplanar network, but the best solutions require $\Omega(n^{1.5} \log n \log k)$ time. We describe a graph matching approach that can decide strict-orthogonal drawability for arbitrary reflex complexity k in $O((nk)^{1.5})$ time, which is faster for constant values of k. In contrast, if the embedding is not fixed, we prove that it is NP-complete to decide whether a planar graph admits a strict-orthogonal drawing with reflex face complexity 4.

1 Introduction

Map schematization is a problem of considerable interest in geography, cartography, information visualization and computational geometry. Rectangular (countries are rectangles) and rectilinear (borders are made of orthogonal line segments) schematizations have been studied for over 80 years; see the comprehensive survey by Tobler [19]. While rectangular schematizations sometimes must distort the topology of the map (e.g., no four mutually neighboring countries can be represented by contact of rectangles), rectilinear schematizations can preserve the topology, at the expense of more complicated country shapes. We consider the problem of rectangular schematization where the "complexity" of each country (as defined by the number of reflex corners) is minimized. We also consider the case where different countries are allowed to have different complexities. We describe efficient algorithms for both of these scenarios.

An *orthogonal drawing* of a planar graph $G = (V, E)$ in $\mathbb{R}^2$ is a planar drawing of G such that each vertex $v \in V$ is drawn as a point and each edge $(u, v) \in E$ is drawn as a rectilinear (axis-aligned) path between the points that correspond to u and v. A *t-bend orthogonal drawing* of G is an orthogonal drawing of G, where each edge is drawn as an orthogonal polyline with at most t bends. An orthogonal drawing is *strict* if it does not contain any bend, i.e., it is a 0-bend orthogonal drawing. In the literature such a drawing is also referred to as *bendless*

© Springer-Verlag Berlin Heidelberg 2016
R.M. Freivalds et al. (Eds.): SOFSEM 2016, LNCS 9587, pp. 121–133, 2016.
DOI: 10.1007/978-3-662-49192-8_10

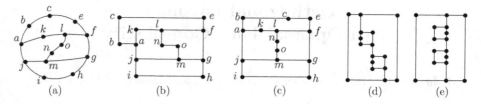

Fig. 1. (a) A plane graph G. (b) A strict-orthogonal drawing of G with reflex face complexity 1. (c) A rectangular drawing of G. (d)–(e) Two strict-orthogonal drawings (0-bend drawings) of the same graph with different reflex face complexities.

or *no-bend* orthogonal drawing [17]. If G is a *plane graph* (i.e., a planar graph with a fixed planar embedding), then an orthogonal drawing of G is additionally constrained to respect the given planar embedding. The *reflex face complexity* of an orthogonal drawing Γ is the smallest integer k such that each inner face of Γ contains at most k reflex angles, and the outer face of Γ contains at most $k + 4$ reflex angles. Thus in an orthogonal drawing of G with reflex face complexity k, each face of G is drawn as an orthogonal polygon with at most $2k + 4$ sides. Figure 1(a)-(c) show a graph G and two strict-orthogonal drawings of G.

From technical drawings and wiring schematics to transportation network layouts, orthogonal drawing (or layout) is one of the most common techniques for visualizing planar graphs [6,11,16] and is also a popular visualization technique provided by most network layout systems (e.g., yEd [21], graphviz [7], and OGDF [3]). Early work on orthogonal layouts was done by Valiant [20] and Leiserson [13] in the context of VLSI design. The input graphs are assumed to be planar and with maximum-degree four, although models incorporating higher degree graphs were introduced later by Tamassia [18] and Fößmeier and Kaufmann [8].

1.1   Optimization Goals and Challenges

The number of reflex corners per face and the number of bends per edge are two important aesthetic criteria in an orthogonal drawing, and a good drawing usually minimizes these two parameters. Note that these two optimization criteria are important not only from the point of view of VLSI complexity and floorplanning, but also due to the readability and aesthetics of the layout. Minimizing the total number of bends over all possible embeddings of the input planar graph is NP-hard [9]. However, for maximum-degree-4 plane graphs, Tamassia [18] uses a maximum-flow based technique to solve the problem in $O(n^{7/4}\sqrt{\log n})$-time. Later, Cornelsen and Karrenbauer [4] proposed a variation of this maximum-flow based approach that improves the running time to $O(n^{3/2})$. Although these algorithms can be adapted to bound the number of bends per edge, there exist more specialized algorithms for such optimizations. For example, Bläsius et al. [1,2] gave efficient algorithms to bound the number of bends per edge, which can also optimize any convex cost associated with the edges of the input graph, even in the variable embedding setting for some specific cost functions.

Note that minimization of the number of total bends, or the number of bends per edge cannot bound the reflex face complexity, see Fig. 1(d)–(e), but a drawing with reflex face complexity k ensures that the number of bends per edge is at most $2k + 4$. Given a plane graph G with four prescribed corner vertices, Miura et $al.$ [15] showed how to decide whether G admits a strict-orthogonal drawing with reflex face complexity 0 (also known as $rectangular$ $drawings$, as shown in Fig. 1(c)), that respects the given corners. They reduced the problem of rectangular drawing to the problem of finding a perfect matching in some graph, which leads to an $O(n^{1.5}/\log n)$-time algorithm. If the four corner vertices are not given, then a trivial solution is to try all possible options for the corner vertices. A variant of Tamassia's [18] flow-based approach can solve this problem in $O(n\log^2 n)$ time, even when the corners are not given in the input.

Tamassia's [18] flow-based approach can be modified to decide strict-orthogonal drawability for arbitrary reflex complexity k by solving a maximum-flow problem. One can adapt many other existing variations of Tamassia's formulation [1,2,4] to decide strict-orthogonal drawability with a given reflex face complexity. Unfortunately, all these modifications require solving some maximum-flow problem in some nonplanar network whose size is linear in the number of vertices n in the input graph. Based on the known complexities for computing a maximum flow, the algorithms in this setting require $\Omega(n^{1.5}\log n\log k)$ running time for solving strict-orthogonal drawability with reflex face complexity k. Thus an interesting question is whether the matching-based approach of Miura et $al.$'s [15] can be generalized to more efficiently decide orthogonal drawability with reflex face complexity k.

1.2   Our Contributions

We study the problem of orthogonal drawing of a biconnected planar graph with a given reflex face complexity k. Note that since every vertex in an orthogonal drawing has degree ≤ 4, we consider only max-degree-4 graphs in this paper. In the fixed embedding setting, we give an algorithm based on bipartite graph matching to compute a strict-orthogonal drawing of a biconnected plane graph G with any given reflex face complexity k (if such a drawing exists). Furthermore, given the nonnegative integers $k_0, k_1, \ldots, k_r$ for the faces $f_0, f_1, \ldots, f_r$ of G, our algorithm can compute a strict-orthogonal drawing of G, with at most k_i reflex corners in each face $f_i, i \in \{0, 1, \ldots, r\}$. For example, one can specify $k_i = k$ for each inner face f_i, and $k_0 = 4$ for the outer face f_0 to compute a complexity-k tessellation of a rectangle.

Although perfect matching problems on bipartite graphs can be solved via maximum flow [14], the matching-based technique we present here does not use this relationship. Based on the best known time-complexity for computing a maximum matching, our matching-based algorithm runs in $O((nk)^{1.5})$ time, where k is largest k_i, which is asymptotically faster than any previous approaches in the practical setting where k is a constant. Our algorithm can also be extended to

124   M. Jawaherul Alam et al.

compute general (non-strict) orthogonal drawings as well as orthogonal drawings with at most t_i bends on each edge e_i, for some nonnegative integer t_i.

Finally, we show that if the embedding of the planar graph G is not given, deciding whether G has a strict-orthogonal drawing with a given reflex face complexity k is NP-complete, even when $k = 4$.

2 Strict-Orthogonal Drawing Algorithms for Plane Graphs

In this section we give our algorithm for deciding strict-orthogonal drawability of planar graphs with a given reflex face complexity, and discuss some subsequent generalizations. We begin with a preliminary result showing that to compute a strict-orthogonal drawing it suffices to specify the angles between pairs of consecutive edges around each vertex. We then describe our algorithm based on a perfect matching in a bipartite graph, proving the following main theorem:

Theorem 1. *Let G be an n-vertex biconnected plane graph with the faces $f_0, \ldots, f_r$. Given the nonnegative integers $k_0, \ldots, k_r$ with $k = \max_i\{k_i\}$, one can decide in $O((nk)^{1.5})$ time whether G has a strict-orthogonal drawing, where each face f_i has at most k_i reflex corners, and construct such a drawing if it exists.*

2.1 Orthogonal Drawing Using Angle Assignment

Tamassia [18] showed that an orthogonal drawing Γ of a biconnected plane graph G can be described by augmenting the embedding of G with the angles at the bends (*bend angles*) and the angles between pairs of consecutive edges around the vertices of G (*vertex angles*). For strict-orthogonal drawings (no bends), we only consider vertex angles. Specifically, an *angle assignment* is a mapping from the set $\{\pi/2, \pi, 3\pi/2\}$ to the angles of G, where each angle is assigned exactly one value. Although an angle assignment of G does not specify edge lengths, it can precisely describe the shape of Γ. Given an angle assignment Φ, one can test if Φ corresponds to a strict-orthogonal drawing by Lemma 1, which is implied from [18]:

Lemma 1. *An angle assignment Φ for a plane graph G corresponds to a strict-orthogonal drawing of G if and only if Φ satisfies the following conditions (P_1–P_2):*

(P_1) The sum of the assigned angles around each vertex v in G is 2π.
(P_2) the total assigned angle of every inner (respectively, outer) face f is $(\gamma-2)\pi$ (respectively, $(\gamma+2)\pi$), where γ is the number of vertices on the boundary of f.

Given an angle assignment Φ satisfying (P_1–P_2), one can obtain a strict-orthogonal drawing of G (i.e., the exact coordinates for the vertices) in linear time.

2.2 Bipartite Graph Matching Formulation

Here we prove Theorem 1 by reducing the drawing problem to the problem of finding a perfect matching in a bipartite graph. We construct a bipartite graph $B(G)$ so that one can compute a strict-orthogonal drawing of G with reflex face complexity k from a perfect matching of $B(G)$, and vice versa. Although our result generalizes the rectangular drawing algorithm by Miura *et al.* [15], the bipartite graph we construct is quite different from the one in [15] and it gives the option of having reflex corners in a face.

Construction of $B(G)$: Let f_0 be the outer face and $f_1, \ldots, f_r$ be the inner faces of G; see Fig. 2. For each inner face $f_i, i \in \{1, \ldots, r\}$ of G we have four vertices $x_i^1, x_i^2, x_i^3, x_i^4$ in $B(G)$, as shown with white squares with thin boundaries. These vertices will correspond to four $\pi/2$ angles in f_i. We also have k_i pairs of vertices $a_i^1, b_i^1, \ldots, a_i^{k_i}, b_i^{k_i}$ associated with f_i, as shown with white and gray squares with bold boundaries. For each $j \in \{1, \ldots, k_i\}$, there is an edge (a_i^j, b_i^j). Later, every a-vertex will correspond to a $\pi/2$ angle, and every b-vertex will correspond to a $3\pi/2$ angle in f_i. In each internal face f_i, there are only k_i pairs of a and b-vertices, which will bound the number of reflex corners of f_i in the final drawing. Observe that by Condition (P_2) of Lemma 1, each internal face of G has exactly four $\pi/2$ angles more than its $3\pi/2$ angles, and hence we have four more white squares than gray squares. Similarly, the outer face f_0 must contain four $3\pi/2$ angles more than its $\pi/2$ angles. Thus for the face f_0, we have four vertices y_0^1, y_0^2, y_0^3 and y_0^4 representing $3\pi/2$ angles, and $p = k_0 - 4$ pairs of vertices $a_0^1, b_0^1, \ldots, a_0^p, b_0^p$. Call the x- and the a-vertices the *convex face-vertices* and the y- and b-vertices the *reflex face-vertices*.

In addition to the face-vertices above, $B(G)$ also has boundary-vertices that correspond to the vertices of G. For each degree-4 vertex v in G, let f_i, f_j, f_k, f_l be the four faces incident to v. For each $h \in \{i, j, k, l\}, B(G)$ has a vertex v_h, which is adjacent to all the convex face-vertices associated with f_h; see vertex h in Fig. 2. We refer to these vertices as *convex boundary-vertices*. Each of these convex boundary-vertices will choose a convex face-vertex ensuring four $\pi/2$ angles around v. For each degree-3 vertex v incident to the faces $f_i, f_j, f_k, B(G)$ has three vertices v_i, v_j, v_k, which are adjacent to all the convex face-vertices of their corresponding faces. We also have an additional vertex v^* in $B(G)$, which is a common neighbor for v_i, v_j, v_k; see vertex n^* in Fig. 2. Again we refer to these vertices v_i, v_j, v_k as *convex boundary-vertices*, and the vertex v^* as the *central-vertex*. Intuitively, v^* will match with one of its incident vertices leaving two vertices among $\{v_i, v_j, v_k\}$, which will choose two $\pi/2$ angles around v. Finally, if v is a degree-2 vertex incident to the faces f_i and f_j, then we have two vertices v' and v'' in $B(G)$ that are adjacent to each other. We call v' a *convex boundary-vertex* (shown as gray circle), and v'' a *reflex boundary-vertex* (shown as white circle). The vertex v' is adjacent to all the convex face-vertices associated with f_i and f_j, and the vertex v'' is adjacent to all the reflex vertices associated with f_i and f_j; see vertex m in Fig. 2. Note that degree-3 and degree-4 vertices of G

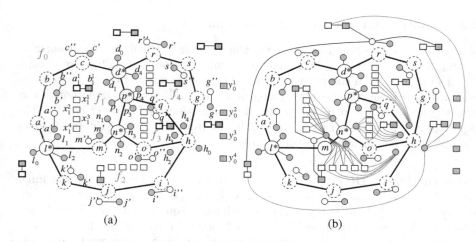

(a)   (b)

Fig. 2. (a) A plane graph G (induced by the bold edges), and the construction of $B(G)$ with $k_0 = 4$, $k_1 = k_2 = k_3 = k_4 = 1$, where only a few edges of $B(G)$ are shown. (b) The remaining edges in $B(G)$: the edges shown are the ones incident to the convex boundary vertices for a degree-4 (red), a degree-3 (green), a degree-2 (blue) vertices and the ones incident to reflex boundary vertices for two degree-2 vertices (black)(Color figure online).

do not have any associated reflex boundary-vertices in $B(G)$, since they cannot induce $3\pi/2$ angles in an orthogonal drawing; see Lemma 1, Condition (P_1).

This completes the construction of $B(G)$. It is bipartite, as shown in gray and white in Fig. 2.

Reduction: The following lemma reduces our problem to the problem of finding a perfect matching in some corresponding graph.

Lemma 2. *There is a perfect matching in $B(G)$ if and only if G has a strict-orthogonal drawing, where each face f_i contains at most k_i reflex corners.*

Proof. Assume that $B(G)$ has a perfect matching M; see Fig. 3(a)–(b). From this matching, we compute an angle assignment Φ for G from the set $\{\pi/2, \pi, 3\pi/2\}$ so that Φ satisfies Conditions $(P_1$–$P_2)$ of Lemma 1.

Consider an arbitrary face f_i of G. We assign an angle inside f_i (at some vertex v) the value $\pi/2$ if the corresponding boundary-vertex in $B(G)$ is matched to some convex face-vertex of f_i. For example, the convex boundary-vertices associated with the vertices b and h in Fig. 3(b) are determining $\pi/2$ angles around b and h in Fig. 3(c). Similarly, a $3\pi/2$ angle is assigned to v when its corresponding boundary-vertex in $B(G)$ is matched with a reflex face-vertex for f_i, e.g., see vertex m in Fig. 3(b). Otherwise, the boundary-vertex is either matched with some central-vertex, or another boundary vertex (e.g., see vertex c). In both cases we assign the corresponding angle the value π.

Note that the above rules may lead to a conflict at some degree-2 vertex, when it has both convex and reflex boundary-vertices matched to the convex and reflex face-vertices of the same face. For example, the vertex q in Fig. 3(b) has its boundary vertices matched with the face-vertices in the same face f_3. In such a case we assign the angle at v a value of π (inside the corresponding face). Since M is a perfect matching, the construction of $B(G)$ implies that each inner face has exactly four more $\pi/2$ angles than $3\pi/2$ angles. Similarly, the outer face f_0 contains exactly four more $3\pi/2$ angles than $\pi/2$ angles. Thus Condition (P_2) of Lemma 1 is satisfied for each face of G.

Consider now the assignment of angles around each vertex v of G. If $\deg(v) = 4$, then all its four convex boundary-vertices are matched to some convex face-vertices, and hence it has exactly four $\pi/2$ angles. If $\deg(v) = 3$, then exactly one of its three convex boundary-vertices is matched with v^*, and hence it has two $\pi/2$ angles and one π angle. Finally, if $\deg(v) = 2$, then it either has two π angles (because v' and v'' are either matched to each other or to the face-vertices in the same face); or it receives exactly one $\pi/2$ angle and exactly one $3\pi/2$ angle. Thus the sum of angles around each vertex is 2π, satisfying Condition (P_1) of Lemma 1. By Lemma 1, this angle assignment gives an orthogonal drawing of G. Since each face f_i can have at most k_i reflex boundary-vertices matched to its k_i reflex face-vertices, the number of reflex corners in the drawing of f_i is at most k_i; see Fig. 3(c).

Conversely, if G has a strict-orthogonal drawing Γ, where each face f_i of G has at most k_i reflex corners, then Γ gives a perfect matching M in G, as follows. For each face f_i of G, traverse around its drawing in Γ, and for each $\pi/2$ (respectively, $3\pi/2$) angle, match the corresponding boundary-vertex to a convex (respectively, reflex) face-vertex of f_i. There are always sufficiently many face-vertices, since each inner face f_i is associated with k_i pairs of convex and reflex face-vertices, and the outer face f_0 has exactly $p = k_0 - 4$ such pairs. It is straightforward to match face-vertices with boundary vertices such that the unmatched face vertices remain in pairs. Hence we can afterwards choose the edges between the unmatched pairs of face-vertices in M. For each degree-2 vertex with two π angles, we take the edge between its boundary-vertices in M. Finally, for each degree-3 vertex v, we match the boundary vertex corresponding to the π angle of v with v^*.   □

Time Complexity: The number of vertices $|V|$ in $B(G)$ is $O(nk)$, where $k = \max_i\{k_i\}$. Since there are $O(n)$ boundary-vertices and for each of the $O(n)$ faces, there are $O(k)$ face-vertices, the number of edges $|E|$ in $B(G)$ is again $O(nk)$. Hence the existence of a perfect matching in $B(G)$ can be tested in $O(\sqrt{|V|}|E|) = O(\sqrt{nk} \times nk) = O((nk)^{1.5})$ time using the Hopcroft-Karp algorithm [10].

2.3   General Orthogonal Drawing with a Given Face-Complexity

Here we extend our algorithm to general (non-strict) orthogonal drawing. Note that each bend in an orthogonal drawing can be thought of as a degree-2 vertex

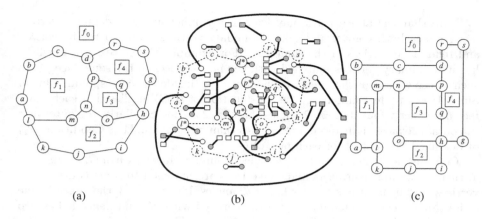

(a)   (b)   (c)

Fig. 3. (a) A biconnected plane graph G with maximum degree four, (b) a perfect matching in $B(G)$, and (c) a strict-orthogonal drawing of G with $k_0 = 4$ and $k_1 = k_2 = k_3 = k_4 = 1$.

on some edge in the graph (e.g., a subdivision of an edge). We use this observation to obtain the following lemma.

Lemma 3. *Let G be a biconnected plane graph with edges $e_1, \ldots, e_m$ and faces $f_0, f_1, \ldots, f_r$. Consider the set of non-negative integers $t_1, \ldots, t_m$ and $k_0, k_1, \ldots, k_r$. Let G_t be a graph obtained from G by subdividing each edge e_i exactly t_i times. Then G has an orthogonal drawing, where each edge e_i has at most t_i bends and each face f_i has at most k_i reflex corners if and only if G_t has a strict-orthogonal drawing where each face f_i has at most k_i reflex corners.*

Proof. Assume that G has a desired orthogonal drawing. For each bend point p, subdivide the corresponding edge at p. In this way each edge e_i is subdivided at most t_i times. For each edge e_i that has not been subdivided t_i times in this process, further subdivide it so that the total number of subdivisions is exactly t_i. Then this corresponds to a strict-orthogonal drawing of G_t, where each face f_i has at most k_i reflex corners.

Conversely, if Γ is a strict-orthogonal drawing of G_t, where each face f_i has at most k_i reflex corners, then a desired orthogonal drawing of G can be obtained from Γ by considering the degree two vertices (with angles $\pi/2$ and $3\pi/2$) of Γ as the bends of the corresponding edges in G. □

It is straightforward to use Lemma 3 to find a polynomial-time algorithm for orthogonal drawing that simultaneously bounds the reflex face complexity and the number of bends per edge. Our goal in this paper is to bound the reflex face complexity and we leave the task of designing fast algorithms optimizing multiple objectives as a future work. There exists specialized algorithms for bounding the number of bends per edge or for optimizing any convex cost associated with the edges of the input graph, even in the variable embedding setting for some specific cost functions [1,2].

3   NP-Hardness for Planar Graphs

In this section we prove that it is NP-complete to decide whether a planar biconnected graph admits a strict-orthogonal drawing with a given reflex face complexity k, even when $k = 4$. Throughout this section we denote this problem by MIN-REFLEX-DRAW.

Garg and Tamassia [9] proved that it is NP-hard to decide whether a maximum-degree-4 planar graph admits a strict-orthogonal drawing. This NP-hardness proof readily implies the NP-hardness of the problem of computing strict-orthogonal drawing with reflex face complexity k, but this proof does not hold if we restrict k to be a constant. On the other hand, our NP-hardness proof holds when $k = 4$, even when it is known that the input graph has a strict-orthogonal drawing.

We prove the NP-completeness with a reduction from the rectilinear monotone planar 3-SAT problem (RMP3SAT), which is NP-hard [5]. The input of an RMP3SAT instance I is a collection C of clauses over a set U of variables such that each clause contains at most three variables, and each clause is either positive or negative (i.e., all its variables are either positive or negative). Moreover, the corresponding *SAT-graph* G_I (i.e., a bipartite graph with vertex set $C \cup U$ and edge set $\{(x,y) | x \in C, y \in U, y \in x\}$) admits a planar drawing Γ satisfying the following properties:

- Each vertex in Γ is drawn as an axis-aligned rectangle. All the vertices representing variables lie along a horizontal line h (known as *backbone*).
- The vertices representing positive (respectively, negative) clauses lie on the top (respectively, bottom) half-plane of h. Each edge is drawn as a vertical line segment that is incident to the drawings of its end vertices.

The RMP3SAT problem asks to decide whether there is a satisfying truth assignment for U satisfying all clauses in C. RMP3SAT remains NP-hard even when each variable appears in at most four clauses [12].

Given an instance $I = (U, C)$ of RMP3SAT, where each variable appears in at least two and at most 4 clauses, we construct a planar graph H so that H has a strict-orthogonal drawing with face complexity 4, if and only if the RMP3SAT instance is satisfiable.

We construct H from the drawing Γ of the SAT-graph G_I; see Fig. 4(a). We first draw a polygon with holes (shown in gray) that represents each edge of Γ as a tunnel; see Fig. 4(b). We then place the drawing onto a regular grid H, where each hole is a collection of grid cells, as shown in Fig. 4(c) using dark regions. Then for each variable and each clause, we assign a corresponding variable cell and a corresponding clause cell in H. Figure 4(c) depicts the cells corresponding to the variable x_1 and clauses c_1, c_2, c_4, respectively. In each variable cell, we create a variable-staircase structure of length three, as shown in the gray region of Fig. 4(f), such that the base of the staircase is adjacent to the bottom side of the cell. Note that this staircase contributes to four reflex corners in the variable cell, which can be transferred to the cell lying below the variable cell by flipping the staircase vertically. For each edge connecting a variable to a clause, we first find a sequence of cells connecting the variable cell to the clause cell, and then add a staircase of length two and a 4×4 grid structure (see Fig. 4(f)) to each of these cells, as described below.

The staircase is added at a corner of the cell that cannot be flipped and contributes to two reflex corners to the cell; we do not show these staircases in the schematic representations of Figs. 4(c)–(e). The grid is added to one side of the cell such that it contributes to two reflex corners to this cell, which can be transferred to the cell adjacent to it by flipping. Since $k = 4$, none of the cells on the path from the variable to the clause cell can contain more than one grid structure. The grid structures are added exploiting this constraint along the variable to clause path such that

- if the clause is positive, then the placement of the variable-staircase in the variable cell eventually forces a grid structure to fall into the corresponding clause cell. On the other hand,
- if the clause is negative, then the placement of the variable-staircase outside of the variable cell will force a grid structure to fall into the corresponding clause cell.

Finally, for each clause c, we add a staircase of length $(6 - 2|c|)$ at the corner of its clause cell, where $|c|$ is the number of variables in c. Such a clause-staircase ensures that at least one of the grid structures incident to the clause cell must lie outside of the clause cell; we do not show these staircases in the schematic representations of Figs. 4(c)–(e).

Let the resulting drawing be Γ'. It is straightforward to carry out the above construction in polynomial time, and one can observe that any strict-orthogonal drawing must respect the axis-alignments of the edges of the underlying graph (up to rotation or reflection).

Theorem 2. *It is NP-complete to decide if a planar graph admits a strict-orthogonal drawing with reflex face complexity 4.*

Proof. Let $I = (U, C)$ be an instance of RMP3SAT, and let H be the corresponding planar graph. We now prove that H admits a strict-orthogonal drawing with face complexity 4, if and only if the RMP3SAT instance is satisfiable.

Given a drawing of H with reflex face complexity 4, we use the above/below (i.e., inside/outside of a variable cell) orientations of a variable staircase to find the truth value of the corresponding variable; see Fig. 4(e). By construction, no clause cell can have all its adjacent grid structures inside it, otherwise it would have at least $(6 - 2|c|) + 2|c| > 4$ reflex corners. Consequently, every clause cell must have one of its grid-structures M outside of the clause cell. Recall that any variable cell that receives a variable staircase obtains at least 4 reflex corners, and hence cannot have any grid structure inside it. Therefore, the grid structure M will force the corresponding variable staircase to lie outside or inside of its variable-cell depending on whether the clause is positive or negative. We assign the outside and inside configurations the values true and false, respectively, which implies that each clause must be satisfied.

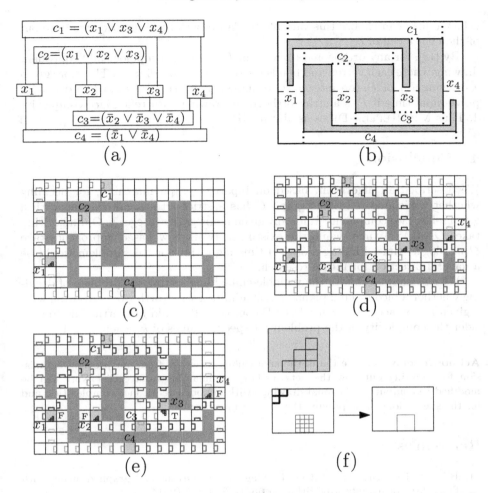

Fig. 4. (a) G_I, (b) Γ', (c)–(d) illustration of the reduction, where the variable and clause cells are shaded, (e) computing truth assignment: $x_1 = x_2 = x_4 = $ false, $x_3 = $ true, (f) insertion of a staircase and a grid.

On the other hand, given a satisfying truth assignment for I, we orient the variable-staircases above/below depending on whether it is false/true. The placement for the grid structures is then straightforward, which is guided by the restrictions on the variable to clause paths. Therefore, to verify that the reflex face complexity is bounded by 4, we only need to examine the clause cells. In the following we show that for any clause cell with more than 4 reflex corners can be fixed without introducing any new bad cells. Let c be a clause that contains all its incident grid-structures inside the cell yielding $(6 - 2|c|) + 2|c| > 4$ reflex face complexity. Without loss of generality assume that the clause is positive. Since c is satisfied, at least one of its variable-staircase must lie outside of its variable cell. We now can choose this variable to clause path to flip a grid structure out of the clause cell of c.

By [18], for any orthogonal drawing of H, there is a topologically equivalent drawing where each vertex and bends are on integer coordinates. Therefore given a drawing Γ_H of H (on integer coordinates), it is straightforward to decide in polynomial time if Γ is a strict-orthogonal drawing with reflex face complexity 4. Thus MIN-REFLEX-DRAW is also in NP.   □

4   Conclusion

We described an algorithm, based on bipartite graph matching, for deciding whether a biconnected plane graph G has a strict-orthogonal drawing with a given reflex face complexity k, for any given nonnegative integer k. Our algorithm takes $O((nk)^{1.5})$ time, while the existing network-flow based approaches take $\Omega(n^{1.5} \log n)$ time. Finding a $o(n^{1.5})$ time algorithm for this problem would be a natural direction for future research.

We also showed that in the variable-embedding setting the problem of deciding whether a biconnected planar graph admits a strict-orthogonal drawing with a given reflex face complexity 4 is NP-complete. It would be worthwhile to consider the complexity of the problem for specific values of k, where $k < 4$.

Acknowledgement. We thank the anonymous reviewers from our previous submission for pointing out how the network-flow formulations from earlier work can be modified to compute orthogonal drawings with bounded reflex face complexities, and for the suggestions on improving the NP-hardness result.

References

1. Bläsius, T., Krug, M., Rutter, I., Wagner, D.: Orthogonal graph drawing with flexibility constraints. Algorithmica **68**(4), 859–885 (2014)
2. Bläsius, T., Rutter, I., Wagner, D.: Optimal orthogonal graph drawing with convex bend costs. In: Fomin, F.V., Freivalds, R., Kwiatkowska, M., Peleg, D. (eds.) ICALP 2013, Part I. LNCS, vol. 7965, pp. 184–195. Springer, Heidelberg (2013)
3. Chimani, M., Gutwenger, C., Jünger, M., Klau, G., Klein, K., Mutzel, P.: The open graph drawing framework. In: Handbook of Graph Drawing and Visualization, pp. 543–571 (2013)
4. Cornelsen, S., Karrenbauer, A.: Acclerated bend minimization. J. Graph Algorithms Appl. **16**(3), 635–650 (2012)
5. de Berg, M., Khosravi, A.: Optimal binary space partitions for segments in the plane. Int. J. Comput. Geom. Appl. **22**(3), 187–206 (2012)
6. Di Battista, G., Eades, P., Tamassia, R., Tollis, I.G.: Graph Drawing: Algorithms for the Visualization of Graphs, 3rd edn. The MIT Press, Cambridge (2009)
7. Ellson, J., Gansner, E.R., Koutsofios, L., North, S.C., Woodhull, G.: Graphviz—open source graph drawing tools. In: Mutzel, P., Jünger, M., Leipert, S. (eds.) GD 2001. LNCS, vol. 2265, p. 483. Springer, Heidelberg (2002)

8. Fößmeier, U., Kaufmann, M.: Drawing high degree graphs with low bend numbers. In: Brandenburg, F.J. (ed.) GD 1995. LNCS, vol. 1027, pp. 254–266. Springer, Heidelberg (1996)

9. Garg, A., Tamassia, R.: On the computational complexity of upward and rectilinear planarity testing. SIAM J. Comput. **31**(2), 601–625 (2001)

10. Hopcroft, J.E., Karp, R.M.: An $n^{5/2}$ algorithm for maximum matchings in bipartite graphs. SIAM J. Comput. **2**(4), 225–231 (1973)

11. Kaufmann, M., Wagner, D.: Drawing Graphs: Methods and Models. LNCS, vol. 2025. Springer, London (2001)

12. Kempe, D.: On the complexity of the "reflections" game (2003). http://www-bcf. usc.edu/dkempe/publications/reflections.pdf

13. Leiserson, C.E.: Area-efficient graph layouts (for VLSI). In: Symposium on Foundations of Computer Science (FOCS), pp. 270–281 (1980)

14. Leiserson, C.E., Cormen, T.H., Stein, C., Rivest, R.: Introduction to Algorithms. Prentice Hall, Englewood Cliffs (1999)

15. Miura, K., Haga, H., Nishizeki, T.: Inner rectangular drawings of plane graphs. Int. J. Comput. Geom. Appl. **16**(2–3), 249–270 (2006)

16. Nishizeki, T., Rahman, M.S.: Planar Graph Drawing. World Scientific, Singapore (2004)

17. Rahman, M.S., Egi, N., Nishizeki, T.: No-bend orthogonal drawings of subdivisions of planar triconnected cubic graphs. IEICE Trans. **88–D**(1), 23–30 (2005)

18. Tamassia, R.: On embedding a graph in the grid with the minimum number of bends. SIAM J. Comput. **16**(3), 421–444 (1987)

19. Tobler, W.: Thirty five years of computer cartograms. Ann. Assoc. Am. Geogr. **94**, 58–73 (2004)

20. Valiant, L.G.: Universality considerations in VLSI circuits. IEEE Trans. Comput. **30**(2), 135–140 (1981)

21. Wiese, R., Eiglsperger, M., Kaufmann, M.: yFiles: visualization and automatic layout of graphs. In: Mutzel, P., Jünger, M., Leipert, S. (eds.) GD 2001. LNCS, vol. 2265, pp. 453–454. Springer, Heidelberg (2002)

L-Drawings of Directed Graphs

Patrizio Angelini[1], Giordano Da Lozzo[2]([✉]), Marco Di Bartolomeo[2],
Valentino Di Donato[2], Maurizio Patrignani[2],
Vincenzo Roselli[2], and Ioannis G. Tollis[3]

[1] Tübingen University, Tübingen, Germany
angelini@informatik.uni-tuebingen.de
[2] Roma Tre University, Rome, Italy
{dalozzo,dibartolomeo,didonato,patrigna,roselli}@dia.uniroma3.it
[3] University of Crete and Institute of Computer Science-FORTH, Heraklion, Greece
tollis@csd.uoc.gr

Abstract. We introduce *L-drawings*, a novel paradigm for representing
directed graphs aiming at combining the readability features of orthogonal drawings with the expressive power of matrix representations. In an
L-drawing, vertices have exclusive x- and y-coordinates and edges consist
of two segments, one exiting the source vertically and one entering the
destination horizontally.

We study the problem of computing L-drawings using minimum
ink. We prove its NP-completeness and provide a heuristic based on
a polynomial-time algorithm that adds a vertex to a drawing using the
minimum additional ink. We performed an experimental analysis of the
heuristic which confirms its effectiveness.

1 Introduction

Drawing directed graphs is a challenging goal to which a vast literature has been
dedicated [15,23]. In fact, most of the theoretical and applied tasks concerning
these graphs turned out to be difficult. To give a few examples, even for planar
and acyclic graphs, it is hard to decide whether they admit an upward planar
drawing [9]; if a directed graph contains directed cycles, it is hard to reverse the
minimum number of edges to make it acyclic [13,17], which is the first step of the
renown Sugiyama approach [23]. From a practical perspective, the more directed
cycles it has, the less a hierarchical drawing of it becomes meaningful, strongly
reducing the possibility of obtaining a clear and unambiguous representation.

In this paper we introduce a novel drawing paradigm specifically conceived
for directed graphs, which combines orthogonal drawings with matrix representations. Namely, we call *L-drawing* a drawing where each vertex has exclusive
x- and y-coordinates and each directed edge has two orthogonal segments, one

Angelini was partially supported by DFG grant Ka812/17-1. Da Lozzo, Di
Bartolomeo, Di Donato, Patrignani, and Roselli were partially supported by MIUR
project "AMANDA – Algorithmics for MAssive and Networked DAta", prot.
2012C4E3KT_001.

R.M. Freivalds et al. (Eds.): SOFSEM 2016, LNCS 9587, pp. 134–147, 2016.
DOI: 10.1007/978-3-662-49192-8_11

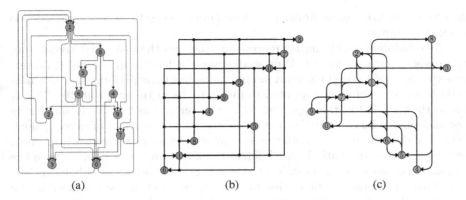

(a)   (b)   (c)

Fig. 1. (a) A hierarchical drawing with "backloop routing" produced by yEd [24], (b) an OOD, and a (c) minimum-ink L-drawing of the same connected random directed graph.

leaving the source vertically and one entering the destination horizontally. Edges are allowed both to overlap and to intersect. Graphically, the joint between the horizontal and the vertical segment of an edge is drawn as a small circular arc, allowing the user to easily identify the edges even in the presence of overlaps and intersections. We remark that L-drawings are strictly related to the popular *confluent drawing* style [7], which also leverages partially collinear edges and smoothened bends to reduce the visual complexity of the representation.

An example of L-drawing is in Fig. 1(c); further examples can be found in [1].

This paradigm is inspired by the *overloaded orthogonal drawings* [19,20] of directed acyclic graphs, in which vertices have exclusive x- and y-coordinates and the edges consist of two segments, one leaving the source from the top and one entering the destination from the left. For graphs that are not acyclic, a minimal set of edges is selected to be drawn backward, leaving the source from the bottom and entering the destination from the right. Edges are hence always drawn with a single bend turning clockwise. As long as the graph has few directed cycles, this model is extremely effective, as also testified by user studies [8]. L-drawings can in fact be seen as a generalization of this model to graphs that may contain many directed cycles, so that edges are allowed both to turn clockwise and counterclockwise. Note that, instead of using small circular arcs, ambiguities are solved in overloaded orthogonal drawings by placing a small dot on each overlapped bend (see Fig. 1(b)).

The relationship of L-drawings with orthogonal drawings is immediate and the benefits are immediate as well, since orthogonal drawings are widely recognized as one of the most readable drawing standards, ensuring a clear readability even in the presence of crossings [3,18]. We remark that a representation very similar to L-drawings was used in [2] as an intermediate step to compute orthogonal drawings of high-degree graphs in the Kandinsky model. However, the main purpose of [2] was to balance edges on the four sides of each vertex, so to reduce the area of the orthogonal drawing obtained once the vertices are expanded into rectangular boxes.

The relationship with matrix representations, and the benefits deriving from it, are also somehow evident. User studies suggest that matrix representations are extremely well suited for many simple tasks, but their performances dramatically decrease when it is requested to follow paths in the graph [8,10]. This is due to the fact that in this representation each vertex has two labels, one for its row and one for its column. Traversing a directed edge consists of moving along the row of the source vertex until the column of the destination vertex is reached. Traversing a directed path, instead, forces the user to repeatedly jump from the column of the vertex that is reached to the row of the same vertex when it is left. L-drawings overcome this limitation by moving the labels inside the matrix. The matrix itself is symbolically represented by the edges, that identify the portions of the rows and columns that have to be followed to connect adjacent vertices. A previous attempt to combine node-link and matrix representations was presented in [16], which introduced the NodeTrix visualization tool.

L-drawings have several strong points: (i) they always exist and are easy to compute; in fact any placement of the vertices such that no two vertices share the same horizontal or vertical grid line yields a valid L-drawing (the placement of the vertices uniquely determines the routing of the edges); (ii) they are not ambiguous, even for very dense graphs; (iii) they are particularly suited for interactive graph drawing, since vertices and edges can be easily added or removed preserving the user's mental map.

Since L-drawings always exist, we are interested in producing readable ones. One of the most desirable features of a graph drawing, especially when the graph is large, is that of having a small size. The classical notion of size of a drawing, namely the area of its bounding box, does not make much sense in this case, due to the requirement of using different x- and y-coordinates. We hence study the problem of minimizing the *ink* of the drawing, which is computed as the sum of the lengths of vertical and horizontal segments, where overlapping portions are counted only once.

We prove in Sect. 3 that this problem is NP-complete. Motivated by this, we describe in Sect. 4 an incremental heuristic, based on adding vertices one at a time using the minimum additional ink. This heuristic is experimentally evaluated in Sect. 5 against the optimal solution (when it was possible to compute one), against overloaded orthogonal drawings, and against a random placement of the vertices. We give definitions in Sect. 2 and conclude in Sect. 6 suggesting future lines of research. Because of space limitations, omitted proofs are deferred to the full version of the paper [1].

2   Preliminaries

In this paper we consider graphs $G = (V, E)$ that are directed. An edge (u, v) is an *outgoing* edge of u and an *incoming* edge of v. We allow G to contain both (u, v) and (v, u), but only a single copy of them; further, we do not allow loops (u, u).

In an *L-drawing* Γ of G each vertex $v \in V$ is assigned an exclusive integer x-coordinate x_v and y-coordinate y_v, and each edge (u, v) is drawn as a 1-bend polyline composed of a vertical segment incident to u and a horizontal segment incident to v. Note that, edges may cross and partially overlap. We resolve the ambiguity among crossings and bends by replacing each bend with a small rounded junction (see Fig. 1(c)).

The *ink* $ink(\Gamma)$ of an L-drawing Γ is the sum of the lengths of vertical and horizontal segments, where overlapping portions are counted only once. Since rounded junctions have all equal size, they are not taken into account when measuring ink.

We are interested in producing L-drawings of minimum cost. Both if the cost is computed in terms of area or in terms of ink, it is immediate that a drawing of minimum cost uses contiguous values for the coordinates of the vertices. Also, since area and ink do not change up to a translation of the whole drawing, in the rest of the paper we assume to use integer x- and y-coordinates in the range $[1 \ldots n]$. With the above assumptions, given a graph $G = (V, E)$, an L-drawing can be immediately obtained by choosing any two orderings π_x and π_y for the vertices in V, where π_x determines x-coordinates and π_y determines y-coordinates. We denote such a drawing by $\Gamma(\pi_x, \pi_y)$, and its ink by $ink(\pi_x, \pi_y)$. For any two orderings π_x and π_y, drawing $\Gamma(\pi_x, \pi_y)$ has area $n \times n$, where $n = |V|$. Hence, we focus on the problem of computing L-drawings with minimum ink. The corresponding decision problem is formally defined as follows.

Problem: MINIMUM-INK-L-DRAWING (MILD)
Instance: A directed graph $G = (V, E)$ and an integer k.
Question: Does G admit an L-drawing Γ such that $ink(\Gamma) \leq k$?

Let Γ be an L-drawing of G and let $ink_x(\Gamma)$ ($ink_y(\Gamma)$, respectively) be the amount of ink used for horizontal (vertical, respectively) segments. Obviously, $ink(\Gamma) = ink_x(\Gamma) + ink_y(\Gamma)$. In the following lemma we prove that $ink_x(\Gamma)$ ($ink_y(\Gamma)$, respectively) only depends on the horizontal (vertical, respectively) permutation of the vertices in Γ, which makes it possible to search for two optimal permutations independently.

Lemma 1. *Let G be a graph and let π_x be any permutation of its vertices. For any two permutations π'_y and π''_y we have that $ink_x(\pi_x, \pi'_y) = ink_x(\pi_x, \pi''_y)$. Symmetrically, $ink_y(\pi'_x, \pi_y) = ink_y(\pi''_x, \pi_y)$ for any two permutations π'_x and π''_x.*

Proof. Each edge (u, v) is composed of two segments, one incident to the source vertex u and one incident to the target vertex v. Hence, if we consider for each vertex only the segments incident to it, then all the segments of the drawing are eventually accounted for. Since overlaps are counted only once, $ink(\Gamma)$ is the sum, for every vertex, of the longest segments exiting it in the four directions North, East, South, and West. Thus, $ink_x(\Gamma)$ is the sum, for every vertex, of the longest segments exiting it in the directions East and West, while $ink_y(\Gamma)$ is the sum of the longest segments exiting it along North and South. Hence, $ink_x(\Gamma)$ only depends on π_x and $ink_y(\Gamma)$ only depends on π_y. □

The *complete graph* K_n is the directed graph $G = (V, E)$, where $|V| = n$ and for all ordered pairs $u, v \in V$, $u \neq v$, we have $(u, v) \in E$. In the following lemma we prove that any placement of the vertices of K_n on the $n \times n$ grid yields an L-drawing whose edges use all the segments of such a grid.

Lemma 2. *Any L-drawing Γ of K_n on the $n \times n$ grid uses $2n(n-1)$ ink.*

Clearly, Lemma 2 implies that any L-drawing of K_n on the $n \times n$ grid is a minimum-ink drawing, since empty rows or columns never reduce the ink. However, when a complete graph K_n is a subgraph of a larger graph, it might make sense to spread its vertices on a larger grid. In Lemma 3 we hence study the cost, in terms of ink, of this operation.

Lemma 3. *Any L-drawing of K_n on the $(n+h) \times (n+k)$ grid uses $2n(n-1) + n(h+k)$ ink.*

Let Γ be an L-drawing of K_n on the $(n+h) \times (n+k)$ grid. If $h > 0$ consider any horizontal grid line l not intersecting any vertex of K_n and such that at least one vertex is above l and at least one vertex is below l. For example, Fig. 2 shows a drawing of K_7 on the 11×10 grid and a possible grid line l in red.

Let p be the number of vertices above l ($n - p$ is the number of vertices below l). Line l is traversed by p vertical segments of Γ exiting the p vertices above l and entering the region below l. Also, l is traversed by $n - p$ vertical segments exiting the $n - p$ vertices below l and entering the region above l. Since vertices have exclusive x-coordinates, these $p + (n - p) = n$ vertical segments use distinct vertical grid lines; thus, removing line l yields an L-drawing Γ' on the $(n + h - 1) \times (n + k)$ grid that saves n ink. Analogous compressions can be performed starting from vertical grid lines that do not inter-

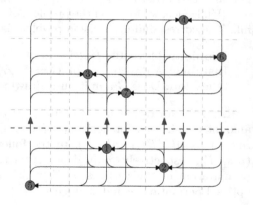

Fig. 2. An L-drawing of K_7 on the 11×10 grid and a removable grid line in red.

sect any vertex. After $h + k$ compressions we produce an L-drawing of K_n of minimum size which, by Lemma 2, uses $2n(n-1)$ ink. It follows that the ink of the original drawing is $2n(n-1) + n(h+k)$, hence the statement.

3   Complexity of the MILD Problem

In order to show the NP-hardness of MILD we reduce the problem PROFILE, which is defined as follows.

Problem: PROFILE
Instance: An undirected graph $G = (V, E)$ and an integer k.
Question: Does there exist an ordering π for the vertices of V such that

$$\sum_{u \in V} \left(\pi(u) - \min_{v \in N(u) \cup \{u\}} \pi(v) \right) \leq k \qquad (1)$$

where $N(u)$ denotes the set of neighbors of u?

It is folklore[1] that PROFILE is equivalent to SUMCUT (see the definition in [1]), which is known to be NP-complete [4,11,21].

Given an instance $I_p = \langle G = (V, E), k \rangle$ of PROFILE, we build an equivalent instance $I_m = \langle G' = (V', E'), k' \rangle$ of MILD as follows. Graph G' contains two subgraphs K^1 and K^2, that are complete graphs on $p = \frac{5}{2}n^2 + \frac{9}{2}n + 1$ vertices, where $n = |V|$. Consider two arbitrary vertices v_1 and v_2 of K^1 and K^2, respectively. For each vertex $v \in V$ we add to V' a vertex u_v with (directed) edges (u_v, v_1), (u_v, v_2), and (v_2, u_v). For each edge $e = (v, w) \in E$ we add to E' edges (u_v, u_w) and (u_w, u_v). We set $k' = k + 4p(p - 1) + \frac{3}{2}n^2 + \frac{9}{2}n$.

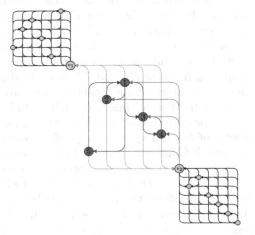

Fig. 3. Instance I_m of PROFILE with $p = 7$ (K^1 and K^2 are drawn smaller for space reasons) (Color figure online).

Lemma 4. *Instance I_p admits a solution if and only if instance I_m does.*

Proof. Suppose instance $I_p = \langle G = (V, E), k \rangle$ of PROFILE admits a solution and let π be the ordering of the vertices of V such that $\sum_{u \in V} (\pi(u) - \min_{v \in N(u) \cup \{u\}} \pi(v)) \leq k$. We show that the corresponding instance $I_m = \langle G' = (V', E'), k' \rangle$ of MILD admits a solution. We draw K^1 and K^2 in such a way that each uses contiguous x- and y-coordinates and the bounding box of K^1 is above and on the left of the bounding box of K^2. In particular, we place v_1 in the bottom right corner of the bounding box of K^1 and v_2 in the top left corner of the bounding box of K^2. We insert between K^1 and K^2 the remaining part of the vertices in V' so that their horizontal ordering corresponds to π and their vertical ordering is arbitrary. See Fig. 3 for an example. We show that the ink is no more than $k' = k + 4p(p - 1) + \frac{3}{2}n^2 + \frac{9}{2}n$. In fact, the ink can be computed as a sum of: (i) the ink used inside the complete subgraphs K^1 and K^2 (black edges of Fig. 3, which by Lemma 2 is $4p(p - 1)$ in total; (ii) the ink used to connect, for each $v \in V$, vertex u_v to v_1 and v_2 (drawn in green in Fig. 3, which is $2n + (n + 1)n$; (iii) the ink of the edges

---

[1] Refer to [6]. A formal proof of the equivalence of the two problems can be found in [12].

(drawn red in Fig. 3 that connect v_2 to u_v, for each $v \in V$, which is $n + \sum_{i=1}^{n} i$; (iv) the ink used for the edges among vertices u_v, with $v \in V$. The vertical ink of the latter contribution is already counted in (ii). The horizontal ink is exactly $\sum_{u \in V} (\pi(u) - \min_{v \in N(u) \cup \{u\}} \pi(v))$. Summing up the contributions (i)–(iii) we have $4p(p - 1) + \frac{3}{2}n^2 + \frac{9}{2}n$. Since $\sum_{u \in V} (\pi(u) - \min_{v \in N(u) \cup \{u\}} \pi(v)) \leq k$ the used ink is at most $k' = k + 4p(p - 1) + \frac{3}{2}n^2 + \frac{9}{2}n$. Conversely, suppose that instance I_m admits an L-drawing using at most k' ink. By Lemmas 2 and 3, any L-drawing of K^1 or K^2 that does not use contiguous x- and y-coordinates uses at least $p = \frac{5}{2}n^2 + \frac{9}{2}n + 1$ ink more than an L-drawing that uses contiguous x- and y-coordinates. Observe that the value on the left side of equation (1) is bounded by n^2, where $n = |V|$. Hence, we can assume $k \leq n^2$ in any non-trivial PROFILE instance. It follows that the additional ink that would be needed to insert grid lines in the drawings of K^1 or K^2 is at least $p = \frac{5}{2}n^2 + \frac{9}{2}n + 1 > k + \frac{3}{2}n^2 + \frac{9}{2}n$. This ensures that in any L-drawing that uses at most k' ink, K^1 and K^2 use contiguous x- and y-coordinates, and vertices u_v, for each $v \in V$, are inserted between the bounding boxes of K^1 and K^2, both in the horizontal and in the vertical order. Hence, by Lemma 2, the total contribution of these two subgraphs is $4p(p-1)$. We can assume that K^1 lies to the left and above K^2 (up to a vertical or horizontal flip of the entire drawing). Also, we can assume that v_1 lies on the bottom-right corner of K^1 and v_2 on the top-left corner of K^2, as they are the only vertices of K^1 and K^2 that are connected to vertices u_v, with $v \in V$. This implies that, for every horizontal and vertical order of vertices $u_v \in V'$, the cost of the green edges in Fig. 3 is $2n + (n + 1)n$, and the cost of the red edges is $n + \sum_{i=1}^{n} i$. Finally, for the blue edges, the vertical contribution is already covered by the green edges and the horizontal contribution is no more than k. Hence, the horizontal order of vertices u_v yields a solution for PROFILE. This concludes the proof.

Theorem 1. MILD *is NP-complete.*

Proof. MILD is trivially in NP by non-deterministically trying all permutations π_x and π_y of the vertices of the graph and computing the ink of $\Gamma(\pi_x, \pi_y)$. Given an instance I_p of PROFILE, the corresponding instance I_m of MILD can be built in polynomial time, and Lemma 4 ensures that the two instances are equivalent. $\square$

4   A Polynomial On-Line Algorithm

Motivated by the NP-completeness result in Theorem 1, we seek in this section an efficient heuristic to construct L-drawings of graphs with reduced ink. In particular, we study the setting in which the drawing is constructed incrementally by adding one vertex at a time to a previously computed drawing; the goal is then to add the new vertex (with all its incident edges) using the minimum additional ink, where the only operation that is allowed on the previous drawing is to insert a row and a column (the cost of elongating the edges traversing the inserted row/column has hence to be taken into account, as well). We prove in Theorem 2 that there exists a polynomial-time algorithm, called `OptAddVertex`, to place the given vertex in the given L-drawing while minimizing the additional ink of the resulting L-drawing with respect to the given one.

We remark that, besides providing a heuristic for the general problem, this incremental approach fits in the framework of streamed graph drawing, in which the graph to be drawn is too large to be stored in the memory and hence comes in the form of a streaming of its elements (vertices, edges, components) that have to be placed in the drawing without a prior knowledge of the elements that are yet to come.

Since, by Lemma 1, the horizontal and vertical coordinates of an L-drawing can be computed independently, we describe Algorithm OptAddVertex by only focusing on how to compute the optimal x-coordinate of the new vertex, adding a column.

Let $G = (V, E)$ be an n-vertex directed graph and let Γ be an L-drawing of it. We assume that the vertices in V have x-coordinates in $\{1, 2, \ldots, n\}$. Vertex v has to be added to the drawing, with its (possibly empty) set of outgoing edges $\{(v, u_1), (v, u_2), \ldots, (v, u_h)\}$ towards vertices of V and its (possibly empty) set of incoming edges $\{(w_1, v), (w_2, v), \ldots, (w_k, v)\}$ from vertices of V.

Algorithm OptAddVertex computes the additional ink needed to insert a vertical grid line l_v for v in each one of the possible $n+1$ positions $\{1, 2, \ldots, n+1\}$, where if l_v is inserted in position i, all vertices of V with x-coordinate greater or equal than i have to be shifted one unit to the right (hence, $i = 1$ and $i = n + 1$ correspond to adding a column to the left and to the right of the drawing, respectively).

We define three integer functions, that we call $StretchInk_x$, $IncomingInk_x$, and $OutgoingInk_x$, in the domain $\{1, 2, \ldots, n + 1\}$ as follows. $StretchInk_x(i)$ is the cost of inserting l_v in position i. This cost is due to the fact that the length of all horizontal segments traversed by l_v is incremented by one. $IncomingInk_x(i)$ is the cost, in terms of horizontal ink, of routing the edges $\{(w_1, v), (w_2, v), \ldots, (w_k, v)\}$ entering v when v is placed in position i. Observe that all these edges will enter v on a horizontal grid line l_h, which is exclusive of v. Hence, the value of $IncomingInk_x(i)$ is the range of the x-coordinates of vertices $\{w_1, w_2, \ldots, w_k\} \cup \{v\}$ after the insertion of v in position i. The computation of function $OutgoingInk_x$ is more complex. Each outgoing edge (v, u_j), $j = 1, \ldots, h$, of v has a vertical segment (which does not contribute to function $OutgoingInk_x$) and a horizontal segment entering u_j at its y-coordinate y_{u_j}. However, u_j may have already horizontal segments entering it at y-coordinate y_{u_j}. Let W_j and E_j be the minimum and the maximum x-coordinate that are used by some horizontal segments at coordinate $y = y_{u_j}$ (if there is no horizontal segment with $y = y_{u_j}$ we set $W_j = E_j = x_{u_j}$). The contribution of edge (v, u_j) to $OutgoingInk_x(i)$ is zero if $W_j \le i \le E_j$ and $\min(|i - W_j|, |i - E_j|)$, otherwise.

Finally, we insert v in a position corresponding to a minimum of function $AddInk_x$ defined as $AddInk_x = StretchInk_x + IncomingInk_x + OutgoingInk_x$.

The heuristic IncrementaLDraw for producing L-drawings of directed graphs works as follows. First, we order the vertices of the graph in such a way that, for any $1 \le j \le n$, the subgraph induced by the first j vertices is connected.

In particular, we consider the vertices in a BFS order. Second, we assign to the first vertex coordinates $(1, 1)$ and add a vertex at a time in the given order using Algorithm OptAddVertex.

We say that a permutation π_1 of the first n positive integers *extends* a permutation π_2 of the first $n - 1$ positive integers if π_2 can be obtained from π_1 by removing element n.

Theorem 2. *Given a directed graph G, a vertex $v \in G$, and an L-drawing $\Gamma'(\pi'_x, \pi'_y)$ of the subgraph $G' = G \setminus v$, algorithm OptAddVertex constructs in linear time an L-drawing $\Gamma^*(\pi^*_x, \pi^*_y)$ of minimum ink among all L-drawings $\Gamma(\pi_x, \pi_y)$ of G such that π_x extends π'_x and π_y extends π'_y.*

Proof. Suppose by contradiction that there exists an L-drawing $\Gamma^\circ(\pi^\circ_x, \pi^\circ_y)$ that uses less ink than $\Gamma^*(\pi^*_x, \pi^*_y)$ and such that π°_x extends π'_x and π°_y extends π'_y. Without loss of generality suppose that $ink_x(\Gamma^\circ) < ink_x(\Gamma^*)$. By removing v we obtain again $\Gamma'(\pi'_x, \pi'_y)$ and we save $AddInk_x(x^\circ_v)$ ink, where x°_v is the x-coordinate of v in Γ^c. Since $ink_x(\Gamma^*) = ink_x(\Gamma') + AddInk_x(x^*_v)$, where x^*_v is the x-coordinate of v in Γ^*, we have that $AddInk_x(x^\circ_v) < AddInk_x(x^*_v)$, contradicting the hypothesis that Γ^* is obtained by inserting v in a minimum of function $AddInk_x$. The complexity of algorithm OptAddVertex is discussed in [1].   □

5   Experimental Evaluation

We implemented Algorithm OptAddVertex and the heuristic IncrementaLDraw, and performed an extensive testing to evaluate the quality of the obtained L-drawings. We compared the performances of our heuristic with the optimum ink, the OOD algorithm of DAGView [20], and random placements. Refer to [1] for details on the time complexity and on running times of IncrementaLDraw.

5.1   An Integer Linear Programming Formulation

In order to compare the heuristic approach with the optimal solution we formulated the problem of finding an L-drawing with minimum ink as an ILP problem. Given an n-vertex graph $G = (V, E)$, in the following we describe only the part to compute its x-coordinates (the computation of y-coordinates is analogous). By definition, the amount ink_x of a drawing Γ of G is obtained by summing up all the horizontal segments of the drawing. Since each y-coordinate is exclusively used for one vertex, there are n (possibly null, if there exists a vertex with no incoming edges) horizontal segments in Γ. The horizontal segment s_i that includes v_i, $i = 1, \ldots, n$ extends from the leftmost to the rightmost bends of the edges entering v_i. We call W_i and E_i the x-coordinates of the endpoints of s_i. Variables:

$$\forall i, j = 1, \ldots, n : \; x_{ij} = \begin{cases} 1 \text{ if vertex } v_i \text{ has } x\text{ -coordinate } j, \\ 0 \text{ otherwise} \end{cases}$$

$$\forall i = 1, \ldots, n : \; E_i, W_i \quad \text{(rightmost and leftmost endpoints of } s_i)$$

Variables x_{ij} are binary, while E_i and W_i are integers. To simplify the description we denote by x_i the x-coordinate of vertex v_i, that is $x_i = \sum_{j=1}^{n} x_{ij} \cdot j$. Constraints:

$$\forall i : \; \sum_{j=1}^{n} x_{ij} = 1 \quad \text{(each vertex has a unique } x \text{ -coordinate)}$$

$$\forall j : \; \sum_{i=1}^{n} x_{ij} \leq 1 \quad \text{(each column contains at most one vertex)}$$

$$\forall i : \; E_i \geq x_i \quad \text{(the rightmost endpoint of } s_i \text{ does not lie to the left of } v_i)$$

$$\forall i : \; W_i \leq x_i \quad \text{(the leftmost endpoint of } s_i \text{ does not lie to the right of } v_i)$$

$$\forall (v_i, v_j) \in E, \; E_j \geq x_i \quad \text{(the rightmost endpoint of } s_j \text{ does not lie to the left of } v_i)$$

$$\forall (v_i, v_j) \in E, \; W_j \leq x_i \quad \text{(the leftmost endpoint of } s_j \text{ does not lie to the right of } v_i)$$

The objective function is: $\min \sum_{i=1}^{n} (E_i - W_i)$.

To compute minimum-ink L-drawings we used Gurobi Optimizer ver. 6.0.4 [14] on a Dual Xeon X5460 Quad Core 3.16 GHz 48GB RAM.

5.2   Random Generation of the Graphs Suites

We generated uniformly at random two graph suites of dense, weakly connected, directed graphs. The first graph suite is meant to compare the performances of Algorithm `IncrementaLDraw` with respect to the optimum. For each number of vertices n in $\{5, 10, 15\}$ and for each percentage p in $\{10, 20, 30, 70\}$ we generated ten graphs whose number of edges m is $p\,\%$ with respect to the maximum possible number of edges, that is, $m = \lfloor n(n-1)p/100 \rfloor$. In particular, we used the procedure `gnm_random_graph` of the NetworkX 1.7 library [22], discarding graphs that were not connected.

The second graph suite is meant to compare `IncrementaLDraw` with a random placement of the vertices and is generated with the same procedure and edge percentages of the first suite, but vertices range in $\{100, 200, 300, 400, 500\}$.

5.3   Results of the Experiments

The results of the experiments are shown in Figs. 4 and 5. Figure 4(a) is devoted to the ten graphs with 15 vertices and 63 edges (30 % of the maximum possible) of the first graph suite. On the x-axis the ten graphs are reported. The curves represent: (i) the ink used by the optimal algorithm; (ii) the average and the standard deviation of the ink used by Algorithm `IncrementaLDraw` over 100

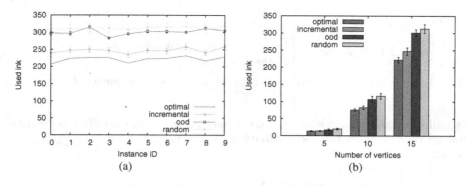

Fig. 4. (a) Ink used for the drawings of the ten graphs of the first graph suite with 15 vertices and 63 edges (corresponding to 30 % of the maximum possible). Optimal, incremental, ODD, and random placement are compared. For the latter three the average and the standard deviation over 100 runs is shown. (b) Ink consumption varying the size of the graphs (fixing at 30 % edge density).

runs, each using a different BFS ordering obtained by starting from a random initial vertex and by shuffling the adjacency lists of the vertices; (iii) the average and the standard deviation of the ink used by Algorithm OOD over 100 runs, each obtained from DAGView [20] by shuffling the adjacency lists of the vertices; and (iv) the average and the standard deviation of the ink used by 100 random placements of the vertices.

From Fig. 4(a) it is apparent that the performances of IncrementaLDraw are always largely better than those of OOD and random placements, and not rarely are close to the optimum. Although this result could be anticipated (OOD was not conceived to reduce ink), we were surprised to note that, even with very small graphs and relatively many runs, the worst case for IncrementaLDraw is always comparable with the best case for OOD and significantly better than the best case of random placement. We found the same pattern in all plots obtained by changing densities and sizes.

Figure 4(b) shows how the size impacts on ink, focusing on 30 % density graphs of the first graph suite. All the points are obtained by averaging ten values (for example, each bar for 15 vertices of Fig. 4(b) is obtained by averaging the ten corresponding values of Fig. 4(a). Figure 5(b) further deepens this analysis showing how much ink each algorithm saves with respect to the maximum theoretical upper bound of $2n \times (n-1)$ for the second graph suite. We observe that, when increasing the number of vertices, both the number of edges and the consumption of ink increase quadratically. At the same time, the ink saved by IncrementaLDraw with respect to OOD and random placement increases linearly.

Figure 5(a) shows how density impacts on ink, focusing on graphs of 15 vertices. Again, each point is the average of ten points obtained for ten different graphs (e.g., the values for density 30 % are obtained by averaging the ten values of Fig. 4(a). For denser graphs, the difference among the alternative algorithms

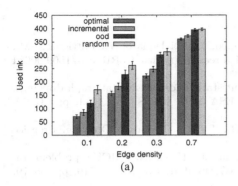
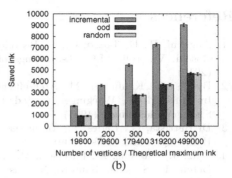

Fig. 5. (a) Ink consumption by varying density (the size of the graphs is fixed at 15 vertices). (b) The difference between the theoretical maximum and the actual ink used by incremental, ODD, and random placement, for the second test-suite (graphs with 30 % of maximum possible edges).

seems to reduce. This could be predicted as Lemma 2 ensures that for any vertex order of a K_n uses the same ink.

Overall, the experiments show that the ink consumption of IncrementaLDraw are closer to the optimum than to those of alternative algorithms and that the heuristic offers a good compromise between effectiveness and running times, even with a naïve implementation of Algorithm OptAddVertex. In [1] we discuss a more efficient version of the algorithm.

6   Conclusions and Open Problems

We introduced L-drawings, a novel paradigm for representing directed graphs. We investigated the problem of producing drawings with minimum ink, which turned out to be NP-complete. Our heuristic, however, proved to produce near-optimal solutions.

Several problems remain open: (i) How much area and ink could be saved if vertices were allowed to share horizontal or vertical grid lines, provided that the drawing is still unambiguous? (ii) Does there exist an ordering of the vertices such that IncrementaLDraw produces a minimum-ink drawing? (iii) Problem PROFILE, which we reduced to show the NP-hardness of MILD, is linear-time solvable for trees [4] and for square grids [5]; what is the complexity of computing minimum-ink L-drawings for these families of graphs? (iv) What is the complexity of minimizing crossings in an L-drawing?

Finally, although in [8] it is shown that overloaded orthogonal drawings are superior to matrix representations under several respects, it would be interesting to contrast both these representations with L-drawings in an extensive user study.

References

1. Angelini, P., Da Lozzo, G., Di Bartolomeo, M., Di Donato, V., Patrignani, M., Roselli, V., Tollis, I.G.: L-drawings of directed graphs. CoRR abs/1509.00684 (2015)
2. Biedl, T.C., Kaufmann, M.: Area-efficient static and incremental graph drawings. In: Burkard, R., Woeginger, G. (eds.) ESA 1997. LNCS, vol. 1284, pp. 37–52. Springer, Heidelberg (1997)
3. Di Battista, G., Eades, P., Tamassia, R., Tollis, I.G.: Graph Drawing. Prentice Hall, Englewood Cliffs (1999)
4. Díaz, J., Gibbons, A., Paterson, M., Toran, J.: The MINSUMCUT problem. In: Dehne, F., Sack, J., Santoro, N. (eds.) WADS 1991. LNCS, vol. 519, pp. 65–79. Springer, Heidelberg (1991)
5. Díaz, J., Penrose, M., Petit, J., Serna, M.: Convergence theorems for some layout measures on random lattice and random geometric graphs. Comb. Prob. Comput. 9(6), 489–511 (2000)
6. Díaz, J., Petit, J., Serna, M.: A survey of graph layout problems. ACM Comput. Surv. 34(3), 313–356 (2002)
7. Dickerson, M., Eppstein, D., Goodrich, M.T., Meng, J.Y.: Confluent drawings: visualizing non-planar diagrams in a planar way. J. Graph Alg. Appl. 9(1), 31–52 (2005)
8. Didimo, W., Montecchiani, F., Pallas, E., Tollis, I.G.: How to visualize directed graphs: a user study. In: IISA 2014, pp. 152–157. IEEE
9. Garg, A., Tamassia, R.: On the computational complexity of upward and rectilinear planarity testing. SIAM J. Comput. 31(2), 601–625 (2001)
10. Ghoniem, M., Fekete, J., Castagliola, P.: On the readability of graphs using node-link and matrix-based representations: a controlled experiment and statistical analysis. Inf. Vis. 4(2), 114–135 (2005)
11. Golovach, P.: The total vertex separation number of a graph. Disk. Mat. 9(4), 86–91 (1997)
12. Golovach, P., Fomin, F.: The total vertex separation number and the profile of graphs. Disk. Mat. 10(1), 87–94 (1998)
13. Grinberg, E., Dambit, J.: Latviiskii Matematicheskii Ezhegodnik 2, 65–70 (1966). in Russian
14. Gurobi Optimization: Gurobi Optimizer. http://www.gurobi.com/
15. Healy, P., Nikolov, N.S.: Hierarchical drawing algorithms. In: Tamassia, R. (ed.) Handbook of Graph Drawing and Visualization. CRC Press, Boca Raton (2013)
16. Henry, N., Fekete, J., McGuffin, M.J.: Nodetrix: a hybrid visualization of social networks. IEEE Trans. Vis. Comput. Graph. 13(6), 1302–1309 (2007)
17. Huang, J., Kang, Z.: A genetic algorithm for the feedback set problems. In: ICPACE 2003 (2003)
18. Huang, W., Hong, S., Eades, P.: Effects of crossing angles. In: PacificVis 2008. IEEE (2008)
19. Kornaropoulos, E.M., Tollis, I.G.: Overloaded orthogonal drawings. In: Speckmann, B. (ed.) GD 2011. LNCS, vol. 7034, pp. 242–253. Springer, Heidelberg (2011)
20. Kornaropoulos, E.M., Tollis, I.G.: DAGView: An Approach for Visualizing Large Graphs. In: Didimo, W., Patrignani, M. (eds.) GD 2012. LNCS, vol. 7704, pp. 499–510. Springer, Heidelberg (2013)
21. Lin, Y., Yuan, J.: Profile minimization problem for matrices and graphs. Acta Mathematicae Applicatae Sinica. English Series. Yingyong. Shuxue Xuebao 10(1), 107–112 (1994)

22. Los Alamos Nat. Lab.: NetworkX. http://networkx.lanl.gov/index.html
23. Sugiyama, K., Tagawa, S., Toda, M.: Methods for visual understanding of hierarchical system structures. IEEE Trans. Syst. Man Cybern. **11**(2), 109–125 (1981)
24. yWorks: yEd Graph Editor. http://www.yworks.com/en/products/yfiles/yed/

A Combinatorial Model of Two-Sided Search

Harout Aydinian[1], Ferdinando Cicalese[2]($\boxtimes$),
Christian Deppe[3], and Vladimir Lebedev[4]

[1] Technische Universität München, Munich, Germany
h.aydinyan@tum.de
[2] University of Verona, Verona, Italy
ferdinando.cicalese@univr.it
[3] University of Bielefeld, Bielefeld, Germany
cdeppe@math.uni-bielefeld.de
[4] Russian Academy of Sciences, Moscow, Russia
lebedev37@mail.ru

Abstract. We study a new model of combinatorial group testing in
a network. An object (the target) occupies an unknown node in the
network. At each time instant, we can test (or query) a subset of the
nodes to learn whether the target occupies any of such nodes. Unlike
the case of conventional group testing problems on graphs, the target in
our model can move immediately after each test to any node adjacent to
each present location. The search finishes when we are able to locate the
object with some predefined accuracy s (a parameter fixed beforehand),
i.e., to indicate a set of s nodes that include the location of the object.

In this paper we study two types of problems related to the above
model: (i) what is the minimum value of the accuracy parameter for
which a search strategy in the above sense exists; (ii) given the accuracy,
what is the minimum number of tests that allow to locate the target. We
study these questions on paths, cycles, and trees as underlying graphs
and provide tight answer for the above questions. We also considered
a restricted variant of the problem, where the number of moves of the
target is bounded.

1 Introduction

Problems involving search arise in various areas of human activity. Search theory
deals with the problem faced by a searcher: finding a hidden object, in a given
"search space", in minimum time. In most of early developments it is assumed
that an object to be searched is stationary and hidden according to a known
distribution or it is moving and its motion is determined, by some known rules.
This model of search is called *one-sided search*. In case the target can attempt

H. Aydinian—Supported by Gottfried Wilhelm Leibniz-Program BO 1734/20-1 and
the DFG project BO 1734/31-1.
V. Vladimir—Supported by the Russian Foundation for Basic Research, project
No.15-01-08051.

to contrast the searcher's activity and react in some intelligent way in order not to be found, the problem is called *two-sided search*.

The first developments in search theory were made by Bernard Koopman and his colleagues in the Anti-Submarine Warfare Operations Research Group of the U.S. Navy during World War II. Their purpose was to provide efficient ways to search for enemy submarines. Their work which was only published later in [15] also mentioned two-sided search (see also [5]).

Here, we consider a *combinatorial* model of two sided search which was proposed by the late Rudolph Ahlswede during the Workshop "Search Methodologies II" (2010). In a *combinatorial search problem* the object(s) to be found live in a discrete space and the tests to be asked satisfy certain specified requirements. The fundamentals of combinatorial search can be found in primary books [1,4,14].

We define our search space $\mathcal{N} = \{1, 2, \ldots, N\}$ as the vertices of an undirected graph $G = (\mathcal{N}, \mathcal{E})$. The object to be found (the *target*) occupies some vertex of $\mathcal{N}$ which is unknown to the searcher. The searcher is able to detect the presence of the target at any subset of $\mathcal{N}$, i.e. for any subset $T \subset \mathcal{N}$, called a *test set*, or *test* for short, the searcher can learn whether the target is located at some node in T or not. The goal is to find the location of the target, with a certain accuracy, in minimum time (number of tests). Note that in case of stationary target, the problem is equivalent to a traditional group testing problem [10] with a single defective item.

In our model, after each test, the target can move to any vertex adjacent to its current location or stay at the same place. For the ease of description we assume that each vertex in our graph G has a loop. Thus, we may formally assume that in each time unit the target moves to an adjacent vertex. For each $j \geq 1$, let d_j be the location of the target at time j. Then, for each $n \geq 1$, the sequence of target positions until time n is given by the vector $(d_1, \ldots, d_n) \in \mathcal{N}^n$ which defines a walk in the graph G, i.e., $(d_i, d_{i+1}) \in \mathcal{E}$ $(i = 1, \ldots, n-1)$. Recall that in case of traditional group testing $d_i = d_j$; $1 \leq i, j \leq n+1$.

For a test T and a node $d \in \mathcal{N}$ we define the test function

$$f_T(d) = \begin{cases} 0 \ \text{(No)} \ , \ \text{if } d \notin T \\ 1 \ \text{(Yes)} \ , \ \text{if } d \in T. \end{cases}$$

Then, $f_T(d)$ represents the result of test T when the target is in position d.

For $j = 1, 2, \ldots$, let T_j be the test performed at time j. We assume that T_j depends on the results of all the previous tests[1]. The sequence $(T_1, T_2, \ldots, T_n)$ is then called a sequential or adaptive strategy of length n.

We denote by $\mathcal{D}_i$ the set of possible positions of the target after the result of the ith test has become available. We have that $\mathcal{D}_0 = \mathcal{N}$ and for each $i \geq 1$, it holds that

$$\mathcal{D}_i = \begin{cases} \Gamma(T_i \cap \mathcal{D}_{i-1}) \ , \ \text{if } f_{T_i}(d_i) = 1 \\ \Gamma(\mathcal{D}_{i-1} \backslash T_i) \ \ , \ \text{if } f_{T_i}(d_i) = 0, \end{cases}$$

---

[1] Formally, T_j is a function mapping the result of the first $j-1$ tests to a subset of $\mathcal{N}$, i.e., $T_j = T_j(f_{T_1}(d_1), \ldots, f_{T_{j-1}}(d_{j-1}))$.

where $\Gamma(\mathcal{A}) := \{j \in \mathcal{N} : \exists i \in \mathcal{A} \text{ with } (i,j) \in \mathcal{E}\}$ is the neighborhood of a subset $\mathcal{A} \subset \mathcal{N}$. Note that $\mathcal{A} \subset \Gamma(\mathcal{A})$. We say that the test $\mathcal{T}_i$ reduces $\mathcal{D}_{i-1}$ to $\mathcal{D}_i$.

Clearly, given the sequence of movements of the target $(d_1, \ldots, d_n)$ and the strategy $\mathcal{T}_1, \ldots, \mathcal{T}_n$, the sequence $\mathcal{D}_1, \ldots, \mathcal{D}_n$ is also determined. With these definitions we can now formalize the concept of a successful strategy as follows.

Given a graph $G = (\mathcal{N}, \mathcal{E})$, and a positive integer s, a sequential strategy of length n, $\mathcal{T}_1, \ldots \mathcal{T}_n$ (as defined above), is called (G, s)–successful if for any possible sequence of the target's movements $(d_1, \ldots, d_n)$, we have that $|\mathcal{D}_i| \leq s$ for some $i \leq n$.

We define $s^*(G)$ as the minimum number s^* such that there exists a (G, s^*)–successful strategy. Given an integer $s \geq s^*(G)$, we denote by $n(G, s)$ the minimum number n such that there exists a (G, s)–successful strategy of length n. We call the corresponding strategy a minimum size or optimal (G, s) strategy.

We also consider a more general problem when the mobility of the target is limited, that is the target can change its position at most t times. We refer to this case as the model with restricted movements of the target. In this case we use the corresponding notation $(G, s; t)$–successful and $n(G, s; t)$.

Since the concept of a successful strategy is defined in terms of worst case scenario, in our analyses we will equivalently assume that the goal of the target is to maximize the length of an (G, s)–successful strategy, or to maximize the size of any $\mathcal{D}_i$, depending on whether our goal is to give a bound on $n(G, s)$ or on $s^*(G)$ respectively. Therefore, the target's movements and the sequence of *test results* $f_{\mathcal{T}_1}(d_1) \ldots, f_{\mathcal{T}_n}(d_n)$, also called *answers*, can be viewed as an *adversarial strategy*, whose goal is to maximize the above mentioned quantities.

Our Results. We study the above quantities for the cases where the underlying graph belongs to the classes of undirected cycles, paths, and trees. In Sect. 2 we give an optimal (C_N, s) strategy for any $s \geq 5$, where C_N denotes a cycle of length N. We also show that $s^*(C_N) = 5$ for $N \geq 5$. For a path P_N on N vertices we give an optimal $(P_N, 4)$ strategy, which is linear in N. For $s \geq 5$ we give an optimal (P_N, s) strategy, which is logarithmic in N. In Sect. 3 we consider the case of trees. For a tree T, we characterize $s^*(T)$ in terms of the maximum degree and the radius of T. In Sect. 4 we consider the variant of the problem where the movement of the target are restricted for the cases of the underlying graph being a cycles or a path. We give optimal $(C_N, 3, t)$ strategies for $t = 1, 2$ and for $t \geq 3$ we give a general strategy. Finally, we give an optimal $(P_N, 3, 1)$ strategy. In Sect. 5 we discuss some directions for future research.

Motivations and Related Work. The model has application to the area of node selection for target tracking in sensor networks. Sensor networks are systems of many small and simple devices. In general, the sensors used have reduced functionalities so that their cost remains low. A sensor may generate as little as one bit of information. Moreover, for energy saving reasons, sensors should not be active continuously but it is important to carefully select at each point in time the set of sensors which should be active to carry on their task. When the task of the network is the tracking of objects, a major initial task is to determine an area where the object to be tracked is surely initially located, and from which the actual tracking procedure can start. This localization together with the minimization of the area of localization is one of the most critical and expensive part of the tracking procedure, as it is typically done by an exhaustive search [12,18]. For the sake of reducing the bandwidth consumption, sensors networks are also hierarchically organized in graph and more specifically tree structures [17]. Therefore, our results can be used to support the localization phase while trying to reduce the area of localization and reducing the number of activations of sensors.

To the best of our knowledge, the search model we are considering in this paper and the related problems stated above have not been studied earlier in the literature. Group testing in graph has been considered both in terms of searching for an edge and for a vertex, and for different models of the test allowed [6,8,13,19]. However, in all these works the basic assumption is that the target is still which makes the problem significantly different.

Another area of research related to the problem studied here is graph searching. Graph searching encompasses a wide variety of combinatorial problems related to the problem of capturing a fugitive residing in a graph using the minimum number of searchers. Although there are many different models of graph searching (the interested reader is referred to the very comprehensive annotated bibliography [11]), none appear to cover the type of two side combinatorial search we are considering here. Models of search closely related to ours appear to be the so called cop and robber game [3], the princess and the monster [2], and domination search games.

For space constraints, some of the proofs are deferred to the extended version.

2   Optimal Strategies for Cycles and Paths

In this section we consider the classes of cycles and path graphs on N vertices, denoted by C_N and P_N respectively. We will focus on the dual of the parameters $n(C_N, s)$ and $n(P_N, s)$. Given integers $n, s \geq 1$, we denote by $N_c(n, s)$ (resp. $N_p(n, s)$) the maximum N, such that there exists an (C_N, s)–successful (resp. (P_N, s)–successful) strategy of length n. Clearly, $n(C_N, s) = \min\{i : N_c(i, s) \geq N\}$ and $n(P_N, s) = \min\{i : N_p(i, s) \geq N\}$.

Cycle Graphs

Let $C_N = (\mathcal{N}, \mathcal{E})$ be an undirected cycle of length N with a loop in each node. Thus, $\mathcal{N} = \{1, \ldots, N\}$ and $\mathcal{E} = \{\{i, i+1\} : 1 \leq i \leq N-1\} \cup \{N, 1\} \cup \{\{i, i\} : i \in \mathcal{N}\}$. We start with the following simple observation.

Proposition 1. *For $N \geq 5$ there does not exist a (C_N, s)–successful strategy with $s \leq 4$, that is $s^*(C_N) \geq 5$.*

Proof. Let $(\mathcal{T}_1, \ldots, \mathcal{T}_n)$ be a (C_N, s)–successful strategy. Since we consider the adversarial setting of our problem, the test results maximally increase the size of s. To prove the statement it is sufficient to show that one can choose the test results in such a way that $|\mathcal{D}_i| \geq 5$ for $i = 1, \ldots, N$. Since $|\mathcal{D}_0| = N \geq 5$ it is easy to see that in the worst case $|\mathcal{D}_i| = \max\{|\Gamma(\mathcal{T}_i \cap \mathcal{D}_{i-1})|, |\Gamma(\mathcal{D}_{i-1} \setminus \mathcal{T}_i)|\} \geq \left\lceil \frac{|\mathcal{D}_{i-1}|}{2} \right\rceil + 2 \geq 5$. $\square$

For $s \geq 5$, we can characterize the size of optimal (C_N, s) strategies.

Theorem 1. *For any $s \geq 5$ and any $n \geq 0$ we have $N_c(n, s) = 2^n(s - 4) + 4$.*

Proof. We proceed by induction on n. The case $n = 0$ is trivial. For the induction step with $n \geq 1$ suppose $N_c(n, s) > 2^n(s - 4) + 4$ with $(\mathcal{T}_1, \ldots, \mathcal{T}_n)$ being an optimal strategy. Skipping the trivial cases: $|\mathcal{T}_1| = 1$ or $|N - 1|$, we notice that $|\mathcal{D}_1| \geq |\mathcal{T}_1| + 2$ if $f_{\mathcal{T}_1}(d_1) = 1$ and $|\mathcal{D}_1| \geq |\mathcal{N} \setminus \mathcal{T}_1| + 2$ if $f_{\mathcal{T}_1}(d_1) = 0$, with equality in both cases if and only if $\mathcal{T}_1$ is connected, that is $\mathcal{T}_1$ is a path in C_N. This implies that $|\mathcal{D}_1| \geq \lceil N_c(n, s)/2 \rceil + 2 > (2^n(s - 4) + 4)/2 + 2 = 2^{n-1}(s - 4) + 4$, a contradiction with the induction hypothesis $|\mathcal{D}_1| \leq N_c(n - 1, s) = 2^{n-1}(s - 4) + 4$. Hence we have $N_c(n, s) \leq 2^n(s - 4) + 4$. On the other hand, in case $N = 2^n(s - 4) + 4$ we take as $\mathcal{T}_1$ a path on $\frac{N}{2}$ vertices, which is sufficient (and necessary) to get an optimal strategy, in view of the induction hypothesis. $\square$

Path Graphs

Let now P_N be a path graph on N vertices, thus $\mathcal{E} = \{\{i, i + 1\} : 1 \leq i < N\} \cup \{\{i, i\} : 1 \leq i \leq N\}$. It turns out that here we have $s^*(P_N) = 4$, as given by the following proposition.

Proposition 2. *For $N \geq 5$ there does not exist a (P_N, s)-successful strategy with $s \leq 3$, that is $s^*(P_N) \geq 4$.*

Our goal now is to find an optimal $(P_N, 4)$ strategy.

Theorem 2. *For $N \geq 4$ we have $n(N, 4) = \left\lceil \dfrac{N}{2} \right\rceil - 2$.*

Proof. The proof consists of two parts. The upper bound is deferred to the extended version of the paper. Here we show the lower bound

Lemma 1. *For $N \geq 4$ we have $n(N, 4) \geq \left\lceil \dfrac{N}{2} \right\rceil - 2$.*

Proof. For the sake of the presentation, let us assume that N is odd. The case where N is even can be dealt with analogously. We shall describe an adversary, that guarantees, for at least $\lceil N/2 \rceil - 3$ rounds, the existence of a set $\mathcal{A}_i \subseteq \mathcal{D}_i$ of 5 adjacent positions among the candidates to be the target position. Let $\mathcal{A}_0 = \{\lceil N/2 \rceil - 2, \ldots, \lceil N/2 \rceil + 2\}$. It is trivially true that $\mathcal{A}_0 \subseteq \mathcal{D}_0 = \{1, \ldots, N\}$.

The adversary strategy is easily described as follows: for each $i = 1, 2, \ldots,$ if $|\mathcal{T}_i \cap \mathcal{A}_{i-1}| < 3$ then the answer will be No, otherwise the answer will be Yes. Let us denote by a_i the central element of $\mathcal{A}_i$, i.e., $a_0 = \lceil N/2 \rceil$. Let us analyze the possible cases. It is not hard to see that when $|\mathcal{T}_i \cap \mathcal{A}_{i-1}| < 3$ (and the answer is No) or $|\mathcal{T}_i \cap \mathcal{A}_{i-1}| > 3$ (and the answer is Yes) or $\mathcal{T}_i \cap \mathcal{A}_{i-1} \in$

$\{\{a_{i-1} - 1, a_{i-1}, a_{i-1} + 1\}, \{a_{i-1} - 2, a_{i-1}, a_{i-1} + 1\}, \{a_{i-1} - 1, a_{i-1}, a_{i-1} + 2\}\}$,
(and the answer is Yes), we have $\mathcal{A}_{i-1} \subseteq \mathcal{D}_i$. Hence $\mathcal{A}_i = \mathcal{A}_{i-1}$ which satisfies
the claim about the existence of a subset of size 5 contiguous positions among
the candidate positions for the target. Finally, we have the two cases given
by $T_i \cap \mathcal{A}_{i-1} \in \{\{a_{i-1} - 2, a_{i-1} - 1, a_{i-1}\}, \{a_{i-1}, a_{i-1} + 1, a_{i-1} + 2\}\}$, where
the answer will be Yes. Then, if $T_i \cap \mathcal{A}_{i-1} = \{a_{i-1} - 2, a_{i-1} - 1, a_{i-1}\}$ and
$a_{i-1} > 3$, we have $\mathcal{A}_i = \{a_{i-1} - 3, \dots, a_{i-1} + 1\} \subseteq \mathcal{D}_i$ with 5 candidate positions
as above. In this case we say that we had a shift towards the left of the 5
candidates. If $T_i \cap \mathcal{A}_{i-1} = \{a_{i-1}, a_{i-1} + 1, a_{i-1} + 2\}$ and $a_{i-1} < N - 2$, we have
$\mathcal{A}_i = \{a_{i-1} - 1, \dots, a_{i-1} + 3\} \subseteq \mathcal{D}_i$ again with 5 possible positions for the target.
In this case we say that we had a shift towards the right of the 5 candidates.

By the above arguments it follows that the existence of the set $\mathcal{A}_i$ can be
guaranteed until a shift towards right or left is possible, i.e., for at least $\lceil N/2 \rceil - 3$
times. This bound is given by the case where the test used always induces a shift
in the same direction of the 5 candidates. At this point there are still 5 candidates
so at least another test is needed, which proves the desired bound $\lceil N/2 \rceil - 2$. $\square$

This completes the proof of Theorem 3. $\square$

Corollary 1. *For $n \geq 0$ we have $N_p(n, 4) = 2n + 4$.*

The next theorem shows that we need much less tests, if we search for a final
target set of size $s \geq 5$.

Theorem 3. *For $n \geq 0$ and $s \geq 4$ we have $N_p(n, s) = (s - 4)2^n + 2n + 4$.*

Proof. The case $s = 4$ follows from Corollary 1, thus we consider the case $s \geq 5$.

Lower bound: For the case $s \geq 5$, we will first prove the lower bound. For this,
we consider two variants of the problem and then reduce the original problem
to them. The first variant (later referred to as variant O for open) arises when
we consider the search space *open* on both sides. More formally, we assume that
$\mathcal{D}_0 = \{a + 1, \dots, a + x\}$, for some $a \in \mathbb{Z}$ and $x \in \mathbb{N}$ and the target can move on
any position in $\mathbb{Z}$, i.e., there is no boundary at a or $a + x$, in the sense that from
position $a + x$ the target can move to position $a + x + 1$ too, and from position
$a + 1$ it can move to position a too. This is different from the problem we fix
at the beginning , since when the target is in 1 (resp. in N), if it moves, it can
only move to 2 (resp. $N - 1$). Let us denote by $N^O(n, s)$ the largest value of x
such that there is an (P_N, s)-successful strategy of length n in this open variant
of the problem over any path $P_N = \{a + 1, \dots, a + x\}$.

Claim 1. $N^O(s, n) \geq 2^n(s - 4) + 4$.

The base case $n = 0$ is trivially true. For the induction step, let $n \geq 1$ and
$\mathcal{D}_0 = \{a + 1, \dots, a + 2^n(s - 4) + 4\}$. Using the first test $T_1 = \{a + 1, a + 2, \dots, a + 2^{n-1}(s - 4) + 2\}$ we have that either $\mathcal{D}_1 = \{a, a + 1, \dots, a + 2^{n-1}(s - 4) + 3\}$ or
$\mathcal{D}_1 = \{a + 2^{n-1}(s - 4) + 2, a + 2^{n-1}(s - 4) + 3, \dots, a + 2^n(s - 4) + 5.\}$ In both cases
we have that $|\mathcal{D}_1| = 2^{n-1}(s - 4) + 4 = N^O(n - 1, s)$. Hence, by the induction
hypothesis, $n - 1$ additional tests are sufficient for a successful strategy starting
from $\mathcal{D}_1$. Thus, n tests are sufficient for a successful strategy starting from $\mathcal{D}_0$,
i.e., we have shown $N^O(n, s) \geq 2^n(s - 4) + 4$ concluding the proof of Claim 1.

As a second variant of the problem (later referred to as variant H for half-open) we consider the case where the search space is *half-open*. We assume that $\mathcal{D}_0 = \{1, \ldots, x\}$, for some $x \in \mathbb{N}$ and the target can move on any position in $\mathbb{N}$, i.e., it can never move to a position to the left of 1 but it can move to a position to the right of x, i.e., positions $x+1, x+2, \ldots$ might become possible candidates later on. Let us denote by $N^H(n, s)$ the largest value of x such that there is an (P_N, s)-successful strategy of length n in this open variant of the problem over any line $P_N = \{1, \ldots, x\}$. We can prove by induction the following

Claim 2. $N^H(n, s) \geq 2^n(s - 4) + n + 4$.

The base case $n = 0$ is trivially true. For the induction step, let $n \geq 1$ and $\mathcal{D}_0 = \{1, \ldots, 2^n(s - 4) + n + 4\}$. Using the first test $T_1 = \{1, 2, \ldots, 2^{n-1}(s - 4) + n + 3\}$ we have that either $\mathcal{D}_1 = \{1, \ldots, 2^{n-1}(s - 4) + n + 4\}$ or $\mathcal{D}_1 = \{2^{n-1}(s - 4) + n + 3, 2^{n-1}(s - 4) + n + 4, \ldots, 2^n(s - 4) + n + 4\}$. In the first case we have that $|\mathcal{D}_1| = 2^{n-1}(s - 4) + (n - 1) + 4 = N^H(n - 1, s)$ hence by induction $n - 1$ additional tests are sufficient for a successful strategy starting from $\mathcal{D}_1$. In the other case, we have $|\mathcal{D}_1| = 2^{n-1}(s - 4) + 4 = N^O(n - 1, s)$ and even allowing the new search space to be half open we can finish the search with a strategy of size $n - 1$, by Claim 1. In both cases n tests are sufficient for a successful strategy starting from $\mathcal{D}_0$, hence the induction step is proved and so is Claim 2.

We are now ready to prove the lower bound $N(n, s) \geq 2^n(s-4)+2n+4$. Again we proceed by induction on n. The base case $n = 0$ is trivially true. For $n \geq 1$, let $\mathcal{D}_0 = \{1, \ldots, 2^n(s-4)+2n+4\}$ and define $T_1 = \{1, \ldots, 2^{n-1}(s-4)+(n-1)+3\}$. As a result of this test we have either $\mathcal{D}_1 = \{1, \ldots, 2^{n-1}(s - 4) + (n - 1) + 4\}$ or $\mathcal{D}_1 = \{2^{n-1}(s - 4) + (n - 1) + 3, \ldots, 2^n(s - 4) + 2n + 4\}$.

In both cases we have $|\mathcal{D}_1| = N^H(n - 1, s)$ and we can finish in $n - 1$ tests by Claim 2, using a strategy for the half-open variant defined above. Notice that in both cases, the resulting set of candidates positions for the target can extend only one direction, like in an instance of the half-open variant. This concludes the proof of the inductive step, hence $N(n, s) \geq 2^n(s - 4) + 2n + 4$ as desired.

Upper bound: Suppose given n, s there always exists an optimal strategy achieving $N(n, s)$ such that all the tests are intervals in $\mathcal{N}$. Then the upper bound $N(n, s) \leq 2^n(s-4)+2n+4$ immediately follows from the inductive proof of Theorem 3.

Next, we show that any strategy using also tests which are not intervals can be transformed into a strategy that only uses intervals and such that the target space is also an interval. Hence the upper bound extends also to such strategies. Let $A \subseteq \mathbb{Z}$. We will say that A is an a-set (shorter for a-interval-set) if A is equal to the union of a intervals and it is not equal to the union of any $a - 1$ intervals. If A is an a-set for some $a > 1$ we will also say that A is a multi-interval set. On the other hand, if A is a 1-set, we will also say that A is a mono-interval set.

In the following, for an a-set we will abuse notation and identify the a-set A with the set of a (maximal) intervals whose union is equal to A. More formally, let A be an a-set, with $A = I_1 \cup \cdots \cup I_a$ where $I_1, \ldots, I_a$ are intervals and for each $1 \leq i < j \leq a$ it holds that $I_i \cap I_j = \emptyset$. Therefore, we will also write $I \in A$ to indicate that $I = I_i$ for some $i = 1, \ldots, a$, i.e., I is one of the a maximal intervals into which A can be partitioned. Let $\mathbf{A}^{(a)}$ denote the family of all a-sets and let $\mathbf{D}^{(k,a)}$ denote the family of all a-sets of cardinality k.

Let us consider a minimum length strategy $\mathcal{S}$. For each test $\mathcal{T}$ used by $\mathcal{S}$ with target space being $\mathcal{D}$ —since the target can only move after the test has been performed—we can assume that $\mathcal{T} \setminus \mathcal{D} = \emptyset$. Moreover, we can also assume that $|\Gamma(\mathcal{T} \cap \mathcal{D})| \geq |\Gamma(\mathcal{D} \setminus \mathcal{T})|$. Otherwise, we can replace $\mathcal{T}$ with $\mathcal{D} \setminus \mathcal{T}$ which satisfies the requirement and only swaps the two possible outcomes of the test. We will use the following two claims, which we prove later.

Claim 3. Fix $a > 1$, $b \geq 1$ and $k \geq 1$. For each $\mathcal{D} \in \mathbf{D}^{(k,a)}$ and $\mathcal{T} \in \mathbf{A}^{(b)}$ such that $\mathcal{T} \setminus \mathcal{D} = \emptyset$, and for each $\mathcal{D}^* \in \mathbf{D}^{(k,1)}$, there exists $\mathcal{T}' \in \mathbf{A}^{(b)}$ such that

$$|\Gamma(\mathcal{D} \cap \mathcal{T})| \geq |\Gamma(\mathcal{T} \cap \mathcal{D}^*)|.$$

Claim 4. Fix $b > 1$. For each $\mathcal{D}, \mathcal{D}^* \in \mathbf{D}^{(k,1)}$ and $\mathcal{T} \in \mathbf{A}^{(b)}$ there exists $\mathcal{T}' \in \mathbf{A}^{(1)}$ such that

$$|\Gamma(\mathcal{D} \cap \mathcal{T})| \geq |\Gamma(\mathcal{T}' \cap \mathcal{D}^*)| \quad \text{and} \quad \Gamma(\mathcal{D}^* \setminus \mathcal{T}') \in \mathbf{A}^{(1)}.$$

Let us take any sequence of target spaces $\mathcal{D}_0, \ldots \mathcal{D}_n$ encountered using the strategy $\mathcal{S}$ and let the corresponding tests be $\mathcal{T}_1, \ldots, \mathcal{T}_n$. We will show that we can map this into a sequence of target spaces $\mathcal{D}_0^*, \ldots \mathcal{D}_n^*$ which are always made of a single interval, and are obtained using tests $\mathcal{T}_1', \ldots \mathcal{T}_n'$ which are always made of a single interval, and such that for each i we have $|\mathcal{D}_i| \geq |\mathcal{D}_i^*|$.

We show this by induction. For $i = 0$, we have $\mathcal{D}_0 = \mathcal{D}_0^* = \{1, \ldots, N\}$ then the statement holds. Let us assume that for some $i \geq 0$, for some k and $a > 1$ we have $|\mathcal{D}_i| = |\mathcal{D}_i^*| = k$ and $\mathcal{D}_i \in \mathbf{D}^{(k,a)}$ and $\mathcal{D}_i^* \in \mathbf{D}^{(k,1)}$. For some $b \geq 1$, let $\mathcal{T}_i \in \mathbf{A}^{(b)}$ be the (possibly multi-interval) test used by the startegy $\mathcal{S}$.

By Claim 3, there is a (possibly multi-interval) test $\tilde{\mathcal{T}}_i \in \mathbf{A}^{(b)}$ such that we can use this test on $\mathcal{D}_i^*$ and the resulting target space is $\mathcal{D}_{i+1}'$ satisfying $|\mathcal{D}_{i+1}| \geq |\mathcal{D}_{i+1}'|$. Let $k' = |\mathcal{D}_{i+1}'|$. Notice that $\mathcal{D}_{i+1}'$ can be multi-interval, i.e., there is some $a' \geq 1$ such that $\mathcal{D}_{i+1}' \in \mathbf{D}^{k',a'}$. Now, by Claim 4, we have that there exists a test $\mathcal{T}_i' \in \mathbf{A}^{(1)}$ such that the target space $\mathcal{D}_{i+1}^*$ resulting from using $\mathcal{T}_i'$ on $\mathcal{D}_i^*$ is not larger than the $\mathcal{D}_{i+1}'$. In addition, $\mathcal{D}_{i+1}^*$ is mono-interval, being the intersection of two intervals, and, by Claim 4, also $\mathcal{D}_i^* \setminus \mathcal{T}_i'$ is also an interval. This means that both the possible outcomes of the test $\mathcal{T}'$ are intervals. Therefore, we have found a (mono-interval) test $\mathcal{T}_i'$ such that using $\mathcal{T}_i'$ on the target space $\mathcal{D}_i^*$ the resulting target space $\mathcal{D}_{i+1}^*$ is mono-interval and $|\mathcal{D}_{i+1}^*| \leq |\mathcal{D}_{i+1}'| \leq |\mathcal{D}_{i+1}|$. This concludes the induction step. A consequence of this is that if there exists a strategy of length n using multi-interval tests then there exists a strategy of length n using only mono-interval tests and such that all the target sets encountered at intermediate steps are mono-intervals.

Our proof of the lower bound also shows that when all tests are to be mono-intervals and such that the intermediate target sets are also intervals, then the lower bound is best possible. It follows that since any multi-interval strategy could be turned into a mono-interval strategy, the lower bound is best possible also for strategies where the use of multi-intervals is allowed. □

Proof of Claim 3. Fix a one-to-one map $f : x \in \mathcal{D} \mapsto f(x) \in \mathcal{D}^*$ such that it preserves the order, i.e., $x < x' \to f(x) < f(x')$. Let f be canonically extended to intervals, i.e., for each interval I, we have $f(I) = \{f(x) \mid x \in I\}$. Therefore, f maps intervals to intervals. Moreover, for each $x, x' \in \mathcal{D}$, such that $x < x'$ we have that $f(x') - f(x) \le x' - x$, since $\mathcal{D}^*$ does not have holes.

Let $\mathcal{D} = I_1^{\mathcal{D}} \cup \cdots \cup I_a^{\mathcal{D}}$ and $\mathcal{T} = T_1^{\mathcal{D}} \cup \cdots \cup T_b^{\mathcal{D}}$. W.l.o.g, assume that in the above expressions, the intervals are taken according to the relative order of their first element. Then, we have $\Gamma(\mathcal{D} \cap \mathcal{T}) = \bigcup\limits_{i=1}^{a} \bigcup\limits_{j=1}^{b} \Gamma(I_i^{\mathcal{D}} \cap T_j^{\mathcal{D}})$.

Let $\mathcal{T}' = \cup_{j=1}^{b} f(T_j)$. Then, $\Gamma(\mathcal{D}^* \cap \mathcal{T}') = \bigcup\limits_{i=1}^{a} \bigcup\limits_{j=1}^{b} \Gamma(f(I_i^{\mathcal{D}}) \cap f(T_j^{\mathcal{D}}))$. It is not hard to see that $\Gamma(f(I_i^{\mathcal{D}}) \cap f(T_j^{\mathcal{D}}))$ is only a translation of $\Gamma(I_i^{\mathcal{D}} \cap T_j^{\mathcal{D}})$. Thus, $|\bigcup\limits_{i=1}^{a} \bigcup\limits_{j=1}^{b} \Gamma(I_i^{\mathcal{D}} \cap T_j^{\mathcal{D}})| \le |\bigcup\limits_{i=1}^{a} \bigcup\limits_{j=1}^{b} \Gamma(f(I_i^{\mathcal{D}}) \cap f(T_j^{\mathcal{D}}))|$. Actually, we could have a strict inequality if some intervals are made adjacent by f. □

Proof of Claim 4. Let $\mathcal{D} = \{u, u+1, \ldots, v\}$, $\mathcal{D}^* = \{w, w+1, \ldots, z\}$ and $\mathcal{T} = T_1^{\mathcal{D}} \cup \cdots \cup T_b^{\mathcal{D}}$, with the intervals being disjoint and included in $\mathcal{D}$. W.l.o.g, assume that in the latter expressions, the intervals are taken according to the relative order of their first element. Fix a one-to-one map $g : x \in \mathcal{D} \mapsto g(x) \in \mathcal{D}^*$ such that it preserves the order, i.e., $x < x' \to g(x) < g(x')$ and such that for each $i = 1, \ldots, b$, it holds that $g(y_i) = w + \sum_{\ell=0}^{i-1} |T_\ell^{\mathcal{D}}| - 1$, where y_i is the smallest element in $T_i^{\mathcal{D}}$. In words, g maps the elements of $T_1^{\mathcal{D}}$ to the first elements of $\mathcal{D}^*$, and for each $i = 2, \ldots, b$ it maps the elements in $T_i^{\mathcal{D}}$ to the first elements of $\mathcal{D}^*$, following the elements of $g(T_{i-1}^{\mathcal{D}})$. Hence, $g(\cup_i T_i^{\mathcal{D}}) = \{w, w+1, \ldots, w-1 + \sum_{j=1}^{b} |T_j^{\mathcal{D}}|\}$ which is an interval beginning at the first element of $\mathcal{D}^*$.

We set $\mathcal{T}' = g(\cup_i T_i^{\mathcal{D}})$. The above observation immediately implies that $\Gamma(\mathcal{D}^* \setminus \mathcal{T}')$ is an interval. Morever, arguing like in the case of Claim 3, we also have that $|\Gamma(\mathcal{D} \cap \mathcal{T})| \ge |\Gamma(\mathcal{T} \cap \mathcal{D}^*)|$. □

3   Trees

For a tree T, we denote by $\Delta(T)$ the *maximum degree* of T and by $r(T)$ the *radius* of T, i.e., $r(T) = \min_{v \in T} \max_{u \in T} d(u, v)$ where $d(u, v)$ denote the length of the unique path between u and v in T. We have the following result.

Theorem 4. *Let T be a tree, and let $r = r(T)$ and $\Delta = \Delta(T) \ge 3$. Then we have*

(i) $s^*(T) \le r(\Delta - 1) + 2$ for $r = 1, 2$.

(ii) $s^*(T) \le \left(\lceil \frac{\Delta - 1}{2} \rceil + 1 \right) (\Delta - 1) + u$ for $r = 3$, where $u = 3$ if $\Delta = 3$ and $u = 2$ for $\Delta \ge 4$.

(iii) $s^*(T) \leq \left(\lceil \frac{\Delta}{2} \rceil + 1\right)(\Delta - 1) + 2$ *for $r \geq 4$.*

Moreover, the above bounds are best possible in the sense that for any $r = 3, 4$ there exists a tree for which a lower bound bound can be shown that differs by 1 and for any $r \notin \{3,4\}$ the lower bound exactly matches the corresponding upper bound in (i) and (iii).

The above theorem is consequence of the following observations which motivate the study of the problem on complete q-ary trees described below.

Given a tree T of radius r and maximum degree Δ we can first root T in a vertex x such that any other vertex of T is at distance at most r. In other words, we choose the root in such a way that the depth of the resulting rooted tree is equal to the radius. In addition, this way, because of the bound on the degree, the tree T can be seen as a super tree of a complete $(\Delta - 1)$-ary tree of depth r and as a subtree of an *expanded complete $(\Delta - 1)$-ary tree of depth r*, which we exactly define later. Therefore, for $q = \Delta - 1$, any strategy that works when the search space is an expanded complete q-ary tree of depth r will also work on T (with the appropriate mapping of the tests). Therefore the results (i)-(iv) follow from Theorem 5 below.

For the lower bound, since a complete $(\Delta - 1)$-ary tree of depth r is a tree of maximum degree Δ and radius r the result also follows from Theorem 5 below.

Complete Trees and Expanded Complete Trees. For the proof of Theorem 4, we study our problem for complete q-ary trees and for a properly defined extension of complete q-ary trees. For integers $k \geq 1$ and $q \geq 1$ we denote by $B_k^q = (\mathcal{N}, \mathcal{E})$ a complete q-ary tree of depth k, where $\mathcal{N} = \{1, 2, \ldots, q^{k+1} - 1\}$ and $|\mathcal{E}| = |\mathcal{N}| - 1$. Such a tree has one vertex of degree q, called the root, $q^k(q - 1) - 2$ vertices of degree $q + 1$, called the inner vertices, and q^k vertices of degree 1, called the leaves. We also define an *expanded complete q-ary tree of depth k* as the tree obtained from joining a B_k^q and a B_{k-1}^q by making the root of the B_{k-1}^q an additional child of the root of the B_k^q. Such an *expanded complete q-ary tree of depth k* is denoted by B_k^{q+}. In such a tree all the internal nodes (including the root) have degree $q + 1$.

Theorem 5. *For $q \geq 2$ we have*

(i) $s^*(B_1^q) = q + 1$, $s^*(B_1^{q+}) = q + 2$

(ii) $s^*(B_2^q) = 2q + 1$, $s^*(B_2^{q+}) = 2q + 2$

(iii) $s^*(B_3^q) = \lceil \frac{q+1}{2} \rceil q + q + j$, where $j = 0$ if $q = 2$ and $j = 1$ if $q \geq 3$; and
$s^*(B_3^{q+}) \leq \lceil \frac{q+1}{2} \rceil q + q + 2$,

(iv) $s^*(B_4^q) = \lceil \frac{q+1}{2} \rceil q + q + i$, where $i = 1$ if q is even and $i = 2$ if q is odd; and
$s^*(B_4^{q+}) \leq s^*(B_4^q) + 1$ if q is even and $s^*(B_4^{q+}) \leq s^*(B_4^q)$ if q is odd

(v) $s^*(B_k^{q+}) = s^*(B_k^q) = \lceil \frac{q+1}{2} \rceil q + q + 2$ for all $k \geq 5$.

4   Optimal Strategies When the Target Is Restricted

We now consider the case of a target that changes its position at most t times. Thus, there are at most t distinct elements in a target walk $(d_1, \ldots, d_n)$. The notations $N_c(n, s, t)$ and $N_p(n, s, t)$ have the same meaning as $N_c(n, s)$ and $N_p(n, s)$ but for the restricted version of problem considered here.

Theorem 6. *For $s \geq 4$ we have*

$$N_c(n, s, t) \geq (s - 4)2^n + 2^{n-t+2} = N_c(n, s) + 4(2^{n-t} - 1).$$

Proof. Let $N = (s-4)2^n + 2^{n-t+2}$. We take $\mathcal{T}_1 = \{1, 2, \ldots, (s-4)2^{n-1} + 2^{n-t+1}\}$ and assume w.l.o.g. that the answer is 1. Since the target can move at most t times, we can assume then that the final position of the target is in some half-open interval. Thus, we have the half-open case of the problem for the lines, defined in the proof of Theorem 3. Recall that for k tests we have $N^H(k, s) = 2^k(s-4) + k + 4$ (keeping the notation introduced earlier). Furthermore, observe that we can choose the test $\mathcal{T}_k$ in such a way that $|\mathcal{D}_k| \leq \frac{N}{2^k} + k + 1$ for $1 \leq k < t$ and $\mathcal{D}_k| \leq \frac{N}{2^k} + t + 1$ for $t \leq k \leq 2^{n-t+1}$. Therefore we have

$$\frac{N_c(n, s, t)}{2^k} + t + 1 \geq N^H(n - k, s) = (s - 4)2^{n-k} + n - k + 4,$$

$$N_c(n, s, t) \geq (s - 4)2^n + 2^k(n - t + 3 - k).$$

We can maximize this quantity choosing $k = n - t + 1$. Then we get $N_c(n, s, t) \geq (s - 4)2^n + 4 + 4(2^{n-t} - 1)$. Furthermore, in view of Theorem 1 we have $N_c(n, s, t) \geq N(n, s) + 4(2^{n-t} - 1)$.   □

Remark 1. *It is easy to see that for paths we have the following bound $N_l(n, s, t) \geq N_c(n, s, t) + 2t$.*

For cases $t = 1, 2$ and $s = 3$ we give optimal strategies.

Theorem 7. *For $n \geq 2$ we have (i) $N_c(n, 3, 1) = 2^n$, (ii) For $n \geq 4$ we have $N_c(n, 3, 2) = 2^{n-2}$.*

Proof. (i): As a first test $\mathcal{T}_1$ we take a path on 2^{n-1} vertices, say $\mathcal{T}_1 = \{1, \ldots, \frac{N}{2}\}$, and assume, w.l.o.g, that the answer is 1. Then clearly $\mathcal{D}_1 = \{N, 1, \ldots, \frac{N}{2} + 1\}$ and we know that the final position of the target is in $\mathcal{D}_1$. Then for each $i = 2, \ldots, n$ we take as a test a path $\mathcal{T}_i \subset \mathcal{D}_{i-1}$ of size $\frac{|\mathcal{D}_{i-1}|}{2}$, which contains one of endpoonts of $\mathcal{D}_{i-1}$. Observe then that $|\mathcal{D}_i| = |\mathcal{D}_{i-1}|/2 + 1$ and $|\mathcal{T}_i| = |\mathcal{D}_{i-1}|/2$, for $i = 2, \ldots, n$. It is easy to see that $|\mathcal{T}_n| = 2$ and $|\mathcal{D}_n| = 3$. This shows that $N_c(n, 3, 1) \geq 2^n$. To see that $N_c(n, 3, 1) \leq 2^n$, suppose that $N \geq 2^n$ and the target moves only after n-th test. Then the well-known information theoretic bound (for a stationary target) says that after $n - 1$ tests there are at least two vertices with unknown status. Hence, $|\mathcal{D}_{n-1}| \geq 4$ for $N \geq 2^n$ and we need at least one more test to find a $\mathcal{D}_n$ of size 3.

(ii): The proof for $t = 2$ is similar. After $n - 2$ tests we get $|\mathcal{D}_{n-2}| \geq 4$ and we need at least two more tests to find a proper set $\mathcal{D}_n$.   □

5   Concluding Remarks

We considered a new model of combinatorial two-sided search and studied the cases where the search space topology belongs to the classes of cycles, paths and trees. We described optimal search strategies for path graphs and cycles. Based on an analysis of complete q-ary tree, for arbitrary tree topologies we characterized the minimum possible size of a target set, in terms of the maximum degree and the radius of the tree. Due to the connections to the field of sensor networks, an interesting question which is left open is about the characterization of strategies of minimum length.

Besides extensions to other network topologies, another direction for future research is the study of probabilistic models of two-sided search, as well as two-sided models of search in the case where some of test results are incorrect [7,16], and the related coding problems.

References

1. Ahlswede, R., Wegener, I.: Suchprobleme. Wiley, New York (1987). Teubner, 1979, (English translation), Search problems
2. Alpern, A., Fokkink, R., Gal, S., Timmer, M.: On search games that include ambush. SIAM J. Control Optim. **51**, 4544–4556 (2013)
3. Alspach, B.: Searching and sweeping graphs: a brief survey. Matematiche (Catania) **59**, 5–37 (2006)
4. Aigner, M.: Combinatorial Search. Wiley, New York (1988)
5. Benkoski, S.J., Monticino, M.G., Weisinger, J.R.: A survey of the search theory literature. Naval Res. Logistics **38**, 469–494 (1991)
6. Cheraghchi, M., Karbasi, A., Mohajer, S., Saligrama, V.: Graph-constrained group testing. IEEE Trans. Inf. Theor. **58**(1), 248–262 (2012)
7. Cicalese, F.: Fault-Tolerant Search Algorithms. Springer, Berlin (2013)
8. Damaschke, P.: A tight upper bound for group testing in graphs. Discrete Appl. Math. **48**, 101–109 (1994)
9. Deppe, C.: Searching with lies and coding with feedback. In: CsiszAr, I., Katona, G.O.H., Tardos, G., Wiener, G. (eds.) Entropy, Search, Complexity, Bolyai Society Mathematical Studies, vol. 16, pp. 27–70 (2007)
10. Du, D., Hwang, F.: Combinatorial group testing and its applications. Series on Applied Mathematics (1993)
11. Fomin, F.V., Thilikos, D.M.: An annotated bibliography on guaranteed graph searching. Theor. Comput. Sci. **399**(3), 236–245 (2008)
12. Kaplan, L.: Global node selection for localization in a distributed sensor network. IEEE Trans. Aerosp. Electron. Syst. **42**, 113–135 (2006)
13. Karbasi, A., Zadimoghaddam, M.: Sequential group testing with graph constraints. In: Proceedings of IEEE Information Theory Workshop (ITW), pp. 292–296 (2012)
14. Knuth, D.E.: The Art of Computer Programming. Combinatorial Algorithms, Part 1, vol. 4A. Addison-Wesley Publishing, Boston (2011)
15. Koopman, B.: Search and Screening. Persimmon Press, New York (1946)
16. Pelc, A.: Searching games with errors - fifty years of coping with liars. Theor. Comput. Sci. **270**, 71–109 (2002)

17. Ramya, K., Kumar, K.P., Rao, V.S.: A survey on target tracking techniques in wireless sensor networks. Int. J. Comput. Sci. Eng. Surv. **3**, 93–108 (2012)
18. Rowaihy, H., Eswaran, S., Johnson, M., Verma, D., Bar-Noy, A., Brown, T., La Porta, T.: A survey of sensor selection schemes in wireless sensor networks. In: Proceedings of SPIE 6562, Unattended Ground, Sea, and Air Sensor Technologies and Applications IX, 65621A (2007)
19. Triesch, E.: A group testing problem for hypergraphs of bounded rank. Discrete Appl. Math. **66**, 185–188 (1996)

On the Power of Laconic Advice
in Communication Complexity

Kfir Barhum$^{(\boxtimes)}$ and Juraj Hromkovič

Department of Computer Science, ETH Zurich, Zurich, Switzerland
{kfir.barhum,juraj.hromkovic}@inf.ethz.ch

Abstract. We continue the study of a recently introduced model of communication complexity with advice, focusing on the power of advice in the context of equality of bitstrings and divisibility of natural numbers. First, we establish that the equality problem admits a protocol of polylogarithmic communication, provided a laconic advice of just one bit. For the divisibility problem, we design a protocol with sublinear communication and advice of roughly $\tilde{O}(\sqrt{n})$. We complement our result on divisibility with a matching lower bound in a restricted setting using a recent result of Chattopadhyay et al. and a reduction from set-disjointness to divisibility.

Keywords: Communication complexity · Communication complexity with advice · Equality · Divisibility

1 Introduction

The research field of communication complexity concerns with the efficiency of *interaction*. The theory of communication complexity quantifies and studies the amount of communication required for different settings of distributed computing between two entities that are allowed to communicate over some channel. In this case, local computations are assumed to be unbounded, i.e., we do not limit the entities with respect to the time and space complexities of their local computations, and are solely interested in the amount of information exchanged during the computation, measured usually by the total number of bits exchanged by the parties. The communication between the parties is guided by a protocol which specifies how each message sent depends on the input and the messages sent previously.

More formally, two computationally unbounded players Bob and Charlie hold partial inputs $x \in \mathcal{X}$ and $y \in \mathcal{Y}$, respectively, to a function $f : \mathcal{X} \times \mathcal{Y} \to \{0,1\}$ known to both of them. The parties interact according to a protocol π, which is modeled as a finite sequence of functions $(M_1, \ldots, M_r)$, where $M_i : \mathcal{X} \times (\{0,1\}^+)^{i-1} \to \{0,1\}^+$ for an odd i and $M_i : \mathcal{Y} \times (\{0,1\}^+)^{i-1} \to \{0,1\}^+$ for an even i, specifying the i'th message of the protocol. The computation proceeds

This work was partially supported by SNF grant 200021-146372.

© Springer-Verlag Berlin Heidelberg 2016
R.M. Freivalds et al. (Eds.): SOFSEM 2016, LNCS 9587, pp. 161–170, 2016.
DOI: 10.1007/978-3-662-49192-8_13

as follows: In the first round, the message $m_1 = M_1(x)$ is sent by Bob, in the second round $m_2 = M_2(y, M_1(x))$ is sent by Charlie, and in general, in the i'th round message $M_i(x, m_1, \ldots, m_{i-1})$ (respectively, $M_i(y, m_1, \ldots, m_{i-1})$) for odd (resp., even) i is sent.

The transcript of a protocol on inputs x and y is $\pi(x, y) \stackrel{\text{def}}{=} (m_1, \ldots, m_r)$. The length of the transcript $|\pi(x, y)|$ is defined as the total length of all the messages exchanged.

A protocol is correct if, for all $x \in \mathcal{X}$ and $y \in \mathcal{Y}$, there exists a referee function EVAL, such that $\text{EVAL}(\pi(x, y)) = f(x, y)$. Put differently, if it is possible to compute $f(x, y)$ only by looking at the transcript.

Finally, the communication complexity of a protocol π is $\max_{x,y} |\pi(x, y)|$ and the deterministic communication complexity of a function $f : \mathcal{X} \times \mathcal{Y} \to \{0, 1\}$ is defined as $\text{CC}(f) \stackrel{\text{def}}{=} \min_\pi \max_{x,y} |\pi(x, y)|$, where the minimum is taken over all correct protocols for f.

Observe that for every function there exists a trivial protocol: Bob sends his entire input to Charlie, who locally computes $f(x, y)$ and announces the output. The question one is usually interested in, is "What is the minimal amount of communication between the parties required to compute f?".

For example, the equality function, $\text{EQ} : \{0, 1\}^n \times \{0, 1\}^n \to \{0, 1\}$ defined by $\text{EQ}(x, y) = 1$ iff $x = y$, can be shown to require a deterministic communication complexity of at least $n + 1$ bits.

In the literature many variations of this model are studied, and in particular various models that involve randomness and non-determinism. For a thorough introduction to communication complexity we refer to the textbooks of Hromkovič [1] and Kushilevitz and Nisan [2].

1.1 Communication Complexity with Advice

First observe that simply adding an advice, which depends on the inputs, of even just one bit to the classical model of communication complexity does not seem to make sense, as the advice bit $f(x, y)$ immediately yields a trivial protocol for the problem.

However, motivated by the problem of proving polynomial lower-bounds on the efficiency of dynamical data structures, Pătraşcu [3] has recently suggested the following model, where Bob has input $x \in \mathcal{X}$ (just as before), but Charlie is given as input k elements $y_1, \ldots, y_k$ from $\mathcal{Y}$. Then, a third party Alice, the advisor, receives both inputs and computes an advice string which she sends to Bob and then remains silent. Finally, Bob and Charlie are presented an index i and are allowed to interact by exchanging messages, where their goal is to compute $f(x, y_i)$.

More formally, a protocol π with m advice bits for the k-instance problem is $\pi = (\pi_a, M_1, \ldots, M_r)$, where $\pi_a : \mathcal{X} \times \mathcal{Y}^k \to \{0, 1\}^m$ is the advice function of the protocol and

$$M_i : \mathcal{X} \times [k] \times \{0, 1\}^m \times (\{0, 1\}^+)^{i-1} \to \{0, 1\}^+$$

for an odd i and

$$M_i : \mathcal{Y}^k \times [k] \times (\{0,1\}^+)^{i-1} \to \{0,1\}^+$$

for an even i.

The computation on inputs $x, y_1, \ldots, y_j, i$ proceeds similarly to before: First, the advice $a = \pi_a(x, y_1, \ldots, y_k)$ is given to Bob, and then the interaction continues as follows: Bob sends message $m_1 = M_1(x, i, a)$ to Charlie, who replies with message $m_2 = M_2(y_1, \ldots, y_k, i, m_1)$ and so forth.

We stress that it is essential that only Bob receives the advice. Otherwise, for example, whenever $m > \log(|\mathcal{X}|)$, the advice could already encode Bob's input, and Charlie could locally compute the answer.

Also, note that the problem is only interesting in the case where $m < k$. Otherwise, a trivial protocol always exists, where the advice just encodes the answer vector $(f(x, y_1), \ldots, f(x, y_k))$.

As before, for $y^{(k)} \stackrel{\text{def}}{=} (y_1, \ldots, y_k)$, the transcript $\pi(x, y^{(k)}, i)$ of π for inputs $x, y^{(k)}$ and i is the list $(m_1, \ldots, m_r)$ of all messages exchanged during its computation on inputs $x, y^{(k)}$. As before, a protocol is correct if it is possible to compute $f(x, y_i)$ only by looking at its transcript. We define the communication complexity of f for k inputs with m bits of advice as

$$\mathsf{CC}^k_m(f) \stackrel{\text{def}}{=} \min_{\pi} \max_{x, y^{(k)}, i} \left| \pi(x, y^{(k)}, i) \right| ,$$

where the minimum is over all protocols π that correctly compute $f(x, y_i)$ for every input $(x, y^{(k)}, i)$.

As mentioned, Pătraşcu offered a plausible approach for lower bounds to a host of dynamic data structure problems via a series of reduction from the problem of set-disjointness in the communication complexity with advice model on which super-polynomial lower bounds are conjectured. In particular, one such problem is subgraph connectivity, where, after a preprocessing of an undirected graph, the data structure supports on/off operation for vertices and queries for pair of vertices u, v asking whether there is a path using only "on" vertices from u to v. Another problem is Langerman's problem, where it is required to maintain updates on an array of length n and support answering the zero-partial-sum question, namely, does there exist a non-empty subset of indices that sum to zero. We refer to Sect. 1.1 in [3] for a complete taxonomy.

Thus, the communication complexity with advice model is well-motivated, whose study offers a promising approach towards polynomial lower bounds on the aforementioned problems.

1.2   Our Contribution

We study two natural problems in the model of communication complexity with advice (CCwA): the equality of two bitstrings, where each of the parties hold a bitstring and the goal is to decide whether they are equal or not, and the problem of divisibility, where each party has a number, and the goal is to decide whether one of them divides the other.

Recall that in the CCwA model, for any function $f : \{0,1\}^n \times \{0,1\}^n \to \{0,1\}$, the problem for k instances with k bits of advice becomes trivial, i.e., $\mathsf{CC}_k^k(f) = 1$. Pătraşcu conjectured that, for all functions f (as before), whenever the number of advice bits m is significantly smaller than k, i.e., for $m \in o(k)$, the communication complexity of f in the CCwA model is linearly related to that in the classical model, i.e., that for all f, $\mathsf{CC}_m^k(f) \in \Omega(\mathsf{CC}(f))$ for $m \in o(k)$.

However, Chattopadhyay et al. [4] refuted his conjecture, and showed that in the CCwA model $\log k$ bits of advice and communication are sufficient to compute equality (recall that $\mathsf{CC}(\mathrm{EQ}) = n + 1$) using the following simple protocol: The advice encodes an index j of a y_j such that $x = y_j$, or 0 for the case that $x \neq y_j$ for all j. After the parties are presented an index i, Bob forwards the advice to Charlie, who then answers with the result of the comarison of y_i and y_j (or with no if the advice was 0).

An important problem in the CCwA model is set-disjointness, where the inputs $x, y \subset \{0,1\}^n$ are interpreted as characteristic vectors of a subset of $[n]$. The vectors x and y are disjoint, if and only if the sets they describe are disjoint. That is, $\mathrm{DISJ} : \{0,1\}^n \times \{0,1\}^n \to \{0,1\}$ is given by $\mathrm{DISJ}(x,y) = \bigwedge_{i=1}^n (\neg((x)_i \wedge (y)_i))$, where for $z \in \{0,1\}^n$, $(z)_i$ denotes its i'th coordinate. Pătraşcu showed that proving a lower bound on set-disjointness in the CCwA model for some specific parameters would imply a polynomial lower bound on many problems of dynamic data-structures. Chattopadhyay et al. studied set-disjointness in the CCwA model, and showed an upper bound of $\tilde{O}(\sqrt{n})$ on its communication complexity, provided the same amount of advice, and a matching lower bound in a more restricted setting.

We first study the power of laconic advice for equality. Somewhat surprisingly, we show that, in the CCwA model, a short communication of only a polylogarithmic number of bits already suffices to deterministically compute equality, provided a laconic advice of just one bit. Our result can be understood as a trade-off between the number of advice bits the protocol utilizes and its communication complexity.

Chattopadhyay et al. observed that for every $k, n \in \mathbb{N}^+$:

$$\mathsf{CC}_{\log(k)+1}^k(\mathrm{EQ}) \in O(\log(k)) \ ,$$

where we prove that for every $k, n \in \mathbb{N}^+$:

$$\mathsf{CC}_1^k(\mathrm{EQ}) \in O(\log(k)(\log(k) + \log(n))) \ .$$

Our second main result is a protocol for divisibility. We give a protocol that uses roughly $\tilde{O}(\sqrt{n})$ bits of advice and communication, which improves on the performance of the trivial protocol in the classical model, where it is optimal. To see that $\mathsf{CC}(\mathrm{DIV}) \in \Omega(n)$, note that any protocol for divisibility could be used to compute equality, since $x = y$ if and only if $x|y$ and $y|x$, and by applying the lower bound on equality. Our protocol here is inspired by that of Chattopadhyay et al. for set-disjointness. Next, we explain how to reduce set-disjointness to divisibility, and employ their lower-bound on set-disjointness to obtain a matching lower-bound for divisibility in a restricted setting.

2   Preliminaries

We denote by $\mathbb{P}$ the set of prime numbers. For a natural number n and a prime number p, we denote by $\nu_p(n)$ the multiplicity of p in n, i.e., the largest exponent i such that $p^i | n$. For a natural number $n = \prod_{i=1}^{k} p_i^{\nu_{p_i}(n)}$, where the p_i's are the different prime factors of n, we denote by $\|n\|_\pi \overset{\text{def}}{=} \sum_{i=1}^{k} \nu_{p_i}(n)$ the total number of its prime factors including repetitions.

We shall use the following fact: Let $a_1, \ldots, a_\ell, b \in \mathbb{N}$. If $b | a_j$ for all $j \in \{1, \ldots, \ell\}$ then $b | \gcd(a_1, \ldots, a_\ell)$.

3   Equality with a Laconic Advice

In this section we continue the study of the Equality problem with advice. We show that, when the advice $A(x, y_1, \ldots, y_k)$ answers the question "Is there an index j such that $x = y_j$?", it is possible to compute equality while exchanging only a polylogarithmic number of bits.

We start with a basic version of our protocol that computes equality using $O(k \log(n))$ bits of communication, which already improves on the trivial protocol for the case $k < n/\log(n)$.

3.1   A Basic Protocol

After receiving the advice bit, Bob forwards it to Charlie, who maintains a set S of inputs that are potentially equal to x. At the beginning, S is just the entire set of inputs. Then, the protocol proceeds in a step-wise manner, where at each step it asserts for at least one of the y_i's that $x \neq y_i$. Eventually, only one y_i remains, and using the advice, it must hold that $x = y$ and the protocol outputs the index i. More precisely, the protocol proceeds as follows:

1. Bob forwards $A(x, y_1, \ldots, y_k)$ to Charlie. Charlie sets $S \leftarrow \{y_1, \ldots, y_k\}$. While $|S| > 1$, repeat the following two steps:
2. Charlie chooses two strings y and $y' \neq y$ from S, and sends Bob an index q of a bit on which y and y' differ, i.e., $(y)_q \neq (y')_q$.
3. Bob answers with $(x)_q$, the q'th bit of x, and Charlie updates $S \leftarrow \{z \in S : (z)_q = (x)_q\}$.
4. There is only one element y left in S, and Charlie outputs *yes* if $y = y_i$ and *no* otherwise.

Correctness follows immediately, since we rule out only elements for which we are sure that $y_j \neq x$, and using the advice, we know that, at step 4, it must hold that $x = y$. By the choice of q, the size of S reduces by at least one in every iteration of steps 2 and 3, and so the protocol terminates after at most k iterations of steps 2 and 3, and in every round $\log(n) + 1$ bits are communicated, amounting to a total communication complexity of $O(k \log n)$ bits.

3.2 A Protocol Using a Polylogarithmic Number of Communication Bits

As in the basic version of the protocol, the protocol now proceeds iteratively, where at each step the protocol asserts for a *constant fraction* of the elements in S that $x \neq y_j$. It follows that after $O(\log k)$ rounds, only a constant number of possible indices remains.

The idea of our protocol is as follows: Think of the messages of Charlie in the basic version of the protocol as describing a predicate from a set of n predicates of the form $p_i : \{0,1\}^n \to \{0,1\}$, where $p_i(s) = 1$ if and only if the i'th bit of s is 1. In each round Charlie chose an appropriate predicate p_i, and Bob's answer was $p_i(x)$.

This allowed in turn to rule out the equality of at least one of the remaining y_i's to x. In contrast, we show next that for every k, there exists a *single* set of $O(n)$ predicates that allows to rule out the equality of x to a *constant fraction* of the remaining y_i's, for any set of $y_1, \ldots, y_k$. This allows to reduce the number of rounds of the protocol to $O(\log(k))$, while essentially maintaining the same number of bits communicated in every round, since describing a particular predicate would require only $\log(n) + O(1)$ bits. The next lemma establishes the existence of the aforementioned predicate set. We shall make use of the following: Let $S \subseteq \{0,1\}^n$. A predicate $p : \{0,1\}^n \to \{0,1\}$ is **good** for S, if $|\{s \in S \mid p(s) = 1\}| > \frac{|S|}{4}$ and $|\{s \in S \mid p(s) = 0\}| > \frac{|S|}{4}$.

Lemma 1. *Let $k > 17$ and $n > 0$. Then there exists a set $P = P(n,k)$ of $30n$ predicates such that, for any pair-wise different $y_1, \ldots, y_k \in \{0,1\}^n$ there exists a predicate $p \in P$, which is good for $\{y_1, \ldots, y_k\}$.*

Proof. Consider first a fixed set $Y \stackrel{\text{def}}{=} \{y_1, \ldots, y_k\}$ and a random subset Z of $\{0,1\}^n$, where each string is chosen independently to Z with probability $1/2$. The expected number of elements in $Z \cap \{y_1, \ldots, y_k\}$ is $k/2$. Furthermore, by the Chernoff bound, with probability at most $2 \cdot e^{-\frac{k}{12}}$ ($< e^{-\frac{k}{24}}$ whenever $k > 17$), any of the "bad" events $B_{Y,Z} \stackrel{\text{def}}{=} \{|Z \cap Y| > \frac{3k}{4} \text{ or } |Z \cap Y| < \frac{k}{4}\}$ happens. Consider now ℓ independent copies of Z, that is, the random subsets $Z_1, \ldots, Z_\ell$, and the corresponding events $B_{Y,Z_i} \stackrel{\text{def}}{=} \{|Z_i \cap Y| > \frac{3k}{4} \text{ or } |Z_i \cap Y| < \frac{k}{4}\}$. Now set $B_Y \stackrel{\text{def}}{=} \cap_{i=1}^{\ell} B_{Y,Z_i}$, i.e., B_Y is the event that for every i, B_{Y,Z_i} happens. Now, by the independence of the B_{Y,Z_i}'s we see that $\Pr[B_Y] < e^{-\frac{k\ell}{24}}$. Finally, applying the union bound we obtain that $\Pr[\exists Y \subset \{0,1\}^n, |Y| = k : B_Y] < 2^{nk} \cdot e^{-\frac{k\ell}{24}}$. Setting $\ell \stackrel{\text{def}}{=} 30n$ ensures that this probability is strictly smaller than 1, which in turn implies the existence of a single choice of $30n$ subsets of $\{0,1\}^n$ which satisfy the condition. By what we have shown, setting the predicate set as the corresponding indicator functions completes the proof. $\square$

We are now ready to describe the protocol:

1. Bob forwards $A(x, y_1, \ldots, y_k)$ to Charlie. Charlie sets $S \leftarrow \{y_1, \ldots, y_k\}$. While $|S| > 17$, repeat the following two steps:

2. Charlie sets $k' = |S|$, computes locally a set $P(n, k')$, chooses a good predicate for S, and sends its index q to Bob along with k'.
3. Bob computes locally $P(n, k')$, and answers with $p_q(x)$. Charlie updates
   $S \leftarrow \{w \in S : p_q(w) = p_q(x)\}$.
4. Continue with the loop of the basic protocol.

Given Lemma 1, the analysis of the protocol follows easily. The lemma guarantees the existence of a good predicate set, which both parties can explicitly find (for example, they can agree on the lexicographically first $30n$ predicates and locally exhaustive search them).

The communication complexity follows readily: In every round of iterating steps 2 and 3, at most $(\log(k) + \log(n) + 1)$ bits are communicated, describing $|S|$ and the index of the predicate. Since the chosen predicate is good, it follows that the updated set in Step 4 has size of at most $\frac{3}{4}|S|$, and therefore after $O(\log(k))$ rounds we continue with at most 17 rounds of the basic protocol, resulting in a total communication complexity of $O(\log(k)(\log(k) + \log(n)))$.

Thus we have shown:

Theorem 1. *For all $k \in \mathbb{N}^+$,*

$$\mathsf{CC}_1^k(\mathrm{EQ}) \in O(\log(k)(\log(k) + \log(n))) \ .$$

4   Divisibility

Here, Bob and Charlie have inputs x and $y_1, \ldots, y_k \in \{0,1\}^n$, respectively, interpreted as natural numbers in the inteval $[1, 2^n]$, identifying 2^n with the bitstring 0^n. After receiving an advice $A(x, y_1, \ldots, y_k)$ they are presented an index i and required to compute $x|y_i$.

We show a protocol that communicates $O(\log(n)(\log(k) + \log(n))\sqrt{n})$ bits, provided an advice of similar size. First we shall describe our protocol using the following simplifying assumption, and later we explain how to remove it. We assume that the prime factors of all $x, y_1, \ldots, y_k$ are either within the interval $[2^{2^\tau}, 2^{2^{\tau+1}})$ for some $\tau \in \{0, \ldots, \log(n)/2 - 1\}$ or in the interval $[2^{2^\tau}, 2^n)$ for $\tau = \log(n)/2$. Observe that this implies that each of the numbers $x, y_1, \ldots, y_k$ has at most $n/2^\tau$ distinct factors. Moreover, if $\tau \in \{0, \ldots, \log(n)/2 - 1\}$, each factor can be described using at most $2^{\tau+1}$ bits.

The main idea in our algorithm is to reveal information about x using $y_1, \ldots, y_k$. The advice consists in a subset of the inputs of Charlie for which $x|y_i$, and it is built by iterating over the inputs. The inputs that are added are those which x divides, and that additionally, given all the indices added so far, contribute enough additional information about x. Roughly speaking, the index of an input y_i is not added to the advice string for one of two reasons: Either x does not divide it, or, it would not have contributed enough information about x, given the previous positive inputs. If later an input not added due to the latter condition is presented, it could be computed using only $\tilde{O}(\sqrt{n})$ bits. In what follows we set

$$t \stackrel{\text{def}}{=} \sqrt{n}/2^\tau \ .$$

More precisely, the advice sent by Alice is a subset S of the indices $\{1, \dots, k\}$, which is constructed as follows:

Start with $S = \emptyset$ and add to S the minimum index j_0 such that $x|y_{j_0}$. Then, loop on the elements $y_{j_0+1}, \dots, y_k$: Let $j_0, \dots, j_\ell$ be the current indices of S. Add the index j of the current element y_j to S if:

- $x|y_j$ and
- $\|\gcd(y_{j_0}, \dots, y_{j_\ell})\|_\pi - \|\gcd(y_{j_0}, \dots, y_{j_\ell}, y_j)\|_\pi \geq t$.

The second condition ensures that the gcd of the elements currently indexed by S contains at least t more prime factors including repetition than the gcd of those elements along with the number y_j.

We claim that during this process, at most $\frac{n}{2^\tau t}$ elements are added to S. Indeed, by our simplifying assumption, y_{j_0} contains at most $\frac{n}{2^\tau}$ different factors and the second condition asserts that the number of prime factors remaining in $\gcd_{j' \in S} a'_j$ after adding an element decreases by at least t.

Substituting for t, we have that $|S| < \sqrt{n}$, and therefore the advice sent by Alice (a description of S) can be encoded using at most $\sqrt{n} \log(k)$ bits.

After Bob receives S from Alice, the parties are presented an index i and need to determine whether $x|y_i$. The protocol between Bob and Charlie continues as follows:

1. If $i \in S$, Bob outputs *yes*.
2. Otherwise, Bob forwards S to Charlie.
3. Charlie computes the set $S' = \{j' \in S \mid j' < i\}$, i.e., the constructed set S as it was just before index i was processed.
4. If $\|\gcd_{j' \in S'}(y_{j'})\|_\pi - \|\gcd(\gcd_{j' \in S'} y_{j'}), y_i)\|_\pi \geq t$, he outputs *no*. Otherwise, there are at most $t - 1$ distinct prime factors appearing in y_i but not in $\gcd_{j' \in S'}(y_{j'})$. For each such factor p, Charlie sends $(p, \nu_p(y_i))$ to Bob.
5. Bob outputs *yes* if, for every received pair $(p, \nu_p(y_i))$, it holds that $\nu_p(x) \leq \nu_p(y_i)$ and otherwise outputs *no*.

Let us now see that the protocol always outputs a correct answer. If it outputs *yes* at step 1, by construction, for all indices in S, it holds that $x|y_j$. If the protocol outputs *no* at step 4, it cannot be the case that $x|y_i$, as otherwise both conditions during the construction of S had been satisfied and i would have been added to S. Lastly, if *no* is output at step 5, it is because the protocol witnesses a factor with higher multiplicity in x than in y_i. When this is not the case, we claim that $x|y_i$. We show that for each prime factor in x, it holds that $\nu_p(x) \leq \nu_p(y_i)$. Let us distinguish two cases: If p is one of the up to $t - 1$ factors not appearing in $\gcd(\gcd_{j' \in S}(y_{j'}), y_i)$, then the inequality is asserted by Bob. For any other such factor p, we have (1) $\nu_p(\gcd_{j' \in S'}(y_{j'})) \leq \nu_p(y_i)$, and additionally, by construction of S it holds that $x|y_{j'}$ for all $j' \in S$ and therefore $x|\gcd_{j' \in S}(y_{j'})$, and, in particular, (2) $\nu_p(x) \leq \nu_p(\gcd_{j' \in S}(y_{j'}))$. The correctness in this case follows from (1) and (2).

Next, we analyze the communication complexity of our protocol. At step 1, S is forwarded and at most $\sqrt{n} \log(k)$ bits are sent, and at step 4 at most $t-1$ pairs are sent. This implies that, for the case that $\tau = \log(n)/2$, it holds that $t = 1$ and therefore no pairs are sent. For $\tau \in \{0, \ldots, \log(n)/2 - 1\}$, using the second part of the assumption, each prime can be described using at most $2^{\tau+1}$ bits. The multiplicity of every prime number is at most n, which can be described using $\log(n)$ bits, and therefore at most $t(2^{\tau+1} + \log(n)) \in O(\sqrt{n}\log(n))$ bits are communicated at this step. Thus, the total number of bits communicated is $O(\sqrt{n}(\log(k) + \log(n)))$.

Finally, in order to get rid of the assumption, note that $x|y_i$ if and only if, $x^{(\tau)}|y_i^{(\tau)}$ for all $\tau \in \{0, \ldots, \log(n)/2\}$, where for $m \in \mathbb{N}$ we define $m^{(\tau)} \stackrel{\text{def}}{=} \prod_{p \in [2^{2^\tau}, 2^{2^{\tau+1}}) \cap \mathbb{P}} p^{\nu_p(m)}$ for $\tau \in \{0, \ldots, \log(n)/2 - 1\}$ and $m^{(\tau)} \stackrel{\text{def}}{=} \prod_{p \in [2^{2^\tau}, 2^n) \cap \mathbb{P}} p^{\nu_p(m)}$ for $\tau = \log(n)/2$.

Thus, for the final protocol we run the protocol in parallel for each of the $\log(n)/2 + 1$ possible values of τ, and output yes in case all runs output yes, and no otherwise. This results in an overhead of an $O(\log(n))$ factor to the original protocol. We summarize: Utilizing $O(\sqrt{n}\log(k)\log(n))$ advice bits our protocol communicates $O(\sqrt{n}\log(n)(\log(k) + \log(n)))$ bits. Thus we have shown:

Theorem 2. *For all $k \in \mathbb{N}^+$,*

$$\mathsf{CC}^k_{\log(n)\log(k)\sqrt{n}}(\mathrm{DIV}) \in O(\sqrt{n}\log(n)(\log(k) + \log(n))) \ .$$

4.1 On the Asymmetry of Divisibility

We note that the communication complexity with advice model is inherently asymmetric, and therefore a protocol for $\mathrm{DIV}(x, y) = x|y$ does not yield a protocol for $\mathrm{DIV}'(x, y) = y|x$, as is the case in the classical model of communication complexity. In this section, we explain the changes needed in our protocol to obtain a protocol for DIV'. In our protocol from the previous section, we used the advice to reveal information about x using the y_i's. In particular, the number encoded by the gcd of the y_i's chosen to the advice could be understood as a relatively tight upper bound on the prime powers of x. The analogous advice information for $y|x$ consists in a lower bound on the prime powers of x. Analogously to before, observe that if $a|x$ and $b|x$ then also $\mathrm{lcm}(a, b)|x$. The advice is generated similarly to before, where the set S first contains the minimal index j_0 such that $y_{j_0}|x$. Then, looping over the indices $j_0 + 1, \ldots, k$, index j is added if and only if $y_j|x$ and $\|\mathrm{lcm}(y_{j_0}, \ldots, y_{j_\ell}, y_j)\|_\pi - \|\mathrm{lcm}(y_{j_0}, \ldots, y_{j_\ell})\|_\pi \geq t$, where $j_0, \ldots, j_\ell$ are the current elements of S. Steps (4) and (5) of the protocol now become:

4. If $\|\mathrm{lcm}(y_{j_0}, \ldots, y_{j_\ell}, y_j)\|_\pi - \|\mathrm{lcm}(y_{j_0}, \ldots, y_{j_\ell})\|_\pi \geq t$ it outputs no. Otherwise, there are at most $t - 1$ distinct prime factors appearing in $\mathrm{lcm}_{j' \in S'}(y_{j'})$ but not in y_i. For each such factor p, Charlie sends $(p, \nu_p(y_i))$ to Bob.
5. Bob outputs yes if, for every received pair $(p, \nu_p(y_i))$, it holds that $\nu_p(x) \geq \nu_p(y_i)$, and outputs no.

The correctness and analysis of the protocol follow analogously to before.

4.2   An Almost Matching Lower Bound in Restricted Settings

The problem of set-disjointness consists in two n-bit inputs x and y, where each is interpreted as the characteristic vector of a subset of a set of n elements. Inputs $x, y \in \{0, 1\}^n$ are disjoint if $(y)_i = 0$ whenever $(x)_i = 1$. Chattopadhyay et al. [4] studied the problem of set-disjointness in the CCwA model, and showed that for $k \geq \sqrt{n}$, any protocol with advice of size $m \leq \alpha\sqrt{n}$ communicates at least $\beta\sqrt{n}$ bits for some constants $0 < \alpha, \beta < 1$. In what follows, we describe a reduction from set-disjointness to divisibility, establishing an analogous lower bound for divisibility.

Given inputs a and b (characteristic vectors of some sets A and B) to set-disjointness, we first observe that A and B are disjoint if and only if $A \subseteq \overline{B}$. Now, let $p_1, \ldots, p_n$ be the first n prime numbers, and set $N_A \stackrel{\text{def}}{=} \prod_{j=1}^{n} p_i{}^{a_i}$ and $N_{\overline{B}} \stackrel{\text{def}}{=} \prod_{j=1}^{n} p_i{}^{1-b_i}$. It follows that $A \cap B = \emptyset$ if and only if $N_A \mid N_{\overline{B}}$. By the prime number theorem, it holds that for all large enough n, the first n prime numbers lie in the interval $[1, 3n\log(n)]$, and therefore both N_A and $N_{\overline{B}}$ are described using at most $n \cdot \log(3n\log(n)) < 2n\log(n)$ bits. Therefore, any protocol in the CCwA model for k inputs of size $2n\log(n)$ and m bits of advice yields a protocol (with the same k and m values) for inputs of size n for divisibility; the parties compute N_x and $N_{\overline{y_1}}, \ldots, N_{\overline{y_k}}$ and run the protocol for divisibility on these inputs. Setting $f(n) = 2n\log(n)$, the lower bound of Chattopadhyay et al. (Theorem 5.2 in [4]) establishes that a protocol for $k \geq \sqrt{f^{-1}(n)}$ inputs of size n with advice of size at most $\alpha\sqrt{f^{-1}(n)}$ communicates at least $\beta\sqrt{f^{-1}(n)}$ bits. In view of our protocol from Sect. 4, it follows that this is best possible (up to a logarithmic factor) with advice of size $\tilde{O}(\sqrt{n})$.

References

1. Hromkovič, J.: Communication Complexity and Parallel Computing. Springer, New York (1997)
2. Kushilevitz, E., Nisan, N.: Communication Complexity. Cambridge University Press, New York (1997)
3. Patrascu, M.: Towards polynomial lower bounds for dynamic problems. In: Schulman, L.J. (ed.) STOC, pp. 603–610. ACM (2010)
4. Chattopadhyay, A., Edmonds, J., Ellen, F., Pitassi, T.: A little advice can be very helpful. In: Rabani, Y. (ed.) SODA, pp. 615–625. SIAM (2012)

Using Attribute Grammars to Model Nested Workflows with Extra Constraints

Roman Barták[✉]

Faculty of Mathematics and Physics, Charles University in Prague,
Malostranské nám. 25, 118 00 Prague 1, Czech Republic
bartak@ktiml.mff.cuni.cz

Abstract. Workflow is a formal description of a process. Nested work-flows were proposed to model processes with a hierarchical structure and they support extra logical and temporal constraints to express relations beyond the hierarchical structure. This workflow model supports scheduling applications with a known number of activities in the process, but it cannot be used to model planning problems, where the number of activities is unknown beforehand. In this paper we propose to model nested workflows using a modified version of attribute grammars. In particular we show that nested workflows with extra constraints can be fully translated to attribute grammars. The major advantage of this novel modeling framework is a support for recursive tasks that can model planning problems in the style of hierarchical task networks.

Keywords: Attribute grammars · Workflows · Modelling · Transformation

1 Introduction

Workflows are used to formally describe processes of various types such as business and manufacturing processes. There exist many formal models to describe workflows [8] that include decision points and conditions for process splitting as well as loops to describe repetition of activities. Hierarchical structure of workflows is in particular interesting for real-life workflows [1] as many workflows are obtained by decompositions of tasks. The paper [2] proposed a hierarchical workflow model called Nested Temporal Networks with Alternatives that was later extended with extra constraints to model a wider range of workflows [3].

An interesting question is whether it would be possible to unify various hierarchical structures of workflows using a single concept. The hierarchical structure naturally resembles the idea of a derivation tree of context-free grammars, but to model constraints beyond the hierarchical structure one needs a stronger concept. In this paper we propose to use attribute grammars [6] as a unifying concept to describe hierarchical workflows. In particular, we will show that nested workflows with extra constraints can be fully translated to attribute grammars.

© Springer-Verlag Berlin Heidelberg 2016
R.M. Freivalds et al. (Eds.): SOFSEM 2016, LNCS 9587, pp. 171–182, 2016.
DOI: 10.1007/978-3-662-49192-8_14

2   Background

2.1   Nested Worfkflows

In this work we use nested workflows from the FlowOpt system [3]. The nested workflows were formally introduced in [4] and for completeness, we will recapitulate the formal definitions here.

The *nested workflow* is obtained from a root task by applying decomposition operations that split the task into subtasks until primitive tasks, corresponding to activities, are obtained. Three decomposition operations are supported, namely parallel, serial, and alternative decompositions. Figure 1 gives an example of a nested workflow that shows how the tasks are decomposed. The root task *Chair* is decomposed serially to two tasks, where the second task is a primitive task filled by activity *Assembly*. The first task *Create Parts* decomposes further to three parallel tasks *Legs*, *Seat*, and *Back Support*. *Back Support* is the only example here of alternative decomposition to two primitive tasks with *Buy* and *Welding* activities (*Welding* is treated as an alternative to *Buy*). Hence the workflow describes two alternative processes. Naturally, the nested workflow can be described as a tree of tasks (Fig. 1 bottom right).

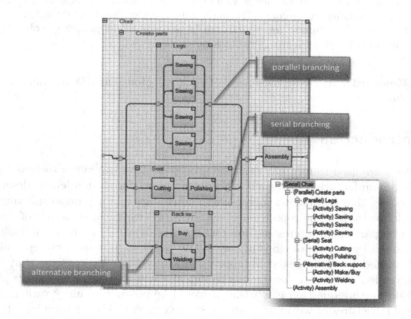

Fig. 1. Example of a nested workflow as it is visualized in the FlowOpt Workflow Editor (from top to down there are parallel, serial, and alternative decompositions)

In this paper we further simplify the formal model from [4] by omitting the serial decomposition and modeling it as a parallel decomposition with extra precedence constraints. For example the serially decomposed task *Chair* can be decomposed in a parallel way with an extra precedence constraints put between the tasks *Create Parts* and *Assembly*. This removes one modeling concept – a serial task – while preserving the expressive power fully (the proof is obvious).

Formally, the nested workflow is a set *Tasks* of tasks that is a union of three disjoint sets: *Parallel*, *Alternative*, and *Primitive*. For each task T (with the exception of the *root* task), function $parent(T)$ denotes the parent task in the hierarchical structure. Similarly for each task T we can define the set $subtasks(T)$ of its child nodes ($subtasks(T) = \{C \in Tasks | parent(C) = T\}$). The tasks from sets Parallel and Alternative are called *compound tasks* and they must decompose to some subtasks:

$$\forall T \in (Parallel \cup Alternative) : subtasks(T) \neq \emptyset, \tag{1}$$

while the *primitive tasks* do not decompose:

$$\forall T \in Primitive : subtasks(T) = \emptyset. \tag{2}$$

The workflow defines one or more processes in the following way. *Process* selected from the workflow is defined as a subset $P \subseteq Tasks$ in the workflow satisfying the following constraints:

$$\forall T \in P, T \neq root : \qquad\qquad parent(T) \in P \tag{3}$$
$$\forall T \in P \cap Parallel : \qquad\qquad subtasks(T) \subseteq P \tag{4}$$
$$\forall T \in P \cap Alternative : \qquad |subtasks(T) \cap P| = 1 \tag{5}$$

Formula (3) says that for each task in the process (except the root) its parent task is also in the process. Formula (4) says that all subtasks of a paralell task in the process must also be in the process. Finally, formula (5) says that exactly one subtask is in the process for each alternative task in the process.

In addition to the hierarchical structure of the nested workflow, the nested structure also defines certain implicit temporal constraints (the arcs in Fig. 1). These temporal relations must hold for all tasks in a single process. Assume that S_i is the start time and E_i is the end time of task T_i. The primitive tasks T_i are filled with activities and each activity has certain duration D_i. Then for tasks in the process P the following relations hold:

$$\forall T_i \in P \cap Primitive : S_i + D_i = E_i \tag{6}$$
$$\forall T_i \in P \cap (Parallel \cup Alternative) :$$
$$S_i = min\{S_j | T_j \in P \cap subtasks(T_i)\}$$
$$E_i = max\{E_j | T_j \in P \cap subtasks(T_i)\}. \tag{7}$$

Notice that the duration of a compound task is defined by the time allocation of its subtasks while the duration of a primitive task is defined by the activity.

A *feasible process* is a process where the time variables S_i and E_i can be instantiated in such a way that they satisfy the above temporal constraints.

It is easy to realize that if there are no additional constraints then any process is feasible. The process defines a partial order of tasks so their start and end times can be set in the left-to-right order while satisfying the constraints (6) and (7).

The nested structure may not be flexible enough to describe naturally some additional relations in real-life processes, for example when an alternative for one task influences the selection of alternatives in other tasks. The following constraints can be added to the nested structure to simplify description of these additional relations between any two tasks T_i and T_j [3]:

precedence constraint $(i \rightarrow j)$: $T_i, T_j \in P \Rightarrow E_i \leq S_j$ (8)

start-start synchronization $(i \; ss \; j)$: $T_i, T_j \in P \Rightarrow S_i = S_j$ (9)

start-end synchronization $(i \; se \; j)$: $T_i, T_j \in P \Rightarrow S_i = E_j$ (10)

end-start synchronization $(i \; es \; j)$: $T_i, T_j \in P \Rightarrow E_i = S_j$ (11)

end-end synchronization $(i \; ee \; j)$: $T_i, T_j \in P \Rightarrow E_i = E_j$ (12)

mutual exclusion constraint $(i \; mutex \; j)$: $T_i \notin P \vee T_j \notin P$ (13)

equivalence constraint $(i \Leftrightarrow j)$: $T_i \in P \Leftrightarrow T_j \in P$ (14)

implication constraint $(i \Rightarrow j)$: $T_i \in P \Rightarrow T_j \in P$ (15)

Note that if extra constraints are used then the existence of a feasible process is no longer guaranteed. For example an equivalence constraint between the tasks *Buy* and *Welding* in Fig. 1 causes no feasible process to exist.

In summary, we can model the nested workflow with extra constraints as a tuple $W = (Parallel, Alternative, Primitive, root, parent, D, C)$, where the *parent* relation defines a tree rooted at the node *root* with leaves *Primitive* and internal nodes *Parallel* $\cup$ *Alternative*. The set C defines the extra constraints (8) – (15) and D maps the tasks in *Primitive* to non-negative integers defining durations of primitive tasks. The process is a subtree of this tree satisfying constraints (3) – (5) and (13) – (15), where each node T_i has assigned two integers S_i and E_i satisfying the constraints (6) – (12).

2.2   Constraint Satisfaction

Constraint satisfaction technology originated in Artificial Intelligence as a technique for declarative modeling and solving of combinatorial optimization problems. A *constraint satisfaction problem* (CSP) is a triple (X, D, C), where X is a finite set of decision variables, for each $x_i \in X$, $D_i \in D$ is a finite set of possible values for the variable x_i (the domain), and C is a finite set of constraints [5]. A *constraint* is a relation over a subset of variables (its scope) that restricts the possible combinations of values to be assigned to the variables in its scope. Constraints can be expressed in extension using a set of compatible value tuples or as a formula. We will be using constraints expressed as arithmetic and logical formulas, such as $S + D = E$. A *solution* to a CSP is a complete instantiation of variables such that the values are taken from respective domains and all constraints are satisfied. We say that a CSP is *consistent* if it has a solution.

2.3   Attribute Grammars

Attribute grammars were introduced by Knuth [6] to add semantics to context-free grammars in a syntax-directed fashion. The primary application area was the design of compilers. Briefly speaking an attribute grammar is a context-free grammar where symbols are associated with sets of attributes and rewriting (production) rules are associated with sets of attribute computation (semantic) rules defining values of certain attributes based on values of other attributes. We will slightly modify the original definition here by using constraints rather than the semantic rules with synthesized attributes and inherited attributes.

We define an *attribute grammar* as a tuple $G = (\Sigma, N, \mathcal{P}, S, A, \mathcal{C})$, where Σ is an alphabet – a finite set of terminal symbols, N is a finite set of non-terminal symbols with S as the start symbol, $\mathcal{P}$ is a set of rewriting (production) rules (see below), A associates each grammar symbol $X \in \Sigma \cup N$ with a set of attributes (variables in terms of a CSP), and $\mathcal{C}$ associates each production $R \in \mathcal{P}$ with a set of constraints over the attributes of symbols used in the rule. $G = (\Sigma, N, \mathcal{P}, S)$ is a classical context-free grammar with the production rules in the form $X \to w$, where $X \in N$ is a non-terminal symbol and $w \in (\Sigma \cup N)^*$ is a finite sequence of terminal and non-terminal symbols. If $\boldsymbol{A}$ are the attributes for symbol X, we will write $X(\boldsymbol{A})$ in the production rule. To separate grammar symbols in a string (a word) we will use the dot (.) notation. We will also include the constraints associated with the rule directly in the rule inside the brackets $[c_1, \dots]$. This is an example of a production rule:

$$Seat(S_S, E_S) \to Cutting(S_C, E_C).Polishing(S_P, E_P)$$
$$[S_S = S_C, E_C \leq S_P, E_P = E_S]$$

Let (w, C) denote a state of the rewriting system, where w is a string of grammar symbols with their attributes and C is a set of constraints over these attributes. We say that (w, C) directly rewrites to (w', C') using the production rule $X(\boldsymbol{A}) \to u[c]$ if and only if $w = u_1.X(\boldsymbol{B}).u_2$, $C' = C \cup \{\boldsymbol{A} = \boldsymbol{B}\} \cup c$, and C' is a consistent CSP over the variables from attributes of w', where $w' = u_1.u.u_2$. We denote this rewriting as $(w, C) \Rightarrow (w', C')$. Briefly speaking, a non-terminal symbol X is substituted by u in the string w and constraints from the production rule are added to the constraints in the state. Note that we assume classical standardization apart of the production rules, that is, each production rule is used with fresh variables/attributes. For a sequence of direct rewritings $(w_1, C_1) \Rightarrow (w_2, C_2) \Rightarrow \cdots \Rightarrow (w_n, C_n)$ we will use classical notation $(w_1, C_1) \Rightarrow^* (w_n, C_n)$. The language generated by an attribute grammar $G = (\Sigma, N, \mathcal{P}, S, A, \mathcal{C})$ is:

$$L(G) = \{w\sigma | (S, \emptyset) \Rightarrow^* (w, C), w \in \Sigma^*, \sigma \text{ is a solution to a CSP } C\}.$$

σ is an instantiation of attributes (substitution of values to variables) and $w\sigma$ means applying substitution σ to w, i.e., a word w, where the attributes are substituted by values defined in σ.

3   Translating Nested Networks to Attribute Grammars

In this paper we propose to model nested workflows with extra constraints using attribute grammars. In particular, we will show that any nested workflow with extra constraints can be translated to an attribute grammar, where the words in its language correspond to feasible processes. We will proceed as follows. First, we will show how the core nested structure is translated to an attribute grammar, with two attributes for each symbol denoting start and end times of corresponding tasks. Then we will show how the extra constraints can be translated by adding some new attributes and constraints between them. Finally, we will prove that this translation is sound, that is, any feasible process corresponds to a word in the language of the grammar and vice versa, each word in the language describes some feasible process.

3.1   Translation of the Nested Structure

Let us assume first a nested workflow without extra constraints as described in Sect. 2.1. Recall that such a workflow can be represented as a tree with the *root* task in its root, with primitive tasks in its leaves, with compound tasks in the inner nodes, and with two types of branching – parallel and alternative. The basic idea of translation is modeling all the tasks as grammar symbols – the compound tasks will be translated to non-terminal symbols with the root task as the start non-terminal symbol, while the primitive tasks will be translated to terminal symbols.

Let $W = (Parallel, Alternative, Primitive, root, parent, D, \emptyset)$ be a nested workflow without extra constraints. Then we define the attribute grammar as $G = (Primitive, Parallel \cup Alternative, \mathcal{P}, root, A, \mathcal{C})$, where each grammar symbol has two attributes S and E. Let $T_i \in Parallel$ be a parallel task and $subtasks(T_i) = \{T_{i_1}, \ldots, T_{i_k}\}$ be all its subtasks. Then in $\mathcal{P}$, we include the following rewriting rule:

$$T_i(S_i, E_i) \rightarrow T_{i_1}(S_{i_1}, E_{i_1}) \ldots T_{i_k}(S_{i_k}, E_{i_k})$$
$$[S_i = min\{S_{i_1}, \ldots, S_{i_k}\}, E_i = max\{E_{i_1}, \ldots, E_{i_k}\}] \qquad (16)$$

Notice that the constraints in the production rule correspond to the constraints (4) and (7) from the definition of a nested workflow. If we want to model a serial decomposition, we should add constraints $E_{i_j} \leq S_{i_{j+1}}$ to the rule, but we will show later how to model precedence constraints in general.

Let $T_i \in Alternative$ be an alternative task and $subtasks(T_i) = \{T_{i_1}, \ldots, T_{i_k}\}$ be all its subtasks. Then in $\mathcal{P}$, we include the following i_k rewriting rules:

$$T_i(S_i, E_i) \rightarrow T_{i_j}(S_{i_j}, E_{i_j}) \; [S_i = S_{i_j}, E_i = E_{i_j}] \qquad (17)$$

Now the constraints in the rule describe the workflow constraints (5) and (7).

Notice that due to the nature of nested workflows, each symbol but the *root* appears exactly once on the right side of some rule. If that symbol is a terminal symbol, that is, a primitive task $T_j \in Primitive$ then we also include the following constraint among the rule constraints:

$$S_j + D_j = E_j \tag{18}$$

to model the workflow constraint (6). D_j is a constant defined by the function D from the workflow (duration of the primitive task).

Figure 2 shows a complete attribute grammar modeling the nested workflow from Fig. 1.

$Chair(S_{chair}, E_{chair}) \rightarrow Parts(S_{parts}, E_{parts}).Assembly(S_{assembly}, E_{assembly})$
$[S_{chair} = min\{S_{parts}, S_{assembly}\}, E_{chair} = max\{E_{parts}, E_{assembly}\},$
$\quad E_{parts} \leq S_{assembly}, S_{assembly} + D_{assembly} = E_{assembly}]$

$Parts(S_{parts}, E_{parts}) \rightarrow Legs(S_{legs}, E_{legs}).Seat(S_{seat}, E_{seat}).Back(S_{back}, E_{back})$
$[S_{parts} = min\{S_{legs}, S_{seat}, S_{back}\}, E_{parts} = max\{E_{legs}, E_{seat}, E_{back}\}]$

$Legs(S_{legs}, E_{legs}) \rightarrow Saw1(S_1, E_1), Saw2(S_2, E_2), Saw3(S_3, E_3), Saw4(S_4, E_4)$
$[S_{legs} = min\{S_1, S_2, S_3, S_4\}, E_{legs} = max\{E_1, E_2, E_3, E_4\},$
$\quad S_1 + D_1 = E_1, S_2 + D_2 = E_2, S_3 + D_3 = E_3, S_4 + D_4 = E_4]$

$Seat(S_{seat}, E_{seat}) \rightarrow Cutting(S_{cut}, E_{cut}).Polishing(S_{polish}, E_{polish})$
$[S_{seat} = min\{S_{cut}, S_{polish}\}, E_{seat} = max\{E_{cut}, E_{polish}\},$
$\quad E_{cut} \leq S_{polish}, S_{cut} + D_{cut} = E_{cut}, S_{polish} + D_{polish} = E_{polish}]$

$Back(S_{back}, E_{back}) \rightarrow Buy(S_{buy}, E_{buy})$
$[S_{back} = S_{buy}, E_{back} = E_{buy}, S_{buy} + D_{buy} = E_{buy}]$

$Back(S_{back}, E_{back}) \rightarrow Weld(S_{weld}, E_{weld})$
$[S_{back} = S_{weld}, E_{back} = E_{weld}, S_{weld} + D_{weld} = E_{weld}]$

Fig. 2. An attribute grammar modeling the nested workflow from Fig. 1

3.2   Translation of Extra Constraints

In the previous section we showed how the nested workflow structure is represented by an attribute grammar. Now, we will add extra constraints (8) – (15). All these constraints are binary and they can be categorized into two groups: temporal constraints (8) – (12) and logical constraints (13) – (15). Recall that the nested workflow can be represented as a tree and each task appears exactly once in this tree. So for two tasks T_i and T_j only two situations may occur as depicted in Fig. 3: either one task is an ancestor of the other one (left part of Fig. 3) or they both share the same ancestor (task A in the right part of Fig. 3).

To represent an extra constraint in the attribute grammar we suggest to add a new attribute M to each grammar symbol representing a task on the path between T_i and T_j in the tree representation of the workflow. Note that this new

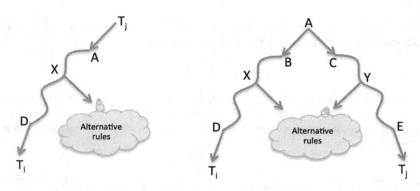

Fig. 3. Two possible positions of tasks T_i and T_j in a tree representing the workflow

attribute is unique for each extra constraint used so we need to introduce as many of these attributes as we have extra constraints. To simplify notation, we allow an attribute to be shared between symbols within a production rule which would otherwise be modeled by an equality constraint between the two attributes (instead of $A(M) \rightarrow B(N)[M = N]$ we will simply write $A(M) \rightarrow B(M)$).

Let us first describe the translation when one task is an ancestor of the other task (left part of Fig. 3). All grammar symbols on the path between the tasks are extended with a new attribute M. For the temporal constraints (8) – (12) we add the constraints to the first and to the last rule in the sequence as shown at Table 1. If T_j is a parent of T_i then we do not need the extra attribute at all and the rule constraint can be expressed using the time attributes of the tasks. Notice that if an alternative rule is used on the path and task T_i is not generated by the grammar then effectively no constraint is imposed on M. However, if both tasks T_i and T_j are included then the respective temporal constraint must hold, which is guaranteed by rule constraints.

Table 1. Translation of temporal constraints when T_j is an ancestor of T_i

Production Rule	$(i \rightarrow j)$	$(j \rightarrow i)$	$(i\ ss\ j)$	$(i\ se\ j)$	$(i\ es\ j)$	$(i\ ee\ j)$
$T_j(S_j, E_j) \rightarrow \ldots, A(\ldots, M), \ldots$	$M \leq S_j$	$E_j \leq M$	$M = S_j$	$M = E_j$	$M = S_j$	$M = E_j$
$D(\ldots, M) \rightarrow \ldots, T_i(S_i, E_i), \ldots$	$E_i \leq M$	$M \leq S_i$	$M = S_i$	$M = S_i$	$M = E_i$	$M = E_i$

We showed translation of all temporal constraints for this situation but obviously, due to constraints (16) – (17), the precedence constraints and the synchronizations se, es can never hold as it is not possible to satisfy the added constraints together with the existing rule constraints. This is formally correct as it implies that tasks T_i and T_j can never be used together if such extra constraint is imposed on them [4].

For the logical constraints (13) – (15) we will use a similar principle, but now we need to assume also the possible alternative rules on the path. In Fig. 3 such a rule is shown as the alternative rule from symbol X. Logical constraints restrict appearance of tasks in the process. So for example to model the equivalence constraint we need to force both tasks T_i and T_j to appear together. This means that no alternative rule on the path between them is allowed, which is modeled by constraint $M = 0$ in these rules. Contrarily, for the *mutex* relation we require an alternative rule to be used, which is forced by using different values of M in rules with T_j and T_i. In fact it means that task T_i can never be used. Note finally, that the constraint $i \Rightarrow j$ is redundant because if T_i is included then all its ancestors are included (see constraint (3)) and T_j is among them. Hence we only show the model for implication $j \Rightarrow i$, which is actually identical to equivalence in this special case. Table 2 shows all these constraints to be added to the rules. The special case where T_j is a parent of T_i can be modeled by combing these constraints. For example for the *mutex* constraint we modify the rule to $T_j \rightarrow \ldots, T_i, \ldots [0 = 1]$, where the constraint can never be satisfied and hence the rule cannot be applied.

Table 2. Translation of logical constraints when T_j is an ancestor of T_i

Production Rule	$(i\ mutex\ j)$	$(i \Leftrightarrow j)$	$(j \Rightarrow i)$
$T_j \rightarrow \ldots, A(\ldots, M), \ldots$	$M = 0$	$M = 1$	$M = 1$
$D(\ldots, M) \rightarrow \ldots, T_i, \ldots$	$M = 1$	$M = 1$	$M = 1$
$X(\ldots, M) \rightarrow \ldots$ (alternative rules)	$M = 0$	$M = 0$	$M = 0$

Assume now the situation from the right part of Fig. 3, where tasks T_i and T_j have a common ancestor A. Again, we introduce a new attribute M to all symbols, but A, on the path between T_i and T_j and we add extra rule constraints. Let us first describe the model for temporal constraints and for the *mutex* constraint. If A is a task with an alternative decomposition then according to rules (17), it is never possible that T_i and T_j will be generated together in a single word. Hence the *mutex* constraint is satisfied by default and also all the temporal constraints (8) – (12) are satisfied. Consequently, no extension of the grammar is necessary to model these constraints. It remains to show the model when A is a parallel task. In this case we use the extra attribute M and we add the rule constraints according to Table 3. Note that no constraint is added to alternative rules so they can be used without any restriction. In fact, the *mutex* constraint enforces one of alternatives to be used as it is not possible to generate both T_i and T_j due to the conflict between constraints $M = 1$ and $M = 0$.

For the equivalence and implication constraints both alternative and parallel decompositions in A need to be handled as well as the alternative rules need to be assumed. Moreover, for the implication constraint the alternative rules in the right branch use different constraints than the alternative rules in the left branch. In Table 4 we summarize all the constraints to be added to rules. Note that we show there also the constraints where A is a parallel task, then there is a single rule for it, and where A is an alternative task, then there are more

Table 3. Translation of temporal and *mutex* constraints when T_j and T_i share a common ancestor

Production Rule	$(i \to j)$	$(i\ ss\ j)$	$(i\ se\ j)$	$(i\ es\ j)$	$(i\ ee\ j)$	$(i\ mutex\ j)$
$D(\ldots, M) \to \ldots, T_i(S_i, E_i), \ldots$	$E_i \leq M$	$M = S_i$	$M = S_i$	$M = E_i$	$M = E_i$	$M = 1$
$E(\ldots, M) \to \ldots, T_j(S_j, E_j), \ldots$	$M \leq S_j$	$M = S_j$	$M = E_j$	$M = S_j$	$M = E_j$	$M = 0$

rules, but only the rules generating B and C are relevant. Briefly speaking, if A is alternative then none of the tasks T_i and T_j can be generated if there is an equivalence constraint between them and only task T_j can be generated if there is implication $(i \Rightarrow j)$. If A is parallel then for the equivalence constraint either the alternative rules are used on both sides or they are not used (then both T_i and T_j are generated). For the implication constraint, an alternative rule can be used on the left side together with an alternative on the right side or with T_j, but it is not possible to use an alternative rule on the right together with T_i (as it would mean that T_j is excluded that would violate the implication constraint).

Table 4. Translation of logical constraints when T_j and T_i share a common ancestor.

Production Rule	$(i \Leftrightarrow j)$	$(i \Rightarrow j)$
$D(\ldots, M) \to \ldots, T_i, \ldots$	$M = 1$	$M = 1$
$E(\ldots, M) \to \ldots, T_j, \ldots$	$M = 1$	$M = 1$
$X(\ldots, M) \to \ldots$ (left alternative)	$M = 0$	$-$
$Y(\ldots, M) \to \ldots$ (right alternative)	$M = 0$	$M = 0$
$A \to \ldots, B(\ldots, M), \ldots, C(\ldots, M), \ldots$ (parallel)	$-$	$-$
$A \to B(\ldots, M)$ (alternative)	$M = 0$	$M = 0$
$A \to C(\ldots, M)$	$M = 0$	$-$

3.3 Soundness of Translation

In the previous sections we showed how to construct an attribute grammar from a nested workflow with extra constraints. To prove correctness of this translation, we need to show that any feasible process corresponds to a word generated by the grammar and vice versa. Due to space limits we only sketch the proofs.

Theorem 1. *A feasible process for a nested workflow corresponds to a word generated by the corresponding attribute grammar.*

Proof. A feasible process defines a subtree of the nested workflow according to constraints $(3) - (5)$. This subtree is a derivation tree of the grammar thanks to production rules $(16) - (17)$. These rules are applicable as the process satisfies the constraints $(6) - (7)$. The only reason, why any rule would not be applicable, is that some constraint presented in the previous section is violated. Note first that the feasible process satisfies all extra constraints. By looking at Tables 1, 2,

3 and 4 one can easily verify that constraints imposed by used rules are satisfied. For example, if there is a precedence $i \rightarrow j$ then constraints $E_i \leq M \wedge M \leq S_j$ are satisfied as $E_i \leq S_j$ is satisfied in the process. Note that if the constrained tasks are not in the process then some rules with attributes M might still be used in the derivation, but they impose restriction only for the logical constraints and again it is easy to check that they are satisfied.

Theorem 2. *Any word generated by an attribute grammar translated from the nested workflow describes a feasible process for that workflow.*

Proof. Let us take the derivation tree of a given word. Then the process P is defined by all symbols (tasks) in the tree. This process satisfies constraints (3) – (7) as they were imposed by production rules used and by the constraints (16) – (18).

One can easily check that if two tasks connected by an extra temporal constraint are in the process then this temporal constraint is satisfied thanks to satisfaction of rule constraints. For logical constraints one can prove their satisfaction by contradiction. Assume for example an extra constraint $i \Leftrightarrow j$, where T_i is in the derivation tree, T_j is not there, and they have a common ancestor A in the workflow. The rule generating T_i introduced constraint $M = 1$. A cannot be an alternative task as $M = 0$ would hold according to Table 4 . So A must be parallel but as T_j is not present then some alternative rule on the path from A to T_j has been used, which again leads to $M = 0$. This is a contradiction as there is not way to satisfy the constraints and having only T_i but not T_j in the derivation tree. Similarly we can explore all other constraints and cases.

The theorems showed that for each feasible process there is a word and for each word there is a feasible process. However, it does not mean that there is a one-to-one mapping between feasible processes and words. For a given process there could be more words as we introduced the auxiliary variables M and there might be more feasible assignments to them (for example to satisfy the constraints $E_i \leq M \wedge M \leq S_j$ there could be several feasible values for M). Nevertheless, if we restrict the attributes in words to variables S_i and E_i only then we get a unique mapping.

4   Conclusions

This paper proposes using attribute grammars to model nested workflows with extra constraints. We first proposed a small extension of attribute grammars where, instead of traditional semantic rules, a constraint satisfaction problem is used to model relations between the attributes. Then we showed how a nested workflow with extra constraints can be translated to an attribute grammar where the language of that grammar corresponds to feasible processes defined by the workflow.

There are two major advantages of the proposed framework. First, the nested workflows do not support recursion so they cannot be used to model planning problems. Other workflow models use loops to describe when some sub-process should be repeated (for example, if the product does not pass the exit test then it goes back to production). This is not possible in nested workflows and due to extra constraints, it is not clear if and how recursion can be added to nested workflows. Attribute grammars provide a natural mechanism to describe recursion and they seem more appropriate to describe a large range of processes including planning problems. The open question is how attribute grammars compare in terms of modeling capabilities and solving efficiency to existing planning domain modeling frameworks, in particular to hierarchical task networks [7]. The second reason for introducing attribute grammars to model workflows is the hope of direct exploitation of existing techniques for attribute grammars also for workflows and in general for planning models. For example, the paper [4] proposed an ad-hoc method to verify nested workflows with extra constraints (ensuring that a feasible process exists). The open question is whether, for example, the methods for reduction of context-free grammars can be exploited for the same purpose. This would be a significant contribution as there are no such verification methods for planning models.

Acknowledgments. Research is supported by the Czech Science Foundation under the project P103-15-19877S.

References

1. Bae, J., Bae, H., Kang, S.-H., Kim, Z.: Automatic control of workflow processes using ECA rules. IEEE Trans. Knowl. Data Eng. **16**(8), 1010–1023 (2004)
2. Barták, R., Čepek, O.: Nested precedence networks with alternatives: recognition, tractability, and models. In: Dochev, D., Pistore, M., Traverso, P. (eds.) AIMSA 2008. LNCS (LNAI), vol. 5253, pp. 235–246. Springer, Heidelberg (2008)
3. Barták, R., Cully, M., Jaška, M., Novák, L., Rovenský, V., Sheahan, C., Skalický, T., Thanh-Tung, D.: Workflow optimization with FlowOpt, on modelling, optimizing, visualizing, and analysing production workflows. In: Proceedings of Conference on Technologies and Applications of Artificial Intelligence (TAAI 2011), pp. 167–172. IEEE Conference Publishing Services (2011)
4. Barták, R., Rovenský, V.: On verification of nested workflows with extra constraints: from theory to practice. Expert Syst. Appl. **41**(3), 904–918 (2014)
5. Dechter, R.: Constraint Processing. Morgan Kaufmann, San Francisco (2003)
6. Knuth, D.E.: Semantics of context-free languages. Math. Syst. Theory **2**(2), 127–145 (1968)
7. Nau, D.S., Au, T.-C., Ilghami, O., Kuter, U., Murdock, J.W., Wu, D., Yaman, F.: SHOP2: an HTN planning system. J. Artif. Intell. Res. (JAIR) **20**, 379–404 (2003)
8. van der Aalst, W., ter Hofstede, A.H.M.: Yawl: yet another workflow language. Inf. Syst. **30**(4), 245–275 (2005)

A Natural Counting of Lambda Terms

Maciej Bendkowski[1]([⊠]), Katarzyna Grygiel[1],
Pierre Lescanne[2], and Marek Zaionc[1]

[1] Faculty of Mathematics and Computer Science,
Theoretical Computer Science Department, Jagiellonian University,
ul. Prof. Łojasiewicza 6, 30–348 Kraków, Poland
{bendkowski,grygiel,zaionc}@tcs.uj.edu.pl
[2] École Normale Supérieure de Lyon,
LIP (UMR 5668 CNRS ENS Lyon UCBL INRIA),
University of Lyon, 46 Allée d'Italie, 69364 Lyon, France
pierre.lescanne@ens-lyon.fr

Abstract. We study the sequence of numbers corresponding to λ-terms of given size in the model based on de Bruijn indices. It turns out that the sequence enumerates also two families of binary trees, i.e. black-white and zigzag-free ones. We provide a constructive proof of this fact by exhibiting appropriate bijections. Moreover, we investigate the asymptotic density of λ-terms containing an arbitrary fixed subterm, showing that strongly normalizing terms are of density 0 among all λ-terms.

Keywords: Lambda calculus · Black-white trees · Zigzag-free trees · Asymptotic density · Functional programming

1 Introduction

Counting combinatorial objects representing entities of logical provenance forms a prominent subject of modern research in computational logic. In recent years, a growing attention has been given to various models of computation, in particular to lambda calculus, which forms the core of functional programming languages, such as Haskell or OCaml.

Different combinatorial models of lambda calculus are known in the literature. In [13] John Tromp introduced a binary encoding of lambda calculus and combinatory logic, which allowed him to construct compact and efficient self-interpreters for both languages. Quantitative aspects of the aforementioned lambda calculus representation were studied in [9]. The authors of [4] considered the model in which the size of every variable, application and abstraction is equal to one. A similar model, in which variables do not contribute to the size of a term, was introduced in [6], where results concerning semantic properties of random λ-terms were stated. In particular, it was proven that in this model

This work was partially supported by the grant 2013/11/B/ST6/00975 founded by the Polish National Science Center.

© Springer-Verlag Berlin Heidelberg 2016
R.M. Freivalds et al. (Eds.): SOFSEM 2016, LNCS 9587, pp. 183–194, 2015.
DOI: 10.1007/978-3-662-49192-8_15

asymptotically almost all λ-terms are strongly normalizing. The same size model was considered in [8], where in order to cope with the infinite number of variables the authors used the de Bruijn notation.

In this paper we consider a natural way of measuring the size of λ-terms represented using the de Bruijn notation. Let us assume that we are given a countable set $\{0, S\,0, S^2\,0, \ldots\}$ of de Bruijn indices. We define λ-terms as follows. Each de Bruijn index is a λ-term. If t and u are λ-terms, then $(\lambda\,t)$ and $(t\,u)$ are λ-terms. Henceforth, we follow standard notational conventions for lambda calculus (see e.g. [9]). In our model we assume the following notion of size:

$$|S^n\,0| = |n| + 1,$$
$$|\lambda\,t| = |t| + 1,$$
$$|t\,u| = |t| + |u| + 1.$$

In other words, each constructor, i.e. 0, S, λ and application, is of size 1.

In Sect. 2 we count the number of λ-terms of size n. Using methods of analytic combinatorics, we give the asymptotic growth of the corresponding sequence. Moreover, we give a holonomic equation associated with this sequence. Next, we study λ-terms with bounded number of free indices and appropriate generating functions. In Sects. 3 and 4 we present mutually inverse bijections among λ-terms, black-white trees, and zigzag-free trees. In addition, each constructed bijection is supported by a corresponding Haskell implementation. In Sect. 5 we focus on the family of λ-terms containing a fixed subterm, showing that in the considered model asymptotically almost all λ-terms are not strongly normalizing.

1.1 Notation

Throughout the paper we use the following notation. We denote combinatorial classes, e.g. λ-terms, by capital calligraphic letters. Given a class $\mathcal{A}$, its corresponding generating function will be denoted as A. By $[z^n]A$ we denote the coefficient standing at z^n in the series expansion of $A(z)$. Whenever such a generating function yields a dominating singularity, we use ρ_A to denote it. Sometimes, when we are interested in the approximate value of ρ_A we write $\rho_A \doteq c$, where c is the approximation. We use the underbar notation to denote de Bruijn indices – $\underline{n}$ stands for the n'th de Bruijn index, i.e. $\underline{n} = S^n\,0$. To denote addition and subtraction operations on combinatorial classes we use $\oplus$ and $\ominus$, respectively. Given two complex functions f and g of the same asymptotic order, i.e. satisfying $\lim_{n\to\infty} f(n)/g(n) = 1$, we write $f \sim g$.

2   Lambda Terms

2.1   Counting λ-terms with Natural Size

In this section we are interested in the generating function for the sequence corresponding to the numbers of λ-terms. Let us start with considering the class of de Bruijn indices.

Lemma 1. *Let D stand for the generating function enumerating de Bruijn indices. Then*

$$D(z) = \frac{z}{1-z}.$$

Proof. Let $n \in \mathbb{N}$. There exists a unique de Bruijn index $\underline{n}$ encoding n. Since application and θ are both of size 1, the size of $\underline{n}$ is equal to $n+1$ and thus $([z^n]D)_{n \in \mathbb{N}} = (0, 1, 1, \ldots)$, which implies $D(z) = \frac{z}{1-z}$. $\square$

Lemma 2. *Let L_∞ stand for the generating function enumerating all λ-terms. Then*

$$L_\infty(z) = \frac{(1-z)^{3/2} - \sqrt{1 - 3z - z^2 - z^3}}{2z\sqrt{1-z}}.$$

Proof. Since λ-terms are either applications, abstractions or de Bruijn indices, the set $\mathcal{L}_\infty$ of lambda terms can be expressed as

$$\mathcal{L}_\infty = \mathcal{L}_\infty \mathcal{L}_\infty \oplus \lambda \mathcal{L}_\infty \oplus \mathcal{D}.$$

Using this presentation, we immediately obtain a corresponding quadratic equation defining the generating function L_∞:

$$L_\infty(z) = zL_\infty^2(z) + zL_\infty(z) + \frac{z}{1-z}. \tag{1}$$

We compute its discriminant $\Delta_{L_\infty} = \frac{1 - 3z - z^2 - z^3}{1 - z}$ and finally solve the above equation:

$$\begin{aligned} L_\infty(z) &= \frac{(1-z) - \sqrt{\Delta_{L_\infty}}}{2z} \\ &= \frac{(1-z)^{3/2} - \sqrt{1 - 3z - z^2 - z^3}}{2z\sqrt{1-z}}. \end{aligned}$$

$\square$

Using the generating function L_∞ we can now find the asymptotic growth of the sequence $([z^n]L_\infty)_{n \in \mathbb{N}}$.

Theorem 1. *The asymptotic approximation of the number of λ-terms of size n is given by*

$$[z^n]L_\infty \sim (3.38298\ldots)^n \frac{C}{n^{3/2}}, \quad where \quad C \doteq 0.60676.$$

Proof. Examining the function L_∞ we note that its dominating singularity ρ_{L_∞} is equal to the root of smallest modulus of $1 - 3z - z^2 - z^3$. Therefore,

$$\rho_{L_\infty} = \frac{1}{3}\left(\sqrt[3]{26 + 6\sqrt{33}} - \frac{4\,2^{2/3}}{\sqrt[3]{13 + 3\sqrt{33}}} - 1\right) \doteq 0.29559774252208393$$

and hence $1/\rho_{L_\infty} \doteq 3.38298$. Let us write L_∞ as

$$L_\infty(z) = \frac{(1-z) - \sqrt{\frac{1-3z-z^2-z^3}{1-z}}}{2z}$$

$$= \frac{(1-z) - \sqrt{\rho_{L_\infty}(1 - \frac{z}{\rho_{L_\infty}})\frac{Q(z)}{1-z}}}{2z}$$

where

$$Q(z) = \frac{R(z)}{\rho_{L_\infty} - z} \qquad \text{and} \qquad R(z) = z^3 + z^2 + 3z - 1.$$

Applying Theorem VI.1 of [7] we obtain

$$[z^n]L_\infty \sim \left(\frac{1}{\rho_{L_\infty}}\right)^n \cdot \frac{n^{-3/2}}{\Gamma(-\frac{1}{2})} \widetilde{C} \qquad \text{with} \qquad \widetilde{C} = \frac{-\sqrt{\rho_{L_\infty}\frac{Q(\rho_{L_\infty})}{1-\rho_{L_\infty}}}}{2\rho_{L_\infty}}.$$

Since $Q(\rho_{L_\infty}) = R'(\rho_{L_\infty}) = 3\rho_{L_\infty}^2 + 2\rho_{L_\infty} + 3$, we finally get

$$C = \frac{\widetilde{C}}{\Gamma(-\frac{1}{2})} \doteq 0.60676.$$

$\square$

The sequence $([z^n]L_\infty)_{n\in\mathbb{N}}$ is known as **A105633** in *Online Encyclopedia of Integer Sequences* [2]. The first 15 values are as follows:

0, 1, 2, 4, 9, 22, 57, 154, 429, 1223, 3550, 10455, 31160, 93802, 284789.

2.2 Holonomic Presentation of L_∞

Using the Maple package gfun [11] we find the following holonomic equation defining L_∞:

$$z^3 + z^2 - 2z + (z^3 + 3z^2 - 3z + 1)L_\infty + (z^5 + 2z^3 - 4z^2 + z)L'_\infty = 0.$$

Such a presentation of L_∞ allows us to derive a simpler, compared to the combinatorial definition, recursive definition of its coefficients. For convenience, let us denote $L_{\infty,n} := [z^n]L_\infty$. Now, we can express the recursive definition of $L_{\infty,n}$ as:

$$L_{\infty,0} = 0, \qquad L_{\infty,1} = 1, \qquad L_{\infty,2} = 2, \qquad L_{\infty,3} = 4,$$

$$L_{\infty,n} = \frac{(4n-1)L_{\infty,n-1} - (2n-1)L_{\infty,n-2} - L_{\infty,n-3} - (n-4)L_{\infty,n-4}}{n+1}.$$

Note that $L_{\infty,n}$ depends on the previous four values $L_{\infty,n-1}, L_{\infty,n-2}, L_{\infty,n-3}$ and $L_{\infty,n-4}$. Exploiting this fact, the above definition allows us to compute the exact value $L_{\infty,n}$ using only linear number of arithmetic operations. Moreover, this holonomic equation could be used to develop a random generator in the spirit of [3].

2.3   Counting Terms with Bounded Number of Free Indices

In this section we are interested in counting terms with bounded number of distinct free de Bruijn indices. We start with giving the generating function associated with the set of first m indices.

Lemma 3. *Let* $\mathcal{D}_m = \{\underline{0}, \underline{1}, \ldots, \underline{m-1}\}$ *where* $m \in \mathbb{N}$. *Then*

$$D_m(z) = \frac{z(1 - z^m)}{1 - z}.$$

Proof. Let us notice that

$$[z^n]D_m = \begin{cases} 1 & \text{if } 1 \le n \le m, \\ 0 & \text{otherwise.} \end{cases}$$

It follows that we can express $D_m(z)$ as $D(z) - z^m D(z)$. Using Lemma 1 we finally obtain $D_m(z) = \frac{z}{1-z} - \frac{z^{m+1}}{1-z} = \frac{z(1-z^m)}{1-z}$. $\qquad\square$

Let $m \in \mathbb{N}$. We denote by $\mathcal{L}_m$ the set of λ-terms whose free indices are elements of $\mathcal{D}_m$. Obviously, for every m we have $\mathcal{L}_m \subseteq \mathcal{L}_{m+1}$.

Lemma 4. *The generating function associated with the set* $\mathcal{L}_m$ *is given by*

$$L_m(z) = \frac{1 - \sqrt{1 - 4z^2\left(L_{m+1}(z) + \frac{1-z^m}{1-z}\right)}}{2z}.$$

Proof. Due to the structure of λ-terms, we can express $\mathcal{L}_m$ in the following way:

$$\mathcal{L}_m = \mathcal{L}_m \mathcal{L}_m \oplus \lambda \mathcal{L}_{m+1} \oplus \mathcal{D}_m,$$

which immediately implies

$$L_m(z) = zL_m(z)^2 - zL_{m+1}(z) + \frac{z(1 - z^m)}{1 - z}.$$

Solving in $L_m(z)$, we obtain

$$L_m(z) = \frac{1 - \sqrt{\Delta_{\mathcal{L}_m}}}{2z} = \frac{1 - \sqrt{1 - 4z^2\left(L_{m+1}(z) + \frac{1-z^m}{1-z}\right)}}{2z}.$$

$\qquad\square$

Notice that L_m, and in particular L_0 – counting the number of closed λ-terms, is defined using L_{m+1}. If this definition is developed, then L_m is expressed by means of infinitely nested radicals – a known phenomenon already observed in other models of λ-calculus (see e.g. [4,9]).

2.4   Counting λ-terms with Another Notion of Size

Assume we take another notion of size in which θ has size 0, applications are of size 2, whereas abstraction and successor keep their original size 1. Formally,

$$|\theta| = 0,$$
$$|S^n\theta| = |n| + 1,$$
$$|\lambda t| = |t| + 1,$$
$$|t\,u| = |t| + |u| + 2.$$

It is easy to verify that the corresponding generating function[1] A_1 fulfills the identity $A_1(z) = z^2 A_1^2(z) - (1 - z)A_1(z) + \frac{1}{1-z}$. The reader may check that

$$L_\infty = z\,A_1 \quad \text{and so} \quad [z^n]A_1 = [z^{n+1}]L_\infty.$$

It follows that both notions of size yield the sequence **A105633**.

3 E-free Black-White Binary Trees

A *black-white binary tree* is a binary tree in which nodes are colored either *black* • or *white* ∘. Let E be a set of edges. An *E-free black-white binary tree* is a black-white binary tree in which edges from the set E are forbidden. For instance, if the set of forbidden edges is $E_1 = \{\, {}^{\circ}\!\nearrow^{\bullet}, {}^{\bullet}\!\searrow_{\circ}, {}^{\bullet}\!\searrow_{\bullet}, {}^{\circ}\!\searrow_{\circ} \,\}$, then the only allowed edges are $A_1 = \{\, {}_{\circ}\!\nearrow^{\bullet}, {}_{\bullet}\!\nearrow^{\bullet}, {}_{\circ}\!\nearrow^{\circ}, {}^{\circ}\!\searrow_{\bullet} \,\}$. The *size* of a black-white tree is the total number of its nodes. For E_1, like for $E_2 = \{\, {}^{\circ}\!\searrow_{\bullet}, {}_{\circ}\!\nearrow^{\bullet}, {}_{\bullet}\!\nearrow^{\bullet}, {}_{\circ}\!\nearrow^{\circ} \,\}$, which is obtained by left-right symmetry, the E-free black-white binary trees are counted by **A105633**, see [10].

Henceforth we consider only the set E_1 and speak rather in terms of allowed edge patterns, i.e. A_1. For convenience, whenever we use the term *black-white trees*, we mean the black-white trees with allowed set of patterns A_1. If not stated otherwise, we assume that black-white trees have black roots.

3.1   Recursive Description

Let $\mathcal{BW}_\bullet$ and $\mathcal{BW}_\circ$ denote the set of black-white trees with a black, respectively white, root. Interpreting the set of allowed edges A_1 combinatorially, we can define both $\mathcal{BW}_\bullet$ and $\mathcal{BW}_\circ$ using the following mutually recursive equations:

---

[1] We write this function A_1 as a reference to the function $A(x, 1)$ described in **A105632** of the *Online Encyclopedia of Integer Sequences* [2].

$$\mathcal{BW}_\bullet = \bullet \oplus \overset{\bullet}{\diagup}_{\mathcal{BW}_\bullet} \oplus \overset{\bullet}{\diagup}_{\mathcal{BW}_\circ}$$

$$\mathcal{BW}_\circ = \circ \oplus \overset{\circ}{\diagup}_{\mathcal{BW}_\circ} \oplus \overset{\circ}{\diagdown}_{\mathcal{BW}_\bullet} \oplus \overset{\circ}{\diagup\diagdown}_{\mathcal{BW}_\circ \quad \mathcal{BW}_\bullet}$$

Such a representation yields the following identities on the corresponding generating functions $BW_\bullet$ and $BW_\circ$:

$$BW_\bullet(z) = z + zBW_\bullet(z) + zBW_\circ(z)$$
$$BW_\circ(z) = z + zBW_\circ(z) + zBW_\bullet(z) + zBW_\circ(z)BW_\bullet(z)$$

Reformulating this system, we obtain

$$BW_\circ(z) = \frac{(1-z)BW_\bullet(z) - z}{z},$$

hence

$$(1-z)zBW_\bullet^2(z) - (1-z)^2 BW_\bullet(z) + z = 0.$$

Notice that the equation defining $BW_\bullet$ is equivalent to the Eq. (1) defining L_∞ up to multiplication by $(1-z)$. It follows that both $([z^n]BW_\bullet)_{n\in\mathbb{N}}$ and $([z^n]L_\infty)_{n\in\mathbb{N}}$ are equal and therefore there exists a bijection between λ-terms and black-white trees.

3.2   Bijection Between λ-terms and Black-White Trees

We are now ready to give the translation LtoBw from λ-terms to black-white trees and the inverse translation BwtoL from black-white trees to λ-terms:

$$0 \xrightarrow{\mathsf{LtoBw}} \bullet \qquad\qquad \bullet \xrightarrow{\mathsf{BwtoL}} 0$$

$$S\,n \xrightarrow{\mathsf{LtoBw}} \overset{\mathsf{LtoBw}(n)}{\bullet\diagup} \qquad\qquad \overset{T}{\bullet\diagup} \xrightarrow{\mathsf{BwtoL}} S\,\mathsf{BwtoL}(T)$$

$$\lambda\,M \xrightarrow{\mathsf{LtoBw}} \overset{\mathsf{LtoBw}(M)}{\circ\diagup} \qquad\qquad \overset{T}{\circ\diagup} \xrightarrow{\mathsf{BwtoL}} \lambda\,\mathsf{BwtoL}(T)$$

$$M_1\,M_2 \xrightarrow{\mathsf{LtoBw}} \circ\overset{\mathsf{LtoBw}(M_2)}{\underset{\mathsf{LtoBw}(M_1)}{\diagup\diagdown}} \qquad \circ\overset{T_2}{\underset{T_1}{\diagup\diagdown}} \xrightarrow{\mathsf{BwtoL}} \mathsf{BwtoL}(T_1)\,\mathsf{BwtoL}(T_2)$$

Proposition 1. *Both* LtoBw *and* BwtoL *are mutually inverse bijections, i.e.*

$$\mathsf{LtoBw} \circ \mathsf{BwtoL} = id_\Lambda \qquad \text{and} \qquad \mathsf{BwtoL} \circ \mathsf{LtoBw} = id_{\mathcal{BW}_\bullet}.$$

In order to translate a given black-white tree t into a corresponding λ-term, we decompose t depending on the type of its leftmost node. If t is a single black node $\bullet$, we translate it into 0. Otherwise, we have to consider three cases based on the set A_1 of allowed edges and map them into λ-abstraction, successor, or application, respectively.

Example 1. Let us give two black-white trees corresponding to:

- $\Omega = (\lambda.xx)(\lambda.xx) = (\lambda(\theta\,\theta))\,\lambda(\theta\,\theta)$, and
- $Y = \lambda f.(\lambda x.f(xx))(\lambda x.f(xx)) = \lambda(\lambda(S\,\theta\,(\theta\,\theta))\,\lambda(S\,\theta\,(\theta\,\theta)))$

LtoBw(Ω)   LtoBw(Y)

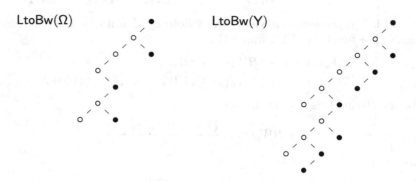

We provide Haskell implementations of LtoBw and BwtoL which can be found at [1]. Our implementations were tested using Quickcheck [5].

4   Binary Trees Without Zigzags

In this section we are interested in zigzag-free binary trees, i.e. trees without a forbidden *zigzag* subtree:

Let us denote the set of such trees as $\mathcal{BZ}_1$. We can define it using the following combinatorial equations:

$$\mathcal{BZ}_1 = \overset{\times}{}{}_{\diagdown \mathcal{BZ}_1} \oplus \mathcal{BZ}_2$$

$$\mathcal{BZ}_2 = \times \;\oplus\; {}_{\mathcal{BZ}_2}\!\diagup^{\times} \oplus\; {}_{\mathcal{BZ}_2}\!\diagup^{\times}\!\diagdown_{\mathcal{BZ}_1}$$

Similarly to L_∞ and $BW_\bullet$, the generating function BZ_1 can be expressed as a solution of the functional equation:

$$z(1-z)BZ_1^2(z) + (1-z)^2 BZ_1(z) + z = 0.$$

It follows that the sequence $([z^n]BZ_1)_{n\in\mathbb{N}}$ is equal to $([z^n]BW_\bullet)_{n\in\mathbb{N}}$ and also to $([z^n]L_\infty)_{n\in\mathbb{N}}$, suggesting that appropriate bijections exist. We note that Sapounakis et al. [12] consider the same sequence defined in terms of constrained Dyck paths and give the following explicit formula:

$$[z^n]BZ_1 = [z^n]L_\infty = \sum_{k=0}^{(n-1)\div 2} \frac{(-1)^k}{n-k}\binom{n-k}{k}\binom{2n-3k}{n-2k-1}.$$

4.1 Bijection Between Black-White Trees and Zigzag-Free Trees

We start with giving the translation BwtoBz from black-white trees to zigzag-free ones. For convenience, we use u_1 and u_2 to denote arbitrary (possibly empty) black-white trees.

Proposition 2. *Let t be a black-white tree. Then trees t and $\mathsf{BwtoBz}(t)$ are of equal size.*

Proof. Let us notice that it suffices to consider the case $\mathsf{BwtoBz}\big(\,{}_t\!\nearrow^{\bullet}\,\big)$, since it results in subtracting one black node. Because the root of t is white, the next translation step is done according to one of the last four rules, which eventually falls into either the fourth or the sixth equation. Since both of them enforce adding one additional $\times$, the total number of nodes is preserved.   □

What remains is to give the inverse translation, which we present as two mutually recursive functions $\mathsf{BztoBw_\bullet}$ and $\mathsf{BztoBw_\circ}$:

Proposition 3. *Let t be a zigzag-free tree. Then trees t and $\mathsf{BztoBw_\bullet}(t)$ are of equal size.*

Proof. The fourth and sixth equations defining $\mathsf{BztoBw_\bullet}$ introduce an additional white node $\circ$, but since both the first and the second equations of $\mathsf{BztoBw_\circ}$ remove one node, the overall tree size is preserved.   □

Proposition 4. *Both $\mathsf{BztoBw_\bullet}$ and BwtoBz are mutually inverse bijections, i.e.*

$$\mathsf{BztoBw_\bullet} \circ \mathsf{BwtoBz} = id_{\mathcal{BW}_\bullet} \qquad and \qquad \mathsf{BwtoBz} \circ \mathsf{BztoBw_\bullet} = id_{\mathcal{BZ}}.$$

Example 2. Let us present the zigzag-free tree corresponding to the aforementioned black-white tree associated with Ω:

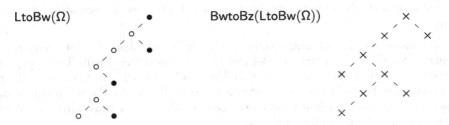

We provide Haskell implementations of BwtoBz, BztoBw$_\bullet$ and BztoBw$_\circ$ which can be found at [1]. Our implementations were tested using Quickcheck [5].

5   Counting λ-terms Containing Fixed Subterms

Let M be an arbitrary λ-term of size p and $\mathcal{T}_M$ denote the set of λ-terms that contain M as a subterm. In this section we focus on the asymptotic density of $\mathcal{T}_M$ in the set of all λ-terms.

Theorem 2. *For a fixed term* M, *the asymptotic density of* $\mathcal{T}_M$ *is equal to* 1. *In other words, asymptotically almost all λ-terms contain* M *as a subterm.*

Proof. Consider an arbitrary $t \in \mathcal{T}_M$. Either t is equal to M, or M is a proper subterm of t. In the latter case we have four additional cases. Either t is an abstraction, or $t = t_1 t_2$ and M is a subterm of t_1, t_2 or both. Combining, we obtain the following equation:

$$\mathcal{T}_M = M \oplus \lambda\, \mathcal{T}_M \oplus \mathcal{T}_M\, \mathcal{L}_\infty \oplus \mathcal{L}_\infty\, \mathcal{T}_M \oplus \mathcal{T}_M\, \mathcal{T}_M.$$

Note that by adding $\mathcal{T}_M \mathcal{L}_\infty$ and $\mathcal{L}_\infty \mathcal{T}_M$ together we count each term $t = t_1 t_2$ containing M in both t_1 and t_1 twice, therefore we have to subtract $\mathcal{T}_M\, \mathcal{T}_M$. Such a presentation yields the following functional quadratic equation involving the corresponding generating function T_M:

$$T_M(z) = z^p + z\, T_M(z) + 2z\, T_M(z)\, L_\infty(z) - z\, T_M^2(z).$$

Since $\sqrt{\Delta_{L_\infty}} = 1 - 2z\, L_\infty(z) - z$ (see Lemma 2), we can express its discriminant as $\Delta_{T_M} = \Delta_{L_\infty} + 4z^{p+1}$. Hence $\Delta_{T_M} > \Delta_{L_\infty}$. It follows that the root ρ_{T_M} of smallest modulus of Δ_{T_M} is strictly larger than the root ρ_{L_∞} of smallest modulus of Δ_{L_∞}, i.e. $\rho_{T_M} > \rho_{L_\infty}$. Moreover, $T_M(z) = \frac{\sqrt{\Delta_{T_M}} - \sqrt{\Delta_{L_\infty}}}{2z}$ and thus the generating function counting the number of λ-terms which do not contain M as a subterm is given by

$$L_\infty(z) - T_M(z) = \frac{(1 - z) - \sqrt{\Delta_{T_M}}}{2z}.$$

Applying Theorem IV.7 of [7] we immediately get that the above set has asymptotic density 0 and thus $\mathcal{T}_M$ has asymptotic density equal to 1. $\square$

Corollary 1. *Asymptotically almost no λ-term is strongly normalizing.*

Proof. Consider the aforementioned Ω. Since it is not normalizing and asymptotically almost all λ-terms contain it as a subterm, we immediately get our claim.   □

Let us notice the striking discrepancy between the density of strongly normalizing terms in the natural model and the corresponding density in the model considered in [6]. In the latter case, variables tend to be arbitrarily far from their binders, since they do not contribute to the overall size. In the natural model, however, increasing an index increases the overall size and thus indices tend to be rather near their binding lambdas.

References

1. Bendkowski, M.: Natural counting of lambda terms - Haskell implementations (2015). https://github.com/maciej-bendkowski/natural-counting-of-lambda-terms
2. Online Encyclopedia of Integer Sequences. http://oeis.org/
3. Bacher, A., Bodini, O., Jacquot, A.: Exact-size sampling for Motzkin trees in linear time via Boltzmann samplers and Holonomic specification. In: Proceedings of the Meeting on Analytic Algorithmics and Combinatorics, pp. 52–61. SIAM (2013)
4. Bodini, O., Gardy, D., Gittenberger, B.: Lambda terms of bounded unary height. In: Proceedings of the Eighth Workshop on Analytic Algorithmics and Combinatorics, pp. 23–32 (2011). http://www.siam.org/proceedings/analco/2011/anl11_03_bodinio.pdf
5. Claessen, K., Hughes, J.: QuickCheck: a lightweight tool for random testing of Haskell programs. In: Proceedings of the Fifth ACM SIGPLAN International Conference on Functional Programming, ICFP 2000, pp. 268–279. ACM, New York (2000)
6. David, R., Grygiel, K., Kozik, J., Raffalli, Ch., Theyssier, G., Zaionc, M.: Asymptotically almost all λ-terms are strongly normalizing. Logical Methods Comput. Sci. **9**, 1–30 (2013)
7. Flajolet, P., Sedgewick, R.: Analytic Combinatorics. Cambridge University Press, New York (2009). ISBN:0521898064, 9780521898065
8. Grygiel, K., Lescanne, P.: Counting and generating lambda terms. J. Funct. Program. **23**(5), 594–628 (2013)
9. Grygiel, K., Lescanne, P.: Counting terms in the binary lambda calculus. In: Proceedings of the 25th International Conference on Probabilistic, Combinatorial and Asymptotic Methods for the Analysis of Algorithms (2014). https://hal.inria.fr/hal-01077251
10. Gu, N.S.S., Li, N.Y., Mansour, T.: 2-Binary trees: bijections and related issues. Discrete Math. **308**(7), 1209–1221 (2008). http://dx.doi.org/10.1016/j.disc.2007.04.007
11. Salvy, B., Zimmermann, P.: Gfun: a Maple package for the manipulation of generating and holonomic functions in one variable. ACM Trans. Math. Softw. **2**, 163–177 (1994). http://dx.doi.org/10.1145/178365.178368
12. Sapounakis, A., Tasoulas, I., Tsikouras, P.: Ordered trees and the inorder traversal. Discrete Math. **306**(15), 1732–1741 (2006). http://dx.doi.org/10.1016/j.disc.2006.03.044
13. Tromp, J.: Binary lambda calculus and combinatory logic. In: Kolmogorov Complexity and Applications (2006)

Online Minimum Spanning Tree with Advice

(Extended Abstract)

Maria Paola Bianchi[1]([✉]), Hans-Joachim Böckenhauer[1], Tatjana Brülisauer[1], Dennis Komm[1], and Beatrice Palano[2]

[1] Department of Computer Science, ETH Zurich, Zürich, Switzerland
{maria.bianchi,hjb,tatjana.bruelisauer,dennis.komm}@inf.ethz.ch
[2] Dipartimento di Informatica, Università Degli Studi di Milano, Milan, Italy
palano@di.unimi.it

Abstract. In the online minimum spanning tree problem, a graph is revealed vertex by vertex; together with every vertex, all edges to vertices that are already known are given, and an online algorithm must irrevocably choose a subset of them as a part of its solution. The advice complexity of an online problem is a means to quantify the information that needs to be extracted from the input to achieve good results. For a graph of size n, we show an asymptotically tight bound of $\Theta(n \log n)$ on the number of advice bits to produce an optimal solution for any given graph. For particular graph classes, e.g., with bounded degree or a restricted edge weight function, we prove that the upper bound can be drastically reduced; e.g., $5(n-1)$ advice bits allow to compute an optimal result if the weight function is the Euclidean distance; if the graph is complete, even a logarithmic number suffices. Some of these results make use of the optimality of Kruskal's algorithm for the offline setting. We also study the trade-off between the number of advice bits and the achievable competitive ratio. To this end, we perform a reduction from another online problem to obtain a linear lower bound on the advice complexity for any near-optimal solution. Using our results from the advice complexity finally allows us to give a lower bound on the expected competitive ratio of any randomized online algorithm for the problem.

1 Introduction

Computing problems are called *online* if the input arrives gradually in consecutive time steps. An *online algorithm* has to create parts of the definite output while only knowing a prefix of the input in the current time step [8]. A broad subclass are online graph problems, i.e., online problems where the input corresponds to some graph that is revealed in an online fashion. In this paper, the mainly studied model reveals the vertices of an underlying graph one after the other; together with every vertex, all edges are shown that connect this vertex to all other vertices that are already known to the online algorithm. Sleator

This work was partially supported by SNF grant 200021–146372 and by MIUR under the project "PRIN: Automi e Linguaggi Formali: Aspetti Matematici e Applicativi."

R.M. Freivalds et al. (Eds.): SOFSEM 2016, LNCS 9587, pp. 195–207, 2016.
DOI: 10.1007/978-3-662-49192-8_16

and Tarjan introduced the concept of *competitive analysis* to measure the performance of an online algorithm [23]. In this worst-case measurement, one compares the cost or gain of the solution produced by the online algorithm to the optimal one that could hypothetically be computed if the whole instance were known from the start. Here, one assumes that the input is produced by a malicious *adversary*. The ratio between the two is called the *competitive ratio* of the online algorithm; a detailed introduction is given by Borodin and El-Yaniv [8].

While competitive analysis is an extremely powerful and widely-used tool to assess the performance of online algorithms, it does not address the question of the essential parts of the input that the online algorithm is missing, i.e., the *information content* of the problem [16]. To be able to answer this question, we study the *advice complexity*. An *online algorithm with advice* has an additional resource available, the so-called *advice tape*, that may contain any kind of information about the instance at hand. The content of this tape is an infinite binary string called the *advice*, and it is written onto the tape before the computation starts by an *oracle* that sees the whole input in advance. The advice complexity then measures the number of *advice bits* that allows to achieve a certain performance.

A first model of online computation with advice was introduced by Dobrev et al. [12]. This model was then refined simultaneously by Fraigniaud et al. [14] and Hromkovič et al. [16]. The latter model, which is the one we use in this paper, was first applied to three different online problems by Böckenhauer et al. [6]. Here, in every time step, the online algorithm can query the advice tape for any number of advice bits (analogously to the model of the random tape of a randomized Turing machine). The advice complexity is the length of a maximum prefix of the advice tape that is read; this length usually depends on the input size n of the given instance.

The advice complexity has been widely applied to a large number of online problems so far including paging [6] and k-server [7,14]. In particular, this model has recently been studied for quite a number of graph problems such as different coloring problems [3,4,15,22], the independent set problem [10], the dominating set problem [9], the Steiner tree problem [2], or graph exploration [11]. Moreover, online computation with advice also has some interesting connections to randomized online computation [7,18]. This line of research tries to answer the question of how well any additional information on the input may be exploited. The crucial part is the generality of the answer that is given: the advice may encode any information, it is not restricted to a specific problem parameter or property of the input. This way, a lower bound on the advice complexity to achieve some given competitive ratio c means that it will never be possible to obtain c-competitiveness with less information, no matter what the information will actually be.

To the best of our knowledge, the advice complexity of the minimum spanning tree problem has not been studied so far. Megow et al. [20] investigated the online minimum spanning tree problem in a model that allows an online algorithm to do some recourse actions, meaning that it can perform a certain amount of edge rearrangements. However, in this model, the algorithm has to compute a feasible spanning tree for the graph that has been presented so far in any time step. For randomly weighted graphs with edge weights that are uniformly distributed over the interval between 0 and 1, the problem was studied by Remy et al. [21]. In their model, both the algorithm and the adversary do not know the edge weights before they are presented. Tsai and Tsang investigated the competitiveness of a certain family of randomized algorithms [24]; they restricted the inputs to graphs in the Euclidean space. The minimum spanning tree problem was also considered in the setting of min-max regret [17], in which the goal is to minimize the maximal possible deviation of a given solution from optimum.

This paper is organized as follows. In Sect. 2, we formally introduce the model of online computation with advice and the online minimum spanning tree problem. In Sect. 3, we study the advice complexity of optimal online algorithms with advice for different graph classes. In Subsect. 3.1, we give an asymptotically tight bound of $\Theta(n \log n)$ to compute an optimal solution for general graphs. In Subsect. 3.2, we present a linear lower bound for graphs that have three different edge weights. Here, we also study a different model of online computation where the structure of the graph is known in advance, but the edge weights appear online. The interesting point of the proof is that the optimality of this online algorithm is a consequence of the optimality of Kruskal's offline algorithm. In Subsect. 3.4, we first study graphs with bounded degree. Furthermore, we prove that there is an optimal online algorithm that uses $5(n-1)$ advice bits to be optimal on graphs with a Euclidean weight function. Section 4 studies the trade-off between the number of advice bits and the competitive ratio that is achievable. We prove a linear lower bound to obtain a near-optimal competitive ratio by giving a reduction from the bit guessing problem. In Sect. 5, we extend this result to randomized algorithms. Due to space limitations, some proofs are omitted in this extended abstract.

2   Preliminaries

In this paper, we only consider the objective to minimize a given cost function.

Definition 1 (Online Minimization Problem). *An online minimization problem consists of a set $\mathcal{I}$ of inputs and a cost function. Every input $I \in \mathcal{I}$ is a sequence $I = (x_1, x_2, \ldots, x_n)$ of requests. Furthermore, a set of feasible outputs (or solutions) is associated with every I; every output is a sequence $O = (y_1, y_2, \ldots, y_n)$ of answers. The cost function assigns a positive real value $\mathrm{cost}(I, O)$ to every input I and any feasible output O. For every input I, we call any feasible output O for I that has smallest possible cost (i.e., that minimizes the cost function) an optimal solution for I.*

In what follows, we will simply write $\mathrm{cost}(I)$ instead of $\mathrm{cost}(I, O)$ as O is always clear from context, and we let $[I]_k = (x_1, \ldots, x_k)$, for $k \leq n$, be the sequence of the first k requests in I. In the settings we study, the input always corresponds to a weighted undirected graph G with a weight function ω. Throughout this paper, the set of vertices of G is denoted by $V(G)$, and $E(G)$ denotes its set of edges; if G is clear from context, we simply write V and E. G is usually revealed to the online algorithm as follows. Let $V = \{v_1, v_2, \ldots, v_n\}$; then v_i is presented in time step i together with all edges $\{v_i, v_j\} \in E$ for which $j < i$, i.e., edges that are connected to vertices that have already been revealed in previous time steps. After every newly revealed vertex, an online algorithm for the *online minimum spanning tree problem* (OMST for short) must choose some of the newly revealed edges that are part of the solution; this decision is final. Note that we do not require the set of chosen edges to be a spanning tree of the already revealed vertices in the intermediate steps. To correctly capture the online environment, the number of vertices is not known to the online algorithm in advance.

Next, we formally define online algorithms with advice.

Definition 2 (Online Algorithm with Advice). *Consider an input I of an online minimization problem. An online algorithm* ALG *with advice computes the output sequence* $\text{ALG}^\phi(I) = (y_1, y_2, \ldots, y_n)$ *such that y_i is computed from $\phi, x_1, x_2, \ldots, x_i$, where ϕ is the content of the advice tape, i.e., an infinite binary sequence.* ALG *is c-competitive with advice complexity $b(n)$ if there exists a non-negative constant α such that, for every n and for any input sequence I of length at most n, there exists some advice string ϕ such that* $\text{cost}(\text{ALG}^\phi(I)) \leq c \cdot \text{cost}(\text{OPT}(I)) + \alpha$ *and at most the first $b(n)$ bits of ϕ have been accessed during the computation of the solution $\text{ALG}^\phi(I)$. If the above inequality holds with $\alpha = 0$, we call* ALG *strictly c-competitive with advice complexity $b(n)$.* ALG *is called optimal if it is strictly 1-competitive.*

For the sake of an easier notation, we omit ϕ as it is always clear from context. Moreover, we denote the binary logarithm of a natural number x simply by $\log x$.

3   Optimality

In this section, we show that any online algorithm with advice that solves the OMST problem optimally on general graphs needs to read $\Omega(n \log n)$ advice bits. We also provide an online algorithm that achieves optimality with $O(n \log n)$ advice bits. We will then discuss the problem for input graphs that are somehow limited (such as special graph classes, bounded edge weights, or bounded degree).

3.1   General Graphs

First, we provide an online algorithm that gets as advice the parent of every newly revealed vertex with respect to an optimal spanning tree.

Theorem 1. *There exists an online algorithm with advice for the OMST problem that uses $n\lceil \log n \rceil + 2\lceil \log(\lceil \log n \rceil + 1) \rceil$ advice bits and that is optimal on every instance of length n.*

Proof. The oracle computes an optimal offline solution T. At each request v, the algorithm asks for the index of the parent of v in T (if the requested vertex is the arbitrarily chosen root of T, then the oracle encodes v itself). This takes $n \lceil \log n \rceil$ bits of advice in total. In order to know how many advice bits it should read after each request, the algorithm needs to ask first for the number $\lceil \log n \rceil$, which is encoded using a prefix code (such as Elias' delta-code [13]) at the beginning of the advice string with $2 \lceil \log(\lceil \log n \rceil + 1) \rceil$ bits.   □

Although the above online algorithm uses a straightforward approach, this upper bound is asymptotically tight; in fact, we can prove the lower bound even on a very restricted class of graphs.

Theorem 2. *Any online algorithm with advice for the OMST problem on bipartite graphs needs to read at least $\log(((n-1)/2)!) \in \Omega(n \log n)$ advice bits to be optimal on every instance of length n.*

Proof. Let $n = 2k + 1$ be an odd number. We consider a bipartite graph G having vertices $\{v_1, v_2, \ldots, v_k, u_1, u_2, \ldots, u_k, w\}$ such that, for each $1 \leq i \leq k$, the vertex u_i is connected to w through an edge of weight 1 and, for $i \leq j \leq k$, u_i is connected to v_j through an edge of weight $k - i + 1$. Clearly, such a graph has a unique minimum spanning tree, as shown in Fig. 1. As set of instances $\mathcal{I}$, we consider all online presentation of G such that first, a permutation of the vertices $v_1, v_2, \ldots, v_k$ is presented, then the vertices $u_1, u_2, \ldots, u_k, w$ are presented in this order. These instances differ only in the order of the first k vertices.

Suppose towards contradiction that there is an algorithm ALG that optimally solves OMST on any instance of $\mathcal{I}$ using less than $\log(((n-1)/2)!)$ bits of advice. This implies that there are two different instances, with two different permutations σ_1 and σ_2 of the vertices $v_1, v_2, \ldots, v_k$, which receive the same advice string. Let v_i be the first vertex that is not at the same position in σ_1 and σ_2, say at position s in σ_1 and position t in σ_2. Up to and including the $(k+i)$-th time step, the input looks the same for ALG. However, in time step $k + i$, ALG

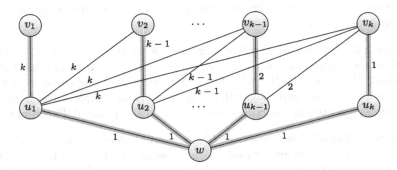

Fig. 1. Graph structure used in the proof of Theorem 2; gray edges denote the (unique) optimal solution.

has to choose the edge that connects u_i to the vertex that was presented in time step s (t, respectively) in the case of σ_1 (σ_2, respectively). But since the input looked exactly the same to the algorithm so far, it cannot distinguish between these two cases, so it will not output the optimal solution for at least one of these instances.   □

3.2   Graphs with Bounded Edge Weights

We now consider graphs with a bounded number of different edge weights. The next theorem shows that, even for only those different weights, still a linear number of advice bits is needed.

Theorem 3. *Any online algorithm with advice that solves the OMST problem on graphs with 3 or more different edge weights needs to read at least $n-2$ advice bits to be optimal on every instance of length n.*   □

Next, we prove a result for graphs with two different edge weights, which will come in handy when considering different graph classes. However, for the following result, the online setting differs in the following sense. The structure of the graph is known to the online algorithm in advance, and only the concrete edge weights are revealed in the respective time steps.

Theorem 4. *Let $G = (V, E, \omega)$ be a connected graph with two different edge weights a and b, where $0 \leq a < b$. For the online problem in which the structure of G is fully known to the algorithm and only the edge weights are presented online, there exists an optimal online algorithm ALG that uses no advice.*

Proof sketch. For each instance I which is an online presentation of the graph $G = (V, E, \omega)$, let $m = |E|$, let e_i be the edge presented to ALG at time step i, and let w_i be the weight of e_i. The algorithm works on I as follows: if $w_i = a$, then ALG includes e_i in the partial solution if and only if e_i does not create a cycle; if $w_i = b$, then it includes $e_i = \{u, v\}$ in the partial solution if and only if, for all the other paths connecting u and v, ALG has already rejected one of the edges composing that path. The resulting graph T_{ALG} is clearly a spanning tree whose optimality is easy to show.   □

Theorem 4 implies logarithmic upper bounds on the advice complexity for optimal online algorithms for complete and complete bipartite graphs. Indeed, with $\max\{2, \lceil \log n \rceil + 2\lceil \log \lceil \log n \rceil \rceil\}$ bits, the input size can be encoded in a self-delimiting way, and an online algorithm can then treat a complete graph instance as if it were presented in the online model used in the proof of Theorem 4.

Corollary 1. *For all complete graphs G of size n with edge weights in $\{a, b\}$, with $0 \leq a < b$, there exists an optimal online algorithm with advice that solves the OMST problem for G and uses $\max\{2, \lceil \log n \rceil + 2\lceil \log \lceil \log n \rceil \rceil\}$ bits of advice.*   □

Next, we show that the bound from Corollary 1 is essentially tight.

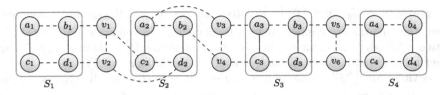

Fig. 2. An instance of the graph class $\mathcal{G}_{22}$ that is used in the proof of Theorem 6. Dashed edges have weight 1, solid edges have weight 2. First, the four isolated squares S_1, S_2, S_3 and S_4 are presented, then the remaining vertices $v_1, v_2, \ldots, v_6$ in this order. To find an optimal minimum spanning tree, ALG has to choose exactly one edge of weight 2 in square S_2, which is oriented horizontally in the ladder, but no edge of weight 2 in the vertically oriented squares S_1, S_3 and S_4.

Theorem 5. *Any online algorithm with advice that solves the OMST problem optimally for complete graphs with at least two different edge weights needs at least $\log(\lfloor n/2 \rfloor)$ bits of advice to be optimal on every input sequence of size at most n.*

Proof sketch. For every even $2 \leq m \leq n$, we consider the complete graph instance $\hat{G}_m$ with edge weights $\{1, 2\}$ defined as follows: $\hat{G}_m$ has m vertices and, by calling v_j the vertex presented at time step j, each v_j with j odd (even, respectively) is connected with all vertices v_i, with $i < j$, by an edge of weight 1 (2, respectively). By using the partition tree technique introduced in [1], we can show that any algorithm needs a different advice string for each $\hat{G}_m$ to be optimal. The intuitive idea is that any algorithm needs to know when the end of the input is reached, as only in the final time step, it has to choose an edge of weight b. $\square$

With a similar technique, we can obtain an analogous upper and lower bounds for the case of complete bipartite graphs.

3.3 Ladders

We now restrict our attention to a special class of bipartite graphs, namely ladders, which can be defined as the Cartesian product of two path graphs, one of which has only one edge. Despite of this simple structure, we show that such graphs still require linear advice, even for only two different edge weights.

Theorem 6. *For ladders with two different edge weights, any online algorithm with advice for the OMST problem needs at least $\lfloor \frac{n+2}{6} \rfloor$ advice bits to be optimal on every input sequence of length n.*

Proof sketch. Let n be an arbitrary natural even number. We now provide a graph class $\mathcal{G}_n$ that contains $2^{\lfloor \frac{n+2}{6} \rfloor}$ graphs of size n and show that, for any two graphs in $\mathcal{G}_n$, ALG needs different advice strings. We define $\mathcal{G}_n$ as follows: For any graph $G \in \mathcal{G}_n$, first $\lfloor \frac{n+2}{6} \rfloor$ isolated squares $S_1, \ldots, S_{\lfloor \frac{n+2}{6} \rfloor}$ are presented. Then, the squares are connected to a ladder with the remaining $n - 4 \cdot \lfloor \frac{n+2}{6} \rfloor$ vertices. All edges incident to these vertices have weight 1. Any square S_i can either be oriented horizontally or vertically in the ladder (see Fig. 2).

As there are $\lfloor \frac{n+2}{6} \rfloor$ squares in every graph $G \in \mathcal{G}_n$ and all squares can be oriented in two ways, there are $2^{\lfloor \frac{n+2}{6} \rfloor}$ graphs in $\mathcal{G}_n$, therefore it suffices to show that ALG needs different advice strings for any two graphs of $\mathcal{G}_n$.   □

Theorem 7. *For ladders with two different edge weights, there exists an online algorithm with advice for the OMST problem that is optimal on every input sequence of even length $n \geq 2$ and reads at most $\lceil \frac{3}{4}n \rceil + 4\lceil \log n \rceil + 2\lceil \log \lceil \log n \rceil \rceil$ advice bits.*

Proof sketch. The algorithm reads the size of the input, the positions of the four corner vertices, and, for each vertex, if it lies on the top or the bottom line of the ladder, if needed.   □

3.4 Further Special Graph Classes

We now analyze our problem on graphs of bounded degree.

Theorem 8. *For graphs with degree at most g, there exists an online algorithm with advice that solves the OMST problem and uses at most $(n-1)\lceil \log g \rceil + \max\{2, \lceil \log n \rceil + 2\lceil \log \lceil \log n \rceil \rceil\}$ advice bits to be optimal on every instance of length n.*   □

For graphs with degree 3 and 4 we obtain asymptotically matching lower bounds.

We complement our results on special graph classes by showing that also in a geometric setting, where the vertices are points in the Euclidean plane and the edge weights are their distance, we can compute an optimal solution using linear advice.

Theorem 9. *For Euclidean graphs, there exists an online algorithm ALG with advice that solves the OMST problem and uses at most $5(n-1)$ advice bits to be optimal on every instance of length n.*   □

4   Competitiveness

In this section, we analyze the trade-off between the advice complexity and the competitiveness of online algorithms for the OMST problem. We recall that, as proved in [20], for general graphs no deterministic online algorithm can be competitive. If we have edge weights bounded by a constant k, then the greedy algorithm is clearly k-competitive on every input, since any spanning tree on a graph with n vertices has weight at least $n-1$ and at most $k(n-1)$. If the degree of the graph is bounded by 3, we can even prove a better competitive ratio without advice for a bounded number of edge weights.

Theorem 10. *Let $G = (V, E)$ be a graph with maximum degree 3 where $\omega \colon E \to W$ is a weight function that maps edges into a bounded set W of weight values. Let a denote the minimum and b the maximum element in W. Then, the greedy algorithm* GREEDY *has a competitive ratio of at most $((a+b)\frac{n}{2} - 2a + b)/(a(n-1))$ for the OMST problem on G.* □

In Subsect. 3.4, we have shown that any online algorithm with advice needs a linear amount of advice to be optimal for the OMST problem on complete graphs with three different edge weights. We now show what an algorithm with logarithmic advice can achieve in this setting. For simplicity, we will only discuss the case with the three edge weights 1, 2 and 3. Note that similar results can be obtained using the same proof idea, and any 3 different edge weights.

Theorem 11. *There exists an online algorithm* ALG *with advice for the OMST problem on complete graphs with edge weights $1, 2$ and 3 that reads $\max\{2, \lceil \log n \rceil + 2\lceil \log \lceil \log n \rceil \rceil\} + 1$ bits of advice on an input of length n and achieves a strict competitive ratio of $(5 - \sqrt{5})/2 \approx 1.382$.*

Proof sketch. For each instance of size n, the oracle encodes n on the advice tape with $\max\{2, \lceil \log n \rceil + 2\lceil \log \lceil \log n \rceil \rceil\}$, then uses an additional bit to suggest one of the following two strategies:

1. in the first $n - 1$ steps, choose greedily all edges of weight 1 which do not close a cycle, then in the last step choose the cheapest edges needed to connect all the currently separated components in the partial solution,
2. in the first $n - 1$ steps choose greedily all edges of weight 1 and 2 without closing cycles, then connect the components in the last step with the cheapest possible edges.

Given the solution T obtained with strategy 2, consider the tree T' constructed in the following way: whenever there exists an edge e_1 of weight 1 in the input graph that would create a cycle in T that contains an edge e_2 of weight 2, replace e_2 with e_1. We call *suboptimal* all the edges of weight 2 removed in this process. The oracle suggests the second strategy if and only if the number of suboptimal edges chosen with this strategy is less than $p(n - 1)$, for a suitable $0 < p < 1$. □

We now provide a linear lower bound that even holds for the case that the maximum degree is 3. To this end, we use a general technique, namely reducing the *bit guessing problem with known history*, which was introduced by Böckenhauer et al. [5], to the OMST problem.

Definition 3 (Bit Guessing with Known History). *The* bit guessing problem with known history *(BGKH) is the following online minimization problem. The input $I = (n, d_1, d_2, \ldots, d_n)$ consists of a natural number n and the bits $d_1, d_2, \ldots, d_n$, that are revealed one by one. The online algorithm* ALG *computes the output sequence* $\mathrm{ALG}(I) = y_1 y_2 \ldots y_n$, *where $y_i = f(n, d_1, \ldots, d_{i-1}) \in \{0, 1\}$, for some computable function f. The algorithm is not required to respond with any output in the last time step. The cost of a solution* $\mathrm{ALG}(I)$ *is the number of wrongly guessed bits, i.e., the Hamming distance* $\mathrm{Ham}(d, \mathrm{ALG}(I))$ *between $d = d_1 d_2 \ldots d_n$ and* $\mathrm{ALG}(I)$.

We start by formally describing the reduction on a specific class of instances for the OMST problem.

Lemma 1. *Let s be any BGKH instance of length n'. Let $\delta \in \mathbb{R}$ with $1/2 \le \delta \le 1$ be such that any online algorithm with advice for BGKH reading b advice bits can guess at most $\delta n'$ bits of s correctly. Then, no online algorithm ALG with advice for the corresponding OMST instance G_s reads b advice bits and achieves*

$$\mathrm{cost}(\mathrm{ALG}(G_s)) < \mathrm{cost}(\mathrm{OPT}(G_s)) + (1 - \delta)n',$$

where $\mathrm{cost}(\mathrm{OPT}(G_s))$ is the cost of an optimal minimum spanning tree of G_s.

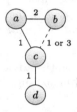

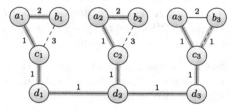

(a) Basic component used to build the graphs

(b) The graph G_s corresponding to the bit string $s = 110$. The dashed edge of the i-th component in G_s has weight 1 if the i-th bit of s has value 0 and weight 3 otherwise.

Fig. 3. Graph construction used in the proof of Theorem 1. Edges that depend on s are dashed.

Proof sketch. For every bit string $s = s_1 s_2 \ldots s_{n'}$ of length n', which is an instance of the BGKH problem, we construct a corresponding instance G_s for the OMST problem as follows: for every bit s_i of s, we build a component as illustrated in Fig. 3a, where the edge $\{b, c\}$ has weight 1 if and only if s_i has value 0 and weight 3 otherwise. Then, we complete these n' components to a connected graph by adding edges of weight 1 between the vertices d_i and d_{i+1}, for all $1 \le i < n'$. As an example, the graph G_{110} is shown in Fig. 3b. The vertex presentation order is $a_1, b_1, c_1, d_1, a_2, \ldots, c_{n'}, d_{n'}$. □

Using Lemma 1, we can now proceed to prove the following lower bound on the advice complexity for any c-competitive algorithm for OMST.

Theorem 12. *There is no online algorithm with advice that is strictly c-competitive, $1 \le c \le 11/10$, for the OMST problem on graphs with maximum degree 3, maximum edge weight 3 and n vertices, $n = 4n'$ for some $n' \in \mathbb{N}$, that reads less than $(1 + (5c - 5) \log(5c - 5) + (6 - 5c) \log(6 - 5c)) \frac{n}{4}$ bits of advice.* □

In the case of unbounded edge weights, we can extend the linear lower bound from Theorem 12 to consider competitive ratios up to $\frac{5}{4}$ as follows: Instead of the edge weights 1, 2, and 3, we choose weights $1, k+1$, and $2k+1$, for arbitrarily large k. Then, Lemma 1 can be proven analogously and we get that ALG has to read at least $\text{cost}(\text{OPT}(G_s)) + k \cdot (1 - \delta)n'$ bits of advice. With the same calculations as above, we can show that the competitive ratio of ALG is $c \geq 1 + (1 - \delta) \cdot \frac{k}{2+2(k+1)} \xrightarrow{k\to\infty} 1 + (1-\delta)\frac{1}{2}$. As a result, we get the following corollary.

Corollary 2. *In the case where edge weights are unbounded, there is no online algorithm with advice that is c-competitive, $1 \leq c \leq \frac{5}{4}$, for the OMST problem on graphs with maximum degree 3 and n vertices, $n = 4n'$ for some $n' \in \mathbb{N}$, that reads less than $(1 + (2c - 2)\log(2c - 2) + (3 - 2c)\log(3 - 2c))\frac{n}{4}$ bits of advice.* □

5   Randomized Online Algorithms

In this section, we give a lower bound on the competitive ratio achievable by any randomized online algorithm. Our proof is based on the following result.

Lemma 2. (Böckenhauer et al. [7]). *Consider an online minimization problem U, and let $\mathcal{I}(n)$ be the set of all possible inputs of length n and $I(n) := |\mathcal{I}(n)|$. Furthermore, suppose that there is a randomized online algorithm for U with worst-case expected competitive ratio at most E. Then, for any fixed $\varepsilon > 0$, it is possible to construct an online algorithm with advice that uses at most*

$$\log n + 2\log\log n + \log\left(\log I(n)/\log\left(1 + \varepsilon\right)\right) + c$$

advice bits, for a constant c, and achieves a competitive ratio of $(1 + \varepsilon)E$.   □

In Theorem 12, we constructed, for each integer n, a set $\mathcal{I}(n)$ of $2^{\frac{n}{4}}$ instances (depicted in Fig. 3) such there is no $\frac{11}{10}$-competitive online algorithm that uses $o(n)$ advice bits. This, together with Lemma 2, implies the following result.

Theorem 13. *For arbitrarily small $\delta > 0$, every randomized algorithm (using an arbitrary number of random bits) for the OMST problem on graphs with maximum degree 3 and maximum edge weight 3 has a worst-case expected competitive ratio of at least $\frac{11}{10}(1 - \delta)$ on sufficiently large instances.*   □

Acknowledgments. The authors would like to thank Juraj Hromkovič for enlightening discussions.

References

1. Barhum, K., Böckenhauer, H.-J., Forišek, M., Gebauer, H., Hromkovič, J., Krug, S., Smula, J., Steffen, B.: On the power of advice and randomization for the disjoint path allocation problem. In: Geffert, V., Preneel, B., Rovan, B., Štuller, J., Tjoa, A.M. (eds.) SOFSEM 2014. LNCS, vol. 8327, pp. 89–101. Springer, Heidelberg (2014)

2. Barhum, K.: Tight bounds for the advice complexity of the online minimum steiner tree problem. In: Geffert, V., Preneel, B., Rovan, B., Štuller, J., Tjoa, A.M. (eds.) SOFSEM 2014. LNCS, vol. 8327, pp. 77–88. Springer, Heidelberg (2014)

3. Bianchi, M.P., Böckenhauer, H.-J., Hromkovič, J., Keller, L.: Online coloring of bipartite graphs with and without advice. In: Gudmundsson, J., Mestre, J., Viglas, T. (eds.) COCOON 2012. LNCS, vol. 7434, pp. 519–530. Springer, Heidelberg (2012)

4. Bianchi, M.P., Böckenhauer, H.-J., Hromkovič, J., Krug, S., Steffen, B.: On the advice complexity of the online $L(2,1)$-coloring problem on paths and cycles. In: Du, D.-Z., Zhang, G. (eds.) COCOON 2013. LNCS, vol. 7936, pp. 53–64. Springer, Heidelberg (2013)

5. Böckenhauer, H.-J., Hromkovič, J., Komm, D., Krug, S., Smula, J., Sprock, A.: The string guessing problem as a method to prove lower bounds on the advice complexity. Theoret. Comput. Sci. **554**, 95–108 (2014)

6. Böckenhauer, H.-J., Komm, D., Královič, R., Královič, R., Mömke, T.: On the advice complexity of online problems. In: Dong, Y., Du, D.-Z., Ibarra, O. (eds.) ISAAC 2009. LNCS, vol. 5878, pp. 331–340. Springer, Heidelberg (2009)

7. Böckenhauer, H.-J., Komm, D., Královič, R., Královič, R.: On the advice complexity of the k-server problem. In: Aceto, L., Henzinger, M., Sgall, J. (eds.) ICALP 2011, Part I. LNCS, vol. 6755, pp. 207–218. Springer, Heidelberg (2011)

8. Borodin, A., El-Yaniv, R.: Online Computation and Competitive Analysis. Cambridge University Press, New York (1998)

9. Boyar, J., Favrholdt, L.M., Kudahl, C., Mikkelsen, J.W.: The advice complexity of a class of hard online problems. CoRR abs/1408.7033 (2014)

10. Dobrev, S., Královič, R., Královič, R.: Independent set with advice: the impact of graph knowledge. In: Erlebach, T., Persiano, G. (eds.) WAOA 2012. LNCS, vol. 7846, pp. 2–15. Springer, Heidelberg (2013)

11. Dobrev, S., Královič, R., Markou, E.: Online graph exploration with advice. In: Even, G., Halldórsson, M.M. (eds.) SIROCCO 2012. LNCS, vol. 7355, pp. 267–278. Springer, Heidelberg (2012)

12. Dobrev, S., Královic, R., Pardubská, D.: Measuring the problem-relevant information in input. RAIRO ITA **43**(3), 585–613 (2009)

13. Elias, P.: Universal codeword sets and representations of the integers. IEEE Trans. Inf. Theory **21**(2), 194–203 (1975)

14. Emek, Y., Fraigniaud, P., Korman, A., Rosén, A.: Online computation with advice. In: Albers, S., Marchetti-Spaccamela, A., Matias, Y., Nikoletseas, S., Thomas, W. (eds.) ICALP 2009, Part I. LNCS, vol. 5555, pp. 427–438. Springer, Heidelberg (2009)

15. Forišek, M., Keller, L., Steinová, M.: Advice complexity of online coloring for paths. In: Dediu, A.-H., Martín-Vide, C. (eds.) LATA 2012. LNCS, vol. 7183, pp. 228–239. Springer, Heidelberg (2012)

16. Hromkovič, J., Královič, R., Královič, R.: Information complexity of online problems. In: Hliněný, P., Kučera, A. (eds.) MFCS 2010. LNCS, vol. 6281, pp. 24–36. Springer, Heidelberg (2010)

17. Kasperski, A.: Discrete Optimization with Interval Data: Minmax Regret and Fuzzy Approach. Springer, Heidelberg (2008)

18. Komm, D., Královič, R.: Advice complexity and barely random algorithms. Theor. Inf. Appl. (RAIRO) **45**(2), 249–267 (2011). IEEE Computer Society

19. Kruskal Jr., J.B.: On the shortest spanning subtree of a graph and the traveling salesman problem. Proc. Am. Math. Soc. **7**(1), 48–50 (1956)

20. Megow, N., Skutella, M., Verschae, J., Wiese, A.: The power of recourse for online MST and TSP. In: Czumaj, A., Mehlhorn, K., Pitts, A., Wattenhofer, R. (eds.) ICALP 2012, Part I. LNCS, vol. 7391, pp. 689–700. Springer, Heidelberg (2012)

21. Remy, J., Souza, A., Steger, A.: On an online spanning tree problem in randomly weighted graphs. Comb. Probab. Comput. **16**(1), 127–144 (2007). Cambridge University Press

22. Seibert, S., Sprock, A., Unger, W.: Advice complexity of the online coloring problem. In: Spirakis, P.G., Serna, M. (eds.) CIAC 2013. LNCS, vol. 7878, pp. 345–357. Springer, Heidelberg (2013)

23. Sleator, D.D., Tarjan, R.E.: Amortized efficiency of list update and paging rules. Commun. ACM **28**(2), 202–208 (1985)

24. Teh Tsai, Y., Yi Tang, C.: The competitiveness of randomized algorithms for on-line Steiner tree and on-line spanning tree problems. Inf. Process. Lett. **48**(4), 177–182 (1993). Elsevier

Subsequence Automata with Default Transitions

Philip Bille, Inge Li Gørtz, and Frederik Rye Skjoldjensen[✉]

Technical University of Denmark, Lyngby, Denmark
{phbi,inge,fskj}@dtu.dk

Abstract. Let S be a string of length n with characters from an alphabet of size σ. The *subsequence automaton* of S (often called the *directed acyclic subsequence graph*) is the minimal deterministic finite automaton accepting all subsequences of S. A straightforward construction shows that the size (number of states and transitions) of the subsequence automaton is $O(n\sigma)$ and that this bound is asymptotically optimal.

In this paper, we consider subsequence automata with *default transitions*, that is, special transitions to be taken only if none of the regular transitions match the current character, and which do not consume the current character. We show that with default transitions, much smaller subsequence automata are possible, and provide a full trade-off between the size of the automaton and the *delay*, i.e., the maximum number of consecutive default transitions followed before consuming a character.

Specifically, given any integer parameter k, $1 < k \leq \sigma$, we present a subsequence automaton with default transitions of size $O(nk\log_k \sigma)$ and delay $O(\log_k \sigma)$. Hence, with $k = 2$ we obtain an automaton of size $O(n\log \sigma)$ and delay $O(\log \sigma)$. On the other extreme, with $k = \sigma$, we obtain an automaton of size $O(n\sigma)$ and delay $O(1)$, thus matching the bound for the standard subsequence automaton construction. The key component of our result is a novel hierarchical automata construction of independent interest.

1 Introduction

Let S be a string of length n with characters from an alphabet of size σ. A *subsequence* of S is any string obtained by deleting zero or more characters from S. The *subsequence automaton* (often called the *directed acyclic subsequence graph*) is the minimal deterministic finite automaton accepting all subsequences of S. Baeza-Yates [1] initiated the study of subsequence automata. He presented a simple construction using $O(n\sigma)$ size (size denotes the total number of states *and* transitions) and showed that this bound is optimal in the sense that there are subsequence automata of size at least $\Omega(n\sigma)$. He also considered variations with encoded input strings and multiple strings. Subsequently, several researchers have further studied subsequence automata (and its variants) [2–9]. See also the surveys by Tronícek [10,11]. The general problem of *subsequence indexing*, not limited to automata based solutions, is investigated by Bille et al. [12].

© Springer-Verlag Berlin Heidelberg 2016
R.M. Freivalds et al. (Eds.): SOFSEM 2016, LNCS 9587, pp. 208–216, 2016.
DOI: 10.1007/978-3-662-49192-8_17

In this paper, we consider subsequence automata in the context of *default transitions*, that is, special transitions to be taken only if none of the regular transitions match the current character, and which do not consume the current character. Each state has at most one default transition and hence the automaton remains deterministic. The key point of default transitions is to reduce the size of standard automata at the cost of introducing a *delay*, i.e., the maximum number of consecutive default transition followed before consuming a character. For instance, given a pattern string of length m the classic Knuth-Morris-Pratt (KMP) [13] string matching algorithm may be viewed as an automaton with default transitions (typically referred to as *failure transitions*). This automaton has size $O(m)$, whereas the standard automaton with no default transition would need $\Theta(m\sigma)$ space. The delay of the automaton in the KMP algorithm is either $O(m)$ or $O(\log m)$ depending on the version. Similarly, the Aho-Corasick string matching algorithm for multiple strings may also be viewed as an automaton with default transitions [14]. More recently, default transitions have also been used extensively to significantly reduce sizes of deterministic automata for regular expression [15, 16]. The main idea is to effectively enable states with large overlapping identical sets of outgoing transitions to "share" outgoing transitions using default transitions.

Surprisingly, no non-trivial bounds for subsequence automata with default transitions are known. Naively, we can immediately obtain an $O(n\sigma)$ size solution with $O(1)$ delay by using the standard subsequence automaton (without default transitions). At the other extreme, we can build an automaton with $n + 1$ states (each corresponding to a prefix of S) with a standard and a default transition from the state corresponding to the ith prefix to the state corresponding to the $i + 1$st prefix (the standard transition is labeled $S[i + 1]$). It is straightforward to show that this leads to an $O(n)$ size solution with $O(n)$ delay. Our main result is a substantially improved trade-off between the size and delay of the subsequence automaton:

Theorem 1. *Let S be a string of n characters from an alphabet of size σ. For any integer parameter k, $1 < k \leq \sigma$, we can construct a subsequence automaton with default transitions of size $O(nk \log_k \sigma)$ and delay $O(\log_k \sigma)$.*

Hence, with $k = 2$ we obtain an automaton of size $O(n \log \sigma)$ and delay $O(\log \sigma)$. On the other extreme, with $k = \sigma$, we obtain an automaton of size $O(n\sigma)$ and delay $O(1)$, thus matching the bound for the standard subsequence automaton construction.

To obtain our result, we first introduce the *level automaton*. Intuitively, this automaton uses the same states as the standard solution, but hierarchically orders them in a tree-like structure and samples a selection of their original transitions based on their position in the tree, and adds a default transition to the next state on a higher level. We show how to do this efficiently leading to a solution with $O(n \log n)$ size and $O(\log n)$ delay. To achieve our full trade-off from Theorem 1 we show how to augment the construction with additional ideas for small alphabets and generalize the level automaton with parameter k,

$1 < k \leq \sigma$, where large k reduces the height of the tree but increases the number of transitions. In the final section we generalize the result to *multiple* strings.

2   Preliminaries

A *deterministic finite automaton* (DFA) is a tuple $A = (Q, \Sigma, \delta, q_0, F)$ where Q is a set of nodes called *states*, δ is a set of labeled directed edges between states, called *transitions*, where each label is a character from the alphabet Σ, $q_0 \in Q$ is the *initial* state and $F \subseteq Q$ is a set of *accepting* states. No two outgoing transitions from the same state have the same label. The DFA is *incomplete* in the sense that every state does not contain transitions for every character in Σ. The *size* of A is the sum of the number of states and transitions.

We can think of A as an *edge-labeled directed graph*. Given a string P and a path p in A we say that p and P match if the concatenation of the labels on the transitions in p is P. We say that A *accepts* a string P if there is a path in A, from q_0 to any state in F, that matches P. Otherwise A *rejects* P.

A *deterministic finite automaton with default transitions* is a deterministic finite automaton AD where each state can have a single unlabeled default transition. Given a string P and a path p in AD we define a match between P and p as before, with the exception that for any default transition d in p the corresponding character in P cannot match any standard transition out of the start state of d. Definition of accepted and rejected strings are as before. The *delay* of AD is the maximum length of any path matching a single character, i.e., if the delay of AD is d then we follow at most $d - 1$ default transitions for every character that is matched in P.

A *subsequence* of S is a string P, obtained by removing zero or more occurrences of characters from S. A *subsequence automaton* constructed from S, is a deterministic finite automaton that accepts string P iff P is a subsequence of S. A subsequence automaton construction is presented in [1]. This construction is often called the *directed acyclic subsequence graph* or DASG, but here we denote it SA. The SA has $n + 1$ states, all accepting, that we identify with the integers $\{0, 1, \ldots, n\}$. For each state s, $0 \leq s \leq n$, we have the following transitions:

- For each unique character α in $S[s + 1, n]$, there is a transition labeled α to the smallest state $s' > s$ such that $S[s'] = \alpha$.

The SA has size $O(n\sigma)$ since every state can have at most σ transitions. An example of an SA is given in Fig. 1.

A *subsequence automaton with default transitions* constructed from S, denoted SAD, is a deterministic finite automaton with default transitions that accepts string P iff P is a subsequence of S.

The next section explores different configurations of transitions and default transitions in SADs.

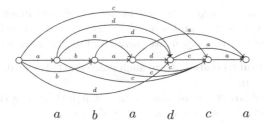

$$a \quad\quad b \quad\quad a \quad\quad d \quad\quad c \quad\quad a$$

Fig. 1. An example of an SA constructed from the string *abadca*

3   New Trade-Offs for Subsequence Automata

We now present a new trade-off for subsequence automata, with default transitions. We will gradually refine our construction until we obtain an automaton that gives the result presented in Theorem 1. In each construction we have $n+1$ states that we identify with the integers $\{0, 1, \ldots, n\}$. Each of these states represents a prefix of the string S and are all accepting states. We first present the *level automaton* that gives the first non-trivial trade-off that exploits default transitions. The general idea is to construct a hierarchy of states, such that every path that only uses default transitions is guaranteed to go through states where the outdegree increases at least exponentially. The level automaton is a SAD of size $O(n \log n)$ and delay $O(\log n)$. By arguing that any path going through a state with outdegree σ will do so by taking a regular transition, we are able to improve both the size and delay of the level automaton. This results in the *alphabet-aware level automaton* which is a SAD of size $O(n \log \sigma)$ and delay $O(\log \sigma)$. Finally we present a generalized construction that gives a trade-off between size and delay by letting parameter k, $1 < k \leq \sigma$, be the base of the exponential increase in outdegree on paths with only default transitions. This SAD has size $O(nk \log_k \sigma)$ and delay $O(\log_k \sigma)$. With $k = 2$ we get an automaton of size $O(n \log \sigma)$ and delay $O(\log \sigma)$. In the other extreme, for $k = \sigma$ we get an automaton of size $O(n\sigma)$ and delay $O(1)$.

3.1   Level Automaton

The level automaton is a SAD with $n+1$ states that we identify with the integers $\{0, 1, \ldots, n\}$. All states are accepting. For each state $i > 0$, we associate a level, $\mathrm{level}(i)$, given by:

$$\mathrm{level}(i) = \max(\{x \mid i \bmod 2^x = 0\})$$

Hence, $\mathrm{level}(i)$ is the exponent of the largest power of two that divides i. The level function is in the literature known as the ruler function. We do not associate any level with state 0. Note that the maximum level of any state is $\log_2 n$. For a nonnegative integer s, we define $\bar{s}$ to be the smallest integer $\bar{s} > s$ such that $\mathrm{level}(\bar{s}) \geq \mathrm{level}(s) + 1$.

The transitions in the level automaton are as follows: From state 0 we have a default transition to state 1 and a regular transition to state 1 with label $S[1]$. For every other state s, $1 \leq s \leq n$, we have the following transitions:

- A default transition to state $\bar{s}$. If no such state exist, the state s does not have a default transition.
- For each unique character α in $S[s+1, \min(\bar{s}, n)]$, there is a transition labeled α to the smallest state $s' > s$ such that $S[s'] = \alpha$.

An example of the level automaton constructed from the string $abacbabcabad$ and alphabet $\{a, b, c, d\}$ is given in Fig. 2. The dashed arrows denote default transitions and the vertical position of the states denotes their level.

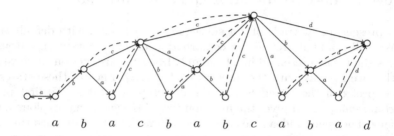

$$a \quad b \quad a \quad c \quad b \quad a \quad b \quad c \quad a \quad b \quad a \quad d$$

Fig. 2. The level automaton constructed from the string $abacbabcabad$

We first show that the level automaton is a SAD for S, i.e., the level automaton accepts a string iff the string is a subsequence of S. To do so suppose that P is a string of length m accepted by the level automaton and let $s_1, s_2, \ldots, s_m$ be the sequence of states visited with regular transitions on the path that accepts P. From the definition of the transition function, we know that if a transition with label α leads to state s', then $S[s'] = \alpha$. This means that $S[s_1]S[s_2]\ldots S[s_m]$ spells out a subsequence of S if the sequence $s_1, s_2, \ldots, s_m$ is strictly increasing. From the definition of the transitions, a state s only have transitions to states s' if $s' > s$. Hence, the sequence is strictly increasing.

For the other direction, we show that the level automaton simulates the SA. At each state s, trying to match character α, we find the smallest state $s' > s$ such that s' has an incoming transition with label α: By the construction, either s has an outgoing transition leading directly to s' *or* we follow default transitions until reaching the first state with a transition to s'. This means that the states visited with standard transitions in the level automaton are the same states that are visited in the SA. Since the SA accepts all subsequences of S this must also hold for the level automaton.

Analysis. The following shows that the number of outgoing transitions increase with a factor two when the level increase by one. For all $s > 0$, we have the following property of $\bar{s}$ and level(s):

$$\overline{s} - s = 2^{\text{level}(s)} \tag{1}$$

By definition, $2^{\text{level}(s)}$ divides s. This means that we can write s as $c \cdot 2^{\text{level}(s)}$, where c is a uneven positive integer. We know that c is uneven because $2^{\text{level}(s)}$ is the largest power of two that divides s. The next integer, larger than s, that $2^{\text{level}(s)}$ divides is $s' = s + 2^{\text{level}(s)}$. This means that $\overline{s} \geq s'$. We can rewrite s' as follows: $s' = s + 2^{\text{level}(s)} = c \cdot 2^{\text{level}(s)} + 2^{\text{level}(s)} = (c+1) \cdot 2^{\text{level}(s)}$. Since c is uneven we know that $c + 1$ is even so we can rewrite s' further: $s' = \frac{(c+1)}{2} \cdot 2^{\text{level}(s+1)}$. This shows that $2^{\text{level}(s+1)}$ divides s' which means that $s' = \overline{s}$ and we conclude that $\overline{s} - s = 2^{\text{level}(s)}$.

Since the maximal level of any state is $\log_2 n$ and the level increase every time we follow a default transition, the delay of the level automaton is $O(\log n)$.

At each level l we have $O(n/2^{l+1})$ states, since every 2^lth state is divided by 2^l, and 2^l is the largest divisor in every second of these cases. Since $\overline{s} - s = 2^{\text{level}(s)}$ each state at level l has at most $2^l + 1$ outgoing transitions. Therefore, each level contribute with size at most $n/2^{l+1} \cdot (2^l + 1) = O(n)$. Since we have at most $O(\log n)$ levels, the total size becomes $O(n \log n)$.

In summary, we have shown the following result.

Lemma 1. *Let S be a string of n characters. We can construct a subsequence automaton with default transitions of size $O(n \log n)$ and delay $O(\log n)$.*

3.2   Alphabet-Aware Level Automaton

We introduce the *Alphabet-aware level automaton*. When the level automaton reaches a state s where $\overline{s} - s \geq \sigma$, then s can have up to σ outgoing transitions without violating the space analysis above. The level automaton only has a transition for each unique character in $S[s + 1, \min(\overline{s}, n)]$. Hence, for all states s in the alphabet-aware level automaton where $\overline{s} - s \geq \sigma$, we let s have a transition for each symbol α in Σ, to the smallest state $s' > s$ such that $S[s'] = \alpha$. No matching path can take a default transition from a state with σ outgoing transitions. Hence, states with σ outgoing transitions do not need default transitions.

We change the level function to reflect this. For each state $1 \leq i \leq n$ we have that:

$$\text{level}(i) = \min(\lceil \log_2 \sigma \rceil, \max(\{x \mid i \bmod 2^x = 0\})) \tag{2}$$

The transitions in the alphabet-aware level automaton is as follows: From state 0 we have a default transition to state 1 and a regular transition to state 1 with label $S[1]$. For every other state s, $1 \leq s \leq n$, we have the following transitions:

- A default transition to state $\overline{s}$. If no such state exist, the state s does not have a default transition.
- If $\overline{s} - s < \sigma$ then for each unique character α in $S[s + 1, \min(\overline{s}, n)]$, there is a transition labeled α to the smallest state $s' > s$ such that $S[s'] = \alpha$.
- If $\overline{s} - s \geq \sigma$ then for each unique character α in $S[s+1, n]$, there is a transition labeled α to the smallest state $s' > s$ such that $S[s'] = \alpha$.

An example of the alphabet-aware level automaton constructed from the string *abacbabcabad* and alphabet $\{a, b, c, d\}$ is given in Fig. 3. The level automaton in Fig. 2 is constructed from the same string and the same alphabet. For comparison, state 4 in Fig. 3 now has outdegree σ and has transitions to the first succeeding occurrence of any unique character and state 8 has been constrained to level $\lceil \log_2 \sigma \rceil$.

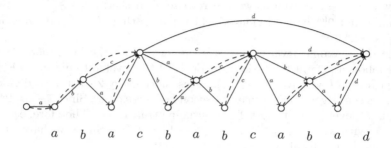

Fig. 3. The alphabet level automaton constructed from the string *abacbabcabad*

The alphabet-aware level automaton is a SAD by the same arguments that led to Lemma 1.

The delay is now bounded by $O(\log \sigma)$ since no state is assigned a level higher than $\lceil \log_2 \sigma \rceil$. The size of each level is still $O(n)$. Hence, the total size becomes $O(n \log \sigma)$

In summary, we have shown the following result.

Lemma 2. *Let S be a string of n characters. We can construct a SAD of S with size $O(n \log \sigma)$ and delay $O(\log \sigma)$.*

3.3   Full Trade-Off

We can generalize the construction above by introducing parameter k, $1 < k \leq \sigma$, which is the base of the exponential increase in outdegree of states on every path that only uses default transitions. Now, when we follow a default transition from s to $\bar{s}$, the number of outgoing transitions increase with a factor k instead of a factor 2. This gives a trade-off between size and delay in the SAD determined by k. Increasing k gives a shorter delay of the SAD but increases the size and vice versa.

Each state, except state 0, is still associated with a level, but we need to generalize the level function to account for the parameter k. For every k and i we have that:

$$\text{level}(i, k) = \min(\lceil \log_k \sigma \rceil, \max(\{x \mid i \bmod k^x = 0\})) \tag{3}$$

Now, the level function gives the largest power of k that divides i.

The transitions in the generalized alphabet-aware level automaton is as follows: From state 0 we have a default transition to state 1 and a regular transition to state 1 with label $S[1]$. For every other state s, $1 \leq s \leq n$, we have the following transitions:

- A default transition to state $\bar{s}$. If no such state exist, the state s does not have a default transition.
- If $\bar{s} - s < \sigma$ then for each unique character α in $S[s + 1, \min(\bar{s}, n)]$, there is a transition labeled α to the smallest state $s' > s$ such that $S[s'] = \alpha$.
- If $\bar{s} - s \geq \sigma$ then for each unique character α in $S[s+1, n]$, there is a transition labeled α to the smallest state $s' > s$ such that $S[s'] = \alpha$.

We can show that the generalized alphabet-aware level automaton is a SAD by the same arguments that led to Lemma 2.

Analysis. The delay is bounded by $O(\log_k \sigma)$ because no state is assigned a level higher than $\lceil \log_k \sigma \rceil$.

With the new definition of the level function we have that

$$\bar{s} - s \leq k^{\text{level}(s,k)+1} \tag{4}$$

for all $s > 0$. This expression bounds the number of outgoing transitions from state s.

At level l we have $O(n(k-1)/(k^{l+1}))$ states each with $O(k^{l+1})$ outgoing transitions such that each level has size $O(nk)$. The size of the automaton becomes $O(nk \log_k \sigma)$ because we have $O(\log_k \sigma)$ levels.

In summary, we have shown Theorem 1.

4   Subsequence Automata for Multiple Strings

Troníček et al. [3] generalizes the simple subsequence automaton to multiple strings: Given a set of strings $\mathcal{S} = \{S_1, S_2, \ldots . S_N\}$, automata are presented to accept a pattern P iff P is a subsequence of *every* string in $\mathcal{S}$ or accept P iff P is a subsequence of *some* string in $\mathcal{S}$. For $\mathcal{S} = \{S_1, S_2\}$ the size, of both automata, is $O(|S_1| \cdot |S_2| \cdot \sigma)$.

It is possible to generalize the alphabet-aware level automaton to multiple strings and for $\mathcal{S} = \{S_1, S_2\}$ we obtain the following result:

Theorem 2. *Let S_1 and S_2 be two strings with characters from an alphabet of size σ. We can construct subsequence automata with default transitions accepting either*

- *string P iff P is a subsequence of at least one of S_1 and S_2.*
- *or string P iff P is a subsequence of both S_1 and S_2.*

For both automata, the size is $O(|S_1| \cdot |S_2| \cdot \log \sigma)$ and the delay is $O(\log \sigma)$.

We present the details in the full version.

References

1. Baeza-Yates, R.A.: Searching subsequences. Theor. Comput. Sci. **78**(2), 363–376 (1991)
2. Troníček, Z., Shinohara, A.: The size of subsequence automaton. Theor. Comput. Sci. **341**(1), 379–384 (2005)
3. Crochemore, M., Melichar, B., Troníček, Z.: Directed acyclic subsequence graph: overview. J. Disc. Algorithms **1**(3–4), 255–280 (2003)
4. Crochemore, M., Troníček, Z.: Directed acyclic subsequence graph for multiple texts. Technical repport, Institut Gaspard-Monge, pp. 99–118. Citeseer (1999)
5. Crochemore, M., Tronicek, Z.: Directed acyclic subsequence graph for multiple texts. Technical Report IGM-99-13, Institut Gaspard-Monge (1999)
6. Hoshino, H., Shinohara, A., Takeda, M., Arikawa, S.: Online construction of subsequence automata for multiple texts. In: Proceedings of the 7th SPIRE, pp. 146–152 (2000)
7. Farhana, E., Ferdous, J., Moosa, T., Rahman, M.S.: Finite automata based algorithms for the generalized constrained longest common subsequence problems. In: Chavez, E., Lonardi, S. (eds.) SPIRE 2010. LNCS, vol. 6393, pp. 243–249. Springer, Heidelberg (2010)
8. Bannai, H., Inenaga, S., Shinohara, A., Takeda, M.: Inferring strings from graphs and arrays. In: Rovan, B., Vojtáš, P. (eds.) MFCS 2003. LNCS, vol. 2747, pp. 208–217. Springer, Heidelberg (2003)
9. Troníček, Z.: Operations on DASG. In: Proceedings of the 4th WIA, pp. 82–91 (1999)
10. Troníček, Z.: Searching subsequences. Department of Computer Science and Engineering, FEE CTU in Prague, Ph.D. thesis (2001)
11. Troníček, Z.: Common subsequence automaton. In: Champarnaud, J.-M., Maurel, D. (eds.) CIAA 2002. LNCS, vol. 2608, pp. 270–275. Springer, Heidelberg (2003)
12. Bille, P., Farach-Colton, M.: Fast and compact regular expression matching. Theoret. Comput. Sci. **409**, 486–496 (2008)
13. Knuth, D.E., Morris Jr., J.H., Pratt, V.R.: Fast pattern matching in strings. SIAM J. Comput. **6**(2), 323–350 (1977)
14. Aho, A.V., Corasick, M.J.: Efficient string matching: an aid to bibliographic search. Commun. ACM **18**(6), 333–340 (1975)
15. Kumar, S., Dharmapurikar, S., Yu, F., Crowley, P., Turner, J.: Algorithms to accelerate multiple regular expressions matching for deep packet inspection. In: Proceedings of the 12th SIGCOMM, pp. 339–350 (2006)
16. Hayes, Ch.L., Luo, Y.: DPICO: a high speed deep packet inspection engine using compact finite automata. In: Proceedings of the 3rd ANCS, pp. 195–203 (2007)

Run-Time Checking Multi-threaded Java Programs

Frank S. de Boer[1,3]($\boxtimes$) and Stijn de Gouw[1,2]

[1] CWI, Amsterdam, The Netherlands
[2] SDL, Amsterdam, The Netherlands
[3] Leiden University, Leiden, The Netherlands
f.s.de.boer@cwi.nl

Abstract. Assertion checking traditionally focused on state-based properties. In a multi-threaded environment, approaches based on shared-state require complex locking mechanisms to ensure that specifications are checked atomically (in the same state). In addition to this increased complexity, locks also negatively affect performance.

In this paper, we extend both the underlying theory and the practical implementation of SAGA, a run-time checker for single-threaded Java programs, to multi-threading, while avoiding locks.

1 Introduction

Runtime assertion checking is an important practical method for finding bugs. However, the scope of run-time assertion checking is restricted mainly to sequential programs. The main problem of run-time assertion checking of parallel shared-variable programs in general is because of interference. Take for example the very simple statement assert x==x;, where x is a field of an object. If after retrieving the value of the first occurrence of x *another* thread modifies x then the assertion may evaluate to false! To prevent this whenever an assertion is checked, the entire parallel execution has in principle to be "frozen". This thus requires control of the underlying execution platform and will in most cases give rise to severe loss in performance.

In [5] we enhance run-time assertion checking with attribute grammars [9] for describing properties of *histories*, e.g., sequences of method calls and returns. This supports strict programming to interfaces because it allows for interface specifications abstracting from the state as represented by the program variables.

The main contribution of this paper is an extension to multi-threaded Java programs which avoids in a natural manner interference problems. As an example of the expressivity and generality of our approach, we show how to detect at runtime deadlocks in multithreaded Java programs.

Related Work. To the best of our knowledge, our extension of run-time assertion checking to multi-threaded programs is the first solution which supports interface specs and does not require fine-grained control of the JVM execution platform to

This research is partly funded by the EU project FP7-610582 Envisage.

© Springer-Verlag Berlin Heidelberg 2016
R.M. Freivalds et al. (Eds.): SOFSEM 2016, LNCS 9587, pp. 217–228, 2016.
DOI: 10.1007/978-3-662-49192-8_18

prevent interference in assertion checking as in the purely state-based approaches of [1,7]. Other state-based approaches like [2,15] require complex atomicity specifications and locking mechanisms. These latter two approaches further require substantial extensions both of the specification language and corresponding tool support for run-time verification.

Besides run-time assertion checking, there are many other tools for run-time verification of multi-threaded programs. In [16] also an automated code instrumentation technique is introduced for the generation of counter-examples of a given property. These counter-examples are given in terms of partial orders of events which record particular modes of access to the state. This state-based approach consequently does not support a design by contract methodology. The run-time verification of multi-threaded programs as described in [10] affects the running system by blocking those threads which would violate the property. Other approaches focus on specific classes of properties, like data-races [3], deadlock [13] and restricted protocol-properties [6].

In [12] a process algebraic approach based on an extension of CSP is introduced which supports a strictly separate run-time verification of properties of data and protocols. In contrast, our approach builts on well-established parser technology and integrates the run-time verification of both data- and protocol-oriented properties of multi-threaded programs.

2   Extending the Framework

Our approach to multi-threading is based on the following perspectives:

Thread view: here we specify the behavior of each thread in isolation.
Object view: here we specify the behavior of objects individually.
Global view: here we specify global properties of a program.

All of the above views can be supported by a single formalism: attribute grammars extended with assertions, but the interpretation of the grammars differs between the various perspectives. We first discuss the required extensions to communication views and grammars. We then illustrate each of the above views with a running 'dining philosophers' example. The behavior of the philosophers is specified by a corresponding thread class. The interfaces of the classes defining the resources, i.e., the forks and the pasta are defined in Fig. 1.

```
interface Fork {
  void get();
  void release();
}

interface Pasta {
    void eat();
}
```

Fig. 1. Interfaces forks and pasta

2.1   Multi-threaded Events

A communication view (as defined in [5]) is a partial mapping which associates a name to each event, e.g., method calls and returns. In multi-threaded programs, due to scheduling and locking, there can be a delay between when a method is called, and when its body starts executing. For synchronized methods, a method call indicates that a lock was requested, whereas the start of the execution of a method body indicates that the lock was acquired successfully. To distinguish these two events, we introduce an 'exec' event, that indicates the start of execution of a method body (and thus, implies acquisition of the lock). See Fig. 4 for a communication view that uses an 'exec' event. Returns of synchronized methods indicate the release of the lock.

In general, our framework incorporates built-in attributes of an event which store the objects involved, e.g., the actual parameters. Here we introduce a new built-in attribute `Long threadId` which stores the identity of the thread in which the event occurred. Thread id's are used in both the object and the global view.

The final addition to our framework are `reset`-actions. Suppose a user desires to check the history only when a specific event occurs, and subsequently reset the history to start a new session. We support this by labeling an event with a `reset(b)` action, where b is a `boolean` expression in which both grammar attributes *of the previous session* (the attributes of the grammar start symbol) and the objects involved in the event can be used. Grammar attributes are prefixed by the keyword 'session'. If the condition b is true, the history is parsed and subsequently reset, with the caveat that attributes values of the previous session are retained. Figures 2 and 4 both depict views with `reset`-actions.

2.2   Grammars and Interference Freedom

The context-free grammar underlying an attribute grammar generates in our approach the valid histories, i.e., traces of events. Event names form the terminal symbols of the grammar, whereas the non-terminal symbols specify the structure of valid sequences of events. In our approach, a communication history is valid if and only if it and all its prefixes are generated by the grammar. The attributes are used in assertions to specify the *data-flow* of the valid histories.

State-based specifications for multi-threaded programs introduce race conditions. During evaluation of an assertion in one thread, another thread can change the state. This implies that different parts of the same assertion are evaluated in different states. For example, `assert o.x==o.x;` is not necessarily true anymore, if `o.x` is changed by another thread. To solve this problem, we must ensure exclusive access to all objects occurring in the assertion. This requires complex locking mechanisms and a fine-grained control over the underlying execution platform. Furthermore, such a complete 'lock-down' of the system can have a severe negative impact on performance. Our history-based approach avoids these problems, provided that a grammar is *well-formed* in the following sense.

Definition 1. *A grammar is well-formed if and only if no built-in attributes (of terminals) are dereferenced.*

This condition is easily checked, and is natural from a conceptual point of view: whenever an event occurs, *only* the actual parameters and return value are communicated between the caller and the callee. Dereferencing an attribute accesses the underlying state of the program, and would mean that the grammar does not only depend on the current history, but potentially on the entire heap! The above condition ensures that the grammar depends only on the history. Well-formed grammars allow an elegant solution to the interference problems mentioned above. See Sect. 4.2 for more details.

3   Multi-threaded Perspectives

Thread View. In the thread perspective, we specify the behavior of each thread in isolation by a corresponding communication view and grammar. We associate a history to each thread, and the grammar generates the set of valid histories of the thread. Semantically, such thread-local histories can be obtained from the global history by projection on the value of the `threadId` attribute[1].

The thread perspective is tailored for the specification of properties that each thread must obey, independently from that of any other threads. Consider for example a client-server scenario. To avoid blocking access to the Server while handling the requests of a Client, the server creates a new thread for each client after accepting an incoming connection, and each client must follow a protocol.

We illustrate the thread view using the running dining philosophers example. Figure 2 presents the communication view for the Phil thread class. It introduces the grammar terminals "start", "get", "release", and "eat" for the corresponding events. Only events from implementations of the `Fork` interface with `synchronized` versions of `get` and `release` are selected.

```
thread view Phil grammar Phil.g {
  return Phil.Phil(Fork left, Fork right) start reset(true),
    call synchronized void Fork.get() get,
    call synchronized void Fork.release() release
              reset(callee==session.right),
    call void Pasta.eat() eat
}
```

Fig. 2. Communication view philosophers

Note that we have included the constructor method of the class Phil in the communication view, which allows us to use its parameters in the attribute

---

[1] There is one subtle technicality: thread id's can be reused in Java. Hence, two events that occur in different threads can share the same thread id. This can be detected by monitoring the `void run()` method of a Thread class in the communication view. A call to `run` signals the creation of the thread, and a return indicates termination.

```
S ::= start
        (S.left=start.left; S.right=start.right;)
    |   gf1=get gf2=get eat rf1=release rf2=release
        (S.left=start.left; S.right=start.right;)
        { assert gf1.callee == S.left && gf2.callee == S.right; }
        { assert rf1.callee == S.left && rf2.callee == S.right; }
```

Fig. 3. Attribute grammar for philosopher session

grammar to specify the behavior of the thread instances of class Phil. This communication view marks each **start** event, and each **release** event for which its **callee** equals the attribute of the previous session, as a reset.

The grammar in Fig. 3 defines the behavior and the attributes of a session[2]. The grammar introduces the aliases **gf1**, **gf2**, **rf1** and **rf2** to distinguish different occurences of the same terminal.

Object View. In the object view of a Java program, we specify the interaction of a single object by means of a corresponding communication view and grammar. The grammar generates the set of all valid traces of events that the object may engage in. In a multi-threaded environment, several threads can be active (executing) in a single object. Intuitively, the local object histories can be obtained from the global history by projection on the values of the built-in attributes **caller** and **callee**.

The object view is convenient for specifying a resource that is shared between different threads. We illustrate this by a specification of the forks in the dining philosophers example. Figure 4 presents the communication view.

```
local view ForkView grammar Fork.g specifies Fork {
     exec synchronized void get() get,
   return synchronized void release() release reset(true)
}
```

Fig. 4. Communication view of a fork

As explained above, the keyword "exec" indicates events which are triggered by the start of the execution of the specified method. The grammar in Fig. 5 then defines the behavior of a session of a Fork object. Note that mutual exclusion here is simply expressed by the identity between the thread id's corresponding to the events which indicate the execution of the get and release methods.

---

[2] Terminals have built-in attributes, which refer to the objects involved. Non-terminals have user-defined attributes defining properties of sequences of terminals.

$$S ::= \text{get release } \{ \text{ assert get.threadid==release.threadid; } \}$$

Fig. 5. Attribute grammar for fork session

Global View. In the global view, we treat the Java program as a single entity that we wish to specify. Traditional approaches based on global invariants [4, 11] are supported in this perspective. The grammar generates the set of all valid *global* traces of the entire program. In particular, the user can specify the desired interleavings between events from different threads.

We illustrate the global view by means of a general method for dynamic detection of deadlocks. We focus on deadlocks that arise from synchronized methods. Given a multi-threaded program we first introduce the communication view in Fig. 6, which includes all events of the synchronized methods.

```
global view DeadlockMyProgram grammar deadlock.g {
    call synchronized T C.m() C_m,
  return synchronized T C.m() ret_C_m,
    exec synchronized T C.m() exec_C_m,
  ...
}
```

Fig. 6. Global communication view

Since unsynchronized methods cannot cause a deadlock, the communication view filters those out. A thread blocks if it calls a synchronized method on an object that is already locked by another thread. The general idea is to build a directed "wait-for" graph to capture such dependencies between threads. A deadlock corresponds to a cycle in the wait-for graph.

In more detail, the nodes of the graph are thread id's, and there is an edge from t_1 to t_2 if t_1 calls a method on some object that is locked by t_2. To build the graph we define three inherited attributes:

- An attribute `reqLock` of type `Map<Long, Object>` that maps a thread id (a `Long`) to the object for which it requested, but has not yet acquired the lock.
- An attribute `hasLock` of type `Map<Long, Map<Object, Integer> >`. Given a thread id and an object, this map returns the number of times the lock on that object has been acquired but not released by the thread[3]. If for a given thread id, the count becomes 0 for an object, the entry is removed.
- An attribute `g` storing a directed graph. Any graph library can be used as long as it has a method `void addEdge(Object node1, Object node2)` for adding an edge between `node1` and `node2` to the graph, and a (pure) method

---

[3] Due to reentrance, locks in Java can be acquired more than once by the same thread.

`boolean noCycle()` for cycle detection. For example, `DirectedGraph` from the `JGraphT` library suffices for our purposes.

The two maps and the graph are initialized to empty. Figure 7 shows how the attributes are updated whenever an event occurs. The graph is built in the last ϵ-production using the `reqLock` and `hasLock` maps. The assertion formalizes the no deadlock requirement.

$S ::= C\_m$
```
    (
      S.reqLock.put(C_m.threadId, C_m.callee);
    )
    S
|   ret_C_m
    (
      Map<Object, Integer> m = S.hasLock.get(ret_C_m.threadId);
      Integer cnt = m.get(ret_C_m.caller);
      if( cnt == 1)m.remove(ret_C_m.caller);
      else m.put(ret_C_m.caller, cnt-1);
    )
    S
|   exec_C_m
    (
      S.reqLock.remove(exec_C_m.threadId);
      Map<Object, Integer> m = S.hasLock.get(exec_C_m.threadId);
      int newCnt = 1;
      if(m == null){
        m = new HashMap<Object, Integer>;
        S.hasLock.put(exec_C_m.threadId, m);
      } else if(m.get(exec_C_m.callee)!= null)
        newCnt = m.get(exec_C_m.callee)+1;
      }
      m.put(exec_C_m.caller, newCnt);
    )
    S
|   ε
    (
      for(Long rl : S.reqLock.keySet())
        for(Long hl : S.hasLock.keySet())
          if(hl.get(rl).containsKey(reqLock.get(rl)))
            g.addEdge(rl, hl);
    )
    { assert g.noCycle(); }
```

Fig. 7. Attribute grammar specifying deadlock

3.1 Combining the Views

Above, we described three different perspectives on a multi-threaded Java program: the thread view, the object view and a global view. Thus, to properly support a separation of concerns, multiple attribute grammars can be present, each grammar focusing on a specific behavioral property of the program. We check all grammars independently. If each grammar is considered to be a formal language (generating the valid histories), this has the same effect as taking the intersection, or conjunction of the formal languages involved. Note that this also provides a simple way to specify the intersection of two context-free languages. Although the intersection of two context-free languages is in general not context-free, we can specify them separately by two different context-free grammars.

Combining the above views on the dining philosophers ensures that

- All philosophers use their (shared) resources correctly (thread view).
- The resources themselves behave properly (object view).
- No deadlock arises during execution (global view).

4   Tool Architecture

In this section we describe the tool architecture of SAGA (Fig. 8). SAGA integrates four different components: a state-based assertion checker, a parser generator, a monitor to intercept events and a general tool for meta-programming. How and when these components are used is explained below by means of a workflow. The tool architecture in a single-threaded setting was previously described in [5], and successfully applied to a large industrial case[4] with 150,000 lines of code and 44 classes, specified by 5 different attribute grammars of 28, 65, 115, 25 en 37 lines, respectively. It is important to note that this application thus involves storage, updates and parsing of different histories. The main difference with respect to multi-threaded programs resides in the additional event "exec" and the different views (i.e., thread, object and global view). Below we describe how the extensions at the implementation level needed for multi-threading can be easily incorporated without affecting the overall tool architecture, which has proven its use in practice, as shown above. Further, we discuss how these extensions avoid the interference problem as explained in the introduction.

4.1 Meta-Program

Suppose that during execution of a multi-threaded Java program, an event listed in a communication view occurs. The corresponding history should be updated to reflect the addition of the new event. Hence, the first question is: *how to represent the history?* A meta-program written in Rascal [8] generates for each event in the communication view a 'token class' with fields for storing: the *caller*

---

[4] This case study has successfully been carried out in the EU HATS project (http://www.hats-project.eu).

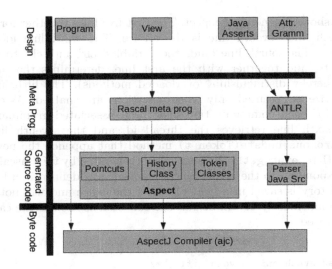

Fig. 8. SAGA tool architecture

and *callee*, the *thread identity*, the actual parameters and (for return events only) the *return value*. We can then represent a history as a Java List of instances of token classes.

4.2   Monitoring and Interference

The monitoring component updates the history whenever a relevant event occurs, and (possibly) triggers the parser.

AspectJ is tailored for monitoring. It can intercept method calls, executions and returns conveniently with pointcuts, and weave in user-defined code (advice) that is executed before or after the intercepted event. Each pointcut corresponds to an event listed in the communication view. The advice is the code that updates the history.

The code for the aspect is generated from the communication view automatically by the Rascal meta-program. Advice is woven into Java source code, byte code or at class load-time fully automatically by AspectJ. We store thread-local histories in a Map<Long, History>, which assigns to each thread id (of type Long) its history, and use the inter-type declarations of AspectJ to store the local history of an object in the object itself in a new field named h. This ensures that whenever the object goes out of scope, so does its history and consequently reduces memory usage. Furthermore, compared to storing a single global history, this method avoids the calculation (by projection upon the global history) of the local object and thread histories. Since in thread-local histories, the threadId is the same for every event, we reduce memory usage by avoiding storing it altogether. We store global histories inside a separate Aspect class.

Figure 9 shows a generated aspect. The first five lines together form a point-cut, the sixth and seventh line is the advice. The third line identifies the method name. The fourth line binds the variables 'clr' and 'cle' to the appropriate objects, and together with the first line, determines the full method signature (thereby distinguishing overloaded methods). The fifth line ensures that the advice is executed only when assertions are enabled. Assertions can be enabled for each communication view (and associated grammar) individually. The sixth line retrieves the thread id, and the seventh line calls a void synchronous update(Token t) method that appends the new event to the history. Here, exec_get is the token class (generated by the Rascal meta program) corresponding to the event. Since the event was defined in a local object view, the history is saved in a field cle.h of the callee and will not persist in the program indefinitely. It will be garbage collected as soon as the callee object itself is destroyed.

```
    /* exec synchronized void get(); */
before(Object clr, Fork cle):
 (execution( synchronized void *.get())
  && this(clr) && target(cle)
  && if(ForkViewAspect.class.desiredAssertionStatus() )) {
    long threadId = Thread.currentThread().getId();
    cle.h.update(new exec_get(clr, cle, threadId));
}
```

Fig. 9. Aspect for the event 'exec synchronized void get()' from Fig. 4

Our approach avoids interference, as a consequence of which concurrently running threads of the program need not be locked, because of the following three main characteristics:

- Whenever the history should be updated with a new event, we create a new instance of the appropriate token class to store the objects involved in the event. This new token class object is not changed (or even visible) in the program under test.
- Assertions in well-formed grammars do not refer to the actual state of the program: they are completely determined by the values of the built-in attributes stored in these newly created objects.
- Event updates that arise from different threads can cause an update to the same history variable: global histories and local object histories are shared between threads. We provide exclusive access to the history with update methods.

4.3   Parser Generator and Assertion Checker

After a history update, SAGA must decide whether it still satisfies the specification given by the grammar. A history can be seen as a sequence of tokens (in our setting: events). Since the grammar together with the assertions generate the set of all valid histories, checking whether a history satisfies the specification reduces to deciding whether the history can be parsed by a parser for the grammar, where moreover during parsing the assertions must evaluate to true. For events labelled with a reset action, if the associated condition is true, the parser is triggered and the history subsequently reset (see below for more information on parsing).

We use ANTLR [14] to create a parser for the given attribute grammar.

During parsing, ANTLR calls a state-based assertion checker to evaluate the assertions in the grammar. We used standard Java assertions in our grammars. This ensures compatibility with all Java compilers. The result of the parsing process is either a parse or assertion error, which indicates that the history violates the specification given by the attribute grammar, or a parse tree decorated with new attribute values.

5   Conclusion

The new version of SAGA can be obtained from https://github.com/cwi-swat/saga. Although we illustrated our framework using synchronized methods, general locks as provided in the package `java.util.concurrent.locks` can be handled just as easily by tracking the methods `lock`, `tryLock` and `unlock` in the communication view. In [5] we have already reported on a successful application of our tool to an industrial case study. This led to the integration of SAGA into the software lifecycle at SDL. Currently, in the context of the European ENVIS-AGE project, we are extending this case to the specification, run-time verification and *monitoring* of a distributed cloud application.

References

1. Aftandilian, E., Guyer, S.Z., Vechev, M.T., Yahav, E.: Asynchronous assertions. In: Proceedings of the 26th Annual ACM SIGPLAN Conference on Object-Oriented Programming, Systems, Languages, and Applications, OOPSLA 2011, Part of SPLASH 2011, Portland, OR, USA, pp. 275–288 (2011)
2. Araujo, W., Briand, L.C., Labiche, Y.: Enabling the runtime assertion checking of concurrent contracts for the java modeling language. In: Proceedings of the 33rd International Conference on Software Engineering, ICSE 2011, Waikiki, Honolulu, HI, USA, pp. 786–795 (2011)
3. Bodden, E., Havelund, K.: Racer: effective race detection using AspectJ. In: Proceedings of the ACM/SIGSOFT International Symposium on Software Testing and Analysis, ISSTA 2008, Seattle, WA, USA, pp. 155–166 (2008)
4. Cohen, E., Moskal, M., Schulte, W., Tobies, S.: Local verification of global invariants in concurrent programs. In: Touili, T., Cook, B., Jackson, P. (eds.) CAV 2010. LNCS, vol. 6174, pp. 480–494. Springer, Heidelberg (2010)
5. de Boer, F.S., de Gouw, S., Johnsen, E.B., Kohn, A., Wong, P.Y.H.: Run-time assertion checking of data- and protocol-oriented properties of java programs: an industrial case study. In: Chiba, S., Tanter, É., Bodden, E., Maoz, S., Kienzle, J. (eds.) Transactions on AOSD XI. LNCS, vol. 8400, pp. 1–26. Springer, Heidelberg (2014)

6. Hurlin, C.: Specifying and checking protocols of multithreaded classes. In: ACM Symposium on Applied Computing (SAC 2009), pp. 587–592. ACM Press (2009)
7. Kandziora, J., Huisman, M., Bockisch, Ch., Zaharieva-Stojanovski, M.: Run-time assertion checking of JML annotations in multithreaded applications with e-OpenJML. In: Proceedings of the 17th Workshop on Formal Techniques for Java-Like Programs, FTfJP 2015, Prague, Czech Republic, pp. 8:1–8:6. ACM, New York (2015)
8. Klint, P., van der Storm, T., Vinju, J.: Rascal: a domain specific language for source code analysis and manipulation. In: Walenstein, A., Schupp, S. (eds.) Proceedings of the IEEE International Working Conference on Source Code Analysis and Manipulation (SCAM 2009), pp. 168–177 (2009)
9. Knuth, D.E.: Semantics of context-free languages. Math. Syst. Theory 2(2), 127–145 (1968)
10. Luo, Q., Rosu, G.: EnforceMOP: a runtime property enforcement system for multithreaded programs. In: International Symposium on Software Testing and Analysis, ISSTA 2013, Lugano, Switzerland, pp. 156–166 (2013)
11. Mizuno, M.: A structured approach for developing concurrent programs in Java. Inf. Process. Lett. 69(5), 233–238 (1999)
12. Möller, M., Olderog, E.-R., Rasch, H., Wehrheim, H.: Integrating a formal method into a software engineering process with UML and Java. Formal Asp. Comput. 20(2), 161–204 (2008)
13. Nonaka, Y., Ushijima, K., Serizawa, H., Murata, S., Cheng, J.: A run-time deadlock detector for concurrent Java programs. In: 8th Asia-Pacific Software Engineering Conference (APSEC 2001), Macau, China, pp. 45–52 (2001)
14. Parr, T.: The Definitive ANTLR Reference. Pragmatic Bookshelf, Lewisville (2007)
15. Rodríguez, E., Dwyer, M.B., Flanagan, C., Hatcliff, J., Leavens, G.T., Robby: Extending JML for modular specification and verification of multi-threaded programs. In: Gao, X.-X. (ed.) ECOOP 2005. LNCS, vol. 3586, pp. 551–576. Springer, Heidelberg (2005)
16. Rosu, G., Sen, K.: An instrumentation technique for online analysis of multithreaded programs. Concurrency Comput. Pract. Experience 19(3), 311–325 (2007)

Online Graph Coloring with Advice and Randomized Adversary

(Extended Abstract)

Elisabet Burjons[1]([✉]), Juraj Hromkovič[2], Xavier Muñoz[1], and Walter Unger[3]

[1] Matemàtica Aplicada IV, Universitat Politècnica de Catalunya, Barcelona, Spain
elisabet.burjons@gmail.com, xml@ma4.upc.edu
[2] Department of Computer Science, ETH Zürich, Zürich, Switzerland
juraj.hromkovic@inf.ethz.ch
[3] Lehrstuhl für Informatik I, RWTH Aachen University, Aachen, Germany
quax@cs.rwth-aachen.de

Abstract. We generalize the model of online computation with three players (algorithm, adversary and an oracle called advisor) by strengthening the power of the adversary by randomization. In our generalized model, the advisor knows everything about the adversary except the random bits the adversary may use.

We examine the expected competitive ratio of online algorithms within this model in order to measure the hardness of online problems in a new way. We start our investigation by proving upper and lower bounds on the competitive ratio for the online graph coloring problem.

Keywords: Online computation · Information · Randomization · Graph coloring

1 Introduction

Advice complexity was introduced in [8] and revised in [5,11] in order to measure the information content of online problems. The question is how many bits about the future are necessary and sufficient for an online algorithm in order to be able to solve a given problem in an optimal way or to guarantee a concrete competitive ratio. Studying online problems from this point of view by getting tradeoffs between the solution quality and the size of advice provided, one obtained a new instrument for measuring the hardness of online problems. Other important conceptual contributions are the development of a powerful method for proving lower bounds on the achievable expected competitive ratios of randomized online algorithms and new insights on the potential power of information.

The investigations in a series of papers (see, e.g.,[1–4,6,7,9,10,13–15]) show very different behaviors of the tradeoff between the solution quality measured

This work has been partially supported by the SNF grant 200021-146372 and the Spanish government under project MTM2011-28800-C02-01.

© Springer-Verlag Berlin Heidelberg 2016
R.M. Freivalds et al. (Eds.): SOFSEM 2016, LNCS 9587, pp. 229–240, 2016.
DOI: 10.1007/978-3-662-49192-8_19

by the competitive ratio and the amount of information of the unknown future. Sometimes this behavior looks really surprising. The typical patterns occurring for different problems are the following ones.

1. The achievable competitive ratio improves continuously with the number of advice bits provided. Sometimes very quickly, sometimes very slowly.
2. The number of advice bits provided does not help at all until a special threshold value is reached. After crossing this threshold, the quality of solutions may jump to a significantly better competitive ratio.
3. The pattern can be a mix of 1. and 2. depending on the interval of the number of advice bits offered.

Additionally to understanding the very different roles and the power of additional information in different situations, we have learned a lot about the possible relations between advice bits and random bits. There are situations in which random bits are as powerful as advice bits and situations in which advice bits are incomparably more powerful than random bits. A byproduct of this research is the fact that, for most "reasonable" online problems, if some competitive ratio is not achievable by a logarithmic number of advice bits, then it is not achievable by any randomized online algorithm with an unrestricted number of random bits. The resulting technique for proving lower bounds on achievable competitive ratios of randomized online algorithms has been quite successful in investigating the limits of randomized online algorithms for concrete online problems.

In our model there are three players, namely an online algorithm, the adversary and the advisor (oracle). As usual in the classical model, the adversary knows everything about a given online algorithm and constructs the hardest problem instance for it with respect to the achievable competitive ratio. The advisor in the extended game is very powerful, it knows everything about the future, i.e., it knows the whole input instance that will be presented request by request. The advisor writes its advice on the oracle tape and the number of bits read by the online algorithm from the oracle tape is the advice complexity on this problem instance. The advice complexity of an online algorithm is in general a function of the input size defined in the worst-case manner as the maximum over all inputs of the same size.

In this paper, we introduce a more general model in which we allow the adversary to use random bits. In spite of the fact that randomization can be very helpful for the design of online algorithms in the classical model of online computations, in the classical model random bits are not helpful for the adversary at all. Since the competitive ratio is defined in a worst-case manner, it is sufficient to produce deterministically one hard input for each online algorithm. In our new model of advice complexity, however, randomization can increase the power of the adversary. If the adversary constructs a set of problem instances from which it can choose one randomly, and the advisor does not know the random bits for this choice in advance, then the advisor can only tell the set of possible problem instances or a probability distribution over the set of problem instances if it is not uniform, but not the exact problem instance that will be presented to the algorithm. In this scenario, we measure the expected competitive ratio over the hardest set of problem instances. A new parameter in the game is the number of random bits of the adversary with respect to the achievable competitive ratio if the advisor is allowed to give an unbounded number of advice bits. A more advanced study may investigate the tradeoff of the number of advice bits, the number of random bits of the adversary, and the achievable competitive ratio.

There is also another interesting view on the scenario. In the classical model without any advisor, the adversary is not allowed to construct one of the hardest problem instances for a given online algorithm. The power of the adversary will be reduced by asking him, for a given positive integer s, to generate a set of 2^s problem instances S. The competitive ratio of a S is the minimum of expected ratios over all online algorithms solving S. Then, the competitive ratio of the problem with respect to s is the maximum of the expected competitive ratios over all sets. This measure for the quality of online algorithms is reasonable, because in the classical model an online algorithm is bad if there exists one hard problem instance for it. Here, investigating the achievable competitive ratio with respect to s (or the size of the set of problem instances) can provide more insight into the hardness of concrete online problems. For instance, if the competitive ratio improves essentially with growing s, then the online problem can look easier than at the first glance in the classical worst-case model.

A formal definition of this scenario can be given as follows. Let, for an $s \in \mathbb{N}$, S be a set of 2^s instances of a considered online problem $\mathcal{P}$. Let $\mathcal{A}$ be an online algorithm solving $\mathcal{P}$. The competitive ratio of $\mathcal{A}$ on S, $\mathrm{comp}_\mathcal{A}(S)$, is the expected competitive ratio of $\mathcal{A}$ on all instances in S with respect to the uniform probability distribution. For any S of 2^s instances we define

$$\mathrm{comp}(S) = \min\{\mathrm{comp}_\mathcal{B}(S) \mid \mathcal{B} \text{ solves the instances in } S\}$$

as the competitive ratio of the "best" online algorithm for S. Finally,

$$\mathrm{comp}_S(\mathcal{P}) = \max\{\mathrm{comp}(S) \mid S \text{ consists of } 2^s \text{ instances in } S\}$$

is the *expected competitive ratio for the problem $\mathcal{P}$ for an adversary with s random bits*. For an online algorithm $\mathcal{A}$ solving $\mathcal{P}$ and any $s \in \mathbb{N}$, we define the *expected competitive ratio of $\mathcal{A}$ with respect to s* as

$$\mathrm{comp}_\mathcal{A}(s) = \max\{\mathrm{comp}_\mathcal{A}(S) \mid S \text{ consists of } 2^s \text{ instances of } \mathcal{P}\}$$

If the corresponding maxima do not exist, one can still prove lower and upper bounds on $\mathrm{comp}_S(\mathcal{P})$ and $\mathrm{comp}_\mathcal{A}(s)$.

In this paper, we apply this concept to the classical online coloring problem, where vertices are presented one after the other with edges leading to the previously revealed vertices only. In the classical model, the competitive ratio is unbounded, i.e., it is a function growing with the number of vertices n. This holds even for trees where this function is in $O(\log_2 n)$ [12]. In contrast to this, we show that, within this new model, for all k-colorable graphs for $k \geq 2$, the expected competitive ratio for sets of 2^s problem instances is at most $(s+2)/2$.

Note that here the sets of problem instances are known to the online algorithm that can choose the appropriate working strategy with respect to the given set of 2^s inputs. If this set is unknown, we have the same unbounded competitive ratio as for the classical online problem since, for each online algorithm, one can take the hardest problem instance and add a few dummy vertices in order to get 2^s problem instances.

For this upper bound of $(s + 2)/2$, we will reach a tight lower bound when restricting to online algorithms using new colors only if necessary.

2   Upper Bound on the Competitive Ratio of the Online Coloring Problem

The competitive ratio of the algorithms in this setting is linear with respect to the number of random bits available to the adversary. A practical way of computing competitive ratios in this setting is to look for the expected number of colors used by the algorithm in order to satisfy a problem instance. Knowing the chromatic number of the problem instance and the expected number of colors used, we obtain the competitive ratio by computing their ratio.

Theorem 1. *Given an adversary that uses s random bits to generate inputs such that all of its 2^s inputs are k-colorable, there exists an online algorithm with unbounded advice that colors the input using $(s + 2)k/2$ colors in expectation.*

Proof. Given the set S of 2^s possible problem instances (defined by their request sequence of vertices) we can construct a rooted tree T_S where a node divides itself into two or more child nodes whenever a next request in the sequence enables to distinguish between two or more groups of different problem instances. The root represents the whole set of 2^s problem instances. For example, the first node never gives any information about a concrete problem instance because it is the same for all 2^s instances. However, the second node may or may not be adjacent to the first node presented. So, looking at the first two nodes of all the request sequences, two groups of problem instances can be derived from that piece of information.

Once constructed, the tree of the set of problem instances (instance tree) will have at most 2^s leaves. The algorithm colors the request sequence given as follows.

First of all, choose one leaf ℓ in the instance tree T_S according to the following criterion. For each node v in the path from the leaf ℓ to the root of the tree, the subtree T_v rooted in the node v has a greater or equal number of leaves than any of its sibling subtrees T_w. Observe that, in a balanced tree, any leaf chosen is good according to the criterion. Thus, there may be more than one possible choice for the leaf. This leaf corresponds to a possible problem instance I_ℓ.

Start coloring the incoming vertices (requests) as if the graph given as a problem instance is the instance I_ℓ corresponding to the selected leaf ℓ. This means that the online coloring strategy used is optimal for I_ℓ.

As long as the actual problem instance I coincides with the problem instance I_ℓ, we use this strategy following the path from the root to ℓ. If a request R comes up that does not coincide with the chosen path, this means that the unknown

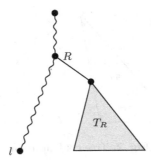

Fig. 1. Subtree T_R

actual problem instance does not correspond to I_ℓ. In this case the problem instance is one that corresponds to a leaf in a subtree T_R of problem instances that coincide with I up to the request R (see Fig. 1). A leaf $\hat\ell$ in T_R is selected according to the same criterion as l was chosen in T_S. Now, k new colors are used to color the new incoming requests according to an optimal coloring of the graph corresponding to $I_{\hat\ell}$. In this context, we speak about a switch from ℓ (I_ℓ) to $\hat\ell$ ($I_{\hat\ell}$) during the execution of our online algorithm. Our online algorithm continues recursively in T_R in the same way as in T_S and makes further switches if necessary.

Coloring in this way, $(t+1)k$ colors are used at most to color the graph where t is the number of switches of leaves until the correct leaf is reached. We have to prove that using this method not more than $s/2$ switches are performed on average. In fact, we will prove in the following lemma that all together not more than $\frac{s2^s}{2}$ switches are required in all 2^s online computations over all 2^s problem sequences considered.

Knowing that, in order to color all 2^s graphs, not more than $s2^s/2$ switches are required, we can easily conclude that on average no more than $s/2$ switches are made, so in expectation no more than $\frac{sk}{2} + k = \frac{(s+2)k}{2}$ colors are used. □

Lemma 1. *Let S be any set of L problem instances, and let T_S be the corresponding tree with L leaves. Let $\#(T_S)$ be the sum of the number of switches over all L online computations on the L input instances performed by the online algorithm described above. Then, $\#(T_S) \le \frac{L \log L}{2}$ for $L \ge 2$.*

Proof. By induction on the number of leaves of the instance tree L. For $L = 2$ the only possible rooted tree is the root v and two leaves v_1, v_2. Either choice that is possible for ℓ is equally valid. Therefore, choosing v_1, there will be no switches for the first problem instance and one switch for the second, making a total number of one switch as expected.

Suppose the lemma is valid for any tree with less than L leaves. For a tree with L leaves, the root will have two or more child vertices, pick the one which has the smallest subtree and name it T_2 and name the rest of the tree with

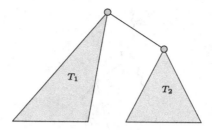

Fig. 2. Tree configuration

the original root T_1 (see Fig. 2). The selected leaf will always be in T_1 (as T_2 is the smallest subtree with a child root of T_S). The total number of switches is $\#(T_S) = \#(T_1) + \ell_2 + \#(T_2)$. This corresponds to the switches inside T_1, the switches inside T_2 and the ℓ_2 first switches in the online computations with problem instances in T_2. Using the induction hypothesis, we have

$$
\begin{aligned}
\#(T_S) &= \#(T_1) + \ell_2 + \#(T_2) \\
&\leq \frac{\ell_1}{2} \log \ell_1 + \frac{\ell_2}{2} + \frac{\ell_2}{2} \log \ell_2 + \frac{\ell_2}{2} \\
&= \frac{\ell_1}{2} \left(\log \ell_1 + \frac{\ell_2}{\ell_1} \right) + \frac{\ell_2}{2} \left(\log \ell_2 + 1 \right) \\
&= \frac{\ell_1}{2} \log \left(2^{\frac{\ell_2}{\ell_1}} \ell_1 \right) + \frac{\ell_2}{2} \log(2\ell_2) .
\end{aligned}
$$

We have rewritten the terms in order to have them as a sum of two logarithms. Now we use the facts that $\ell_2 \leq \ell_1$ and $2\ell_2 \leq L$, and the inequality $2^x \leq 1 + x$ for $0 \leq x \leq 1$ for $x = \frac{\ell_2}{\ell_1} \leq 1$ in the first term and we conclude

$$
\begin{aligned}
\#(T_S) &\leq \frac{\ell_1}{2} \log \left(2^{\frac{\ell_2}{\ell_1}} \ell_1 \right) + \frac{\ell_2}{2} \log(2\ell_2) \\
&\leq \frac{\ell_1}{2} \log \left(\left(1 + \frac{\ell_2}{\ell_1} \right) \ell_1 \right) + \frac{\ell_2}{2} \log L \\
&= \frac{\ell_1}{2} \log L + \frac{\ell_2}{2} \log L \\
&= \frac{L}{2} \log L.
\end{aligned}
$$

Observe that, if all of the child subtrees of the root are equal in number of leaves, we can still choose the selected leaf amongst T_1 or in the other way around, select a subtree T_2 where the chosen leaf is not included.   □

Following Lemma 1 the expected number of switches is at most

$$
\frac{\frac{L \log L}{2}}{L} = \frac{\log L}{2}.
$$

For $L = 2^s$, the expected number of switches is at most $s/2$, and so the expected number of colors used over all 2^s input instances is $k + \frac{sk}{2}$. If the optimal coloring of the instances uses k colors, then the expected competitive ratio is $1 + \frac{s}{2}$.

3   Lower Bound

We were not able to prove matching lower bounds to the upper bounds proved above. So we present lower bounds only for the subclass of so-called *minimalistic online algorithms* having the following property: They never use a new color if the presented vertex can be colored with one of the colors used up to now.

3.1   Idea of the Proof

We start by proving the lower bound $(s+2)/2$ on the expected competitive ratio for bipartite graphs, i.e., for 2-colorable graphs, if the adversary has s random bits. The idea is to construct a concrete graph $G_s = (V_s, E_s)$ for each integer $s \geq 1$ by a recursion with respect to s. Then, for each s, the 2^s problem instances of the adversary are generated by presenting G_s vertex by vertex in 2^s different orders. Hence, at the very end the graph G_s is colored in all 2^s cases and the difficulty is in recognizing which one of the vertices is the currently presented vertex.

To get some first intuition about our proof strategy consider the graph G_1 in Fig. 3.

At the beginning, the four vertices labeled by strings of length 2, namely 00, 01, 10, and 11, are presented. An online algorithm sees only two pairs of vertices, but does not know their labels. The algorithm has to use two colors c_1 and c_2 in order to color them. If it uses the same color c_1 (or c_2) for 00 and 10 and the color c_2 (or c_1) for 01 and 11, then after presenting vertices 0 and 1 the online algorithm can use c_2 for 0 and c_1 for 1 and G_1 is colored in an optimal way. If the online algorithm chooses the same color c_1 (or c_2) for 00 and 11 and the other color for 01 and 10, then one needs two more colors to color 0 and 1 because c_1 and c_2 are forbidden for both, and 0 and 1 are adjacent. The corresponding problem instances are the following sequences of vertices

$$00, 01, 10, 11, 0, 1 \quad \text{and} \quad 00, 01, 11, 10, 0, 1.$$

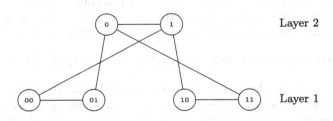

Fig. 3. Graph G_1

After getting the first four vertices, no online algorithm can distinguish between these two problem instances and so both will be colored in the same way with respect to the order of the vertices. In this way, G_1 will be colored optimally with 2 colors for one instance and suboptimally with 4 colors for the other instance. The expected number of colors is 3 and the expected competitive ratio is 1.5.

One immediately observes that one random bit is sufficient to randomly choose one of these two problem instances. Note, that we denoted the vertices $00, 01, 10,$ and 11 in such a way that, for an optimal coloring, the vertices with the same last bit have to be colored by the same color.

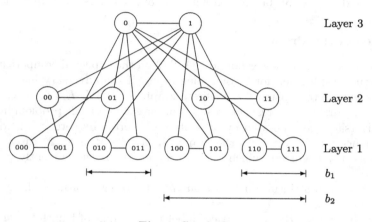

Fig. 4. Graph G_2

Now, we need to exploit this idea for an arbitrary number s of random bits. To get G_2, one adds a new layer of $2^3 = 8$ vertices labeled by strings of length 3 as depicted in Fig. 4. The edges connecting these new vertices of the third layer with the vertices in the first layer and in the second layer are chosen in such a way that G_2 can be colored optimally only if all vertices ending with the bit 0 have the same color. The 4 instances are presented in three phases: First the vertices labeled by strings of length 3, then the vertices labeled by strings of length 2, and finally the vertices 0 and 1 are presented. The 4 instances are not distinguishable when presenting the 4 pairs of vertices of the first layer. If they are not colored in an optimal way, then the first two colors cannot be used either for coloring any of the vertices in the third layer and not for coloring at least one pair of vertices in the second layer. The basic problem instance is

$$000, 001, 010, 011, 100, 101, 110, 101, 111, 00, 01, 10, 11, 0, 1.$$

One random bit b_1 can be used to exchange the order of $010 \leftrightarrow 011$ and $110 \leftrightarrow 111$. The second random bit b_2 can be used to exchange the order of $10 \leftrightarrow 11$ and correspondingly resume or reorder the four vertices $100, 101, 110, 111$.

3.2 Construction of G_s and the Corresponding 2^s Problem Instances

In general, one can define G_s recursively from G_{s-1} as follows. Let G^0_{s-1} (G^1_{s-1}) be the same graph as G_{s-1} except for the labeling of the vertices. In G^0_{s-1} (G^1_{s-1}) each label is prolonged by one additional first bit 0 (1), i.e., a vertex labeled by $x_1 x_2 \ldots x_{s-1}$ is now labeled by $0 x_1 x_2 \ldots x_{s-1}$. First, we take G^0_{s-1} and G^1_{s-1} as two components. Then a pair of connected vertices 0 and 1 is added and the vertex 0 is connected to all vertices in G^0_{s-1} and G^1_{s-1} where the last bit of the label is 1 and the vertex 1 is connected to all vertices where the last bit is 0. If, for $i = 0, 1$, one denotes by $G^i_{s-1}(0)$ and $G^i_{s-1}(1)$ the subgraphs of G^i_{s-1} induced by the vertices whose labeling ends with 0 and 1, respectively, then one can realize our recursive construction as depicted in Fig. 5.

Explicitly, $G_s = (V_s, E_s)$ can be described as follows.

$$V_s = \{x_0 x_1 \ldots x_m \mid x_i \in \{0,1\} \text{ for } i \in \{1, \ldots, m\}, 1 \le m \le s\}$$
$$E_s = \{\{x_0 x_1 \ldots x_r, y_0 y_1 \ldots y_t\} \mid \text{if } (r < t \text{ and } x_r = \bar{y}_t) \text{ or } (r = t \text{ and}$$
$$x_0 x_1 \ldots x_{r-1} = y_0 y_1 \ldots y_{t-1} \text{ and } x_r = \bar{y}_t)\}$$

We say that the vertex $x_0 x_1 \ldots x_m$ is in the $(s - m + 1)$-th layer of G_s.

Now, we have to describe the 2^s problem instances corresponding to G_s. We consider 2^s bit strings $b_1 b_2 \ldots b_s$ for $b_i \in \{0,1\}$, for $i = 1, \ldots, s$ and assign one order of vertices to each string. For the string consisting of zeros only, one can consider the order of vertices starting with the lowest labels of length $s + 1$ and finishing with labels 0 and 1. Inside of the same length, the vertices are ordered lexicographically (see Fig. 4 for G_2 when reading the labels from left to right in each layer). The last bit b_s is responsible for the order of 10 and 11. If $b_s = 1$, then one exchanges the order of 10 and 11 to 11, 10. If $b_s = 1$, it immediately means that in the layer $s - 2$ the order 100, 101, 110, 111 has to be changed to 110, 111, 100, 101 to still have the same structure of G_s (see Fig. 5). If $b_1 = 1$

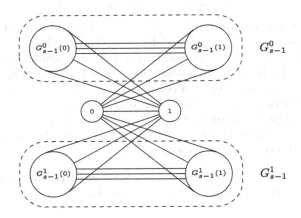

Fig. 5. Explicit construction of the graph G_s

and $b_2 = 0$, then this is already the corresponding instance. If $b_1 = b_2 = 1$, then one has still to exchange the order $010, 011$ for $011, 010$ and the order $100, 101$ for $101, 100$ in order to get the corresponding problem instance. The last case is that $b_1 = 0$ and $b_2 = 1$. In this case, the layers 2 and 3 remain unchanged and the order $110, 111$ is exchanged for $111, 110$.

If one wants to describe these changes for each G_s in general, it is sufficient to say, for each binary string $b_1 \ldots b_s$, which position in the same layer is taken by a vertex labeled by $x_0 x_1 \ldots x_m$ for $0 \leq m \leq s$. This position can be described by the function $Z_{b_1 b_2 \ldots b_s}$:

$$Z_{b_1 b_2 \ldots b_s}(x_0 x_1 \ldots x_m) = z_0 z_1 \ldots z_m \quad \text{with } z_i \equiv x_i + \sum_{j=0}^{i-1} x_j b_{s-j} \mod 2$$

for all $x_0 x_1 \ldots x_m$ and $0 \leq m \leq s$.

Let $\text{Set}(G_s)$ be the set of 2^s problem instances specified by the 2^s binary strings $b_1 \ldots b_s$.

3.3 Lower Bounds

The following fact can be proved easily by induction on the number of layers. Let $G'_m(a)$ be the union of graphs $G^1_{m-1}(a)$ and $G^0_{m-1}(a)$ (see Fig. 5).

Observation 1. *Let $1 \leq m \leq s$ and let G_m be any subgraph of G_s induced by presenting the vertices of the first $m + 1$ layers of G_s. Let A be a coloring of G_s obtained by an arbitrary minimalistic online algorithm, and let $\text{Col}_A(H)$ denote the set of the colors used to color a subgraph H of G_m. Then*

$$\text{Col}_A(G'_m(0)) = \text{Col}_A(G'_m(1))$$

or

$$|\text{Col}_A(G'_m(0)) - \text{Col}_A(G'_m(1))| = 1.$$

Note that the situation $\text{Col}_A(G_m(0)) = \text{Col}_A(G_m(1))$ forces to take two new colors when coloring the vertices 0 and 1 in G_m. The graphs G_s are constructed in such a way that until the very end it is not clear which vertices have labels finishing with 0 (1). If one takes a wrong guess in some subgraph, then one uses the colors from $\text{Col}_A(G_{m-1}(a))$ to color vertices in $G^1_{m-1}(a)$ in one subgraph and to color vertices in $G^1_{m-1}(\bar{a})$ in the second subgraph.

We call an online algorithm on $\text{Set}(G_s)$ *symmetric* if it uses the same strategy to color different copies (G^0_m or G^1_m) of the same subgraph with respect to the order in which the vertices are presented.

Lemma 2. *For each $s \in \mathbb{N}$, any minimalistic symmetric online algorithm working on $\text{Set}(G_s)$ uses at least $s + 2$ colors on average.*

Proof (Sketch). Let us prove this fact by induction. Let $\mathcal{A}$ be a minimalistic online algorithm for coloring bipartite graphs.

Let $s = 0$. To color $G_0 = K_2$ one needs 2 colors for the single input instance in $\text{Set}(G_0)$, and so we are done.

In general, consider the two subgraphs G_{s-1}^0 and G_{s-1}^1. Following the induction hypothesis, each one of them uses at least $(s-1) + 2 = s + 1$ colors on average over its 2^{s-1} problem instances when colored by $\mathcal{A}$. We have to prove that $\mathcal{A}$ uses $s + 2$ colors on average over $\text{Set}(G_s)$ when coloring G_s.

When the vertices of the $(s-1)$-th layer labeled by $00, 01, 10, 11$ are presented, $\mathcal{A}$ cannot distinguish between 10 and 11. If $\mathcal{A}$ makes the right decision, no new color is required for 0 and 1. If $\mathcal{A}$ makes a wrong decision, one can show that $\text{Col}_{\mathcal{A}}(G_s'(0)) = \text{Col}_{\mathcal{A}}(G_s'(1))$, and so none of the colors used for G_{s-1} can be used to color 0 and 1. Thus, two new colors are necessary.

This means that for 2^{s-1} problem instances $\mathcal{A}$ does not need a new color and for 2^{s-1} problem instances $\mathcal{A}$ needs two new colors. On average $\mathcal{A}$ needs at least one new color which concludes the proof. □

We are not completely happy with forcing the online algorithms to be symmetric because this looks like a strong restriction. We can remove this requirement by using the full power of the adversary. For a given minimalistic online algorithm $\mathcal{A}$, the adversary renames the vertices in G_s (i.e., reorders the sequence of vertices) in such a way that $\mathcal{A}$ behaves on this concrete problem instance $\alpha_{\mathcal{A}}$ as if it were a symmetric algorithm. Then this instance $\alpha_{\mathcal{A}}$ is used to generate the set of 2^s instances $\text{Set}(G_s(\alpha_{\mathcal{A}}))$ in the same way as $\text{Set}(G_s)$ was constructed with respect to 2^s binary strings. This finally offers the following result.

Theorem 2. *For each $s \in \mathbb{N}$ and each minimalistic online algorithm $\mathcal{A}$ there exists a set of instances, namely $\text{Set}(G_s(\alpha_{\mathcal{A}}))$, such that the expected competitive ratio of $\mathcal{A}$ is at least $(s+2)/2$.*

Making more copies of G_s and connecting them by taking edges from each vertex of one copy to all vertices of all other copies one gets a structure that is a starting point for proving the following theorem.

Theorem 3. *For any $k, s \in \mathbb{N}$, and any minimalistic online coloring algorithm $\mathcal{A}$, there exists an adversary using s random bits to construct 2^s problem instances that are k-colorable such that the competitive ratio of $\mathcal{A}$ is at least $(s+2)/2$.*

References

1. Barhum, K., Böckenhauer, H.-J., Forišek, M., Gebauer, H., Hromkovič, J., Krug, S., Smula, J., Steffen, B.: On the power of advice and randomization for the disjoint path allocation problem. In: Geffert, V., Preneel, B., Rovan, B., Štuller, J., Tjoa, A.M. (eds.) SOFSEM 2014. LNCS, vol. 8327, pp. 89–101. Springer, Heidelberg (2014)
2. Bianchi, M.P., Böckenhauer, H.-J., Hromkovič, J., Keller, L.: Online coloring of bipartite graphs with and without advice. Algorithmica **70**(1), 92–111 (2014)

3. Böckenhauer, H.-J., Hromkovič, J., Komm, D., Krug, S., Smula, J., Sprock, A.: The string guessing problem as a method to prove lower bounds on the advice complexity. Theor. Comput. Sci. **554**, 95–108 (2014)
4. Böckenhauer, H.-J., Komm, D., Královič, R., Královič, R.: On the advice complexity of the k-server problem. In: Aceto, L., Henzinger, M., Sgall, J. (eds.) ICALP 2011, Part I. LNCS, vol. 6755, pp. 207–218. Springer, Heidelberg (2011)
5. Böckenhauer, H.-J., Komm, D., Královič, R., Královič, R., Mömke, T.: On the advice complexity of online problems. In: Dong, Y., Du, D.-Z., Ibarra, O. (eds.) ISAAC 2009. LNCS, vol. 5878, pp. 331–340. Springer, Heidelberg (2009)
6. Böckenhauer, H.-J., Komm, D., Královič, R., Rossmanith, P.: The online knapsack problem: advice and randomization. Theor. Comput. Sci. **527**, 61–72 (2014)
7. Dobrev, S., Královič, R., Markou, E.: Online graph exploration with advice. In: Even, G., Halldórsson, M.M. (eds.) SIROCCO 2012. LNCS, vol. 7355, pp. 267–278. Springer, Heidelberg (2012)
8. Dobrev, S., Královič, R., Pardubská, D.: Measuring the problem-relevant information in input. RAIRO Theor. Inform. Appl. **43**(3), 585–613 (2009)
9. Forišek, M., Keller, L., Steinová, M.: Advice complexity of online coloring for paths. In: Dediu, A.-H., Martín-Vide, C. (eds.) LATA 2012. LNCS, vol. 7183, pp. 228–239. Springer, Heidelberg (2012)
10. Gebauer, H., Komm, D., Královič, R., Královič, R., Smula, J.: Disjoint path allocation with sublinear advice. In: Xu, D., Du, D., Du, D. (eds.) COCOON 2015. LNCS, vol. 9198, pp. 417–429. Springer, Heidelberg (2015)
11. Hromkovič, J., Královič, R., Královič, R.: Information complexity of online problems. In: Hliněný, P., Kučera, A. (eds.) MFCS 2010. LNCS, vol. 6281, pp. 24–36. Springer, Heidelberg (2010)
12. Keller, L.: Complexity of optimization problems: advice and approximation. Ph.D. thesis, ETH Zürich (2014)
13. Komm, D., Královič, R.: Advice complexity and barely random algorithms. In: Černá, I., Gyimóthy, T., Hromkovič, J., Jefferey, K., Královič, R., Vukolić, M., Wolf, S. (eds.) SOFSEM 2011. LNCS, vol. 6543, pp. 332–343. Springer, Heidelberg (2011)
14. Komm, D., Královič, R., Mömke, T.: On the advice complexity of the set cover problem. In: Hirsch, E.A., Karhumäki, J., Lepistö, A., Prilutskii, M. (eds.) CSR 2012. LNCS, vol. 7353, pp. 241–252. Springer, Heidelberg (2012)
15. Wehner, D.: Advice complexity of fine-grained job shop scheduling. In: Paschos, V.T., Widmayer, P. (eds.) CIAC 2015. LNCS, vol. 9079, pp. 416–428. Springer, Heidelberg (2015)

Pseudoknot-Generating Operation

Da-Jung Cho[1], Yo-Sub Han[1(✉)], Timothy Ng[2], and Kai Salomaa[2]

[1] Department of Computer Science, Yonsei University,
50, Yonsei-Ro, Seodaemun-Gu, Seoul 120-749, Republic of Korea
{dajung,emmous}@cs.yonsei.ac.kr
[2] School of Computing, Queen's University, Kingston, ON K7L 3N6, Canada
{ng,ksalomaa}@cs.queensu.ca

Abstract. A pseudoknot is an intra-molecular structure formed primarily in RNA strands and much research has been done to predict efficiently pseudoknot structures in RNA. We define an operation that generates all pseudoknots from a given sequence and consider algorithmic and language theoretic properties of the operation. We give an efficient algorithm to decide whether a given string is a pseudoknot of a regular language L—the runtime is linear if L is given by a deterministic finite automaton. We consider closure and decision properties of the pseudoknot-generating operation. For DNA encoding applications, pseudoknot structures are undesirable. We give polynomial-time algorithms to decide whether a regular language L contains a pseudoknot or a pseudoknot generated by some string of L. Furthermore, we show that the corresponding questions for context-free languages are undecidable.

Keywords: Pseudoknots · Pseudoknot-generating operation · Closure and decision properties · Formal languages

1   Introduction

A ribonucleic acid (RNA) often forms secondary structures according to the base-pairing with Adenine (A), Uracil (U), Guanine (G) and Cytosine (C) [5]. These bases A, G, C and U complementarily bind and form a double helix called *stem*, and double helix with unpaired loop known as *stem-loop*. A RNA structure generally has stems and various kinds of loops as a structural motif, which then gives rise to well-known structures such as hairpin or pseudoknot. RNA structures play an important role in cells and give insights to molecular evolution and function of RNA molecule [19]. Therefore, in bioinformatics, it is one of the most important and fundamental problems to predict RNA structures made up of a set of stems with optimal thermodynamic energy. Note that stabilized optimal foldings of a RNA sequence are closely related to the minimum free energy of RNA secondary structures based on the theory of thermodynamics.

A pseudoknot structure contains at least two stem-loops that occur in RNA with intramolecular base-pairing: Second half of one stem is embedded

© Springer-Verlag Berlin Heidelberg 2016
R.M. Freivalds et al. (Eds.): SOFSEM 2016, LNCS 9587, pp. 241–252, 2016.
DOI: 10.1007/978-3-662-49192-8_20

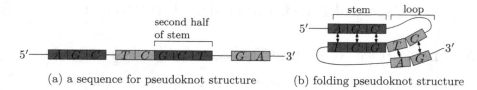

(a) a sequence for pseudoknot structure   (b) folding pseudoknot structure

Fig. 1. A pseudoknot structure example: (a) A sequence contains a pseudoknot structure in which the second half of stem (blue box) exists between the two halves of another stem (green boxes) (b) A sequence folds into a pseudoknot (Color figure online).

in between the two halves of another stem. (See Fig. 1 for an example of pseudoknot structure.) Pseudoknot structures appear in many natural RNA molecules and are closely related with the ribosomal frameshifting that allows viruses to create many protein structures from a relatively small genome [9]. Since the ribosomal frameshifting affects on encoding protein and the pseudoknot structure gives a tertiary structure of molecule, it is a major topic of biomolecular computing to predict pseudoknot structures [3,9]. This led researchers to study efficient methods that predict pseudoknot structures [2,4]. From a formal language viewpoint, several researchers [8,15,17,18] characterized the pseudoknot structure and suggested pseudoknot predicting algorithms. Given an input, the problem of predicting or aligning arbitrary pseudoknot structures is NP-hard [2,11]. Möhl et al. [17] presented an algorithm that computes the edit-distance of two RNA structures with arbitrary pseudoknots and showed that the algorithm is applicable in practice. Kari and Seki [15] formalized particular case of pseudoknot structures under formal language theory and investigated its properties. Evans [8] proposed the first polynomial-time algorithm for finding maximum common substructures that include pseudoknots. Rinaudo et al. [18] generalized several RNA structures and presented an alignment algorithm based on tree decomposition approach.

While most researchers considered a problem of predicting pseudoknot structures from a (long) sequence, we consider pseudoknot structures from a different angle: A sequence may be expanded (namely, append a new sequence itself) to form a pseudoknot structure. We consider this process and define a new operation *pseudoknot-generating* operation that generates all pseudoknot structures (from now we just call *pseudoknots* in short) from a given sequence. Then the resulting sequences fold itself into pseudoknots. Thus the input string becomes a seed string to generate pseudoknots. We establish the closure properties of the pseudoknot-generating operation on a string and present an algorithm that determines whether or not a string is a pseudoknot. Since pseudoknot structures are related to some biological mutations, they are crucial for detecting mutational patterns of a DNA sequence. We also study the closure properties of pseudoknot-generating operation on languages and examine several questions related to pseudoknots with respect to languages. From a biological view point, we can think of the pseudoknot-generating operation on a language as a procedure to generate all possible pseudoknots that may cause a mutation from a set of subsequences. In particular, we theoretically demonstrate that one can check whether or not two sets of DNA sequences contain common mutational seed sequences. Furthermore, we define the pseudoknot-freeness and investigate the decidability problem for pseudoknot-freeness for regular and context-free languages.

In Sect. 2, we recall some notation and define the pseudoknot-generating operation. We consider the pseudoknot-generating operation and design several algorithms for recognizing generated pseudoknots from strings and finite automata in Sect. 3. Then, we study closure and decision properties of the pseudoknot-generating operation, and, investigate the pseudoknot-free languages in Sect. 4.

2   Preliminaries

Let Σ denote a finite alphabet of characters and Σ^* denote the set of all strings over Σ. The size $|\Sigma|$ of Σ is the number of characters in Σ. A language over Σ is a subset of Σ^*. The symbol $\emptyset$ denotes the empty language and the symbol λ denotes the null string. Given a string $x = x_1 \cdots x_n$, $|x|$ is the number of characters in x, $x(i)$ denotes the ith character x_i of x and $x(i,j) = x_i x_{i+1} \cdots x_j$ is the substring of x from position i to position j, where $i \leq j$. Given two strings x and y in Σ^*, x is a *prefix* of y if there exists $z \in \Sigma^*$ such that $xz = y$ and x is a *suffix* of y if there exists $z \in \Sigma^*$ such that $zx = y$. Furthermore, x is said to be a *substring* or an *infix* of y if there are two strings u and v such that $uxv = y$.

An FA A is specified by a tuple $(Q, \Sigma, \delta, s, F)$, where Q is a finite set of states, Σ is an input alphabet, $\delta : Q \times \Sigma \to 2^Q$ is a transition function, $s \in Q$ is the start state and $F \subseteq Q$ is a set of final states. If F consists of a single state f, then we use f instead of $\{f\}$ for simplicity. Let $|Q|$ be the number of states in Q and $|\delta|$ be the number of transitions in δ. Then, the size of A is $|A| = |Q| + |\delta|$. For a transition $\delta(p, a) = q$ in A, we say that p has an *out-transition* and q has an *in-transition*. If $\delta(q, a)$ has a single element q', then we denote $\delta(q, a) = q'$ instead of $\delta(q, a) = \{q'\}$ for simplicity.

A string x over Σ is accepted by A if there is a labeled path from s to a final state such that this path spells out x. We call this path an *accepting path*. Then, the language $L(A)$ of A is the set of all strings spelled out by accepting paths in A. We say that a state of A is *useful* if it appears in an accepting path in A; otherwise, it is *useless*. Unless otherwise mentioned, in the following we assume that all states of A are useful.

Given a string x, we say that x has a pseudoknot if there exists a substring w of x such that $w = w_1 w_2 w_3 w_4 w_1^R w_5 w_3^R$ for some strings $w_1, w_2, w_3, w_5 \in \Sigma^+$ and $w_4 \in \Sigma^*$. We call the string w *pseudoknot string* (or *pseudoknot* in short). We consider a restricted pseudoknot in which $w_4 = \lambda$, which means that half of one stem is adjacent to half of another stem. (See Fig. 2 for an example of restricted pseudoknot.)

Given a string x, we define the restricted pseudoknot-generating operation

$$\mathbb{PK}_{\mathbb{R}}(x) = \{x_1 x_2 x_3 x_1^R x_4 x_3^R \mid x = x_1 x_2 x_3 \text{ and } x_1, x_2, x_3, x_4 \in \Sigma^+\}.$$

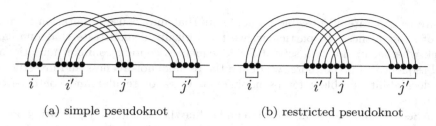

(a) simple pseudoknot   (b) restricted pseudoknot

Fig. 2. An example of two types of pseudoknot on a sequence of length n: (a) a simple pseudoknot with two base-pairings (i, j) and (i', j') (b) a restrict version of pseudoknot with two base-pairings (i, j) and (i', j') such that the first half of (i', j') is immediately followed by the second half of (i, j), where $0 \leq i < i' < j < j' \leq n$.

For a language L,

$$\mathbb{PK}_\mathbb{R}(L) = \bigcup_{x \in L} \mathbb{PK}_\mathbb{R}(w).$$

We define the iterated operation of $\mathbb{PK}_\mathbb{R}$ to be, for $i \geq 0$,

$$\mathbb{PK}_\mathbb{R}^{(0)}(L) = L, \quad \mathbb{PK}_\mathbb{R}^{(i+1)}(L) = \mathbb{PK}_\mathbb{R}(\mathbb{PK}_\mathbb{R}^i(L)), \quad \mathbb{PK}_\mathbb{R}^*(L) = \bigcup_{i=0}^{\infty} \mathbb{PK}_\mathbb{R}^i(L).$$

In the following, we only consider restricted pseudoknots and call them simply pseudoknots unless we need to distinguish restricted pseudoknots from general pseudoknots.

3   Algorithms for Recognizing Generated Pseudoknots

We first study the problem for checking whether or not a string $w = w_1 w_2 \cdots w_n$ is a pseudoknot; namely, is $w = x_1 x_2 x_3 x_1^R x_4 x_3^R$ for some $x_1, x_2, x_3, x_4 \in \Sigma^+$. The main idea of our approach is to check if there exists a substring $x_3 x_1^R$ of w such that x_1 is a prefix and x_3^R is a suffix of w. A naive approach is to consider all possible substrings and check this condition. We design a better algorithm that checks the required condition more efficiently based on the Aho-Corasick algorithm [1].

Fig. 3. A naive approach for checking whether or not w is a pseudoknot. For each substring $w(i, j)$, we check whether or not $w(i, j)$ is a catenation of x_3 and x_1^R for a prefix x_1 and a suffix x_3 of w—$w(i, j) = x_3 x_1^R$—by comparing characters from both directions.

Before we describe the whole algorithm, we first present an algorithm that finds the shortest length of the matching prefix of the input pattern string with respect to the input for each index of the input. This algorithm is crucial for checking whether or not w is a pseudoknot.

Procedure. ShortestMatchingLength(w, T)

/* w **is a length** m **pattern and** T **is a length** n **text**   */
Construct a DFA $A = (Q, \Sigma, \delta, 0, Q \setminus \{0\})$ for w, where $Q = \{0, 1, \ldots, m\}$

/* **construct the goto function** $\mathbb{G}$ */
$\mathbb{G}(0, a \neq w_1 \in \Sigma) \leftarrow 0$
for $i \leftarrow 0$ **to** $m - 1$ **do**
$\quad \lfloor \ \mathbb{G}(i, w_{i+1}) \leftarrow i + 1$

/* **construct the failure function** $\mathbb{F}$ **and the output function** $\mathbb{O}$ */
$\mathbb{F}(1) \leftarrow 0$
for $i \leftarrow 1$ **to** m **do**
$\quad \lfloor$ **if** $\mathbb{G}(i, a) = i + 1$ **then**
$\qquad \lfloor \ v \leftarrow \mathbb{F}(i)$
$\qquad\quad$ **while** $\mathbb{G}(v, a) \neq \emptyset$ **do**
$\qquad\qquad \lfloor \ v \leftarrow \mathbb{F}(v)$
$\qquad\quad \mathbb{F}(i + 1) \leftarrow \mathbb{G}(v, a)$
$\qquad\quad \mathbb{O}(i + 1) \leftarrow \min(\mathbb{O}(i + 1), \mathbb{O}(\mathbb{F}(i + 1)))$

/* **read** T **using** $\mathbb{G}, \mathbb{F}, \mathbb{O}$ */
$q \leftarrow 0$
for $i \leftarrow 1$ **to** n **do**
$\quad \lfloor$ **while** $\mathbb{G}(q, T(i)) \neq \emptyset$ **do**
$\qquad \lfloor \ q \leftarrow \mathbb{F}(q)$
$\qquad q = \mathbb{G}(q, T(i))$
$\qquad$ **if** $\mathbb{O}(q) \neq \emptyset$ **then**
$\qquad \lfloor \ \mathrm{SML}[q] \leftarrow \mathbb{O}(q)$

return SML

Given an input pattern string w and a text T, **Proc.** ShortestMatchin-gLength is a modified Aho-Corasick algorithm that finds the shortest length of the matching prefix of w at each index of T; if u the shortest matching prefix of w, then the reversal u^R of u appears as an infix of T. The two main differences from the original Aho-Corasick algorithm are

1. it receives only one string w as an input pattern and regards all prefixes of w as matching patterns

2. the output function $\mathbb{O}$ returns the shortest length of the matching pattern instead of reporting all matching patterns: $\mathbb{O}(i+i) \leftarrow \min(\mathbb{O}(i+1), \mathbb{O}(\mathbb{F}(i+i)))$.

It is easy to verify that **Proc.** ShortestMatchingLength runs in $O(m + n)$ time, where $m = |w|$ and $n = |T|$.

Now we design the whole algorithm that determines whether or not w is a pseudoknot using **Proc.** ShortestMatchingLength. First, we consider all prefixes of w up-to length $\frac{n}{2}$—candidates for being w_1 in the pseudoknot—and compute the set $w_p[i]$ of the shortest length of the matching prefix of each index using **Proc.** ShortestMatchingLength with $w = w_1 w_2 \cdots w_{\frac{n}{2}}$ and $T = w^R$. Next, we similarly consider all suffixes of w up-to length $\frac{n}{2}$—candidates for being w_3^R in the pseudoknot—and compute the set $w_s[i]$ of the shortest length of the matching suffix for each index $1 \le i \le n$ using **Proc.** ShortestMatchingLength with $w = w_{\frac{n}{2}+1} \cdots w_{n-1} w_n$ and $T = w$.

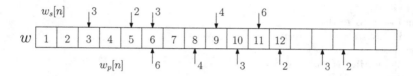

Fig. 4. An example of running **Proc.** ShortestMatchingLength for checking whether or not w is a pseudoknot.

Figure 4 is an example of running **Proc.** ShortestMatchingLength for a string w and obtain $w_p[n]$ and $w_s[n]$. In this example, because of $w_s[9]$ and $w_p[10]$, we know that w is a pseudoknot. However, a pair of $w_s[5]$ and $w_p[6]$ is invalid since, at index 6, w cannot have a prefix of size 6 ($=w_p[6]$). Similarly, a pair of $w_s[11]$ and $w_p[12]$ is invalid for checking the pseudoknot structure for w since, after index 11, w cannot have a $w_1^R w_4 w_3^R$, where $|w_3^R| = 6$.

Lemma 1. *Given a string w of length n, we can determine whether or not w is a pseudoknot in $O(n)$ time.*

Note that if w is a pseudoknot, then we can find all indices i of w such that $w(1, i) = x_1 x_2 x_3$ and $w(i + 1, n) = x_1^R x_4 x_3^R$ from the algorithm. Let $I_{pk}(w)$ be the set of such indices.

Corollary 1. *Given two pseudoknot strings x and y, we can determine whether or not both $x, y \in \mathbb{PK}_\mathbb{R}(w)$ for a string w in linear-time in the size of x and y. We can also identify such w using $I_{pk}(x)$ and $I_{pk}(y)$ within the same runtime.*

We next consider a problem of determining whether or not w is in $\mathbb{PK}_\mathbb{R}(L(A))$ of a given FA A. Our approach is simple: we read w character by character with A and find all indices j of w when we enter a final state of A while reading w. Namely, $w(1, j) \in L(A)$. Let $I_p(w, A)$ be the set of such indices.

Lemma 2. *Given a string w and an FA A,*

$$w \in \mathbb{PK}_{\mathbb{R}}(L(A)) \text{ iff } I_{pk}(w) \cap I_p(w, A) \neq \emptyset.$$

Note that a pseudoknot string may have several different pseudoknots. Therefore, even if $w(1, i) = x_1 x_2 x_3$ and $w(i + 1, n) = x_1^R x_4 x_3^R$ for an index i of w, $w(1, i)$ may not be accepted by A. This is why we have considered all possible indices in $I_{pk}(w)$. We now establish the following result based on pseudoknot checking algorithm and Lemma 2.

Theorem 1. *Given a string w of size n and an FA A of size $m = |A|$, we can determine whether or not $w \in \mathbb{PK}_{\mathbb{R}}(L(A))$ in $O(mn)$ time. If A is a DFA, then the runtime becomes $O(n)$.*

Proof. It takes $O(n)$ time to compute $I_{pk}(w)$ and $O(mn)$ time to compute $I_p(w, A)$. If A is a DFA, then we can compute $I_p(w, A)$ in $O(n)$ time.   □

Note that pseudoknots of RNA are closely related with the frameshifting mutation of protein expressions and commonly found in viral genomes, in particular influenza virus [7]. This leads researchers to consider the structural comparison among several sequences to find the common mutational pattern, in particular, pseudoknots [6]. Here, we investigate a necessary condition of $\mathbb{PK}_{\mathbb{R}}(x) \cap \mathbb{PK}_{\mathbb{R}}(y) \neq \emptyset$ for two strings x and y and show that it is decidable to check whether or not two strings have a common pseudoknot in $\mathbb{PK}_{\mathbb{R}}(x)$ and $\mathbb{PK}_{\mathbb{R}}(y)$.

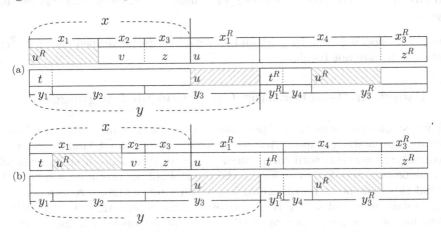

Fig. 5. An example of two strings x and y such that $\mathbb{PK}_{\mathbb{R}}(x) \cap \mathbb{PK}_{\mathbb{R}}(y) \neq \emptyset$. First, x is a prefix of y. Second, the longer part u (slanted line box in the figure) of y appears as (a) a prefix of x or (b) an infix (but not prefix) of x in the reversed form u^R, where $t, v, z \in \Sigma^+$.

Let x and y be two strings, where $|x| < |y|$. Figure 5 shows that $\mathbb{PK}_{\mathbb{R}}(x) \cap \mathbb{PK}_{\mathbb{R}}(y) \neq \emptyset$ if and only if x is a prefix of y, and $y(|x| + 1, |y|)$ $(= u$ in the figure) appears as an infix of $x(1, |x| - 2)$—we consider $x(1, |x| - 2)$ to ensure $t, v, z \in \Sigma^+$, if exists. There are two possible cases for being an infix as follows:

(a) Since u^R appears as a prefix of $x(1, |x|-2)$, we can select an arbitrary prefix t of $x(1, |x|-2)$ for y_1 as depicted in Fig. 5(a).

(b) Since u^R appears an infix (but not a prefix) of $x(1, |x|-2)$, we can select the prefix t of x_1 such that $x_1 = tu^R$ for y_1 as depicted in Fig. 5(b).

We can check this in $O(|x|)$ time using the KMP algorithm [16] and, therefore, obtain the following result.

Lemma 3. *Given two strings x and y such that $|x| < |y|$, we can determine whether or not $\mathbb{PK}_{\mathbb{R}}(x) \cap \mathbb{PK}_{\mathbb{R}}(y) \neq \emptyset$ in $O(|x|)$ time.*

Proof. The proof is immediate from Fig. 5. □

We know if two strings have a common pseudoknot in $\mathbb{PK}_{\mathbb{R}}(x)$ and $\mathbb{PK}_{\mathbb{R}}(y)$. We next investigate the inclusion between $\mathbb{PK}_{\mathbb{R}}(x)$ and $\mathbb{PK}_{\mathbb{R}}(x)$ when $|x| < |y|$.

Lemma 4. *Given two strings x and y, if $|x| < |y|$, then it is impossible that $\mathbb{PK}_{\mathbb{R}}(x) \subset \mathbb{PK}_{\mathbb{R}}(y)$.*

Given a string z, it is straightforward to verify that $\mathbb{PK}_{\mathbb{R}}(z)$ is regular and $\mathbb{PK}_{\mathbb{R}}(z)$ is infinite from the definition of the operation. We consider the case of applying the $\mathbb{PK}_{\mathbb{R}}$ operation on the resulting language several times, and prove that the iterated $\mathbb{PK}_{\mathbb{R}}$ does not preserve the regularity.

Theorem 2. *There exists a string z such that $\mathbb{PK}_{\mathbb{R}}^2(z)$ and $\mathbb{PK}_{\mathbb{R}}^*(z)$ are not regular.*

Proof. By choosing $\Sigma = \{a, \textcent, \$\}$ and $z = \textcent a\$$, it can be verified that $\mathbb{PK}_{\mathbb{R}}^2(z)$ and $\mathbb{PK}_{\mathbb{R}}^*(z)$ are not regular. □

4   Pseudoknot-Generating Operation on Languages

We investigate the properties of pseudoknot on a set of strings. The pseudoknot operation on a string implies that we generate a pseudoknot family related by a common structural motif for pseudoknot. Note that a given string expands and becomes a pseudoknot string by the pseudoknot operation on a string. We extend this view point into languages in which a set of strings represents a set of all subsequences of a long RNA sequence. This implies that the pseudoknot operation on a language generates all possible pseudoknots that may partially occur as a mutation.

4.1   Closure and Decision Properties of the Pseudoknot-Generating Operation

We first consider the closure property of the pseudoknot operation and determine whether or not two sets of pseudoknots on languages contain a common string.

Theorem 3. *Regular and context-free languages are not closed under $\mathbb{PK}_{\mathbb{R}}$.*

Next, given regular languages L and R we consider the problem of checking whether or not there is a pseudoknot generated both by a string of L and a string of R. Note that, by Theorem 3, we know that $\mathbb{PK}_{\mathbb{R}}(L)$ need not be even context-free in general. This means that we cannot simply first produce a representation of the languages $\mathbb{PK}_{\mathbb{R}}(L)$ and $\mathbb{PK}_{\mathbb{R}}(R)$, respectively, and then check whether or not they have a non-empty intersection. Instead, our algorithm is based directly on finite automata for the original language L and R.

Let $A = (Q_A, \Sigma, \delta_A, s_A, F_A)$ and $B = (Q_B, \Sigma, \delta_B, s_B, F_B)$ be two FAs for L and R. Then, we first construct an FA $C = (Q_A \times Q_B, \Sigma, \delta_C, s_A \times s_B, F_A \times F_B)$ for $L(A) \cap L(B)$ by the standard Cartesian product, where

$$\delta_C((p,q),a) = \{(p',q') \mid p' \in \delta_A(p,a) \text{ and } q' \in \delta_B(q,a)\}.$$

Our algorithm is similar to the idea illustrated in Fig. 5: We check if there exists a pair of strings—say $x \in L(A)$, $y \in L(B)$ and $|x| < |y|$ (the other case is symmetric)—such that $\mathbb{PK}_{\mathbb{R}}(x) \cap \mathbb{PK}_{\mathbb{R}}(y) \neq \emptyset$. Since x is a prefix of y, there exists a path from s_B to a nonfinal state q that spells out x in B. We search for such paths in C. For each state (f,q) of C, where $f \in F_A, q \in Q_B$, we define two FAs as follows:

- $\overleftarrow{C_{f,q}} = (Q_A \times Q_B, \Sigma, \delta_C, s_A \times s_B, \{(f,q)\})$; in other words, (f,q) is the only final state of C.
- $\overrightarrow{B_q} = (Q_B, \Sigma, \delta_B, q, F_B)$; in other words, q is the new start state of B.

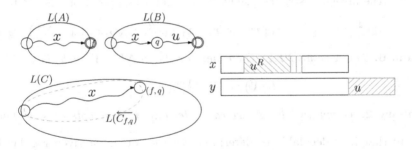

Fig. 6. An example of two FAs A and B such that $\mathbb{PK}_{\mathbb{R}}(L(A)) \cap \mathbb{PK}_{\mathbb{R}}(L(B)) \neq \emptyset$. Note that u^R is an infix of the string $x(2, |x| - 2)$.

Lemma 5. *There exists a state* (f,q) *of* C *such that*

$$L(\overleftarrow{C_{f,q}}) \cap \Sigma^* \cdot (L(\overrightarrow{B_q})^R \cdot \Sigma^2) \cdot \Sigma^* \neq \emptyset$$

or a state (p, f') *of* C *such that*

$$L(\overleftarrow{C_{p,f'}}) \cap \Sigma^* \cdot (L(\overrightarrow{A_p})^R \cdot \Sigma^2) \cdot \Sigma^* \neq \emptyset$$

if and only if $\mathbb{PK}_{\mathbb{R}}(L(A)) \cap \mathbb{PK}_{\mathbb{R}}(L(B)) \neq \emptyset$.

Once we have an intersection FA C, there are at most $|Q_A||Q_B|$ states in the form of (f,q) or (p,f'). Then, for each state, say (f,q), we need to check whether or not $L(\overleftarrow{C_{f,q}}) \cap \Sigma^* \cdot (L(\overrightarrow{B_q})^R \cdot \Sigma^2) \cdot \Sigma^*$ is empty. Since the size of $C_{f,q}$ is at most $|A||B|$ and the size of $\overrightarrow{B_q}$ is at most $|B|$, it takes $O(|A||B|^2)$ time. Therefore, in the worst-case, the total runtime is

$$O(n^2) \text{ (the number of states) } \times O(n^6) \text{ (intersection test) } = O(n^8),$$

where n is the maximum number of states between A and B.

Theorem 4. *Given two FAs A and B, we can determine whether or not*

$$\mathbb{PK}_\mathbb{R}(L(A)) \cap \mathbb{PK}_\mathbb{R}(L(B)) \neq \emptyset$$

in polynomial-time.

Often we need to verify if there exists a pseudoknot in the input set—a set of strings. In biology, a RNA sequence might first fold into non-pseudoknot, and then form a more complex structure including pseudoknots. According to this phenomenon, Jabbari and Condon [10] considered non-pseudoknots for predicting pseudoknots capturing all possible pre-structures of pseudoknots.

When an input set has a finite number of elements, we may check them one by one. However, if the set is infinite, then we need a better algorithm. We consider this problem when the set is a regular language. Before we present an algorithm, we define the inverse restricted pseudoknot-generating operation $\mathbb{PK}_\mathbb{R}^{-1}$ to be

$$\mathbb{PK}_\mathbb{R}^{-1}(w) = \{x_1 x_2 x_3 \mid w \in x_1 x_2 x_3 x_1^R x_4 x_3^R, x_1, x_2, x_3, x_4 \in \Sigma^+\}.$$

Lemma 6. *Let A be an NFA. Then there exists an NFA A' such that*

$$L(A') = \mathbb{PK}_\mathbb{R}^{-1}(L(A)).$$

Corollary 2. *Given an NFA A, we can determine if A accepts a pseudoknot.*

Note that it is decidable to determine whether or not a given regular language contains a pseudoknot. Here, we contrast the result of Corollary 2 by showing that it is undecidable whether or not a context-free language contains a pseudoknot. We use a reduction from the *Post Correspondence Problem* (PCP) [20]. Recall that an instance of PCP consists of two lists of strings $((u_1, \ldots, u_n), (v_1, \ldots, v_n))$, $u_i, v_i \in \Sigma^*$, $1 \leq i \leq n$, and a solution of this instance is a sequence of integers $i_1, \ldots, i_k \in \{1, \ldots, n\}$ such that $u_{i_1} \cdots u_{i_k} = v_{i_1} \cdots v_{i_k}$. It is well known that PCP is unsolvable [20].

Proposition 1. *For a given context-free language L, it is undecidable whether or not L contains a pseudoknot.*

4.2   Pseudoknot-Free Languages

Analogously with the definition of restricted code classes, such as prefix- or suffix-codes [12], we define that a language L is $\mathbb{PK}_\mathbb{R}$-free (informally just pseudoknot-free) if no string of L is a pseudoknot generated by another string of L.

Definition 1. *We say that a language L is $\mathbb{PK}_\mathbb{R}$-free if $L \cap \mathbb{PK}_\mathbb{R}(L) = \emptyset$.*

In DNA coding applications, pseudoknots are in general undesirable because they can result in undesired bonds in DNA sequences [13,14]. This means that if we can efficiently check the property of $\mathbb{PK}_\mathbb{R}$-freeness, it might be worthwhile to add a preprocessing stage for predicting pseudoknot-structures. Note that some approaches for prediction pseudoknots consider also "non-pseudoknots" because an RNA sequence can fold a non-pseudoknot to form a pseudoknot.

We consider the case when L is regular and show that we can decide whether or not L is $\mathbb{PK}_\mathbb{R}$-free. We use a similar construction for constructing an FA for $\mathbb{PK}_\mathbb{R}^{-1}$.

Lemma 7. *Let A be an FA. Then there exists an NFA A' that accepts a set of strings $u = u_1 u_2 u_3$ such that $u_1^R u_4 u_3^R u_1 u_2 u_3 \in L(A)$.*

Theorem 5. *Given an FA A, we can determine whether or not $L(A)$ is $\mathbb{PK}_\mathbb{R}$-free in polynomial-time.*

Here, we observe that deciding $\mathbb{PK}_\mathbb{R}$-freeness of a context-free language is undecidable based on Proposition 1.

Theorem 6. *For a given context-free language L it is undecidable whether or not L is $\mathbb{PK}_\mathbb{R}$-free.*

5   Conclusions

We have considered one of RNA structures called pseudoknot and specific phenomenon in which a sequence expands itself and forms a pseudoknot. We have defined the restrict version of the pseudoknot-generating operation: For a string x, $\mathbb{PK}_\mathbb{R}(x)$, roughly speaking, consists of all possible continuations of x that can fold back onto x to form a pseudoknot.

We have investigated (closure-)properties of pseudoknot-generating operation on a string and designed linear-time algorithm for determining whether or not given string is a pseudoknot. We have shown that for two strings x and y, it is decidable whether or not $\mathbb{PK}_\mathbb{R}(x) \cap \mathbb{PK}_\mathbb{R}(y) \neq \emptyset$. Moreover, we have examined the pseudoknot-generating operation on languages, and showed that regular and context-free languages are not closed under pseudoknot-generating. On the other hand, we have established that given two FAs A and B, it is decidable whether or not $\mathbb{PK}_\mathbb{R}(L(A)) \cap \mathbb{PK}_\mathbb{R}(L(B)) \neq \emptyset$ in polynomial-time in the size of A and B. Furthermore, we have shown that it is decidable whether or not a given regular language is $\mathbb{PK}_\mathbb{R}$-free in polynomial-time. However, it is undecidable to determine whether or not a given context-free language is $\mathbb{PK}_\mathbb{R}$-free.

References

1. Aho, A.V., Corasick, M.J.: Efficient string matching: an aid to bibliographic search. Commun. ACM **18**(6), 333–340 (1975)
2. Akutsu, T.: Dynamic programming algorithms for RNA secondary structure prediction with pseudoknots. Discrete Appl. Math. **104**(1), 45–62 (2000)
3. Brierley, I., Digard, P., Inglis, S.C.: Characterization of an efficient coronavirus ribosomal frameshifting signal: requirement for an RNA pseudoknot. Cell **57**(4), 537–547 (1989)
4. Condon, A., Davy, B., Rastegari, B., Zhao, S., Tarrant, F.: Classifying RNA pseudoknotted structures. Theor. Comput. Sci. **320**(1), 35–50 (2004)
5. Dirks, R.M., Lin, M., Winfree, E., Pierce, N.A.: Paradigms for computational nucleic acid design. Nucleic Acids Res. **32**(4), 1392–1403 (2004)
6. Doose, G., Metzler, D.: Bayesian sampling of evolutionarily conserved RNA secondary structures with pseudoknots. Bioinformatics **28**(17), 2242–2248 (2012)
7. Du, Z., Hoffman, D.W.: An NMR and mutational study of the pseudoknot within the gene 32 mRNA of bacteriophage T2: insights into a family of structurally related RNA pseudoknots. Nucleic Acids Res. **25**(6), 1130–1135 (1997)
8. Evans, P.A.: Finding common RNA pseudoknot structures in polynomial time. J. Discrete Algorithms **9**(4), 335–343 (2011)
9. Giedroc, D.P., Theimer, C.A., Nixon, P.L.: Structure, stability and function of RNA pseudoknots involved in stimulating ribosomal frameshifting. J. Mol. Biol. **298**(2), 167–185 (2000)
10. Jabbari, H., Condon, A.: A fast and robust iterative algorithm for prediction of RNA pseudoknotted secondary structures. BMC Bioinform. **15**(1), 147 (2014)
11. Jiang, T., Lin, G., Ma, B., Zhang, K.: A general edit distance between RNA structures. J. Comput. Biol. **9**(2), 371–388 (2002)
12. Jürgensen, H., Konstantinidis, S.: Codes. In: Word, Language, Grammar. Handbook of Formal Languages, vol. 1, pp. 511–607 (1997)
13. Kari, L., Konstantinidis, S., Kopecki, S.: Transducer Descriptions of DNA Code Properties and Undecidability of Antimorphic Problems. arXiv:1503.00035 (2015)
14. Kari, L., Mahalingam, K.: Watson-Crick palindromes in DNA computing. Nat. Comput. **9**(2), 297–316 (2010)
15. Kari, L., Seki, S.: On pseudoknot-bordered words and their properties. J. Comput. Syst. Sci. **75**(2), 113–121 (2009)
16. Knuth, D.E., Morris Jr, J.H., Pratt, V.R.: Fast pattern matching in strings. SIAM J. Comput. **6**(2), 323–350 (1977)
17. Möhl, M., Will, S., Backofen, R.: Fixed parameter tractable alignment of RNA structures including arbitrary pseudoknots. In: Ferragina, P., Landau, G.M. (eds.) CPM 2008. LNCS, vol. 5029, pp. 69–81. Springer, Heidelberg (2008)
18. Rinaudo, P., Ponty, Y., Barth, D., Denise, A.: Tree decomposition and parameterized algorithms for RNA structure-sequence alignment including tertiary interactions and pseudoknots. In: Raphael, B., Tang, J. (eds.) WABI 2012. LNCS, vol. 7534, pp. 149–164. Springer, Heidelberg (2012)
19. Saraiya, A.A., Lamichhane, T.N., Chow, C.S., SantaLucia Jr, J., Cunningham, P.R.: Identification and role of functionally important motifs in the 970 loop of Escherichia coli 16S ribosomal RNA. J. Mol. Biol. **376**(3), 645–657 (2008)
20. Shallit, J.: A Second Course in Formal Languages and Automata Theory, vol. 179. Cambridge University Press, Cambridge (2009)

Capabilities of Ultrametric Automata with One, Two, and Three States

Maksims Dimitrijevs[(✉)]

Faculty of Computing, University of Latvia, Raiņa bulvāris 19, Riga LV-1586, Latvia
`mvdmaks@inbox.lv`

Abstract. Ultrametric automata use p-adic numbers to describe the random branching of the process of computation. Previous research has shown that ultrametric automata can have a significant decrease in computing complexity. In this paper we consider the languages that can be recognized by one-way ultrametric automata with one, two, and three states. We also show an example of a promise problem that can be solved by ultrametric integral automaton with three states.

1 Introduction

Rūsiņš Freivalds has recently introduced the idea of using p-adic numbers in Turing machines and finite automata to describe the random branching of the process of computation [1]. He proved that the use of p-adic numbers exposes new possibilities which do not inhere in deterministic or probabilistic approaches. Moreover, in 1916 Alexander Ostrowski proved that any non-trivial absolute value of the rational numbers $\mathbb{Q}$ is equivalent to either the usual real absolute value or a p-adic absolute value. So using p-adic numbers was the only remaining possibility not yet explored [1].

Ultrametric automata are similar to probabilistic automata but research has shown that the capabilities of these types of automata can differ very much. Ultrametric automata are able to recognize nonrecursive languages [1] and can have significant state complexity advantages over other types of automata [2,3]. Ultrametric automata can also solve the tasks of Turing machines with various requirements for computing complexity [4,5].

In this paper we look at ultrametric automata with one, two and three states to explore the capabilities of ultrametric automata with a small number of states. We work with general ultrametric automata and ultrametric automata with restricted definitions, which has the effect of reducing the power of ultrametric automata. We will see that, even with restrictions, ultrametric automata with a small number of states have great language recognition power.

2 p-adic Numbers

p-adic numbers are used in different sciences, including chemistry and physics [6,7], and are also used in ultrametric automata and Turing machines [1].

© Springer-Verlag Berlin Heidelberg 2016
R.M. Freivalds et al. (Eds.): SOFSEM 2016, LNCS 9587, pp. 253–264, 2016.
DOI: 10.1007/978-3-662-49192-8_21

A p-adic digit is a natural number between 0 and $p-1$ where p is an arbitrary prime number. A p-adic integer $(a_i)_{i \in N}$ is an infinite sequence of p-adic digits written from right to left. A p-adic integer can be written like a decimal number $...a_i...a_2 a_1 a_0$.

For each natural number, there exists its p-adic representation and only a finite number of p-adic digits are not zeroes. Negative integers have a different representation in p-adic numbers, namely, they have an infinite sequence of digits $p-1$ to the left. If all digits of a p-adic integer are $p-1$ then we have the p-adic number -1. We can add, subtract and multiply p-adic integers in the same way as natural numbers in base p. The only division that is not possible in p-adic integers is division by p. For example, if we want to have p-adic integer $1/p$, equation $p * x = 1$ should have a solution, but multiplication by p-adic integer p gives zero in the right-most p-adic digit. That being said, p-adic integers can represent any integer and most of the rational numbers, except for those having a positive integral power of p in the divisor.

p-adic float numbers can have a decimal point and are infinite to the left side but finite to the right side. For example, p-adic number $1/p$ can be written as $...0000.1$. The field of p-adic numbers is denoted as $\mathbb{Q}_p$. For the curious reader, David A. Madore has written extensively about p-adic numbers and further information on the subject can be found in [8].

To measure p-adic number we require the absolute value of a p-adic number. If p is a prime number, then the p-adic ordinal of the rational number a, denoted by $ord_p a$, is the largest m such that p^m divides a.

Definition 1. *For any rational number x its p-norm (p-adic absolute value) is*

$$\|x\|_p = \begin{cases} 1/p^{ord_p x}, & \text{if } x \neq 0 \\ 0, & \text{if } x = 0. \end{cases}$$

For example, if $p = 5$, $\|50\|_5 = \|2 * 5^2\|_5 = \|5^2\|_5 = 1/5^2 = 1/25$, if $p = 2$, $\|50\|_2 = \|2\|_2 = 1/2$, but for any other prime number p, $\|50\|_p = 1$.

3   Definitions of Ultrametric Automata

Ultrametric automata are similar to probabilistic automata. Probabilistic finite automata were introduced by Michael O. Rabin, and the reader can refer to [9] for more details about probabilistic automata.

Definition 2. *A one-way p-ultrametric finite automaton is a tuple $(Q, S, \delta, q_0, F, \Lambda)$ where*

- *Q is the finite set of states,*
- *S is the input alphabet,*
- *$\delta : Q \times S \times Q \to \mathbb{Q}_p$ is the transition function,*
- *$q_0 : Q \to \mathbb{Q}_p$ is the initial amplitude distribution,*
- *$F \subseteq Q$ is the set of accepting states,*

– $\Lambda = (\lambda, \Diamond)$ *is the acceptance condition where* $\lambda \in \mathbb{R}$ *is the acceptance threshold and* $\Diamond \in \{\geq, \leq\}$.

A probabilistic automaton has transition probabilities that are real numbers. In the case of p-ultrametric automaton the transitions have amplitudes, which are p-adic numbers. Therefore, we can assume that, for a p-ultrametric automaton, prime number p is also a parameter. Probabilistic automata have their beginning distribution of probabilities among the states and transitions are performed with probabilities. In ultrametric automata every state has a beginning amplitude, and by reading input word, transitions are done with amplitudes. This means that final amplitudes of the states are calculated the same way as probabilities in probabilistic automata. To get the result after reading the input word, the amplitude of every accepting state is transformed into p-norm and the word is accepted if and only if p-norm sum of accepting states satisfies the acceptance condition.

We allow usage of all possible p-adic numbers in p-ultrametric automata. We allow this in the first definition of ultrametric automata because Paavo Turakainen defined probabilistic finite automata where the "probabilities" can be arbitrary real numbers and he has proven that languages recognizable by these probabilistic finite automata are the same as for ordinary probabilistic finite automata. Ultrametric automata defined in this way have great capabilities, for example, they are able to recognize nonrecursive languages [1]. This is also the reason why more restricted versions of ultrametric automata were introduced.

Definition 3. *Finite p-ultrametric automaton is called integral if all the p-adic numbers in its initial distribution and transition function are p-adic integers.*

There are no examples of ultrametric integral automata recognizing nonrecursive languages. The next definition is even more restrictive.

Definition 4. *A state of a p-ultrametric automaton is called regulated if there exist constants* λ, c *such that for every input word the p-norm of amplitude* γ *of this state is bounded by* $\lambda - c < \|\gamma\|_p < \lambda + c$. *A finite p-ultrametric automaton is called regulated if all of its states are regulated.*

Ultrametric regulated automata can recognize only regular languages [1]. In this paper we don't research capabilities of ultrametric regulated automata with a small number of states because their capabilities are limited to the recognition of regular languages. However, ultrametric regulated automata can have many fewer states than deterministic and even probabilistic automata [1,2].

4   Ultrametric Automata with One State

In this section we will show that the capabilities of ultrametric automata with one state can be described with the limitations of known types of automata. Mentioned capabilities are not so small.

Theorem 1. *For every prime number p there exists a p-ultrametric automaton with one state that can recognize a non-regular language.*

Proof. We will take the following non-regular language: $L_1 = \{x|x \in \{0,1\}^*$ and $|x|_0 \geq |x|_1\}$, where $|x|_a$ denotes the number of symbols a in x. So L_1 contains all binary words in which the number of zeroes is not less than the number of ones. L_1 can be easily shown nonregular based on the argument that the difference between the number of zeroes and the number of ones can increase infinitely. A deterministic finite automaton cannot recognize this language because it is nonregular.

We can construct a p-ultrametric automaton with one accepting state with initial amplitude 1. When the automaton reads symbol one, it will multiply the amplitude by p. When the symbol zero is read the amplitude is then multiplied by p^{-1}. After reading the input word the amplitude of the state will be $p^{|x|_1 - |x|_0}$. The p-norm of this number is equal to $1/p^{|x|_1 - |x|_0} = p^{|x|_0 - |x|_1}$. The automaton will accept the input word x if and only if the p-norm of the amplitude is at least 1, and this is possible only when $|x|_0 \geq |x|_1$. $\square$

Theorem 2. *For every prime number p, the languages recognizable by p-ultrametric automaton with one state form the proper subset of languages recognizable by one-way deterministic pushdown automata.*

Proof. If an ultrametric automaton has one state, it has an initial amplitude and ability to multiply this amplitude by some number. The p-norm of the state's amplitude determines the acceptance's condition. In [3] it is shown that after receiving any input symbol the p-norm of one state's amplitude monotonically increases, decreases or does not change. That being said, we can see that in this case one state works like a counter and the value of this counter is determined by p-norm.

Now we can replace the ultrametric automaton with one state with a deterministic pushdown automaton. This automaton has two symbols in the stack's alphabet, denoted by α and β. The number of symbols α in the stack will represent the p-norm $p^{|\alpha|}$, where $|\alpha|$ denotes the number of symbols α in the stack. The number of symbols β in the stack will represent the p-norm $p^{-|\beta|}$. The empty stack will represent the p-norm 1. At the beginning of work we will put the required number of α or β in the stack of deterministic pushdown automaton to represent the beginning p-norm. By reading input symbols the stack will change its symbols to keep up with the changes of p-norm of ultrametric automaton. It is possible to do this deterministically because it is known which symbols to put into the stack or to take from it. After reading the input word we have to check if p-norm satisfies the accepting condition. Therefore we have to check that the stack contains the required number of symbols α or β. This can also be done deterministically. After that, we can easily decide whether to go to the accepting or rejecting state.

The remaining situation is when the amplitude equals zero. If the amplitude of the only state of the ultrametric automaton is zero, it cannot be changed because multiplying by zero always gives zero. By knowing the acceptance condition for p-norm, we can deterministically go to accepting or rejecting state and not leave it.

The described method allows us to replace an ultrametric automaton with one state with a one-way deterministic pushdown automaton. This method works equally for all prime numbers p. This means that all languages recognizable by ultrametric automaton with one state are recognizable by one-way deterministic pushdown automata. There exist regular languages that are not recognized by ultrametric automaton with one state. For example, we can take the one-letter alphabet language containing only one word with a length greater than zero. This language requires at least two states [3].   □

We can improve this result by looking at a model that is weaker than one-way deterministic pushdown automata.

Theorem 3. *For every prime number p, the languages recognizable by a p-ultrametric automaton with one state form the proper subset of languages recognizable by one-way deterministic counter automata.*

Proof. We can improve the proof of Theorem 2 by replacing the deterministic pushdown automaton with a deterministic counter automaton. We will use the value of a counter instead of the number of symbols in stack memory. The classic definition of counter automata allow the counter to have only nonnegative integral values [10]. In our case, the counter automaton will have two sets of states: if the automaton is in the first set of states, then the value of counter c represents the p-norm p^c; if the automaton is in the second set of states, then the value of counter c represents the p-norm p^{-c}. By doing so we solve the problem of representing p-norm being negative powers of p. Other principles remain like in the proof of Theorem 2. Multiplication of p-norm by p will be represented by increasing the value of counter by 1 if the automaton is in the first set of states and decreasing the value of counter by 1 if the automaton is in the second set of states. Multiplication of p-norm by p^{-1} will be represented by doing the opposite operations. If p-norm becomes greater than 1, the automaton will go to the first set of states. The automaton will go to the second set of states if p-norm becomes less than 1.

Like in the proof of Theorem 2, this method will work equally for all prime numbers p and situations with amplitude zero will be handled in the same way. To finish the proof we can again use one-letter alphabet language containing only one word with the length greater than zero to show that the sets of recognizable languages are not equal.   □

The aforementioned three theorems are valid for all ultrametric automata with one state. Now we will show that the capabilities of ultrametric integral automata with one state are even smaller.

Theorem 4. *For every prime number p, a p-ultrametric integral automata with one state can recognize only regular languages.*

Proof. In the case of ultrametric integral automata multiplication by negative degrees of p is not allowed because these numbers are not p-adic integers. This means that p-norm of one single state of the ultrametric integral automaton can only decrease or remain the same. If the ultrametric automaton has an accepting threshold and accepting condition $\leq \lambda$, then the input word will become acceptable after a finite number of input symbols that increase the amplitude. If the ultrametric automaton has an accepting threshold and accepting condition $\geq \lambda$ we can see the opposite situation - the input word will become rejectable after a finite number of input symbols that increase the amplitude. That being said, we can see that the only thing that an ultrametric integral automaton with one state can do is count specific input symbols to a finite value to decide whether the input word is acceptable or not. The same task is also solvable by deterministic finite automata, which means that languages recognizable by ultrametric integral automaton with one state can only be regular.   □

To conclude the results of Theorem 4 we should also mention that some regular languages are not recognizable by ultrametric integral automata with one state, because one-letter alphabet languages containing only one word with a length greater than zero require at least two states [3].

5   Ultrametric Automata with Two States

In this section, we will see that a second state significantly increases the capabilities of ultrametric automata.

Theorem 5. *For every prime number p there exists a p-ultrametric automaton with two states that can recognize non-context-free language.*

Proof. In this proof, we will use the following language, which is not context-free: $L_2 = \{x | x \in \{0, 1, 2\}^* \text{ and } |x|_0 < |x|_1 \text{ and } |x|_1 < |x|_2\}$. We will construct an ultrametric automaton with two states to recognize the language L_2 (see Fig. 1).

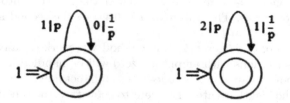

Fig. 1. Ultrametric automaton recognizing L_2

The following illustrations of ultrametric automata have some details that we need to describe, so it will be easier for the reader to understand the depicted automata. The big arrows with numbers denote the beginning amplitudes of the states. The transitions are shown as lines with arrows that connect two states (note they can also connect a state with itself). The symbols before the vertical line are input symbols and the symbol after the vertical line is the amplitude of transition.

Both states of the automaton in Fig. 1 act like independent counters. After reading the input word x, the amplitude of the left state becomes equal to $p^{|x|_1-|x|_0}$ and the amplitude of the right state becomes equal to $p^{|x|_2-|x|_1}$. That gives a p-norm of the left state's amplitude equal to $p^{|x|_0-|x|_1}$ and a p-norm of the right state's amplitude equal to $p^{|x|_1-|x|_2}$. If the input word belongs to L_2, both p-norms will not exceed $1/p$. The smallest prime number is 2, so the largest possible p-norm sum of the word that belongs to L_2 does not exceed $1/2 + 1/2 = 1$. Otherwise at least one state will have p-norm of at least 1, and this gives us a p-norm sum greater than 1. Therefore constructed p-ultrametric automaton will recognize L_2 for all prime numbers p. □

This result shows us the power of the ability of ultrametric automata to have independent amplitudes making independent counters. This example will be followed by better results about ultrametric automata with two states, and these results will show that we do not require different states to have multiple counters - they can be "hidden" in one common state.

Theorem 6. *For every prime number p there exists a p-ultrametric integral automaton with two states that has only one accepting state and can recognize non-context-free language.*

Proof. We will use the following language, which is not context-free: $L_3 = \{x | x = \{a, b, c, d\}^*$ *and* $|x|_a = |x|_b = |x|_c$ *before the first symbol* $d\}$. This time letter d will be like an end-marker after the part of the word that we are interested in. We will construct an ultrametric integral automaton with two states to recognize the language L_3 (see Fig. 2).

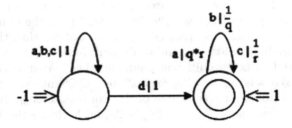

Fig. 2. Ultrametric automaton recognizing L_3

The accepting state has beginning amplitude 1. Now we take two distinct prime numbers q and r, with conditions $q \neq p$ and $r \neq p$. This allows us to divide by q and r in p-adic integers. After reading an input word x before the first symbol d, accepting state will have amplitude $q^{|x|_a-|x|_b} * r^{|x|_a-|x|_c}$. This number will be equal to 1 if and only if $|x|_a = |x|_b = |x|_c$. Then with the help of the second state, we subtract 1 from the amplitude of the accepting state after reading the first symbol d. After that, the amplitude of the non-accepting state will be zero, so the amplitude of the second state will remain zero if it was so after reading the first symbol d, or it will remain as a positive number otherwise. The automaton will accept the input word if and only if the p-norm of the accepting state does not exceed zero. The automaton will work similarly for any chosen prime number p. □

We can see that for ultrametric automata it is enough to have two states, only one being accepting and having a restriction of ultrametric integral automata to recognize a language that cannot be recognized by one-way nondeterministic pushdown automata.

6   Ultrametric Automata with Three States

Ultrametric automata with two states have shown great capabilities and it is interesting to consider what we might get if we add one more state. Rūsiņš Freivalds has shown that for every prime number $p \geq 3$ it is possible to construct an ultrametric automaton with three states which will recognize nonrecursive languages. This ultrametric automaton used p-adic numbers that are not p-adic integers. By having three states the ultrametric automaton surpassed the capabilities of Turing machines. We will demonstrate the theorem with the proof from [1]. After that, we will improve these results.

Theorem 7. *There is a continuum of languages recognizable by finite ultrametric automata [1].*

Proof. Let $\beta = ...2a_3 2a_2 2a_1 2a_0 2$ be an arbitrary p-adic number (not a p-adic integer), where $p \geq 3$ and all $a_i = \{0,1\}$. Denote by B the set of all possible such β. Consider an automaton A_β with 3 states, the initial amplitudes of the states being $(\beta, -1, -1)$. The automaton is constructed to have the following property: if the input word is $2a_0 2a_1 2a_2 2a_3 2...2a_n 2$ then the amplitude of the first state becomes $...2a_{n+4} 2a_{n+3} 2a_{n+2} 2a_{n+1} 2$. To achieve this, the automaton adds -2, multiplies by p, adds $-a_n$ and again multiplies by p.

Now let β_1 and β_2 be two different p-adic numbers. Assume that they have the same first symbols $a_m...2a_3 2a_2 2a_1 2a_0 2$ but different symbols a_{m+1} and b_{m+1}. Then the automaton accepts one of the words $a_{m+1} 2a_m...2a_3 2a_2 2a_1 2a_0 2$ and rejects the other one $b_{m+1} 2a_m...2a_3 2a_2 2a_1 2a_0 2$. Hence the languages are distinct [1]. □

After this proof we can make two conclusions. First, ultrametric automata can recognize nonrecursive languages. Second, this is possible for ultrametric automata with three states for every prime number $p \geq 3$. The natural next step to improve this result would be to prove that ultrametric automata with three states can recognize nonrecursive languages when $p = 2$. Although we had expected to achieve the latter result, in the process we also found that we were able to reduce the state complexity.

Theorem 8. *For every prime number p there exists a p-ultrametric automaton with two states that can recognize nonrecursive languages.*

Proof. Let $\beta = ...a_3a_2a_1a_0$ be an arbitrary p-adic number, which is not a p-adic integer, p be an arbitrary prime number and all $a_i = \{0,1\}$. We define a language L_β in the following way: a binary sequence belongs to L_β if and only if it is equal to the last digits of β. p-adic numbers can only be finite to the right side from a decimal point. Assume that number β has k p-adic digits after the decimal point. Now we can construct an ultrametric automaton to recognize L_β (see Fig. 3).

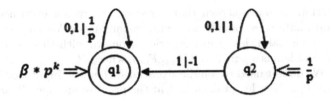

Fig. 3. Ultrametric automaton recognizing L_β

The accepting state $q1$ has a beginning amplitude $\beta * p^k$, where k is the number of p-adic digits after the decimal point in β. This gives us a p-adic number with the p-adic digits after the decimal point being zeroes. State $q2$ has a beginning amplitude p^{-1}. Assume that the input word is $c_0c_1c_2c_3...c_n$. If $c_i = 1$, p^{-1} will be subtracted from the amplitude of state $q1$, otherwise nothing will be subtracted $(0 * p^{-1})$. If i-th p-adic digit from the right in β was equal to c_i $(a_i = c_i)$, the remaining amplitude of $q1$ will have i-th p-adic digit equal to zero, and it will remain zero for all the time of the work of an automaton. If i-th p-adic digit from the right in β was not equal to c_i, $q1$ will have i-th p-adic digit different from zero (it will be 1 if β had $a_i = 1$ and the input word had the digit $c_i = 0$, or it will be $p - 1$ if β had the digit $a_i = 0$ and the input word had the digit $c_i = 1$).

When the whole input word $c_0c_1c_2c_3...c_n$ is read, the accepting state $q1$ will have an amplitude followed by $b_nb_{n-1}...b_0$ after the decimal point, where all the digits in $b_nb_{n-1}...b_0$ will be zeroes if and only if the input word belongs to L_β. If the input word does not belong to L_β, the amplitude of the state $q1$ will have at least one digit after the decimal point which is not zero. The p-norm of this number will be at least p. If the p-adic number does not have nonzero digits after the decimal point, its p-norm will not be greater than 1. An ultrametric automaton will accept the input word if the p-norm is less than or equal to 1. Otherwise, the input word will be rejected.

A constructed ultrametric automaton will accept the word $a_0a_1a_2a_3...$ a_ma_{m+1} and reject the word $a_0a_1a_2a_3...a_mb_{m+1}$ if all symbols $a_0a_1a_2a_3...a_m$ are equal and a_{m+1} and b_{m+1} are different. Hence, all possible languages L_β are distinct. The automaton will work in the same manner for all prime numbers p and is capable of recognizing nonrecursive languages L_β.   □

The results of the proven theorem show that ultrametric automata do not need even three states to recognize nonrecursive languages; it is enough to have two states. Moreover, we do not require both of them to be accepting. Recognizable nonrecursive language can also be depicted like a search on an infinite binary tree. We choose one infinite branch on the binary tree. Then we get as an input one particular path on the tree, which goes from the root to the depth of this binary tree. The automaton will accept the input path if and only if it belongs to the chosen path, in other words, if the path is equal to the prefix of the chosen branch.

In this section, we have shown that three states are not even necessary for ultrametric automata to recognize nonrecursive languages. We can provide other results about the capabilities of ultrametric automata with three states.

A promise problem is a pair $P = (P_{yes}, P_{no})$, where $P_{yes}, P_{no} \subseteq \Sigma^*$ and $P_{yes} \cap P_{no} = \emptyset$ [11]. We do not need to handle situations when the input word is outside of the set $P_{yes} \cup P_{no}$. Now let's take a look at the promise problem $PromiseEQ = (PromiseEQ_{yes}, PromiseEQ_{no})$, where $PromiseEQ_{yes} = \{a^m ba^m ba^n | m \neq n\}$ and $PromiseEQ_{no} = \{a^m ba^n ba^m | m \neq n\}$. To distinguish $PromiseEQ_{yes}$ from $PromiseEQ_{no}$ it is enough to check whether the length of the first block of symbols a is equal to the length of the second block of symbols a.

Theorem 9. *There exists a promise problem that cannot be solved by any bounded-error $o(\log\log n)$-space probabilistic Turing machines in sub-exponential expected time, but for every prime number p this problem can be solved by a one-way p-ultrametric integral automaton with three states [12].*

Proof. Promise problem $PromiseEQ$ cannot be solved by any bounded-error $o(\log\log n)$-space probabilistic Turing machines in sub-exponential expected

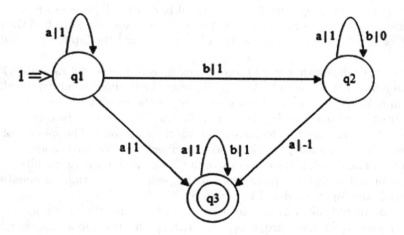

Fig. 4. Integral ultrametric automaton solving the problem $PromiseEQ$

time [13]. We can construct a p-ultrametric integral automaton to solve the problem *PromiseEQ* by comparing the lengths of the first two blocks of symbols a (see Fig. 4).

The illustrated automaton just adds 1 to the amplitude of state $q3$ after reading each symbol a in the first fragment and subtracts 1 from the amplitude of state $q3$ after reading each symbol a in the second fragment. The amplitude will be zero if and only if both fragments have an equal number of symbols a. All amplitudes are p-adic integers and the automaton works identically for all prime numbers p [12].   □

7 Summary

Previous research has found many examples of ultrametric automata having state complexity advantages over other types of automata. In this paper we have shown the capabilities of ultrametric automata with a small number of states. Ultrametric automata with one state are able to recognize non-regular languages, but cannot recognize languages that are not recognizable by one-way deterministic pushdown or counter automata. Ultrametric integral automata with one state cannot recognize non-regular languages.

Ultrametric automata with two states can recognize languages that are not context-free. This is also true for ultrametric integral automata with two states with just one state being accepting. Ultrametric automata with two states are also able to recognize nonrecursive languages and a third state is not necessary. Furthermore, ultrametric integral automata with three states can solve a problem that cannot be solved by any bounded-error $o(loglogn)$-space probabilistic Turing machines in sub-exponential expected time. All the proven results are true for p-ultrametric automata for every prime number p.

References

1. Freivalds, R.: Ultrametric finite automata and turing machines. In: Béal, M.-P., Carton, O. (eds.) DLT 2013. LNCS, vol. 7907, pp. 1–11. Springer, Heidelberg (2013)
2. Balodis, K., Berina, A., Cīpola, K., Dimitrijevs, M., Iraids, J., et al.: On the State Complexity of Ultrametric Finite Automata. In: SOFSEM 2013, Proceedings, vol. 2, pp. 1–9, Špindlerův Mlýn (2013)
3. Balodis, K.: Counting with probabilistic and ultrametric finite automata. In: Calude, C.S., Freivalds, R., Kazuo, I. (eds.) Gruska Festschrift. LNCS, vol. 8808, pp. 3–16. Springer, Heidelberg (2014)
4. Dimitrijevs, M., Ščeguļnaja, I., Freivalds, R.: Complexity Advantages of Ultrametric Machines. In: SOFSEM 2014, Proceedings, vol. 2, pp. 21–29, Nový Smokovec (2014)
5. Krišlauks, R., Rukšāne, I., Balodis, K., Kucevalovs, I., Freivalds, R., Nāgele, I.: Ultrametric Turing Machines with Limited Reversal Complexity. In: SOFSEM 2013. Proceedings, vol. 2, pp. 87–94, Špindlerův Mlýn (2013)
6. Kozyrev, S.V.: Ultrametric analysis and interbasin kinetics. In: 2nd International Conference on p-Adic Mathematical Physics, American Institute of Physics, pp. 121–128 (2006)

7. Vladimirov, V.S., Volovich, I.V., Zelenov, E.I.: P-Adic Analysis and Mathematical Physics. World Scientific (1995)
8. Madore, D.A.: A first introduction to p-adic numbers. http://www.madore.org/david/math/padics.pdf
9. Rabin, M.O.: Probabilistic Automata. Inf. Control **6**(3), 230–245 (1963)
10. Kravtsev, M.: Quantum finite one-counter automata. In: Bartosek, M., Tel, G., Pavelka, J. (eds.) SOFSEM 1999. LNCS, vol. 1725, pp. 431–440. Springer, Heidelberg (1999)
11. Watrous, J.: Quantum computational complexity. In: Meyers, R.A. (ed.) Encyclopedia of Complexity and Systems Science, pp. 7174–7201. Springer, New York (2009)
12. Dimitrijevs, M.: Ultrametric Finite Automata for Turing Machine Tasks of Various Complexity. Submitted (2015)
13. Rashid, J., Yakaryılmaz, A.: Implications of quantum automata for contextuality. In: Holzer, M., Kutrib, M. (eds.) CIAA 2014. LNCS, vol. 8587, pp. 318–331. Springer, Heidelberg (2014)

The Complexity of Paging Against a Probabilistic Adversary

Stefan Dobrev[1], Juraj Hromkovič[2], Dennis Komm[2(✉)], Richard Královič[3],
Rastislav Královič[4], and Tobias Mömke[5]

[1] Mathematical Institute, Slovak Academy of Sciences, Bratislava, Slovakia
stefan.dobrev@savba.sk
[2] Department of Computer Science, ETH Zürich, Zurich, Switzerland
{juraj.hromkovic,dennis.komm}@inf.ethz.ch
[3] Google Inc., Zurich, Switzerland
richard.kralovic@dcs.fmph.uniba.sk
[4] Department of Computer Science, Comenius University, Bratislava, Slovakia
kralovic@dcs.fmph.uniba.sk
[5] Saarland University, Saarbrücken, Germany
moemke@cs.uni-saarland.de

Abstract. We consider deterministic online algorithms for paging. The
offline version of the paging problem, in which the whole input is given
in advance, is known to be easily solvable. If the input is random,
chosen according to some known probability distribution, an $\mathcal{O}(\log k)$-
competitive algorithm exists. Moreover, there are distributions, where
no algorithm can be better than $\Omega(\log k)$-competitive.

In this paper, we ask the question of what happens if it is known
that the input is one from a set of ℓ potential candidates, chosen according to some probability distribution. We present an $\mathcal{O}(\log \ell)$-competitive
algorithm, and show a matching lower bound.

1 Introduction

In algorithmics, that is, the "study of algorithms" [12], one is concerned with
constructing and analyzing algorithms for given computing problems that perform well with respect to some given constraints, for instance, being efficient or
obtaining some specific solution quality. In a classical setup, an input is given
to an algorithm, and some particular information needs to be extracted. This
is usually done while having full knowledge about the input. For instance, the
information that needs to be extracted may be a cheapest Hamiltonian cycle
that is "hidden" in an instance that corresponds to a complete weighted graph.
Many computing problems, however, are what is called "intrinsically online"
which means that the whole input is not known in advance, but arrives gradually in consecutive time steps while a part of the definite output already needs

The research is partially funded by SNF grant 200021–146372, VEGA grant
1/0979/12, and Deutsche Forschungsgemeinschaft grant BL511/10-1.

R.M. Freivalds et al. (Eds.): SOFSEM 2016, LNCS 9587, pp. 265–276, 2016.
DOI: 10.1007/978-3-662-49192-8_22

to be created. Typical members of this class are scheduling or packing problems, and various kinds of resource management problems.

Such problems are called "online problems" [1,6,15,22], and they are found in many real-world situations. One of the most prominent and well-understood online problems is the paging (caching) problem. Here, an online algorithm maintains a cache containing up to k logical pages out of m possible ones. The input is a sequence of n requests for logical pages, and the algorithm has to process each request: if the requested page is in the cache (this is called a cache hit), nothing happens; if not (which is called a cache miss, or page fault), the algorithm has to evict one page from the cache, and replace it with the requested one. The goal of the algorithm is to process the input while minimizing the number of page faults. Let us give a formal definition; for the ease of presentation, we identify pages with their indices.

Definition 1 (Paging Problem). *An instance of the paging problem is a sequence of integers representing requests to logical pages $I = (x_1, x_2, \ldots, x_n)$, $x_i > 0$. An online algorithm ALG maintains a buffer (content of the physical memory) $B = \{b_1, b_2, \ldots, b_k\}$ of k integers, where k is a fixed constant known to ALG. Before processing the first request, the buffer gets initialized as $B = \{1, 2, \ldots, k\}$. Upon receiving a request x_i, if $x_i \in B$, then ALG creates the partial output $y_i = 0$. If $x_i \notin B$, then a page fault occurs, and ALG has to find some victim b_j, that is, $B := B \setminus \{b_j\} \cup \{x_i\}$, and $y_i = b_j$. The cost of the solution $\mathrm{ALG}(I)$ is the number of page faults, that is, $\mathrm{cost}(\mathrm{ALG}(I)) = |\{i \mid y_i > 0\}|$.*

In this paper, we consider deterministic online algorithms for the paging problem. The performance of an algorithm is measured by the *competitive ratio* (which basically is the online counterpart of the *approximation ratio*[1] for offline problems) where the cost of the algorithm on a particular input I is compared to an optimal solution for I. In a very general setting, one may consider the inputs to come from a probability distribution ρ over all possible inputs. The quality of a solution depends on the *data model* that specifies two orthogonal aspects of the setting. First, the data model specifies the class of possible input distributions $\mathcal{P}$; an algorithm ALG is called *c-competitive* if there is a constant α, such that

$$\forall \rho \in \mathcal{P} \colon \mathbb{E}_\rho \left[\frac{\mathrm{cost}(\mathrm{ALG}(I)) - \alpha}{\mathrm{cost}(\mathrm{OPT}(I))} \right] \leq c \tag{1}$$

where the instance I is taken according to the distribution ρ. Second, the data model specifies what information about the distribution ρ is known to the algorithm.

When $\mathcal{P}$ is the class of all point mass distributions (that is, distributions where one particular input has probability 1), the impact of the algorithm's knowledge has been widely studied. One extreme case is when the whole point mass distribution (that is, the input) is known to the algorithm; this corresponds to the

---

[1] Note that unlike offline algorithms, in an online setting we usually ignore the running time of the algorithm.

offline deterministic case, which is easily solvable by a greedy algorithm (LFD, longest forward distance, called MIN by Bélády [2] who first proved its optimality). The other extreme case, when the point mass distribution is unknown, corresponds to the online deterministic worst-case scenario, and it is known [21] that no deterministic online algorithm can be better than k-competitive. The spectrum between these two extremes has been studied by means of advice complexity [4,5,11,14], which was first studied for paging by Dobrev et al. [10].

Orthogonally, in what is known as *distributional* approach [6] to the analysis of online algorithms, it is supposed that the inputs come from a fixed distribution (that is, $\mathcal{P}$ is a singleton, in our setting), which usually is a uniform distribution. Franaszek and Wagner [13] studied the particular input distribution where each request is selected independently from a given distribution. Note that this was done before competitive analysis was introduced by Sleator and Tarjan [21]. In the *Markov paging* model by Shedler and Tung [20], the inputs are generated by a Markov chain. Pandurangan and Upfal [18] studied a setting where the page requests come from a stochastic process with given entropy.

There are also approaches where the worst case from several distributions is analyzed. Notably, in the *access graph* model by Borodin et al. [7], $\mathcal{P}$ is the set of point mass distributions corresponding to walks in a (known) graph G and the particular distribution ρ is, of course, unknown. The *statistical adversary* introduced by Raghavan [19] considers (in the role of $\mathcal{P}$ in our notation) the class of all point mass distributions that fulfill certain statistical properties, for instance, each page is requested the same number of times. This model was later sucessfully applied to two-way currency trading by Chou et al. [9].

Our setting follows the general notion of the *diffuse adversary* introduced by Koutsoupias and Papadimitriou [17] with the only distinction that in the diffuse adversary model, the particular distribution is always unknown. The diffuse adversary is a generalization of the statistical adversary (and other models that have been introduced in the past). In 1998, Young further studied the diffuse adversary [25]; he gave both tight bounds (up to a factor of 2) for deterministic and randomized online algorithms in this setting.

As pointed out before [1,3,8,15,17,23], we argue that both extreme cases, that is, either knowing *nothing* about the input, or knowing *everything* about it, are unrealistic. As the requested pages depend on the user's actions, it is clearly impossible for any operating system to completely foresee which pages will be accessed in the future. On the other hand, it seems equally unrealistic to assume that *every* input sequence is possible.

In Sect. 2, we briefly state our contribution, and put it into context. We formally state and prove our results in Sect. 3; we conclude in Sect. 4. Throughout this paper, log denotes the logarithm with base 2.

2   Our Contribution

On one hand, a known point mass distribution corresponds to the offline case, and it is easily solvable as already mentioned [6]. Also, it follows from Yao's principle [24] and the results on randomized paging, that, for any known point mass distribution, there is an $\mathcal{O}(\log k)$-competitive algorithm, and there are such distributions where any algorithm is at least $\Omega(\log k)$-competitive. Moreover, as shown by Komm and Královič [16], it follows that, for any point mass distribution, there is an $\mathcal{O}(\log k)$-bit long binary string from which the $\mathcal{O}(\log k)$-competitive algorithm can be efficiently decoded.

Motivated by this, we analyze known distributions that are somewhat "close" to being point mass: instead of the probability mass being concentrated in one input, it can be distributed arbitrarily among ℓ inputs.

Definition 2 (Class $\mathcal{P}_\ell$ of Distributions). *By $\mathcal{P}_\ell$ we denote the class of ℓ-point distributions, that is, probability distributions over inputs such that at most ℓ inputs have non-zero probability.*

An online algorithm for paging that only evicts pages in case of a page fault is called a "demand paging" (for k-server, a generalization of paging, the term "lazy" is more common) algorithm. The laziness requirement comes with no loss of generality [6], hence we shall consider only lazy algorithms to obtain easier arguments. Note that we incorporated this already in our formal definition of paging that only allows an algorithm to evict a page if a page fault occurs.

We analyze the expected competitive ratio of deterministic paging algorithms over a known ℓ-point distribution. We show that, if ℓ is constant, there is an "almost optimal" (that is, 1-competitive) algorithm. Basically, we prove that, since ℓ is fixed, the overhead the algorithm pays until it realizes which input it is working on, can be hidden in the additive constant α from the definition of the competitive ratio. For the *strict* competitive ratio (that is, when demanding that $\alpha = 0$) we show an $\mathcal{O}(\log \ell)$-competitive algorithm for $\ell < k$ (for $\ell \geq k$, one can use the $\mathcal{O}(\log k)$-competitive algorithm). We complement the result by a matching lower bound stating that no online algorithm can be better than $(\log \ell)/2$-competitive.

3   Results

We start with the non-strict case, that is, the case where the additive constant α from the definition of the competitive ratio may be strictly positive. From the order of the quantifiers in (1), it follows that α may depend on the cache size k, and the class of possible distributions (in our case parametrized by ℓ). In this case, we can prove the following theorem.

Theorem 1. *There is a 1-competitive paging algorithm for any known distribution from the class $\mathcal{P}_\ell$, for any constant ℓ.*

Before presenting an online algorithm that obtains this bound, let us make the following observation.

Lemma 1. *Consider two inputs I_1 and I_2. Let OPT_1 and OPT_2 be the optimal (LFD) algorithms for I_1 and I_2, respectively. Then OPT_1 and OPT_2 make the same number of page faults on the common prefix of I_1 and I_2.*

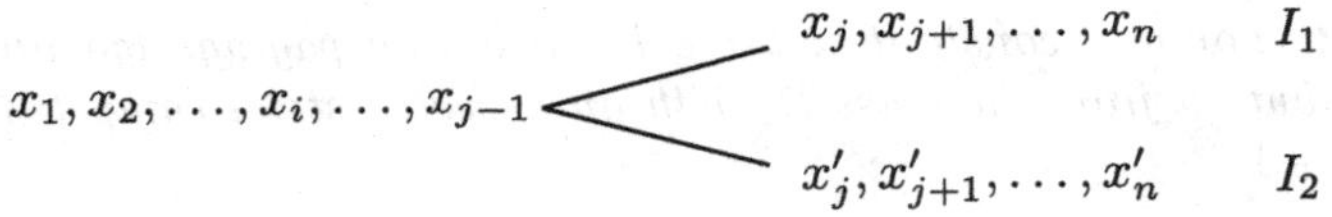

$$x_1, x_2, \ldots, x_i, \ldots, x_{j-1} \left< \begin{array}{ll} x_j, x_{j+1}, \ldots, x_n & I_1 \\ \\ x'_j, x'_{j+1}, \ldots, x'_n & I_2 \end{array} \right.$$

Fig. 1. I_1, I_2, and the branching point

Proof. Let j be the first position where I_1, and I_2 differ (see Fig. 1). To prove the claim by contradiction, suppose that there is a page requested in the prefix $x_1, x_2, \ldots, x_{j-1}$ such that it causes a page fault for OPT_1 but not for OPT_2; let x_i be the first page with this property. OPT_1 evicted this page in some preceding time step as it was requested farthest in the future, namely in time step i. But since this happened on the common prefix of I_1 and I_2 (that is, $i \leq j-1$), OPT_2 would have taken the same action.   □

We now prove Theorem 1 using Lemma 1.

Proof (of Theorem 1). Let $\rho \in \mathcal{P}_\ell$ be the input distribution, and let $I_1, I_2, \ldots, I_\ell$ denote the inputs that are chosen with non-zero probability sorted in non-increasing order $\rho(I_1) \geq \rho(I_2) \geq \cdots \geq \rho(I_\ell)$. Let OPT_i be the optimum (LFD) solution for I_i, $1 \leq i \leq \ell$.

The algorithm ALG starts by simulating OPT_1. If the actual input is I_1, ALG is optimal. If, at some point, ALG realizes that the instance is not I_1, ALG switches to OPT_2: it replaces the cache by the pages that would have been in the cache of OPT_2 at this moment, and continues as OPT_2 (actually, since we are considering lazy algorithms exclusively, ALG only remembers the state, and replaces the pages as needed). Let j be the time step where I_1 and I_2 differ for the first time; and thus, if ALG is not optimal on I_2, the actions of OPT_1 and OPT_2 differ in some time step $i < j$.

From Lemma 1 it follows that, after ALG switched from OPT_1 to OPT_2, the number of page faults so far was the same as in OPT_2 plus the at most k faults needed to change the cache content to be consistent with OPT_2.

This process can be iterated: ALG simulates OPT_2 until it sees a difference in the input, switches to OPT_3, and so on. After any switch to OPT_d, the cost of ALG so far is bounded by the cost of OPT_d plus at most kd page faults needed to replace the cache after each switch. Overall, since there are at most ℓ switches, we have

$$\mathrm{cost}(\mathrm{ALG}) \leq \mathrm{cost}(\mathrm{OPT}) + k\ell.$$

Since this inequality holds for any execution, (1) holds for $\alpha = k\ell$, which finishes the proof.   □

The previous theorem asserts that the price of each switch of the algorithm ALG is at most k. If we consider the *strict* competitive ratio, in which $\alpha = 0$ in (1), this price is too high, since for any known distribution there is a $\mathcal{O}(\log k)$-competitive algorithm. In the following theorem, we present a more detailed accounting of the expected price of switching. In the proof, we again make use of Lemma 1.

Theorem 2. *For any constant ℓ, there is an online paging algorithm for any known distribution from the class $\mathcal{P}_\ell$ with an expected strict competitive ratio of at most $\ln \ell + 1$.*

Proof. Let $\rho \in \mathcal{P}_\ell$, and let $\mathcal{I} := \{I_1, I_2, \ldots, I_\ell\}$ be the instances that are chosen with non-zero probability. ALG computes the prefix tree T of $\mathcal{I}$ with root r. Note that in T there is exactly one node for each distinct prefix in any of the request strings in $\mathcal{I}$, and any instance is represented by a path from r to a distinct leaf of T. Each node of T that is not a leaf is labeled by its requested page, that is, the page that distinguishes its prefix from its predecessor's prefix. The root and the leaves obtain an empty label. We define $N^+(v)$ to be the out-neighborhood of a node v, that is, the children of v.

For each node v of T, we define a probability $p(v)$ inductively. If v is a leaf, then there is a exactly one instance $I_v \in \mathcal{I}$ that corresponds to a path from r to v and we set $p(v) = \rho(I_v)$. If v is no leaf but all vertices in $N^+(v)$ already have been assigned probabilities, we set

$$p(v) = \sum_{w \in N^+(v)} p(w).$$

Clearly, eventually all nodes are labeled and $p(r) = 1$.

We now identify a subgraph $\mathcal{T}$ of T as follows. Initially, $\mathcal{T}$ has all nodes of T but no arcs. Then for each node v, we introduce exactly one arc (v, w), where

$$w = \arg\max_{w' \in N^+(v)} p(w'),$$

breaking ties arbitrarily. Note that $\mathcal{T}$ is a collection of directed paths (possibly of length 0) that end in leaves of T. For each node v, denote by P_v the suffix of the path in $\mathcal{T}$ that contains v where v is the start vertex of P_v. Then the strategy of ALG is to move within $\mathcal{T}$ according to the requests and to follow the LFD strategy of the instance defined by the labels of P_v, where v is the currently visited node. Note that after requests the strategy may change. For a leaf w, OPT_w is the optimal LFD solution for the instance defined by the path from r to w. To estimate the impact of strategy changes, we use the following claim that follows by applying Lemma 1 to all pairs of vertices that are reachable from a given vertex.

Claim. Let v be a node of T, and let S be the set of leaves reachable from v. Then the number of page faults on the prefix of instances defined by the path from r to v are identical for all solutions OPT_w with $w \in S$.

The claim allows us to estimate the cost of changing strategies. For each pair of leaves w, w', there is a vertex v such that the paths from r to w and from r to w' fork at v. There is a *critical phase* from where OPT_w and $\text{OPT}_{w'}$ differ first until v, and there is some number κ of differences between the two solutions within the critical phase. Therefore, changing the strategy from OPT_w to $\text{OPT}_{w'}$ causes at most κ page faults and $\kappa \leq \text{cost}(\text{OPT}_{w'})$. As a consequence, the number of different strategies is an upper bound on the attained competitive ratio. In the remaining analysis, we give an upper bound on the expected number of strategy changes.

Let us fix an internal node v with $s := |N^+(v)| > 0$, that is, v is not a leaf. Furthermore, let w be the subsequent node in P_v. For each $w' \in N^+(v)$ we define

$$p'(w') = \frac{p(w')}{\sum_{w'' \in N^+(v)} p(w'')}.$$

In other words, $p'(w')$ is the probability that the given path continues with w', provided that it contains v. Then the probability to change the strategy after v is $1 - p'(w)$. We now give an upper bound on the fraction of reachable leaves left after leaving v.

For any node w', let leaves(w') be the set of leaves reachable from w'. Then, for each $w' \in N^+(v)$, we define the fraction of leaves reachable from v via w' by

$$\gamma(v, w') := \frac{|\text{leaves}(w')|}{\sum_{w'' \in N^+(v)} |\text{leaves}(w'')|}.$$

Then the expected fraction of leaves left after leaving v is

$$\sum_{w' \in N^+(v)} p'(w')\gamma(v, w').$$

This number is maximized if $\gamma(v, w) = 1$ and $\gamma(v, w') = 0$ for all $w' \neq w$. We obtain the upper bound $\gamma(v, w) \leq p'(w)$.

Therefore, we obtain an upper bound on the total number of strategy changes if we consider an integer t and a sequence $(p_i)_{i=1}^t$ of probabilities such that

$$\sum_{i=1}^{t}(1 - p_i)$$

is maximized subject to

$$\prod_i p_i \geq 1/\ell.$$

The meaning of the objective is that there are t nodes with probabilistic decisions. The probability of the selected strategy at the i-th node is p_i and thus there is a strategy change with probability $1 - p_i$. Intuitively, the maximum is attained when both the length t of the sequence is long and the probabilities to change strategies are large. However, for increasing values of t, the constraints enforce that most of the $1 - p_i$ are small. The constraints stem from the fact that there is no leaf left if the fraction of remaining leaves is smaller than $1/\ell$.

We claim that the maximum can be attained with $p_i = p_{i'}$ for all pairs of indices i, i'. Suppose towards contradiction that there is no maximal solution with this property. Let $p_1, p_2, \ldots, p_t$ be a solution attaining the maximum such that

$$\mu := \max_{i, i'} |p_i - p_{i'}|$$

is minimal and among these solutions one where

$$|\{(i, i') \mid |p_i - p_{i'}| = \mu\}|$$

is minimal. Let us fix two indices $\hat{i}, \hat{i}'$ such that $|p_{\hat{i}} - p_{\hat{i}'}| = \mu$. Now let us consider the solution $p'_1, p'_2, \ldots, p'_t$ where $p'_i = p_i$ for all indices except $\hat{i}$ and $\hat{i}'$ and where

$$p'_{\hat{i}} = p'_{\hat{i}'} = \frac{p_{\hat{i}} + p_{\hat{i}'}}{2}.$$

Clearly, still $\max_{i,i'} |p'_i - p'_{i'}| \le \mu$. Also, the value of the objective function did not change since $p_{\hat{i}} + p_{\hat{i}'} = p'_{\hat{i}} + p'_{\hat{i}'}$. To show that the constraints are satisfied, we claim that

$$p_{\hat{i}} p_{\hat{i}'} \le p'_{\hat{i}} p'_{\hat{i}'}.$$

By renaming the indices, we assume without loss of generality that $p_{\hat{i}} \le p_{\hat{i}'}$. Then we have

$$
\begin{aligned}
p'_{\hat{i}} p'_{\hat{i}'} &= \left(\frac{p_{\hat{i}} + p_{\hat{i}'}}{2} \right)^2 \\
&= \frac{p_{\hat{i}}^2 + 2 p_{\hat{i}} p_{\hat{i}'} + p_{\hat{i}'}^2}{4} \\
&= p_{\hat{i}} p_{\hat{i}'} + \frac{p_{\hat{i}}^2}{4} - \frac{p_{\hat{i}}(p_{\hat{i}} + \delta)}{2} + \frac{(p_{\hat{i}} + \delta)^2}{4} \\
&= p_{\hat{i}} p_{\hat{i}'} + \frac{\delta^2}{4}
\end{aligned}
$$

where $\delta = p_{\hat{i}'} - p_{\hat{i}}$. Therefore, unless $\delta = 0$,

$$|\{(i, i') \mid |p'_i - p'_{i'}| = \mu\}| < |\{(i, i') \mid |p_i - p_{i'}| = \mu\}|,$$

which is a contradiction to the minimality of $|\{(i, i') \mid |p_i - p_{i'}| = \mu\}|$. Hence, from now on we may assume that $p_i = p$ for some probability p and all indices i. We obtain

$$t = \log_p p^t = \log_p(1/\ell) = \frac{\ln \ell}{\ln(1/p)}.$$

As a consequence, we have to find

$$\max_p \frac{(1 - p) \ln \ell}{\ln(1/p)}.$$

We have

$$\frac{\partial}{\partial p} \frac{(1 - p) \ln \ell}{\ln(1/p)} = \ln \ell \frac{\ln p - 1 + 1/p}{\ln^2 p}.$$

For $p \in (0, 1)$, the derivative is always positive since $1/p > 1$. Thus,

$$\frac{(1 - p) \ln \ell}{\ln(1/p)}$$

is monotonously increasing in p and the maximum is attained for $p \to 1$. On a high level this means that t is large whereas the probability of strategy changes, $1 - p$, is small. Now the number of strategy changes is bounded from above by

$$\lim_{p \to 1} \frac{(1 - p) \ln \ell}{\ln(1/p)} = \lim_{p \to 1} \frac{-\ln \ell}{-1/p} = \ln \ell \quad .$$

where we used l'Hôspital's rule. With our previous discussion, the statement of the theorem follows.   □

Finally, we argue that the algorithm from Theorem 2 is in a sense best possible by proving the following lower bound.

Theorem 3. *Any online algorithm* ALG *on the class* $\mathcal{P}_\ell$ *with* $\ell \leq k$ *has an expected strict competitive ratio of at least* $(\log \ell)/2$.

Proof. First, assume that ℓ is a power of 2; without loss of generality, let ALG be a demand paging (that is, lazy) algorithm. We assume that the cache is always organized such that the page indices of all pages residing in the cache at any given point in time are in increasing order in every time step; we do this without loss of generality and to keep our arguments simple and not being forced to argue about permutations of the cache cells. We describe a class $\mathcal{I}$ of ℓ instances, and the probability distribution ρ will be a uniform distribution over $\mathcal{I}$. Let us assume that initially the cache contains pages $1, 2, \ldots, k$.

All instances in $\mathcal{I}$ start by introducing a page $k + 1$, and particular instances can be described by a number i, such that in I_i, the optimal algorithm evicts page i from the cache in the first time step. For all I_i, the optimal cost will be one, so this is the only page fault the optimal algorithm incurs. In each instance, the first request is followed by $\log \ell$ rounds, and we show that any algorithm causes a page fault in every round with probability at least $1/2$. Together with the one page fault in the first time step, this gives the expected number of faults, and also the expected competitive ratio.

Consider any algorithm ALG running on an instance I_i. Every round starts and ends by a sequence requesting pages $\ell + 1, \ell + 2, \ldots, k$; if these pages are not in the cache of ALG at the beginning of the round, ALG makes a page fault with probability 1, and we are done. Hence, we can assume that ALG never evicts pages from the range $\ell + 1, \ell + 2, \ldots, k$ from the cache. Let us call the pages $1, 2, \ldots, \ell$ *active*.

For round 2, the active pages are partitioned into two halves, and all pages of that half that does not contain i are requested consecutively (see Fig. 2). Clearly, ALG makes another page fault with probability at least $1/2$. This procedure is applied recursively to the half that contains i until, in round $\log \ell + 1$, there are only two halves that contain single pages (out of which one is the page i). Moreover, in every round, all pages that were requested in the previous round are requested again.

·

Initial cache:	17	1	2	3	4	5	6	7	8	9	10	11	12	13	14	15	16
Round 1:	17																
Round 2:	17									9	10	11	12	13	14	15	16
Round 3:	17	1	2	3	4					9	10	11	12	13	14	15	16
Round 4:	17	1	2	3	4			7	8	9	10	11	12	13	14	15	16
Round 5:	17	1	2	3	4		6	7	8	9	10	11	12	13	14	15	16

Fig. 2. Example for $k = \ell = 16$; the optimal solution removes page 5 in time step one, when page 17 is requested; after that, there are 4 rounds that each consist of requests that were in the cache of OPT at the beginning

It follows that, in each round r, $2 \leq r \leq \log \ell + 1$, ALG makes a page fault with probability at least $1/2$ (note that the corresponding events are all independent) and an additional page fault in round 1 with probability 1. At the same time, OPT makes exactly 1 page fault in total. Summing up, the ratio of the costs is $(\log \ell)/2 + 1$.

Finally, assume that ℓ is not a power of 2. Let ℓ' denote the largest power of 2 that is smaller than ℓ. We follow the exact same strategy as above with ℓ' instead of ℓ; in particular, we request all pages $\ell' + 1, \ell' + 2, \ldots, k$ in every round. By definition, we have $\ell' > \ell/2$, thus there is at most one less round, and consequently the lower bound decreases by at most 1.   □

4   Conclusion

We studied the case of paging against a known distribution. On one hand, there are distributions where no algorithm can perform better than being $\Omega(\log k)$-competitive; on the other hand, for a point mass distribution, an easy optimal algorithm exists. We addressed the general question of characterizing the distributions in terms of the complexity of paging algorithms for them. We showed that if the distribution has at most ℓ inputs that are chosen with non-zero probability each, there is an $(\ln \ell + 1)$-competitive online algorithm. Complementing, by constructing a class of hard instances, we showed that this bound is tight up to a small factor when $\ell \leq k$. In the case that the additive constant α from the competitive ratio is allowed to be positive, there is a simple 1-competitive online algorithm where $\alpha = k\ell$.

Acknowledgement. The authors would like to thank Hans-Joachim Böckenhauer for very valuable discussions.

References

1. Albers, S.: Online algorithms: a survey. Math. Program. **97**(1), 3–26 (2003)
2. Bélády, L.A.: A study of replacement algorithms for virtual-storage computer. IBM Syst. J. **5**(2), 78–101 (1966)
3. Ben-David, S., Borodin, A.: A new measure for the study of on-line algorithms. Algorithmica **11**(1), 73–91 (1994)
4. Böckenhauer, H.-J., Komm, D., Královič, R., Královič, R., Mömke, T.: On the advice complexity of online problems. In: Dong, Y., Du, D.-Z., Ibarra, O. (eds.) ISAAC 2009. LNCS, vol. 5878, pp. 331–340. Springer, Heidelberg (2009)
5. Böckenhauer, H.-J., Komm, D., Královič, R., Královič, R.: On the advice complexity of the k-server problem. In: Aceto, L., Henzinger, M., Sgall, J. (eds.) Automata, Languages and Programming. LNCS, vol. 6755, pp. 207–218. Springer, Heidelberg (2011)
6. Borodin, A., El-Yaniv, R.: Online Computation and Competitive Analysis. Cambridge University Press, New York (1998)
7. Borodin, A., Irani, S., Raghavan, P., Schieber, B.: Competitive paging with locality of reference. J. Comput. Syst. Sci. **50**(2), 244–258 (1995)
8. Boyar, J., Larsen, K.S., Nielsen, M.N.: The accommodating function: a generalization of the competitive ratio. SIAM J. Comput. **31**(1), 233–258 (2001)
9. Chou, A., Cooperstock, J., El-Yaniv, R., Klugerman, M., Leighton, T.: The statistical adversary allows optimal money-making trading strategies. In: Proceeding of SODA 1995, pp. 467–476. Society for Industrial and Applied Mathematics (1995)
10. Dobrev, S., Královič, R., Pardubská, D.: How much information about the future is needed? In: Geffert, V., Karhumäki, J., Bertoni, A., Preneel, B., Návrat, P., Bieliková, M. (eds.) SOFSEM 2008. LNCS, vol. 4910, pp. 247–258. Springer, Heidelberg (2008)
11. Emek, Y., Fraigniaud, P., Korman, A., Rosén, A.: Online computation with advice. Theor. Comput. Sci. **412**(24), 2642–2656 (2011)
12. Harel, D., Feldman, Y.: Algorithmics: The Spirit of Computing. Addison-Wesley, 3rd edn (2004)
13. Franaszek, P.A., Wagner, T.J.: Some distribution-free aspects of paging algorithm performance. J. ACM **21**(1), 31–39 (1974)
14. Hromkovič, J., Královič, R., Královič, R.: Information complexity of online problems. In: Hliněný, P., Kučera, A. (eds.) MFCS 2010. LNCS, vol. 6281, pp. 24–36. Springer, Heidelberg (2010)
15. Irani, S., Karlin, A.R.: On online computation. In: Hochbaum, D.S. (ed.) Approximation Algorithms for NP-hard Problems, pp. 521–564. PWS Publishing Company (1997)
16. Komm, D., Královič, R.: Advice complexity and barely random algorithms. Theor. Inf. Appl. (RAIRO) **45**(2), 249–267 (2011)
17. Koutsoupias, E., Papadimitriou, C.H.: Beyond competitive analysis. SIAM J. Comput. **30**(1), 300–317 (2000)
18. Pandurangan, G., Upfal, E.: Entropy-based bounds for online algorithms. ACM Trans. Algorithms **3**(1), 1–19 (2007)
19. Raghavan, P.: A statistical adversary for on-line algorithms. DIMACS **7**, 79–83 (1991)
20. Shedler, G.S., Tung, C.: Locality in page reference strings. SIAM J. Comput. **1**, 218–241 (1972)

21. Sleator, D.D., Tarjan, R.E.: Amortized efficiency of list update and paging rules. Communications of the ACM **28**(2), 202–208 (1985)
22. Fiat, A. (ed.): Online Algorithms 1996. LNCS, vol. 1442. Springer, Heidelberg (1998)
23. Fiat, A., Woeginger, G.J.: Competitive odds and ends. In: Fiat, A., Woeginger, G.J. (eds.) Online Algorithms 1996. LNCS, vol. 1442, pp. 385–394. Springer, Heidelberg (1998)
24. Yao, A.C.-C.: Probabilistic computations: Toward a unified measure of complexity (extended abstract). In: Proceeding of FOCS 1977, pp. 222–227. IEEE Computer Society (1977)
25. Young, N.E.: Bounding the diffuse adversary. In: Proceeding of SODA 1998, pp. 420–425. Society for Industrial and Applied Mathematics (1998)

On Parity Game Preorders
and the Logic of Matching Plays

M.W. Gazda and T.A.C. Willemse[⊠]

Eindhoven University of Technology, Eindhoven, The Netherlands
t.a.c.willemse@tue.nl

Abstract. Parity games can be used to solve satisfiability, verification
and controller synthesis problems. As part of an effort to better understand their nature, or the nature of the problems they solve, preorders on
parity games have been studied. Defining these relations, and in particular proving their transitivity, has proven quite difficult on occasion. We
propose a uniform way of lifting certain preorders on Kripke structures
to parity games and study the resulting preorders. We explore their relation with parity game preorders from the literature and we study new
relations. Finally, we investigate whether these preorders can also be
obtained via modal characterisations of the preorders.

1 Introduction

Parity games [6,15] are two player games played on a directed graph. These
games are interesting as they underpin, *e.g.* solutions to verification, satisfiability and synthesis problems, see [2,7] and they appear as solution to fundamental problems such as complementing tree automata [6]. The problem of solving
a parity game (computing the set of vertices won by each player) is one of those
rare problems that are in NP∩coNP, and for which no polynomial time algorithm
has yet been found.

In an effort to increase the general understanding of the parity game solving problem or of those problems mapped to parity game solving, preorders on
parity games have been studied on various occasions and for different purposes.
For instance, in [13], Namjoshi investigated simulation in the context of abstraction using a variant of parity games called *model checking games*, leading to
a framework that was *complete* (in the sense of having finite abstract objects)
for the μ-calculus; in [8], Fritz and Wilke [8] defined and studied *delayed simulation*, an adaptation of simulation; Cranen *et al.* [3,4] studied variations of
stuttering bisimulation for parity games; Kissig and Venema [12] defined *basic
game bisimulation* for studying complementation. Dawar and Grädel [5] defined
yet two other forms of bisimulation on parity games by viewing these as relational structures; their purpose is to analyse the descriptive complexity of parity
games.

For the most part, the preorders on parity games are inspired by similar relations on computational models such as Kripke structures or Labelled Transition

© Springer-Verlag Berlin Heidelberg 2016
R.M. Freivalds et al. (Eds.): SOFSEM 2016, LNCS 9587, pp. 277–289, 2016.
DOI: 10.1007/978-3-662-49192-8_23

Systems. However, there seems to be no systematic method by which the afore-mentioned parity game relation have been obtained from a relation on a computational model, as testified by the existence of several variations of bisimulation on parity games. Moreover, defining new relations on parity games, showing transitivity and proving that they approximate the winning set of some player can be quite cumbersome; for instance, proving transitivity of delayed simulation required analysing 24 different cases, see [8], and in [4], the proof that governed stuttering bisimilarity is an equivalence relation is technically involved, requiring a step-wise rephrasing of the definition and intricate arguments.

The contributions of this paper are as follows. We propose a novel, more generic method for obtaining a parity game relation from a relation on a computational model. It is based on the notion of matching paths, which we lift naturally to matching plays. In this approach, we can lift any preorder on a computational model that can be specified using the matching paths to a corresponding preorder in the parity game setting. Moreover, we identify conditions that guarantee that the resulting relation is a preorder and that it approximates the winning set of a particular player.

We exemplify our approach using a number of well-known and some less-known relations on Kripke structures and show that some of the thus obtained relations coincide with existing parity game relations. Finally, for all the relations we study in detail, we provide logical characterisations, by identifying sound and complete fragments of an alternating-time temporal logic for the respective relations. The logical characterisations reveal interesting differences between the behavioural relations and the induced parity game relations.

Structure. In Sect. 2, we recall the basics of parity games. Then, in Sect. 3, we introduce our generic scheme for obtaining parity game relations. In Sect. 4, we show how this theory can be applied to recover existing and define new parity game preorders; in Sect. 5, we provide modal characterisations of these relations. We finish with conclusions in Sect. 6. Proofs for all results can be found in [10].

2   Preliminaries

A parity game is an infinite duration game, played by players *odd*, denoted by $\square$ and *even*, denoted by $\lozenge$, on a directed, finite graph.

Definition 1. *A parity game is a tuple $\langle V, E, \Omega, (V_\lozenge, V_\square) \rangle$, where*

- *V is a set of vertices, partitioned in a set $V_\lozenge$ of vertices owned by player $\lozenge$, and a set of vertices $V_\square$ owned by player $\square$,*
- *$E \subseteq V \times V$ is a total edge relation, i.e. for all v, $(v, w) \in E$ for some w,*
- *$\Omega : V \to \mathbb{N}$ is a priority function that assigns priorities to vertices.*

We depict parity games as graphs in which diamond-shaped vertices represent vertices owned by player $\lozenge$, box-shaped vertices represent vertices owned by player $\square$ and priorities, associated with vertices, are written inside vertices; see Fig. 1(a) and (b) for examples.

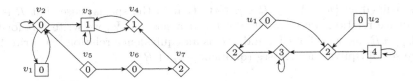

Fig. 1. (a) Left: A parity game with three different priorities. (b) Right: A parity game with four different priorities.

We use the following notational conventions: we write $v \to w$ instead of $(v, w) \in E$, and we write $v^\bullet$ for the set $\{w \in V \mid v \to w\}$. Henceforth, $\bigcirc$ denotes an arbitrary player. We write $\bar{\bigcirc}$ for $\bigcirc$'s opponent; *i.e.* $\bar{\Diamond} = \Box$ and $\bar{\Box} = \Diamond$. Finally, given the set of vertices V, the subset of vertices of V with priority n is denoted V_n: we have $V_n = \{v \in V \mid \Omega(v) = n\}$.

A play starts by placing a token on some vertex $v \in V$. Players $\Diamond$ and $\Box$ move the token indefinitely according to a single simple rule: if the token is on some vertex that belongs to player $\bigcirc$, that player gets to move the token to an adjacent vertex. The *parity* of the highest priority that occurs infinitely often on a play defines the *winner* of the play: player $\Diamond$ wins if, and only if this priority is even. This is known as the *parity condition*.

A strategy for player $\bigcirc$ is a partial function $\sigma:V^* \to V$ satisfying that for all sequences of vertices $u_1 \cdots u_n \in V^*$ on which σ is defined, both $u_n \in V_\bigcirc$ and $\sigma(u_1 \cdots u_n) \in u_n^\bullet$. The set of all strategies for player $\bigcirc$ is denoted $\mathbb{S}_\bigcirc^*$.

Let $\pi = v_1\, v_2\, v_3 \cdots$, with $v_1 = v$ be a play starting in v. We denote the i-th vertex on π by π_i; that is, $\pi_i = v_i$. Play π is *consistent* with a strategy $\sigma \in \mathbb{S}_\bigcirc^*$ if all prefixes $v_1 \cdots v_n$ of π for which $\sigma(v_1 \cdots v_n)$ is defined, we have $v_{n+1} = \sigma(v_1 \cdots v_n)$. The set of plays consistent with strategy σ, starting in v is denoted $\mathsf{Plays}(\sigma, v)$; we sometimes refer to this set as the set of σ-plays. A strategy σ is winning for player $\bigcirc$ from vertex v iff all plays consistent with σ are won by $\bigcirc$. Vertex v is won by player $\bigcirc$ whenever she has a winning strategy for vertex v.

Example 1. In the parity game depicted in Fig. 1(a), v_1, v_2, v_5 are won by player $\Diamond$, and player $\Box$ wins the remaining vertices.

Let $C, D \subseteq V$ be sets of vertices. We generalise the one-step reachability relation E to (forced) reachability, where reachability is confined to a set of vertices. Let $v \in V$.

$$v_\bigcirc{\to} D = \begin{cases} v^\bullet \cap D \neq \emptyset & \text{if } v \in V_\bigcirc \\ v^\bullet \subseteq D & \text{otherwise} \end{cases}$$

$$v_\bigcirc{\mapsto}_C D = \exists \sigma \in \mathbb{S}_\bigcirc^* : \forall \pi \in \mathsf{Plays}(\sigma, v) : \exists k : \pi_k \in D \wedge \forall j < k : \pi_j \in C$$

$$v_\bigcirc{\mapsto}_C = \exists \sigma \in \mathbb{S}_\bigcirc^* : \forall \pi \in \mathsf{Plays}(\sigma, v) : \forall k : \pi_k \in C$$

Finally, let $R \subseteq V \times V$ be a relation on V. The set of vertices *below* some vertex v, denoted Rv, is the set $\{w \in V \mid w\, R\, v\}$; the set of vertices *above* v, denoted

vR, is defined as $\{w \in V \mid v \; R \; w\}$. Note that in the special case that R is an equivalence relation, we have $Rv = vR$. For a set $U \subseteq V$, we write UR for the set $\bigcup_{v \in U} vR$; likewise for RU, and if R is an equivalence relation, we write $V_{/R}$ for the set of equivalence classes (quotient set) of R.

3   Inducing Parity Game Preorders and Equivalences

In the past, behavioural relations for transition systems, such as the simulation preorder and bisimulation, have been ported to parity games, see [3–5, 8, 9, 12]. These efforts, however, did not appear to have followed a general guiding principle. In this section, we propose a scheme for lifting particular types of behavioural preorders for Kripke structures to parity game preorders. More specifically, the Kripke structure preorders and equivalences we consider are those that can be phrased in terms of *matching paths*.

Let $K = \langle S, T, \mathsf{AP}, \mathcal{L} \rangle$ be an arbitrary Kripke structure, where S is a (possibly infinite) set of states, $T \subseteq S \times S$ is a total transition relation, AP is a set of atomic propositions and $\mathcal{L}\!:\!S \to \mathcal{P}(\mathsf{AP})$ is a state labelling function. A *path* through K, starting in some state $s_1 \in S$, is an infinite sequence of states $s_1 \; s_2 \; s_3 \; \ldots$ for which $(s_i, s_{i+1}) \in T$ for all i, and $\mathsf{Paths}(s)$ denotes the set of paths starting in s.

Let, within the context of some Kripke structure K, $\mathsf{Rel}(R)$ denote a predicate on relations $R \subseteq S \times S$ on K's states. Think, for instance, of the predicate that a relation R is a simulation relation. For given predicate Rel and relation R, a *matching predicate* is a predicate $\mathsf{Rel}\!-\!\mathsf{match}^{\mathcal{L}}_R(\pi_t, \pi_s)$ on *paths* π_t and π_s in K, and K's labelling $\mathcal{L}$ and the relation R on K's states satisfying:

$$\mathsf{Rel}(R) \text{ iff } \forall (s,t) \in R : \; \forall \pi_s \in \mathsf{Paths}(s) : \; \exists \pi_t \in \mathsf{Paths}(t) : \; \mathsf{Rel}\!-\!\mathsf{match}^{\mathcal{L}}_R(\pi_t, \pi_s)$$

Note that matching predicates do *not* use K's transition relation T.

Example 2. A typical instance of Rel is the simulation predicate Sim: for $R \subseteq S \times S$, $\mathsf{Sim}(R)$ holds iff for all $(s,t) \in R$ we have $\mathcal{L}(s) = \mathcal{L}(t)$ and for any $s' \in S$ with $(s, s') \in T$, there is a $t' \in S$ for which $(t, t') \in T$ and $(s', t') \in R$. An associated matching predicate $\mathsf{Sim}\!-\!\mathsf{match}^{\mathcal{L}}_R(\pi, \pi')$ is $\forall i : \; \mathcal{L}(\pi'_i) = \mathcal{L}(\pi_i) \wedge \pi'_i \; R \; \pi_i$.

In case a matching predicate is to be interpreted on a parity game, we use the priority function Ω as the state labelling function, writing $\mathsf{Rel}\!-\!\mathsf{match}^{\Omega}_R$. A matching predicate $\mathsf{Rel}\!-\!\mathsf{match}^{\mathcal{L}}$ is *monotonic* if $\mathsf{Rel}\!-\!\mathsf{match}^{\mathcal{L}}_R(\pi_t, \pi_s)$ implies $\mathsf{Rel}\!-\!\mathsf{match}^{\mathcal{L}}_{R'}(\pi_t, \pi_s)$ for all $R \subseteq R'$.

Definition 2. *Let $G = \langle V, E, \Omega, (V_\diamond, V_\square) \rangle$ be a parity game and let $\mathsf{Rel}\!-\!\mathsf{match}^{\mathcal{L}}_R$ be a matching predicate for a predicate Rel on Kripke structure relations. A relation $R \subseteq V \times V$ is a parity game Rel-relation whenever $v \; R \; w$ implies that for all strategies $\sigma_v \in \mathbb{S}^*_\diamond$, there is a strategy $\sigma_w \in \mathbb{S}^*_\diamond$ such that*

$$\forall \pi_w \in \mathsf{Plays}(\sigma_w, w) : \exists \pi_v \in \mathsf{Plays}(\sigma_v, v) : \; \mathsf{Rel}\!-\!\mathsf{match}^{\Omega}_R(\pi_w, \pi_v)$$

We write $v \sqsubseteq_{\mathsf{Rel}} w$ iff for some parity game Rel-relation R we have $v \; R \; w$.

Example 3. A relation $R \subseteq V \times V$ is a *parity game* Sim-*relation* whenever $v \, R \, w$ implies that for all strategies $\sigma_v \in \mathbb{S}_\Diamond^*$, there is a strategy $\sigma_w \in \mathbb{S}_\Diamond^*$ such that

$$\forall \pi_w \in \mathsf{Plays}(\sigma_w, w) : \exists \pi_v \in \mathsf{Plays}(\sigma_v, v) : \forall i : \; \Omega(\pi_{v,i}) = \Omega(\pi_{w,i}) \wedge \pi_{v,i} \, R \, \pi_{w,i}$$

In Sects. 4.1 and 4.2, we study further instances of Definition 2, showing that the theory of this section can be used to recover existing parity game relations (Sect. 4.1) and how it can be used to obtain new parity game relations (Sect. 4.2). For the remainder of this section, we focus on establishing under which conditions one can prove the resulting parity game relations are preorders and when they can be used to approximate the winning partition for player $\Diamond$. The theorem below shows that an induced parity game Rel-relation is a preorder whenever a simple monotonicity criterion for the matching predicate holds.

Theorem 1. *Assume that for all preorders R for which Rel holds, Rel$-$match$_R^{\mathcal{L}}$ is a preorder, too. If Rel$-$match$^{\mathcal{L}}$ is monotonic, then $\sqsubseteq_{\mathsf{Rel}}$ is a preorder.*

Under similar conditions, one can prove that $\sqsubseteq_{\mathsf{Rel}}$ is an equivalence relation. The next theorem states that we can conclude that the parity game relations *under-approximate* the winning partition for player $\Diamond$ from a simple condition on the matching predicate.

Theorem 2. *Let R be a parity game Rel-relation. Assume $v \, R \, w$ and suppose that for all π_v, π_w, if Rel$-$match$_R^{\Omega}(\pi_w, \pi_v)$ and π_v is won by $\Diamond$ then so is π_w. If v is won by $\Diamond$, then w is won by $\Diamond$.*

4 Applications

We first illustrate how the theory we developed in the previous section can be put to use to recover preorders and equivalences already present in the literature. More specifically, we show that *governed simulation* [11], also known as *direct simulation* [8,9], *governed bisimulation* [11] and *governed stuttering bisimulation* [4] are all instances of our general theory. We then proceed to show that we can also obtain relations that did not appear in the literature.

4.1 Existing Parity Game Relations

Consider the Kripke structure Rel predicates for simulation, bisimulation and stuttering bisimulation (aka stuttering equivalence), listed in Table 1. For lack of space, we refrain from giving the standard definition of these predicates, given that these can be found in most standard textbooks, and since these predicates are essentially also defined via their matching predicates next to them. Note that also the latter can be found in the literature (although less commonly).

Using the Rel-predicates of Table 1 and Definition 2, we immediately obtain parity game simulation, parity game bisimulation and parity game stuttering bisimulation. One can check with little effort that the matching predicates for simulation, bisimulation and stuttering bisimulation meet the conditions of Theorems 1 and 2. As a result, we can claim the following:

Table 1. Matching predicates for well-known behavioural relations; R is a relation on states, $\mathcal{L}$ is a state labelling function, and π, π' are infinite sequences of states

Rel	Rel$-$match$_R^{\mathcal{L}}(\pi, \pi')$
Simulation	*For all i, $\mathcal{L}(\pi_i') = \mathcal{L}(\pi_i)$ and $\pi_i' \, R \, \pi_i$.*
Bisimulation	*For all i, $\mathcal{L}(\pi_i') = \mathcal{L}(\pi_i)$, $\pi_i' \, R \, \pi_i$ and $\pi_i \, R \, \pi_i'$.*
Stuttering bisimulation	*There is a non-decreasing, unbounded function $f : \omega \to \omega$ with $f(1) = 1$ such that for all i and all $j \in [f(i), f(i+1))$, $\mathcal{L}(\pi_i') = \mathcal{L}(\pi_j)$, $\pi_i' \, R \, \pi_j$ and $\pi_j \, R \, \pi_i'$.*

Proposition 1. *Relation $\sqsubseteq_{simulation}$ is a preorder and relations $\sqsubseteq_{bisimulation}$ and $\sqsubseteq_{stuttering \, bisimulation}$ are equivalences. Moreover, all three relations underapproximate the winning set for player $\diamondsuit$.*

Definition 3. *A relation R is a* governed simulation *iff for all $v \, R \, w$:*

1. *$\Omega(v) = \Omega(w)$,*
2. *if $v \in V_\diamondsuit$, then for each $v \to v'$ we have $w \diamond\!\to v' R$,*
3. *if $v \in V_\square$, then $w \diamond\!\to v^\bullet R$.*

We write $v \leq w$ iff there is a governed simulation R such that $v \, R \, w$. A relation $R \subseteq V \times V$ is a governed bisimulation *if both R and R^{-1} are governed simulations. We write $v \leftrightarrow w$ iff for some governed bisimulation R, $v \, R \, w$.*

The example below illustrates governed simulation and governed bisimulation.

Example 4. Consider the parity game of Fig. 1(a). We have $v_3 \leftrightarrow v_4$, because, even though both vertices belong to different players, neither player can force play to vertices of different priorities. On the other hand, $v_1 \leftrightarrow v_2$ does not hold.

The theorem below confirms that governed similarity and governed bisimilarity coincide with the preorder and equivalence induced by Definition 2 using the simulation and bisimulation matching predicates of Table 1.

Theorem 3. *We have $\leq = \sqsubseteq_{simulation}$ and $\leftrightarrow = \sqsubseteq_{bisimulation}$.*

Next, we focus on the notion of governed stuttering bisimulation [4].

Definition 4. *An equivalence relation $R \subseteq V \times V$ is a* governed stuttering bisimulation *iff $v \, R \, w$ then*

1. *$\Omega(v) = \Omega(w)$,*
2. *for any $v \to \mathcal{C}$ with $\mathcal{C} \in V_{/R} \setminus \{vR\}$ and $v \in V_\bigcirc$, then $w \bigcirc\!\mapsto_{vR} \mathcal{C}$,*
3. *for any player $\bigcirc$, we have $v \bigcirc\!\mapsto_{vR}$ iff $w \bigcirc\!\mapsto_{vR}$.*

We write $v \leftrightarrow_s w$ iff there is a governed stuttering bisimulation R such that $v \, R \, w$.

Example 5. Reconsider the parity game from Fig. 1(a). Observe that we did not have $v_1 \leftrightarrow v_2$. However, we do have $v_1 \leftrightarrow_s v_2$: player $\diamond$ is capable of enforcing divergent plays (plays that only pass through vertices with the same priority), and it is capable of enforcing plays to reach vertices with priority 1.   □

Governed stuttering bisimulation coincides with parity game stuttering bisimulation; this confirms that governed stuttering bisimulation naturally generalises Kripke structure stuttering bisimulation to the parity game setting.

Theorem 4. *We have* $\leftrightarrow_s = \sqsubseteq_{stuttering\ bisimulation}$.

4.2 Two New Parity Game Relations

So far, we have illustrated that Definition 2 can be used to recover parity game relations from the literature, and that Theorems 1 and 2 are instrumental in establishing that the resulting parity game relations are preorders and that they approximate the winning set for player $\diamond$. In this section, we show that Definition 2 immediately gives us definitions for *parity game trace inclusion* and *parity game stuttering simulation*; these relations have so far not appeared in the literature. Consider the matching predicates listed in Table 2.

Table 2. Matching predicates for behavioural relations; R is a relation on states, $\mathcal{L}$ is a state labelling function, and π, π' are infinite sequences of states

Rel	Rel$-$match$_R^{\mathcal{L}}(\pi, \pi')$
Trace inclusion	*For all i, $\mathcal{L}(\pi'_i) = \mathcal{L}(\pi_i)$.*
Stuttering simulation	*There is a non-decreasing, unbounded function $f : \omega \twoheadrightarrow \omega$ with* $f(1) = 1$ *such that for all i and all $j \in [f(i), f(i+1))$,* $\mathcal{L}(\pi'_i) = \mathcal{L}(\pi_j)$ *and* $\pi'_i \, R \, \pi_j$

Stuttering simulation for Kripke structures is *coarser* than simulation: it allows for abstracting from finite computations through states with the same information and computational branching structure. Trace inclusion is even coarser than stuttering simulation. Both stuttering similarity and trace inclusion for Kripke structures are known to be preorders. Theorem 1 allows us to establish that the parity game relations they induce are preorders too. Theorem 2 again allows us to conclude that both preorders under-approximate the winning partition for player $\diamond$.

Proposition 2. *Relations* $\sqsubseteq_{stuttering\ simulation}$ *and* $\sqsubseteq_{trace\ inclusion}$ *are preorders. Both preorders under-approximate the winning partition for player* $\diamond$.

We next focus on giving a coinductive definition for parity game stuttering simulation. Apart from providing a deeper understanding of this relation, and understanding how it compares to the ones from the previous section, the coinductive definition gives rise to a polynomial time algorithm for deciding this relations.

Definition 5. *A preorder $R \subseteq V \times V$ is a* governed stuttering simulation *iff $v \, R \, w$ implies:*

1. $\Omega(v) = \Omega(w)$,
2. *left-to-right even transfer:*
   (a) *if* $v \in V_\Diamond$, *then for all* $v \to v_s$ *we have* $w \,_\Diamond\!\!\mapsto_{vR} v_s R$,
   (b) *if* $v \in V_\Box$, *then* $w \,_\Diamond\!\!\mapsto_{vR} v^\bullet R$,
3. *right-to-left odd transfer:*
   (a) *if* $w \in V_\Diamond$, *then* $v \,_\Box\!\!\mapsto_{Rw} Rw^\bullet$
   (b) *if* $w \in V_\Box$, *then for all* $w \to w_s$, *we have* $v \,_\Box\!\!\mapsto_{Rw} Rw_s$

We write $v \leftrightarroweq_s w$ *iff there is a governed stuttering bisimulation* R *such that* $v\, R\, w$.

For the above coinductive definition of our parity game stuttering simulation relation one can deduce that a symmetric relation that meets its properties is a governed stuttering bisimulation relation—a basic sanity check for the correctness of the definition. The link with governed stuttering bisimulation is, however, not obvious. We have the following theorem relating governed stuttering simulation to parity game stuttering simulation.

Theorem 5. *We have* $\leq_s = \sqsubseteq_{stuttering\ simulation}$.

Governed stuttering simulation can be computed in polynomial time; we sketch a naive algorithm based on fixpoint iteration. We start with a trivial relation R that relates all states with the same priorities. Upon every iteration, every pair $(s, t) \in R$ is checked as to whether the conditions of Definition 5 hold; if it is not the case, the pair is removed from R; thus every iteration a monotonic transformer is applied that after at most a quadratic number of steps will reach a fixpoint. As for checking that s, t and R meet the conditions of Definition 5, the main source of the complexity is in computing the $_\Diamond\!\!\mapsto$ relation. The latter can be done in $\mathcal{O}(|V| + |E|)$ time using a modified attractor computation [4] and such a computation is performed for $O(|V|)$ successors. This means that deciding governed stuttering simulation can be done in at most $\mathcal{O}(|V|^5 \cdot (|V| + |E|))$.

We remark that we did not strive to have optimal running times for deciding the preorders in this section. We leave it for future research to tighten our bounds. For deciding governed stuttering simulation, it may be fruitful to incorporate ideas from [14], which describes, as far as we could trace, the first algorithm for stuttering simulation in the Kripke structure setting.

5   Logical Characterisations of Parity Game Relations

An alternative approach to defining a behavioural preorder is by identifying an appropriate fragment of a modal logic. A natural question is thus whether, given a fragment of a modal logic for Kripke structures that coincides with a given preorder, there is a uniform way of obtaining a fragment of a modal logic for parity games that coincides with the parity game relation. While a logical characterisation of a behavioural relation offers an alternative angle for understanding it, and, as such, is interesting to study in its own right, our results in this section suggest it is unlikely such a uniform method exists.

5.1 A Modal Logic for Parity Games

The logic we consider is called the Alternating-time Hennessy-Milner logic with Until. It is essentially based on the alternating-time temporal logic of [1], but its syntax is inspired by Hennessy-Milner logic for Labelled Transition Systems. Our syntax facilitates characterising all relations we study in this paper by imposing restrictions on our base grammar.

Definition 6. *The* Alternating-time Hennessy-Milner logic with Until, *henceforth referred to as the logic AHML, is defined as follows:*

$$\phi, \psi := \bot \mid \top \mid \neg\phi \mid \langle n \rangle_\bigcirc \phi \mid \phi \wedge \psi \mid \phi \vee \psi \mid \phi \langle\!\langle n \rangle\!\rangle_\bigcirc \psi \mid \phi \langle\!\langle n \rangle\!\rangle_\bigcirc^\infty \psi$$

where $n \in \mathbb{N}$ and $\bigcirc \in \{\Diamond, \Box\}$. The semantics of AHML formulae is defined inductively in the context of a parity game $G = \langle V, E, \Omega, (V_\Diamond, V_\Box) \rangle$:

$$
\begin{aligned}
[\![\bot]\!] &= \emptyset \\
[\![\top]\!] &= V \\
[\![\neg\phi]\!] &= V \setminus [\![\phi]\!] \\
[\![\langle n \rangle_\bigcirc \phi]\!] &= \langle n \cdot \rangle_\bigcirc [\![\phi]\!] \\
[\![\phi \wedge \psi]\!] &= [\![\phi]\!] \cap [\![\psi]\!] \\
[\![\phi \vee \psi]\!] &= [\![\phi]\!] \cup [\![\psi]\!] \\
[\![\phi \langle\!\langle n \rangle\!\rangle_\bigcirc \psi]\!] &= (V_n \cap [\![\psi]\!]) \cup \mu V' \subseteq V.([\![\phi]\!] \cap (\langle n \cdot \rangle_\bigcirc [\![\psi]\!] \cup \langle n \cdot \rangle_\bigcirc V')) \\
[\![\phi \langle\!\langle n \rangle\!\rangle_\bigcirc^\infty \psi]\!] &= (V_n \cap [\![\psi]\!]) \cup \nu V' \subseteq V.([\![\phi]\!] \cap (\langle n \cdot \rangle_\bigcirc [\![\psi]\!] \cup \langle n \cdot \rangle_\bigcirc V'))
\end{aligned}
$$

where, for $W \subseteq V$ and $n \in \mathbb{N}$, operator $\langle n \cdot \rangle_\bigcirc W$ yields the set $\{v \in V_n \mid v_\bigcirc \rightarrow W\}$. We write $v \models \phi$ iff $v \in [\![\phi]\!]$.

Intuitively, $\langle n \rangle_\bigcirc \phi$ holds in vertices with priority n (*i.e.* those from the set V_n) for which $\bigcirc$ can force play to vertices satisfying ϕ. The strong until operator $\phi \langle\!\langle n \rangle\!\rangle_\bigcirc \psi$ holds in vertices with priority n for which $\bigcirc$ can govern the plays through ϕ vertices, ultimately reaching ψ vertices. The weak until operator $\phi \langle\!\langle n \rangle\!\rangle_\bigcirc^\infty \psi$ is more or less the same but also holds whenever $\bigcirc$ governs plays through ϕ-invariant vertices. Observe that our use of fixpoints in the semantics is permitted as the associated predicate transformers to which they are applied are monotonic and the set $(2^V, \subseteq)$ is a complete lattice. The example below illustrates typical properties one can express using AHML.

Example 6. Reconsider the parity game depicted in Fig. 1(a). Observe that $v_1 \models \langle 0 \rangle_\Diamond \top$ and $v_1 \models \langle 0 \rangle_\Box \top$. We have $v_3 \models \top \langle\!\langle 1 \rangle\!\rangle_\Diamond^\infty \bot$ and $v_3 \models \top \langle\!\langle 1 \rangle\!\rangle_\Box^\infty \bot$; we also have $v_2 \models \top \langle\!\langle 0 \rangle\!\rangle_\Diamond^\infty \bot$ but $v_2 \not\models \top \langle\!\langle 0 \rangle\!\rangle_\Box^\infty \bot$. Moreover, we have $v_7 \models (\langle 1 \rangle_\Diamond \top) \langle\!\langle 2 \rangle\!\rangle_\Diamond \neg(\langle 0 \rangle_\Diamond \top)$ because v_7 has (1) priority 2 as demanded by the until operator and (2) satisfies the goal formula $\neg(\langle 0 \rangle_\Diamond \top)$. □

In general, we are interested in comparing the "observations" that we can make in different vertices in a parity game; that is, we wish to compare the set of modal formulae satisfied by different vertices. We formalise observations as follows.

Definition 7. *Let L be a fragment of AHML. We write $\mathcal{O}^L(v)$ to denote the set of formulae $\phi \in L$ for which $v \models \phi$.*

5.2   Characterising Preorders Using AHML

Throughout this section, we assume that $G = \langle V, E, \Omega, (V_\Diamond, V_\Box) \rangle$ is an arbitrary parity game. In Table 3, we list the sound and complete fragments of the modal logic of Sect. 5.1 and the relations of Sects. 4.1 and 4.2.

Table 3. Parity game preorders and equivalences and their corresponding sound and complete fragments of AHML with their grammars. In these grammars, $n \in \mathbb{N}$

Relation	Fragment	Grammar
$\leq$	AHML$^\leq$	$\phi, \psi ::= \top \mid \langle n \rangle_\Diamond \phi \mid \phi \wedge \psi \mid \phi \vee \psi$
$\leftrightarrow$	AHML$^\leftrightarrow$	$\phi, \psi ::= \top \mid \neg \phi \mid \langle n \rangle_\Diamond \phi \mid \phi \wedge \psi \mid \phi \vee \psi$
$\leq_s$	AHML$^{\leq_s}$	$\phi, \psi ::= \top \mid \phi \wedge \psi \mid \phi \vee \psi \mid \phi \langle\!\langle n \rangle\!\rangle_\Diamond \psi \mid \phi \langle\!\langle n \rangle\!\rangle_\Diamond^\infty \psi$
$\leftrightarrow_s$	AHML$^{\leftrightarrow_s}$	$\phi, \psi ::= \top \mid \neg \phi \mid \phi \wedge \psi \mid \phi \vee \psi \mid \phi \langle\!\langle n \rangle\!\rangle_\Diamond \psi \mid \phi \langle\!\langle n \rangle\!\rangle_\Diamond^\infty \psi$
$\leq_t$	AHML$^{\leq_t}$	$\phi ::= \top \mid \bigvee\limits_{n \in N} \langle n \rangle_\Diamond \phi_n \quad (\emptyset \neq N \subset \mathbb{N}$ is a finite set of priorities$)$

Before we address the soundness and completeness of the fragments of the modal logic listed in Table 3, we first point out that there are interesting and fundamental differences with the modal characterisations for the preorders on computational models such as Kripke structures. For instance, disjunction is a necessary part of the logic for the simulation relations in the parity game setting: without it, one cannot show that $w \leq v$ does not hold in the following game:

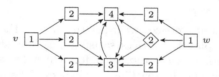

While disjunction *can* be part of the characteristic logic for the corresponding Kripke structure preorders, there, it is redundant: it does not add to the distinguishing power of the modal logic. This can be explained by the phenomenon that in the Kripke structure setting, it suffices to describe one *fixed* behaviour. In the parity game setting, one must be able to express that a player can *guarantee* (*i.e.* regardless of the strategies of her opponent) a certain *set* of behaviours; this requires disjunctions. As we will show in Example 9, it is also not the case that disjunction can be added harmlessly to the characteristic logic for a parity game preorder. This suggests there is no easy way to obtain a modal characterisation for a parity game preorder from a modal characterisation of a Kripke structure preorder.

We next state the relation between the fragments identified in Table 3 and the studied preorders, and we illustrate these correspondences using small examples.

Theorem 6. *Let* v, w *be arbitrary vertices in* G. *Then:*

1. $v \leq w$ *iff* $\mathcal{O}^{AHML^\leq}(v) \subseteq \mathcal{O}^{AHML^\leq}(w)$;

2. $v \leftrightarrow w$ iff $\mathcal{O}^{AHML^{\leftrightarrow}}(v) = \mathcal{O}^{AHML^{\leftrightarrow}}(w)$;

Example 7. Consider the parity game of Fig. 1(a). Recall that $v_1 \leq v_2$, see Example 4. We thus have $\mathcal{O}^{AHML^{\leq}}(v_1) \subseteq \mathcal{O}^{AHML^{\leq}}(v_2)$. For instance, both v_1 and v_2 satisfy $\langle 0 \rangle_\Diamond \langle 0 \rangle_\Diamond \top$. However, we do not have $v_2 \leq v_1$. This follows from the fact that $\langle 0 \rangle_\Diamond \langle 1 \rangle_\Diamond \top$ is a distinguishing formula that holds in v_2, but fails for v_1. $\square$

Theorem 7. *Let v, w be arbitrary vertices in G. Then*

1. $v \leq_s w$ iff $\mathcal{O}^{AHML^{\leq_s}}(v) \subseteq \mathcal{O}^{AHML^{\leq_s}}(w)$.
2. $v \leftrightarrow_s w$ iff $\mathcal{O}^{AHML^{\leftrightarrow_s}}(v) = \mathcal{O}^{AHML^{\leftrightarrow_s}}(w)$.

Example 8. One can check that in the parity game of Fig. 1(a), we do not have $v_5 \leq_s v_2$. This is confirmed by the formula $(\top \langle\!\langle 0 \rangle\!\rangle_\Diamond \top) \langle\!\langle 0 \rangle\!\rangle_\Diamond (\top \langle\!\langle 2 \rangle\!\rangle_\Diamond \top)$ that expresses that through vertices with priority 0, a vertex with priority 2 can be reached. It holds in v_5, but not in v_2. $\square$

The relation between parity game trace inclusion and $AHML^{\leq_t}$ is given below.

Theorem 8. *For all $v, w \in V$, we have $v \sqsubseteq_{trace\ inclusion} w$ iff $\mathcal{O}^{AHML^{\leq_t}}(v) \subseteq \mathcal{O}^{AHML^{\leq_t}}(w)$.*

The fragment of AHML needed to characterise the parity game trace inclusion preorder is non-obvious. In particular, the restriction on $AHML^{\leq_t}$ to at all depths of the formulae only allow for $\langle n \rangle_\Diamond\_$ for which the priorities are distinct is needed to reduce player $\Diamond$'s powers. Omitting this constraint and allowing for arbitrary disjunctions will lead to a finer relation, as shown by the example below.

Example 9. Consider the parity game depicted below.

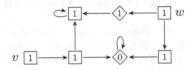

Clearly, we have $v \leq_t w$. By Theorem 8 we have $\mathcal{O}^{AHML^{\leq_t}}(v) \subseteq \mathcal{O}^{AHML^{\leq_t}}(w)$. Note that we also have $w \leq_t v$. However, we have $w \models \langle 1 \rangle_\Diamond (\langle 1 \rangle_\Diamond \langle 1 \rangle_\Diamond \top \vee \langle 1 \rangle_\Diamond \langle 0 \rangle_\Diamond \top)$ but $v \not\models \langle 1 \rangle_\Diamond (\langle 1 \rangle_\Diamond \langle 1 \rangle_\Diamond \top \vee \langle 1 \rangle_\Diamond \langle 0 \rangle_\Diamond \top)$. Omitting the constraint on the disjunctions would therefore lead to incorrect distinguishing formulae. $\square$

6   Conclusions

We proposed a scheme for lifting preorders on Kripke structures that can be defined through *matching paths* to preorders on parity games. We showed that our scheme can be used to recover preorders for parity games that have already been defined in the literature. Moreover, we demonstrated that we can easily construct new ones, such as parity game trace inclusion and parity game stuttering simulation, and prove such new relations are preorders (or equivalences) and that they approximate the winning partition of player $\Diamond$.

Our scheme for obtaining parity game relations from existing behavioural relations also extends to other relations by choosing different parameters for the matching predicate. For instance, the bisimulation of [5] on parity games with a finite number of priorities can be recovered using the matching predicate bisimulation$-$match$_R^{\mathcal{L}}$, where $\mathcal{L}(v) = \{\Omega(v), \bigcirc \mid v \in V_\bigcirc\}$; that is, the vertex labelling is extended with information which player owns the vertex. Observe that the resulting relation is finer than the ones studied in this paper.

Lastly, we provided modal characterisations of all parity game relations studied. Given the fundamental differences between these modal characterisations and their Kripke structure counterparts, we deem it highly unlikely that a logical approach to a systematic way of obtaining parity game relations from Kripke structure relations will be successful.

References

1. Alur, R., Henzinger, T.A., Kupferman, O.: Alternating-time temporal logic. J. ACM **49**(5), 672–713 (2002)
2. Arnold, A., Vincent, A., Walukiewicz, I.: Games for synthesis of controllers with partial observation. TCS **303**(1), 7–34 (2003)
3. Cranen, S., Keiren, J.J.A., Willemse, T.A.C.: Stuttering mostly speeds up solving parity games. In: Bobaru, M., Havelund, K., Holzmann, G.J., Joshi, R. (eds.) NFM 2011. LNCS, vol. 6617, pp. 207–221. Springer, Heidelberg (2011)
4. Cranen, S., Keiren, J.J.A., Willemse, T.A.C.: A cure for stuttering parity games. In: Roychoudhury, A., D'Souza, M. (eds.) ICTAC 2012. LNCS, vol. 7521, pp. 198–212. Springer, Heidelberg (2012)
5. Dawar, A., Grädel, E.: The descriptive complexity of parity games. In: Kaminski, M., Martini, S. (eds.) CSL 2008. LNCS, vol. 5213, pp. 354–368. Springer, Heidelberg (2008)
6. Emerson, E.A., Jutla, C.S.: Tree automata, Mu-Calculus and determinacy. In: FOCS 1991, pp. 368–377. IEEE Computer Society (1991)
7. Friedmann, O., Lange, M.: The modal μ-calculus caught off guard. In: Brünnler, K., Metcalfe, G. (eds.) TABLEAUX 2011. LNCS, vol. 6793, pp. 149–163. Springer, Heidelberg (2011)
8. Fritz, C., Wilke, T.: Simulation relations for alternating parity automata and parity games. In: Ibarra, O.H., Dang, Z. (eds.) DLT 2006. LNCS, vol. 4036, pp. 59–70. Springer, Heidelberg (2006)
9. Gazda, M.W., Willemse, T.A.C.: Consistent consequence for boolean equation systems. In: Bieliková, M., Friedrich, G., Gottlob, G., Katzenbeisser, S., Turán, G. (eds.) SOFSEM 2012. LNCS, vol. 7147, pp. 277–288. Springer, Heidelberg (2012)
10. Gazda, M.W.: Parity Games, Fixpoint Logic and Relations of Consequence. Eindhoven University of Technology, Forthcoming (2016)
11. Keiren, J.J.A.: Advanced Reduction Techniques for Model Checking. Eindhoven University of Technology (2013)
12. Kissig, C., Venema, Y.: Complementation of coalgebra automata. In: Kurz, A., Lenisa, M., Tarlecki, A. (eds.) CALCO 2009. LNCS, vol. 5728, pp. 81–96. Springer, Heidelberg (2009)

13. Namjoshi, K.S.: Abstraction for branching time properties. In: Hunt Jr., W.A., Somenzi, F. (eds.) CAV 2003. LNCS, vol. 2725, pp. 288–300. Springer, Heidelberg (2003)
14. Ranzato, F., Tapparo, F.: Computing stuttering simulations. In: Bravetti, M., Zavattaro, G. (eds.) CONCUR 2009. LNCS, vol. 5710, pp. 542–556. Springer, Heidelberg (2009)
15. Zielonka, W.: Infinite games on finitely coloured graphs with applications to automata on infinite trees. TCS **200**(1–2), 135–183 (1998)

A PTAS for Scheduling Unrelated Machines of Few Different Types

Jan Clemens Gehrke, Klaus Jansen, Stefan E.J. Kraft$^{(\boxtimes)}$, and Jakob Schikowski

Department of Computer Science, Kiel University, 24098 Kiel, Germany
{jcg,kj,stkr,schi}@informatik.uni-kiel.de

Abstract. Scheduling on Unrelated Machines is a classical optimization problem where n jobs have to be distributed to m machines. Each of the jobs $j \in \{1, \ldots, n\}$ has on machine $i \in \{1, \ldots, m\}$ a processing time $p_{ij} \geq 0$. The goal is to minimize the makespan, i.e. the maximum completion time of the longest-running machine. Unless P = NP, this problem does not allow for a polynomial-time approximation algorithm with a ratio better than $\frac{3}{2}$. A natural scenario is however that many machines are of the same type, like a CPU and GPU cluster: for each of the K machine types, the machines $i \neq i'$ of the same type k satisfy $p_{ij} = p_{i'j}$ for all jobs j. For the case where the number K of machine types is constant, this paper presents an approximation scheme, i.e. an algorithm of approximation ratio $1 + \varepsilon$ for $\varepsilon > 0$, with an improved running time only single exponential in $\frac{1}{\varepsilon}$.

1 Introduction

Scheduling is a classical optimization problem. Jobs—e.g. computing tasks—have to be distributed to machines such that one objective is minimized, normally the maximum completion time of the jobs. One example is a cluster of processors that has to perform a large amount of computing tasks. In general, the machines may be heterogeneous: a processor may have been designed to perform a certain type of calculations very fast, but may not be suited for other ones. However, the number of different machine types may indeed be limited, as can be the case for e.g. a cluster of CPUs and GPUs.

Formally, an instance I consists of a set $\mathcal{J} = \mathcal{J}(I)$ of n jobs and a set $\mathcal{M} = \mathcal{M}(I)$ of m machines. Every job j has a processing time on machine i of $p_{ij} \geq 0$ for $i \in \{1, \ldots, m\}$ and $j \in \{1, \ldots, n\}$. A non-preemptive schedule is a distribution of the jobs to the machines such that every job is processed by exactly one machine. Formally, it is a mapping $\sigma : \mathcal{J} \to \mathcal{M}$ of each job j to a machine i. The objective is to find a schedule σ that minimizes the makespan $\max_{i \in \mathcal{M}} \sum_{j:\sigma(j)=i} p_{ij}$, i.e. the maximum completion time of all jobs. Thus, even

Research supported by DFG project JA612/14-2, "Entwicklung und Analyse von effizienten polynomiellen Approximationsschemata für Scheduling- und verwandte Optimierungsprobleme".

© Springer-Verlag Berlin Heidelberg 2016
R.M. Freivalds et al. (Eds.): SOFSEM 2016, LNCS 9587, pp. 290–301, 2016.
DOI: 10.1007/978-3-662-49192-8_24

the longest-running machine shall finish the processing as soon as possible. This classical problem is called Scheduling on Unrelated Machines and is denoted by $R | \, | C_{\max}$ in the 3-field notation [11].

As suggested above, we consider a variant where the machines are only of K different types, where K is seen as constant: for two different machines i and i' (with $i \neq i'$) of the same type, we have $p_{ij} = p_{i'j}$ for all jobs $j \in \{1, \ldots, n\}$. The machines of type k are denoted by $\mathcal{M}_k$ such that the sets $\mathcal{M}_1, \ldots, \mathcal{M}_K$ are a disjoint partition of $\mathcal{M}$. The number of machines of one type is $m_k := |\mathcal{M}_k|$ for $k \in \{1, \ldots, K\}$. Hence, $m_1 + \cdots + m_K = m$ holds. The problem is denoted by $(Pm_1, \ldots, Pm_K) | \, | C_{\max}$. We can assume without loss of generality that $m_k \leq n$ for all $k \in \{1, \ldots, K\}$ and therefore $m = m_1 + \cdots + m_K \leq n \cdot K$. In fact, a solution cannot use more than n machines of type k because there are only n jobs. For one type, machines whose number exceeds n can therefore be discarded.

1.1 Known Results

Even Scheduling on Identical Machines $P | \, | C_{\max}$ (where, as the name suggests, all machines are of the same type) is NP-complete [9]. Thus, finding the optimum objective value $\text{OPT}(I)$ and a corresponding schedule efficiently (i.e. in polynomial time in the input length $|I|$) seems unlikely for the general case. We are therefore looking for efficient approximation algorithms. The approximation ratio of an algorithm A is $\sup_I \frac{A(I)}{\text{OPT}(I)}$, where $A(I)$ is the objective value of the solution found by A. List Scheduling is a well-known heuristic with the approximation ratio $2 - \frac{1}{m}$ for $P | \, | C_{\max}$. Hochbaum and Shmoys [12] presented the first polynomial time approximation scheme (PTAS) for $P | \, | C_{\max}$, a family of algorithms $(A_\varepsilon)_{\varepsilon > 0}$ where A_ε has an approximation ratio of $1 + \varepsilon$ for $\varepsilon > 0$. The running time is polynomial in $|I|$, but the degree of the polynomial may depend exponentially (or worse) on $\frac{1}{\varepsilon}$.

Unfortunately, $R | \, | C_{\max}$ does not allow for a PTAS unless $P = NP$: a polynomial algorithm cannot in general have an approximation ratio $c < \frac{3}{2}$ as shown by Lenstra et al. [19]. Approximation algorithms with a ratio of 2 were presented by Lenstra et al. [19], by Shmoys and Tardos [22], and by Gairing, Monien, and Woclaw [8]. A $2 - \frac{1}{m}$ approximation algorithm was found by Shchepin and Vakhania [21]. These algorithms are based on solving a linear program (LP) and rounding the solution to an integer one, with the exception of the purely combinatorial algorithm in [8]. Recently, Arad et al. [1] have presented a new algorithm that decides that a schedule σ with a makespan of at most T and an average machine load $L = \Sigma_{i \in \mathcal{M}} \Sigma_{j:\sigma(j)=i} p_{ij}/m$ does not exist, or it finds one with a makespan of at most $\min\{T + \frac{L}{h}, 2T\}$, where $h = h(T)$ is the so-called feasibility factor.

No algorithm is known for the general problem with a ratio better than 2. For a long time, this was even true for the Restricted Assignment Problem, a special case where $p_{ij} \in \{p_j, \infty\}$. A breakthrough was the estimation algorithm by Svensson [23]. The algorithm does not return an actual solution, but it can estimate the optimal makespan within $\frac{33}{17} + \varepsilon \approx 1.9412 + \varepsilon$, i.e. with a ratio better than 2. Chakrabarty et al. [5] have presented for a constant $\delta^* > 0$ a $(2 - \delta^*)$-approximation algorithm (that also returns a solution) for the $(1, \bar{\varepsilon})$-Restricted Assignment Problem. In this case of Restricted Assignment, the finite processing times are additionally either $p_j = 1$ or $p_j = \bar{\varepsilon}$ for constant $\bar{\varepsilon} > 0$.

Bhaskara et al. [2] studied the matrix $P = (p_{ij})_{m \times n}$ of the processing times, more precisely the influence of its rank on the non-approximability of $R \mid \mid C_{\max}$. Rank 1 is the case of identical or uniform machines (where the processing times are of the form $p_{ij} = \frac{p_j}{s_i}$), which allows for PTAS (see above for identical and e.g. [13, 17] for uniform machines). Unless $P = NP$, rank 4 is already APX-hard (i.e. a PTAS cannot exist), and rank 7 cannot be better approximated than $\frac{3}{2}$, as in the general case (see above). This was improved by Chen et al. [6]: rank 4 does already not allow for a polynomial-time approximation algorithm better than $\frac{3}{2}$ unless $P = NP$.

If the number m of machines is constant (i.e. $Rm \mid \mid C_{\max}$ is considered), the problem has a PTAS [19] and a fully polynomial time approximation scheme (FPTAS) [14]. An FPTAS is a PTAS where the running time is polynomial in $|I|$ and also $\frac{1}{\varepsilon}$. Faster FPTAS were successively found [7, 16], and the fastest known has a running time in $O(n) + (\frac{m}{\varepsilon})^{O(m)} \leq O(n) + (\frac{\log m}{\varepsilon})^{O(m \log m)}$ [18]. It should be noted that the algorithm by Lenstra et al. [19], while "only" being a PTAS, has a space complexity only polynomial in m, $\log \frac{1}{\varepsilon}$, and the input length. Interestingly, the special case of Scheduling on a constant number of m identical machines ($Pm \mid \mid C_{\max}$) has a lower bound of $n^{O(1)} + (\frac{1}{\varepsilon})^{O(m)}$ on the running time unless the Exponential Time Hypothesis fails [6]. For ε small enough, e.g. $\varepsilon \leq \frac{1}{m}$, the running time of the algorithm in [18] can be bounded by $O(n) + (\frac{1}{\varepsilon})^{O(m)}$ and therefore attains this lower bound.

Finally, Imreh [15] considered the Scheduling problem on $K = 2$ types. He presented heuristic algorithms with ratios $2 + \frac{m-1}{k}$ and $4 - \frac{2}{m}$, where m is the number of processors of the first and k the number of processors of the second type. Bleuse et al. [3] described an algorithm with the approximation ratio $\frac{4}{3} + \frac{1}{3k} + \varepsilon$ for scheduling on m cores (CPUs) and k GPUs. If all jobs are accelerated when executed on a GPU, the algorithm has the ratio $\frac{3}{2} + \varepsilon$. Wiese et al. [24] presented a PTAS for $(Pm_1, \ldots, Pm_K) \mid \mid C_{\max}$ (where $K = O(1)$). It has to solve $m^{O(K \cdot ((1/\varepsilon)^{1/\varepsilon \log 1/\varepsilon}))}$ linear programs, which is therefore a lower bound on the overall running time. It is double exponential in $\frac{1}{\varepsilon}$. (An earlier paper [4] with a more sophisticated rounding explained an algorithm for Δ-dimensional jobs.) Raravi and Nélis [20] presented for $K = 2$ a PTAS single exponential in $\frac{1}{\varepsilon}$.

1.2   Our Result

Theorem 1. *For general* $K = O(1)$, *there is a PTAS for* $(Pm_1, \ldots, Pm_K) \mid \mid C_{\max}$ *whose running time is bounded by* $O(K \cdot n) + m^{O(K/\varepsilon^2)} \cdot (\frac{\log m}{\varepsilon})^{O(K^2)}$: *it is only single exponential in* $\frac{1}{\varepsilon}$.

1.3   Techniques

Our algorithm first preprocesses the instance I with a method presented in [7,18] to get a new instance I^{merge} whose set of jobs $\mathcal{J}(I^{\text{merge}})$ has a bounded cardinality. Then, the well-known dual approximation approach [12,13,19] is employed: it uses an oracle that either returns for the makespan T a solution of value at most $(1 + \Theta(\varepsilon))T$ or does not return a solution if T is too small. The dual approach then finds a solution close the optimal makespan of I^{merge} (and therefore an approximate solution to I) by a binary search.

For a given makespan T, our oracle first partitions the jobs into large and small jobs for every machine type k. The processing times of the jobs are then rounded so that they have discrete values. Our main contribution in this paper is the dynamic program of the oracle: every feasible schedule has a *profile*, and one profile represents several (real) schedules. As the job processing times are discrete, so are the profiles.

After the rounding, the dynamic program of the oracle constructs all possible profiles. A simple condition is checked for every constructed profile to see whether the small jobs can be greedily assigned to the machines. If yes, a real schedule for the large jobs is found by backtracking, and the small jobs are greedily scheduled, which yields a solution of makespan at most $(1 + \Theta(\varepsilon))T$. If no profile has been generated by the dynamic program or no profile allows for a distribution of the small jobs, the value T is too small. The binary search adapts T according to the output of the oracle until the optimum has been approximated.

Note that the principle of our algorithm is similar to [7,20], but it was found independently of [20]. We think that the algorithm presented here can be considered to be less complicated and to have an easier analysis than the algorithm in [20].

1.4   General Remarks and Notation

The processing time of a job j on the machine type $k \in \{1, \ldots, K\}$ is denoted by p_{kj}. The value $k(i)$ is the type of a machine $i \in \mathcal{M}$. We suppose that K is constant, that $0 < \varepsilon \leq \frac{1}{2}$ and that computing the logarithm needs time in $O(1)$. Note that all results and proofs are also valid for non-integral p_{kj} etc., especially if they are rounded to non-integral values. The complete paper (with the missing proofs and additional explanations) has been published as a technical report [10].

2   Preprocessing of the Instance

As a first step, we preprocess the items with a technique from [7,18] to get an instance I^{merge} with a smaller number of jobs.

Let $0 < \varepsilon' \leq \frac{1}{3}$ with $\varepsilon' = \Theta(\varepsilon)$. The actual value of ε' will be determined later. First, let $d_j := \min_{k \in \{1,\ldots,K\}} p_{kj}$ be the smallest processing time of a job j, and let $D := \sum_{j \in \mathcal{J}} d_j$. It is easy to see that $\frac{D}{m} \leq \text{OPT}(I) \leq D$.

Hence, we can divide all processing times p_{kj} by $\frac{D}{m}$ such that we get the following:

Assumption 1. *Without loss of generality, the jobs are scaled such that* $1 \leq$ $\mathrm{OPT}(I) \leq m$ *and* $D = \sum_{j \in \mathcal{J}} d_j = m$.

The jobs are now partitioned into fast and slow ones for each type k. A job is slow on type k if $p_{kj} \geq \frac{m}{\varepsilon'} d_j$, otherwise it is fast on type k. Should j be slow on type k, we set $p_{kj}^{\mathrm{round}} := \infty$ (or to a sufficiently large value like $2m$) such that a reasonable algorithm will not schedule j on such a machine. If j is fast on type k, we round it down to the nearest lower value $p_{kj}^{\mathrm{round}} := d_j(1 + \varepsilon')^h$ for $h \in \mathbb{N}$ (where in fact $h = \lfloor \log_{1+\varepsilon'} \frac{p_{kj}}{d_j} \rfloor$). The new instance of scaled and rounded jobs $\mathcal{J}^{\mathrm{round}}$ together with the (unchanged) machines is called I^{round}.

Remark 2. We will sometimes directly refer to "fast jobs" and "slow jobs" if we mean e.g. "jobs scheduled on machines where they are fast."

Jobs will later on also be called "large" and "small" such that similar expressions will be used.

Lemma 3 [18]. *We have* $\mathrm{OPT}(I^{\mathrm{round}}) \leq (1 + \varepsilon') \mathrm{OPT}(I)$.

Lemma 4 [18]. *If there is an approximation algorithm A_r for I^{round} such that $A_r(I^{\mathrm{round}}) \leq \alpha \mathrm{OPT}(I) + \beta$, then there is also an approximation algorithm A for I with $A(I) \leq \alpha(1 + \varepsilon')^2 \mathrm{OPT}(I) + \beta(1 + \varepsilon') \leq (\alpha(1 + \varepsilon')^2 + \beta(1 + \varepsilon')) \mathrm{OPT}(I)$. (The proof of this lemma is constructive.)*

A rounded job $j \in \mathcal{J}^{\mathrm{round}}$ has the profile $(\Pi_{1,j}, \ldots, \Pi_{k,j})$, where $\Pi_{k,j} \in \mathbb{N}$ is the exponent such that $p_{kj}^{\mathrm{round}} = d_j(1 + \varepsilon')^{\Pi_{k,j}}$. We set $\Pi_{k,j} = \infty$ if $p_{kj}^{\mathrm{round}} = \infty$.

Lemma 5 [7]. *The jobs in I^{round} have $l \leq (2 + \log_{1+\varepsilon'}(\frac{m}{\varepsilon'}))^K$ profiles.*

We derive from I^{round} the modified instance I^{merge}. Let $\nu := \frac{1}{\lceil m/\varepsilon' \rceil}$. The jobs are partitioned into large jobs $L := \{j \mid d_j > \nu\}$ and small jobs $S := \{j \mid d_j \leq \nu\}$. Take an enumeration of the l profiles such that we can denote a profile directly by its number $\varsigma \in \{1, \ldots, l\}$. The set S is then further partitioned into the subsets $S_\varsigma = \{j \in \mathcal{J} \mid j \text{ has the profile } \varsigma\}$ for $\varsigma \in \{1, \ldots, l\}$. Two jobs j_a and j_b with the same profile ς and for which $d_{j_a}, d_{j_b} \leq \frac{\nu}{2}$ holds are now grouped together to a new composed job j_c with $p_{kj_c} := p_{kj_a} + p_{kj_b}$ for every k. The composing is repeated until there is at most one job $j \in S_\varsigma$ with $d_j \leq \frac{\nu}{2}$ for every profile ς. The other jobs (including the jobs in L) now have all a processing time of at least $\frac{\nu}{2}$. The set of all jobs is called $\mathcal{J}^{\mathrm{merge}}$, which yields together with the (unchanged) machines the instance I^{merge}.

Lemma 6. *If two jobs j_a and j_b are grouped together to j_c, then j_c has the same profile as j_a and j_b.*

Lemma 7. *After composing the items, we still have $1 \leq \mathrm{OPT}(I^{\mathrm{merge}}) \leq m$.*

Theorem 8. *We have reduced the number of jobs: the cardinality of $\mathcal{J}^{\mathrm{merge}}$ (i.e. of jobs in I^{merge}) is bounded by $\min\{n, \frac{2D}{\nu} + l\} = \min\{n, O(\frac{m^2}{\varepsilon'}) + (\frac{\log m}{\varepsilon'})^{O(K)}\}$.*

Theorem 9 [18]. *We have $\mathrm{OPT}(I^{\mathrm{round}}) \leq \mathrm{OPT}(I^{\mathrm{merge}}) \leq \mathrm{OPT}(I^{\mathrm{round}}) + \varepsilon'$.*

Theorem 10. *I^{merge} can be constructed from I in time $O(n \cdot K)$.*

3   The Main Algorithm

Let $0 < \delta \leq \varepsilon' \leq \frac{1}{3}$ with $\delta = \Theta(\varepsilon')$. We present our algorithm for an instance I with n' items and $1 \leq \mathrm{OPT}(I) \leq m$: it finds a solution of value at most $(1 + \delta)\mathrm{OPT}(I)$. As I is in fact I^{merge}, we make the following assumption:

Assumption 2. I has $n' \leq O(\frac{m^2}{\varepsilon'}) + (\frac{\log m}{\varepsilon'})^{O(K)} = O(\frac{m^2}{\delta}) + (\frac{\log m}{\delta})^{O(K)}$ items.

3.1   Approximating the Optimum by Binary Search

We introduce $\delta' = \Theta(\delta)$ with $0 < \delta' \leq \delta \leq \varepsilon' \leq \frac{1}{3}$. Suppose that we have an oracle $\mathtt{Oracle}(I, T)$ that returns for a given makespan T and a constant $C > 0$ either a solution of value at most $(1 + C\delta')T$ or $\perp$ (false). The answer $\perp$ implies that there is not a solution of value at most T, i.e. $T < \mathrm{OPT}(I)$. We use the famous dual approximation approach [12,13,19]: we start with the lower bound $LB = 1$ and the upper bound $UB = m$. Then, a binary search with the oracle approximates the optimum $\mathrm{OPT}(I)$ up to the desired approximation ratio.

Lemma 11. *Suppose that the* $\mathtt{Oracle}$ *function has the properties above. Then, the dual approximation approach, i.e. the binary search, finds a schedule with a makespan of at most* $\left(1 + (C+1)\delta' + C(\delta')^2\right) \mathrm{OPT}(I)$ *and needs* $O(\log(\frac{m}{\delta'}))$ *calls of the oracle.*

3.2   The Oracle

We now describe the principle of the oracle. As a first step, the processing times of the jobs in I are divided by T. We get a new instance I^{scale} with $\mathrm{OPT}(I^{\mathrm{scale}}) \leq 1$ (if $\mathrm{OPT}(I) \leq T$). Then, the jobs are rounded to get the instance I^{r} with $\mathrm{OPT}(I^{\mathrm{r}}) \leq (1+\delta')$. A dynamic program $\mathtt{DynProg}$ is used to iteratively construct the sets $TS_0, \ldots, TS_{n'}$ of *profiles*, where each profile represents several (real) schedules. The profiles in TS_j consider the first j jobs $\{1, \ldots, j\}$. At the end, a function $\mathtt{CreateSchedule}$ tries to construct a discrete schedule σ for the instance I^{p} (which is similar to I^{r}) from each profile $t \in TS_{n'}$, where the profiles in $TS_{n'}$ consider all n' jobs. If there is a solution to I with a makespan of at most T, one discrete schedule σ for I with a makespan of at most $(1 + C\delta')T$ will be found by $\mathtt{CreateSchedule}$. We first have the following obvious lemma:

Lemma 12. *The set* I^{scale} *can be constructed in* $O(n' \cdot K)$.
In a slight abuse of notation, we still denote the scaled processing times by p_{kj}.

Definition 13. *A (scaled) job j is large on a machine type k if $p_{kj} \geq \delta'$. Otherwise, it is small.*

Take one job j. If its processing time is large on a machine type k, it is rounded up to the next $\gamma \cdot (\delta')^2$ for $\gamma \in \mathbb{N}$. If the processing time is small, i.e. $p_{kj} < \delta'$, the processing time is rounded *down* to the next multiple of $\frac{m_k \cdot \delta'}{n'}$. This new instance with processing times p^{r}_{kj} is denoted by I^{r}. Note that jobs are large (or small) on a machine type in I^{r} if they are large (or small) in I^{scale}, and vice versa.

Lemma 14. *If I^{scale} has a schedule with a makespan of at most 1, then I^r has a schedule with a makespan of at most $1 + \delta'$. I^r can be constructed in $O(n' \cdot K)$.*

Take one schedule with a makespan of at most $1 + \delta'$. If only large jobs of I^r are scheduled on a machine i, its total processing time is a multiple of $(\delta')^2$. In fact, it must be one of the values $\{0\} \cup \{\gamma \cdot (\delta')^2 \mid \gamma \in \mathbb{N} \text{ and } \delta' \leq \gamma \cdot (\delta')^2 \leq 1 + \delta'\}$. These processing times can be numbered with $\gamma = 0$ (for the total processing time 0) and $\gamma \in \{\gamma_0 := \lceil \frac{1}{\delta'} \rceil, \gamma_0 + 1, \ldots, \gamma_1 - 1, \gamma_1 := \lfloor \frac{1+\delta'}{(\delta')^2} \rfloor\}$. If j is large, the value $\gamma(k,j)$ is the factor such that $p^r_{kj} = \gamma(k,j) \cdot (\delta')^2$.

Lemma 15. *For I^r, there are $O(\frac{1}{(\delta')^2})$ processing times of the form $\gamma \cdot (\delta')^2$ of large jobs on a machine.*

Similarly, take all small jobs assigned to a machine type k. Their total processing time is at most $m_k \cdot (1 + \delta')$ because we consider a schedule with a makespan of at most $1 + \delta'$. Moreover, the total processing time is also a multiple of $\frac{m_k \cdot \delta'}{n'}$ because of the rounding, i.e. it is one of the values in $\Sigma_k := \{\tau \cdot \frac{m_k \cdot \delta'}{n'} \mid \tau \in \mathbb{N} \text{ and } 0 \leq \tau \cdot \frac{m_k \cdot \delta'}{n'} \leq m_k \cdot (1 + \delta')\}$.

Lemma 16. *For one machine type k of I^r, there are $O((1+\delta') \cdot m_k / \frac{m_k \cdot \delta'}{n'}) = O(\frac{n'}{\delta'}(1 + \delta')) = O(\frac{n'}{\delta'})$ possible total processing times of small jobs in Σ_k.*

Based on the observations above, we introduce several useful definitions.

Definition 17. *Let I' be a sub-instance of I^r, i.e. an instance whose jobs $\mathcal{J}'$ are a subset of the jobs in I^r, and which has the same machines as I^r. Let $\sigma : \mathcal{J}' \to \mathcal{M}$ be a feasible schedule with a makespan of at most $1 + \delta'$. Let b_i be the total processing time of the large jobs assigned to machine i, i.e. $b_i = b_i(\sigma) := \sum_{j:\sigma(j)=i, p^r_{k(i)j} \geq \delta'} p^r_{k(i)j}$. The remaining processing time (or remaining machine capacity) of every machine type k for the makespan $1 + \delta'$ is $r_k = r_k(\sigma) := \sum_{i \in \mathcal{M}_k} (1 + \delta' - b_i) = m_k \cdot (1 + \delta') - \sum_{i \in \mathcal{M}_k} b_i$. Moreover, the value $ab_k(\sigma, \gamma)$ denotes the number of machines of type k where $b_i = \gamma \cdot (\delta')^2$, i.e. $ab_k(\sigma, \gamma) := |\{i \in \mathcal{M}_k \mid b_i = \gamma \cdot (\delta')^2\}|$ for $\gamma \in \{0, \gamma_0, \ldots, \gamma_1\}$. Furthermore, $as_k(\sigma)$ is the total processing time of all small jobs assigned to machine type k, i.e. $as_k(\sigma) := \sum_{j:\sigma(j)\in\mathcal{M}_k, p^r_{kj} < \delta'} p^r_{kj}$. As seen above, we have $as_k(\sigma) \in \Sigma_k$. Since the small jobs have to fit into the remaining processing time, $as_k(\sigma) \leq r_k(\sigma)$ holds for all $k \in \{1, \ldots, K\}$. The values $ab_k(\sigma) := (ab_k(\sigma, \gamma))_{\gamma=0, \gamma_0, \ldots, \gamma_1}$ and $as_k(\sigma)$ for $k \in \{1, \ldots, K\}$ form the profile of σ.*

3.3 Dynamic Programming

The dynamic program `DynProg` determines all possible profiles for I^r. It is together with the definitions above the main contribution of this paper. One profile t for a sub-instance I' is represented like above: it is a tuple of K tuples, one for each machine type: $t = ((AB_1, AS_1), \ldots, (AB_k, AS_k), \ldots, (AB_K, AS_K))$. For each $k \in \{1, \ldots, K\}$, the entry AS_k denotes the total processing time of all small jobs that are assigned to the machines of type k. One AB_k is again a tuple $AB_k = (q_0, q_{\gamma_0}, \ldots, q_\gamma, \ldots, q_{\gamma_1})$. Each entry q_γ denotes the number of machines of type k where the large jobs have the total processing time $\gamma \cdot (\delta')^2$. Obviously, $q_0 + \sum_{\gamma = \gamma_0}^{\gamma_1} q_\gamma = m_k$ holds. For convenience, $AS_k(t)$ denotes the entry AS_k of a profile t, and $AB_k(t)$ stands for the tuple AB_k of profile t. Additionally, $(AB_k(t))_\gamma$ is the entry q_γ in the tuple $AB_k(t)$.

Lemma 18. *One profile t has $O(\frac{K}{(\delta')^2})$ entries.*

The dynamic program `DynProg` starts with the profile set that represents the empty schedule: $TS_0 = \{(((m_1, 0, \ldots, 0), 0), \ldots, ((m_K, 0, \ldots, 0), 0))\}$. For every machine type k, small jobs have not been assigned ($AS_k = 0$), and m_k machines (i.e. all machines of type k) have a total processing time of large jobs equal to 0.

Suppose that the set TS_{j-1} has been determined: it contains all profiles that can be obtained for the first $j-1$ jobs $\{1, \ldots, j-1\}$. The profiles for $\{1, \ldots, j\}$ are constructed by considering for each $t \in TS_{j-1}$ all possibilities to add j to t. Fix one $t \in TS_{j-1}$. We go over all k. If j is small on type k, then $AS_k(t)$ is simply increased by p_{kj}^r: we have a new profile $t' \in TS_j$ where additionally j is assigned to the machine type k. If j is large on type k, all $q_\gamma = (AB_k(t))_\gamma > 0$ are taken into account: there are q_γ machines of type k in t where each has the total processing time $\gamma \cdot (\delta')^2$. If we add j to one of these machines, there is one machine less with the processing time $\gamma \cdot (\delta')^2$ and one machine more with the processing time $\gamma \cdot (\delta')^2 + p_{kj}^r = \gamma \cdot (\delta')^2 + \gamma(k, j) \cdot (\delta')^2$. Hence, q_γ decreases and $q_{\gamma + \gamma(k,j)}$ increases by one. Thus, each $q_\gamma > 0$ generates a new profile $t' \in TS_j$.

If t' is new, we save t' together with the corresponding backtracking information to later construct a schedule. Otherwise, t' has been derived in another way (e.g. from another $t \in TS_{j-1}$), and we only keep the old backtracking information. The check whether t' is new can be done in the size of the profile $O(\frac{K}{(\delta')^2})$: all profiles $t'' \in TS_j$ can be saved in one array with the position of t'' given by its values $(AB_k(t''))_\gamma$ and $AS_k(t'')$. Note that the dynamic program only adds a large item if $\gamma \cdot (\delta')^2 + p_{kj}^r \le (1 + \delta')$, i.e. $\gamma + \gamma(k, j) \le \gamma_1$: we only want to find the profiles representing schedules with makespans of at most $1 + \delta'$. Similarly, the bound $AS_k(t') + p_{kj}^r \le m_k \cdot (1 + \delta')$ for AS_k is checked. It is therefore possible that there is not any profile $t \in TS_{j-1}$ such that j can be assigned: the value T is too small. Then, `DynProg` returns the empty set, and `Oracle` will return $\perp$. It is obvious that a schedule for I^r with a makespan of at most $(1 + \delta')$ corresponds to at least one profile.

Lemma 19. *Let σ be a schedule for I^r with a makespan of at most $1 + \delta'$. Then* `DynProg` *generates a profile t for I^r where $(AB_k(t))_\gamma = ab_k(\sigma, \gamma)$ for all $\gamma \in \{0, \gamma_0, \ldots, \gamma_1\}$ and $k \in \{1, \ldots, K\}$, and $AS_k(t) = as_k(\sigma)$ for all $k \in \{1, \ldots, K\}$.*

Lemma 20. *One TS_j has at most $\kappa \le (m^{O(1/(\delta')^2)} \cdot O(\frac{n'}{\delta'}))^K \le m^{O(K/(\delta')^2)} \cdot (\frac{\log m}{\delta'})^{O(K^2)}$ profiles.*

Lemma 21. *The dynamic program needs $O(\frac{K}{(\delta')^2} \cdot m \cdot \kappa \cdot n') = m^{O(K/(\delta')^2)} \cdot (\frac{\log m}{\delta'})^{O(K^2)}$ for the construction of all profiles $TS_0, \ldots, TS_{n'}$.*

Remark 22. Profiles can of course be stored in a more compact form by only saving the strictly positive $ab_k(\sigma, \gamma)$ and $(AB_k(t))_\gamma$. However, it can be shown that the number of profiles κ and the asymptotic running time do not change.

3.4   Construction of a Schedule

The goal of this section is the construction of a schedule from a suitable profile.

Definition 23. *For a given profile t', the remaining total processing time for every machine type k is defined by $R_k = R_k(t') := m_k \cdot (1 + \delta') - \sum_\gamma (AB_k(t))_\gamma \cdot \gamma \cdot (\delta')^2$. The definition corresponds to the one of r_k in Definition 17.*

Definition 24. *Let I^{p} be an instance where the processing time is $p_{kj}^{\mathrm{p}} := p_{kj}^{\mathrm{r}}$ if j is large on type k and $p_{kj}^{\mathrm{p}} := p_{kj}$ if j is small on k.*

After `DynProg`, `Oracle` calls the function `CreateSchedule` for every $t' \in TS_{n'}$. First, `CreateSchedule` (t') checks whether enough processing time is left for the small jobs by controlling whether $R_k(t') \leq AS_k(t')$ holds for all k. If yes, the Function `Backtracking` is called to construct a schedule for the large jobs and to find for every k the set of small jobs $\mathcal{J}_k$ assigned to $\mathcal{M}_k$. First, it finds with the backtracking information the tuples $t_0' \in TS_0, t_1' \in TS_1, \ldots, t_{n'-1}' \in TS_{n'-1}, t' = t_{n'}' \in TS_{n'}$ from which t' has been constructed. Starting from t_0' (which is the profile for the empty schedule), we have two cases. Either t_j' has been constructed from t_{j-1}' by adding j to a machine type k where it is large, and in $\mathcal{M}_k$ to a machine with the current processing time $\gamma \cdot (\delta')^2$. Then `Backtracking` assigns j to one machine in $\mathcal{M}_k$ with a current processing time of $\gamma \cdot (\delta')^2$. Otherwise, j is small on k, and j is added to $\mathcal{J}_k$. When all t_j' have been processed by `Backtracking`, the small jobs $\mathcal{J}_k$ are greedily added by `CreateSchedule`: a machine in $\mathcal{M}_k$ gets assigned jobs in $\mathcal{J}_k$ until the total processing time of the machine exceeds $1 + 2\delta'$. Then, the next machine in $\mathcal{M}_k$ is processed in the same way. `Oracle` returns $\perp$ if it cannot construct a schedule (because $TS_{n'} = \emptyset$ or no t' satisfies $R_k(t') \leq AS_k(t')$ for all k).

Lemma 25. *Let $t \in TS_{n'}$ be a profile for which $AS_k(t) \leq R_k(t)$ holds for all k. Then, `CreateSchedule` and `Backtracking` return a schedule σ for I^{p} with a makespan of at most $1 + 3\delta'$, i.e. a schedule for I of value at most $(1 + 3\delta')T$.*

Proof (Sketch). The proof is an indirect one. It can be shown that the constructed schedule σ for I^{p} satisfies $ab_k(\sigma, \gamma) = (AB_k(t))_\gamma$ for all k and γ. This yields $r_k(\sigma) = R_k(t)$ for all k. Moreover, $p_{kj}^{\mathrm{p}} \leq p_{kj}^{\mathrm{r}} + \frac{m_k \cdot \delta'}{n'}$ holds for small jobs. Since we have $AS_k(t) \leq R_k(t)$ by assumption, $\sum_{j \in \mathcal{J}_k} p_{kj}^{\mathrm{p}} \leq R_k(t) + m_k \cdot \delta'$ holds: the small jobs in $\mathcal{J}_k$ with their processing times p_{kj}^{p} only slightly exceed the remaining capacity of the machines of type k. This would be contradicted if there were a machine type k on which the greedy assignment yielded a processing time larger than $1 + 2\delta'$ for every machine of this type and some small jobs were still left. All small jobs can therefore be greedily scheduled. As we have $p_{kj}^{\mathrm{p}} < \delta'$ for small jobs on type k, the makespan of σ is bounded by $1 + 3\delta'$ because at

most one small job is assigned to a machine i such that the processing time of i exceeds $1 + 2\delta'$. The schedule σ for I^{P} yields one for I of value at most $(1 + 3\delta')T$.
□

Lemma 26. *One call of* Backtracking *needs time in* $O(K + n' \cdot m)$. *Therefore, the total running time for all calls of* CreateSchedule *is in* $O(\kappa \cdot \frac{K}{(\delta')^2} + n' \cdot m) = m^{O(K/(\delta')^2)} \cdot \left(\frac{\log m}{\delta'}\right)^{O(K^2)}$.

Theorem 27. *Let I be an instance with $1 \leq \mathrm{OPT}(I) \leq m$. The dual approximation approach with the* Oracle *function is a PTAS: it finds for given $\delta > 0$ and $\delta' := \frac{\delta}{5}$ a solution of value at most $(1 + \delta)\mathrm{OPT}(I)$. The running time is in* $m^{O(K/\delta^2)} \cdot \left(\frac{\log m}{\delta}\right)^{O(K^2)}$.

Proof. Lemma 11 states that the binary search returns a solution of value at most $(1 + (C + 1)\delta' + C(\delta')^2)\mathrm{OPT}(I)$ if Oracle has the stated properties. Suppose first that Oracle returns a solution σ for T and input I. This is only the case if there is a profile $t \in TS_{n'}$ for which $AS_k(t) \leq R_k(t)$ holds. Hence, the returned schedule σ has a makespan of at most $1 + 3\delta'$ for I^{P} as seen in Lemma 25, and it is also a schedule for I with a makespan of at most $(1 + 3\delta')T$.

It remains to prove that there is not a solution to I of value (at most) T if Oracle (I, T) returns $\bot$. We show this by demonstrating that the oracle will always return a solution if there is a solution of value (at most) T: the instance I satisfies $\mathrm{OPT}(I) \leq T$, which implies that the instance I^{scale} has an optimum $\mathrm{OPT}(I^{\mathrm{scale}}) \leq 1$, which again implies that the instance I^{r} satisfies $\mathrm{OPT}(I^{\mathrm{r}}) \leq 1 + \delta'$ (see Lemma 14). Let σ be an optimal schedule for I^{r}. Then, DynProg, CreateSchedule and Backtracking will find a schedule: Lemma 19 states that there is a profile t_σ for which we have $ab_k(\sigma, \gamma) = (AB_k(t_\sigma))_\gamma$ and $as_k(\sigma) = AS_k(t_\sigma)$ for all k and γ. As in the proof of Lemma 25, the identity $ab_k(\sigma, \gamma) = (AB_k(t_\sigma))_\gamma$ implies $r_k(\sigma) = R_k(t_\sigma)$. Since the small jobs assigned by σ to the machine type k have to fit into the remaining machine capacity $r_k(\sigma)$, we have $AS_k(t_\sigma) = as_k(\sigma) \leq r_k(\sigma) = R_k(t_\sigma)$. By Lemma 25, CreateSchedule (t_σ) will therefore construct a schedule σ' for I with a makespan of at most $(1 + 3\delta')T$. Hence, the oracle has the desired properties with $C = 3$. We now want that $(1 + (C + 1)\delta' + C(\delta')^2)\mathrm{OPT}(I) \leq (1 + \delta)\mathrm{OPT}(I)$. Set $\delta' := \frac{\delta}{5}$ such that we have $(C + 1)\delta' + C(\delta')^2 = 4\delta' + 3(\delta')^2 \leq \delta$ because $\delta \leq \frac{1}{3}$.

The overall running time follows from Lemmas 11, 12, 14, 21, and 26.   □

4   The General PTAS

We set $\varepsilon' := \frac{\varepsilon}{5}$ and $\delta := \varepsilon'$. The PTAS first constructs the instance I^{merge} from I (see Sect. 2). Then, it calls the binary search (with Oracle) for I^{merge} with $\delta' = \frac{\delta}{5} = \frac{\varepsilon}{25}$. When it has found a solution, the combination of the items in I^{merge} and then the rounding of I^{round} are undone to get a schedule σ for I. The following two lemmas show Theorem 1.

Lemma 28. *For the values of $\varepsilon' = \frac{\varepsilon}{5}$, $\delta = \varepsilon'$ and $\delta' = \frac{\varepsilon}{25}$, the algorithm returns a solution of value $A_\varepsilon(I) \leq (1 + \varepsilon)\mathrm{OPT}(I)$.*

Proof. Let $A_m(I^{\mathrm{merge}})$ be the makespan of the solution returned by the binary search. Transforming I^{round} into I^{merge}, using the binary search and then undoing the combination of items is an algorithm A_r for I^{round} with $A_r(I^{\mathrm{round}}) = A_m(I^{\mathrm{merge}}) \leq (1+\delta)\mathrm{OPT}(I^{\mathrm{merge}}) = (1+\varepsilon')\mathrm{OPT}(I^{\mathrm{merge}})$ (see Theorem 27). Since $\mathrm{OPT}(I^{\mathrm{round}}) \leq \mathrm{OPT}(I^{\mathrm{merge}}) \leq \mathrm{OPT}(I^{\mathrm{round}}) + \varepsilon'$ (see Theorem 9), we have an algorithm with $A_r(I^{\mathrm{round}}) \leq (1+\varepsilon')\mathrm{OPT}(I^{\mathrm{merge}}) \leq (1+\varepsilon')(\mathrm{OPT}(I^{\mathrm{round}})+\varepsilon') = (1+\varepsilon')\mathrm{OPT}(I^{\mathrm{round}}) + \varepsilon'(1+\varepsilon')$. By Lemma 4, this implies an algorithm for I with $A_\varepsilon(I) \leq ((1+\varepsilon')^3 + \varepsilon' \cdot (1+\varepsilon')^2)\mathrm{OPT}(I) \leq (1+\varepsilon)\mathrm{OPT}(I)$. This upper bound holds because of $\varepsilon' = \frac{\varepsilon}{5}$. (Main parts of the proof are taken from [18].) □

Lemma 29. *The PTAS has a running time in* $O(K \cdot n) + m^{O(K/\varepsilon^2)} \cdot (\frac{\log m}{\varepsilon})^{O(K^2)}$.

5   Concluding Remarks

We have described a PTAS for $(Pm_1, \ldots, Pm_K)||C_{\max}$ that is single exponential in $\frac{1}{\varepsilon}$. Natural questions are improvements of the running time and the existence of an efficient PTAS (a PTAS with a running time of the form $f(\frac{1}{\varepsilon}) \cdot |I|^{O(1)}$, i.e. where the degree of the polynomial is independent of $\frac{1}{\varepsilon}$). Another interesting task is the generalization of this algorithm to jobs with Δ dimensions, the general case considered by Bonifaci and Wiese [4].

We mention in closing the question whether the FPTAS for a constant number of machines m presented by Jansen and Mastrolilli [18] can also be adapted to $(Pm_1, \ldots, Pm_K)||C_{\max}$ and yield an FPTAS for a constant number of machine types K.

Acknowledgements. We would like to thank the anonymous reviewers for their comments, especially for the reference to the algorithm by Raravi and Nélis [20]. The bibliography contains information from the DBLP database (www.dblp.org), which is made available under the ODC Attribution License.

References

1. Arad, D., Mordechai, Y., Shachnai, H.: Tighter Bounds for Makespan Minimization on Unrelated Machines (2014). arXiv:1405.2530
2. Bhaskara, A., Krishnaswamy, R., Talwar, K., Wieder, U.: Minimum makespan scheduling with low rank processing times. In: Proceedings of the ACM-SIAM Symposium on Discrete Algorithms, SODA 2013, pp. 937–947. SIAM (2013)
3. Bleuse, R., Kedad-Sidhoum, S., Monna, F., Mounié, G., Trystram, D.: Scheduling independent tasks on multi-cores with GPU accelerators. Concurr. Comput. **27**(6), 1625–1638 (2015)
4. Bonifaci, V., Wiese, A.: Scheduling Unrelated Machines of Few Different Types (2012). arXiv:1205.0974
5. Chakrabarty, D., Khanna, S., Li, S.: On $(1, \varepsilon)$-restricted assignment makespan minimization. In: Proceedings of the ACM-SIAM Symposium on Discrete Algorithms, SODA 2015, pp. 1087–1101. SIAM (2015)

6. Chen, L., Ye, D., Zhang, G.: An improved lower bound for rank four scheduling. Oper. Res. Lett. **42**(5), 348–350 (2014)
7. Fishkin, A.V., Jansen, K., Mastrolilli, M.: Grouping techniques for scheduling problems: simpler and faster. Algorithmica **51**(2), 183–199 (2008)
8. Gairing, M., Monien, B., Woclaw, A.: A faster combinatorial approximation algorithm for scheduling unrelated parallel machines. Theor. Comput. Sci. **380**(1–2), 87–99 (2007)
9. Garey, M.R., Johnson, D.S.: Computers and Intractability. A Guide to the Theory of NP-Completeness. W.H. Freeman and Company, New York (1979)
10. Gehrke, J.C., Jansen, K., Kraft, S.E.J., Schikowski, J.: A PTAS for Scheduling Unrelated Machines of Few Different Types. Technical report, No. 1506, Christian-Albrechts-Universität zu Kiel (2015). ISSN 2192-6247
11. Graham, R.L., Lawler, E.L., Lenstra, J.K., Rinnooy Kan, A.H.G.: Optimization and approximation in deterministic sequencing and scheduling: a survey. In: Hammer, P.L., Johnson, E.L., Korte, B.H. (eds.) Discrete Optimization II: Proceedings of the Advanced Research Institute on Discrete Optimization and Systems Applications of the Systems Science Panel of NATO and of the Discrete Optimization Symposium. Annals of Discrete Mathematics, vol. 5, pp. 287–326. Elsevier (1979)
12. Hochbaum, D.S., Shmoys, D.B.: Using dual approximation algorithms for scheduling problems: theoretical and practical results. J. ACM **34**(1), 144–162 (1987)
13. Hochbaum, D.S., Shmoys, D.B.: A polynomial approximation scheme for scheduling on uniform processors: using the dual approximation approach. SIAM J. Comput. **17**(3), 539–551 (1988)
14. Horowitz, E., Sahni, S.: Exact and approximate algorithms for scheduling nonidentical processors. J. ACM **23**(2), 317–327 (1976)
15. Imreh, C.: Scheduling problems on two sets of identical machines. Computing **70**(4), 277–294 (2003)
16. Jansen, K., Porkolab, L.: Improved approximation schemes for scheduling unrelated parallel machines. Math. Oper. Res. **26**(2), 324–338 (2001)
17. Jansen, K., Robenek, C.: Scheduling jobs on identical and uniform processors revisited. In: Solis-Oba, R., Persiano, G. (eds.) WAOA 2011. LNCS, vol. 7164, pp. 109–122. Springer, Heidelberg (2012)
18. Jansen, K., Mastrolilli, M.: Scheduling unrelated parallel machines: linear rogramming strikes back. Christian-Albrechts-Universität zu Kiel (2010). ISSN: 2192-6247, No. 1004
19. Lenstra, J.K., Shmoys, D.B., Tardos, E.: Approximation algorithms for scheduling unrelated parallel machines. Math. Program. **46**, 259–271 (1990)
20. Raravi, G., Nélis, V.: A PTAS for assigning sporadic tasks on two-type heterogeneous multiprocessors. In: Proceedings of the 33rd IEEE Real-Time Systems Symposium, RTSS 2012, pp. 117–126. IEEE Computer Society (2012)
21. Shchepin, E.V., Vakhania, N.: An optimal rounding gives a better approximation for scheduling unrelated machines. Oper. Res. Lett. **33**(2), 127–133 (2005)
22. Shmoys, D.B., Tardos, E.: An approximation algorithm for the generalized assignment problem. Math. Program. **62**, 461–474 (1993)
23. Svensson, O.: Santa claus schedules jobs on unrelated machines. SIAM J. Comput. **41**(5), 1318–1341 (2012)
24. Wiese, A., Bonifaci, V., Baruah, S.K.: Partitioned EDF scheduling on a few types of unrelated multiprocessors. Real-Time Syst. **49**(2), 219–238 (2013)

Compacting a Dynamic Edit Distance Table by RLE Compression

Heikki Hyyrö[1]($\boxtimes$) and Shunsuke Inenaga[2]

[1] Department of Computer Sciences, University of Tampere, Tampere, Finland
heikki.hyyro@uta.fi
[2] Department of Informatics, Kyushu University, Fukuoka, Japan
inenaga@inf.kyushu-u.ac.jp

Abstract. Kim and Park [A dynamic edit distance table, J. Disc. Algo., 2:302–312, 2004] proposed a method (KP) based on a "dynamic edit distance table" that allows one to efficiently maintain edit distance information between two strings A of length m and B of length n when the strings can be modified by single-character edits to their left or right ends. This type of computation is useful e.g. in cyclic string comparison. KP uses linear time, $O(m + n)$, to update the distance representation after each single edit. As noted in a recent extension of KP by Hyyrö et al. [Incremental string comparison, J. Disc. Algo., 34:2-17, 2015], a practical bottleneck is that the method needs $\Theta(mn)$ space to store a representation of a complete $m \times n$ edit distance table. In this paper we take the first steps towards reducing the space usage by RLE compressing A and B. Let M and N be the lengths of RLE compressed versions of A and B, respectively. We propose how to store the edit distance table using $\Theta(mN + Mn)$ space while maintaining the same time complexity as the original method that does not use compression.

1 Introduction

Edit distance is a classic and widely used similarity measure between two strings A and B. In this paper we concentrate on Levenshtein distance that is defined as the minimum number of single-character insertions, deletions, and/or substitutions needed in order to transform A into B (or vice versa).

Let m and n denote the lengths of A and B, respectively, and let $ed(A, B)$ denote the Levenshtein distance between A and B. The fundamental $\Theta(mn)$ time dynamic programming edit distance algorithm computes information in a "directional" manner. Without loss of generality we assume the typical left-to-right direction that allows one to efficiently cope with changes to the *right* ends of A or B: the solution to $ed(A, B)$ can for example be updated into a solution to $ed(Ac, B)$ or $ed(A, Bc)$, where c is a character appended to the right end of A or B, in $\Theta(n)$ or $\Theta(m)$ additional time, respectively.[1] Changes to the *left* end

---

[1] The case of right-to-left direction is symmetric and would within the context of this paper only result in interchanging the notions of "left" and "right" ends of a string.

© Springer-Verlag Berlin Heidelberg 2016
R.M. Freivalds et al. (Eds.): SOFSEM 2016, LNCS 9587, pp. 302–313, 2016.
DOI: 10.1007/978-3-662-49192-8_25

are much more costly: e.g. updating a standard dynamic programming solution to $ed(A, B)$ into a solution to $ed(cA, B)$ or $ed(A, cB)$, where c is prepended to the left end of A or B, takes $\Theta(mn)$ worst-case time.

There are several solutions (mainly [4,6]) that can handle left-end modifications in linear $O(m + n)$ time. Efficient support for left-end modifications is important in many applications, such as e.g. cyclic string comparison and computing approximate periods (see [2,3,5–7] for more details). The key to overcome the $\Theta(mn)$ limit of the basic dynamic programming algorithm is to use some kind of an indirect representation of the edit distance information. In this paper we concentrate on the difference-representation used by the "dynamic edit distance table" that was first introduced by Kim and Park [5] for Levenshtein edit distance and recently extended by Hyyrö et al. [4] to weighted edit distance.

It was noted in [4] that the main practical limitation of the dynamic edit distance table is its $\Theta(mn)$ space requirement. In this paper we take the first steps towards reducing the space usage by compressing A and B with run-length encoding (RLE). Let M and N be the lengths of RLE compressed versions of A and B, respectively. We propose how to store the dynamic edit distance table only partially using $\Theta(mN + Mn)$ space while still supporting left-end modifications in linear $O(m + n)$ time. Our method builds on Arbell et al.'s edit distance computation algorithm for RLE compressed strings [1].

2 Preliminaries

We use the following notation with strings. Let A be a string consisting of m characters. For $1 \leq i \leq m$, $A[i]$ denotes the ith character of A, and for $1 \leq i \leq j \leq m$, $A[i : j]$ denotes the substring of A that starts at its ith character and ends at its jth character. If $i > j$, we define $A[i : j] = \varepsilon$, where ε means an empty string. For any two characters x and y, we also define a character (mis)match function $\delta(x, y)$ whose value is 0 iff $x = y$ and 1 otherwise.

Run length encoding (RLE) is a string compression method that compresses a string A by replacing each maximally long substring $A[i : j]$, where $A[i] = A[k]$ for all $k \in [i .. j]$, by the pair $(a, j - i + 1)$, where $a = A[i]$. That is, each maximally long run of equal characters is replaced by a value-pair that describes the character and the length of the run. It is usual to express such pairs (a, x) in the form a^x. For example if $A = \text{aaaabbacccbbaabbb}$, the RLE compressed representation of A may be written as $a^4 b^2\ a^1 c^3 b^2 a^2\ b^3$. We define the length of an RLE compressed string (or the RLE length of a string) as the number of maximal runs in it. E.g. the length of the preceding example string A is 17 and its RLE length is 7. RLE compression is effective if the strings contain long runs of equal characters.

The edit distance between two strings A and B is denoted by $ed(A, B)$ and is defined as the minimum number of edit operations that are required in order to transform A into B or vice versa. We concentrate on the classic Levenshtein edit distance, which permits the following three edit operations for a string A:

1. Insert a character c after position i of A. If $i = 0$, insert it to the left end.

2. Delete the character a_i from position i of A.
3. Substitute the character a_i at position i of A by a character c.

For example $ed(\text{apple}, \text{carpe}) = 3$ and an optimal three-operation way to transform $A = \text{apple}$ into $B = \text{carpe}$ is to delete $A[4] = \text{l}$, substitute $A[2] = \text{p}$ by r and insert the character c to the front.

Throughout the paper we will let m denote the length of A and n denote the length of B. The fundamental $\Theta(mn)$ time solution for computing $ed(A, B)$ fills an $(m + 1) \times (n + 1)$ dynamic programming table D with values $D[i, j] = ed(A[1 : i], B[1 : j])$ for $0 \leq i \leq m$ and $0 \leq j \leq n$. Note that $D[m, n]$ will tell the desired result $ed(A, B)$. Each value $D[i, j]$ is computed using the following well-known recurrence (1).

$$D[i, 0] = i \text{ for } 0 \leq i \leq m, D[0, j] = j \text{ for } 0 \leq j \leq n, \text{ and}$$
$$D[i, j] = \min\{D[i, j-1] + 1, D[i-1, j] + 1, D[i-1, j-1] + \delta(A[i], B[j])\}, \quad (1)$$
$$\text{for } 1 \leq i \leq m \text{ and } 1 \leq j \leq n.$$

Let us recall a well-known observation about neighboring cell values in D.

Observation 1. *The following properties hold for D:*

1. $D[i, j] - D[i - 1, j - 1]$ *is 0 or 1, for $i \in [1 .. m]$ and $j \in [1 .. n]$.*
2. $D[i, j] - D[i - 1, j]$ *is -1, 0 or 1, for $i \in [1 .. m]$ and $j \in [0 .. n]$.*
3. $D[i, j] - D[i, j - 1]$ *is -1, 0 or 1, for $i \in [0 .. m]$ and $j \in [1 .. n]$.*

For the rest of the paper we assume that we are given edit distance information, for example the table D, that corresponds to computing $ed(A, B)$, and that the string B will then be subjected to an edit operation at its left or right end.[2] Let B' denote B after the operation. The goal is to update the edit distance information, for example D, so that it corresponds to $ed(A, B')$.

Let D' denote D after it has been updated to correspond to $ed(A, B')$. If the operation to B is done at its right end, in which case either $B' = Bc$ (insertion), $B' = B[1 : n - 1]$ (deletion) or $B' = B[1 : n - 1]c$ (substitution), D may be updated into D' in $O(m)$ time by computing a single column at index $j = n$ or $j = n + 1$ using recurrence (1). It is well-known (see e.g. [5]) that any of the analogous left end modifications, corresponding to either $B' = cB$ (insertion), $B' = B[2 : n]$ (deletion) or $B' = cB[2 : n]$ (substitution), may lead to up to $\Theta(mn)$ differences between D and D'. This gives a worst-case bound of $\Theta(mn)$ for updating D into D'.

3   The Dynamic Edit Distance Table

The "dynamic edit distance table" proposed by Kim and Park [5] avoids the $\Theta(mn)$ bound of updating D into D' by maintaining a difference representation DR of D (instead of the original D). Each cell $DR[i, j]$ of the difference table

---

[2] Without loss of generality; the case of editing A is symmetric.

has two fields: a vertical (upper) difference $DR[i,j].U$ and a horizontal (left) difference $DR[i,j].L$. These difference values are defined as

$$DR[i,j].U = D[i,j] - D[i-1,j] \text{ and}$$
$$DR[i,j].L = D[i,j] - D[i,j-1], \text{ for } i = 1,\ldots,m \text{ and } j = 1,\ldots,n.$$

That is, $DR[i,j].U$ tells the difference between $D[i,j]$ and its upper neighbor $D[i-1,j]$ and $DR[i,j].L$ tells the difference between $D[i,j]$ and its left neighbor $D[i,j-1]$. Figure 1 shows an example of D and the corresponding U- and L- values in DR. Note that if we have only DR available, computing an arbitrary $D[i,j]$ value requires $O(\min\{m,n\})$ time since we need to backtrack $\min\{i,j\}$ cells from $DR[i,j]$. However, the computation of DR can be set to keep track of a constant number of specific values of interest, such as $D[m,n] = ed(A,B)$, without causing asymptotic (or practically significant) overhead. This makes DR sufficient for many applications.

		c	a	r	p	e
	0	1	2	3	4	5
a	1	1	1	2	3	4
p	2	2	2	2	2	3
p	3	3	3	3	2	3
l	4	4	4	4	3	3
e	5	5	5	5	4	3

		c	a	r	p	e
a	1	0	-1	-1	-1	-1
p	1	1	1	0	-1	-1
p	1	1	1	1	0	0
l	1	1	1	1	1	0
e	1	1	1	1	1	0

	c	a	r	p	e
	1	1	1	1	1
a	0	0	1	1	1
p	0	0	0	0	1
p	0	0	0	-1	1
l	0	0	0	-1	0
e	0	0	0	-1	-1

$$D \qquad\qquad DR.U \qquad\qquad DR.L$$

Fig. 1. The tables D and DR for $A =$ apple and $B =$ carpe

Let DR' denote DR after it has been updated to correspond to $ed(A, B')$. Modifying the left end of B may shift column indices within B and DR. E.g. if a character is deleted from the left end of B, then for $j = 2,\ldots,n$ the equality $B[j-1] = B'[j]$ holds and column $j-1$ in DR corresponds to column j in DR'. We define ℓ as a correcting offset: $\ell = -1$ if a character was deleted from the left end, $\ell = 1$ if a character was inserted to the left end of B, and $\ell = 0$ otherwise. Now $B[j - \ell] = B'[j]$ and column $j - \ell$ in DR corresponds to column j in DR'.

The important benefit from using DR instead of D is that DR' can differ from DR in at most $O(m+n)$ positions. The following Theorem 1 can be derived from the proof of Theorem 8 in [4].

Theorem 1 [4]. *Any single row $i \in [1 \mathrel{..} m]$ of DR' contains at most $O(1)$ columns j where $DR'[i,j].L \neq DR[i,j-\ell].L$. Any single column $j \in [1 \mathrel{..} n - \ell]$ of DR' contains at most $O(1)$ rows i where $DR'[i,j].U \neq DR[i,j-\ell].U$. Overall the table DR' contains at most $O(m+n)$ positions where $DR'[i,j] \neq DR[i,j-\ell]$.*

Kim and Park [5] presented the first algorithm that updates DR into DR' in $O(m+n)$ time. Later Hyyrö et al. [4] gave a more general and simple algorithm that achieves the same result.

Theorem 2 [5]. *DR can be updated to DR' in $O(m+n)$ time.*

4   Edit Distance of RLE Compressed Strings

For the rest of the paper we assume that A and B have been RLE compressed and denote their RLE lengths by M and N, respectively. In this section we review the algorithm of Arbell et al. [1] that computes $ed(A, B)$ in $\Theta(mN + Mn)$ time. Note that the time complexity holds even if A and B are given in uncompressed form: in that case A and B can first be RLE compressed in $O(m + n)$ time.

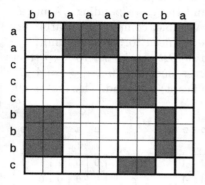

Fig. 2. Matching (black) and non-matching (white) blocks at intersecting RLE runs

The key idea is to divide the dynamic programming table D into "boxes" that are defined by intersections of maximal runs of A and B (see Fig. 2). D contains $M \times N$ such boxes. Let M^I denote the length of the Ith run in A, N^J the length of the Jth run in B, and $\mathcal{B}^{I,J}$ the box that corresponds to the Ith run of A and the Jth run of B. The box $\mathcal{B}^{I,J}$ consists of the two dimensional index interval that spans the rows $i = i_T^I \ldots i_B^I$ and the columns $j = j_L^J \ldots j_R^J$, where the index bounds are $i_T^I = 1 + \Sigma_{k=1}^{I-1} M^k$, $j_L^J = 1 + \Sigma_{k=1}^{J-1} N^k$, $i_B^I = i_T^I + M^I - 1$ and $j_R^J = j_L^J + N^J - 1$.

For convenience we also define boxes for row and/or column 0: $\mathcal{B}^{I,0}$ spans the rows $i = i_T^I \ldots i_B^I$ in column 0 and $\mathcal{B}^{0,J}$ the columns $j = j_L^J \ldots j_R^J$ in row 0. Since the box $\mathcal{B}^{I,J}$ is an index interval instead of a concrete sub-table of D, we may refer to a box $\mathcal{B}^{I,J}$ also in alternative representations of D, such as DR.

As depicted in Fig. 2, each box may be classified as a matching (black) or a non-matching (white) box, depending on whether the runs of A and B that define $\mathcal{B}^{I,J}$ are runs of the same character or not.

The table D is processed one box at a time, and in each box $\mathcal{B}^{I,J}$ only the cells on its right/bottom boundary (in rightmost column j_R^J and/or bottom row i_B^I) are filled. It is convenient to define the left/up boundary to consist of those cells that are immediate left/up neighbours of $\mathcal{B}^{I,J}$ (on column $j_L^J - 1$ and/or row $i_T^I - 1$). The boxes are processed in such a manner that the left/up neighboring boxes $\mathcal{B}^{I-1,J-1}$, $\mathcal{B}^{I,J-1}$ and $\mathcal{B}^{I-1,J}$ are processed before the box $\mathcal{B}^{I,J}$. This guarantees that the cells in the left/up boundary have been computed before $\mathcal{B}^{I,J}$. The values in the right/bottom boundary can be computed from the values in the left/up boundary.

The diagonal d of D consists of those cells $D[i, j]$ where $d = j - i$. A first observation, based on the fact that $D[i, j] = D[i - 1, j - 1]$ if $A[i] = B[j]$, is that black boxes are very easy to handle. Consider a cell $D[i, j]$ on the right/bottom boundary of a black box and let $h = 1 + \min\{i - i_T^I, j - j_L^J\}$. The value h tells the (minimum) distance between $D[i, j]$ and the left/up boundary. Then the cell $D[i - h, j - h]$ is on the up/left boundary, resides on the same diagonal as $D[i, j]$ and the equality $A[i - h + k] = B[j - h + k]$ holds for $k = 1, \ldots, h$.

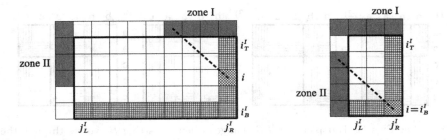

Fig. 3. Processing a white box. The filled cells on the right/bottom boundary are highlighted with a grid-pattern. The cells of zone I and zone II are shown in dark grey. The dashed diagonal lines go from $D[i-h, j_R^J - h]$ to $D[i, j_R^J]$.

The above-mentioned fact then tells that we can set $D[i, j] = D[i-1, j-1] = \cdots = D[i-h, j-h]$.

Now we consider the more complex case of updating white boxes. Figure 3 depicts the processing of a white box $\mathcal{B}^{I,J}$. We discuss only the computation of values in column j_R^J, as row i_B^I is processed in a symmetric manner. If (i, j) is a position inside a white box, then $A[i] \neq B[j]$ and recurrence (1) reduces into $D[i, j] = 1 + \min\{D[i-1, j-1], D[i, j-1], D[i-1, j]\}$. This can be interpreted as increasing the distance by 1 for each diagonal, horizontal or vertical step that a single use of the recurrence "moves" inside the white box. Now if a cell $D[i, j_R^J]$ along the right boundary of $\mathcal{B}^{I,J}$ inherits its value (via repeated uses of the previous rule) from a cell $D[i^*, j^*]$ on the left/up boundary, the equality $D[i, j_R^J] = D[i^*, j^*] + \max\{i - i^*, j_R^J - j^*\}$ holds. Here the term $\max\{i - i^*, j_R^J - j^*\}$ gives the minimum number of repeated uses of recurrence (1) we need in order to reach cell $D[i, j_R^J]$ from the cell $D[i^*, j^*]$. The remaining problem is how to determine which cell $D[i^*, j^*]$ on the left/up boundary gives the optimal (minimal) value for $D[i, j_R^J]$. Let $h = 1 + \min\{i - i_T^I, j_R^J - j_L^J\}$ again be the distance to the left/up boundary. Arbell et al. showed that the optimal source cell $D[i^*, j^*]$ must reside in either "zone I" or "zone II", where zone I consists of the cells $D[i_T^I - 1, j^*]$ for $j^* = j_R^J - h, \ldots, j_R^J$ and zone II of the cells $D[i^*, j_L^J - 1]$ for $i^* = i - h, \ldots, i$ (see Fig. 3). Note that both zones have length $h+1$. A crucial observation for computing $D[i, j_R^J]$ is that the term $\max\{i - i^*, j_R^J - j^*\}$ has the fixed value $i - i_T^I + 1$ for any cell in zone I and the fixed value $j_R^J - j_L^J + 1$ for any cell in zone II. Now $D[i, j_R^J] = \min\{i - i_T^I + 1 + Z_1, j_R^J - j_L^J + 1 + Z_2\}$, where $Z_1 = \min\{D[i_T^I - 1, j^*] \mid j^* \in [j_R^J - h \mathinner{\ldotp\ldotp} j_R^J]\}$ and $Z_2 = \min\{D[i^*, j_L^J - 1] \mid i^* \in [i - h \ldots i]\}$. Computing $D[i, j_R^J]$ is then reduced to finding Z_1 and Z_2, which are simply the minimum values in zone I and zone II, respectively. Each of them can be found in $O(1)$ additional time per each row i while processing the rows of column j_R^J in the order $i = i_T^I, \ldots, i_B^I$.

First consider Z_1. At the first row $i = i_T^I$ we have $h = 1 + \min\{i - i_T^I, j_R^J - j_L^J\} = 1$ and $Z_1 = \min\{D[i_T^I - 1, j_R^J - 1], D[i_T^I - 1, j_R^J]\}$. Then whenever i is incremented, h either grows by one or has reached the limit $j_R^J - j_L^J$. Using the

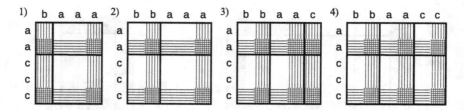

Fig. 4. The Figs. (1–4) depict how DS changes when the string B evolves through the strings `baaa`, `bbaaa`, `bbaac` and `bbaaacc` by modifications to its left or right end. Cells stored in DS are shaded with a pattern: vertical pattern shows cells with U-fields, horizontal pattern cells with L-fields and grid-pattern cells with both U- and L-fields.

update rule $Z_1 := \min\{Z_1, D[i_T^I - 1, j_R^J - h]\}$, we can maintain the correct value $Z_1 = \min\{D[i_T^I - 1, j^*] \mid j^* \in [j_R^J - h \mathinner{..} j_R^J]\}$ for the new row i.

Now consider Z_2. This case is slightly more complicated due to how the top row of zone II starts moving down when $h < i - i_T^I + 1$ (e.g. on the right side of Fig. 3 the top row of zone II has moved two steps down to row $i_T^I + 1$). As long as $h = i - i_T^I + 1$, zone II can be handled in similar manner to zone I. We have $h = 1$ and $Z_2 = \min\{D[i_T^I - 1, j_L^J - 1], D[i_T^I, j_L^J - 1]\}$ at the first row $i = i_T^I$, and the update rule in the following rows is $Z_2 := \min\{Z_2, D[i, j_L^J - 1]\}$. If zone II starts moving, the update also needs to reflect that the previous top row cell $D[i - h - 1, j_L^J - 1]$ leaves zone II. Let C be a counter array such that for $0 \le x \le \max\{m, n + \ell\}$, $C[x]$ tells how many times the value x appears in the cells of zone II. C can be maintained by always setting $C[D[i, j_L^J - 1]] := C[D[i, j_L^J - 1]] + 1$ when updating Z_2, and then also setting $C[D[i - h - 1, j_L^J - 1]] := C[D[i - h - 1, j_L^J - 1]] - 1$ if zone II moves one step down. If currently $Z_2 = D[i - h - 1, j_L^J - 1]$ and the latter update results in $C[D[i - h - 1, j_L^J - 1]] = 0$, the new minimum in zone II is $> Z_2$. The correct new minimum is $Z_2 := Z_2 + 1$, because Observation 1 guarantees that the new top row value $D[i - h, j_L^J - 1] \le D[i - h - 1, j_L^J - 1] + 1 = Z_2 + 1$.

Since filling a single cell in a right/bottom boundary of a box can be done in $O(1)$ time and D has altogether $\Theta(mN + Mn)$ such cells, the overall time for computing $ed(A, B)$ is $\Theta(mN + Mn)$.

5   Dynamic Edit Distance Table for RLE Strings

Let us now turn to the main topic (and contribution) of this paper: handling left (and right) end modifications efficiently when the strings A and B are RLE compressed. Instead of the full difference table DR, we will maintain a "sparse" table DS that contains only those columns and rows that coincide with the right/bottom boundaries of the boxes $\mathcal{B}^{I,J}$. To be more precise, DS stores the values $\{DR.U[i, j_R^J] \mid i \in [1 \cdots m], J \in [1 \cdots N]\}$ and $\{DR.L[i_B^I, j] \mid I \in [1 \cdots M], j \in [1 \mathinner{..} n]\}$. Note that the stored columns contain only the U-fields and the stored rows only the L-fields. The cells at intersections of these columns and rows contain both fields. See Fig. 4 for an example. Assume that DS corresponds to $ed(A, B)$ and that B has been changed into B' by a modification to its left or right end. Let DS' denote DS after it has been updated to correspond to $ed(A, B')$. Our goal is to find an efficient way to update DS into DS'.

First we note that even though we discuss only the case of editing B, the goal is to also allow left or right end edits to A. This means, among other things, that we should be able to efficiently add/remove rows or columns to/from DS when updating it to correspond to the DS'. A suitable solution (like in [5]) is to store DS as a linked structure where each cell $DS[i, j]$ has a pointer to its four neighbours (left, up, right and down). Here we define a "neighbour" to be the nearest cell that actually exists in DS, effectively hopping over those cells of the boxes $\mathcal{B}^{I,J}$ that do not reside on the right/bottom boundary of any $\mathcal{B}^{I,J}$. Such a linked sparse table DS can be stored using $\Theta(mN + Mn)$ space and adding or removing a column or row can be done in $\Theta(n)$ or $\Theta(m)$ time, respectively.

Figure 4 shows examples of how the form of DS (which cells are stored in it) may change when the left or right end of B is modified. For example if a character is inserted, it either expands the current boxes (step $1 \to 2$ in Fig. 4) or adds completely new boxes (imagine the situation of step 3 without the last character c in B). Performing this kind of changes to DS is straight-forward in $O(m)$ time. We assume that when we start to update DS into DS', the preprocessing step of changing the form of DS, if necessary, has already been done. For convenience, we will already refer to this preprocessed (but not yet fully updated) table as DS' (or DR', as the two tables differ only in that the former is a partial representation of the latter).

We will concentrate on the case where the left end of B has been modified. The case of right end modifications is relatively simple to handle in $O(m)$ time by essentially (re)computing right boundaries of the boxes in at most two rightmost box columns. The procedure is a straight-forward modification of Arbell et al.'s method for computing distances on right boundaries. We omit further details due to lack of space

Our method for updating DS processes DS as if it contained all values of DR. The modified algorithm will process (roughly) the same set of values of DR as the previous dynamic edit distance table algorithms of Kim and Park [5] or Hyyrö et al. [4], but since we are now working with DS instead of DR, needed values of DR that are not in DS will be computed on the fly and forgotten once they are no longer needed.

Let us first briefly review how the algorithm of Hyyrö et al. [4] (we will from now on call it HNI) works. It is based on using Lemma 1 which states those cells in DR' (and DS') that need to be recomputed. Recall that the value ℓ referred to in Lemma 1 is a correcting offset that keeps the indices aligned correctly when comparing values in DR and DR'. This same ℓ is valid also for DS and DS'.

Lemma 1 [4]. *Assume that the values $DR'[i^*, j^*]$ are correct for all cells where $i^* < i$ or $j^* < j$. The entry $DR'[i, j]$ needs to be recomputed if and only if $DR'[i - 1, j].L \neq DR[i - 1, j - \ell].L$ or $DR'[i, j - 1].U \neq DR[i, j - 1 - \ell].U$.*

Assume that HNI is currently processing column j. The basic principle of the algorithm is to maintain a list $prev\Delta$, in ascending row order, of those rows i that should be recomputed in column j. That is, $prev\Delta$ will hold those indices i for which the inequality $DR'[i, j-1].U \neq DR[i, j-1-\ell].U$ was true while processing the previous column $j-1$.[3] This enforces the second condition in Lemma 1. HNI processes the column j rows listed in $prev\Delta$ in increasing row order. Each such cell $DR'[i, j]$ is recomputed, and the U- and L-fields of the new value are compared with the old ones (which corresponded to $DR[i, j-\ell]$). If the U-fields do not match, the row i of the next column $j+1$ is added to a second list, $curr\Delta$, that will later become $prev\Delta$ for column $j+1$. If the L-fields do not match, the first rule of Lemma 1 is enforced: also the row $i+1$ in column j will be computed (regardless of whether row $i+1$ is present in $prev\Delta$ or not). The computation can be stopped if $curr\Delta$ remains empty or j was the last column of DR.

In the case of RLE strings and DS', we will process DS' one box at a time. As we are concentrating on the case where the left end of B is modified, the first column affected by the modification is $j = 1$. This is the column where the left boundaries $j_L^0 = 1$ of the boxes $\mathcal{B}^{I,1}$ reside.

The boxes will be processed in a column-wise manner: first the boxes $\mathcal{B}^{1,J'}$, ..., $\mathcal{B}^{M,J'}$, then the boxes $\mathcal{B}^{1,J'+1}, \ldots, \mathcal{B}^{M,J'+1}$ (if the algorithm did not decide to stop already), and so on. During the computation we maintain two lists for each box: $\Delta_B^{I,J}$ and $\Delta_R^{I,J}$. The list $\Delta_B^{I,J}$ records the position-value pairs $(j, D'[i, j])$ for all cells $DR'[i_B^I, j]$ on the bottom boundary of $\mathcal{B}^{I,J}$ where the inequality $DR'[i_B^I, j].L \neq DR[i_B^I, j-\ell].L$ holds. In similar manner, each list $\Delta_R^{I,J}$ records the position-value pairs $(i, D'[i, j])$ for all cells $DR'[i, j_R^J]$ on the right boundary of $\mathcal{B}^{I,J}$ where the inequality $DR'[i, j_R^J].U \neq DR[i, j_R^J - \ell].U$ holds. The positions may be accompanied by pointers to allow direct reference in a linked DS. We will also keep the bottom-left values $D'[i_B^I, j_L^J - 1]$ and $D[i_B^I, j_L^J - 1 - \ell]$ together with the list $\Delta_B^{I,J}$ and the top-right values $D'[i_T^I - 1, j_R^J]$ and $D[i_T^I - 1, j_R^J - \ell]$ together with the list $\Delta_R^{I,J}$. The reason for keeping such concrete distance values from D' and D will become clear later. It is important to note that Theorem 1 guarantees that the lists $\Delta_B^{I,J}$ and $\Delta_R^{I,J}$ contain at most $O(1)$ positions.

The lists $\Delta_B^{I,J}$ and $\Delta_R^{I,J}$ serve a similar purpose as $prev\Delta$ in HNI: $\Delta_B^{I-1,J}$ lists the cells on the top row i_T^I of $\mathcal{B}^{I,J}$ and $\Delta_R^{I,J-1}$ lists the cells on the left column j_L^J of $\mathcal{B}^{I,J}$ that need to be recomputed due to Lemma 1. In the beginning, the positions and values in column 1 where the condition $DR'[i, 1].U \neq DR[i, 1 - \ell].U$ holds are recorded in the corresponding lists. This can be done in $O(m)$ time e.g. by temporarily computing the whole column 1 of DR and the whole column $1 - \ell$ of DR'. Each such $\Delta_R^{I,0}$ is also supplied with the top-right value $D'[i_T - 1, 0] = i_T - 1$, using the fact that $D'[i, 0] = i$ for $i \in [0 .. m]$.

Processing a box $\mathcal{B}^{I,J}$ means to recompute and record the position-value pairs at all changed difference values on the right/bottom boundary, compose the lists $\Delta_R^{I,J}$ and $\Delta_B^{I,J}$, and provide the bottom-left and/or top-right values for them. Note that a box $\mathcal{B}^{I,J}$ is processed only if the list $\Delta_B^{I-1,J}$ or the list $\Delta_R^{I,J-1}$ is not empty.

---

[3] In case of using a linked representation of DR or DS, the list should also contain pointers to the corresponding entries in column j.

5.1 Processing Black Boxes

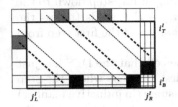

Fig. 5. Updating differences in a black box

The matching black boxes are fairly simple to update. The differences between cells in the left/up boundary are transferred as such diagonally to the right/bottom boundary. The lists $\Delta_B{}^{I-1,J}$ and $\Delta_R{}^{I,J-1}$ tell in which positions along the left/up boundary the inequality $DR'[i,j] \neq DR[i, j - \ell]$ was true, that is, the difference representation changed. Figure 5 shows an example. The grey cells mark difference changes and the black cells are positions on the right/bottom boundary where the changes need to be reflected. Let $DR'[i,j]$ be a changed left/up cell, $DR'[i',j']$ be the affected right/bottom cell, and set $k = \min\{i' - i, j' - j\}$. As the values propagate diagonally, there are three cases (also shown in Fig. 5): the cell $DR'[i',j']$ is $DR'[i' + k, j' + k]$, its right, or down neighbour. We can traverse from $DR'[i',j']$ to $DR'[i'+k,j'+k]$ in the linked DS' structure in $\Theta(k)$ steps and then update $DR'[i',j']$ in $O(1)$ time. If both cells are inside vertical or horizontal boundaries (the middle case in Fig. 5), then we can first follow a single link that hops from row $i_T^I - 1$ to row i_B^I and then directly walk the remaining k steps. If one cell is on a vertical side and the other on a horizontal side (left or right case in Fig. 5), then walking from one to the other via the nearest top right or low left corner takes $2k = \Theta(k)$ steps. Updating $DR'[i',j']$ is a simple matter of copying the source difference from $DR'[i,j]$ in appropriate manner. It is also simple to construct the lists $\Delta_B{}^{I,J}$ and $\Delta_R{}^{I,J}$. The distance value $D'[i',j']$ may be computed in $O(1)$ time from the already known values $D'[i,j] = D'[i + k, j + k]$.

Let $\#^{I,J}$ be the number of cells in box $\mathcal{B}^{I,J}$ of DR' that differ from DR. The preceding procedure for updating a black box requires $O(\#^{I,J})$ work. Each top/left change $DR'[i,j]$ is propagated to each cell along the $\Theta(k)$ long diagonal path to $DR'[i + k, j + k]$, so there are at least $\Theta(k)$ distinct changed differences in $\mathcal{B}^{I,J}$ for each $\Theta(k)$-work phase done by the algorithm. The bottom-left and/or top-right values that are required by the lists $\Delta_B{}^{I,J}$ and/or $\Delta_R{}^{I,J}$ may be updated within this time limit in similar manner as with white boxes (see below). We omit further details due to lack of space.

5.2 Processing White Boxes

Now consider updating a white box. The initial situation is shown in Fig. 6. The grey cells again mark positions on the left/up boundary, given by lists $\Delta_B{}^{I-1,J}$ and $\Delta_R{}^{I,J-1}$, where the difference has changed. Our strategy for a white box is to traverse difference changes by depth-first-search (DFS). This partially resembles how the KP algorithm [5] traces changes in DR. We start a separate DFS from each of the $O(1)$ positions listed in $\Delta_B{}^{I-1,J}$ or $\Delta_R{}^{I,J-1}$. Each search traces neighboring cells with difference changes as long as possible while still remaining inside the current box. If the search is currently in cell

$DR[i,j]$, it may proceed one step right to the cell $DR[i,j+1]$ if the condition $DR'[i,j].U \neq DR[i,j-\ell].U$ (from Lemma 1) holds, and/or one step down to the cell $DR[i+1,j]$ if the condition $DR'[i,j].L \neq DR[i,j-\ell].L$ holds (unless the step would leave the current box). Figure 6 shows examples of the first step from the left column or top row.

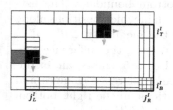

Fig. 6. Updating differences in a white box

Let $DR'[i,j]$ be the cell that the DFS has just moved to. Consider first the case where the step was to the right. We assume (in inductive manner) that the previous column values $D'[i-1,j-1]$, $D'[i,j-1]$, $D[i-1,j-1-\ell]$ and $D[i,j-1-\ell]$ are available in $O(1)$ time. The assumption holds in the first step from the left boundary, and the further steps always maintain this property. We will then compute the current column values $D'[i-1,j]$, $D'[i,j]$, $D[i-1,j-\ell]$, and $D[i,j-\ell]$. Once all these are known, it is trivial to determine both $DR'[i,j]$ and $DR[i,j-\ell]$. If the step was down, we assume (again in inductive manner) that the above row values $D'[i-1,j-1]$, $D'[i-1,j]$, $D[i-1,j-1-\ell]$ and $D[i-1,j-\ell]$ are available, compute the current row values $D'[i,j-1]$, $D'[i,j]$, $D[i,j-1-\ell]$, and $D[i,j-\ell]$ and then derive the differences $DR'[i,j]$ and $DR[i,j-\ell]$. Once $DR'[i,j]$ and $DR[i,j-\ell]$ are known, the DFS compares them to decide if it should next go right and/or down or backtrack.

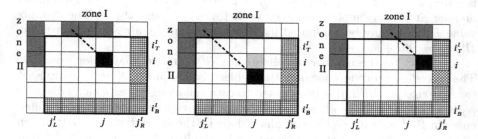

Fig. 7. An example of how the zones change upon a down or right step

Due to lack of space, we sketch only briefly how to compute a certain $D'[i,j]$. The other values of D' and D are computed in similar manner. The method is similar to Arbell et al.'s algorithm. Figure 7 depicts the cell $D'[i,j]$ (in black) and the corresponding zones, illustrating how now also zone I may move when the DFS makes a step right. Now both zones have their own counter array C that is used for maintaining the minimum value. The only delicacy is that we need to know the actual D'- or D-values inside the zones. We note that when a DFS makes its first step, both zones have two cells, and both of them touch the top-left cell $D'[i_T^I-1,j_L-1]$ and/or the start cell of the DFS. The list $\Delta_B^{I-1,J}$ and/or $\Delta_R^{I,J-1}$ provides the values $D'[i_T^I-1,j_L-1]$ and $D[i_T^I-1,j_L-1-\ell]$, and the values of D' and D at the starting cell of the DFS are available in $O(1)$ time. Therefore both zones have a known base value from which the rest of the values in the zone can be computed incrementally using the difference information stored by DS' along the left/up boundary. This poses no problems as the zones always move/change only one step/cell at a time.

The main remaining question is how to build the lists $\Delta_B{}^{I,J}$ and $\Delta_R{}^{I,J}$. If a DFS reaches a cell in the right/bottom boundary, the position and the distance (maintained by the DFS) is recorded in the respective list. Since there can be only $O(1)$ such positions along the right/bottom boundary and there are only $O(1)$ different DFS searches, the lists are easy to combine in $O(1)$ time. Recall that we also need to compute the bottom-left values $D'[i_B^I, j_L^J - 1]$ and $D[i_B^I, j_R^J - 1 - \ell]$, if $\Delta_B{}^{I,J}$ is not empty, and the top-right values $D'[i_T^I - 1, j_R^J]$ and $D[i_T^I - 1, j_R^J - \ell]$, if $\Delta_R{}^{I,J-1}$ is not empty. We only describe the first case, as the second case is symmetric. If $\Delta_B{}^{I,J}$ is not empty, then a DFS has traversed from some source cell on the left/up boundary to the bottom row i_B^I. Let i be the row of the DFS source cell. That DFS made at least $i_B^I - i$ steps. If the source cell resides on the left boundary, we compute $D'[i_B^I, j_L^J - 1]$ and $D[i_B^I, j_R^J - 1 - \ell]$ in $\Theta(i_B^I - i)$ time by starting with the values $D'[i, j_L^J - 1]$ and $D[i, j_R^J - 1 - \ell]$ that the source cell has available, and then traverse the difference values stored on the left boundary. If the source cell is on the top boundary, then the process is otherwise the same but now we start from the known top-left values $D'[i_T^I - 1, j_L - 1]$ and $D[i_T^I - 1, j_L - 1 - \ell]$. In both cases the $\Theta(i_B^I - i)$ time can be charged to the work of a DFS that walked at least $i_B - i$ steps.

The number of cells visited by each DFS is at most $O(\#^{I,J})$, and each step takes $O(1)$ time. Since there are $O(1)$ different DFS-searches per box, the overall work for a white box is $O(\#^{I,J})$.

Theorem 3. *DS can be updated to DS' in $O(m+n)$ time.*

Proof. It follows from Theorem 1 and the fact that both black and white boxes can be processed in $O(\#^{I,J})$ time. $\square$

References

1. Arbell, O., Landau, G.M., Mitchell, J.S.: Edit distance of run-length encoded strings. Inf. Process. Lett. **83**(6), 307–314 (2002)
2. Barton, C., Iliopoulos, C.S., Pissis, S.P.: Average-case optimal approximate circular string matching. In: Dediu, A.-H., Formenti, E., Martín-Vide, C., Truthe, B. (eds.) LATA 2015. LNCS, vol. 8977, pp. 85–96. Springer, Heidelberg (2015)
3. Hsu, P., Chen, K., Chao, K.: Finding all approximate gapped palindromes. Int. J. Found. Comput. Sci. **21**(6), 925–939 (2010)
4. Hyyrö, H., Narisawa, K., Inenaga, S.: Dynamic edit distance table under a general weighted cost function. J. Disc. Algorithms **34**, 2–17 (2015)
5. Kim, S.R., Park, K.: A dynamic edit distance table. J. Disc. Algorithms **2**, 302–312 (2004)
6. Landau, G.M., Myers, E.W., Schmidt, J.P.: Incremental string comparison. SIAM J. Comp. **27**(2), 557–582 (1998)
7. Schmidt, J.P.: All highest scoring paths in weighted grid graphs and their application in finding all approximate repeats in strings. SIAM J. Comp. **27**(4), 972–992 (1998)

Walking Automata in Free Inverse Monoids

David Janin[✉]

UMR CNRS LaBRI, Inria Bordeaux, Bordeaux INP,
University of Bordeaux, 33405 Bordeaux, Talence, France
janin@labri.fr

Abstract. Walking automata, be they running over words, trees or even graphs, possibly extended with pebbles that can be dropped and lifted on vertices, have long been defined and studied in Computer Science. However, questions concerning walking automata are surprisingly complex to solve. In this paper, we study a generic notion of walking automata over graphs whose semantics naturally lays within inverse semigroup theory. Then, from the simplest notion of walking automata on birooted trees, that is, elements of free inverse monoids, to the more general cases of walking automata on birooted finite subgraphs of Cayley's graphs of groups, that is, elements of free E-unitary inverse monoids, we provide a robust algebraic framework in which various classes of recognizable or regular languages of birooted graphs can uniformly be defined and related one with the other.

1 Introduction

General Context. Walking automata, be they running over words, trees or even graphs, possibly extended with pebbles, have long been defined and studied in Computer Science [7,8]. For instance, tree walking automata with pebbles have been an important subject of study the last decades since they are natural abstract models of machine for XML query languages such as XPATH, or XML transformation languages such as XSL [6].

Although based on well studied computation models: finite state machines or pushdown automata, questions about walking automata are often surprisingly complex to solve and, to a lesser extent, quite dependent on such or such details in automata's definition.

For instance, in the case of tree languages, bounding the number of pebbles an automaton leads to defining classes of recognizable languages. Various logical characterizations of these classes have been obtained [8] and difficult separation results have also been proved [1–3]. However, for separation results, proof arguments apply to the case of pebbles that are marked and visible [2,3], leaving open the cases of unmarked and/or invisible pebbles.

Even though walking automata are sequential machines much like string automata, the classical algebraic tools that have been developed to study word automata are not easily applicable to tree walking automata. Despite numerous results, little is known about the underlying mathematical framework, say in algebra, that walking automata may induce.

R.M. Freivalds et al. (Eds.): SOFSEM 2016, LNCS 9587, pp. 314–328, 2016.
DOI: 10.1007/978-3-662-49192-8_26

Contribution of the Paper. In this paper, we initiate the development of an algebraic framework, within inverse semigroup theory, for walking automata. We provide a generic notion of automata walking on edge-labeled graphs. They act as some kind of observers of their input graphs much in the same way observational semantics has been defined in concurrency theory by Hennessy and Milner [17].

Unlike most classical definitions, we do not require walking automata to start and end in the same vertex, neither do we require the complete traversal of input structures. Moreover, the capacity given to a walking automaton to check or not the absence of an (incoming or outgoing) edge labeled by a given letter induces two possible semantics for walking automata, much in the same way there are various observational semantics in [17], with or without observable failures.

The languages recognized by our walking automata are languages of birooted graphs, that is, graphs extended with an input root: the vertex where the run starts, and an output root: the vertex where the run ends. These languages are shown to be closed under root preserving graph morphisms (Lemma 12). Moreover, the sequential composition of partial runs of walking automata is shown to induce a composition of the traversed birooted graphs, yielding an inverse monoid structure (see Remark 3 and Theorem 22).

Then we prove (Theorem 16) that, in many cases, the stronger semantics (with observable reading failures) can be uniformly encoded in the weaker semantics (with unobservable reading failures).

As particular cases, walking automata in Cayley's graphs of finitely generated groups are considered in Sect. 4. The recognized languages are subsets of monoids known as freest E-unitary inverse monoids [15,16]. Then, based on the underlying monoid structure, an extension of regular expressions is defined and shown (Theorem 25) to characterize the classes of recognizable languages induced by limiting numbers of allowed pebbles.

2   Graphs

It is very likely that most concepts and properties detailed here have already appeared in the literature. However, for the sake of completeness we provide our own presentation.

Graphs and Morphisms. Let $A = \{a, b, c, \cdots\}$ be a finite alphabet. A *(relational) graph* on the edge alphabet A is a pair $G = \langle V, E \rangle$ with a set of vertex V and sets of a-labeled (directed) edges $E(a) \subseteq V \times V$ for all $a \in A$. A *graph morphism*, or simply morphism, from $G_1 = \langle V_1, E_1 \rangle$ to $G_2 = \langle V_2, E_2 \rangle$ is a mapping $f : V_1 \to V_2$ such that $f(E_1(a)) \subseteq E_2(a)$, for every $a \in A$. Such a morphism is denoted by $f : G_1 \to G_2$.

Walking Paths. Let $\bar{A} = \{\bar{a}, \bar{b}, \bar{c}, \cdots\}$ be a copy of the alphabet A. Let $(A + \bar{A})^*$ be the free monoid generated by $A + \bar{A}$ with the unit (empty word) denoted by 1 and the concatenation of two words $u, v \in (A + \bar{A})^*$ simply denoted by uv.

A (back and forth) walking *path* on the graph G from a vertex x to a vertex y is an alternating sequence of vertices of V and letters of $A + \bar{A}$ of the form

$$\pi = x_0 z_1 x_1 z_2 x_2 \cdots x_{n-1} z_n x_n$$

such that $x = x_0$, $y = x_n$ and, for every $1 \leq i \leq n$, we have $(x_{i-1}, x_i) \in E(z_i)$ where, for every $a \in A$, the relation $E(\bar{a})$ denotes the inverse relation $E^{-1}(a) = \{(x, y) \in V \times V : (y, x) \in E(a)\}$. The vertex x is the *source* of such a path. It is denoted by $\boldsymbol{sr}(\pi)$. The vertex y is the *target* of such a path. It is denoted by $\boldsymbol{tg}(\pi)$.

In such path, a letter $a \in A$ models a *forward traversal* of an a-labeled edge and a letter $\bar{a} \in \bar{A}$ models a *backward traversal* of an a-labeled edge. The *inverse path* π^{-1} of the path π is defined by

$$\pi^{-1} = x_n z_{n-1}^{-1} x_{n-1} \cdots x_2 z_1^{-1} x_1 z_0^{-1} x_0$$

with $(a)^{-1} = \bar{a}$ and $(\bar{a})^{-1} = a$ for every $a \in A$. We easily observe that π^{-1} is indeed a walking path in the graph G from x_n to x_0.

As a particular case, the graph G is *bideterministic* when for every $z \in A + \bar{A}$, for every $(p, q), (p', q') \in E(z)$, if $p = p'$ then $q = q'$. In this case, every path π as above, emanating from a given vertex x, is completely determined by its source x and the path label $\lambda(\pi) = z_1 z_2 \cdots z_n \in (A + \bar{A})^*$ obtained from π by deleting all vertices.

Path-induced Birooted Subgraph. Let $\pi = x_0 z_1 x_1 z_2 x_2 \cdots x_{n-1} z_n x_n$ be a path on the graph G. The *subgraph* $G|\pi$ of graph G induced by the path π is defined by $G|\pi = \langle V|\pi, E|\pi \rangle$ with the set of vertices $V|\pi = \{x_0, x_1, \cdots, x_n\}$ and, for every $a \in A$, the set of a-labeled edges $(E|\pi)(a)$ defined as the set of pairs $(x, y) \in V' \times V'$ such that either xay or $y\bar{a}x$ occurs as a subsequence in the path π. Then, the following lemma is immediate.

Lemma 1. *The graph $G|\pi$ induced by the path π is finite and the inclusion mapping $\iota : V|\pi \to V$ that maps every vertex to itself is a one-to-one morphism, i.e. graph $G|\pi$ is a finite subgraph of graph G.*

Definition 2 (Birooted induced subgraphs). Let π be a path of G. The triple $\theta_G(\pi) = (G|\pi, \boldsymbol{sr}(\pi), \boldsymbol{tg}(\pi))$ defined by distinguishing the source and the target of the path π in the subgraph $G|\pi$, is called the *birooted subgraph* of G induced by the path π.

Remark 3. It is an easy observation that the trivial birooted subgraph product $\theta_G(\pi_1) \cdot \theta_G(\pi_2)$ defined to be the graph $(G|\pi_1 \cup G|\pi_2, \boldsymbol{sr}(\pi_1), \boldsymbol{tg}(\pi_2))$ (with union defined over subsets of vertices and edges) when $\boldsymbol{tg}(\pi_1) = \boldsymbol{sr}(\pi_2)$ and 0 otherwise (with 0 an additional zero element) yields an inverse semigroup: for every element x there is a unique element x^{-1} such that $xx^{-1}x = x$ and $x^{-1}xx^{-1} = x^{-1}$. Indeed, we have $0^{-1} = 0$ and $\theta_G(\pi)^{-1} = \theta_G(\pi^{-1})$, and, additionally, it can be shown that the non zero idempotent elements exactly correspond to the birooted graphs induced by cyclic paths.

The much more interesting case when the birooted subgraphs are invariant under translation (in Cayley's graphs) is detailed in Sect. 4.

Vertex-Labeled Graphs. So far, the graphs we consider have no vertex label. The following definition and lemma shows that this fact does not reduce the generality of our language theoretical study.

Definition 4 (Induced vertex label). The *vertex label* of a vertex $x \in V$ in a graph $G = \langle V, E \rangle$ is defined to be the set $\lambda_V(x) = \{a \in A : (x,x) \in E(a)\}$.

The following lemma, whose proof is immediate, emphasizes the relevance of this notion.

Lemma 5. *Let $G = \langle V, E \rangle$ be a graph on the edge alphabet A. Let $g : V \to \mathcal{P}(B)$ be a vertex labeling function with some new alphabet B disjoint from A. Let $\langle G, g \rangle$ be the resulting vertex-labeled graph and let $\varphi(\langle G, g \rangle) = \langle V', E' \rangle$ be the edge-labeled graph defined by $V' = V$, by $E'(a) = E(a)$ for every $a \in A$, and $E'(b) = \{(x,x) \in V' \times V' : b \in g(v)\}$ for every $b \in B$.*

Then, the vertex identity mapping from V into V' is a one-to-one and onto graph morphism from G into $\varphi(\langle G, g \rangle)$. Moreover, φ is a one-to-one mapping from the class of graphs with A-labeled edge and $\mathcal{P}(B)$-labeled vertices into the class of $(A \cup B)$-labeled edges such that, given the vertex labeling $\lambda'_V : V' \to \mathcal{P}(A + B)$ as defined above, then, for every $v \in V' = V$, we have $g(v) = \lambda'_V(v) \cap B$.

In other words, graphs with vertex labels are easily encoded into graphs without vertex labels. Moreover, in the case both A and B are finite, such a mapping induces a fairly simple MSO-transduction (see [4] Chap. 7). It follows that every MSO-definable language of graphs with A-labeled edges and $\mathcal{P}(B)$-labeled vertices can be encoded into an MSO-definable language of graphs with $A \cup B$-labeled edges. Since graphs with unlabeled vertices are particular case of graphs with labeled vertices this really says that, up to MSO definable languages, studying languages of edge-labeled graphs or languages of edge-and-vertex-labeled graphs is essentially the same.

3   Walking on Graphs

In this paper, a walking automata is sort of a graph observer that traverses the input graph possibly dropping and lifting (in the reverse order) some pebbles. Since walking automata cannot jump between disconnected graphs, all graphs considered from this point are assumed to be connected via walking paths.

Definition 6 (Walking automata with pebbles). A *walking automata with pebbles* on the alphabet A is a tuple $\mathcal{A} = \langle Q, I, T, \delta, \Delta \rangle$ with set of states Q, initial states $I \subseteq Q$, terminal states $T \subseteq Q$, edge transitions $\delta(z) \subseteq Q \times Q$ for every $z \in A + \bar{A}$, and pebble transitions $\Delta((r,s)) \subseteq Q \times Q$ for every $(r,s) \in Q \times Q$.

Informally, from a given vertex, the automaton can traverse forward any outgoing a-labeled edge (reading a) or it can traverse backward any incoming a-labeled edge (reading $\bar{a}$). In both cases, the automaton state is updated according to the first-order transition function δ applied to the traversed edge label a or $\bar{a}$. Additionally, the automaton may drop a pebble on that vertex, interrupting the current run and starting a new subrun. It may also lift a pebble, ending the current subrun and resuming the former run. When ending a subrun, the automaton state is updated according to the second-order transition function Δ applied to the pair of states resulting from the start state and the stop state of the subrun.

Definition 7 (Automaton configuration). Let $\mathcal{A} = \langle Q, I, T, \delta, \Delta \rangle$ be a walking automaton. Let $G = \langle V, E \rangle$ be a graph on the alphabet A. An *automaton configuration* $\Gamma \in (Q \times Q \times V)^+$ is a non-empty stack of (dot separated) triples over $Q \times Q \times V$.

In a stack of the form $\Gamma.(p, q, x)$ with $p, q \in Q$ and $x \in V$, the triple (p, q, x) describes the current run configuration: *from state p, the automaton $\mathcal{A}$ walked to the current vertex x reaching current state q*. The additional stack Γ, possibly empty, contains the configurations of formerly interrupted runs.

As formalized in the next definition, when dropping a pebble on a vertex x, the automaton *interrupts* the current run, *pushes* its configuration (p_1, q_1, x) on the stack, and *starts a subrun* in a configuration (p_2, p_2, x).

On the contrary, when lifting a pebble from a vertex x, the automaton *terminates* a subrun in a configuration (p_1, q_1, x), *pops* the saved configuration (p_2, s, x), and *resumes* the former run in an updated configuration (p_2, q_2, x), chosen according to the (second-order) transition condition $(s, q_2) \in \Delta((p_1, q_1))$.

Definition 8 (Automaton transition and run). On a graph $G = \langle V, E \rangle$, a *transition step* from a configuration $\Gamma_1.(p_1, q_1, x)$ to a configuration $\Gamma_2.(p_2, q_2, y)$ reading $z \in \{1\} \cup A \cup \bar{A}$ is defined according to one of the following three cases:

(1) edge traversal: $z \in A \cup \bar{A}$, $\Gamma_1 = \Gamma_2$, $p_1 = p_2$, $(q_1, q_2) \in \delta(z)$ and $(x, y) \in E(z)$,
(2) pebble drop: $z = 1$, $y = x$, $\Gamma_2 = \Gamma_1 \cdot (p_1, q_1, x)$ and $q_2 = p_2$,
(3) pebble lift: $z = 1$, $y = x$, $\Gamma_1 = \Gamma_2 \cdot (p_2, s, x)$ and $(s, q_2) \in \Delta((p_1, q_1))$.

Such a transition step is denoted by $\Gamma_1.(p_1, q_1, x) \vdash_z \Gamma_2.(p_2, q_2, y)$.

An *run* of the automaton $\mathcal{A}$ on the graph G from a vertex x to a vertex y is then defined as a sequence of transition steps

$$\rho = \Gamma_0.(p_0, q_0, x_0) \vdash_{z_1} \Gamma_1.(p_1, q_1, x_1) \cdots \vdash_{z_n} \Gamma_n \cdot (p_n, q_n, x_n)$$

with $x_0 = x$ and $x_n = y$, also denoted by $\rho = \Gamma_0 \cdot (p_0, q_0, x_0) \vdash_u^* \Gamma_n \cdot (p_n, q_n, x_n)$ with $u = z_1 z_2 \cdots z_n$. The *path* $\pi(\rho)$ induced by run ρ is defined by

$$\pi(\rho) = x_0 z_1 x_1 z_2 x_2 \cdots x_{n-1} z_n x_n$$

Such a run ρ is an *accepting run* from x to y when Γ_0 is the empty stack with $p_0 = q_0 \in I$ (the first configuration is initial), and Γ_n is the empty stack with $p_n \in I$ and $q_n \in T$ (the last configuration is terminal).

Given an integer $k \geq 0$, the run ρ is *k-accepting run* when it is accepting and $|\Gamma_i| \leq k$ for every $0 \leq i \leq n$, where $|\Gamma_i|$ is the length of the sequence Γ_i. For notational purpose, an accepting run with no bound on the number of allowed pebbles is also called an ∞-*accepting run*.

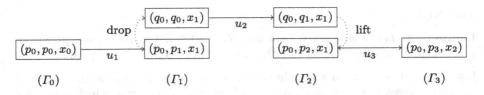

Fig. 1. A run $\Gamma_0 \vdash^*_{u_1} \Gamma_1 \vdash^*_{u_2} \Gamma_2 \vdash^*_{u_3} \Gamma_3$

Example 9. An example of run is depicted Fig. 1 where configuration stacks are depicted vertically. In this run, when no other pebble but the one depicted above is used, then first order transition conditions imply that $(p_0, p_1) \in \delta(u_1)$, $(q_0, q_1) \in \delta(u_2)$ and $(p'_1, p_2) \in \delta(u_3)$, with an obvious extension of δ to $(A + \bar{A})^*$, and second order transition conditions imply that $(p_1, p'_1) \in \Delta((q_0, q_1))$.

Definition 10 (Recognized languages). Given a class of graphs $\mathcal{G}$ (possibly omitted when clear from the context), the language *recognized* (resp. *k*-recognized) by the automaton $\mathcal{A}$ in the class of graph $\mathcal{G}$ is the set $L^\infty_{\mathcal{G}}(\mathcal{A})$ (resp. $L^k_{\mathcal{G}}(\mathcal{A})$) of birooted graphs (G, x, y) with $G \in \mathcal{G}$ and x, y two vertices of G, such that there is an accepting run (resp. a *k*-accepting run) of the automaton $\mathcal{A}$ over G from x to y.

Remark 11. The walking automata defined here are walking automata with unmarked and invisible pebbles in the sense of [6]. However, generalizing Pécuchet's study of two-way automata on strings [18] (see also [5,11]), we do not require that accepting runs starts and ends in the same vertex of the input structures. Moreover, our definition also differs from the definition proposed in [1,6] in the sense that, a priori, the absence of edges *cannot* be detected by the automata and the walking automaton is not required to traverse the entire structure. The consequences of these facts are discussed below.

Lemma 12. *Let $\mathcal{A}$ be a walking automaton on the alphabet A. Let $G_1 = \langle V_1, E_1 \rangle$ and $G_2 = \langle V_2, E_2 \rangle$ be two graphs on the same alphabet. Assume that there is a graph morphism $f : G_1 \rightarrow G_2$. Then, for every $0 \leq k \leq \infty$ and $x, y \in V_1$ if $(G_1, x, y) \in L^k(\mathcal{A})$ then $(G_2, f(x), f(y)) \in L^k(\mathcal{A})$.*

For every environment Γ (run ρ) of the automaton $\mathcal{A}$ on G_1, let $f(\Gamma)$ (resp. $f(\rho)$) be the environment (resp. the run) of the automaton $\mathcal{A}$ on G_2 obtained from Γ (from ρ) by replacing all vertices $x \in V_1$ by their images $f(x) \in V_2$. Then, a simple induction shows that for every run $\rho = \Gamma_1 \vdash^*_u \Gamma_2$ of $\mathcal{A}$ from x to y in graph G_1, $f(\rho) = (\Gamma_1) \vdash^*_u f(\Gamma_2)$ is a well defined run of $\mathcal{A}$ from $f(x)$ to $f(y)$ in graph G_2. The rest of the proof is the routine. In general, the converse does not hold. However, as detailed below, the converse holds in the case the graph G_1 is the subgraph of G_2 induced by an accepting run.

Definition 13 (Graphs induced by a run). Let $\rho : \Gamma_1 \vdash \Gamma_2$ be a run in a graph G from x to y. The *graph induced by a run* ρ is defined to be the subgraph $G_\rho = G|\pi(\rho)$ induced by the path $\pi(\rho)$ traversed by $\mathcal{A}$ in G.

Then we have:

Lemma 14. *Let G be a graph such that $(G, x, y) \in L^k(\mathcal{A})$ via an accepting run ρ. Let G_ρ be the graph induced by the run ρ. Then $(G_\rho, x, y) \in L^k(\mathcal{A})$.*

Let ρ be a run of $\mathcal{A}$ from x to y in G. Then a simple induction on the length of the run ρ shows that ρ is also a run of $\mathcal{A}$ from x to y in $G|\pi(\rho)$ which concludes the proof. The more classical notion of accepting runs defined by complete traversals of the input structure can be related with ours as follows.

Definition 15 (Strict recognizability). A birooted graph (G, x, y) is *strictly recognized* (resp. *strictly k-recognized*) by an automaton $\mathcal{A}$ when there is an accepting (resp. k-accepting) run ρ of $\mathcal{A}$ over G from x to y such that (G_ρ, x, y) and (G, x, y) are isomorphic.

As an immediate corollary of Lemmas 1, 12 and 14, we thus have:

Theorem 16. *Let $\mathcal{A}$ be a walking automaton. For every $k \geq 0$, let $L_S^k(\mathcal{A})$ be the class of birooted graphs strictly k-recognized by $\mathcal{A}$ and let $L^k(\mathcal{A})$ the class of birooted graphs k-recognized by $\mathcal{A}$. Then $L^k(\mathcal{A})$ is the morphism closure of the language $L_S^k(\mathcal{A})$, that is, $(G, x, y) \in L^k(\mathcal{A})$ if, and only if, there exists $(G', x', y') \in L_S^k(\mathcal{A})$ and a graph morphism $f : G' \to G$ such that $f(x') = x$ and $f(y') = y$.*

In particular, when the birooted structures cannot be related by morphisms (as with end markers in two-way word automata [5]), studying strict recognizability just amounts to study recognizability.

4   Walking in Cayley's Graphs of Groups

So far, we have not much used the fact that walking automata recognize sets of birooted graphs. When the underlying graph G is the Cayley's graph of a (presented) group, then the (isomorphic classes of) finite birooted subgraphs induced by paths form a inverse monoid (see [15] for more details and also [16] for a general presentation). Based on the underlying monoid structure, various classes of languages can then be defined and characterized by means of certain restriction of walking automata.

Definition 17 (The Cayley graph of a presented group). Let G be a group generated by $A \subseteq G$ and let $\varphi : (A + \bar{A})^* \to G$ be the corresponding inverse-preserving monoid morphism[1]. Then, the *Cayley graph* of the presented group G is defined to be the graph $C_G = \langle V, E \rangle$ with vertex set defined by $V = G$ and, for every $a \in A$, edge set defined by $E(a) = \{(x, y) \in V \times V : x \cdot \varphi(a) = y\}$.

For convenience, we extend the edge relation function E to $(A + \bar{A})^*$ by taking $E(u) = \{(x, y) \in G \times G : x \cdot \varphi(u) = y\}$. As a particular case, since φ is inverse-preserving and G is a group, we indeed have $E(\bar{a}) = E(a)^{-1}$ which is consistent with our previous extension of the edge relations.

---

[1] The group G is *presented* by the morphism φ.

Remark 18. Clearly, the Cayley graph C_G of the group G is a (possibly infinite) bideterministic graph as well as its (finite) birooted subgraphs. Depending on the chosen group, various interesting examples can be defined (see [16]).

For instance, taking the free group $FG(A)$ we have *birooted trees*. Taking the group defined from $A = \{a, b, c, d\}$ by $cc = 1$, $dd = 1$ and $cd = 0$, we obtain *vertex-labeled birooted trees* with edges labeled by a or b and, following Lemma 5, vertices labeled over $\mathcal{P}(\{c, d\})$. Thanks to the axiom $cd = 0$, only the birooted graph encoding zero has a vertex labeled by both c and d. The language theory of these vertex-labeled birooted graphs has been studied in [10, 14].

Another example, taking the group defined from $A = \{a, b, c, d\}$ by $ab = ba$, $cc = 1$ and $dd = 1$ and $cd = 0$, we obtain *vertex-labeled birooted grids* with (say) horizontal edges labeled by a, vertical edges labeled by b, and, similarly, vertices labeled over $\mathcal{P}(\{c, d\})$.

In other words, this group-theoretic based approach to graphs leads to a vast variety of classes of birooted graphs.

Definition 19 (Induced graphs revisited). Let G be a group presented by a morphism φ, and let C_G be its Cayley graph. For every $u \in (A + \bar{A})^*$, let $C_G|u$ be the graph induced by u defined by $C_G|u = \langle V, E \rangle$ with set of vertices $V = \varphi(Pref(u))$ where $Pref(u) = \{v \in (A + \bar{A})^* : \exists w \in (A + \bar{A})^*, u = vw\}$ is the set of word prefixes of u, and sets of edges $E(a)$ defined as the union of $\{(\varphi(v_1), \varphi(v_2)) \in V \times V : v_1 a = v_2\}$ and $\{(\varphi(v_2), \varphi(v_1)) \in V \times V : v_1 a^{-1} = v_2\}$.

The next lemma, whose proof is immediate, relates our two definitions of induced subgraphs.

Lemma 20. Let π be a path in C_G. Let $\lambda(\pi) \in (A + \bar{A})^*$ be the word of $(A + \bar{A})^*$ obtained from π by deleting all vertices. The birooted subgraph $\theta_{C_G}(\pi)$ induced by the path π isn the graph C_G is isomorphic to the birooted graph $(C_G|\lambda(\pi), 1, \varphi(u))$.

This leads us to the following definition:

Definition 21 (Birooted subgraphs and their product). A birooted finite subgraph of the Cayley graph C_G of the presented group G is a quadruple $B = (V, E, 1, x)$ where $V \subseteq G$ is a finite subset of G such that $1, x \in V$, $E(a) \subseteq \{(x, y) \in V \times V : x \cdot \varphi(a) = y\}$ for every $a \in A$, and such that, the resulting subgraph $\langle V, E \rangle$ is connected. The set of such finite birooted subgraphs of C_G is denoted by $BSG(G)$. Then, the product of two birooted finite subgraphs $B_1 = (V_1, E_1, 1, x_1)$ and $B_2 = (V_2, E_2, 1, x_2)$ is defined by

$$B_1 \cdot B_2 = (V_1 \cup x_1 \cdot V_2, E, 1, x_1 \cdot x_2) \quad \text{with} \quad E(a) = E_1(a) \cup x_1 \cdot E_2(a)$$

with the notation $x_1 \cdot E_2(a) = \{(x, y) \in (x_1 \cdot V_2, x_1 \cdot V_2) : (x, y) \in E_2(a)\}$ for every $a \in A$.

Theorem 22 (Margolis, Meakin [15]). *The set $BSG(G)$ with birooted graph product is an inverse monoid. The mapping $\theta_G : (A + \bar{A})^* \to BSG(G)$ is an onto monoid morphism, and, for every $u \in (A + \bar{A})^*$, we have $\theta_G(u)^{-1} = \theta_G(u^{-1})$ and $\theta_G(u)$ is idempotent if and only if $\varphi(u) = 1$.*

Remark 23. In [15], it is proved that $BSG(G)$ is the freest inverse monoid generated by A whose group image is the group G. This result is much stronger than Theorem 22.

A subset $X \subseteq BSG(C_G)$ of the monoid $BSG(G)$ is called a *G-language*. Following our previous definitions, given $0 \leq k \leq \infty$, the G-language X is k-recognized (resp. strictly k-recognized) by a walking automaton $\mathcal{A}$ when $X = L^k(\mathcal{A}) \cap BSG(C_G)$ (resp. $X = L_S^k(\mathcal{A}) \cap BSG(C_G)$). Then, let k-PWA (resp. k-PWA^S) be the class of G-languages k-recognized (resp. strictly k-recognized) by a finite state walking automaton.

We aim at providing a Kleene-like characterization of these classes of languages by means of regular expressions. For such a purpose, the following operations are defined over G-languages:

(1) sum : $X_1 + X_2 = X_1 \cup X_2$,
(2) product: $X_1 \cdot X_2 = \{x_1 \cdot x_1 \in BSG(G) : x_1 \in X, x_2 \in X\}$,
(3) star: $X^* = \bigcup X^n$,
(4) inverse: $X^{-1} = \{x^{-1} \in BSG(G) : x \in X\}$,
(5) idempotent projection: $X^E = \{x \in X : xx = x\}$,

for all languages $X, X_1, X_2 \subseteq BSG(G)$.

A *k-regular expression* is defined to be any finite expression built over the alphabet $A \cup \bar{A} \cup \{1\}$, combined with sum, product, star and idempotent restriction operators such that the nesting depth of idempotent projection is at most k. A language $X \subseteq BSG(G)$ is a *k-regular language* when it can be defined by a k-regular expressions, mapping 1 to $\theta_G(1)$ and every letter $z \in A + \bar{A}$ to its birooted image $\theta_G(z) \in BSG(G)$.

The class of k-regular languages is denoted by k-REG. The class of languages recognizable by finite monoids M and morphisms from $BSG(G)$ onto M is denoted by REC. Observe that, by definition, the usual class REG of languages definable by finite Kleene regular expressions equals 0-REG. Last, for every class of languages X, let $X^{\downarrow}$ be the class of closure of the languages of X under root-preserving graph morphisms within $BSG(G)$.

Remark 24. Observe that the notion of k-recognizability is not necessarily preserved under (inverse) monoid morphisms. Indeed, given an inverse monoid morphism $\varphi : M \to N$, we certainly have $\varphi(X^E) \subseteq \varphi(X)^E$ for every $X \subseteq M$. However, the reverse inclusion may be false as illustrated by the expression $ab\bar{a}\bar{b}$. Indeed, it induces a non-idempotent birooted tree in the free inverse semigroup but a cycle in any E-unitary inverse semigroup induced by a group in over which the equation $ab = ba$ is satisfied.

Theorem 25 (Hierarchy). *For every presented group G generated by A, the following equalities and inequalities holds. In this figure, strict inequalities $\subset$ are only known to hold in the free inverse monoid, that is, when $G = FG(A)$: the free group generated by A. They have to be read non-strict in all other cases.*

$$0\text{-}PWA^S \subset 1\text{-}PWA^S \subseteq \cdots k\text{-}PWA^S \cdots \subseteq \omega\text{-}PWA^S$$

$$REC \subset 0\text{-}REG \subset 1\text{-}REG \subseteq \cdots k\text{-}REG \cdots \subseteq \bigcup_k k\text{-}REG$$

$$0\text{-}REG^\downarrow \subset 1\text{-}REG^\downarrow \subseteq \cdots k\text{-}REG^\downarrow \cdots \subseteq \bigcup_k k\text{-}REG^\downarrow$$

$$0\text{-}PWA \subset 1\text{-}PWA \subseteq \cdots k\text{-}PWA \cdots \subseteq \omega\text{-}PWA$$

Proof. Each horizontal inclusion follows from the definition. The separation result $REC \subset 0\text{-}REG$ is known over languages of birooted trees [19]. The separation $0\text{-}PWA^S \subset 1\text{-}PWA^S$ follows from the language example of idempotent birooted trees that cannot, by a simple pumping argument, be recognized by an automaton without pebble but that can easily be recognized with a single pebble (see below).

The first row of (vertical) equalities follows from Lemma 26, proven below. Over birooted trees, that is, in the case $G = FG(A)$, these equalities imply the separation result $0\text{-}REG \subset 1\text{-}REG$. Indeed, over birooted trees the language of idempotent trees is recognizable by a one-pebble automaton while a simple argument shows that is cannot be recognized without pebble.

The second row of (vertical) inclusions follows from the known fact [20] (see also [13]) that, for all birooted graphs $x, y \in BSG(G)$, there is a root-preserving graph morphism $f : y \to x$ if and only if $x \leq y$ in the natural order defined by $x \leq y$ when $x = e \cdot y$ for some idempotent element e.

It follows that, we have $X^\downarrow = E(BSG(G)) \cdot X$ for all language $X \subseteq BSG(G)$, and the language $(BSG(G))^E$ of all idempotent elements of $BSG(G)$ belongs to $1\text{-}REG$ as shown by the one-state automaton $\mathcal{A} = \langle \{p, q\}, \{p\}, \{q\}, \delta, \Delta \rangle$ with $\delta(z) = \{(p, p)\}$ for every $z \in A + \bar{A}$ and $\Delta((r, s)) = \{(p, q)\}$ when $r = s = p$ and $\Delta((r, s)) = \emptyset$ otherwise. The fact these inclusions are strict follows, for language of birooted trees, from the fact that the language $\{\theta_G(a)\}$ is not closed under morphisms since it is not closed under natural order.

The last row of (vertical) equalities follows from the first row of vertical equalities and Theorem 16.   □

Lemma 26. *For every ≥ 0, we have $k\text{-}PWA^S = k\text{-}REG$.*

Proof (sketch of). Direct inclusion ($\subseteq$). Let $\mathcal{A} = \langle Q, I, F, \delta, \Delta \rangle$ be a finite-state walking automaton. For every pair of states $p, q \in Q$, let $L_S^k(p, q) \subseteq BSG(G)$ be the class of languages strictly k-recognized by the automaton $\mathcal{A}$ from an initial configuration of the form (p, p, x) to a terminal configuration of the form (p, q, y) for some vertices x, y. Let $E_S^k(p, q)$ be restriction of that language to the case $x = y$, or, equivalently, $E_S^k(p, q) = (L_S^k(p, q))^E$. Then, much like in the proof of Kleene's theorem for regular languages of strings, by mimicking walking automata transition rules, we can define a system of equations relating the languages $L_S^k(p, q)$ and $E_S^k(p, q)$ which resolution yields the expected regular expressions.

Reverse inclusion ($\supseteq$). This can be proved by induction on the syntactic complexity of regular expressions. More precisely, we first prove that the singleton languages $\{\theta_G(1)\}$ and $\{\theta_G(z)\}$ for every $z \in A + \bar{A}$, are strictly 0-recognizable by finite automata. Then, it suffices to show that the class of languages strictly k-recognized by finite walking automata is closed under sum, product and star, and that, if X is strictly k-recognized by a finite walking automaton, then X^E is $(k+1)$-recognized by a finite walking automaton.

It must be noticed that the existence of second order transitions makes these constructions slightly more complex than in the case of string languages. In particular, building an automaton $\mathcal{A}^*$ such that $L_S^k(\mathcal{A}^*) = (L_S^k(\mathcal{A}))^*$ is done from $k+1$ copies of the automaton $\mathcal{A}$. Indeed, this allows to count in any state the number of pebbles that have been dropped and to ensure, between two runs of the automaton $\mathcal{A}$ simulated in the automaton $\mathcal{A}^*$, that all pebbles have been lifted. $\square$

We first prove the direct inclusion ($\subseteq$). Let $\mathcal{A} = \langle Q, I, F, \delta, \Delta \rangle$ be a finite state walking automaton. For every pair of state $p, q \in Q$, let $L_S^k(p, q) \subseteq BSG(G)$ be the class of language strictly k-recognized by the automaton $\mathcal{A}$ from an initial configuration of the form (p, p, x) to a terminal configuration of the form (p, q, y) for some vertices x, y. Let $E_S^k(p, q)$ be restriction of that language to the case $x = y$, or, equivalently, $E_S^k(p, q) = (L_S^k(p, q))^E$.

In the case $k = 0$, for every $p, q \in Q$, we have,

$$L_S^0(p, q) = \delta_{p,q} + \sum_{\substack{z \in A + \bar{A} \\ (p, r) \in \delta(z)}} \theta_G(z) \cdot L_S^0(r, q)$$

with $\delta_{p,q} = \theta_G(1)$ if $p = q$ and $\emptyset$ otherwise. By applying standard argument, this proves that $L_S^0(p, q)$ is 0-REG fo revery $p, q \in Q$ hence $L_S^0(\mathcal{A})$ as well.

Assume that the claim is true for every $k' < k + 1$. From the definition of runs, we easily check that

$$L_S^{k+1}(p, q) = \delta_{p,q} + \sum_{\substack{z \in A + \bar{A} \\ (p, r) \in \delta(z)}} \theta_G(z) \cdot L_S^{k+1}(r, q)$$

$$+ \sum_{\substack{(p', q') \in Q \times Q \\ (p, r) \in \Delta(p', q')}} E_S^k(p', q') \cdot L_S^{k+1}(r, q)$$

since we just enumerate the two cases in which the run may start: either it reads an edge, or it drops a pebble. Then, by induction hypothesis, for every $p', q' \in Q$, the language $E_S^k(p', q')$ is $(k+1)$-regular hence, by application standard reduction techniques, for every $p, q \in Q$, the language $L_S^{k+1}(p, q)$ is also $(k+1)$-regular and thus so is $L_S^{k+1}(\mathcal{A})$.

We prove now the reverse inclusion case ($\supseteq$). This is proved by induction on the syntactic complexity of k-regular expressions.

For the basic cases, let $\mathcal{A}_1$ the automaton a single state, both initial and terminal, and with empty transition functions. Clearly, $L_S^0(\mathcal{A}_1) = \{\theta_G(1)\}$.

Similarly for every $a \in A$. Let $\mathcal{A}_a = \langle D, I, T, \delta, \Delta \rangle$ be the automaton defined by $Q = \{1, \varphi(a)\}, I = \{1\}, T = \{\varphi(a)\}, \delta(a) = \{(1, \varphi(a))\}, \delta(a^{-1}) = \{(\varphi(a), 1)\}$, and $\delta(z') = \emptyset$ for every $z' \notin \{a, a^{-1}\}$, and $\Delta((t, s)) = \emptyset$ for every $r, s \in Q$. Clearly, we have $L_S^0(\mathcal{A}_a) = \{\theta_G(a)\}$.

Then, it suffices to show that the class of languages strictly k-recognized by finite walking automata is closed under sum, product and star, that, if X is strictly k-recognized by a finite walking automaton, then X^E is $(k+1)$-recognized by a finite walking automaton. Let then $X_1, X_2 \subseteq BSG(G)$ be two languages strictly k-recognizable by the automaton $\mathcal{A}_1 = \langle Q_1, I_1, T_1, \delta_1, \Delta_1 \rangle$ and the automaton $\mathcal{A}_2 = \langle Q_2, I_2, T_2, \delta_2, \Delta_2 \rangle$.

Clearly, the sum $X_1 + X_2$ is strictly k-recognizable by the disjoint sum automaton $\mathcal{A}_1 \uplus \mathcal{A}_2$.

For the product, possibly splitting both languages into a disjoint sum with $\{\theta_G(1)\}$ and applying classical identities, we may assume, without loss of generality, that none of these languages contains $\theta_G(1)$ henceforth $I_1 \cap T_1 = \emptyset$ and $I_2 \cap T_2 = \emptyset$. Let then $\mathcal{A}_1 \cdot \mathcal{A}_2 = \langle Q, I, F, \delta, \Delta \rangle$ be the automaton defined by $Q = Q_1 \uplus Q_2$, $I = I_1$, $T = T_2$, for every $z \in A + \bar{A}$,

$$\delta(z) = \delta_1(z) \cup \delta_2(z) \cup \{(p, q) \in Q \times Q : \exists q' \in T_1, (p, q') \in \delta_1(z), q \in I_2\}$$

and, for every $r, s \in Q$, the second order transitions $\Delta((r, s))$ defined by

$$\Delta((r, s)) = \Delta_1((r, s)) \cup \{(p, q) : q \in I_2, \exists q' \in T_1, (p, q') \in \Delta_1((r, s))\}$$

when $r, s \in Q_1$, $\Delta((r, s)) = \Delta_2((r, s))$ when $r, s \in Q_2$, and $\Delta((r, s)) = \emptyset$ otherwise. Then, we can check by induction on the length of runs, that $L_S^k(\mathcal{A} \cdot \mathcal{A}_2) = X_1 \cdot X_2$. The sensitive point, the direct inclusion, is handled by the fact that no runs can goes back and forth between the copy of $\mathcal{A}_1$ and the copy of $\mathcal{A}_2$, and, in an accepting run, moving from a state of $\mathcal{A}_1$ to a state of $\mathcal{A}_2$ is necessarily done with all pebble lifted.

The case of the star is slightly more complex because we must make sure that, at each iteration, no pebble is left on the input.

Still, this can be done by first normalizing the automaton $\mathcal{A}$ (with disjoint set of initial and terminal states) into $\mathcal{A}' = \langle Q', I', T', \delta', \Delta' \rangle$ built out of $k+1$ disjoint copy of the automaton $\mathcal{A}$ by taking $Q' = Q \times \{0, 1, \cdots, k\}$, $I' = I \times \{0\}$, $T' = T \times \{0\}$, for every $z \in A + \bar{A}$, $\delta'(z) = \{(p, i), (q, i)) \in Q' \times Q' : (p, q) \in \delta(z), 0 \leq i \leq k\}$, and for every $r, s \in Q$ and $0 \leq i_1, i_2 \leq k$, taking

$$\Delta((r, i_1), (s, \dot{k}_2)) = \{((p, i), (q, i)) \in Q' \times Q' : (p, q) \in \delta((r, s))\}$$

in the case $i_1 = i_2 = i + 1$ for $i \geq 0$, and $\Delta((r, i_1), (s, k_2)) = \emptyset$ otherwise.

Then, given the automaton $\mathcal{A}'$ as above, we build an automaton $\mathcal{A}^+$, by adding a new initial state q_0, and by duplicating any first order transition from $(q_1, 0)$ to $(q_2, 0)$ to a transition from q_0 to $(q_2, 0)$ in the case $q_1 \in I$, and a transition from $(q_1, 0)$ to q_0 in the case $q_2 \in T$. Then, it is routine to check that $\mathcal{A}^+$ strictly k-recognized X^+. Adding $\theta_G(1)$ poses no difficulty.

The case of the idempotent projection is done as follows. Let $\mathcal{A} = \langle Q, I, F, \delta, \Delta \langle$ that strictly k-recognized $X \subseteq BSG(G)$ with $\theta_G(1) \notin X$. Then we define $\mathcal{A}^E = \langle Q \uplus \{q_i, q_f\}, \{q_i\}, \{q_f\}, \delta', \Delta' \rangle$ by defining for every $z \in A + \bar{A}$, the first order transitions

$$\delta'(z) = \delta(z) \cup \{(q_i, q) : \exists p \in I, (p, q) \in \delta(z)\} \cup \{(p, q_f) : \exists q \in T, (p, q) \in \delta(z)\}$$

and, for every $r, s \in Q \uplus \{q_i, q_f\}$ we put

$$\Delta'((r, s)) = \{(q_i, q_f)\} \cup \Delta((r, s))$$

in the case $r \in I$ and $s \in T$,

$$\Delta'((r, s)) = \Delta((r, s))$$

in the case $r, s \in Q$ with $r \notin I$ or $s \notin T$, and $\Delta'(r, s) = \emptyset$ otherwise.

An accepting run with $k + 1$-pebble of the automaton $\mathcal{A}^E$, necessarily starts in the state q_i and ends in the state q_f. By construction, from the state initial q_i, a pebble must be dropped. Then an accepting run with at most k-pebble is ran in $\mathcal{A}$ and that initial pebble can be lifted, reaching the terminal state q_f. This proves that $L_S^{k+1}(\mathcal{A}^E) \subseteq X^E$. Since the converse inclusion is immediate from the definition, this concludes the idempotent projection case and the proof of the lemma.

5   Conclusion

We have defined walking automata on graphs. By allowing automata to start and stop in arbitrary graph vertices, we have defined the language recognized by a walking automaton in terms of birooted graphs that form inverse semigroups.

Although we do not require walking automata to perform complete traversal of their input structures, thanks to the preorder relation induced by root preserving graph homomorphisms, we eventually provide a correspondence between our notion of recognizability and the more classical one.

In the particular case of Cayley's graphs of groups, we obtain a rather rich array of classes of recognizable languages of birooted graphs, and a notion of k-regular expressions that characterizes the number of allowed pebbles in accepting runs (Theorem 25). How these induced hierarchies of languages of trees or graphs may be related is left as an intriguing open problem.

We conjecture that the hierarchy induced by the number of pebbles is strict for languages of birooted trees. The strictness of the hierarchies for languages of birooted graphs induced by other groups than the free group is also an open problem. Ideally, the algebraic framework proposed here may provide simpler arguments than in [3] for solving these questions.

It has already been observed that an adequate algebraic theory for inverse monoid morphisms can be developed by means of certain kind of premorphisms instead of morphisms [9,10,12,14]. As a matter of fact, transition monoids of walking automata induce a different type of premorphisms that could also be investigated as new language recognizers.

References

1. Bojańczyk, M.: Tree-walking automata. In: Martín-Vide, C., Otto, F., Fernau, H. (eds.) LATA 2008. LNCS, vol. 5196, pp. 1–2. Springer, Heidelberg (2008)
2. Bojańczyk, M., Colcombet, T.: Tree-walking automata do not recognize all regular languages. In: STOC, ACM (2005)
3. Bojańczyk, M., Samuelides, M., Schwentick, T., Segoufin, L.: Expressive power of pebble automata. In: Bugliesi, M., Preneel, B., Sassone, V., Wegener, I. (eds.) ICALP 2006. LNCS, vol. 4051, pp. 157–168. Springer, Heidelberg (2006)
4. Courcelle, B., Engelfriet, J.: Graph structure and monadic second-order logic, a language theoretic approach, vol. 138 of Encyclopedia of mathematics and its applications. Cambridge University Press (2012)
5. Dicky, A., Janin, D.: Two-way automata and regular languages of overlapping tiles. Fundamenta Informaticae **142**, 1–33f (2015)
6. Engelfriet, J., Hoogeboom, H.J., Samwel, B.: XML transformation by tree-walking transducers with invisible pebbles. In: Principles of Database System (PODS). ACM (2007)
7. Engelfriet, J., Hoogeboom, H.J.: Tree-walking pebble automata. In: Karhumäki, J., Maurer, H., Paun, G., Rozenberg, G. (eds.) Jewels are Forever: Contributions to Theoretical Computer Science in Honor of Arto Salomaa, pp. 72–83. Springer, Heidelberg (1999)
8. Engelfriet, J., Hoogeboom, H.J.: Nested pebbles and transitive closure. In: Durand, B., Thomas, W. (eds.) STACS 2006. LNCS, vol. 3884, pp. 477–488. Springer, Heidelberg (2006)
9. Janin, D.: Quasi-recognizable vs MSO definable languages of one-dimensional overlapping tiles. In: Rovan, B., Sassone, V., Widmayer, P. (eds.) MFCS 2012. LNCS, vol. 7464, pp. 516–528. Springer, Heidelberg (2012)
10. Janin, D.: Algebras, automata and logic for languages of labeled birooted trees. In: Kwiatkowska, M., Peleg, D., Fomin, F.V., Freivalds, R. (eds.) ICALP 2013, Part II. LNCS, vol. 7966, pp. 312–323. Springer, Heidelberg (2013)
11. Janin, D.: On languages of one-dimensional overlapping tiles. In: van Emde Boas, P., Groen, F.C.A., Italiano, G.F., Nawrocki, J., Sack, H. (eds.) SOFSEM 2013. LNCS, vol. 7741, pp. 244–256. Springer, Heidelberg (2013)
12. Janin, D.: Towards a higher-dimensional string theory for the modeling of computerized systems. In: Geffert, V., Preneel, B., Rovan, B., Štuller, J., Tjoa, A.M. (eds.) SOFSEM 2014. LNCS, vol. 8327, pp. 7–20. Springer, Heidelberg (2014)
13. Janin, D.: Inverse monoids of higher-dimensional strings. In: Leucker, M., et al. (eds.) ICTAC 2015. LNCS, vol. 9399, pp. 126–143. Springer, Heidelberg (2015)
14. Janin, D.: On labeled birooted trees languages: Algebras, automata and logic. Inf. Comput. **243**, 222–248 (2015)
15. Margolis, S.W., Meakin, J.C.: E-unitary inverse monoids and the Cayley graph of a group presentation. J. Pure and Appl. Algebra **58**, 46–76 (1989)
16. Meakin, J.: Groups and semigroups: connections and contrasts. In: Groups St Andrews 2005, vol. 2. London Mathematical Society, Lecture Note Series 340. Cambridge University Press (2007)
17. Milner, R.: Communication and concurrency. Prentice-Hall, Upper Saddle River (1989)

18. Pécuchet, J.-P.: Automates boustrophedon, semi-groupe de Birget et monoide inversif libre. ITA **19**(1), 71–100 (1985)
19. Silva, P.V.: On free inverse monoid languages. ITA **30**(4), 349–378 (1996)
20. Stephen, J.B.: Presentations of inverse monoids. J. Pure Appl. Algebra **63**, 81–112 (1990)

Precedence Scheduling with Unit Execution Time is Equivalent to Parametrized Biclique

Klaus Jansen, Felix Land, and Maren Kaluza[(⊠)]

Department of Computer Science, University to Kiel, Kiel, Germany
{kj,fku,mkal}@informatik.uni-kiel.de
https://www.algo.informatik.uni-kiel.de

Abstract. We consider the following scheduling problem. Given m machines and n jobs with unit execution times and a precedence relation between the jobs, the problem is to assign each job to one of the machines. The objective is to find a schedule that minimizes the makespan (i.e. the length of the schedule).

We reduce 3-CNF-SAT to this problem and obtain a new lower bound for the running time of $2^{o(\sqrt{n \log n})}$ assuming the Expontential Time Hypothesis (ETH). This improves the previous lower bound of $2^{o(\sqrt{n})}$ also due to the ETH and a reduction by Ullman [13] or, alternatively, a reduction from the k-Clique problem by Lenstra and Rinnooy Kan [10].

For the corresponding decision problem of whether there is a schedule with target makespan $\mathbf{T} = 3$ or not, we further show the equivalence to a classical graph problem, the parametrized Biclique problem. The equivalence also holds for the same scheduling problem with the additional restriction that no job has both a predecessor and a successor. By this we show that an improved lower bound for the running time for the Biclique problem will lead to an improved lower bound for the running time for our scheduling problem and vice versa. Moreover a transfered lower bound for the running time from the Biclique problem would also hold for the running time of approximation algorithms with ratio better than $\frac{4}{3}$OPT. That is, if for example there was no algorithm solving Biclique in $2^{o(n)}$ and U was the set of vertices in the Biclique problem, then there would be no approximation algorithm finding a solution for the introduced scheduling problem with $\Theta(|U|)$ jobs, that finds a solution with a target makespan smaller than $\frac{4}{3}$ times the optimal makespan in time $2^{o(n)}$.

1 Introduction

Complexity theory mainly deals with the question if there exists a polynomial time algorithm for a problem or not. The Exponential Time Hypothesis (ETH), first disposed by Impagliazzo and Paturi, addresses the more explicit question whether NP-hard problems can have algorithms that run in "subexponential time" [3] or not. More precisely, the hypothesis asserts that a truth assignment for a 3-CNF-SAT formular with n_{SAT} variables and m_{SAT} clauses cannot be computed in time $2^{o(n_{\text{SAT}})}$. An important result used in plenty of reductions is implied

© Springer-Verlag Berlin Heidelberg 2016
R.M. Freivalds et al. (Eds.): SOFSEM 2016, LNCS 9587, pp. 329–343, 2016.
DOI: 10.1007/978-3-662-49192-8_27

by the Sparsification Lemma due to Impagliazzo et al. [6]: under assumption of the ETH there is no algorithm that decides 3-CNF-SAT in time $2^{o(m_{SAT})}$. This enables parametrization by the number of clauses. Our paper mainly focuses on reductions to and from our scheduling problem and the consequences the ETH has on its solvability.

Notation. The scheduling problem with precedence constraints will be referred to as $P \mid \text{prec} \mid C_{max}$. If we further add unit execution time as a condition to the problem we will write $P \mid \text{prec}, p_j = 1 \mid C_{max}$. If the target makespan is $\mathbf{T} = 3$ we call the problem $P \mid \text{prec}, p_j = 1 \mid C_{max} = 3$ and if moreover no job has both a successor and a predecessor it is $P \mid \text{prec}, p_j = 1, \text{cl} \leq 2 \mid C_{max} = 3$.
If the number of machines is 1 and the target is to minimize $\sum p_j C_j$, where C_j is the time the job j is finished, we call the problem $P1 \mid \text{prec} \mid \sum p_j C_j$.

Known Results. The problem $P \mid \text{prec} \mid C_{max}$ asks for the best possible schedule with respect to the total running time on a fixed number of identical machines. The instance consists of a number of machines, a set of jobs of different length and of precedence constraints.

This problem arises as an intuitive approach to handle parallel computing, and it is actually already NP-hard for the special case of precedence constraints where only chains are allowed and there exist just two machines [5]. According to Chen et al. [4], $P \mid \text{prec} \mid C_{max}$ on m machines cannot be decided in time $f(m) \cdot |I|^{o(m)}$ where $|I|$ is the size of the scheduling instance and f is an arbitrary function.

In 2011 Svensson gave a reduction from $P1 \mid \text{prec} \mid \sum p_j C_j$ to the scheduling problem $P \mid \text{prec} \mid C_{max}$. This result provides the possibility to transfer bounds for the running times from one scheduling problem to another. For instance this implies that if for any $\varepsilon > 0$ there is no approximation algorithm with ratio $(2 - \varepsilon)$ for $P1 \mid \text{prec} \mid \sum p_j C_j$, then for any $\varepsilon > 0$ there is no $(2 - \varepsilon)$ approximation algorithm for $P \mid \text{prec} \mid C_{max}$ [12]. Assuming a strong version of the Unit Game Conjecture (UGC [9]), Bansal and Khot showed the scheduling problem $P \mid \text{prec} \mid C_{max}$ to be NP-hard to approximate within a factor of $(2 - \varepsilon)$ [1]. With Svenssons result this bound on the running time assuming the UGC can be transferred to $P1 \mid \text{prec} \mid \sum p_j C_j$.

For precedence scheduling with unit execution times ($P \mid \text{prec}, p_j = 1 \mid C_{max}$) the best known lower bound with the ETH is $2^{o(\sqrt{n})}$ and can be achieved both with a reduction from 3-CNF-SAT due to Ullman [13] and a reduction from Clique due to Lenstra and Rinnooy Kan [10]. These results are described in detail in Sect. 2. Dynamic programming over subsets and time slots provides the asymptotically best known running time of $2^{\mathcal{O}(n)}$ for $P \mid \text{prec}, p_j = 1 \mid C_{max}$.

The $k_1 \times k_2$-Biclique problem is closely related to the k-Clique problem. The instance consists of a bipartite graph which is a graph $G = (V \dot\cup W, E)$ with $\{v, w\} \notin E$ if $v \in V$ and $w \in W$, and two parameters k_1 and k_2. The goal is to find a subset $V' \subseteq V$ of size k_1 and a subset $W' \subseteq W$ of size k_2 and for each $v \in V'$ and each $w \in W'$ the edge $\{v, w\}$ is in E.

There is no exact algorithm that decides if there is a $k_1 \times k_2$-Biclique in a bipartite graph in time $2^{o(\sqrt{|V \dot\cup W|})}$, unless the ETH fails, which was already known via a reduction by Johnson [7]. In fact, this reduction transforms a k-Clique instance into a balanced $k \times k$-Biclique problem which is a special case of our problem. On the other hand, no algorithm has been found that solves the $k_1 \times k_2$-Biclique problem faster than $2^{o(|V \dot\cup W|)}$ and it is unknown if there is a better reduction which confirms a tighter bound.

Recently, it has been shown that the parametrized $k \times k$-Biclique problem with parameter k is not fixed parameter tractable by Bingkai Lin [11] via a reduction from k-Clique to $k \times k$-Biclique. That is, if G is the instance, then there is no algorithm deciding wether there is a $k \times k$-Biclique in G or not in running time $f(k) \cdot |G|^{\mathcal{O}(1)}$ for an arbitrary function f depending only on k. Nevertheless the Biclique problem remains one of the simpliest graph problems without exact matching upper and lower bounds for the running time of exact algorithms assuming the ETH [2].

New Results and Organization. In Sect. 2 we will improve the result developed by Ullman in a way it only needs $\mathcal{O}(\frac{m_{\text{SAT}}^2}{\log(m_{\text{SAT}})})$ jobs and show that this leads to a new bound for the running time of $2^{o(\sqrt{n \log n})}$:

Theorem. Assuming ETH there is no algorithm for $P \,|\, \text{prec}, p_j = 1 \,|\, C_{\max}$ in time $2^{o(\sqrt{n \log(n)})}$.

In Sect. 3 we will further show the equivalence between the precedence scheduling with unit execution time and target makespan 3 and the $k_1 \times k_2$-Biclique problem via a linear reduction forth and back:

Theorem. $P \,|\, \text{prec}, p_j = 1 \,|\, C_{\max} = 3$ is equivalent to the $k_1 \times k_2$-Biclique problem.

Our reduction from the Biclique Problem to the $P \,|\, \text{prec}, p_j = 1 \,|\, C_{\max} = 3$ problem produces a scheduling instance for each Biclique instance with the additional property that there is no chain length greater than two, i.e. no job has a predecessor and a sucessor at the same time. By this the equivalence of the $k_1 \times k_2$-Biclique problem to the scheduling instance also holds with this additional restriction on the chain length.

Theorem. $P \,|\, \text{prec}, p_j = 1, \text{cl} \leq 2 \,|\, C_{\max} = 3$ is equivalent to the $k_1 \times k_2$-Biclique problem.

Our reduction shows a new connection between the Biclique Problem and the $P \,|\, \text{prec}, p_j = 1 \,|\, C_{\max}$ and provides the possibility of improving the lower bound for the scheduling problem as soon as there will be an improvement for the parametrized Biclique Problem. In our reduction the number of jobs is linear in the number of only the vertices in V and W and the target makespan is $\mathbf{T} = 3$. By this a new reduction improving the bound for Biclique parametrized by the number of vertices would lead to a new lower bound for an approximation algorithm with ratio better than $\frac{4}{3}$. Our reduction already generalizes the reduction by Lenstra and Rinnooy Kan regarding the extra condition of restricting the chain length to two. Concluding our result presents a new viewpoint to the parametrized Biclique problem and an improvement of the lower bound of the scheduling problem would also improve a lower bound for algorithms solving the parametrized Biclique problem.

2   Lower Bounds for Precedence Constrained Scheduling

First we will introduce a slightly modified version of the reduction by Ullman [13] and derive the lower bound $2^{o(\sqrt{n})}$ from it. Then we will improve the reduction to obtain the new lower bound of $2^{o(\sqrt{n \log n})}$. Both lower bounds are based on the assumption that the ETH holds.

The instance of a 3-CNF-SAT problem is given by a formula φ in CNF with n_{SAT} variables x_i, $i \in [n_{\text{SAT}}]$ and m_{SAT} clauses. The goal is to decide whether there is a truth assignment that satisfies φ.

Let φ be a 3-CNF-SAT instance with n_{SAT} variables and m_{SAT} clauses. The main idea is to create two equal gadgets for each variable x_i. The instance is constructed in a way that there are two possible starting times for the two gadgets in a schedule not exceeding a target makespan $\mathbf{T}$. Exactly one can be started in the first time slot and one in the second. Thus either the one meaning "x_i is true" starts and finishes first and then the one meaning "x_i is false" or vice versa.

For each clause there are 7 jobs representing all possible assignments of its three literals that make the clause true. Each of the jobs is dependent on the last job of the gadgets with the corresponding assignment. For example if there is a clause $(x_i \vee x_j \vee \neg x_k)$ and we choose the corresponding clause job that stands for x_i being false and x_j and x_k being true then it is allowed to be scheduled earlier than all other clause jobs for this clause if all the jobs of the gadget meaning "x_i is false", "x_j is true", and "x_k is true" are scheduled before. That is, if all the corresponding gadgets were started in the first time slot, a clause job representing a partial assignment without contradiction to the assignment represented by the gadgets can be started earlier. This leads to the insight that the formula has a satisfying truth assignment if and only if for each clause there is one job that can be scheduled earlier than the other jobs of the clause. This is the main idea of both reductions.

More precisely let $m = \max\{2n_{\text{SAT}} + 2, 6m_{\text{SAT}} + 1, n_{\text{SAT}} + m_{\text{SAT}} + 1\}$ be the number of machines and let the target makespan $\mathbf{T} = 2n_{\text{SAT}} + 3$. Let there be sets of dummy jobs J_i^{dummy} for $i \in [\mathbf{T}]$. For each pair $j \in J_i^{\text{dummy}}$, $j' \in J_{i'}^{\text{dummy}}$ let $j \prec j'$ iff $i < i'$, for all $i, i' \in [\mathbf{T}]$. Let

$$|J_i^{\text{dummy}}| = \begin{cases} m - n_{\text{SAT}} & \text{, if } i = 1 \text{ or } i = 2, \\ m - (2n_{\text{SAT}} + 1) & \text{, if } 3 \le i \le 2n_{\text{SAT}} + 1, \\ m - (n_{\text{SAT}} + m_{\text{SAT}}) & \text{, if } i = 2n_{\text{SAT}} + 2, \\ m - 6m_{\text{SAT}} & \text{, if } i = 2n_{\text{SAT}} + 3. \end{cases}$$

There is only one possible assignment with target makespan $\mathbf{T}$ of the dummy jobs to the time slots, because the number of machines is chosen in such a way that there is at least one dummy job in each time slot. This leads to a schedule where there is space left for n_{SAT} jobs in the first time slot, in the second up to the $(2n_{\mathrm{SAT}}+1)$st there is space left for $2n_{\mathrm{SAT}}+1$ jobs, and so on.

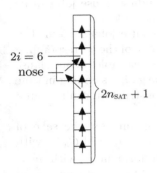

$2i = 6$ —

nose —

$2n_{\mathrm{SAT}}+1$

Fig. 1. A gadget for the variable x_i, where $i = 3$ and $n_{\mathrm{SAT}} = 4$. The arrows represent the precedences between the jobs.

For each variable x_i we create two equal gadgets $g_{x_i=b}$, for $b \in \{0,1\}$, consisting of $2n_{\mathrm{SAT}}+1$ chain jobs and one job we will call nose job: $j^{\mathrm{chain}}_{x_i=b,0} \prec \cdots \prec j^{\mathrm{chain}}_{x_i=b,2n_{\mathrm{SAT}}}$ and $j^{\mathrm{nose}}_{x_i=b}$, $j^{\mathrm{chain}}_{x_i=b,2(i-1)} \prec j^{\mathrm{nose}}_{x_i=b} \prec j^{\mathrm{chain}}_{x_i=b,2i}$. The idea is that the chain jobs must be scheduled in consecutive time slots and the nose job has to be scheduled in the time slot before the job $j^{\mathrm{chain}}_{x_i=b,2i}$ and is some kind of extension of the chain, see Fig. 1 for an illustration.

Next, for each clause c_j let $a_{j,1},\ldots,a_{j,7}$ represent the seven different truth assignments for the three literals in c_j that make c_j true. Let $j^{\mathrm{clause}}_{j,d}$ be a job for each c_j and each $a_{j,d}$. Let $j^{\mathrm{chain}}_{x_i=b,2n_{\mathrm{SAT}}} \prec j^{\mathrm{clause}}_{j,d}$ if and only if $x_i = b$ is included in $a_{j,d}$. The total number of jobs is $n = m \cdot \mathbf{T} = m(2n_{\mathrm{SAT}}+3) \in \mathcal{O}(\max\{m_{\mathrm{SAT}}, n_{\mathrm{SAT}}\}^2)$.

So far, this is a slight modification of the reduction by Ullman, and we will now see how to prove the correctness of the reduction to help the understanding of our own reduction. First of all, each gadget contains a chain of $2n_{\mathrm{SAT}}+1$ jobs and there is at least one other clause job dependent on the last job of each gadget. The clause jobs are not dependent on each other, so each job of each gadget must have been scheduled until the second to last time slot. It is $\mathbf{T} = 2n_{\mathrm{SAT}}+3$, so the gadgets must start in the first or in the second time slot and not any later. The clause jobs can then only be scheduled in the last and in the second to last time slot and not earlier. The space left by the dummy jobs in the last time slot allows only $6m_{\mathrm{SAT}}$ clause jobs to be scheduled. Therefore the other m_{SAT} clause jobs must be scheduled in the second to last time slot because each clause job is dependent on a gadget. And so there is only space for n_{SAT} gadget jobs left in the second to last time slot.

We say a gadget ends (or starts) in time slot t if the last job or the last jobs of the gadgets are scheduled in time slot t (or the first job or the first jobs are scheduled in time slot t, respectively). So only n_{SAT} gadgets end in the second to last time slot while the other n_{SAT} gadgets end in the third to last time slot and therefore start in the first. Also, only n_{SAT} gadgets can start in the first, so the other n_{SAT} start in the second time slot.

In the second time slot up to the $(2n_{\mathrm{SAT}}+1)$-st there is space left for one nose job per time slot. This means, no two gadgets associated to the same variable can be started in the same time slot. Again this means for each variable x_i either $j^{\mathrm{chain}}_{x_i=1,2n_{\mathrm{SAT}}}$ is scheduled in time slot $2n_{\mathrm{SAT}}+1$ and $j^{\mathrm{chain}}_{x_i=0,2n_{\mathrm{SAT}}}$ is scheduled in time slot $2n_{SAT}+2$ or vice versa, and so the gadgets ending in time slot $2n_{SAT}$ represent an assignment for the variables.

If we have a satisfying truth assignment for φ, we can schedule the gadgets corresponding to the true literals (according to the assignment) first. Then, for each clause, one of the possible assignments of the three variables is without contradiction to the truth assignment and the corresponding clause job can be scheduled in time slot $2n_{SAT}+2$. So there is a feasible schedule.

If we have a feasible schedule then there are m_{SAT} clause jobs scheduled in time slot $2n_{SAT}+2$. For each clause there must be exactly one of the corresponding clause jobs, because two different assignments for the three variables of one clause contradict each other. The truth assignment given by the gadgets ending in time slot $2m_{SAT}+1$ therefore satisfies φ. An example of the construction can be found in Appendix A.

To prove the lower bound on the running time, assume for the sake of contradiction that there is an algorithm solving $P\mid prec, p_j = 1 \mid C_{max}$ with n jobs in time $2^{o(\sqrt{n})}$. Then we can transform each formula φ with m_{SAT} clauses and $n_{SAT} \leq 3m_{SAT}$ variables into a $P\mid prec, p_j = 1 \mid C_{max}$ instance with $\mathcal{O}(\max\{m_{SAT}, n_{SAT}\}^2) = \mathcal{O}(m_{SAT}^2)$ jobs and solve it in $2^{o(\sqrt{n})} = 2^{o(m_{SAT})}$. Using the Sparsification Lemma, this contradicts the ETH. So there is no algorithm that solves $P\mid prec, p_j = 1 \mid C_{max}$ with n jobs in time $2^{o(\sqrt{n})}$, unless the ETH fails.

Alternatively, Lenstra and Rinnooy Kan gave a reduction from the k-Clique problem to precedence scheduling where each node and each edge is represented by a job. The standard reduction from 3-SAT to the k-Clique problem [8] produces three nodes per clause and the parameter k matches the number m_{SAT} of clauses. In a clique each node is connected with each other node, so there are at least $\frac{m_{SAT}(m_{SAT}-1)}{2}$ edges in a k-clique instance reduced from a 3-CNF-SAT instance with m_{SAT} clauses. The schedule length of an optimal schedule in the reduction is 3. So the reduction also leads to the same bound on the running time for approximation algorithms with ratio better than $\frac{4}{3}$. To our knowledge, no reduction from 3-CNF-SAT to Clique is known where the number of edges in the constructed instance is $o(\max\{n_{SAT}, m_{SAT}\}^2)$.

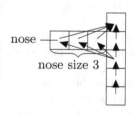

nose

nose size 3

Fig. 2. A gadget for a variable with nose size 3

The main idea of our improvement is to create gadgets with differently sized noses on the same nose height for different variables. The space available for the noses obviously must be increased—that increases the number of machines—but the total height of the schedule is shorter. Differently sized noses are a group of jobs per gadget that only can be scheduled next to one particular job in the chain (see Fig. 2). The size of a nose is defined by the number of jobs in such a group.

We will again create two equal gadgets for each variable x_i, one meaning "x_i is true" and one meaning "x_i is false". According to which gadget starts earlier we decide how the truth assignment is chosen. Everything else is very much the same as before.

Assume we have $2k$ gadgets for k variables $x_1, \ldots, x_k$ with its noses on the same height. Let us say $f_k : \mathbb{N}_{\leq k} \longrightarrow \mathbb{N}$ is the function, that assigns the nose sizes to the gadgets corresponding to the k variables. The number of machines must then be at least $2n_{\text{SAT}} + \sum_{i=1}^{k} f_k(i)$. The function f must have the following property:

Property 1. For all $A, B \subseteq f_k(\mathbb{N}_{\leq k})$ where $|A| = |B|$ and $A \cap B = \emptyset$, it is $\sum_{a \in A} a \neq \sum_{b \in B} b$.

This implies f_k to be injective, because otherwise, if there were two gadgets for x_i and two for x_j, that had the same nose size on the same height, we could start both gadgets for x_i first instead of one of the two assigned to x_i and one of the two assigned to x_j and we had no feasable truth assignment any more.

The identity does not fulfill Property 1. If $k = 4$ then $1 + 4 = 2 + 3$. The quality of such an f_k with Property 1 is measured by how minimal the sum $\sum_{a \in f_k(\mathbb{N}_{\leq k})} a$ is compared to all other injective function with this property. One possibility of successfully constructing a function with Property 1 is to recursively define the function independently from k: $f(1) = 1$ and $f(a) = 1 + \sum_{i=1}^{a-1} a$. If $A, B \subseteq f(\mathbb{N}_{\leq k})$, $A \cap B = \emptyset$, then either A or B contains the largest element of $A \cup B$, and this element is greater than the sum of all elements smaller than the element in $f(\mathbb{N}_{\leq k})$, so in particular Property 1 holds for f. This is actually an iterative way to express $f(a) = 2^{a-1}$.

One question we do not yet have an answer for is whether f is a function with best possible quality. An asymptotically improved value of $\sum_{a \in f_k(\mathbb{N}_{\leq k})} a$ would improve the result of our following reduction.

The exponential function with basis 2 will be the base for the construction in the proof of the next Theorem. Note that $\sum_{i=1}^{k} 2^{i-1} = 2^k - 1$. So next to the $2n_{\text{SAT}}$ chain jobs must be space left for $2^k - 1$ nose jobs.

Theorem 1. *Assuming ETH there is no algorithm for* $\mathrm{P} \,|\, \text{prec}, p_j = 1 \,|\, C_{\max}$ *in time* $2^{o(\sqrt{n \log(n)})}$.

Proof. Let again φ be a formula with n_{SAT} variables and m_{SAT} clauses. We start with an arbitrary factor $k \in \{a \in \mathbb{N} \mid a | n_{\text{SAT}}\}$, by which we want to decrease $\mathbf{T}$. Let

$$m = \max\{2n_{\text{SAT}} + (2^k - 1) + 1, n_{\text{SAT}} + m_{\text{SAT}} + 1, 6m_{\text{SAT}} + 1\}$$

and $\mathbf{T} = \frac{2n_{\text{SAT}}}{k} + 3$. The dummy jobs now build a similar structure as before with updated heights. Let J_i^{dummy} for $i \in [\mathbf{T}]$ be sets of dummy jobs and again for each pair $j \in J_i^{\text{dummy}}$, $j' \in J_{i'}^{\text{dummy}}$ let $j \prec j'$ iff $i < i'$ for all $i, i' \in [\mathbf{T}]$. Let

$$|J_i^{\text{dummy}}| = \begin{cases} m - n_{\text{SAT}} & \text{, if } i = 0, \\ m - (2n_{\text{SAT}} + 2^k - 1) & \text{, if } i = 1, \\ m - (2n_{\text{SAT}} + 2^k - 1) & \text{, if } i \in \{2, \ldots, \frac{2n_{\text{SAT}}}{k}\} \\ m - (n_{\text{SAT}} + m_{\text{SAT}}) & \text{, if } i = \frac{2n_{\text{SAT}}}{k} + 1, \text{ and} \\ m - 6m_{\text{SAT}} & \text{, if } i = \frac{2n_{\text{SAT}}}{k} + 2. \end{cases}$$

Next to the chain jobs we have space left for $2^k - 1$ nose jobs in the second time slot up to the $\frac{2n_{\text{SAT}}}{k}$-st time slot, which is the new length of the gadgets, and space for the clause jobs in the last two time slots as before.

The variables will be partitioned into $\frac{n}{k}$ groups of size k, where each job from the same group has its nose on the same height: For each x_i we create two identical gadgets, one with $b = 0$ and one with $b = 1$:

$$j^{\text{chain}}_{x_i=b,0} \prec \cdots \prec j^{\text{chain}}_{x_i=b,\frac{2n_{\text{SAT}}}{k}} \text{ and } j^{\text{nose}}_{x_i=b,1}, \ldots, j^{\text{nose}}_{x_i=b,2^{(i-1)\bmod k}},$$

with $j^{\text{chain}}_{x_i=b,2\lceil \frac{i}{k}\rceil-1} \prec j^{\text{nose}}_{x_i=b,j} \prec j^{\text{chain}}_{x_i=b,2\lceil \frac{i}{k}\rceil}$ for all $j \in \{1, \ldots, 2^{(i-1)\bmod k}\}$.

The term $\lceil \frac{i}{k}\rceil$ achieved the partitioning of the gadgets into groups and the term $2^{(i-1)\bmod k}$ is the nose size of the gadget corresponding to the variable x_i.

We can again argue that n_{SAT} gadgets end in the second to last time slot and the other n_{SAT} end in the third to last time slot. So at least n_{SAT} gadgets must start in the bottom time slot to be able to end in the third to last, the other n_{SAT} must start the first time slot.

Then k noses of the gadgets of the first group of variables $x_1, \ldots, x_k$ with a total size of $2^k - 1$ must be scheduled in the first time slot as well as in the second. Since $2^k - 1$ can only be represented by nose sizes of gadgets associated with k pairwise different variables of the same group, in the bottom time slot can not start two gadgets associated to the same variable of the first group. This holds for each pair of gadgets associated with the same variables in every group. Again we can conclude: In the bottom layer only gadgets corresponding to pairwise different variables start.

For each clause c_j and the seven different truth assignments $a_{j,1}, \ldots, a_{j,7}$ let $j^{\text{clause}}_{j,d}$ again be a job for each c_j and each $a_{j,d}$. Let $j^{\text{chain}}_{x_i=b,\frac{2n_{\text{SAT}}}{k}} \prec j^{\text{clause}}_{j,d}$ if and only if $x_i = b$ is included in $a_{j,d}$. The argument that m_{SAT} clause jobs can to be scheduled in the second to last time slot if and only if there is a satisfying truth assignment for φ is completely the same as before (Fig. 3).

If $\max\{2n_{\text{SAT}} + (2^k - 1) + 1, n_{\text{SAT}} + m_{\text{SAT}} + 1, 6m_{\text{SAT}} + 1\} = 2n_{\text{SAT}} + (2^k - 1) + 1$ (which is the interesting case, because otherwise the number of jobs would be linear in m_{SAT} and n_{SAT}) the total number of jobs is $m \cdot \mathbf{T} = (2n_{\text{SAT}} + 2^k - 1)\left(\frac{2n_{\text{SAT}}}{k} + 3\right) = \frac{4n^2_{\text{SAT}} + 2^k n_{\text{SAT}} - n_{\text{SAT}}}{k} + 6n_{\text{SAT}} + 3 \cdot 2^k - 3$.

Choosing $k = \log(n_{\text{SAT}})$ seems to be a good choice. Then the number of jobs is $\mathcal{O}(\frac{n^2_{\text{SAT}}}{\log(n_{\text{SAT}})})$. If there was an algorithm solving $P \,|\, \text{prec}, p_j = 1 \,|\, C_{\max}$ with n jobs in time $2^{o(\sqrt{n \log n})}$, then we could transform each 3-SAT formula φ with n_{SAT} variables into an instances of $P \,|\, \text{prec}, p_j = 1 \,|\, C_{\max}$ with $\mathcal{O}(\frac{n^2_{\text{SAT}}}{\log n_{\text{SAT}}})$ jobs and solve it in time

$$2^{o\left(\sqrt{n \log(n)}\right)} = 2^{o\left(\sqrt{\frac{n^2_{\text{SAT}}}{\log(n_{\text{SAT}})}\log\left(\frac{n^2_{\text{SAT}}}{\log(n_{\text{SAT}})}\right)}\right)} = 2^{o\left(\sqrt{\frac{n^2_{\text{SAT}}}{\log(n_{\text{SAT}})}(2\log(n_{\text{SAT}})-\log(\log(n_{\text{SAT}})))}\right)}$$

$$= 2^{o\left(\sqrt{2n^2_{\text{SAT}}-\frac{n^2_{\text{SAT}}\log(\log(n_{\text{SAT}}))}{\log(n_{\text{SAT}})}}\right)} \leq 2^{o(n_{\text{SAT}})},$$

a contradiction to the ETH.

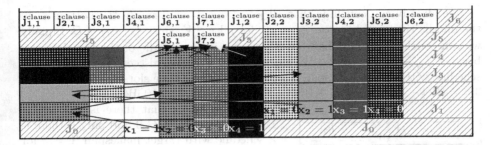

Fig. 3. Example 1 for the new reduction. The formular is again $\varphi = (x_1 \vee x_2 \vee \neg x_3) \wedge (x_1 \vee \neg x_2 \vee x_4)$. For reasons of clarity only a few precendences are presented. Jobs of the same color are corresponding to the same variable. Jobs of the same color and pattern are assigned to the same gadget. The dottet gadgets represent the negative assignments and the gadgets without pattern represent the positive assignments.

3 Precedence Constrained Scheduling and Biclique

This Section concentrates on a reduction from the $k_1 \times k_2$-Biclique problem to $P \mid \text{prec}, p_j = 1 \mid C_{\max}$ with target makespan $\mathbf{T} = 3$ and back. The same result also holds with the additional constraint to the chain length in the scheduling problem not to be greater than two, due to the properties of our reduction. In particular this means for all instances $G = (V \dot\cup W, E)$ of $k_1 \times k_2$-Biclique there is a $P \mid \text{prec}, p_j = 1 \mid C_{\max}$ instance with $\mathcal{O}(|V \dot\cup W|)$ jobs and the Biclique instance has a Biclique of size $k_1 \times k_2$ if and only if there is a schedule with target makespan $\mathbf{T} = 3$ for the corresponding $P \mid \text{prec}, p_j = 1 \mid C_{\max}$ instance. Also for each $P \mid \text{prec}, p_j = 1 \mid C_{\max}$ instance with $n \in \mathcal{O}(m)$ jobs there is a $k_1 \times k_2$-Biclique instance $G = (V \dot\cup W, E)$ with $\mathcal{O}(n)$ vertices and there is a schedule for the scheduling instance with target makespan $\mathbf{T} = 3$ if and only if the Biclique instance has a Biclique of size $k_1 \times k_2$.

We will start with the reduction from the parametrized Biclique problem to the $P \mid \text{prec}, p_j = 1 \mid C_{\max} = 3$ (Fig. 4).

Theorem 2. *There is a linear reduction from the $k_1 \times k_2$-Biclique problem to the $P \mid \text{prec}, p_j = 1 \mid C_{\max} = 3$ and maximum chain length of at most two, that is, for all Jobs j there are no two jobs j_1, j_2 in a way that $j \prec j_1 \prec j_2$.*

Proof. We start our reduction from a $k_1 \times k_2$-Biclique instance $G = (V \dot\cup W, E)$ where $|V| - k_1 = k_1 + k_2 = |W| - k_2$ and there is a vertex $v_d \in V$ and a vertex $w_d \in W$ which have no edges to any other node. This is no loss of generality because otherwise we could modify our instance in an easy way it suits our conditions by adding jobs and edges and increasing the parameters. The modification in detail can be viewed in the Appendix B.

Let $m = k_1 + k_2$. For each node $v \in V$ we create a job j_v^- and for each $w \in W$ we create a job j_w^+. Let $j_v^- \prec j_w^+$ iff $\{v, w\} \notin E$. Let us further define $J_A^- = \{j_v^- \mid v \in A\}$ and $J_B^+ = \{j_w^+ \mid w \in B\}$ for all $A \subseteq V$, $B \subseteq W$.

We already can conclude that all jobs from J_V^- will run in the bottom or in the middle time slot because $j_v^- \prec j_{w_d}^+$ for all $v \in V$, since $\{v, w_d\} \notin E$. Equally we can conclude for all jobs from J_W^+ to run in the middle or the top time slot because $j_{v_d}^- \prec j_w^+$ for all $w \in W$, since $\{v_d, w\} \notin E$. We also can say if there is a schedule with target makespan $\mathbf{T} = 3$ then all three time slots are filled because $3m = (|V| - k_1) + k_1 + k_2 + (|W| - k_2) = |V| + |W| = |V \mathbin{\dot\cup} W|$.

Fig. 4. Visualization of a reduced schedule from the Biclique Instance. The arrows denote that there possibly are precedences between the jobs in the particular sets.

Now we show that there is a $k_1 \times k_2$-Biclique instance in G if and only if there is a schedule with target makespan $\mathbf{T} = 3$ for the scheduling instance. If there is a schedule of target makespan $\mathbf{T} = 3$, then there are $|V| - k_1$ jobs from J_V^- in the bottom time slot and k_1 from J_V^- in the middle time slot. There also are $|W| - k_2$ jobs from J_W^+ in the top time slot and k_2 from J_W^+ in the middle time slot. The jobs from J_V^- and J_W^+ that run in the middle time slot together therefore have no precedences. Let us call the corresponding vertex sets V' and W'. We can conclude, that each node from V' has edges to all vertices in W' and vice versa, because otherwise there would be precendeces between the jobs from $J_{V'}^-$ and $J_{W'}^+$. So V' and W' form a Biclique of the desired size.

If on the other hand we start knowing that there is a Biclique of size $k_1 \times k_2$ in the Biclique instance, we schedule the jobs $J_{V \setminus V'}^-$ in the bottom time slot, the jobs from $J_{W \setminus W'}^+$ in the top time slot and those from $J_{V'}^-$ and from $J_{W'}^+$ together in the middle time slot, which have no precedences because the corresponding vertices form a Biclique.

Theorem 3. *There is a linear reduction from* $\mathrm{P} \mid \mathrm{prec}, p_j = 1 \mid C_{\max} = 3$ *to the* $k_1 \times k_2$*-Biclique problem. In particular the same reduction works for the restricted version of* $\mathrm{P} \mid \mathrm{prec}, p_j = 1, cl \leq 2 \mid C_{\max} = 3$, *where the chain length is limited by* 2.

Proof. Let us now assume a $\mathrm{P} \mid \mathrm{prec}, p_j = 1 \mid C_{\max}$ instance. We suggest the number of jobs to be smaller or equal to $3m$, because if it would be larger, there would be no schedule.

We know there are no precedence chains of length larger than 3, otherwise $\mathbf{T} > 3$. If there is a precedence chain of length 3 we already can conclude which of the jobs in that chain has to be scheduled in which of the three time slots. So, we part our instance in the sets

$$J_{\text{bottom}} = \{j \in J \mid \exists j_1, j_2 \in J : j \prec j_1 \prec j_2\},$$
$$J_{\text{middle}} = \{j \in J \mid \exists j_1, j_2 \in J : j_1 \prec j \prec j_2\},$$
$$J_{\text{top}} = \{j \in J \mid \exists j_1, j_2 \in J : j_1 \prec j_2 \prec j\},$$
$$J_V^- = \{j \in J \mid \exists j' \in J : j \prec j'\} \setminus (J_{\text{bottom}} \cup J_{\text{middle}}),$$
$$J_W^+ = \{j \in J \mid \exists j' \in J : j' \prec j\} \setminus (J_{\text{middle}} \cup J_{\text{top}}) \text{ and}$$
$$J_{\not\prec} = \{j \in J \mid \not\exists j' \in J : j' \prec j \text{ or } j \prec j'\}.$$

First of all, all these sets are disjoint because $\mathbf{T} = 3$ as argued before. Further all jobs from J_{bottom} are already determined to run in the bottom time slot, the jobs from J_{middle} in the middle time slot and those from J_{top} in the top time slot. The jobs from $J_{\not\prec}$ are allowed to run anywhere. The difficulty of the problem lies in choosing jobs from J_V^- and from J_W^+ that can be scheduled in the middle time slot together. The more jobs from J_V^- can be started in the bottom time slot and the more jobs of J_W^+ can be startet in the top time slot the easier the problem becomes. If the middle time slot can be filled with Jobs from $J_{\not\prec}$ the problem is solved. So, we schedule all jobs from $J_{\not\prec}$ in time slot 2 without loss of generality.

Let $k_1 = |J_V^-| - (m - |J_{\text{bottom}}|)$ and $k_2 = |J_W^+| - (m - |J_{\text{top}}|)$. These are the numbers of jobs from J_V^- and J_W^+ respectively that has to run in the middle time slot together. Let $V = \{v_j \mid j \in J_V^-\}$ and $W = \{w_j \mid j \in J_W^+\}$ be the set of vertices of the $k_1 \times k_2$-Biclique instance. Let $\{v_{j^-}, w_{j^+}\} \in E$ iff $j^- \not\prec j^+$ for all $v_{j^-} \in V$, $w_{j^+} \in W$. Then we claim that there is a $k_1 \times k_2$-Biclique if and only if the precedence scheduling instance with unit execution time has a schedule with makespan 3. The proof is simular to the end of the proof of Theorem 2.

Both reductions lead to the two main Theorems:

Theorem 4. $P \mid \text{prec}, p_j = 1 \mid C_{\max} = 3$ *is equivalent to the* $k_1 \times k_2$-*Biclique problem.*

Theorem 5. $P \mid \text{prec}, p_j = 1, cl \leq 2 \mid C_{\max} = 3$ *is equivalent to the* $k_1 \times k_2$-*Biclique problem.*

Acknowledgements. We want to thank Matthias Schulte-Althoff and Eike Lurz for helpful discussions.

A   Example for the Modified Ullman Reduction

Example 1. Figure 5 visualizes the reduction using an example where $\varphi = (x_1 \vee x_2 \vee \neg x_3) \wedge (x_1 \vee \neg x_2 \vee x_4)$. This leads to gadgets of length 9, $m = 13$ and $\mathbf{T} = 11$. We choose $x_1 = 1$, $x_2 = 0$, $x_3 = 0$ and $x_4 = 1$ as the truth assignment that makes φ true.

The truth assignments for the clauses, $a_{j,d}$, are numbered in order to the binary output of the clauses: $a_{1,1}$ makes x_1, x_2 and x_3 false so the output of the first clause would be $0 \vee 0 \vee 1$. $a_{1,2}$ makes x_1 false but x_2 and x_3 true and so on.

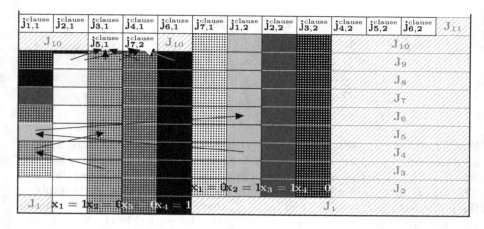

Fig. 5. An example for the modified Ulman reduction. In this example the formula is
$\varphi = (x_1 \vee x_2 \vee \neg x_3) \wedge (x_1 \vee \neg x_2 \vee x_4)$. For reasons of clarity only a few precedences
are presented. Jobs of the same color are corresponding to the same variable. Jobs of
the same color and pattern are assigned to the same gadget. The dottet gadgets represent the negative assignments and the gadgets without pattern represent the positive
assignments.

B Modifiing the Biclique Instance to One with the Needed Property

Given a $k_1 \times k_2$-Biclique instance $G = (V \dot\cup W, E)$ we want to obtain an equivalent
$\widehat{k_1} \times \widehat{k_2}$-Biclique instance $\widehat{G} = ((V) \dot\cup \widehat{W}, \widehat{E})$ with the property, that there is a
$v_d \in \widehat{V}$ and a $w_d \in \widehat{W}$ without any edges and $|\widehat{V}| - \widehat{k_1} = \widehat{k_1} + \widehat{k_2} = |\widehat{W}| - \widehat{k_2}$:
First we add a dummy node v_d to V and a dummy node w_d to W without any
edges. Then we take the maximum of $k_1 + k_2$, $|V \cup \{v_d\}| - k_1$ and $|W \cup \{w_d\}| - k_2$
and take this number as number of machines m. We modify each instance in a
way that $\widehat{k_1} + \widehat{k_2} = |\overline{V}| - \overline{k_1} = |\overline{W} - \overline{k_2}| = m$:
We obtain a set $\overline{V}$ by adding v_d and other nodes without any edges to V until
$|\overline{V}| - k_1 = m$ and we do the same with W until $|\overline{W}| - k_2 = m$.
Then we add vertices to $\overline{V}$ with edges to every node in $\overline{W}$ and at the same time
increase k_1 by one for every vertex we add, until we get a set $\widehat{V}$ and a parameter
$\widehat{k_1}$ where now $\widehat{k_1} + k_2 = m$ and $|\widehat{V} - |\widehat{k_1} = m$ still holds and define $\widehat{k_2} = k_2$ and
$\widehat{W} = \overline{W}$.
In this modification we find a $\widehat{k_1} \times \widehat{k_2}$-Biclique if and only if we found a $k_1 \times k_2$-
Biclique in the old version: The added nodes without any edges do not touch
Bicliques in the instance and those with edges to every node increase all Bicliques
by the number of added nodes of that kind.
So without loss of generality we can assume a $k_1 \times k_2$-Biclique instances $G =
(V \dot\cup W)$ where $|V| - k_1 = k_1 + k_2 = |W| - k_2$ and dummy nodes $v_d \in V$, $w_d \in W$
with no edges.

C   Alternative Reduction from Biclique to the Scheduling Problem Without Chains

There is an alternative for a reduction from $k_1 \times k_2$-Biclique to $P \mid \text{prec}, p_j = 1 \mid C_{\max}$ with target makespan $\mathbf{T} = 3$, where the Biclique instance is not manipulated in the beginning. Further this reduction provides a possibility to reduce the instance without being important from which set, V or W, the k_1 nodes come from and from which the k_2 come from. This could also be done with the old reduction from the Biclique problem to $P \mid \text{prec}, p_j = 1 \, \text{cl} \leq 2 \mid C_{\max}$ by testing both. This reduction however may reveal some new ideas for techniques.

Lemma 1. $P \mid \text{prec}, p_j = 1 \mid C_{\max}$ with target makespan $\mathbf{T} = 3$ can be reduced from the $k_1 \times k_2$-Biclique Problem.

Proof. Let us assume a $k_1 \times k_2$-Biclique instance $G = (V \dot\cup W)$. Let $U = V \dot\cup W$. For each node $v \in U$ we create two jobs j_v^- and j_v^+. For each $v, w \in U$ let $j_v^- \prec j_w^+$ iff $\{v, w\} \notin E$. Let $m = \max\{|U| - k_1 + 1, k_1 + k_2 + 1, |U| - k_2 + 1\}$ and let there be sets of dummy jobs J_{bottom}, J_{middle} and J_{top}, where $|J_{\text{bottom}}| = m - (|U| - k_1)$, $|J_{\text{top}}| = m - (|U| - k_2)$ and $|J_{\text{middle}}| = m - (k_1 + k_2)$. Let $j_{\text{bottom}} \prec j_{\text{middle}} \prec j_{\text{top}}$ for all $j_{\text{bottom}} \in J_{\text{bottom}}, j_{\text{middle}} \in J_{\text{middle}}, j_{\text{top}} \in J_{\text{top}}$.
Let further $j_v^- \prec j_{\text{top}}$ and $j_{\text{bottom}} \prec j_v^+$ for all $v \in U$, and at least one $j_{\text{bottom}} \in J_{\text{bottom}}, j_{\text{top}} \in J_{\text{top}}$.
The dummy job structure is to assure if the schedule has length 3, then all jobs from J_{bottom} are scheduled in the first (bottom) time slot, all jobs J_{middle} in the second (middle) and all jobs from J_{top} in the third (top) time slot. Further all j_v^- are scheduled in the first two time slots, $|U| - k_1$ of them in the first one, k_1 in the second. Likewise all j_v^+ jobs are scheduled in the second and third time slot, k_2 of them in the second and $|U| - k_2$ of them in the third.
Let us denote $J_A^- = \{j_v^- \mid v \in A\}$ and $J_A^+ = \{j_v^+ \mid v \in A\}$ for all $A \subseteq U$.
Note that there are no precedences between two jobs j_v^- and j_w^- aswell as no precedences between j_v^+ and j_w^+ for any $v, w \in U$.
If it is part of the instance, that the k_1 vertices are supposed to come from V and the k_2 vertices from W, we add the condition $j_v^- \prec j_{\text{top}}$ for all $v \in V$, and at least one $j_{\text{top}} \in J_{\text{top}}$.
If there is a solution of this instance with target makespan $\mathbf{T} = 3$, then there are subsets $U^-, U^+ \subseteq U$, $|U^-| = k_1$ and $|U^+| = k_2$ and the corresponding jobs $j_v^-, j_w^+, v \in U^-, w \in U^+$ are scheduled in the middle time slot together while all other jobs $j_v^-, j_w^+, v, w \in U \setminus (U^- \cup U^+)$ are in the bottom or top time slot. So for all $v \in U^-$ and $w \in U^+$ it is $j_v^- \not\prec j_w^+$, that means $\{v, w\} \in E$.

First we know by this, v and w are not both in V or both in W, because otherwise $\{v, w\} \notin E$. So we know $U^- \subseteq V$ and $U^+ \subseteq W$ or vice versa. With the extra condition we would know $U^- \subseteq V$, which induces $U^+ \subseteq W$. Second we know that for each $v \in U^-$ and each $w \in U^+$ it is $\{v, w\} \in E$ and so U^- and U^+ form a $k_1 \times k_2$-Biclique (Fig. 6).

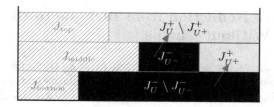

Fig. 6. Visualization of a reduced schedule from the Biclique Instance. The arrows denote that there possibly are precedences between the jobs in the particular sets. All dark gray areas represent the jobs from the set J^- and the light gray area represent the jobs from J^+.

Now assume there is a $k_1 \times k_2$-Biclique. Let again $U^-, U^+ \subseteq U$ be the sets of nodes, that form the biclique, where $|U^-| = k_1$ and $|U^+| = k_2$ and $U^- \subseteq V$ and $U^+ \subseteq W$ or vice versa. For all $v \in U$ we now schedule all corresponding jobs j_v^- in the bottom time slot exept the j_v^- with $v \in U^-$, which we schedule in the middle time slot. For all $v \in U$ we schedule all corresponding jobs j_v^+ in the top time slot except the j_v^+ where $v \in U^+$, which we also schedule in the middle time slot. The schedule is feasable, because there are no precedences in between the jobs corresponding to U^- or in between those of U^+ and there are none between the jobs j_v^- and j_w^+ in the middle time slot, because their corresponding vertices form a biclique by assumption.

References

1. Bansal, N., Khot, S.: Optimal long code test with one free bit. In: 50th Annual IEEE Symposium on Foundations of Computer Science (FOCS 2009), pp. 453–462. IEEE Computer Society (2009)
2. Bulatov, A.A., Marx, D.: Constraint satisfaction parameterized by solution size. SIAM J. Comput. **43**(2), 573–616 (2014)
3. Calabro, C., Impagliazzo, R., Kabanets, V., Paturi, R.: The complexity of unique k-SAT: An isolation lemma for k-CNFs. J. Comput. Syst. Sci. **74**(3), 386–393 (2008)
4. Chen, J., Huang, X., Kanj, I.A., Xia, G.: On the computational hardness based on linear fpt-reductions. J. Comb. Optim. **11**(2), 231–247 (2006)
5. Du, J., Leung, J.Y., Young, G.H.: Scheduling chain-structured tasks to minimize makespan and mean flow time. Inf. Comput. **92**(2), 219–236 (1991)
6. Impagliazzo, R., Paturi, R., Zane, F.: Which problems have strongly exponential complexity? J. Comput. Syst. Sci. **63**(4), 512–530 (2001)
7. Johnson, D.S.: The NP-completeness column: an ongoing guide. J. Algorithms **8**, 438–448 (1987)
8. Karp, R.M.: Reducibility among combinatorial problems. In: Miller, R.E., Thatcher, J.W. (eds.) Proceedings of a Symposium on the Complexity of Computer Computations. The IBM Research Symposia Series, pp. 85–103. Plenum Press, New York (1972)

9. Khot, S.: On the power of unique 2-prover 1-round games. In: Proceedings of the 17th Annual IEEE Conference on Computational Complexity, p. 25. IEEE Computer Society (2002)

10. Lenstra, J.K., Kan, A.H.G.R.: Complexity of scheduling under precedence constraints. Oper. Res. **26**(1), 22–35 (1978)

11. Lin, B.: The parameterized complexity of k-biclique. In: Indyk, P. (ed.) Proceedings of the Twenty-Sixth Annual ACM-SIAM Symposium on Discrete Algorithms, SODA 2015, pp. 605–615. SIAM (2015)

12. Svensson, O.: Conditional hardness of precedence constrained scheduling on identical machines. In: Schulman, L.J. (ed.) Proceedings of the 42nd ACM Symposium on Theory of Computing, STOC 2010, pp. 745–754. ACM (2010)

13. Ullman, J.D.: Np-complete scheduling problems. J. Comput. Syst. Sci. **10**(3), 384–393 (1975)

Grover's Search with Faults on Some Marked Elements

Dmitry Kravchenko, Nikolajs Nahimovs$^{(\boxtimes)}$, and Alexander Rivosh

Faculty of Computing, University of Latvia, Riga, Latvia
nikolajs.nahimovs@lu.lv

Abstract. Grover's algorithm is a quantum query algorithm solving the unstructured search problem of size N using $O(\sqrt{N})$ queries. It provides a significant speed-up over any classical algorithm [2].

The running time of the algorithm, however, is very sensitive to errors in queries. Multiple authors have analysed the algorithm using different models of query errors and showed the loss of quantum speed-up [1,4].

We study the behavior of Grover's algorithm in the model where the search space contains both faulty and non-faulty marked elements. We show that in this setting it is indeed possible to find one of marked elements in $O(\sqrt{N})$ queries.

1 Introduction

Grover's algorithm is a quantum query algorithm solving the unstructured search problem of size N using $O(\sqrt{N})$ queries. It is known that any deterministic or randomized algorithm needs linear time (number of queries) to solve the above problem. Thus, Grover's algorithm provides a significant speed-up over any classical algorithm.

The running time of the algorithm (number of queries), however, is very sensitive to errors in queries. Regev and Schiff [4] have shown that if query has a small probability of failing (reporting that *none* of the elements are marked), then quantum speed-up disappears: no quantum algorithm can be faster than a classical exhaustive search by more than a constant factor. Ambainis et al. [1] have studied Grover's algorithm in the model where each marked element has its own probability to be reported as unmarked, independent of probabilities of other marked elements. Similarly to the result of [4], they have shown that if all marked elements are faulty (have non-zero probability of failure) then the algorithm needs $\Omega(N)$ queries to find a marked element.

Although, technically the model of [1] allows one non-faulty marked element (element with zero probability of failure) this case was not included into the analysis[1].

This research was supported by EU FP7 project QALGO (Dmitry Kravchenko, Nikolajs Nahimovs) and ERC project MQC (Alexander Rivosh).

[1] The limitation of at most one non-faulty marked element comes from the probability independence assumption – two or more marked elements with zero error probability of failure would not be independent.

R.M. Freivalds et al. (Eds.): SOFSEM 2016, LNCS 9587, pp. 344–355, 2016.
DOI: 10.1007/978-3-662-49192-8_28

We study the behavior of the algorithm in the model where the search space contains both faulty and non-faulty marked items. Specifically, we focus on the case where the search space contains multiple non-faulty and one faulty marked element. We analyze the effect of a fault on the state of the algorithm and show that in this setting it is indeed possible to find one of non-faulty marked elements in $O(\sqrt{N})$ queries.

Up to the best our knowledge, this is the first demonstration of query fault modes which can be tolerated by the Grover's algorithm.

2   Technical Preliminaries

In this paper we use standard notions of quantum states, density matrices etc., as described in [3]. Description of Grover's algorithm can be found in [2].

Spherical Trigonometry

Spherical trigonometry is a branch of geometry which deals with the relationships between trigonometric functions of the sides and angles of the spherical polygons. Trigonometry on a sphere differs from the traditional planar trigonometry. For example, all distances are measured as angular distances (Fig. 1).

In the context of this paper we need only a few basic formula for right spherical triangles. Let ABC be a right spherical triangle with a right angle C. Then the following set of rules (known as Napier's rules) applies:

$$
\begin{aligned}
\cos c &= \cos a \cos b & (R1) & & \tan b &= \cos A \tan c & (R6) \\
\sin a &= \sin A \sin c & (R2) & & \tan a &= \cos B \tan c & (R7) \\
\sin b &= \sin B \sin c & (R3) & & \cos A &= \sin B \cos a & (R8) \\
\tan a &= \tan A \sin b & (R4) & & \cos B &= \sin A \cos b & (R9) \\
\tan b &= \tan B \sin a & (R5) & & \cos c &= \cot A \cot B & (R10)
\end{aligned}
$$

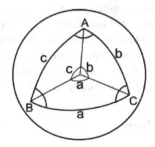

Fig. 1. Spherical trigonometry basic triangle

For more detailed introduction into spherical trigonometry see [5].

3   Model and Results

Error Model

Suppose we have a search space of size N containing k marked elements $i_1, i_2, \ldots,$ i_k. First $k-1$ marked elements are *non-faulty* – the query always returns them as marked. Last marked element is *faulty* – the query might return it as unmarked.

More formally, on each step, instead of the correct query Q, we apply a faulty query Q' defined as follows:

○ $Q'|i_k\rangle = |i_k\rangle$ with probability p;
○ $Q'|i_k\rangle = -|i_k\rangle$ with probability $1 - p$;
○ $Q'|j\rangle = Q|j\rangle$ if $j \neq i_k$.

Summary of Results

We show that if there is at least one non-faulty marked element, then it is still possible to find a non-faulty marked element in $O(\sqrt{N})$ queries with $\Theta(1)$ probability.

Theorem 1. *Let $k \geq 3$, then we can choose $t = O(\sqrt{N/k})$ so that, if we run Grover's algorithm for t steps and measure the final state, the probability of finding a marked element is at least $\cos^2 \frac{\pi}{8} = 0.85 \ldots$.*

For $k = 2$, the probability of finding a marked element is at least $\cos^2 \frac{\pi}{8} = 0.85 \ldots$ under a promise that at most one fault occurs and at least $0.74 \ldots$ in the general case.

We conjecture that, for $k = 2$, the probability is at least $0.85 \ldots$ even in the general case.

4   Analysis of the Algorithm

In this section we analyze the evolution of the state of Grover's algorithm in presence of multiple non-faulty and one faulty marked item. First we review the original Grover's algorithm, then we describe the effect of faults on the state of the algorithm. We derive upper bounds on the effect of faults and provide a modification of the original Grover's search algorithm which finds one of non-faulty marked items with $\Theta(1)$ probability in $O(\sqrt{N})$ queries.

4.1   No Faulty Marked Items

Let us first consider the very basic search problem of Grover's algorithm. Namely, we have N items among which k are marked[2].

---

[2] It is usually considered that $k \ll N$, as for $\frac{k}{N} \geq \lambda$ with sufficiently large λ the search problem can be trivially solved by a probabilistic algorithm in time $O\left(\lambda^{-1}\right)$.

Operator D is symmetric w.r.t. permutations of amplitudes of all items, and operator Q is symmetric w.r.t. permutations of amplitudes of marked items, as well as permutations of amplitudes of non-marked items. So, on any step t amplitudes of all marked items are equal to each other and amplitudes of all non-marked items are equal to each other. Thus, we can represent $|\psi_t\rangle$ as:

$$|\psi_t\rangle = \sum_{i \in U} \alpha_t|i\rangle + \sum_{j \in M} \beta_t|j\rangle,$$

where U stands for the set of indexes of non-marked items and M stands for the set of indexes of k marked items. α_t and β_t denote the amplitudes of respectively a non-marked item and a marked item on step t. At each step of the algorithm we shall take care of two numbers only:

$$\alpha_t\sqrt{N-k} \qquad \text{and} \qquad \beta_t\sqrt{k}. \tag{1}$$

Since $|\psi_t\rangle$ is a unit vector, we have

$$\sum_{i \in U} \alpha_t^2 + \sum_{j \in M} \beta_t^2 = 1.$$

Thus, values (1) meet the equality $\left(\alpha_t\sqrt{N-k}\right)^2 + \left(\beta_t\sqrt{k}\right)^2 = 1$ and correspond to a point on the unit circle.

Initially all amplitudes are equal, so $\alpha = \beta = \frac{1}{\sqrt{N}}$, and the numbers (1) are

$$\alpha_0\sqrt{N-k} = \frac{\sqrt{N-k}}{\sqrt{N}} \qquad \text{and} \qquad \beta_0\sqrt{k} = \frac{\sqrt{k}}{\sqrt{N}}. \tag{2}$$

During the first step of the algorithm operator D does not change amplitudes of the state $|\psi_0\rangle$, and operator Q negates amplitudes of all marked items: $\beta_1 = -\beta_0 = -\frac{1}{\sqrt{N}}$. So the numbers (1) are

$$\alpha_1\sqrt{N-k} = \frac{\sqrt{N-k}}{\sqrt{N}} \qquad \text{and} \qquad \beta_1\sqrt{k} = -\frac{\sqrt{k}}{\sqrt{N}}. \tag{3}$$

According to (2) and (3), transformation $|\psi_0\rangle \xrightarrow{DQ} |\psi_1\rangle$ can be represented on the unit circle as shown on Fig. 2. As before, we assume $k \ll N$, so that $\sqrt{k} \ll \sqrt{N-k}$, and the angle between $|\psi_0\rangle$ and $|\psi_1\rangle$ is

$$2\arcsin\left(\frac{\sqrt{k}}{\sqrt{N}}\right) \approx 2\frac{\sqrt{k}}{\sqrt{N}} \tag{4}$$

(this approximation holds for small-valued $\frac{\sqrt{k}}{\sqrt{N}}$).

Similarly, all further applications of operator DQ are nothing but clockwise rotations by angle $\sim 2\frac{\sqrt{k}}{\sqrt{N}}$. After $\sim \frac{\pi/2}{2\sqrt{k}/\sqrt{N}} = \frac{0.785...\sqrt{N}}{\sqrt{k}}$ such rotations the resulting state $|\psi_{\lfloor 0.785...\sqrt{N/k}\rfloor}\rangle$ reaches the neighborhood of the point $(0, -1)$. Measuring $|\psi_{\lfloor 0.785...\sqrt{N/k}\rfloor}\rangle$ results in getting index of a marked item, with probability almost 1.

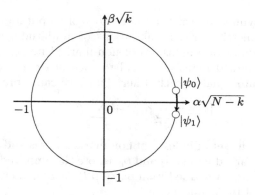

Fig. 2. The first step of Grover's algorithm

4.2 One Faulty Marked Item

Let us now consider the case with

- $N - k$ non-marked items,
- $k - 1$ marked items, and
- 1 *faulty* marked item.

To simplify the analysis we shall interpret the step of the algorithm as consequent application of three operators: ordinary diffusion D and ordinary query Q, and – with probability ϵ – error E, which negates back the amplitude of the faulty marked item.

As the operation E is probabilistic one must deal not with a pure state $|\psi_t\rangle$, but with a mixed state ρ_t (probabilistic mixture of pure states). We shall denote components of the mixture after t steps as $|\psi_t^w\rangle$, where $w \in \{0,1\}^t$ stands for the sequence of t events: 0 – the query has negated the amplitude of the faulty marked item (DQ), and 1 – the query has left that amplitude of the faulty marked item unchanged (DQE). So the mixture ρ_t looks as follows:

$$\rho_t = \sum_{w \in \{0,1\}^t} \epsilon^{|w|} \left(1 - \epsilon\right)^{t-|w|} |\psi_t^w\rangle\langle\psi_t^w|.$$

Transformations D, Q and E are symmetric w.r.t. permutations of amplitudes of non-faulty marked items, as well as permutation of amplitudes of non-marked items. So, in any state $|\psi_t^w\rangle$ of the mixture $|\psi_t^*\rangle$, amplitudes of all non-faulty marked items are equal to each other and amplitudes of all non-marked items are equal to each other. Thus, we can represent $|\psi_t^w\rangle$ as:

$$|\psi_t^w\rangle = \sum_{i \in U} \alpha_t^w |i\rangle + \sum_{\substack{j \in M, \\ j \neq i_k}} \beta_t^w |j\rangle + \gamma_t^w |i_k\rangle, \tag{5}$$

where U stands for the set of indexes of non-marked items and M stands for the set of indexes of k marked items. α_t^w, β_t^w and γ_t^w denote the amplitudes of respectively a non-marked item, a non-faulty marked item and the faulty marked item.

At each step of the algorithm for each of 2^t scenarios w we shall take care of three numbers:

$$\alpha_t^w \sqrt{N-k}, \qquad \beta_t^w \sqrt{k-1} \qquad \text{and} \qquad \gamma_t^w. \tag{6}$$

Since $|\psi_t^w\rangle$ are unit vectors we have

$$\sum_{i \in U} (\alpha_t^w)^2 + \sum_{\substack{j \in M, \\ j \neq f}} (\beta_t^w)^2 + (\gamma_t^w)^2 = 1.$$

Thus, values (6) meet the equality $\left(\alpha_t^w \sqrt{N-k}\right)^2 + \left(\beta_t^w \sqrt{k-1}\right)^2 + (\gamma_t^w)^2 = 1$ and correspond to a points on the unit sphere.

Initially the mixture consists of state $|\psi_0\rangle$ with amplitudes of all items being equal, so $\alpha = \beta = \gamma = \frac{1}{\sqrt{N}}$, and the numbers (6) for $t = 0$ are

$$\alpha_0 \sqrt{N-k} = \frac{\sqrt{N-k}}{\sqrt{N}}, \qquad \beta_0 \sqrt{k} = \frac{\sqrt{k-1}}{\sqrt{N}} \qquad \text{and} \qquad \gamma_0 = \frac{1}{\sqrt{N}}. \tag{7}$$

During the first step of the algorithm

- D does not change amplitudes of the state $|\psi_0\rangle$;
- Q negates the amplitudes of all marked items: $\beta_1^w = \gamma_1^0 = -\frac{1}{\sqrt{N}}$;
- E negates back the amplitude of the faulty marked item: $\gamma_1^1 = -\gamma_1^0 = \frac{1}{\sqrt{N}}$.

So the numbers (6) for $t = 1$ are as follows:

$$\begin{array}{llll}
\alpha_1^0 \sqrt{N-k} = \frac{\sqrt{N-k}}{\sqrt{N}}, & \beta_1^0 \sqrt{k-1} = -\frac{\sqrt{k-1}}{\sqrt{N}}, & \gamma_1^0 = -\frac{1}{\sqrt{N}} & \text{for } w = 0; \\
\alpha_1^1 \sqrt{N-k} = \frac{\sqrt{N-k}}{\sqrt{N}}, & \beta_1^1 \sqrt{k-1} = -\frac{\sqrt{k-1}}{\sqrt{N}}, & \gamma_1^1 = \frac{1}{\sqrt{N}} & \text{for } w = 1.
\end{array} \tag{8}$$

According to (7) and (8), transformation $|\psi_0\rangle\langle\psi_0| \xrightarrow{DQ(E)} (1 - \epsilon) |\psi_1^0\rangle\langle\psi_1^0| + \epsilon |\psi_1^1\rangle\langle\psi_1^1|$ can be represented on the unit sphere as shown on Fig. 3.

Note that if $\epsilon = 0$ the state of the algorithm travels clockwise along the slanted orthodrome[3] which contains points $|\psi_0\rangle$ and $(1, 0, 0)$. The travel lasts until the state reaches the neighborhood of the vertical orthodrome (hereafter we call it *meridian*) where $\alpha = 0$ after $t \approx 0.785\ldots \sqrt{N/k}$ steps.

If $k \ll N$ the rotation angle (4) is sufficiently small. The run of the algorithm can be viewed as clockwise rotation of a state in parallel to the slanted orthodrome (DQ-movement) with occasional (ϵ-probable) up-and-down jumps around the horizontal orthodrome (hereafter we call it *equator*), which correspond to operator E, as shown on Fig. 4. During the first $0.785\ldots \sqrt{N/k}$ steps, the state cannot go out of the area which is covered with arrows on the figure.

---

[3] Orthodrome, also known as a great circle, of a sphere is the intersection of the sphere with a plane which passes through the center point of the sphere.

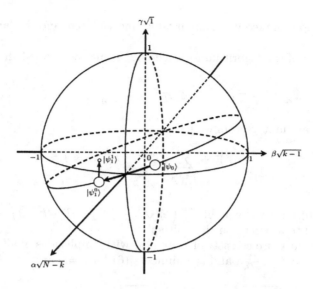

Fig. 3. The first step of Grover's algorithm with a faulty marked item. Size of a ball corresponds to a probability of the state in the mixture. $|\psi_0^*\rangle\langle\psi_0^*| = 1|\psi_0\rangle$ and $|\psi_1^*\rangle\langle\psi_1^*| = (1 - \epsilon)\,|\psi_1^0\rangle\langle\psi_1^0| + \epsilon|\psi_1^1\rangle\langle\psi_1^1|$

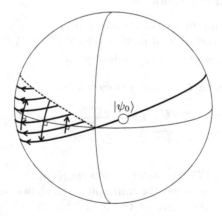

Fig. 4. Probable routes of the initial state $|\psi_0\rangle$ in the run of Grover's algorithm with a faulty marked item

As we already mentioned, $0.785\ldots\sqrt{N/k}$ steps are necessary to reach the desired plane where $\alpha = 0$, given that no fault occurs on the way (in our notation: in expression (5), $\alpha^{00\ldots0}_{\lfloor 0.785\ldots\sqrt{N/k}\rfloor} \approx 0$).

But what could the length of the route be if some faults occur on the state's way to the desired plane? In the two following subsections we will derive upper bounds for the effect of these faults.

At Most One Fault. First, let us assume that the total number of faults is at most one. Although this assumption seems to be rather implausible, we have some arguments for it:

o for sufficiently small ϵ, we have ... $\ll \epsilon^3 \ll \epsilon^2 \ll \epsilon$, so that probability of more than one fault $\sum_{f=2}^{t} \binom{t}{f} \epsilon^f (1-\epsilon)^{t-f} < t^2 \epsilon^2$ could be neglected for number of steps $t \in o\left(\frac{1}{\epsilon}\right)$;

o as we shall see later, the second and all subsequent faults have smaller effect and even have great chances to drive the state closer to the desired meridian.

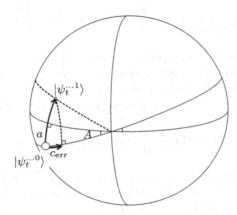

Fig. 5. Metrics for a fault

Let us calculate the effect of a fault in the sense of its projection onto the "no-faults" route. On the Fig. 5 we illustrate ϵ-probable transformation $|\psi_i^{\cdots 0}\rangle \xrightarrow{E} |\psi_i^{\cdots 1}\rangle$, which happened on some step t. The fault increases the angular distance to the desired meridian ($\alpha = 0$) by c_{err}. Using rules of spherical trigonometry ((R8) and (R7)) we have:

$$c_{\text{err}} = \arctan\left(\tan 2a \, \cos\arcsin\frac{\cos A}{\cos a}\right) = \arctan\left(\tan 2a \, \sqrt{1 - \frac{\cos^2 A}{\cos^2 a}}\right), \quad (9)$$

where $A = \arctan\frac{1}{\sqrt{k-1}}$ is angle between the two equators, and a is the distance between $|\psi_i^{\cdots 0}\rangle$ and the horizontal equator. Note that a is at most A ($a = A$ only when $|\psi_t^w\rangle$ reaches the desired meridian), so $1 - \frac{\cos^2 A}{\cos^2 a} \geq 0$.

In equation (9) we assumed that $|\psi_i^{\cdots 0}\rangle$ is located on the bottom-margin of the arrow-filled area of the Fig. 4 (which always holds for one fault case). For $|\psi_i^{\cdots 0}\rangle$ located above the bottom margin, we should calculate (9) for a smaller value of A, which will result in smaller value of c_{err}. For $|\psi_i^{\cdots 0}\rangle$ located above the horizontal equator, the fault-effect c_{err} is negative, i.e. the resulting state $|\psi_i^{\cdots 1}\rangle$ is closer to the desired meridian w.r.t. the direction in parallel to the slanted orthodrome.

Relaxing the above-mentioned assumption, we can conclude the following rough bound:

$$c_{\text{err}} \leq \arctan\left(\tan 2a \ \sqrt{1 - \frac{\cos^2 A}{\cos^2 a}}\right). \tag{10}$$

Now, based on the inequality (10) we shall derive more precise bounds.

If the number of non-faulty marked items $k - 1 \geq 1$, then the angle $A = \arctan\frac{1}{\sqrt{k-1}} \in \left(\arctan\frac{1}{\infty}; \arctan\frac{1}{1}\right] = \left(0; \frac{\pi}{4}\right]$. So we have $0 \leq a \leq A \leq \frac{\pi}{4}$.

Now let us consider different values of a. If $0 \leq a \leq \frac{\pi}{6}$, then (9) is bounded by

$$c_{\text{err}} \leq \max_{\substack{0 \leq a \leq A, \\ a \leq \frac{\pi}{6}}} \arctan\left(\tan 2a \ \sqrt{1 - \frac{\cos^2 A}{\cos^2 a}}\right) \leq \frac{\pi}{4}, \tag{11}$$

where the inequalities become equalities for $A = \frac{\pi}{4}$ and $a = \frac{\pi}{6}$.

If $\frac{\pi}{6} < a \leq \frac{\pi}{4}$, then we can follow that the state $|\psi_i^{\cdots 0}\rangle$ is gone far away from the point "$\alpha\sqrt{N - k} = 1$" of the unit sphere. This distance between the point $(1, 0, 0)$ and the state $|\psi_i^{\cdots 0}\rangle$ can be derived from the rule (R2):

$$c = \arcsin\frac{\sin a}{\sin A} \geq \arcsin\frac{\sin \pi/6}{\sin \pi/4} = \frac{\pi}{4} \tag{12}$$

Since the total distance between the point "$\alpha\sqrt{N - k} = 1$" and any point of the meridian "$\alpha\sqrt{N - k} = 0$" is exactly $\frac{\pi}{2}$, we follow that the state $|\psi_i^{\cdots 0}\rangle$ is at most $\frac{\pi}{2} - \frac{\pi}{4} = \frac{\pi}{4}$ far from the desired meridian.

From (11) and (12) we formulate the following joint conclusion:

Corollary 1. *At least one of the following claims holds for any state $|\psi_i^{\cdots 0}\rangle$ with $0 \leq a \leq A \leq \frac{\pi}{4}$:*

- *either the fault-effect $c_{\text{err}} \leq \frac{\pi}{4}$,*
- *or the state $|\psi_i^{\cdots 0}\rangle$ is already at most $\frac{\pi}{4}$ far from the desired meridian "$\alpha = 0$".*

Any Number of Faults. Now let us use another approach to study the evolution of a state in the considered settings. Transformation $|\psi_t^w\rangle \xrightarrow{DQ} |\psi_{t+1}^{w,0}\rangle$ drives the state $|\psi_t^w\rangle$ clockwise in parallel to the slanted orthodrome by a distance, which depends on the position of this state on the unit sphere.

On Fig. 6 we show a "speed" for each possible position of the state $|\psi_t^w\rangle$. On the "no-faults" route (i.e. the slanted orthodrome) the speed coincides with that of the original Grover's algorithm: $v_G = 2 \arcsin\frac{\sqrt{k}}{\sqrt{N}} \approx 2\sqrt{\frac{k}{N}}$. After a state jumps up, its speed decreases depending on its distance to the slanted orthodrome: e.g. on the parallel circle which contains point "$\beta\sqrt{k - 1} = 1$" of the units sphere, its speed is $v_G \cos A$.

We can also calculate the speed of a state w.r.t. the direction in parallel to the horizontal equator, i.e. the speed of its projection onto the horizontal equator v_π.

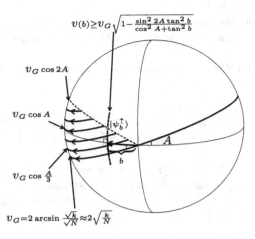

Fig. 6. "Speed" of a state located on some meridian b

Obviously, among all states on some meridian b, the uppermost state $|\psi_b^{\uparrow}\rangle$ has the least speed. Distance between $|\psi_b^{\uparrow}\rangle$ and the point "$\alpha\sqrt{N} = 1$" of the unit sphere, can be derived from angle A, distance b and rule (R6): $c = \arctan\frac{\tan b}{\cos A}$. Distance between $|\psi_b^{\uparrow}\rangle$ and the slanted orthodrome can be derived from distance c, angle $2A$ and rule (R2): $a' = \arcsin(\sin 2A \sin c) = \arcsin\left(\sin 2A \sin \arctan\frac{\tan b}{\cos A}\right)$. And the speed of $|\psi_b^{\uparrow}\rangle$

$$v^{\uparrow}(b) = v_G \cos(a') = v_G \sqrt{1 - \frac{\sin^2 2A \, \tan^2 b}{\cos^2 A + \tan^2 b}}$$

serves as a natural lower bound for the speed of any state located on the same meridian b:

$$v(b) \geq v^{\uparrow}(b) = v_G \sqrt{1 - \frac{\sin^2 2A \, \tan^2 b}{\cos^2 A + \tan^2 b}}.$$

Projection of the speed $v(b)$ onto the horizontal equator might be slightly less:

$$v_{\Pi}(b) \geq v(b) \cos A \geq v_G \sqrt{1 - \frac{\sin^2 2A \, \tan^2 b}{\cos^2 A + \tan^2 b}} \cos A. \qquad (13)$$

Vertical jumps leave states on the same meridians, so faults does not affect value b. Moving at least at speed (13), a state will pass distance $\frac{\pi}{2}$ in at most $\int_0^{\frac{\pi}{2}} \frac{1}{v_{\Pi}(b)}\, db$ steps. From (13) we obtain an upper bound for the number of steps until some meridian b^* (for arbitrary number of faults, i.e. for any ϵ):

$$t_{b^*} \leq \int_0^{b^*} \frac{1}{v_{\Pi}(b)}\, db \leq \int_0^{b^*} \frac{1}{v_G \sqrt{1 - \frac{\sin^2 2A \, \tan^2 b}{\cos^2 A + \tan^2 b}} \cos A}\, db. \qquad (14)$$

We note that on the fastest "no-fault" route a state will travel exactly at speed v_G, so that in the same many t_{b^*} steps it can reach at most $(v_G t_{b^*})^{\text{th}}$ meridian.

From (14) we can derive the upper bound for the distance between the two meridians b^* and $v_G t_{b^*}$:

$$v_G t_{b^*} - b^* \leq \cancel{\times} \int_0^{b^*} \frac{1}{\cancel{\times}\sqrt{1 - \frac{\sin^2 2A \, \tan^2 b}{\cos^2 A + \tan^2 b} \cos A}} \, db - b^* \tag{15}$$

For example, if we fix $b^* = \frac{3}{8}\pi$, then for any $0 \leq A \leq 0.1953\ldots\pi$, value (15) does not exceed $\frac{\pi}{4}$. That is, when the fastest state of a quantum ensemble reaches meridian $b^* + \frac{\pi}{4} = \frac{5}{8}\pi$, the slowest state of the ensemble with certainty reaches at least meridian $b^* = \frac{3}{8}\pi$.

4.3   Proof of Theorem 1

We run standard Grover's algorithm $\frac{5}{4}$ times longer than usually, and then perform a measurement.

- If at most one fault is promised, and $A \leq 0.25\pi$,[4] then we use Corollary 1 and follow that any component of the resulting mixture $|\psi^*\rangle$ is at most $\frac{\pi}{8}$ far from the meridian "$\alpha = 0$".
- If $A \leq 0.1953\ldots\pi$,[5] then we substitute $b^* = \frac{3}{8}\pi$ in (15) and follow exactly the same.

Measurement of such $|\psi^*\rangle$ results in finding a marked item (the faulty or a non-faulty one) with probability at least $\cos^2 \frac{\pi}{8} = 0.853553\ldots$

- Otherwise, if $0.1953\ldots\pi < A \leq 0.25\pi$ and there is no promise on the number of faults, it means that there is exactly 1 non-faulty marked element (so $A = 0.25\pi$).

In this specific case we run standard Grover's algorithm ≈ 1.34 times longer than usually, and then perform a measurement. We substitute $A = 0.25\pi, b^* \approx 0.33\pi$ in (15) and follow that value (15) does not exceed $\approx 0.34\pi$. That is, when the fastest state of a quantum ensemble reaches meridian $b^* + 0.34\pi = 0.67\pi(= 0.5 \times 1.34\pi)$, the slowest state of the ensemble with certainty reaches at least meridian $b^* = 0.33\pi$. Measurement of such $|\psi^*\rangle$ results in finding a marked item (the faulty or a non-faulty one) with probability at least $\cos^2 0.17\pi = 0.74\ldots$ □

---

[4] $A \leq 0.25\pi$ means that there is at least as many non-faulty marked items as faulty marked items. Since we limit our considerations with only one faulty marked item, it suffices with only one non-faulty marked item.

[5] For one faulty marked item, it means existence of at least $\lceil \text{arccot}\, 0.1953\ldots\pi \rceil = \lceil 1.02047 \rceil = 2$ non-faulty marked items.

5   Summary and Open Problems

In this paper we focus on the case where search space contains multiple non-faulty and one faulty marked element. We show that if there are at least two non-faulty marked elements or it there is at most one fault, then it is still possible to find a marked element in $O(\sqrt{N})$ queries with $\Theta(1)$ probability.

It is an open question to generalize the Theorem 1 to more than one faulty marked element. This, however, might be tricky as one needs to deal with hyper-spherical geometry.

References

1. Ambainis, A., Bačkurs, A., Nahimovs, N., Rivosh, A.: Grover's algorithm with errors. In: Kučera, A., Henzinger, T.A., Nešetřil, J., Vojnar, T., Antoš, D. (eds.) MEMICS 2012. LNCS, vol. 7721, pp. 180–189. Springer, Heidelberg (2013)

2. Grover, L.: A fast quantum mechanical algorithm for database search. In: Proceedings of the 28th ACM STOC, pp. 212–219 (1996)

3. Kaye, P., Laflamme, R.: An Introduction to Quantum Computing. Cambridge University Press, Cambridge (2007)

4. Regev, O., Schiff, L.: Impossibility of a quantum speed-up with a faulty oracle. In: Aceto, L., Damgård, I., Goldberg, L.A., Halldórsson, M.M., Ingólfsdóttir, A., Walukiewicz, I. (eds.) ICALP 2008, Part I. LNCS, vol. 5125, pp. 773–781. Springer, Heidelberg (2008)

5. Todhunter, I.: Spherical Trigonometry, 5th edn. MacMillan, London (1886)

Reachability Problems for PAMs

Oleksiy Kurganskyy[1] and Igor Potapov[2]([⊠])

[1] Institute of Applied Mathematics and Mechanics,
NAS of Ukraine, Donetsk, Ukraine
[2] Department of Computer Science, University of Liverpool, Liverpool, UK
potapov@liverpool.ac.uk

Abstract. Piecewise affine maps (PAMs) are frequently used as a reference model to show the openness of the reachability questions in other systems. The reachability problem for one-dimensional PAM is still open even if we define it with only two intervals. As the main contribution of this paper we introduce new techniques for solving reachability problems based on p-adic norms and weights as well as showing decidability for two classes of maps. Then we show the connections between topological properties for PAM's orbits, reachability problems and representation of numbers in a rational base system. Finally we show a particular instance where the uniform distribution of the original orbit may not remain uniform or even dense after making regular shifts and taking a fractional part in that sequence.

1 Introduction

The simplification of real programs shows that there is a number of quite basic models/fragments for which we have fundamental difficulties in the design of verification tools. One of them is the model of iterative map that appears in many different contexts, including discrete-event/discrete-time/hybrid systems, qualitative biological models, chaos-based cryptography, etc. [4,9,12,19].

The one-dimensional affine piecewise iterative map is a very rich mathematical object and at the same time one of the simplest dynamical system producing very complex and sensitive effects. A function $f \colon \mathbb{Q} \to \mathbb{Q}$ is a one-dimensional piecewise-affine map (PAM) if f is of the form $f(x) = a_i x + b_i$ for $x \in X_i$ where all coefficients a_i, b_i and the extremities of a finite number of bounded intervals X_i are rational numbers. Let us consider the sequence of iterations starting from a rational point $x : x, f(x), f^2(x) = f(f(x))$, and so on. The reachability in PAM is a problem to decide for a given f and two rational points x and y whether y is reachable from x. In other words, is there an $n \in \mathbb{N}$ such that $f^n(x) = y$?

The decidability of the reachability problem for one dimensional piecewise-affine map is still an open problem, which is related to other challenging questions in the theory of computation, number theory and linear algebra [8,15,16]. This model plays a crucial role in the recent research about verification of hybrid

This research is supported by EPSRC grant "Reachability problems for words, matrices and maps" (EP/M00077X/1).

© Springer-Verlag Berlin Heidelberg 2016
R.M. Freivalds et al. (Eds.): SOFSEM 2016, LNCS 9587, pp. 356–368, 2016.
DOI: 10.1007/978-3-662-49192-8_29

systems [2,3], timed automata [2] control systems [10,11], representation of numbers in a rational base (β-expansions) [21,23], discounted sum automata [13]. In particular PAM is often used as a reference model to show the openness of the reachability questions in other systems. It also has a very natural geometrical interpretation as pseudo-billiard system [17] and Hierarchical Piecewise Constant Derivative (HPCD) system [3]. The reachability problem for one-dimensional PAM is still open even if we define it with only two intervals [2,3,5,6].

The primary goal of this paper is to demonstrate new approaches for solving reachability problem in PAMs, connecting reachability questions with topological properties of maps and widening connections with other important theoretical computer science problems. First, we show new techniques for decidability of the reachability problem in PAMs based on p-adic norms and weights. We illustrate these techniques showing decidability of two classes of PAMs. The algorithm in Theorem 1 solves point to point reachability problem for two-interval injective PAM under the assumption that a PAM has bounded invariant densities. While our numerical experiments shows that the sequence of invariant densities converge to smooth functions it is not yet clear whether it holds for all PAMs or if not whether this property can be algorithmically checked.

Following the proposed approach based on p-adic weights in Theorem 2 we define another fragment of PAMs for which the reachability problem is decidable. In particularly we remove the condition on bounded invariant densities and injectivity of piecewise-affine map and consider a PAM f with a constraint on linear coefficients in affine maps. This class of PAMs is also related to encoding of rational numbers in the rational base (β-expansions). The decidability of the point-to-point problem for this class is shown in Theorem 2 and decidability of point-to-set problem for the same class can give someone an answer to the open problem related to β-expansions.

Then we establish the connections of topological properties for PAM's orbits with reachability problems and representation of numbers in a rational base system. We show that the reachability problems for above objects tightly connected to questions about distribution of the fractional parts in the generated sequences and moreover about distribution of the fractional part after regular shifts.

2   Preliminaries and Notations

In what follows we use traditional denotations $\mathbb{N}, \mathbb{Z}, \mathbb{Z}^+ = \{0, 1, 2, \ldots\}, \mathbb{P}, \mathbb{Q}$ and $\mathbb{R}$ for sets of natural, integers, positive integers, primes, rational and real numbers, respectively. Let us denote by $S^1 = \mathbb{Q}/\mathbb{Z}$ the unit circle which consists only rational numbers. By $\{x\}$, $\lfloor x \rfloor$ and $\lceil x \rceil$ we denote the fractional part[1] of a number, floor and ceiling functions.

Let Y be a set of numbers and x is a single number, then we define their addition and multiplication as follows: $Y + x = x + Y = \{x + y | y \in Y\}$ and $xY = Yx = \{xy | y \in Y\}$. The application of a function $f : X \to Y$ to a set

---

[1] It will be clear from the context if brackets are used in other conventional ways, for example, to indicate a set of numbers.

$X' \subseteq X$ is defined as $f(X') = \{f(x)|x \in X'\}$. If $f \subseteq X \times Y$ is a nondeterministic map, i.e. $f : X \to 2^Y$ and $x \in X$, $X' \subseteq X$, we define $f(x) = \{y|(x,y) \in f\}$ and $f(X') = \bigcup_{x \in X'} f(x)$.

p-adic norms and weights: Let us consider an arbitrary finite set of prime numbers $\mathbb{F} = \{p_1, p_2, \ldots, p_k\} \subset \mathbb{P}$ in ascending order and define the product of prime numbers from $\mathbb{F}$ by $m = p_1 p_2 \ldots p_k$. Let x be a positive rational number that can be represented by primes from a set $\mathbb{F}$. Then its prime factorization is $x = \prod_{p \in \mathbb{F}} p^{\alpha_p}$, where $\alpha_p \in \mathbb{Z}$, $p \in \mathbb{F}$.

Any nonzero rational number x can be represented by $x = (p^{\alpha_p} r)/s$, where p is a prime number, r and s are integers not divisible by p, and α_p is a unique integer. The p-adic norm of x is then defined by $|x|_p = p^{(-\alpha_p)}$. The p-adic weight of x is defined as $\|x\|_p = \log_p(|x|_p)$, i.e. $\|x\|_p = -\alpha_p$. The following properties of p-adic weights are directly follows from the properties of p-adic norm:

$$\|x\|_p = \|y\|_p \Rightarrow \|x + y\|_p \leq \|x\|_p. \tag{1}$$

$$\|x\|_p < \|y\|_p \Rightarrow \|x + y\|_p = \|y\|_p, \tag{2}$$

$$\|x \cdot y\|_p = \|x\|_p + \|y\|_p, \tag{3}$$

$$\|x^r\|_p = r\|x\|_p, \tag{4}$$

If there is a prime $p \notin \mathbb{F}$ such that $\|x\|_p > 0$, then we define $\|x\|_m = +\infty$, otherwise $\|x\|_m = \max_{p \in F} \|x\|_p$.

By m-weight and m-vector-weight of x in respect to a set $\mathbb{F}$ we denote $\|x\|_m$ and $(\|x\|)_m = (\|x\|_{p_1}, \|x\|_{p_2}, \ldots, \|x\|_{p_k})^T$ respectively. Informally speaking the m-weight of a number x (if $\|x\|_m > 0$) is the number of digits after the decimal point in the representation of x in base m, i.e. x can be written as $x = y \cdot m^{-\|x\|_m}$, where y is an integer which the last digit in the m-ary representation is non zero. If $\|x\|_m \leq 0$, then x is an integer number.

Without loss of generality let us consider from now on only such $x \in X$ for which $\|x\|_m < +\infty$. Alternatively if $\|x\|_m = +\infty$, it is enough to change the set $\mathbb{F} = \{p_1, \ldots, p_k\}$ in order to fulfill the requirements of $\|x\|_m < +\infty$.

Lemma 1. *For all rational $x \in [0, 1]$ with an upper bound $a \in \mathbb{Z}^+$ on $\|x\|_m$, i.e. $\|x\|_m < a$, there is a lower bound $b \in \mathbb{Z}$ based on a and $\mathbb{F}$ such that $\|x\|_p \geq b$ for all $p \in \mathbb{P}$.*

Proof. Let us denote two sums of weights: $\alpha = -\sum_{p \in \mathbb{P}, \|x\|_p \leq 0} \|x\|_p$ and $\beta = \sum_{p \in \mathbb{F}, \|x\|_p \geq 1} \|x\|_p$. Assuming that 2 is the smallest possible prime number and p_k is the largest number in $\mathbb{F}$, we have the following inequality: $\frac{2^\alpha}{p_k^{ka}} \leq \frac{2^\alpha}{p_k^\beta} \leq x \leq 1$. Then $-\alpha \geq -ka\log_2 p_k$ and if $b = -\alpha$ we have $\|x\|_p \geq b$ for all $p \in \mathbb{P}$. $\qquad \square$

Corollary 1. *For any $a \in \mathbb{Z}$ there is only a finite number of rational $x \in [0, 1]$ for which $\|x\|_m < a$.*

Reachability problem for PAMs: We say that $f : S^1 \to S^1$ is a one-dimensional piecewise affine map (PAM) whenever f is of the form $f(x) = a_i x + b_i$, where $a_i, b_i \in \mathbb{Q} \Leftrightarrow x \in X_i$, $S^1 = X_1 \cup X_2 \cup \ldots \cup X_l$ and where $\{X_1, \ldots, X_l\}$ is a finite family of disjoint (rational) intervals. If the intervals are not disjoint we call it *non-deterministic piecewise affine map* and by default a piecewise-affine mapping is understood to be deterministic. The derivative f' of a PAM f we define as $f'(x) = a_i$ for $x \in X_i$, $1 \le i \le l$.

If f is deterministic PAM, an orbit (trajectory) of a point x is denoted by $O_f(x)$ and will be understood either as a set $O_f(x) = \{f^i(x) | i \in \mathbb{Z}^+\}$ or as a sequence, i.e. $O_f(x) : \mathbb{Z}^+ \to S^1$, $O_f(x)(i) = f^i(x)$. We also define that a point y is reachable from x if $y \in O_f(x)$.

In general the reachability problem for PAM can be defined as follows. Given a PAM f, $x \in S^1$ and $Y \subseteq S^1$, decide whether the intersection $Y \cap O_f(x)$ is empty. If Y is a finite union of intervals, we name the reachability problem as *point-to-set (interval)* problem. If Y is a one element set (i.e. a single point), the reachability problem is known as *point-to-point* reachability.[2]

In this paper we only consider one-dimensional PAMs and by the *reachability problem for PAM* we understand point-to-point reachability and explicitly state the type of the problem when we need to refer to other reachability questions. Note that the point-to-interval reachability can be reduced to point-to-point reachability problem by extending a map with a few intervals in which the current value is just sequentially deleted. It works for all PAMs but may not preserve the properties and the form of the original map.

Generally speaking the piecewise-affine mapping does not need to be defined as $f : X \to X$, where $X = S^1 = [0, 1)$. However if the set X is a union of any finite number of bounded intervals we always can scale it to S^1. If $f : X \to X$ is such that $X \neq [0, 1)$, and $X \subseteq [a, b)$, then by applying conjugation $h(x) = \frac{x-a}{b-a}$ the original reachability problem for f is reduced to the reachability problem for the mapping $g = h \circ f \circ h^{-1}$ from $[0, 1)$ to $[0, 1)$. Moreover the interest to PAMs as $f : S^1 \to S^1$ is also motivated by their use in the research of chaotic systems.

3   Decidability Using p-Adic Norms

It is well know in dynamical systems research that due to complexity of orbits in iterative maps it is less useful, and perhaps misleading, to compute the orbit of a single point and it is more reasonable to approximate the statistics of the under-lying dynamics [14, 22]. This information is encoded in the so-called *invariant measures*, which specify the probability to observe a typical trajectory within a certain region of state space and their corresponding *invariant densities*.

Let us consider a density as an ensemble of initial starting points (i.e. initial conditions). The action of the dynamical system on this ensemble is described by the Perron-Frobenius operator. The ensembles which are fixed under the linear

---

[2] Also in a similar way it is possible to define set-to-point and set-to-set reachability problems.

Perron-Frobenius operator is known as invariant densities or in other words, they are eigenfunctions with eigenvalue 1 [14].

Formally under an ensemble A we understand an enumerated set (sequence) of points in phase space. With ensemble we can associate the distribution function and the density function. Let I be a set of points. We denote by $F_I^A(n) = |\{i \in \mathbb{Z}^+ | i \leq n, A(i) \in I\}|$ the number of elements in the sequence A which belong to the set I and which indexes are less or equal n. The distribution function of the ensemble A is defined as $\Phi_A(x) = \lim_{n \to \infty} \frac{F_{(-\infty, x)}^A(n)}{n}$, if the limits exist. The density function ϕ_A of the ensemble A is defined as $\phi_A(x) = \Phi'_A(x)$.

Suppose given an ensemble A_0 with density ϕ_0. If we apply PAM f to each point of the ensemble, we get a new ensemble A_1 with some density distribution ϕ_1. We say that the function ϕ_1 is obtained from ϕ_0 using the Frobenius-Perron or transfer operator, which we denote by L_f. It is known that

$$\phi_1(x) = L_f(\phi_0)(x) = \sum_{y \in f^{-1}(x)} \frac{\phi_0(y)}{|f'(y)|}.$$

If $\phi_1 = \phi_0$ we say that ϕ_0 is a f-invariant density function or an eigenfunction of the transfer operator L_f.

We prove that if for an injective PAM f there exists an invariant bounded density function then the reachability problem for f is decidable.

Lemma 2. *Let f be an injective PAM, and ϕ be a f-invariant density function. If there are $K_{min} > 0$ and $K_{max} < +\infty$ such that for any x from the domain of f the following inequality holds: $K_{min} < \phi(x) < K_{max}$, then for an arbitrary segment of the orbit $x_1, x_2, \ldots, x_{n+1}$, where $x_{i+1} = f(x_i)$, we have $\frac{K_{min}}{K_{max}} \leq |c_1 \cdot c_2 \cdot \ldots \cdot c_n| \leq \frac{K_{max}}{K_{min}}$, where $c_i = a_j$ if $x_i \in X_j$.*

Proof. Let ϕ be an eigenfunction of the Perron-Frobenius operator for an injective PAM f. Then injectivity of f and the fact that $y = f(x)$ implies that $\phi(y) = \frac{\phi(x)}{|f'(x)|}$. We denote $f'(x_i)$ by c_i. Then $\phi(x_{n+1}) = \frac{\phi(x_1)}{|c_1 \cdot c_2 \cdot \ldots \cdot c_n|}$ and $|c_1 \cdot c_2 \cdot \ldots \cdot c_n| = \frac{\phi(x_1)}{\phi(x_{n+1})}$. Now we can bound $|c_1 \cdot c_2 \cdot \ldots \cdot c_n|$ by $\frac{K_{min}}{K_{max}}$ and $\frac{K_{max}}{K_{min}}$. □

Theorem 1. *Given an injective PAM f with two intervals and the existence of a f-invariant density function ϕ such that there are $K_{min} > 0$ and $K_{max} < +\infty$ and the following inequality holds $K_{min} < \phi(x) < K_{max}$ for all x from the domain of f. Then the reachability problem for f is decidable.*

A complete proof of the theorem is available at http://arxiv.org/abs/1510.04121

The theorem can be applied for a larger class of PAMs if more information would be known about the convergence of density functions under the action of the Perron-Frobenius operator. Let us call an ensemble A to be statistically fixed with respect to f, if $\phi_A = L_f(\phi_A)$. E.g. if someone can show that in injective

PAM all statistically fixed ensembles have identical distribution functions then Theorem 1 can be applied to show decidability in injective PAMs.

Following the proposed approach based on p-adic weights we define another fragment of PAMs for which the reachability problem is decidable. In particularly we remove the condition on eigenfunction of the transfer operator and injectivity of piecewise-affine map and consider a PAM f with only constraints on linear coefficients in affine maps. More specifically we require that the powers of prime numbers from prime factorizations of linear coefficients should have the same signs (i.e. two sets of prime numbers used in nominator and denominator are disjoint). Let us denote for a PAM f a matrix A_f with values (a_{ji}), where $a_{ji} = \|a_i\|_{p_j}$, $1 \leq i \leq l$, $1 \leq j \leq k$. The rank of A_f is denoted by $\mathrm{rank}(A_f)$.

Theorem 2. *The reachability problem for a PAM f is decidable if every row of a matrix A_f contains values of the same sign, (i.e. $a_{ji} \cdot a_{j'i} \geq 0$, for all i, j such that $1 \leq i \leq l$, $1 \leq j, j' \leq k$).*

Proof. Let us consider a PAM f of the form $f(x) = a_i x + b_i$ for $x \in X_i$ where all coefficients a_i, b_i and the extremities of a finite number of bounded intervals X_i are rational numbers. Let us define $h = \max\{\|b_1\|_m, \|b_2\|_m, \ldots, \|b_l\|_m\}$. The condition of the theorem means that for any prime $p \in \mathbb{F}$ all linear coefficients of the map f have non-zero p-adic weights of the same sign.

In this case, if p-adic weights of linear coefficients of f are non-negative, then for any $x \in X$ from $\|x\|_p > h$ follows that $\|f(x)\|_p \geq \|x\|_p$ and therefore $\|f(x)\|_m \geq \|x\|_m$ (i.e. m-adic weight does not decrease). If p-adic weight of linear coefficients of the mapping are negative, then for any $x \in X$ we have $\|f(x)\|_p \leq \max\{\|x\|_p, h\}$.

Thus, in the sequence of reachable points for an orbit of a map f either all points of the orbit have m-adic weights bounded from above by h, then we have a cyclic orbit, or from some moment when m-adic weight of a reachable point exceeds h it does not decrease and again, either orbit loops or m-adic weight increases indefinitely.

Thus, in order to decide whether y is reachable, i.e. $y \in O_f(x)$, it is sufficient to start generating a sequence of reachable points in the orbit $O_f(x)$ and wait for one of the events, where either (1) a point in the orbit is equal to y (y is reachable), or (2) the orbit will loop and $y \notin O_f(x)$ (y is not reachable), or (3) a point x' is reachable, such that $\|x'\|_m > \max\{h, \|y\|_m\}$, and then $y \notin O_f(x)$ (y is not reachable). $\qquad\square$

Definition 1. *A piecewise affine mapping $f : S^1 \to S^1$ is complete if for a set of disjoint intervals $S^1 = X_1 \cup X_2 \cup \ldots \cup X_n$, $f(X_i) = S^1$ for any $i = 1..n$.*

Definition 2. *Let be $F : \mathbb{R} \to \mathbb{R}$ is the lifting of a continuous map $f : S^1 \to S^1$ on $\mathbb{R}$, i.e. $f(\{x\}) = \{F(x)\}$. Then by the degree $\deg(f)$ of a map f we denote the number $F(x + 1) - F(x)$, which is independent from the choice of the point x and the lifting F.*

Corollary 2. *The reachability problem for complete piecewise affine mappings with two intervals[3] is decidable.*

Proof. The condition of a piecewise affine map with two intervals $f : S^1 \to S^1$ to be complete means that $S^1 = X_1 \cup X_2$ and $f(X_1) = f(X_2) = S^1$. Thus, if $X_1 = [0, \frac{m}{n}]$ and $X_2 = [\frac{m}{n}, 1)$, then $f(x) = a_1 x + b_1$, where $a_1 = \pm \frac{n}{m}$, when $x \in X_1$, and $f(x) = a_2 x + b_2$, where $a_2 = \pm \frac{n}{n-m}$ at $x \in X_2$, $m, n \in \mathbb{N}$, $\gcd(m, n) = 1$. It is clear that n, m, $n - m$ are relatively prime. So the conditions of Theorem 2 are satisfied. $\qquad \square$

4   PAM Representation of β-Expansions

Given a rational non-integer $\beta > 1$ and the number $x \in [0, 1]$. The target discounted-sum 0-1 problem [13,20] is defined as follows: *Is there a sequence $w : \mathbb{N} \to \{0, 1\}$ of zeros and ones such that $x = \sum_{i=1}^{\infty} w(i) \frac{1}{\beta^i}$.*

For any $x \in S^1$, there exists β-expansion $w : \mathbb{N} \to \{0, 1, \ldots, \lceil \beta \rceil - 1\}$ such that $x = \sum_{i=1}^{\infty} w(i) \frac{1}{\beta^i}$. If $w(i) \in \{0, 1\}$ we call it $(0, 1) - \beta$-**expansion**. Therefore, when $\beta \leq 2$ the answer to the target discounted-sum problem is always positive. Therefore, the only interesting case is when $\beta > 2$. We denote $D = \{0, 1, \ldots, \lceil \beta \rceil - 1\}$. Then the minimal and maximal numbers, which are representable in the basis β with digits from the alphabet D, are $min = \sum_{i=1}^{\infty} 0 \frac{1}{\beta^i} = 0$ and $max = \sum_{i=1}^{\infty} (\lceil \beta \rceil - 1) \frac{1}{\beta^i} = \frac{\lceil \beta \rceil - 1}{\beta - 1}$. When $\beta > 2$ then max is always less then two. Let us denote by X_d the interval $[\frac{min+d}{\beta}, \frac{max+d}{\beta})$ for each $d \in D$. If β is not an integer number then two intervals X_d and X_{d+1} intersect. Also taking into account that $max < 2$, then the intervals X_d and X_{d+2} have no common points. Finally from the above construction we get the next lemma:

Lemma 3. *If $\beta > 2$ and β is rational/non-integer number: $X_d \cap X_{d+1} \neq \emptyset$, $d < \lceil \beta \rceil - 1$; $X_d \cap X_{d+2} = \emptyset$, $d < \lceil \beta \rceil - 2$; $[min, max) = \cup_{d \in A} X_d$.*

Proposition 1. *For any β-expansion there is a non-deterministic PAM where a symbolic dynamic of visited intervals (i.e. a sequence of symbols associated with intervals) from an initial point x_0 corresponds to its representation in base β.*

Proof. Let us define the piecewise affine mapping $f \subseteq [min, max) \times [min, max)$ as follows $f = \{(x, \beta x - d) | x \in X_d, d \in D\}$. It directly follows from this definition that $f(X_d) = [min, max)$.

Let us consider an orbit $f^i(x) = x(i)$, $i \in \mathbb{Z}^+$. We say that $d_i \in D$ such that $x(i) \in X_{d_i}$. Then for any $n \in \mathbb{N}$ $min < |\beta^n x - d_1 \beta^{n-1} - d_2 \beta^{n-2} - \ldots - d_n \beta^0| < max$, and in other form $\frac{min}{\beta^n} < |x - \sum_{i=1}^{n}(d_i \frac{1}{\beta^i})| < \frac{max}{\beta^n}$. So $|x - \sum_{i=1}^{n}(d_i \frac{1}{\beta^i})| \to 0$, $n \to \infty$, and therefore $x = \sum_{i=1}^{\infty}(d_i \frac{1}{\beta^i})$. Let us consider it in other direction. Let $x = \sum_{i=1}^{\infty}(d_i \frac{1}{\beta^i})$, then the sequence $\mathbf{x}(i)$, where $\mathbf{x}(0) = x$, $\mathbf{x}(i+1) = \beta \mathbf{x}(i) - d_i$, is the orbit of x in PAM f. Let us name the constructed map as the β-**expansion PAM**. $\qquad \square$

---

[3] In particularly the continuous piecewise affine mapping of degree two.

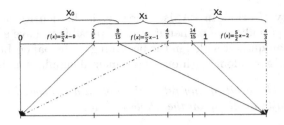

Fig. 1. A non-deterministic PAM for $\frac{5}{2}$-expansion

The nondeterministic β-expansion can be translated into deterministic maps corresponding to *greedy* and *lazy* expansions as follows:

Definition 3. *A function* $f : [min, max) \to [min, max)$ *is the greedy* β-*expansion PAM if the domain* $[min, max)$ *is divided on intervals* X'_d, $d \in \{0, 1, \ldots, \lceil \beta \rceil - 1\}$ *such that* $X'_{\lceil \beta \rceil - 1} = X_{\lceil \beta \rceil - 1}$, $X'_{d-1} = X_{d-1} - X_d$, $d \in \{1, 2, \ldots, \lceil \beta \rceil - 1\}$ *and* $f(x) = \beta x - d$ *iff* $x \in X'_d$.

Since $X_d = [\frac{min+d}{\beta}, \frac{max+d}{\beta})$ then $X'_d = [\frac{min+d}{\beta}, \frac{min+d+1}{\beta}) = [\frac{d}{\beta}, \frac{d+1}{\beta})$, $d \in \{0, 1, \ldots, \lceil \beta \rceil - 2\}$ and the length of the interval X'_d is equal to $\frac{1}{\beta}$, $d < \lceil \beta \rceil - 1$.

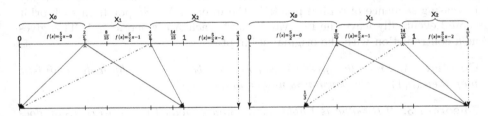

Fig. 2. Deterministic greedy (on the left) and lazy (on the right) $\frac{5}{2}$-expansion PAM

Definition 4. *A function* $f : [min, max) \to [min, max)$ *is the lazy* β-*expansion PAM, if the domain* $[min, max)$ *is divided into intervals* X''_d, $d \in \{0, 1, \ldots, \lceil \beta \rceil - 1\}$, *such that* $X''_0 = X_0$, $X''_d = X_d - X_{d-1}$, $d \in \{1, 2, \ldots, \lceil \beta \rceil - 1\}$, *and* $f(x) = \beta x - d$ *iff* $x \in X''_d$.

Proposition 2. *Let* f *and* g *are greedy and lazy* β-*expansion PAM's respectively.* f *and* g *are (topologically) conjugate by the homeomorphism* $h : h(x) = h^{-1}(x) = max - x$, *i.e.* $f = h \circ g \circ h$.

Proof. The statement holds since $X'_d = max - X''_{\lceil \beta \rceil - 1 - d}$, $d \in \{0, 1, \ldots, \lceil \beta \rceil - 1\}$

We would like to highlight that the questions about reachability as well as representation of numbers in rational bases are tightly connected with questions about the density of orbits in PAMs. Moreover if the density of orbits are the same for all non-periodic points then it may be possible to have a wider application of p-adic techniques provided in the beginning of the paper. Let us formulate a hypothesis that goes along with our experimental simulations in PAMs:

Hypothesis 1. *The orbit of any rational point in any expanding deterministic PAM is either finite or dense on the whole domain.*

Lemma 4. *Any $(0,1) - \beta$-expansion is greedy.*

Proof. Let f be a β-expansion PAM. Assume that there is a point x and the orbit $\mathbf{x}(i)$, where $\mathbf{x}(0) = x$, $\mathbf{x}(i+1) = \beta\mathbf{x}(i) - d_i$ in the map f such that $d_i \in \{0, 1\}$ for all $i \in \mathbb{N}$ and the orbit does not correspond to the β greedy expansion of x.

The intersection of intervals X_0 and X_1 is an interval $X_{01} = [\frac{\min + 1}{\beta}, \frac{\max}{\beta})$. Applying a map $y = \beta x$ to X_{01} we see that X_{01} is scaled into $[\min + 1, \max) = [1, \frac{\lceil \beta \rceil - 1}{\beta - 1})$. The interval $[1, \frac{\lceil \beta \rceil - 1}{\beta - 1})$ does not have any common points with X_0 as the point 1 lies on the right side of the left border of the interval $X_2 = [\frac{\min + 2}{\beta}, \frac{\max + 2}{\beta})$ and by Lemma 3 $X_i \cap X_{i+2} = \emptyset$.

Note that when $x > \frac{1}{\beta - 1}$ we have $\beta x - 1 > x$. Let us assume that for some i $\mathbf{x}(i) \in X_{01}$ and $\mathbf{x}(i+1) = \beta\mathbf{x}(i) - 0$, i.e. we did not followed a greedy expansion and therefore $\mathbf{x}(i+1) \in [1, \frac{\lceil \beta \rceil - 1}{\beta - 1})$. Then $\mathbf{x}(i+2) = \beta\mathbf{x}(i+1) - 1 > \mathbf{x}(i+1)$ and $\mathbf{x}(i+2) \notin X_0$, etc. In this case starting from $\mathbf{x}(i+1)$ there is monotonically increasing sequence of orbital points in the interval X_1. So points in such orbit should eventually leave the interval X_1 and reach X_d, where $d > 1$. This gives us a contradiction with the original assumption. □

Corollary 3. *Since the greedy expansion can be expressed by a deterministic map then $(0,1) - \beta$-expansion is unique and greedy.*

Theorem 3. *If Hypothesis 1 holds then a non-periodic $(0,1) - \beta$-expansion does not exist.*

Proof. Any $(0,1) - \beta$-expansion can be constructed by expanding[4] deterministic greedy β-expansion PAM. If the orbit of a rational point in greedy β-expansion PAM is non-periodic, then by Hypothesis 1 it should be dense and therefore should intersect all intervals and cannot provide $(0,1) - \beta$-expansion. □

Theorem 4. *If Hypothesis 1 holds then for any rational number its deterministic β-expansion is either eventually periodic or it contains all possible patterns (finite subsequences of digits) from $\{0, 1, \ldots, \lceil \beta \rceil - 1\}$.*

Proof. The statement is obvious as Hypothesis 1 implies that the orbit is either periodic or it is dense and the dense orbit is visiting all intervals. □

---

[4] I.e. with linear coefficients that are greater than 1.

It looks that the point-to-interval problem is harder than the point-to-point reachability problem for the expanding PAMs, as for example Theorem 2 gives an algorithm for the point-to-point reachability problem in the β-expansion PAMs, but not for the point-to-interval reachability that is equivalent to the β-expansion problem.

Note that in the β-expansion PAMs all linear coefficients are the same, so the density of the orbit correspond to the density of the following sequence $\mathbf{x}(n) = f^n(x_0)$, where $f(x) = \{\beta x\}$. For example when $\beta = \frac{5}{2}$ and $x_0 = 1$ we get the sequence:

$$\{\tfrac{5}{2}\}, \{\tfrac{5}{2}\{\tfrac{5}{2}\}\}, \{\tfrac{5}{2}\{\tfrac{5}{2}\{\tfrac{5}{2}\}\}\}, \cdots$$

The question about the distribution of a similar sequence $\{\frac{3}{2}\}, \{\frac{3^2}{2^2}\}, \{\frac{3^3}{2^3}\}, \ldots$, where the integer part is removed once after taking a power of a fraction (for example $3/2$) is known as Mahler's $3/2$ problem, that is a long standing open problem in analytic number theory.

5 Density of Orbits and its Geometric Interpretation

It is well known that $\mathbf{x}(n) = \{\alpha n\}$, where α is an irrational number, has an uniform distribution. Let us give some geometric interpretation of the orbit density. Consider the Cartesian plane with the y-axis x and the x-axis y (just swapping their places). Now let us divide the set of lines $x = n$, $n \in \mathbb{N}$, by integer points on the segments of the unit length. The set of points (y, x), where $m \le y < m+1$, $x = n$, i.e. the interval $[m, m+1] \times n$ on the line $x = n$, will be denoted by $S_{m,n}$, $m \in \mathbb{Z}^+$, $n \in \mathbb{N}$. In other words, $S_{m,n} = (m + [0, 1)) \times n$. Let I be an interval such that $I \subseteq [0, 1)$ and by $I_{m,n}$ let us denote the set $(m + I) \times n$.

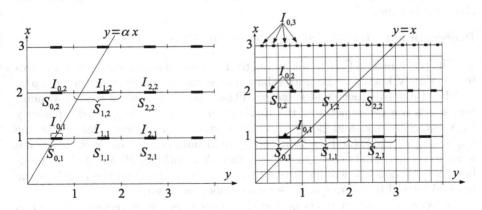

Fig. 3. Left: An example for two sets $S_{m,n}$ and $I_{m,n}$; Right: A dynamic interval $I(n)$

Two points of the plane are defined to be equivalent if they belong to a same line passing through the origin. We call α as homogeneous coordinate of a point (y, x) if $y = \alpha x$. By $H(I)$ we denote the set of homogeneous coordinates of all points from $\bigcup\limits_{m \in \mathbb{Z}^+, n \in \mathbb{N}} I_{m,n}$. The sequence $\mathbf{x}(n) = \{\alpha n\}$ is dense in $[0, 1)$ if and only if for any interval $I \subseteq [0, 1)$ there are m and n, such that the line $y = \alpha x$ intersects the set $I_{m,n}$. It is known that $[0, 1) - H(I) \subseteq \mathbb{Q}$ for any interval $I \subseteq [0, 1)$, i.e. for any irrational $\alpha > 0$ the line $y = \alpha x$ intersect the set $\bigcup\limits_{m \in \mathbb{Z}^+, n \in \mathbb{N}} I_{m,n}$. Moreover in the case of irrational factors it is known that the frequency of occurrence of $\mathbf{x}(n) = \{\alpha n\}$ in the interval I is equal to its length.

In some sense the interval I, in the above example, can be named as *static* because it does not change in time n. However in order to study and describe previously mentioned problems such as the target discounted-sum problem, PAMs reachability problems, the Mahler's 3/2 problem we require the notion of "dynamic intervals".

Let $\mathbf{x}$ be a sequence of numbers from $[0, 1)$. What is the distribution of a sequence $\mathbf{x}'(n) = \{p^{\mathbf{k}(n)}\mathbf{x}(n)\}$, where $\mathbf{k} : \mathbb{Z}^+ \to \mathbb{Z}^+$ is a non-decreasing sequence? For example, if $\mathbf{k}(n) = n - 1$ and the number $\mathbf{x}(n)$ has in the base p the following form $\mathbf{x}(n) = 0.a_{n1}a_{n2} \dots a_{nn}a_{n,n+1} \dots$, then $\mathbf{x}'(n) = 0.a_{nn}a_{n,n+1} \dots$.

Let us assume that $I \subseteq S^1$ and $\mathbf{k} : \mathbb{Z}^+ \to \mathbb{Z}^+$ is a non-decreasing sequence, $p \in \mathbb{N}$. Now we define "dynamical intervals" as an evolving infinite sequence $\mathbf{I}(1), \mathbf{I}(2), \mathbf{I}(3), \dots$:

$$\mathbf{I}(1) = I, \mathbf{I}(n) = \bigcup_{j=0}^{p^{\mathbf{k}(n)}-1} \frac{I + j}{p^{\mathbf{k}(n)}}.$$

By $F_{\mathbf{I}}^{\mathbf{x}}(n) = |\{i \in \mathbb{Z}^+ | i \le n, \mathbf{x}(i) \in \mathbf{I}(i)\}|$ we denote a function representing a frequency of hitting dynamical interval $\mathbf{I}$ by the sequence $\mathbf{x}$. In contrast to $F_I^{\mathbf{x}}(n)$ which only counts the number of hittings to a fixed interval I, our new function $F_{\mathbf{I}}^{\mathbf{x}}$ counts the number of hittings when both points and intervals are changing in time.

Proposition 3. *The following equation holds:* $F_I^{\mathbf{x}'}(n) = F_{\mathbf{I}}^{\mathbf{x}}(n)$.

The phenomenon that significant digit distribution in real data are not accruing randomly known as Benford's Law. For example the sequence $p^1, p^2, p^3, \dots$ satisfies Benford's Law, under the condition that $\log_{10} p$ is an irrational number, which is a consequence of the *Equidistribution theorem* (proved separately by Weyl, Sierpinski and Bohl). The Equidistribution theorem states that the sequence $\{\alpha\}, \{2\alpha\}, \{3\alpha\}, \dots$ is uniformly distributed on the circle $\mathbb{R}/\mathbb{Z}$, when a is an irrational number. It gives us the fact that each significant digit of numbers in (p^n) sequence will correspond to the interval $\mathbb{R}/\mathbb{Z}$ and the length of the interval related to the frequency for each appearing digit.

However the question about the distribution of the sequence $\{(3/2)^n\}$ is different in the way that it is not about the distribution of the first digits of 3^n in base 2, i.e. not about the distribution of the sequence $\frac{3^n}{2^{\lceil n \log_2 3 \rceil}}$, but related to the sequence of digits after some shift of the number $\frac{3^n}{2^{\lceil n \log_2 3 \rceil}}$ corresponding to the multiplication by a power of 2.

So in the above notations the distribution of numbers in the sequence $\mathbf{x}'(n) = \{(3/2)^n\}$ corresponds to the $F_I^{\mathbf{x}'}(n)$ for the logarithmic (Benford's law) distributed sequence $\mathbf{x}(n) = \frac{3^n}{2^{\lceil n \log_2 3 \rceil}}$, $p = 2$ and $\mathbf{k}(n) = \lceil n \log_2 3 \rceil - n$.

Now we will show that even if the sequence $\{\alpha\}, \{2\alpha\}, \{3\alpha\}, \dots$ is uniformly distributed on the circle $\mathbb{R}/\mathbb{Z}$, the irrationality of α is not enough to guarantee uniform distribution or even density of the sequence $\mathbf{x}'(n)$ on the circle corresponding to the linear shifts $\mathbf{k}(n) = n$.

Theorem 5. *Let us define* $\alpha = \sum_{i=1}^{\infty} \frac{1}{2^{\Delta_i}}$ *where* $\Delta_1 = 1$, $\Delta_{i+1} = 2^{\Delta_i} + \Delta_i$, $i \geq 1$ *(http://oeis.org/A034797). Then for all* $n \in \mathbb{N} \cup \{0\}$ *the sequence* $\{2^n n\alpha\}$ *is not dense in the interval* $[0,1]$ *and* $\{2^n n\alpha\} < \frac{1}{2}$.

While the question about the distributions for PAM orbits remains open we have unexpectedly shown that in a very similar system, operating with irrational numbers, the uniform distribution of original orbits in maps may not remain uniform or even dense when taking the fractional part after regular shifts. This makes the questions about PAMs even more "mysterious" as it is not clear whether such property may hold for a sequence of points generated by PAMs, β-expansion and Mahler's problem.

References

1. Asarin, E., Maler, O., Pnueli, A.: Reachability analysis of dynamical systems having piecewise-constant derivatives. Theor. Comput. Sci. **138**, 35–66 (1995)
2. Asarin, E., Mysore, V., Pnueli, A., Schneider, G.: Low dimensional hybrid systems - decidable, undecidable, don't know. Inf. Comput. **211**, 138–159 (2012)
3. Asarin, E., Schneider, G.: Widening the boundary between decidable and undecidable hybrid systems. In: Brim, L., Jančar, P., Křetínský, M., Kučera, A. (eds.) CONCUR 2002. LNCS, vol. 2421, pp. 193–208. Springer, Heidelberg (2002)
4. Aswani, A., Tomlin, C.J.: Reachability algorithm for biological piecewise-affine hybrid systems. In: Bemporad, A., Bicchi, A., Buttazzo, G. (eds.) HSCC 2007. LNCS, vol. 4416, pp. 633–636. Springer, Heidelberg (2007)
5. Bazille, H., Bournez, O., Gomaa, W., Pouly, A.: On the complexity of bounded time reachability for piecewise affine systems. In: RP 2014, pp. 20–31 (2014)
6. Bell, P.C., Chen, S., Jackson, L.: Reachability and mortality problems for restricted hierarchical piecewise constant derivatives. In: RP 2014, pp. 32–44 (2014)
7. Bell, P., Potapov, I.: On undecidability bounds for matrix decision problems. Theor. Comput. Sci. **391**(1), 3–13 (2008)
8. Ben-Amram, A.M.: Mortality of iterated piecewise affine functions over the integers: decidability and complexity. Computability **4**(1), 19–56 (2015)
9. Blank, M., Bunimovich, L.: Switched flow systems: pseudo billiard dynamics. Dyn. Syst. **19**(4), 359–370 (2004)
10. Blondel, V., Bournez, O., Koiran, P., Papadimitriou, C., Tsitsiklis, J.: Deciding stability and mortality of piecewise affine dynamical systems. Theor. Comput. Sci. **255**(1–2), 687–696 (2001)
11. Blondel, V., Bournez, O., Koiran, P., Tsitsiklis, J.: The stability of saturated linear dynamical systems is undecidable. J. Comput. Syst. Sci. **62**(3), 442–462 (2001)

12. Blondel, V., Tsitsiklis, J.: A survey of computational complexity results in systems and control. Automatica **36**, 1249–1274 (2004)
13. Boker, U., Henzinger, T.A., Otop, J.: The target discounted-sum problem. In: LICS 2015, pp. 750–761 (2015)
14. Dellnitz, M., Froyland, G., Sertl, S.: On the isolated spectrum of the Perron-Frobenius operator. Nonlinearity **13**(4), 1171–1188 (2000)
15. Koiran, P., Cosnard, M., Garzon, M.: Computability with low-dimensional dynamical systems. Theor. Comput. Sci. **132**, 113–128 (1994)
16. Koiran, P.: The topological entropy of iterated piecewise affine maps is uncomputable. Discrete Math. Theor. Comput. Sci. **4**(2), 351–356 (2001)
17. Kurganskyy, O., Potapov, I., Sancho-Caparrini, F.: Reachability problems in low-dimensional iterative maps. Int. J. Found. Comput. Sci. **19**(4), 935–951 (2008)
18. Kurganskyy, O., Potapov, I.: Computation in one-dimensional piecewise maps and planar pseudo-billiard systems. In: Calude, C.S., Dinneen, M.J., Păun, G., Jesús Pérez-Jímenez, M., Rozenberg, G. (eds.) UC 2005. LNCS, vol. 3699, pp. 169–175. Springer, Heidelberg (2005)
19. Ouaknine, J., Sousa Pinto, J., Worrell, J.: On termination of integer linear loops. In: SODA 2015, pp. 957–969 (2015)
20. Randour, M., Raskin, J.-F., Sankur, O.: Percentile queries in multi-dimensional Markov decision processes. In: Dagstuhl Seminar "Non-zero-sum Games and Control" (2015)
21. Renyi, A.: Representations for real numbers and their ergodic properties. Acta Mathematica Academiae Scientiarum Hungarica **8**(3–4), 477–493 (1957)
22. Setti, G., Mazzini, G., Rovatti, R., Callegari, S.: Statistical modeling of discrete-time chaotic processes: basic finite dimensional tools and applications. Proc. IEEE **90**(5), 662–690 (2002)
23. Sidorov, N.: Almost every number has a continuum of β-expansions. Am. Math. Mon. **110**(9), 838–842 (2003)

On the Effects of Nondeterminism
on Ordered Restarting Automata

Kent Kwee and Friedrich Otto[✉]

Fachbereich Elektrotechnik/Informatik, Universität Kassel, 34109 Kassel, Germany
{kwee,otto}@theory.informatik.uni-kassel.de

Abstract. While (stateless) deterministic ordered restarting automata accept exactly the regular languages, it is known that nondeterministic ordered restarting automata accept some languages that are not context-free. Here we show that, in fact, the class of languages accepted by these automata is an abstract family of languages that is incomparable to the linear languages, the context-free languages, and the growing context-sensitive languages with respect to inclusion, and that the emptiness problem is decidable for these languages. In addition, it is shown that stateless ordered restarting automata just accept regular languages, and we present an infinite family of regular languages C_n such that C_n is accepted by a stateless ordered restarting automaton with an alphabet of size $O(n)$, but each stateless deterministic ordered restarting automaton for C_n needs $2^{O(n)}$ letters.

Keywords: Restarting automaton · Ordered rewriting · Abstract family of languages · Descriptional complexity

1 Introduction

The *ordered restarting automaton* (ORWW-automaton[1] for short) was introduced in [9], where it was extended into a device for recognizing picture languages. An ORWW-automaton (for strings) has a finite-state control, a tape with end markers that initially contains the input, and a window of size three. Based on its state and the content of its window, the automaton can either perform a *move-right step*, a *rewrite/restart step*, or an *accept step*. While the deterministic variant of the ORWW-automaton characterizes the regular languages, it has been observed that the nondeterministic variant accepts some languages that are not context-free. However, the nondeterministic ORWW-automaton and the languages it accepts have not yet been studied in detail.

Here we present such a study. First we prove that the class of languages accepted by the ORWW-automaton forms an abstract family of languages, that is, it is closed under union, intersection (with regular sets), product, Kleene star, inverse morphisms, and non-erasing morphisms (see, e.g., [5]). However,

---

[1] Following the notation that is used for restarting automata in general (see, e.g., [10]), the suffix -WW says that this automaton can rewrite using non-input symbols.

© Springer-Verlag Berlin Heidelberg 2016
R.M. Freivalds et al. (Eds.): SOFSEM 2016, LNCS 9587, pp. 369–380, 2016.
DOI: 10.1007/978-3-662-49192-8_30

it is neither closed under complementation nor under reversal. Further, it is incomparable to the linear, the context-free, and the growing context-sensitive languages [3,5] with respect to inclusion, as it contains a language that is not even growing context-sensitive, while on the other hand, it does not even include all linear languages. In addition, we show that the emptiness problem is decidable for ORWW-automata. Several of these proofs are based on a Cut-and-Paste Lemma for ORWW-automata that is derived from Higman's Theorem [4].

In [11] an investigation of the descriptional complexity of the deterministic ORWW-automaton was initiated. Each deterministic ORWW-automaton can be simulated by an automaton of the same type that has only a single state, which means that for these automata, states are actually not needed. Accordingly, such an automaton is called a *stateless* det-ORWW-automaton. For these automata, the size of their working alphabets can be taken as a measure for their descriptional complexity. For $n \geq 1$, there exists a regular language that is accepted by a stateless det-ORWW-automaton of size $O(n)$ such that each nondeterministic finite-state acceptor (NFA) for this language has at least 2^n states. On the other hand, each stateless det-ORWW-automaton of size n can be simulated by an unambiguous NFA with $2^{O(n)}$ states [7]. Thus, the stateless det-ORWW-automaton is exponentially more succinct than NFAs.

Here we are also interested in the computational capacity of stateless ORWW-automata and in their descriptional complexity. We prove that stateless ORWW-automata only accept regular languages, which shows that states are actually useful for nondeterministic ORWW-automata. Finally, we present a family of example languages $(C_n)_{n \geq 1}$ such that the language C_n is accepted by a stateless ORWW-automaton with an alphabet of size $O(n)$, while each stateless deterministic ORWW-automaton for C_n needs an alphabet of size 2^n.

This paper is structured as follows. In Sect. 2, we introduce the ORWW-automaton, we present an example of a language that is not even growing context-sensitive, but that is accepted by an ORWW-automaton, and we derive the Cut-and-Paste Lemma. In Sect. 3 we study the closure properties of the class of languages that are accepted by ORWW-automata. Then, in Sect. 4, we show that stateless ORWW-automata only accept regular languages, and we present the announced result on the descriptional complexity of stateless ORWW-automata. The paper closes with Sect. 5, which summarizes our results in short and states a number of open problems for future work.

2   Ordered Restarting Automata

An *ORWW-automaton* is a one-tape machine that is described by an 8-tuple $M = (Q, \Sigma, \Gamma, \triangleright, \triangleleft, q_0, \delta, >)$, where Q is a finite set of states containing the initial state q_0, Σ is a finite input alphabet, Γ is a finite tape alphabet such that $\Sigma \subseteq \Gamma$, the symbols $\triangleright, \triangleleft \notin \Gamma$ serve as markers for the left and right border of the work space, respectively,

$$\delta : (Q \times ((\Gamma \cup \{\triangleright\}) \cdot \Gamma \cdot (\Gamma \cup \{\triangleleft\}) \cup \{\triangleright\triangleleft\})) \rightarrow 2^{(Q \times \{\mathsf{MVR}\}) \cup \Gamma \cup \{\mathsf{Accept}\}}$$

is the *transition relation*, and $>$ is a *partial ordering* on Γ. The transition relation describes three different types of transition steps:

(1) A *move-right step* has the form $(q', \mathsf{MVR}) \in \delta(q, a_1 a_2 a_3)$, where $q, q' \in Q$, $a_1 \in \Gamma \cup \{\triangleright\}$, and $a_2, a_3 \in \Gamma$. It causes M to shift the window one position to the right and to change from state q into state q'. Observe that no move-right step is possible, if the window contains the symbol $\triangleleft$.
(2) A *rewrite/restart step* has the form $b \in \delta(q, a_1 a_2 a_3)$, where $q \in Q$, $a_1 \in \Gamma \cup \{\triangleright\}$, $a_2, b \in \Gamma$, and $a_3 \in \Gamma \cup \{\triangleleft\}$ such that $a_2 > b$ holds. It causes M to replace the symbol a_2 in the middle of its window by the symbol b and to restart, that is, the window is moved back to the left end of the tape, and M reenters its initial state q_0.
(3) An *accept step* has the form $\mathsf{Accept} \in \delta(q, a_1 a_2 a_3)$, where $q \in Q$, $a_1 \in \Gamma \cup \{\triangleright\}$, $a_2 \in \Gamma$, and $a_3 \in \Gamma \cup \{\triangleleft\}$. It causes M to halt and accept. In addition, we allow an accept step of the form $\delta(q_0, \triangleright \triangleleft) = \{\mathsf{Accept}\}$.

If $\delta(q, u) = \emptyset$ for some state q and a word u, then M necessarily halts, when it is in state q seeing u in its window, and we say that M *rejects* in this situation. Further, the letters in $\Gamma \smallsetminus \Sigma$ are called *auxiliary symbols*.

If $|\delta(q, u)| \leq 1$ for all q and u, then M is a *deterministic ORWW-automaton* (det-ORWW-automaton), and if $Q = \{q_0\}$, that is, if the initial state is the only state of M, then we call M a *stateless ORWW-automaton* (stl-ORWW-automaton) or a *stateless deterministic ORWW-automaton* (stl-det-ORWW-automaton), as in this case the state is actually not needed. Accordingly, for stateless ORWW-automata, we will drop the components that refer to states to simplify the notation.

A *configuration* of an ORWW-automaton M is a word $\alpha q \beta$, where $q \in Q$ is the current state, $|\beta| \geq 3$, and either $\alpha = \lambda$ (the empty word) and $\beta \in \{\triangleright\} \cdot \Gamma^+ \cdot \{\triangleleft\}$ or $\alpha \in \{\triangleright\} \cdot \Gamma^*$ and $\beta \in \Gamma \cdot \Gamma^+ \cdot \{\triangleleft\}$; here $\alpha \beta$ is the current content of the tape, and it is understood that the window contains the first three symbols of β. In addition, we admit the configuration $q_0 \triangleright \triangleleft$. A *restarting configuration* has the form $q_0 \triangleright w \triangleleft$; if $w \in \Sigma^*$, then $q_0 \triangleright w \triangleleft$ is also called an *initial configuration*. Further, we use Accept to denote the *accepting configurations*, which are those configurations that M reaches by an accept step.

Any computation of an ORWW-automaton M consists of certain phases. A phase, called a *cycle*, starts in a restarting configuration, the head is moved along the tape by MVR steps until a rewrite/restart step is performed and thus, a new restarting configuration is reached. If no further rewrite operation is performed, any computation necessarily finishes in a halting configuration – such a phase is called a *tail*. By $\vdash_M^c$ we denote the execution of a complete cycle, and $\vdash_M^{c*}$ is the reflexive transitive closure of this relation. It can be seen as the *rewrite relation* that is realized by M on the set of restarting configurations.

An input $w \in \Sigma^*$ is accepted by M, if there is a computation of M which starts with the initial configuration $q_0 \triangleright w \triangleleft$ and ends with an accept step. The language consisting of all words that are accepted by M is denoted by $L(M)$.

As each cycle ends with a rewrite operation, which replaces a symbol a by a symbol b that is strictly smaller than a with respect to the given ordering $>$, each computation of M on an input of length n consists of at most $(|\Gamma| - 1) \cdot n$ cycles. Thus, M can be simulated by a nondeterministic single-tape Turing machine in time $O(n^2)$. While nondeterministic ORWW-automata are quite expressive as we will see below, the deterministic variants are fairly weak.

Theorem 1. [7,11]

(a) *For each det-ORWW-automaton* $M = (Q, \Sigma, \Gamma, \rhd, \lhd, q_0, \delta, >)$, *there exists a stateless det-ORWW-automaton* $M' = (\Sigma, \Gamma', \rhd, \lhd, \delta', >')$ *such that* $L(M') = L(M)$ *and* $|\Gamma'| = |Q| \cdot |\Gamma|^2 + 2 \cdot |\Gamma|$.

(b) *For each DFA* $A = (Q, \Sigma, q_0, F, \varphi)$, *there is a stl-det-ORWW-automaton* $M = (\Sigma, \Gamma, \rhd, \lhd, \delta, >)$ *such that* $L(M) = L(A)$ *and* $|\Gamma| = |Q| + |\Sigma|$.

(c) *For each stl-det-ORWW-automaton* M *with an alphabet of size* n, *there exists an NFA* A *of size* $2^{O(n)}$ *such that* $L(A) = L(M)$ *holds.*

(d) *For each* $n \geq 1$, *there exists a regular language* $B_n \subseteq \{0, 1, \#, \$\}^*$ *such that* B_n *is accepted by a stl-det-ORWW-automaton over an alphabet of size* $O(n)$, *but each NFA for accepting* B_n *has at least* 2^n *states.*

Let $\Sigma = \{a, b, \$\}$, and let

$$L'_{\text{copy}} = \{ w\$u \mid w, u \in \{a, b\}^*, |w|, |u| \geq 2, u \text{ is a scattered subword of } w \}.$$

Lemma 2. *The language* L'_{copy} *is not growing context-sensitive, but there exists an ORWW-automaton* M *such that* $L(M) = L'_{\text{copy}}$.

Proof. Let M be the ORWW-automaton on $\Sigma = \{a, b, \$\}$ and $\Gamma = \{a, a_1, a_2, b, b_1, b_2, \$\}$ that is given by the following meta-instructions[2] using the ordering $\$ > a > b > a_1 > b_1 > a_2 > b_2$, where $c, d, e \in \{a, b\}$:

(1) $(\lambda, \rhd cd \rightarrow \rhd c_1 d)$,
(2) $(\lambda, \rhd c_1 d \rightarrow \rhd c_2 d)$,
(3) $(\rhd \cdot \{a_2, b_2\}^*, c_2 de \rightarrow c_2 d_1 e)$,
(4) $(\rhd \cdot \{a_2, b_2\}^*, c_2 d_1 e \rightarrow c_2 d_2 e)$,
(5) $(\rhd \cdot \{a_2, b_2\}^*, c_2 d\$ \rightarrow c_2 d_1 \$)$,
(6) $(\rhd \cdot \{a_2, b_2\}^*, c_2 d_1 \$ \rightarrow c_2 d_2 \$)$,
(7) $(\rhd \cdot \{a_2, b_2\}^* \cdot c_1 \cdot \{a, b\}^+, \$cd \rightarrow \$c_1 d)$,
(8) $(\rhd \cdot \{a_2, b_2\}^* \cdot c_2 \cdot \{a, b\}^+, \$c_1 d \rightarrow \$c_2 d)$,
(9) $(\rhd \cdot \{a_2, b_2\}^* \cdot c_1 \cdot \{a, b\}^+ \cdot \$ \cdot \{a_2, b_2\}^*, d_2 ce \rightarrow d_2 c_1 e)$,
(10) $(\rhd \cdot \{a_2, b_2\}^* \cdot c_2 \cdot \{a, b\}^+ \cdot \$ \cdot \{a_2, b_2\}^*, d_2 c_1 e \rightarrow d_2 c_2 e)$,
(11) $(\rhd \cdot \{a_2, b_2\}^* \cdot c_1 \cdot \{a, b\}^+ \cdot \$ \cdot \{a_2, b_2\}^*, d_2 c\lhd \rightarrow d_2 c_1 \lhd)$,
(12) $(\rhd \cdot \{a_2, b_2\}^* \cdot c_2 \cdot \{a, b\}^+ \cdot \$ \cdot \{a_2, b_2\}^*, d_2 c_1 \lhd \rightarrow d_2 c_2 \lhd)$,
(13) $(\rhd \cdot \{a_2, b_2\}^+ \cdot c_1 \cdot \$ \cdot \{a_2, b_2\}^*, d_2 c\lhd \rightarrow d_2 c_1 \lhd)$,
(14) $(\rhd \cdot \{a_2, b_2\}^+ \cdot c_2 \cdot \$ \cdot \{a_2, b_2\}^*, d_2 c_1 \lhd \rightarrow d_2 c_2 \lhd)$,
(15) $(\rhd \cdot \{a_2, b_2\}^+ \cdot \$ \cdot \{a_2, b_2\}^+ \cdot \lhd, \text{Accept})$.

---

[2] A meta-instruction $(E, u \rightarrow v)$ is applicable to a restarting configuration $q_0 \rhd w\lhd$, if w can be factored as $w = w_1 u w_2$ such that $\rhd w_1 \in E$, which would give the cycle $q_0 \rhd w\lhd \vdash^c_M q_0 \rhd w_1 v w_2 \lhd$, and a meta-instruction (E, Accept) allows M to accept from any restarting configuration $q_0 \rhd w\lhd$ such that $\rhd w\lhd \in E$ (see, e.g., [10]).

Given an input of the form $w\$u$, where $w, u \in \{a, b\}^*$, it is easily seen from rules (15), (1), and (7) that $|w|, |u| \geq 2$. By rules (1) to (6), the prefix w is rewritten from left to right, where each symbol is first replaced by its copy with index 1, and then this is replaced by the corresponding letter with index 2. Also the suffix u is rewritten in this way by rules (7) to (14); however, here the first letter from $\{a, b\}$ from the left, say c, can only be rewritten to c_1, if at that moment the rightmost already rewritten letter in w happens to be the letter c_1, and analogously, c_1 can further be rewritten to c_2 only if at that moment the rightmost already rewritten letter in w happens to be the letter c_2. Thus, it follows that u is a scattered subword of w, that is, $L(M) = L'_{copy}$.

In [2] it is shown that each growing context-sensitive language is accepted by a one-way auxiliary pushdown automaton with a logarithmic space bound, that is, the class GCSL of growing context-sensitive languages is contained in $\mathcal{L}(\mathsf{OW\text{-}auxPDA}(\log))$. On the other hand, Lautemann has shown in [8] that the language $L_{copy} = \{\, ww \mid w \in \{a, b\}^* \,\}$ is not accepted by any OW-auxPDA with a logarithmic space bound, and his argument immediately extends to the language $L^\$_{copy} = \{\, w\$w \mid w \in \{a, b\}^* \,\}$.

Now assume that the language L'_{copy} is growing context-sensitive. Then there is a OW-auxPDA A that accepts this language with a logarithmic space bound. By using an extra track of the auxiliary tape to implement a binary counter, we can extend A into a OW-auxPDA B for the language $L^\$_{copy}$, which contradicts the statement above. Hence, L'_{copy} is not growing context-sensitive. $\square$

Thus, we see that the ORWW-automata are quite expressive in contrast to their deterministic variants. We conclude this section with a Cut-and-Paste Lemma that will be of importance later.

Lemma 3 (Cut-and-Paste Lemma).
For each ORWW-automaton M, there exists a constant $N(M) > 0$ such that each word $w \in L(M)$, $|w| \geq N(M)$, has a factorization $w = xyz$ satisfying all of the following conditions:

$$\text{(a)}\ |yz| \leq N(M), \text{ (b)}\ |y| > 0, and\,\text{(c)}\ xz \in L(M).$$

Proof. Let $M = (Q, \Sigma, \Gamma, \rhd, \lhd, q_0, \delta, >)$ be an ORWW-automaton, where $Q = \{q_0, q_1, \ldots, q_k\}$ and $\Gamma = \{s_1, s_2, \ldots, s_n\}$. Without loss of generality we may assume that M accepts at the left end of its tape, that is, it accepts in state q_0 with its window containing the left sentinel $\rhd$ and the first two symbols of the proper tape inscription. The transition relation δ of M can be represented by a finite set of five-tuples of the form (q, a, b, c, r), where $q \in Q$, $a \in \Gamma \cup \{\rhd\}$, $b \in \Gamma$, $c \in \Gamma \cup \{\lhd\}$, and $r \in Q \cup \Gamma \cup \{\mathsf{Accept}\}$. Here a five-tuple (q, a, b, c, q') with $q' \in Q$ represents the move-right operation $(q', \mathsf{MVR}) \in \delta(q, abc)$, and a five-tuple (q, a, b, c, d) with $d \in \Gamma$ represents the rewrite/restart operation $d \in \delta(q, abc)$, and analogously, for $(q, a, b, c, \mathsf{Accept})$. As $|Q| = k + 1$ and $|\Gamma| = n$, we see that this set consists of $K \leq (k+1) \cdot n \cdot (n+1)^2 \cdot (k+n+2)$ five-tuples. We introduce a new alphabet $\Omega = \{t_1, t_2, \ldots, t_K\}$ the symbols of which are in 1-to-1 correspondence to these five-tuples.

We now consider a shortest accepting computation C of M on an input $w_m \in \Sigma^m$, where m is sufficiently large. To each number $j = 1, 2, \ldots, m - 1$, we associate a word $x_j \in \Omega^*$ such that x_j describes the sequence of operations that are performed within the computation C at position $m + 1 - j$. Thus, we obtain a sequence of words $X_C = (x_1, x_2, \ldots, x_{m-1})$ over Ω.

Claim. $|x_1| \leq n - 1$, and for all $j = 2, \ldots, m - 1$, $|x_{j-1}| \leq |x_j| \leq j \cdot (n - 1)$.

Proof. We proceed by induction on j. For $j = 1$, x_j is the sequence of operations that are performed within C at the right-most position. Hence, x_1 only consists of rewrite/restart operations, and as $|\Gamma| = n$, it follows that $|x_1| \leq (n - 1)$.

Now assume that $|x_{j-1}| \leq (j - 1) \cdot (n - 1)$ has been established for some $j \geq 2$. We consider the sequence of operations that is described by the word x_j. This word describes all the rewrite/restart operations and all the move-right operations that are performed within C at position $m + 1 - j$. Each move-right operation executed at this position leads to an operation that is performed at its right neighbour, that is, at position $m + 2 - j$. Hence, the number of these move-right operations is exactly $|x_{j-1}|$, which implies that $|x_{j-1}| \leq |x_j| \leq |x_{j-1}| + (n - 1) \leq j \cdot (n - 1)$.

Finally, we extend each word x_j into $a_j x_j s_j$, where a_j is the input letter at position $m + 1 - j$ and s_j is the letter from Γ into which the input symbol a_j is being rewritten by the sequence of operations x_j. Now we consider the sequence $(a_1 x_1 s_1, a_2 x_2 s_2, \ldots, a_{m-1} x_{m-1} s_{m-1})$ over $\Omega \cup \Gamma$.

To determine the constant $N(M)$ we use Higman's theorem [4] and the corresponding *Length function H* from [6] (see also [13]). Let $H(2, n + 1, \Omega \cup \Gamma)$ be the maximal positive integer N such that there exists a sequence $\sigma_1, \sigma_2, \ldots, \sigma_N$ of words over $\Omega \cup \Gamma$ such that $|\sigma_j| \leq j \cdot (n + 1)$ for all $j \geq 1$ and σ_{j_1} is not a scattered subsequence of σ_{j_2} for any indices $1 \leq j_1 < j_2 \leq N$. It is shown in [6] that H is a total recursive function.

Now we choose $N(M) = H(2, n + 1, \Omega \cup \Gamma) + 2$. We see from Claim 1 that $|a_j x_j s_j| = |x_j| + 2 \leq j \cdot (n - 1) + 2 \leq j \cdot (n + 1)$ for all $j \geq 1$. If $m \geq N(M)$, then we see from the definition of the function H that there are indices $1 \leq j_1 < j_2 \leq m - 1$ such that $a_{j_1} x_{j_1} s_{j_1}$ is a scattered subsequence of $a_{j_2} x_{j_2} s_{j_2}$. Thus, $a_{j_1} = a_{j_2}$, $s_{j_1} = s_{j_2}$, and if $x_{j_1} = t_1 t_2 \ldots t_r$ for some $r \geq 1$ and $t_1, t_2, \ldots, t_r \in \Omega$, then x_{j_2} can be written as $x_{j_2} = y_0 t_1 y_1 t_2 y_2 \cdots y_{r-1} t_r y_r$ for some $y_0, y_1, y_2, \ldots, y_r \in \Omega^*$.

The subsequence of rewrite operations of x_{j_1} rewrites the input letter a_{j_1} into the letter s_{j_1}, and the subsequence of rewrite operations of x_{j_2} rewrites the input letter $a_{j_2} = a_{j_1}$ into the letter $s_{j_2} = s_{j_1}$. Hence, as the former is a scattered subsequence of the latter, it follows that actually the same rewrite operations occur in x_{j_2} and in x_{j_1}, and they occur in the same order. In particular, this means that the factors $y_0, y_1, y_2, \ldots, y_r$ only consist of move-right operations.

Finally we take x to be the prefix of w_m up to position $m + 1 - j_2$, y to be the factor of w_m from positions $m + 2 - j_2$ to $m + 1 - j_1$, and z to be the remaining suffix of w_m. Then $w_m = xyz$, $|yz| \leq N(M)$ and $|y| = j_2 - j_1 > 0$. Finally, let $w' = xz$. We will show that M has an accepting computation C' for input w'. This computation is obtained from the computation C as follows.

Let $(C_1, C_2, \ldots, C_\mu)$ be the sequence of cycles of the computation C. For $j = 1, 2, \ldots, \mu$, let C_j be the cycle currently considered.

1. If the rewrite step in C_j is performed on the prefix of length $m + 1 - j_2$ of the current tape, then we append C_j to C'. This includes in particular all those cycles that include a rewrite operation at position $m + 1 - j_2$, that is, the rewrite operations encoded within the word x_{j_2}.
2. If C_j includes a move-right step at position $m + 1 - j_2$ that contributes a letter to one of the factors $y_0, y_1, \ldots, y_r$ of x_{j_2}, then C_j is not appended to C'.
3. Finally, if C_j includes a move-right operation at position $m + 1 - j_2$ that corresponds to a letter t_l of the word x_{j_2} for some $1 \leq l \leq r$, then we combine the initial part of this cycle, up to the point where the operation t_l is executed at position $m + 1 - j_2$, with the final part of the cycle which starts with this very operation at position $m + 1 - j_1$. The resulting cycle C_j' is appended to C'. As $x_{j_1} = t_1 t_2 \ldots t_r$ is a subsequence of x_{j_2}, C_j' is indeed a valid cycle of M.

The computation C' is completed by appending the accepting tail of C to it. Then it is easily checked that C' is indeed an accepting computation of M on input w'. This completes the proof of the Cut-and-Paste Lemma. □

3 Closure Properties

Here we first show that $\mathcal{L}(\text{ORWW})$ is an abstract family of languages.

Theorem 4. $\mathcal{L}(\text{ORWW})$ *is closed under union, intersection, product, Kleene star, inverse morphisms, and non-erasing morphisms.*

Proof. In [11] it is shown that $\mathcal{L}(\text{stl-det-ORWW})$ is closed under union and intersection. The same proof idea can be used here.

Closure under product: Let $M_1 = (Q_1, \Sigma, \Gamma_1, \triangleright, \triangleleft, q^{(1)}, \delta_1, >_1)$ and $M_2 = (Q_2, \Sigma, \Gamma_2, \triangleright, \triangleleft, q^{(2)}, \delta_2, >_2)$ be two ORWW-automata. Without loss of generality we may assume that both these automata accept at the right end of their tapes. We present an ORWW-automaton M for the language $L(M_1) \cdot L(M_2)$. It proceeds as follows:

1. Given a word $w \in \Sigma^*$ as input, M rewrites w from right to left, letter by letter, such that each letter of a suffix v of w is marked by an index 2, and then each letter of the corresponding prefix u is marked by an index 1. In this way $w \in \Sigma^*$ is (nondeterministically) split into $w = uv$ with the idea that $u \in L(M_1)$ and $v \in L(M_2)$ are to be checked.
2. Then M simulates M_1 on the prefix u. During this process, the leftmost occurrence of a letter with index 2 is interpreted as the right delimiter $\triangleleft$.
3. When the simulated computation of M_1 on u accepts, then M realizes this with either the right delimiter $\triangleleft$ or with the leftmost letter with index 2 in its window. In the former case, it accepts iff $\lambda \in L(M_2)$, while in the latter case it rewrites all the letters from the prefix u, from right to left, by the special symbol □.

4. When the first letter of w has been rewritten by the letter $\square$, then M simulates M_2. During this process it simply ignores the prefix of $\square$-symbols on the tape, simulating M_2 on the suffix v. Now M accepts iff this computation of M_2 accepts.

5. If in step 1, all letters are marked with an index 2, that is, $v = w$ and $u = \lambda$ are chosen, then M simply simulates M_2 on v, provided $\lambda \in L(M_1)$; otherwise, it simply halts without acceptance.

Closure under Kleene star: Here the idea is essentially the same as for the operation of product. Given a word $w \in \Sigma^*$ as input, M rewrites the word from right to left, letter by letter, attaching indices 1 or 2 to these letters. In this way a factorization $w = u_1 u_2 \ldots u_m$ is chosen nondeterministically, and it remains to check that $u_1, u_2, \ldots, u_m \in L(M_1)$ hold. This can be done as above, using two copies of the automaton M_1.

Closure under inverse morphisms: Let M be an ORWW-automaton on Σ and let $f : \Sigma'^* \to \Sigma^*$ be a morphism. We present an ORWW-automaton M' such that $L(M') = f^{-1}(L(M))$.

Basically M' proceeds as follows. Given an input $w = a_1 a_2 \ldots a_n \in \Sigma'^*$, it rewrites each letter $a_i \in \Sigma'$, from right to left, by its image $f(a_i) \in \Sigma^*$, and then it simulates the computation of M on input $f(w)$. However, there are two problems that we need to overcome. First, the length of a word $|f(a_i)|$ may be larger than one. Accordingly, the tape alphabet of M' will contain block symbols of the form $[u]$ that represent a word from $u \in \Sigma^*$ of length up to $\mu = \max\{ |f(a)| \mid a \in \Sigma' \}$. Secondly, it may happen that $f(a) = \lambda$ for some letters $a \in \Sigma^*$. In this situation, a will be rewritten into a symbol $[u]^c$ that represents a copy of its right-hand neighbour $[u]$. Of course, there can be several of these copy symbols in a row. In the course of the computation they will always be updated from right to left. As these copy symbols may separate the block symbols on the tape from each other, M' must carry information on the last block symbol it has seen when moving to the right.

Closure under non-erasing morphisms: Let M be an ORWW-automaton on Σ, and let $f : \Sigma^* \to \Omega^*$ be a non-erasing morphism. We present an ORWW-automaton M' such that $L(M') = f(L(M))$, where M' proceeds as follows.

Given a word $w \in \Omega^*$ as input, M' first guesses a factorization $u_1 u_2 \ldots u_m$ of w such that, for all $i = 1, 2, \ldots, m$, $|u_i| \leq \mu = \max\{ |f(a)| \mid a \in \Sigma \}$. This is done by marking the letters of w, one by one, from right to left, by indices 1 and 2 (see the proof for the closure under Kleene star above). Each factor u_i is a candidate for an image of a letter under the morphism f. Then, processing the factors u_i from right to left, M' checks whether $u_i = f(a)$ for some letter $a \in \Sigma$. In the negative, it halts immediately without accepting, while in the affirmative it nondeterministically chooses a letter $a_i \in \Sigma$ satisfying $f(a_i) = u_i$ and rewrites u_i into the word $[a_i]^c \ldots [a_i]^c [a_i]$, that is, the last letter of u_i is rewritten into a block symbol that encodes the letter $a_i \in \Sigma$, and all the other letters of u_i (if any) are rewritten into corresponding copy symbols. Thereafter, M' simulates a computation of M on the input $a_1 a_2 \ldots a_m$ using the technique from the above proof of closure under inverse morphisms. It follows that M' accepts on input $w \in \Omega^*$ iff there exists a word $u \in \Sigma^*$ such that $f(u) = w$ and $u \in L(M)$.   □

In order to establish some non-closure properties we consider the example language $L_{\leq} = \{ a^m b^n \mid 1 \leq m \leq n \}$. Clearly, $L_{\leq}$ is a linear language [5].

Theorem 5. $L_{\leq} \notin \mathcal{L}(\mathsf{ORWW})$.

Proof. Assume to the contrary that there exists an ORWW-automaton $M = (Q, \Sigma, \Gamma, \rhd, \lhd, q_0, \delta, >)$ such that $L(M) = L_{\leq}$, and let $N(M)$ be the corresponding constant from the Cut-and-Paste Lemma (Lemma 3). We consider the word $w = a^{N(M)} b^{N(M)} \in L_{\leq}$. Then $w = xyz$ such that $|yz| \leq N(M)$, $|y| > 0$, and $w' = xz \in L_{\leq}$. From the two factorizations of w we see that $y = b^i$ for some $i > 0$, which implies that $w' = a^{N(M)} b^{N(M)-i} \notin L_{\leq}$, a contradiction. Thus, it follows that the language $L_{\leq}$ is not accepted by any ORWW-automaton.   □

From Lemma 2 and Theorem 5 we obtain the following result.

Corollary 6. *The language class* $\mathcal{L}(\mathsf{ORWW})$ *is incomparable to the language classes* LIN, CFL, *and* GCSL *with respect to inclusion.*

Finally, Theorem 5 allows us to derive the following non-closure properties.

Theorem 7. *The language class* $\mathcal{L}(\mathsf{ORWW})$ *is neither closed under the operation of reversal nor under complementation.*

Proof. Using the technique from the proof of Lemma 2 it can be shown that $L_{\geq} = \{ b^m a^n \mid m \geq n \geq 1 \}$ is accepted by some ORWW-automaton. As $L_{\leq} = L_{\geq}^R$, Theorem 5 implies that $\mathcal{L}(\mathsf{ORWW})$ is not closed under the operation of reversal.

Obviously, also the language $L'_{\geq} = \{ a^m b^n \mid m \geq n \geq 1 \}$ is accepted by some ORWW-automaton. Assume that its complement $(L'_{\geq})^c$ is accepted by some ORWW-automaton. Then also the language $(L'_{\geq})^c \cap (a^+ \cdot b^+)$ is acceped by some ORWW-automaton, since all regular languages are accepted by these automata and $\mathcal{L}(\mathsf{ORWW})$ is closed under intersection. However, $(L'_{\geq})^c \cap (a^+ \cdot b^+) = \{ a^m b^n \mid 1 \leq m < n \}$, and in analogy to the proof of Theorem 5 it can be shown that this language is not accepted by any ORWW-automaton, either. Thus, it follows that $\mathcal{L}(\mathsf{ORWW})$ is not closed under complementation.   □

Although $\mathcal{L}(\mathsf{ORWW})$ is an abstract family of languages that is incomparable to LIN, CFL, and GCSL, we have the following decidability result.

Theorem 8. *The emptiness problem for ORWW-automata is decidable.*

Proof. Let $M = (Q, \Sigma, \Gamma, \rhd, \lhd, q_0, \delta, >)$ be an ORWW-automaton, and let $N(M)$ be the corresponding constant from the Cut-and-Paste Lemma (Lemma 3). As shown in the proof of that lemma, $N(M) = H(2, n + 1, \Omega \cup \Gamma) + 2$, where H is the Length function corresponding to Higman's theorem, which is a recursive function [6]. It now follows that $L(M) \neq \emptyset$ iff $L(M)$ contains a word of length at most $N(M)$.   □

Notice that the Length function H grows very fast, which means that the algorithm for the emptiness problem described above is by no means practical.

4   Stateless ORWW-Automata

In [7] it is shown that each stateless det-ORWW-automaton can be turned into an unambiguous NFA that accepts the same language. This construction can easily be carried over to stateless ORWW-automata.

Theorem 9. Let $M = (\Sigma, \Gamma, \rhd, \lhd, \delta_M, >)$ be a stl-ORWW-automaton. Then an NFA $A = (Q, \Sigma, \Delta_A, q_0, F)$ can be constructed from M such that $L(A) = L(M)$ and $|Q| \in 2^{O(|\Gamma|)}$.

For all $n \geq 3$, the language $U_n = \{a^{2^n}\}$ can be shown to be accepted by a stl-det-ORWW-automaton with an alphabet of $3n - 1$ letters, while each NFA for U_n needs at least $2^n + 1$ states. Hence, the bound given in Theorem 9 is sharp up to the O-notation. In addition, we have the following consequence.

Corollary 10. $\mathcal{L}(\mathsf{stl\text{-}ORWW}) = \mathsf{REG}$.

Finally, we show that stateless ORWW-automata can describe some (regular) languages much more succinctly than stateless det-ORWW-automata.

Let $\Sigma = \{a, b, \#\}$, and let, for $n \geq 1$, C_n denote the following language over Σ:

$$C_n = \{ u_1 \# u_2 \# \ldots \# u_m \mid m \geq 2, u_1, u_2, \ldots, u_m \in \{a, b\}^n, \exists i < j : u_i = u_j \}.$$

Lemma 11. The language C_n is accepted by a stl-ORWW-automaton M_n with a tape alphabet of size $70n - 21$.

Proof. In [12] a stl-det-ORWW-automaton A_n with $68n - 22$ letters is given that accepts the language

$$B_n = \{ v_1 \# v_2 \# \ldots \# v_m \$ u \mid m \geq 1, v_1, v_2, \ldots, v_m, u \in \{a, b\}^n, \exists i : v_i = u \}.$$

The stl-ORWW-automaton M_n is obtained from A_n by adding the $2n + 1$ letters $a_1', a_2', \ldots, a_n', b_1', b_2', \ldots, b_n'$, and $\#'$. It proceeds as follows.

It first marks the letters from right to left. In a syllable $v = v^{(1)} v^{(2)} \ldots v^{(n)}$, the last letter is replaced by $v_n^{(n)'}$, $v^{(n-1)}$ is replaced by $v_{n-1}^{(n-1)'}$, and so forth until finally $v^{(1)}$ is replaced by $v_1^{(1)'}$, and then the preceding letter $\#$ is replaced by $\#'$. In this way it is ensured that all $\{a, b\}$-syllables have length n. This continues until M_n nondeterministically chooses to replace a symbol $\#$ by the symbol $\$$. In this way the input of the form $u_1 \# u_2 \# \ldots \# u_k \# u_{k+1} \# \ldots \# u_m$ is transformed into a word of the form $u_1 \# u_2 \# \ldots \# u_k \$ \hat{u}_{k+1} \#' \ldots \#' \hat{u}_m$, where $\hat{u}_i$ $(k + 1 \leq i \leq m)$ denotes the word that is obtained from $u_i \in \{a, b\}^n$ by the above marking process. Now on the prefix $u_1 \# u_2 \# \ldots \# u_k \$ \hat{u}_{k+1}$, M_n simulates the stl-det-ORWW-automaton A_n, and it accepts if the computation of A_n being simulated accepts. It is obvious now that $L(M_n) = C_n$. □

Now we claim that every stl-det-ORWW-automaton for C_n needs at least $2^{O(n)}$ letters, that is, the above presentation by a stl-ORWW-automaton is exponentially more succinct than any presentation by a stl-det-ORWW-automaton.

Proposition 12. *Let $s : \mathbb{N} \to \mathbb{N}$ be a function such that, for each $n \geq 1$, there exists a stl-det-ORWW-automaton $D_n = (\Sigma, \Gamma_n, \triangleright, \triangleleft, \delta_n, >)$ such that $L(D_n) = C_n$ and $|\Gamma_n| \leq s(n)$. Then $s(n) \notin o(2^n)$.*

Proof. Let $n \geq 1$, and let $D_n = (\Sigma, \Gamma_n, \triangleright, \triangleleft, \delta_n, >)$ be a stl-det-ORWW-automaton such that $L(D_n) = C_n$ and $|\Gamma_n| \leq s(n)$. As D_n is deterministic, we obtain a stl-det-ORWW-automaton $E_n = (\Sigma, \Gamma_n, \triangleright, \triangleleft, \eta_n, >)$ such that $L(E_n) = C_n^c = \Sigma^* \setminus D_n$ simply by interchanging accept steps with undefined steps (see [11] Theorem 10). From E_n we can construct an NFA F_n of size $2^{r \cdot s(n)}$ for C_n^c [7], where $r \in \mathbb{N}_+$ is a constant.

We now present a large fooling set for C_n^c. Let A be a subset of $\{a, b\}^n$ of size 2^{n-1}. Then also the set $\bar{A} = \{a, b\}^n \setminus A$ has size 2^{n-1}. With these sets we associate the following languages:

$$P_A = \{ u_1 \# u_2 \# \ldots \# u_{2^{n-1}} \mid A = \{u_1, u_2, \ldots, u_{2^{n-1}}\} \} \text{ and}$$
$$Q_A = \{ \# v_1 \# v_2 \# \ldots \# v_{2^{n-1}} \mid \bar{A} = \{v_1, v_2, \ldots, v_{2^{n-1}}\} \}.$$

Then $uv \in C_n^c$ for all $u \in P_A$ and all $v \in Q_A$. On the other hand, if B is a subset of $\{a, b\}^n$ of size 2^{n-1} such that $A \neq B$, then $uv \in C_n$ for all $u \in P_A$ and all $v \in Q_B$, because there is a word $x \in \{a, b\}^n$ such that $x \in A$ and $x \in \bar{B}$. Hence, by choosing a pair $(u_A, v_A) \in P_A \times Q_A$ for all subsets A of $\{a, b\}^n$ of size 2^{n-1}, we obtain a fooling set for F_n of size $\binom{2^n}{2^{n-1}} = \frac{(2^n)!}{(2^{n-1})! \cdot (2^{n-1})!} > 2^{2^{n-1}}$. This implies by [1] that the number $2^{r \cdot s(n)}$ of states of the NFA F_n satisfies the inequality $2^{r \cdot s(n)} \geq 2^{2^{n-1}}$, that is, $r \cdot s(n) \geq 2^{n-1}$. Hence, $\frac{s(n)}{2^n} \geq \frac{1}{2r}$, which clearly shows that $\liminf_{n \to \infty} \frac{s(n)}{2^n} \geq \frac{1}{2r} > 0$, that is, $s(n) \notin o(2^n)$.   □

From the proof above we can also derive the following complexity result.

Corollary 13. *Let $s : \mathbb{N} \to \mathbb{N}$ be a function such that, for each $n \geq 1$, there exists a stl-ORWW-automaton $E_n = (\Sigma, \Gamma_n, \triangleright, \triangleleft, \delta_n, >)$ such that $L(E_n) = C_n^c$ and $|\Gamma_n| \leq s(n)$. Then $s(n) \notin o(2^n)$.*

Proof. By Theorem 9 we can construct an NFA F_n of size $2^{r \cdot s(n)}$ for C_n^c from E_n. Now the proof of Proposition 12 shows that $s(n) \notin o(2^n)$.   □

As by Lemma 11 the language C_n is accepted by a stl-ORWW-automaton M_n with a tape alphabet of size $70n - 21$, Corollary 13 shows that the conversion of a stl-ORWW-automaton of size n into a stl-ORWW-automaton for the complement of $L(M)$ can actually increase the alphabet size exponentially.

5   Concluding Remarks

We have seen that the class of languages accepted by nondeterministic ORWW-automata forms an abstract class of languages that is incomparable to the linear, the context-free, and the growing context-sensitive languages with respect to inclusion. However, we don't know yet whether this class is closed under arbitrary morphisms. In addition, we have shown that the emptiness problem is decidable for these languages, but it remains to find an algorithm for this problem that is more efficient. Also it is still open whether finiteness, inclusion, and equivalence are decidable. Further, we have seen that stateless ORWW-automata provide another characterization for the regular languages, which provides exponentially more succinct representations than stateless deterministic ORWW-automata. However, we are still missing an algorithm for turning a stl-ORWW-automaton into an equivalent stl-det-ORWW-automaton without constructing an equivalent NFA as an intermediate step.

References

1. Birget, J.-C.: Intersection and union of regular languages and state complexity. Inform. Proc. Lett. **43**, 185–190 (1992)
2. Buntrock, G., Otto, F.: Growing context-sensitive languages and Church-Rosser languages. Inform. Comp. **141**, 1–36 (1998)
3. Dahlhaus, E., Warmuth, M.: Membership for growing context-sensitive grammars is polynomial. J. Comput. Syst. Sci. **33**, 456–472 (1986)
4. Higman, G.: Ordering by divisibility in abstract algebras. Proc. Lond. Math. Soc. **2**, 326–336 (1952)
5. Hopcroft, J.E., Ullman, J.D.: Introduction to Automata Theory, Languages, and Computation. Addison-Wesley, Reading (1979)
6. Karandikar, P., Schnoebelen, Ph.: Generalized Post embedding problems. Theory Comput. Syst. **56**, 697–716 (2015)
7. Kwee, K., Otto, F.: On some decision problems for stateless deterministic ordered restarting automata. In: Shallit, J., Okhotin, A. (eds.) DCFS 2015. LNCS, vol. 9118, pp. 165–176. Springer, Heidelberg (2015)
8. Lautemann, C.: One pushdown and a small tape. In: Wagner, K.W. (ed.) Dirk Siefkes zum 50. Geburtstag, pp. 42–47. Technische Universität Berlin and Universität Augsburg (1988)
9. Mráz, F., Otto, F.: Ordered restarting automata for picture languages. In: Geffert, V., Preneel, B., Rovan, B., Štuller, J., Tjoa, A.M. (eds.) SOFSEM 2014. LNCS, vol. 8327, pp. 431–442. Springer, Heidelberg (2014)
10. Otto, F.: Restarting automata. In: Ésik, Z., Martín-Vide, C., Mitrana, V. (eds.) Recent Advances in Formal Languages and Applications. SCI, vol. 25, pp. 269–303. Springer, Heidelberg (2006)
11. Otto, F.: On the descriptional complexity of deterministic ordered restarting automata. In: Jürgensen, H., Karhumäki, J., Okhotin, A. (eds.) DCFS 2014. LNCS, vol. 8614, pp. 318–329. Springer, Heidelberg (2014)
12. Otto, F., Wendlandt, M., Kwee, K.: Reversible ordered restarting automata. In: Krevine, J., Stefani, J.-B. (eds.) RC 2015. LNCS, vol. 9138, pp. 60–75. Springer, Heidelberg (2015)
13. Schmitz, S., Schnoebelen, Ph.: Multiply-recursive upper bounds with Higman's lemma. In: Aceto, L., Henzinger, M., Sgall, J. (eds.) ICALP 2011, Part II. LNCS, vol. 6756, pp. 441–452. Springer, Heidelberg (2011)

Quantum Walks on Two-Dimensional Grids with Multiple Marked Locations

Nikolajs Nahimovs[✉] and Alexander Rivosh

Faculty of Computing, University of Latvia, Raina bulv. 19, Riga 1586, Latvia
nikolajs.nahimovs@lu.lv

Abstract. The running time of a quantum walk search algorithm depends on both the structure of the search space (graph) and the configuration of marked locations. While the first dependence has been studied in a number of papers, the second dependence remains mostly unstudied. We study search by quantum walks on the two-dimensional grid using the algorithm of Ambainis, Kempe and Rivosh [AKR05]. The original paper analyses one and two marked locations only. We move beyond two marked locations and study the behaviour of the algorithm for an arbitrary configuration of marked locations.

In this paper, we prove two results showing the importance of how the marked locations are arranged. First, we present two placements of k marked locations for which the number of steps of the algorithm differs by a factor of $\Omega(\sqrt{k})$. Second, we present two configurations of k and $\sqrt{k}$ marked locations having the same number of steps and probability to find a marked location.

1 Introduction

Quantum walks are quantum counterparts of classical random walks [11]. They have been useful to design quantum algorithms for a variety of problems [2,3,5, 10]. In many of those applications, quantum walks are used as a tool for search.

To solve a search problem using quantum walks, we introduce the notion of marked locations. Marked locations correspond to elements of the search space that we want to find. We then perform a quantum walk on the search space with one transition rule at unmarked locations and another transition rule at marked locations. If this process is set up properly, it leads to a quantum state in which marked locations have higher probability than the unmarked ones. This state can then be measured, finding a marked location with a sufficiently high probability. This method of search using quantum walks was first introduced in [12] and has been used many times since then.

The running time of a quantum walk search algorithm depends on both the structure of the search space and the configuration — the number and the placement — of marked locations. There have been a number of papers studying the

NN is supported by EU FP7 project QALGO, AR is supported by ERC project MQC.

© Springer-Verlag Berlin Heidelberg 2016
R.M. Freivalds et al. (Eds.): SOFSEM 2016, LNCS 9587, pp. 381–391, 2016.
DOI: 10.1007/978-3-662-49192-8_31

dependence of the running time on the structure of the graph. Krovi [7] has studied symmetries of a graph and explained fast hitting times using the concept of symmetry. Janmark et al. [8] show that global symmetry of the graph is not necessary for fast quantum search. They demonstrate graphs with automorphism group consisting of an identity mapping only that still achieve the $\Theta(\sqrt{N})$ quantum speed-up. Meyer and Wong [9] have studied connectivity of the graph and have shown that it is also a poor indicator of fast quantum search: there exists graphs with low connectivity but fast search, and graphs with high connectivity but slow search. So, despite of significant progress in the field the overall picture is still far from being complete.

On the other hand, the dependence on the number and the placement of marked locations remains mostly unstudied. Up to our best knowledge, the only such paper is [14], which studies the continuous-time quantum walk on the "simplex of complete graphs" and shows that rearranging the marked elements can cause the parameters of the walk and running time to vary significantly. Most of papers on quantum walk algorithms [3,4] prove their results for one or two marked locations only.

We study search by quantum walks on a finite two-dimensional grid using the algorithm of Ambainis, Kempe and Rivosh (AKR). The original [3] paper analyses the behaviour of the algorithm for one or two marked locations. We move beyond two marked locations and study the behaviour of the algorithm for an arbitrary configuration of marked locations. We show that the placement of marked locations has at least the same effect on the number of steps of the algorithm as the number of marked locations.

First, we present two placements of k marked locations for which the number of steps of the algorithm differs by $\Omega(\sqrt{k})$ factor. Here the first configuration is a block of $\sqrt{k} \times \sqrt{k}$ marked locations and the second configuration is k uniformly distributed marked locations (placed at $\sqrt{N/k}$ distance from each other). We prove that the number of steps of the algorithm for the grouped placement is $\widetilde{\Omega}(\sqrt{N} - \sqrt{k})$, while for the distributed placement is $\widetilde{O}(\sqrt{N/k})$.

Second, we present two configurations of k and $\sqrt{k}$ marked locations, respectively, having the same number of steps and probability to find a marked location. Here, the first configuration is a block of $\sqrt{k} \times \sqrt{k}$ marked locations and the second configuration is the perimeter of a $\sqrt{k} \times \sqrt{k}$ block (all internal locations are not marked).

The dependence of the number of steps on the placement of marked locations makes quantum walks different from Grover's search algorithm, where the number of steps have exact dependence on the number of marked locations. In the case of quantum walks on non-complete graphs, even if the number of marked locations is known, the number of steps can vary depending on a placement of marked locations. On the other hand, for all configurations studied in this paper, if the number of marked locations is in $[1, k]$ then the number of steps of the algorithm is still in $[\widetilde{O}(\sqrt{N/k}), \widetilde{O}(\sqrt{N})]$ — the same as it is for Grover's algorithm.

2 Quantum Walks in Two Dimensions

Suppose we have N items arranged on a two dimensional grid of size $\sqrt{N} \times \sqrt{N}$. We denote $n = \sqrt{N}$. The locations on the grid are labelled by their x and y coordinate as (x, y) for $x, y \in \{0, \ldots, n-1\}$. We assume that the grid has periodic boundary conditions. For example, going right from a location $(n-1, y)$ on the right edge of the grid leads to the location $(0, y)$ on the left edge of the grid.

To introduce a quantum version of a random walk, we define a location register with basis states $|i, j\rangle$ for $i, j \in \{0, \ldots, n-1\}$. Additionally, to allow non-trivial walks, we define a direction or coin register with four basis states, one for each direction: $|\Uparrow\rangle$, $|\Downarrow\rangle$, $|\Leftarrow\rangle$ and $|\Rightarrow\rangle$. Thus, the basis states of quantum walk are $|i, j, d\rangle$ for $i, j \in \{0, \ldots, n-1\}$ and $d \in \{\Uparrow, \Downarrow, \Leftarrow, \Rightarrow\}$. The state of the quantum walk is given by:

$$|\psi(t)\rangle = \sum_{i,j} (\alpha_{i,j,\Uparrow}|i, j, \Uparrow\rangle + \alpha_{i,j,\Downarrow}|i, j, \Downarrow\rangle + \alpha_{i,j,\Leftarrow}|i, j, \Leftarrow\rangle + \alpha_{i,j,\Rightarrow}|i, j, \Rightarrow\rangle).$$

A step of the quantum walk is performed by first applying $I \otimes C$, where C is unitary transform on the coin register. The most often used transformation on the coin register is the Grover's diffusion transformation D:

$$D = \frac{1}{2} \begin{pmatrix} -1 & 1 & 1 & 1 \\ 1 & -1 & 1 & 1 \\ 1 & 1 & -1 & 1 \\ 1 & 1 & 1 & -1 \end{pmatrix}.$$

Then, we apply the shift transformation S:

$$\begin{aligned} |i, j, \Uparrow\rangle &\rightarrow |i, j-1, \Downarrow\rangle \\ |i, j, \Downarrow\rangle &\rightarrow |i, j+1, \Uparrow\rangle \\ |i, j, \Leftarrow\rangle &\rightarrow |i-1, j, \Rightarrow\rangle \\ |i, j, \Rightarrow\rangle &\rightarrow |i+1, j, \Leftarrow\rangle \end{aligned}$$

Notice that after moving to an adjacent location we change the value of the direction register to the opposite. This is necessary for the quantum walk algorithm of [3] to work.

We start quantum walk in the state

$$|\psi_0\rangle = \frac{1}{\sqrt{4N}} \sum_{i,j} (|i, j, \Uparrow\rangle + |i, j, \Downarrow\rangle + |i, j, \Leftarrow\rangle + |i, j, \Rightarrow\rangle).$$

It can be easily verified that the state of the walk stays unchanged, regardless of the number of steps. To use the quantum walk as a tool for search, we "mark" some locations. For unmarked locations, we apply the same transformations as above. For marked locations, we apply $-I$ instead of D as the coin flip transformation. The shift transformation remains the same in both marked and unmarked locations.

Another way to look at a step of the algorithm is that we first perform a query Q transformation, which flips signs of amplitudes of marked locations, then conditionally perform the coin transformation (I or D depending on whether the location is marked or not) and then perform the shift transformation S.

If there are marked locations, the state of the algorithm starts to deviate from $|\psi(0)\rangle$. It has been shown [3] that after $O(\sqrt{N \log N})$ steps the inner product $\langle\psi(t)|\psi(0)\rangle$ becomes close to 0.

In case of one or two marked locations AKR algorithm finds a marked location with $O(1/\log N)$ probability. The probability is small, thus, the algorithm uses amplitude amplification to get $\Theta(1)$ probability. The amplitude amplification adds an additional $O(\sqrt{\log N})$ factor to the number of steps. Thus, the total running time of the algorithm is $O(\sqrt{N} \log N)$.

3   Results

3.1   Grouped and Distributed Placements of Marked Locations

In this subsection we show that the number of steps of the algorithm for two placements of k marked locations can differ by $\Omega(\sqrt{k})$ factor.

Consider two configurations (placements) of k marked locations. The first configuration is a block of $\sqrt{k} \times \sqrt{k}$ marked locations. The second configuration is k uniformly distributed marked locations (placed at $\sqrt{N/k}$ distance from each other) (see Fig. 1). We will refer them as grouped and distributed placements respectively.

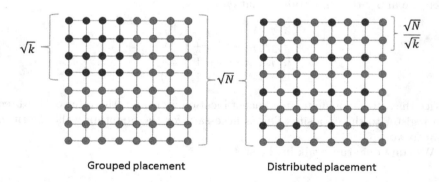

Grouped placement   Distributed placement

Fig. 1. Grouped and distributed placements of k marked locations

Lemma 1 (Grouped placement). *Let k be a full square and let k marked locations be placed as a $\sqrt{k} \times \sqrt{k}$ square on $\sqrt{N} \times \sqrt{N}$ grid. Then AKR algorithm needs $\Omega(\sqrt{N} - \sqrt{k})$ steps.*

Proof. This follows from the fact that the average distance from a location on the grid to a marked location is $\Omega(\sqrt{N} - \sqrt{k})$. Thus, the algorithm needs at least this number of steps to achieve a constant probability of finding a marked location. $\square$

Lemma 2 (Distributed placement). *Let k be a full square and let k marked locations be uniformly distributed on $\sqrt{N} \times \sqrt{N}$ grid (placed at $\sqrt{N/k}$ distance from each other). Then AKR algorithm needs $O(\sqrt{N/k} \cdot \log(N/k))$ steps and finds a marked location with $O(1/\log(N/k))$ probability.*

Proof. By symmetry each of $\sqrt{N/k} \times \sqrt{N/k}$ regions of the grid is experiencing the same evolution (here by region we mean a part of the grid with a marked location in its top-left corner).

More formally, consider basis states corresponding to locations with $\sqrt{N/k}$ distance from each other pointing to the same direction. Initially amplitudes of all such pairs of basis states are equal. For each pair of basis states the step of the algorithm applies the same transformations to the same amplitudes. Thus, after a step of the algorithm amplitudes of a pair of basis states are also equal. Therefore, the evolution of each of the $\sqrt{N/k} \times \sqrt{N/k}$ regions of the grid is essentially the same.

We have k copies of quantum walk on $\sqrt{N/k} \times \sqrt{N/k}$ grid with a single marked location. Therefore, after $O(\sqrt{N/k} \cdot \log(N/k))$ steps — the number of steps for the $\sqrt{N/k} \times \sqrt{N/k}$ grid with a single marked location — overlap of the current and the initial states of the algorithm becomes close to 0. If we measure the state at this point the probability to get one of basis states corresponding to a marked location is $O(1/\log(N/k))$. $\square$

We have shown that the number of steps for the grouped and the distributed placements differ by an $\Omega(\sqrt{k})$ factor. The grouped and the distributed placements are two extreme cases, therefore, we believe that $O(\sqrt{k})$ is the maximal possible gap for any two placements of k marked locations. We conjecture

Conjecture 1. *Let P_1 and P_2 be two placements of k marked locations on the $\sqrt{N} \times \sqrt{N}$ grid. Then the number of steps of AKR algorithm for P_1 and P_2 can differ by at most a $O(\sqrt{k})$ factor.*

3.2 Evolution of Amplitudes of Near-By Marked Locations

In the previous subsection we showed that AKR algorithm is inefficient for grouped marked locations. The reason for this is the coin transformation, which does not rearrange amplitudes within a marked location. Therefore, marked locations inside the group have almost no effect on the number of steps and the probability to find a marked location of the algorithm.

In this subsection, we explore grouped marked locations in more details. We analyse the evolution of amplitudes of two near-by marked locations. We show, that a step of AKR algorithm does not change absolute values of adjoint amplitudes of near-by marked locations.

Theorem 1. *Let $|\psi(t)\rangle$ be a state of AKR algorithm after t steps and let locations (i, j) and $(i, j + 1)$ be marked. Then for any t we have*

$$\langle\psi(t)|i, j, \Rightarrow\rangle = \langle\psi(t)|i, j + 1, \Leftarrow\rangle = (-1)^t/\sqrt{4N}.$$

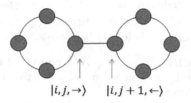

$$|i, j, \rightarrow\rangle \qquad |i, j + 1, \leftarrow\rangle$$

Fig. 2. Amplitudes of near-by marked locations.

Proof. Consider the effect of the step of the algorithm on amplitudes of $|i, j, \Rightarrow\rangle$ and $|i, j + 1, \Leftarrow\rangle$ (see Fig. 2). The query changes signs of both amplitudes (both locations are marked); the coin flip does nothing (both locations are marked); and the shift swaps the amplitudes. More formally,

$$\begin{aligned}
Q|i, j, \Rightarrow\rangle &= -|i, j, \Rightarrow\rangle & Q|i, j + 1, \Leftarrow\rangle &= -|i, j + 1, \Leftarrow\rangle \\
C|i, j, \Rightarrow\rangle &= |i, j, \Rightarrow\rangle & C|i, j + 1, \Leftarrow\rangle &= |i, j + 1, \Leftarrow\rangle \\
S|i, j, \Rightarrow\rangle &= |i, j + 1, \Leftarrow\rangle & S|i, j + 1, \Leftarrow\rangle &= |i, j, \Rightarrow\rangle.
\end{aligned}$$

Therefore, the step of the algorithm changes signs of the amplitudes and swaps their values.

Initially all amplitudes are equal to $1/\sqrt{4N}$. Thus, the values of the amplitudes will be $1/\sqrt{4N}$ after an even number steps and $-1/\sqrt{4N}$ after an odd number steps. □

3.3 Filled and Perimeter Configurations of Marked Locations

In this subsection we present two configurations of k and $\sqrt{k}$ marked locations, respectively, having the same number of steps and probability to find a marked location.

Consider two configurations of marked locations: k marked locations placed as a $\sqrt{k} \times \sqrt{k}$ square and $4(\sqrt{k} - 1)$ marked locations placed as the perimeter of a $\sqrt{k} \times \sqrt{k}$ square (Fig. 3). We will refer them as filled and perimeter configuration respectively.

Let $|\psi(t)\rangle$ be the state of AKR algorithm after t steps for the filled configuration and $|\phi(t)\rangle$ be the state of AKR algorithm after t steps for the perimeter configuration. For further analysis, we split $|\psi(t)\rangle$ into three parts (Fig. 4):

– $|\psi_{out}(t)\rangle$ consisting of basis states of the outer part of the square as well as basis states of the perimeter pointing to the outer part

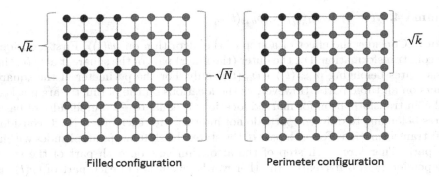

Filled configuration   Perimeter configuration

Fig. 3. Filled and perimeter configurations of marked locations

- $|\psi_{in}(t)\rangle$ consisting of basis states of the inner part of the square as well as basis states of the square pointing to the inner part
- $|\psi_{per}(t)\rangle$ consisting of basis states of the perimeter pointing to other locations on the perimeter.

Similarly we define $|\phi_{out}(t)\rangle$, $|\phi_{in}(t)\rangle$ and $|\phi_{per}(t)\rangle$.

Lemma 3. $\forall t \geq 0 : |\psi_{per}(t)\rangle = |\phi_{per}(t)\rangle$.

Proof. According to Theorem 1, all amplitudes of basis states of $|\psi_{per}(t)\rangle$ and $|\phi_{per}(t)\rangle$ are equal to $(-1)^t/\sqrt{4N}$.   □

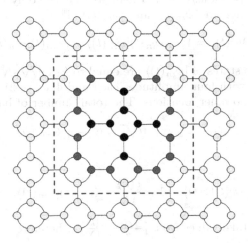

Fig. 4. Group of 3×3 marked locations (in the green dashed box). Basis states of $|\psi_{out}(t)\rangle$ are coloured with light gray, basis states of $|\psi_{per}(t)\rangle$ are coloured with dark gray, basis states of $|\psi_{in}(t)\rangle$ are coloured with black (Color figure online).

Lemma 4. $\forall t \geq 0 : |\psi_{out}(t)\rangle = |\phi_{out}(t)\rangle.$

Proof. Consider the effect of a step of the algorithm on $|\psi(t)\rangle$. First, consider the coin transformation. For the outer (the inner) part of the square it acts on the basis states belonging $|\psi_{out}(t)\rangle$ ($|\psi_{in}(t)\rangle$) only. For the perimeter of the square it acts on all three parts. However, as the locations on the perimeter are marked and coin transformation for marked locations is equal to $-I$, amplitudes of basis states belonging to different parts do not interact with each other. Next, consider shift transformation. For each part of the state, the shift swaps amplitudes within the part. Therefore, each step of the algorithm acts on each part of the state independently of other parts. In other words, evolution of each part of $|\psi(t)\rangle$ is independent on evolutions of other parts. The above argument holds for $|\phi(t)\rangle$ without any changes.

Initially $|\psi_{out}(0)\rangle = |\phi_{out}(0)\rangle$. There are no marked locations in the outer part of the square. Thus, the transformation applied to $|\psi_{out}(t)\rangle$ and $|\phi_{out}(t)\rangle$ are the same. Therefore, $|\psi_{out}(t)\rangle = |\phi_{out}(t)\rangle$ will hold for all t.   □

The next theorem estimates the overlap between the state of the algorithm after t steps for the filled and the perimeter configurations.

Theorem 2. $\forall t \geq 0 : \langle\psi(t)|\phi(t)\rangle \geq 1 - \Theta(k/N).$

Proof.

$$\langle\psi(t)|\phi(t)\rangle = \langle\psi_{out}(t)|\phi_{out}(t)\rangle + \langle\psi_{in}(t)|\phi_{in}(t)\rangle + \langle\psi_{per}(t)|\phi_{per}(t)\rangle.$$

It follows from the previously proved lemmas that the only parts of $|\psi(t)\rangle$ and $|\phi(t)\rangle$ which may differ are $|\psi_{in}(t)\rangle$ and $|\phi_{in}(t)\rangle$. Thus,

$$\langle\psi(t)|\phi(t)\rangle = 1 - \langle\psi_{in}(t)|\psi_{in}(t)\rangle + \langle\psi_{in}(t)|\phi_{in}(t)\rangle.$$

Amplitudes of basis states of $|\psi_{in}(t)\rangle$ are equal to $(-1)^t/\sqrt{4N}$. There are $(\sqrt{k} - 2)^2$ inner locations with four amplitudes each and $4(\sqrt{k} - 2)$ amplitudes of the perimeter pointing to inner locations. The total number of basis states in $|\psi_{in}\rangle$ is

$$c(k) = 4(\sqrt{k} - 2)^2 + 4(\sqrt{k} - 2) = 4(k - 3\sqrt{k} + 2)$$

and, thus, we have

$$\langle\psi(t)|\phi(t)\rangle = 1 - \frac{c(k)}{4N} + \langle\psi_{in}(t)|\phi_{in}(t)\rangle.$$

$\langle\psi_{in}(t)|\phi_{in}(t)\rangle$ can take values from $[-\frac{c(k)}{4N}, \frac{c(k)}{4N}]$. Therefore,

$$\langle\psi(t)|\phi(t)\rangle \geq 1 - 2 \cdot \frac{c(k)}{4N} = 1 - \Theta\left(\frac{k}{N}\right).$$

□

Now we give a corollary of the above theorem which bounds the maximal difference in the number of steps of the algorithm for the configurations. Note that we are interested in the case $k = o(N)$. Otherwise, if k is of the same order as N, then the trivial "measure on the first step" approach finds a marked location with constant probability.

Corollary 1. *Let $k = o(N)$. Then the number of steps of AKR algorithm for the filled and the perimeter placements can differ by at most one.*

Proof. It follows from [13] that the number of steps of AKR algorithm can not increase if we mark a previously unmarked location. Therefore, the total number of steps for k marked locations is at most $O(\sqrt{N \log N})$ — the number of steps of the algorithm for a single marked location. The angle between the state for the filled and the perimeter configurations is less than the angle to which the state is rotated by the step of the algorithm. Thus, the number of steps of the algorithm for the configurations can differ by at most one.   □

The next theorem estimates the maximal difference in the probability to find a marked location after t steps for the filled and the perimeter configurations.

Theorem 3. $\forall t \geq 0$: *the probability of finding a marked location for $|\psi(t)\rangle$ and $|\phi(t)\rangle$ differs by at most $\Theta(k/N)$.*

Proof. It follows from the previously proved lemmas that the only parts of $|\psi(t)\rangle$ and $|\phi(t)\rangle$ which may differ are $|\psi_{in}(t)\rangle$ and $|\phi_{in}(t)\rangle$. For the filled configuration all amplitudes of $|\psi_{in}(t)\rangle$ are equal to $(-1)^t/\sqrt{4N}$. For the perimeter configuration inner part is not marked. Additionally, amplitudes of the perimeter pointing to the inner part might become zero. Thus, the maximal possible difference in probability to measure a marked location is $\frac{1}{4N} \cdot c(k) = \Theta\left(\frac{k}{N}\right)$.   □

A typical probability of finding a marked location for AKR algorithm is $\Omega(1/\log N)$. Thus, the probability of finding a marked location for the configurations differs by an insignificant factor.

We have shown that for the filled and the perimeter configurations of marked locations AKR algorithm has the same number of steps and probability to find a marked location. However, the filled configuration has a quadratically larger number of marked locations than the perimeter configuration.

4   Conclusions and Discussion

In this paper we analysed AKR quantum walk search algorithm for the two-dimensional grid with multiple marked locations. First, we showed that the placement of k marked locations can change the number of steps of the algorithm by an $\Omega(\sqrt{k})$ factor. Namely, we showed that the number of steps of the algorithm for the grouped placement (k marked locations are placed as $\sqrt{k} \times \sqrt{k}$ group) is $\widetilde{\Omega}(\sqrt{N} - \sqrt{k})$, while for the distributed placement (marked locations are placed at $\sqrt{N/k}$ distance from each other) it is $\widetilde{O}(\sqrt{N/k})$.

The proved result shows that the number of steps for k marked locations can be in range $[\widetilde{O}(\sqrt{N/k}), \widetilde{\Omega}(\sqrt{N})]$. We conjecture that this is the maximal possible gap and the number of steps of the AKR algorithm for two placements of k marked locations can differ by at most $O(\sqrt{k})$.

It would be interesting to extend the analysis to three and more-dimensional grids. While our argument for the distributed placement still holds for higher dimensions, the argument for the grouped placement is bound to the two-dimensional case.

Second, we presented two configurations of k and $\sqrt{k}$ marked locations, respectively, having the same number of steps and probability to find a marked location. Here, the first configuration is a block of $\sqrt{k} \times \sqrt{k}$ marked locations and the second configuration is the perimeter of a $\sqrt{k} \times \sqrt{k}$ block (all internal locations are not marked). We showed that marked locations inside the block have almost no effect on the number of steps of the algorithm or the probability to find a marked location. More formally, we showed that internal locations of the block do not contribute to the growth of probability to find a marked location as well as do not affect the number of steps of the algorithm. Thus, the proved result holds not just for square blocks, but for any block of marked locations. Our analysis includes a number of supporting theorems and observations that might be of independent interest.

References

1. Ambainis, A., Bačkurs, A., Nahimovs, N., Ozols, R., Rivosh, A.: Search by quantum walks on two-dimensional grid without amplitude amplification. In: Kawano, Y. (ed.) TQC 2012. LNCS, vol. 7582, pp. 87–97. Springer, Heidelberg (2012)
2. Ambainis, A.: Quantum walk algorithm for element distinctness. SIAM J. Comput. **37**, 210–239 (2007)
3. Ambainis, A., Kempe, J., Rivosh, A.: Coins make quantum walks faster. In: Proceedings of SODA 2005, pp. 1099–1108 (2005)
4. Ambainis, A., Portugal, R., Nahimov, N.: Spatial search on grids with minimum memory (2014). arXiv:1312.0172
5. Buhrman, H., Spalek, R.: Quantum verification of matrix products. In: Proceedings SODA 2006, pp. 880–889 (2006)
6. Krovi, H., Magniez, F., Ozols, M., Roland, J.: Finding is as easy as detecting for quantum walks. In: Abramsky, S., Gavoille, C., Kirchner, C., Meyer auf der Heide, F., Spirakis, P.G. (eds.) ICALP 2010. LNCS, vol. 6198, pp. 540–551. Springer, Heidelberg (2010)
7. Krov, H.: Symmetry in quantum walks (Ph.D thesis) (2007). arXiv:0711.1694
8. Janmark, J., Meyer, D.A., Wong, T.G.: Global symmetry is unnecessary for fast quantum search. Phys. Rev. Lett. **112**, 210502 (2014). arXiv:1403.2228
9. Meyer, D.A., Wong, T.G.: Connectivity is a poor indicator of fast quantum search. Phys. Rev. Lett. **114**, 110503 (2015). arXiv:1409.5876
10. Magniez, F., Santha, M., Szegedy, M.: An $O(n^{1.3})$ quantum algorithm for the triangle problem. In: Proceedings of SODA 2005, pp. 413–424 (2005)
11. Portugal, R.: Quantum Walks and Search Algorithms. Springer, New York (2013)
12. Shenvi, N., Kempe, J., Whaley, K.B.: A quantum random walk search algorithm. Phys. Rev. A **67**(5), 052307 (2003)

13. Szegedy, M: Quantum speed-up of markov chain based algorithms. In: Proceedings of FOCS 2004, pp. 32–41 (2004)
14. Wong, T.G.: On the Breakdown of Quantum Search with Spatially Distributed Marked Vertices vol. 1501, p. 07071 (2015)

How to Smooth Entropy?

Maciej Skorski[✉]

Cryptology and Data Security Group, University of Warsaw, Warsaw, Poland
maciej.skorski@mimuw.edu.pl

Abstract. Smooth entropy of X is defined as possibly biggest entropy of a distribution Y close to X. It has found many applications including privacy amplification, information reconciliation, quantum information theory and even constructing random number generators. However the basic question about the *optimal shape* for the distribution Y has not been answered yet. In this paper we solve this problem for Renyi entropies in *non-quantum settings*, giving a formal treatment to an approach suggested at TCC'05 and ASIACRYPT'05. The main difference is that we use a *threshold cut* instead of a *quantile cut* to rearrange probability masses of X. As an example of application, we derive tight lower bounds on the number of bits extractable from Shannon memoryless sources.

Keywords: Smooth Renyi entropy · Randomness extractors · Asymptotic equipartition property

1 Introduction

1.1 Entropy Smoothing

Security Based on Statistical Closeness. In most of cryptographic applications, probability distributions which are close enough in the variational (statistical) distance are considered indistinguishable. More informally, they have similar cryptographic "quality", when used as randomness sources (randomness extracting) [15] or secure keys (in the context of key derivation [1,6]).

Entropy Notions do not See Statistical Closeness. Unfortunately, standard entropy notions (including important min-entropy and collision entropy which are widely used as randomness measures in cryptography), are not robust with respect to small probability perturbations. Consider the AES cipher with a 256-bit key which is $\epsilon = 2^{-80}$-close to uniform. While such a key is considered secure nowadays, it may happen that it has no more than 81 bits of min-entropy (more precisely, fix $x \in \{0,1\}^{256}$ and consider the key X which is x_0 with probability $2^{-256} + 2^{-80}$ and uniform for $x \in \{0,1\}^{256} \setminus \{x_0\}$). This is a mismatch with respect to our intuitive understanding of min-entropy as a measure of how many almost random bits can be extracted.

M. Skorski—This work was partly supported by the WELCOME/2010-4/2 grant founded within the framework of the EU Innovative Economy Operational Programme.

R.M. Freivalds et al. (Eds.): SOFSEM 2016, LNCS 9587, pp. 392–403, 2016.
DOI: 10.1007/978-3-662-49192-8_32

Smooth Entropy takes Probability Perturbations into Account. To fix the issue described above, the concept of *smooth min-entropy* has been proposed [4,15]. Smooth entropy is defined as the maximal possible entropy within a certain distance to a given distribution. More precisely, for a given entropy notion $H(\cdot)$ (which is usually Renyi entropy, see Sect. 2 for its formal definition) we define the ϵ-smooth entropy of X as the value of the following optimization program

$$\begin{aligned} \text{maximize} \quad & H(Y) \\ \text{s.t.} \quad & \mathrm{SD}(X;Y) \leqslant \epsilon \end{aligned} \tag{1}$$

where $\mathrm{SD}()$ stands for statistical (variational) distance (see Sect. 2) for a formal definition). This definition is now well-suited for cryptographic applications, because does not depend anymore on negligible variations of the probability distribution. In particular, setting $\epsilon = 2^{-80}$ for our AES example we obtain the "correct" result of 256 bits of (smooth) entropy.

Importance of Smooth Entropy. Smooth Renyi entropy, formally introduced by Renner and Wolf in [15], found many applications including privacy amplification [13,15,19], information reconciliation [15] and quantum information theory [17,18]. The technique of perturbing a distribution to get more-entropy was actually known before. For example, entropy smoothing is implicitly used to prove the Asymptotic Equipartition Property [10] or more concretely in the construction of a pseudorandom generator from one-way functions [7–9]. However the simple question

Question 1. How does the shape of optimal Y depend on X?

has not been fully understood so far. In this paper we answer Question 1 by explicitly characterizing the shape of Y depending on X, and give some applications of the derived characterization.

1.2 Related Works and Our Contribution

Related Works. The problem of finding the optimal shape for Y has been addressed in [13,15]. The authors argued intuitively that for min-entropy (which is a special case of Renyi entropy, particularly useful in randomness extraction) the optimal solution cuts down the biggest probabilities of X.

Our Contribution. We show that this characterization is not true, and the problem is more subtle: the optimal solution uses a threshold not a quantile cut (see Fig. 1).

The precise answer to Question 1 is given in Theorem 1. We provide an intuitive explanation, as a three-step algorithm, in Fig. 2.

Theorem 1 (Optimal Renyi entropy smoothing). *Let $\alpha > 1$ be fixed, X be an arbitrary distribution over a finite set and $\epsilon \in (0,1)$. Let $t \in (0,1)$ be such that*

$$\sum_x \max\left(\mathbf{P}_X(x) - t, 0\right) = \epsilon, \tag{2}$$

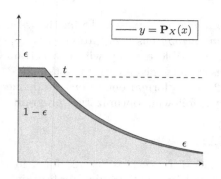
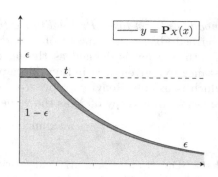

(a) **Quantile Cut** - a folklore solution (TCC'05,ASIACRYPT'05), far from optimal.

(b) **Threshold Cut** - our idea, nearly optimal.

Fig. 1. Our result - the optimal shape for entropy smoothing

and Y be distributed according to

$$\mathbf{P}_Y(x) = \frac{\min\left(t, \mathbf{P}_X(x)\right)}{1 - \epsilon}. \tag{3}$$

Then Y is nearly optimal, that is we have

$$\mathrm{SD}(X; Y) \leqslant \epsilon \tag{4}$$

and

$$\mathbf{H}_\alpha(Y) \leqslant \mathbf{H}_\alpha^\epsilon(X) \leqslant \mathbf{H}_\alpha(Y) + \frac{\alpha}{\alpha - 1} \log\left(\frac{1}{1 - \epsilon}\right) \tag{5}$$

Corollary 1 (Tightness of Theorem 1). *Note that for fixed $\alpha > 1$ we have $\frac{\alpha}{\alpha-1} \log\left(\frac{1}{1-\epsilon}\right) = O(\epsilon)$. Thus, our solution differs from the ideal one by only a negligible additive constant in the entropy amount, which is good enough for almost all applications.*

1.3 Tight No-Go Results for Extracting from Stateless Shannon Sources

A *stateless source* (called also *memoryless*) is a source which produces consecutive samples independently. While this is a restriction, it is often assumed by practitioners working on random number generators (cf. [2,3,5,11]) and argued to be reasonable under some circumstances (so called *restart mode* which enforces fresh samples, see [3,5]). An important result is obtained from a more general fact called Asymptotic Equipartition Property (AEP). Namely, for a stateless source the min-entropy rate (min-entropy per sample) is close to its Shannon entropy per bit. The convergence holds *in probability*, for large number of samples.

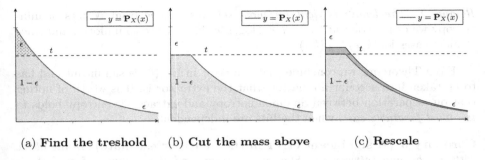

(a) **Find the treshold**   (b) **Cut the mass above**   (c) **Rescale**

Fig. 2. Our result - details of optimal entropy smoothing

A variant of the AEP: The min entropy per bit in a sequence $X_1, \ldots, X_n$ of i.id. samples from X *converges*, when $n \to \infty$, to the Shannon entropy of X. More precisely

$$\frac{-\log \mathbf{P}_{X_1,\ldots,X_n}(\cdot)}{n} \xrightarrow{\text{in probability}} H(X), \tag{6}$$

where the probability is taken over $X_1, \ldots, X_n$.

Thus, the AEP is a bridge connecting the heuristic use of Shannon entropy as a measure of extractable randomness (practice) and the provable security (randomness extractors theory). The best known quantitative form of Eq. (6) appears in [9].

Lemma 1 (Asymptotic Equipartition Property [9]**).** *Let $X_1, \ldots, X_n$ be i.i.d. samples from a distribution X of Shannon entropy k. Then the sequence $(X_1, \ldots, X_n)$ is ϵ-close to a distribution of min entropy $kn - O\left(\sqrt{kn \log(1/\epsilon)}\right)$.*

Corollary 2. *In particular, one can extract $kn - O\left(\sqrt{kn \log(1/\epsilon)}\right) - 2\log(1/\epsilon)$ bits which are ϵ-close to uniform (e.g. using independent hash functions [8] as an extractor).*

Based on Theorem 1 we reprove the following result which matches the bound in [9]. Our result can be understood as the lower bound on the convergence speed in the Asymptotic Equipartition Property given in Lemma 1.

Theorem 2 (An upper bound on the extraction rate [16]**).** *Let $X_1, X_2, \ldots$ be of i.i.d. random variables, each of Shannon entropy k. Then from the sequence $(X_1, X_2, \ldots, X_n)$ no extractor can get more than*

$$N = kn - \Theta(\sqrt{kn \log(1/\epsilon)}) \tag{7}$$

bits which are ϵ-close (in the variation distance) to uniform (the constant under $\Theta(\cdot)$ depends on the source).

Remark 1 (The bound is tight for most settings). Since from N bits of min-entropy we can extract at least $N - 2\log(1/\epsilon)$ bits ϵ-close to uniform, and since in most cases $\log(1/\epsilon) = o(kn)$

From Theorem 2 we conclude that the error in Eq. (6) is significant and has to be taken into account no matter what the extractor is. It is worth of noting that our separation between Shannon entropy and extractable entropy holds *in the most favorable case*, when the bits are independent.

Corollary 3 (A significant error in the heuristic estimate). *In the above setting, the gap between the Shannon entropy and the number of extractable bits ϵ-close to uniform equals at least $\Theta(kn - \sqrt{\log(1/\epsilon)})$. In particular, for the recommended security level ($\epsilon = 2^{-80}$) we obtain the loss of $kn - N \approx \sqrt{80kn}$ bits, no matter what an extractor we use.*

1.4 Organization

Notions we use, as well as some auxiliary technical facts, are explained in Sect. 2. We prove our main result, that is Theorem 1, in Sect. 3. The proof of Theorem 2 appears in Sect. 4.

2 Preliminaries

2.1 Basic Definitions

The most popular way of measuring how two distributions are close is the statistical distance.

Definition 1 (Statistical Distance). *The statistical (or total variation) distance of two distributions X, Y over the same finite set is defined as*

$$\mathrm{SD}\,(X;Y) = \sum_x |\Pr[X = x] - \Pr[Y = x]| \tag{8}$$

We also say that X and Y are ϵ-close.

Below we recall the definition of Renyi entropy of order α. The logarithms are taken at base 2.

Definition 2 (Renyi Entropy). *The Renyi entropy of order α of a distribution X equals $\mathbf{H}_\alpha(X) = \frac{1}{1-\alpha}\log\left(\sum_x \Pr[X = x]^\alpha\right)$.*

Choosing $\alpha \to 1$ and $\alpha \to \infty$ we recover two important notions: Shannon entropy and min entropy.

Definition 3 (Shannon Entropy). *The Shannon Entropy of a distribution X equals $H(X) = -\sum_x \Pr[X = x]\log\Pr[X = x]$.*

Definition 4 (Min Entropy). *The min entropy of a distribution X equals* $H_\infty(X) = -\max_x \log \Pr[X = x]$.

Smooth Renyi Entropy is defined as the value of the program (1).

Definition 5 (Smooth Renyi Entropy, [4]). *The ϵ-smooth Renyi entropy of order α of a distribution X equals $H_\alpha^\epsilon(X) = \max_Y H_\alpha(Y)$ where the maximum is taken over Y satisfying the constraint $\mathrm{SD}(X;Y) \leqslant \epsilon$.*

Definition 6 (Extractable Entropy, [14]). *We say that X has k extractable bits within distance ϵ, denoted $H_{\mathrm{ext}}^\epsilon(X) \geqslant k$, if for some randomized function Ext we have $\mathrm{SD}\left(\mathrm{Ext}(X,S); U_k, S\right) \leqslant \epsilon$, where U_k is a uniform k-bit string and S is an independent uniform string.*

2.2   Technical Facts

We will need the following simple fact on convex functions

Proposition 1. *Let f be a strictly convex differentiable real-valued function and $x < y$. Then for any $\delta > 0$ we have*

$$f(x) - f(x - \delta) \leqslant f(y) - f(y - \delta).$$

Our proof uses the following characterization of "extractable" distributions.

Theorem 3 (An Upper Bound on Extractable Entropy, [14]). *If $H_{\mathrm{ext}}^\epsilon(X) \geqslant k$ then X is ϵ-close to Y such that $H_\infty(Y) \geqslant k$.*

Another important fact we use is the sharp bound on binomial tails.

Theorem 4 (Tight Binomial Tails [12]). *Let $B(n,p)$ be a sum of independent Bernoulli trials with success probability p. Then for $\gamma \leqslant \frac{3}{4}q$ we have*

$$\Pr\left[B(n,p) \geqslant pn + \gamma n\right] = Q\left(\sqrt{\frac{n\gamma^2}{pq}}\right) \cdot \psi\left(p, q, n, \gamma\right) \tag{9}$$

with the error term satisfies

$$\psi\left(p, q, n, \gamma\right) =$$
$$\exp\left(\frac{n\gamma^2}{2pq} - n\mathrm{KL}\left(p + \gamma \,\|\, p\right) + \frac{1}{2}\log\left(\frac{p+\gamma}{p} \cdot \frac{q}{q-\gamma}\right) + O_{p,q}\left(n^{-\frac{1}{2}}\right)\right) \tag{10}$$

where $\mathrm{KL}\left(a \,\|\, b\right) = a\log(a/b) + (1-a)\log((1-a)/(1-b))$ is the Kullback-Leibler divergence, and Q is the complement of the cumulative distribution function of the standard normal distribution.

3   Proof of Theorem 1

We start by rewriting Eq. (1) in the following way

$$\text{minimize} \quad \left(\sum_x (\mu(x) + \epsilon(x))^\alpha\right)^{\frac{1}{\alpha-1}}$$

$$\text{s.t.} \quad \begin{cases} \sum_x \epsilon(x) = 0 \\ \sum_x |\epsilon(x)| = 2\epsilon \\ \forall x \quad 0 \leqslant \mu(x) + \epsilon(x) \leqslant 1 \end{cases} \tag{11}$$

where for the sake of clarity we replace $\mathbf{P}_X$ by μ_X.

Claim. Let $\epsilon(x)$ be optimal for Eq. (11). Define $S^+ = \{x : \epsilon(x) < 0\}$. Then $\mu(x) + \epsilon(x) = \mu(y) + \epsilon(y)$ for all $x, y \in S^+$. We will show the optimality by a mass-shifting argument.

Proof (of Claim). Suppose that

$$\epsilon(x_1), \epsilon(x_2) < 0 \text{ and } \mu(x_1) + \epsilon(x_1) > \mu(x_2) + \epsilon(x_2) > 0 \tag{12}$$

for two different points x_1, x_2 (the statement is trivially true when there is only one point). Take a number δ such that

$$0 < \delta < \min\left(-\epsilon(x_2), \frac{(\mu(x_1) + \epsilon(x_1)) - (\mu(x_2) + \epsilon(x_2))}{2}\right) \tag{13}$$

and modify μ by shifting the mass from x_2 to x_1 in the following way

$$\epsilon'(x) = \begin{cases} \epsilon(x), & x \notin \{x_1, x_2\} \\ \epsilon(x) - \delta, & x = x_1 \\ \epsilon(x) + \delta, & x = x_2 \end{cases}$$

that is shifting the mass from the biggest point to the smallest point. Note that from Eqs. (11) and (13) it follows that the constraints in (11) are satisfied with $\epsilon(x)$ replaced by $\epsilon'(x)$. Let Y be a random variable distributed according to $\mathbf{P}_Y(x) = \mu(x) + \epsilon(x)$ and let Y' be distributed as $\mathbf{P}_{Y'}(x) = \mu(x) + \epsilon'(x)$. Note that we have

$$\sum_x (\mu(x) + \epsilon'(x))^\alpha - \sum_x (\mu(x) + \epsilon(x))^\alpha =$$

$$((\mathbf{P}_Y(x_2) + \delta)^\alpha - (\mathbf{P}_Y(x_2))^\alpha) - ((\mathbf{P}_Y(x_1))^\alpha - (\mathbf{P}_Y(x_1) - \delta)^\alpha)$$

Note that we have

$$\mathbf{P}_Y(x_2) < \mathbf{P}_Y(x_2) + \delta < \mathbf{P}_Y(x_1) - \delta < \mathbf{P}_Y(x_1)$$

by Eqs. (12) and (13). Now from Proposition 1 applied to $f(u) = u^\alpha$, $x = \mathbf{P}_Y(x_2) + \delta$, $y = \mathbf{P}_Y(x_2)$ and δ (here we also use the assumption $\alpha > 1$), it follows that

$$\sum_x (\mu(x) + \epsilon'(x))^\alpha - \sum_x (\mu(x) + \epsilon(x))^\alpha < 0$$

which means $\mathbf{H}_\alpha(Y') > \mathbf{H}_\alpha(Y)$. In other words, Y is not optimal. $\qquad\square$

By the last claim it is clear that there is a number $t \in (0,1)$ such that the set $S^+ = \{x : \mathbf{P}_{Y^*}(x) < \mu(x)\}$ is contained in $\{x : \mu(x) \geqslant t\}$ and that $\mathbf{P}_{Y^*}(x) \geqslant t$ for $x \in S^+$. Therefore

$$\sum_x (\mathbf{P}_{Y^*}(x))^\alpha \geqslant \# \{x : \mu(x) \geqslant t\} \cdot t^\alpha + \sum_{x:\ \mu(x)<t} (\mu(x))^\alpha$$

$$= \sum_x \min(\mu(x), t)^\alpha$$

$$\geqslant (1 - \epsilon)^\alpha \sum_x (\mathbf{P}_Y(x))^\alpha \qquad (14)$$

which, since $\mathbf{H}_\alpha^\epsilon(X) = \mathbf{H}_\alpha(Y^*)$, proves the second inequality in Eq. (5). To prove the first inequality in Eq. (5) note that

$$\mathrm{SD}(X;Y) = \sum_{x:\ \mathbf{P}_X(x)>\mathbf{P}_Y(x)} (\mathbf{P}_X(x) - \mathbf{P}_Y(x)) = \sum_{x:\ \mu(x)>t/(1-\epsilon)} \left(\mu(x) - \frac{t}{1-\epsilon}\right).$$

Since $\frac{t}{1-\epsilon} > t$ we have

$$\mathrm{SD}(X;Y) = \sum_{x:\ \mu(x)>t/(1-\epsilon)} \left(\mu(x) - \frac{t}{1-\epsilon}\right) \leqslant \sum_{x:\ \mu(x)>t} (\mu(x) - t)$$

and therefore by Eq. (2) we obtain

$$\mathrm{SD}(X;Y) < \sum_{x:\ \mu(x)>t} (\mu(x) - t) = \epsilon.$$

which finishes the proof.

4 Proof of Theorem 2

4.1 Characterizing Extractable Entropy

We state the following fact with an explanation in Fig. 3.

Lemma 2 (Lower bound on the extractable entropy). *Let X be a distribution. Then for every distribution Y which is ϵ-close to X, we have $H_\infty(Y) \leqslant -\log t$ where t satisfies*

$$\sum_x \max(\mathbf{P}_X(x) - t, 0) = \epsilon. \qquad (15)$$

The proof follows by Theorem 1.

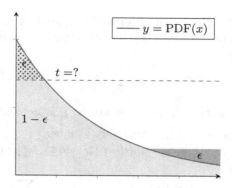

Fig. 3. Entropy Smoothing Problem. For a given probability density function, we want to cut a total mass of up to ϵ above a possibly highest threshold (in dotted red) and rearrange it (in green), to keep the upper bound smallest possible (Color figure online)

Without losing generality, we assume from now that $X \in \{0,1\}$ where $\Pr[X = 1] = p, q = 1 - p$. Define $X^n = (X_1, \ldots, X_n)$. For any $x \in \{0,1\}^n$ we have

$$\Pr[X^n = x] = p^{\|x\|}q^{n-\|x\|}. \tag{16}$$

According to the last lemma and Theorem 3, we have

$$H^{\epsilon}_{\mathrm{ext}}(X^n) \leqslant -\log t \tag{17}$$

where

$$\sum_x \max\left(\mathbf{P}_{X^n}(x) - t, 0\right) = \epsilon. \tag{18}$$

From now we assume that

$$t = p^{pn+\gamma n}q^{qn-\gamma n}. \tag{19}$$

4.2 Determining the Threshold t

The next key observation is that t is actually small and can be omitted. That is, we can simply cut the $(1 - \epsilon)$-quantile. This is stated in the lemma below.

Lemma 3 (Replacing the threshold by the quantile). *Let $x_0 \in \{0,1\}^n$ be a point such that $\|x_0\| = pn + \gamma n$. Then we have*

$$\sum_{x:\ \|x\|\geqslant\|x_0\|} \max\left(\mathbf{P}_{X^n}(x) - \mathbf{P}_{X^n}(x_0)\right) \geqslant \frac{1}{2} \sum_{x:\ \|x\|\geqslant\|x_0\|} \mathbf{P}_{X^n}(x) \tag{20}$$

To prove the lemma, note that from Theorem 4 it follows that setting

$$\gamma' = \gamma + n^{-1} \log\left(\frac{p}{q}\right) \tag{21}$$

we obtain

$$\sum_{j \geqslant pn+\gamma'n} \binom{n}{j} \geqslant \frac{3}{4} \cdot \sum_{j \geqslant pn+\gamma n} \binom{n}{j} \tag{22}$$

when γ is sufficiently small comparing to p and q (formally this is justified by calculating the derivative with respect to γ and noticing that it is bigger by at most a factor of $1 + \frac{\gamma}{\sqrt{npq}}$). But we also have

$$p^j q^{n-j} \geqslant 2 \cdot p^{(p+\gamma)n} q^{(q-\gamma)n} \quad \text{for } j \geqslant \gamma' n \tag{23}$$

Therefore,

$$\sum_{j \geqslant pn+\gamma n} \binom{n}{j} p^j q^{n-j} \geqslant \sum_{j \geqslant pn+\gamma'n} \binom{n}{j} p^j q^{n-j}$$

$$\geqslant 2 \cdot p^{(p+\gamma)n} q^{(q-\gamma)n} \cdot \sum_{j \geqslant pn+\gamma'n} \binom{n}{j}$$

$$\geqslant 2 \cdot \frac{3}{4} \cdot p^{(p+\gamma)n} q^{(q-\gamma)n} \cdot \sum_{j \geqslant pn+\gamma n} \binom{n}{j} \tag{24}$$

which finishes the proof.

4.3 Putting This All Together

Now, by combining Lemmas 2 and 3 and the estimate $Q(x) \approx x^{-1} \exp(-x^2/2)$ for $x \gg 0$ we obtain

$$\epsilon \geqslant \exp\left(-n\text{KL}\left(p + \gamma \parallel p\right) - \log\left(\frac{n\gamma^2}{2pq}\right) + O_{p,q}(1)\right) \tag{25}$$

which, because of the Taylor expansion $\text{KL}\left(p + \gamma \parallel p\right) = \frac{\gamma^2}{2pq} + O_{p,q}(\gamma^3)$, gives us

$$\gamma \geqslant \Omega\left(\sqrt{\frac{\log(1/\epsilon)}{pqn}}\right) \tag{26}$$

Setting $\gamma = c \cdot \sqrt{\frac{\log(1/\epsilon)}{pqn}}$, with sufficiently big c, we obtain the claimed result.

References

1. Barak, B., Dodis, Y., Krawczyk, H., Pereira, O., Pietrzak, K., Standaert, F.-X., Yu, Y.: Leftover hash lemma, revisited. Cryptology ePrint Archive, Report 2011/088 (2011). http://eprint.iacr.org/
2. Bouda, J., Krhovjak, J., Matyas, V., Svenda, P.: Towards true random number generation in mobile environments. In: Jøsang, A., Maseng, T., Knapskog, S.J. (eds.) NordSec 2009. LNCS, vol. 5838, pp. 179–189. Springer, Heidelberg (2009). http://dx.doi.org/10.1007/978-3-642-04766-4_13
3. Bucci, M., Luzzi, R.: Design of testable random bit generators. In: Rao, J.R., Sunar, B. (eds.) CHES 2005. LNCS, vol. 3659, pp. 147–156. Springer, Heidelberg (2005)
4. Cachin, C.: Smooth entropy and Rényi entropy. In: Fumy, W. (ed.) EUROCRYPT 1997. LNCS, vol. 1233, pp. 193–208. Springer, Heidelberg (1997)
5. Dichtl, M., Golić, J.D.: High-speed true random number generation with logic gates only. In: Paillier, P., Verbauwhede, I. (eds.) CHES 2007. LNCS, vol. 4727, pp. 45–62. Springer, Heidelberg (2007)
6. Dodis, Y., Pietrzak, K., Wichs, D.: Key derivation without entropy waste. In: Nguyen, P.Q., Oswald, E. (eds.) EUROCRYPT 2014. LNCS, vol. 8441, pp. 93–110. Springer, Heidelberg (2014)
7. Hastad, J., Impagliazzo, R., Levin, L.A., Luby, M.: Pseudo-random generation from one-way functions. In: Proceedings of the 20th STOC, pp. 12–24 (1988)
8. Hastad, J., Impagliazzo, R., Levin, L.A., Luby, M.: A pseudorandom generator from any one-way function. SIAM J. Comput. **28**(4), 1364–1396 (1999). http://citeseerx.ist.psu.edu/viewdoc/summary?doi=10.1.1.35.3930
9. Holenstein, T.: Pseudorandom generators from one-way functions: a simple construction for any hardness. In: Halevi, S., Rabin, T. (eds.) TCC 2006. LNCS, vol. 3876, pp. 443–461. Springer, Heidelberg (2006)
10. Holenstein, T., Renner, R.: On the randomness of independent experiments. IEEE Trans. Inf. Theory **57**(4), 1865–1871 (2011)
11. Lacharme, P., Röck, A., Strubel, V., Videau, M.: The linux pseudorandom number generator revisited. Cryptology ePrint Archive, Report 2012/251 (2012). http://eprint.iacr.org/
12. McKay, B.D.: On Littlewood's estimate for the binomial distribution. Adv. Appl. Probab. **21**(2), 475–478 (1989)
13. Renner, R.S., König, R.: Universally composable privacy amplification against quantum adversaries. In: Kilian, J. (ed.) TCC 2005. LNCS, vol. 3378, pp. 407–425. Springer, Heidelberg (2005). http://dx.doi.org/10.1007/978-3-540-30576-7_22
14. Renner, R., Wolf, S.: Smooth Renyi entropy and applications. In: ISIT 2004, Chicago, Illinois, USA, p. 232 (2004)
15. Renner, R.S., Wolf, S.: Simple and tight bounds for information reconciliation and privacy amplification. In: Roy, B. (ed.) ASIACRYPT 2005. LNCS, vol. 3788, pp. 199–216. Springer, Heidelberg (2005)
16. Skorski, M.: How much randomness can be extracted from memoryless shannon entropy sources. In: WISA 2015 (2015)
17. Schoenmakers, B., Tjoelker, J., Tuyls, P., Verbitskiy, E.: Smooth Renyi entropy of ergodic quantum information sources. In: 2007 IEEE International Symposium on Information Theory. ISIT 2007, pp. 256–260 (2007)

18. Tomamichel, M.: A framework for non-asymptotic quantum information theory. Ph.D. thesis, ETH Zurich (2012)
19. Watanabe, S., Hayashi, M.: Non-asymptotic analysis of privacy amplification via Renyi entropy and inf-spectral entropy. In: 2013 IEEE International Symposium on Information Theory Proceedings (ISIT), pp. 2715–2719 (2013)

Bounded TSO-to-SC Linearizability Is Decidable

Chao Wang[1,2]([⊠]), Yi Lv[1], and Peng Wu[1]

[1] State Key Laboratory of Computer Science, Institute of Software,
CAS, Beijing, China
wangch@ios.ac.cn
[2] University of Chinese Academy of Sciences, Beijing, China

Abstract. TSO-to-SC linearizability is a variant of linearizability for concurrent libraries on the Total Store Order (TSO) memory model. In this paper we propose the notion of k-bounded TSO-to-SC linearizability, a subclass of TSO-to-SC linearizability that concerns only bounded histories. This subclass is non-trivial in that it does not restrict the number of write, flush and *cas* (compare-and-swap) actions, nor the size of a store buffer, to be bounded. We prove that the decision problem of k-bounded TSO-to-SC linearizability is decidable for a bounded number of processes. We first reduce this decision problem to a marked violation problem of k-bounded TSO-to-SC linearizability, where specific *cas* actions are introduced to mark call and return actions. Then, we further reduce the marked violation problem to a control state reachability problem of a lossy channel machine, which is already known to be decidable. Moreover, we can show that the decision problem of k-bounded TSO-to-SC linearizability has non-primitive recursive complexity.

1 Introduction

Linearizability [9] has been accepted as a *de facto* correctness condition for a concurrent library with respect to its sequential specification on the sequential consistency (SC) memory model [10]. However, modern multiprocessors (e.g., x86 [12], POWER [13]) and programming languages (e.g., Java [11], C11/C++11 [3]) do not comply with the SC memory model. Instead, they provide *relaxed memory models* that allow non-SC behaviors due to hardware or compiler optimization. For instance, in a multiprocessor system implementing the TSO memory model [12], each processor is equipped with a FIFO store buffer. Any written action performed by a processor will append an item into its store buffer before the item is eventually flushed into the memory. The TSO memory model requires that all processes in a concurrent system observe the same order of write and *cas* actions, which is referred to as a total store order.

Accordingly, linearizability has been extended for relaxed memory models, e.g., *TSO-to-TSO linearizability* [7] and *TSO-to-SC linearizability* [8] for the

---

This work is partially supported by the National Natural Science Foundation of China under Grants No. 60721061, No. 60833001, No. 61272135, No. 61572478, No. 61700073, No. 61100069, No. 61472405, and No. 61161130530.

R.M. Freivalds et al. (Eds.): SOFSEM 2016, LNCS 9587, pp. 404–417, 2016.
DOI: 10.1007/978-3-662-49192-8_33

TSO memory model and two variants of linearizability [3] for the C++ memory model. TSO-to-SC linearizability has been proposed for reasoning about the correctness of a concurrent library, which is native to the TSO memory model but is used with a concurrent program that needs to be protected from the relaxed semantics [8].

It is well known that the linearizability of a concurrent library on the SC memory model is decidable for a bounded number of processes [1], but undecidable for an unbounded number of processes [4]. However, to our knowledge, there are only a few decidability results about linearizability on relaxed memory models. We have recently proved that the decision problem of *TSO-to-TSO linearizability* is undecidable for a bounded number of processes [16]. But the decision problem of TSO-to-SC linearizability still remains open for a bounded number of processes.

We propose a decidable subclass of TSO-to-SC linearizability for a bounded number of processes, which is referred to as *k-bounded TSO-to-SC linearizability*. It concerns only k-traces, which are traces with at most k call and return actions, and hence it defined over k-bounded histories of TSO libraries. Note that k-traces may still contain arbitrarily many write, flush and *cas* actions, and store buffers may still contain arbitrarily many items along k-traces. Hence, the k-boundedness on the number of call and return actions does not necessarily restrict the behaviors of a concurrent program to be finite-state. As we prove in this paper, the decision problem of this non-trivial subclass of TSO-to-SC linearizability is decidable for a bounded number of processes.

As in [6,16], we first show that history inclusion is an equivalent characterization of k-bounded TSO-to-SC linearizability. Then, as inspired by [2], we consider to reduce the history inclusion problem to a control state reachability problem of a lossy channel machine. Thus, the decidability of k-bounded TSO-to-SC linearizability follows from the fact that a control state reachability problem of a lossy channel machine is decidable [2]. However, the reduction method in [2] does not directly apply to linearizability of concurrent libraries. This is because that the call and return actions concerned by linearizability are beyond the scope of the TSO memory model, while the reduction method in [2] ensures only the total store orders among write/*cas* actions.

We extend the reduction method in [2] to effectively handle call and return actions. Suppose a concurrent system that contains n client processes running independently and interacting with a shared library. We introduce a new process that keeps launching the specific *cas* actions nondeterministically. These specific *cas* actions are used to mark the possible occurrences of the call and return actions along a trace of the concurrent system. Then, a correctly marked trace of this new process replicates the history of the trace of the concurrent system with only specific *cas* actions. Correspondingly, a counterexample trace of TSO-to-SC linearizability in the original concurrent system (of n processes) can be witnessed by a marked trace of the extended concurrent system (of $n+1$ processes) with the call and return actions bypassed. This marked trace is called a marked violation of TSO-to-SC linearizability. In this way, the complement problem of TSO-to-SC linearizability on the original concurrent system can be characterized by the marked violation problem of the extended concurrent system, to which the reduction method in [2] can be applied.

A lossy channel machine M_i^k ($1 \leq i \leq n+1$) is then constructed such that its traces contain at most k call and return actions and can simulate the k-bounded behaviors of the extended concurrent system from the perspective of each process P_i. Each M_i^k contains only one channel to store the pending written items according to the total store orders under the original concurrent system. Thus, the marked violation problem of k-bounded TSO-to-SC linearizability can be reduced to a control state reachability problem of the product of $M_1^{k-w}, \ldots, M_{n+1}^{k-w}$. Each M_i^{k-w} is resulted from M_i^k by replacing its all but write and *cas* transitions with internal transitions. The reduction is achieved by requiring that each written item in a channel contains a run-time snapshot of the memory, while always keeping bounded the amount of information that needs to be stored as in a perfect channel. With these specialized lossy channels, missing some intermediate channel contents would not break the reachability between control states under perfect channels.

Furthermore, we can show that the decision problem of k-bounded TSO-to-SC linearizability has non-primitive recursive complexity. This is proved in the technical report version of this paper [15] by a reduction from a reachability problem of a lossy single-channel machine, which is known to have non-primitive recursive complexity [14]. Besides, the decision problem of TSO-to-SC linearizability can be reduced to a control state reachability problem of a perfect channel machine in a similar way. This opens a potential way towards determining the decidability of TSO-to-SC linearizability itself.

Related Work. Efforts have been devoted on verification of linearizability on the SC memory model [1,4–6]. A similar reduction method was applied to verify the linearizablility of certain concurrent data structures for an unbounded number of processes on the SC memory model [5]. However, relaxed memory models remain a great challenge for linearizability verification. Our previous work [16] revealed the first undecidability result on TSO-to-TSO linearizability for a bounded number of processes. In [16], the trace inclusion problem of a classic-lossy single-channel system, which has been known to be undecidable, was reduced to the TSO-to-TSO linearizability problem. The closest work to ours is [2] by Atig *et al.*, where a state reachability problem of a concurrent system is reduced to a control state reachability problem of a lossy channel machine.

2   Concurrent Systems

In this section, we first present the notations of libraries, client programs, most general clients and concurrent systems. We then introduce their operational semantics on the TSO and SC memory models.

2.1 Notations

In general, a finite sequence on an alphabet Σ is denoted $l = \alpha_1 \cdot \alpha_2 \cdot \ldots \cdot \alpha_k$, where $\cdot$ is the concatenation symbol and $\alpha_i \in \Sigma$ for each $1 \leq i \leq k$. Let $|l|$ and $l(i)$ denote the length and the i-th element of l, respectively, i.e., $|l| = k$ and $l(i) = \alpha_i$ for $1 \leq i \leq k$. Let $l \uparrow_\Sigma$ denote the projection of l to Σ. Given a function f, let $f[x : y]$ be the function that is the same as f everywhere, except for x, where it has the value y. Let _ denote an item, of which the value is irrelevant, and ϵ the empty word.

A *labelled transition system* (*LTS*) is a tuple $\mathcal{A} = (Q, \Sigma, \rightarrow, q_0)$, where Q is a set of states (a.k.a. configurations), Σ is an alphabet of transition labels, $\rightarrow \subseteq Q \times \Sigma \times Q$ is a transition relation and q_0 is the initial state. A path of $\mathcal{A}$ is a finite transition sequence $q_0 \xrightarrow{\beta_1} q_1 \xrightarrow{\beta_2} \ldots \xrightarrow{\beta_k} q_k$ with $k \geq 0$. A trace of $\mathcal{A}$ is a finite sequence $t = \beta_1 \cdot \beta_2 \cdot \ldots \cdot \beta_k$ with $k \geq 0$ if there exists a path $q_0 \xrightarrow{\beta_1} q_1 \xrightarrow{\beta_2} \ldots \xrightarrow{\beta_k} q_k$ of $\mathcal{A}$.

2.2 Libraries and Client Programs

A library implementing a concurrent data structure provides a number of methods for accessing the data structure. A client program is a program that interacts with libraries. Libraries and client programs may contain private memory locations for their own uses. For simplicity of notations, we assume that a method has just one argument and one return value (if it returns).

Given a finite set $\mathcal{X}$ of memory locations, a finite set $\mathcal{M}$ of method names and a finite data domain $\mathcal{D}$, the set $PCom$ of primitive commands has the forms below:

$$PCom ::= \tau \mid read(x, a) \mid write(x, a) \mid cas\_suc(x, a, b) \mid cas\_fail(x, a, b) \mid call(m, a)$$

where $a, b \in \mathcal{D}, x \in \mathcal{X}$ and $m \in \mathcal{M}$. Herein, τ is the internal command. A *cas* (compare-and-swap) command compresses a read and a write commands into a single one, which is meant to be executed atomically. A successful *cas* command $cas\_suc(x, a, b)$ changes the value of x from a to b, while a failed *cas* command $cas\_fail(x, a, b)$ does nothing and happens only when the value of x is not a.

A library $\mathcal{L}$ can then be defined as a tuple $\mathcal{L} = (\mathcal{X}_\mathcal{L}, \mathcal{M}_\mathcal{L}, \mathcal{D}_\mathcal{L}, Q_\mathcal{L}, \rightarrow_\mathcal{L})$, where $\mathcal{X}_\mathcal{L}, \mathcal{M}_\mathcal{L}$ and $\mathcal{D}_\mathcal{L}$ are a finite memory location set, a finite method name set and a finite data domain of $\mathcal{L}$ respectively; $Q_\mathcal{L} = \bigcup_{m \in \mathcal{M}_\mathcal{L}} Q_m$ is the union of disjoint finite sets Q_m of program positions of each method $m \in \mathcal{M}_\mathcal{L}$; $\rightarrow_\mathcal{L} = \bigcup_{m \in \mathcal{M}_\mathcal{L}} \rightarrow_m$ is the union of disjoint transition relations of each method $m \in \mathcal{M}_\mathcal{L}$. Let $PCom_\mathcal{L}$ be the set of primitive commands (except call commands) upon $\mathcal{X}_\mathcal{L}, \mathcal{M}_\mathcal{L}$ and $\mathcal{D}_\mathcal{L}$. Then, for each $m \in \mathcal{M}_\mathcal{L}, \rightarrow_m \subseteq Q_m \times PCom_\mathcal{L} \times Q_m$; while for each $a \in \mathcal{D}_\mathcal{L}$ there exists an initial state $is_{(m,a)}$ and a final state $fs_{(m,a)}$ in Q_m such that there are neither incoming transitions to $is_{(m,a)}$ nor outgoing transitions from $fs_{(m,a)}$ in $\rightarrow_m$. Similarly, a client program $\mathcal{C}$ can then be defined as a tuple $\mathcal{C} = (\mathcal{X}_\mathcal{C}, \mathcal{M}_\mathcal{C}, \mathcal{D}_\mathcal{C}, Q_\mathcal{C}, \rightarrow_\mathcal{C})$ where $\mathcal{X}_\mathcal{C}, \mathcal{M}_\mathcal{C}, \mathcal{D}_\mathcal{C}$ and $Q_\mathcal{C}$ are a finite memory location set, a finite method name set and a final data domain of $\mathcal{C}$ and a finite program position set, respectively. Let $PCom_\mathcal{C}$ be the set of primitive commands upon $\mathcal{X}_\mathcal{C}$, $\mathcal{M}_\mathcal{C}$ and $\mathcal{D}_\mathcal{C}$. Then, $\rightarrow_\mathcal{C} \subseteq Q_\mathcal{C} \times PCom_\mathcal{C} \times Q_\mathcal{C}$ is a transition relation of $\mathcal{C}$.

A most general client is a special client program that is designed to exhibit all the possible behaviors of a library. A most general client $\mathcal{MGC}$ can be formally defined as a client $(\mathcal{X}_\mathcal{C}, \mathcal{M}_\mathcal{C}, \mathcal{D}_\mathcal{C}, \{q_c\}, \rightarrow_{mgc})$, where q_c is a program position and $\rightarrow_{mgc} = \{(q_c, call(m, a), q_c) | m \in \mathcal{M}_\mathcal{C}, a \in \mathcal{D}_\mathcal{C}\}$ is a transition relation. Intuitively, a most general client simply repeatedly calls an arbitrary method with an arbitrary argument for arbitrarily many times. It does not access any memory location in $\mathcal{X}_\mathcal{C}$, so $\mathcal{X}_\mathcal{C}$ does not influence the behavior of a most general client.

2.3 TSO Operational Semantics

Suppose a concurrent system $C(\mathcal{L})$ that consists of n processes, each of which runs a client program $\mathcal{C}_i = (\mathcal{X}_\mathcal{C}, \mathcal{M}, \mathcal{D}_\mathcal{C}, Q_{\mathcal{C}_i}, \rightarrow_{\mathcal{C}_i})$ on a separate processor for $1 \leq i \leq n$, and all the client programs interact with the same library $\mathcal{L} = (\mathcal{X}_\mathcal{L}, \mathcal{M}, \mathcal{D}_\mathcal{L}, Q_\mathcal{L}, \rightarrow_\mathcal{L})$. The library and client programs have disjoint memory locations, i.e., $\mathcal{X}_\mathcal{L} \cap \mathcal{X}_\mathcal{C} = \emptyset$. The operational semantics of the concurrent system $C(\mathcal{L})$ on the TSO memory model is defined as an LTS $[\![C(\mathcal{L}), n]\!]_{tso} = (Conf_{tso}, \Sigma_{tso}, \rightarrow_{tso}, InitConf_{tso})$, where $Conf_{tso}, \Sigma_{tso}, \rightarrow_{tso}, InitConf_{tso}$ are defined as follows.

Each configuration of $Conf_{tso}$ is a tuple (p, d, u), where p represents current control states of each process, d is the current valuation of the memory locations and u represents the contents of the store buffers for each process. Σ_{tso} is a set of actions in the following forms:

$$\Sigma_{tso} ::= \tau(i) \mid read(i, x, a) \mid write(i, x, a) \mid cas(i, x, a, b) \mid$$
$$flush(i, x, a) \mid call(i, m, a) \mid return(i, m, a)$$

where $1 \leq i \leq n, m \in \mathcal{M}$ and either $x \in \mathcal{X}_\mathcal{L}$ and $a, b \in \mathcal{D}_\mathcal{L}$, or $x \in \mathcal{X}_\mathcal{C}$ and $a, b \in \mathcal{D}_\mathcal{C}$. Informally, $\rightarrow_{tso}$ is a minimum transition relation such that for process P_i $(1 \leq i \leq n)$, a write action $write(i, x, a)$ appends an item (x, a) to $u(i)$; a read action $read(i, x, a)$ obtains the current value a of x, either from the latest pending item (x, a) in $u(i)$ (if exists), or from the current valuation in the memory; a flush action $flush(i, x, a)$ flushes the item (x, a) at the head of $u(i)$ into the memory; a $cas(i, x, a, b)$ action clears $u(i)$ and intends to replace the current value a of x with b in the memory directly; while an call action $call(i, m, a)$ starts to execute a method m with an argument a, and a return action $return(i, m, a)$ returns from the method m with a return value a.

The initial configuration $InitConf_{tso} \in Conf_{tso}$ is a tuple $(p_{init}, d_{init}, \epsilon^n)$, where ϵ^n initializes each process with an empty buffer. If each client program C_i is a most general client, $[\![C(\mathcal{L}), n]\!]_{tso}$ can be abbreviated as $[\![\mathcal{L}, n]\!]_{tso}$.

According to [8], the operational semantics of the concurrent system on the SC memory model can be defined as an LTS $[\![C(\mathcal{L}), n]\!]_{sc} = (Conf_{sc}, \Sigma_{sc}, \rightarrow_{sc}, InitConf_{sc})$ similar to $[\![C(\mathcal{L}), n]\!]_{tso}$, except that each configuration of $Conf_{sc}$ has only the empty store buffer for each process and all write actions are flushed into the memory immediately. The detailed definition of $[\![C(\mathcal{L}), n]\!]_{tso}$ and $[\![C(\mathcal{L}), n]\!]_{sc}$ can be found in [15].

3 Correctness Conditions and Equivalent Characterization

The behavior of a library is typically represented by histories of interactions between the library and the client programs calling it (through call and return actions). Let Σ_{cal} and Σ_{ret} represent the sets of all call and return actions, respectively. A finite sequence $h \in (\Sigma_{cal} \cup \Sigma_{ret})^*$ is a history of an LTS $\mathcal{A}$ if there exists a trace t of $\mathcal{A}$ such that $t \upharpoonright_{(\Sigma_{cal} \cup \Sigma_{ret})} = h$. Let $history(t)$ be the history along trace t, i.e., $history(t) = t \upharpoonright_{(\Sigma_{cal} \cup \Sigma_{ret})}$, and $history(\mathcal{A})$ the set of all histories of $\mathcal{A}$. Moreover, let $h|_i$ denote the projection of history h to the call and return actions of process P_i.

TSO-to-SC linearizability is a variant of linearizability on the TSO memory model. It is used to reason about the interoperability between a high-level data race free client and a low-level library native to the TSO memory model. Hence, it concerns only call and return actions.

Definition 1 (*TSO-to-SC linearizability* [8]). *For histories $h_1, h_2 \in (\Sigma_{cal} \cup \Sigma_{ret})^*$, h_1 is linearizable to h_2, if*

- *for each process P_i, $h_1|_i = h_2|_i$.*
- *there is a bijection $\pi : \{1, \ldots, |h_1|\} \rightarrow \{1, \ldots, |h_2|\}$ such that for any $1 \leq i \leq |h_1|$, $h_1(i) = h_2(\pi(i))$ and for any $1 \leq i < j \leq |h_1|$, if $h_1(i) \in \Sigma_{ret} \wedge h_1(j) \in \Sigma_{cal}$, then $\pi(i) < \pi(j)$.*

For two libraries $\mathcal{L}$ and $\mathcal{L}'$, $\mathcal{L}'$ TSO-to-SC linearizes $\mathcal{L}$ for n processes, if for any history $h_1 \in history(\llbracket \mathcal{L}, n \rrbracket_{tso})$, there exists history $h_2 \in history(\llbracket \mathcal{L}', n \rrbracket_{sc})$, such that h_1 is linearizable to h_2.

The following lemma shows that history inclusion is an equivalent characterization of TSO-to-SC linearizability.

Lemma 1. *Library $\mathcal{L}'$ TSO-to-SC linearizes library $\mathcal{L}$ for n processes if and only if $history(\llbracket \mathcal{L}, n \rrbracket_{tso}) \subseteq history(\llbracket \mathcal{L}', n \rrbracket_{sc})$.*

For an LTS $\mathcal{A}$, a k-trace $t \in trace(\mathcal{A})$ is a trace that contains at most k call and return actions. Let k-$trace(\mathcal{A})$ denote all the k-traces of $\mathcal{A}$.

Definition 2 (k-bounded TSO-to-SC linearizability). *Library $\mathcal{L}'$ k-bounded TSO-to-SC linearizes library $\mathcal{L}$ for n processes, if for each k-trace $t \in k$-$trace (\llbracket \mathcal{L}, n \rrbracket_{tso})$, there exists a history $h \in history(\llbracket \mathcal{L}', n \rrbracket_{sc})$, such that $history(t)$ is linearizable to h.*

For two libraries $\mathcal{L}$, $\mathcal{L}'$ and $n, k \geq 0$, the decision problem of (k-bounded) TSO-to-SC linearizability is to determine whether $\mathcal{L}'$ (k-bounded) TSO-to-SC linearizes $\mathcal{L}$ for n processes.

4   Perfect/Lossy Channel Machines

A classical channel machine is a finite control machine equipped with channels of unbounded sizes. It can perform send and receive operations on its channels. A lossy channel machine is a channel machine where arbitrary many items in its channels may be lost nondeterministically at any time without any notification. In this section we sketch our definition of (S, K)-channel machines, which slightly differs from the definition of channel machines in [2].

The channel machines defined in [2] extend classical channel machines in the following aspects:

- Each transition is guarded by a condition about whether the content of a channel is in a regular language.
- A substitution to the content of a channel may be performed before a send operation on the channel.
- A set of specific symbols, called "strong symbols", are introduced that are not allowed to be lost, but the number of strong symbols in a channel is always bounded.

In this paper, we extend the channel machines defined in [2] with multiple sets of strong symbols, while the number of strong symbols in a channel from the same strong symbol set is separately bounded.

A *channel machine* is formally defined as a tuple $M = (Q, \mathcal{CH}, \Sigma_{\mathcal{CH}}, \Lambda, \Delta)$, where (1) Q is a finite set of states, (2) $\mathcal{CH}$ is a finite set of channel names, (3) $\Sigma_{\mathcal{CH}}$ is an alphabet for channel contents, (4) Λ is a finite set of transition labels, and (5) $\Delta \subseteq Q \times (\Lambda \cup \{\epsilon\}) \times Guard(\mathcal{CH}) \times Op(\mathcal{CH}) \times Q$ is a finite set of transitions with $Guard(\mathcal{CH})$ and $Op(\mathcal{CH})$ being the sets of guards and operations over $\mathcal{CH}$, respectively.

Let $S = \langle s_1, \ldots, s_m \rangle$ be a vector of sets with $s_i \subseteq \Sigma_{\mathcal{CH}}$ for $1 \leq i \leq m$, and $K = \langle k_1, \ldots, k_m \rangle$ be a vector of nature numbers or ∞. A (S, K)-channel machine, abbreviated as (S, K)-*CM*, is a channel machine M with the strong symbol restriction (S, K), i.e., each of M's configuration is a tuple (q, u), where q is a control state and u maps contents to each channel such that for each $1 \leq j \leq m$ and channel c, $|u(c) \uparrow s_j| \leq k_j$. As in [2], a (S, K)-channel machine M with lossy channels is referred to as a (S, K)-lossy channel machine, abbreviated as (S, K)-*LCM*. This is achieved by interpreting a (S, K)-channel machine with a lossy transition relation, instead of a usual (perfect) one. A formal definition of a (S, K)-$(L)CM$ can be found in [15].

For states $q, q' \in Q$, let $T_{q,q'}^{S,K}(M)$ (respectively, $LT_{q,q'}^{S,K}(M)$) denote the sets of traces of (S, K)-*CM* (respectively (S, K)-*LCM*) M from the configuration (q, ϵ^n) to the configuration (q', ϵ^n). For channel machines $M_1 = (Q_1, \mathcal{CH}_1, \Sigma_{\mathcal{CH}}, \Lambda, \Delta_1)$ and $M_2 = (Q_2, \mathcal{CH}_2, \Sigma_{\mathcal{CH}}, \Lambda, \Delta_2)$ such that $\mathcal{CH}_1 \cap \mathcal{CH}_2 = \emptyset$, the product of M_1 and M_2 is also a channel machine $M_1 \otimes M_2 = (Q_1 \times Q_2, \mathcal{CH}_1 \cup \mathcal{CH}_2, \Sigma_{\mathcal{CH}}, \Lambda, \Delta_{12})$, where Δ_{12} is defined by synchronizing transitions sharing the same label in Λ under the conjunction of their guards, and letting other transitions asynchronous. The following lemma holds as in [2].

Lemma 2. *For* (S,K)*-CM* $M_1 = (Q_1, \mathcal{CH}_1, \Sigma_{\mathcal{CH}}, \Lambda, \Delta_1)$ *and* $M_2 = (Q_2, \mathcal{CH}_2,$ $\Sigma_{\mathcal{CH}}, \Lambda, \Delta_2)$, $q_1, q_1' \in Q_1$ *and* $q_2, q_2' \in Q_2$, *let* $q = (q_1, q_2)$, $q' = (q_1', q_2')$, *then* $LT_{q,q'}^{S,K}(M_1 \otimes M_2) = LT_{q_1,q_1'}^{S,K}(M_1) \cap LT_{q_2,q_2'}^{S,K}(M_2)$ *and* $T_{q,q'}^{S,K}(M_1 \otimes M_2) = T_{q_1,q_1'}^{S,K}(M_1) \cap$ $T_{q_2,q_2'}^{S,K}(M_2)$.

Given a (S,K)-*CM* (respectively, (S,K)-*LCM*) M and two states $q, q' \in Q$, a control state reachability problem of M is to determine whether $T_{q,q'}^{S,K}(M) \neq \emptyset$ (respectively, $LT_{q,q'}^{S,K}(M) \neq \emptyset$). As in [2], it can be shown that the control state reachability problem is decidable for (S,K)-*LCM*.

5 Verification of k-Bounded TSO-to-SC Linearizability

In this section we show the proof idea about the decidability of k-bounded TSO-to-SC linearizability for a bounded number of processes. The main theme is to reduce its complement problem to a control state reachability problem of a (S,K)-lossy channel machine. In the same way, we can reduce the decision problem of TSO-to-SC linearizability to a control state reachability problem of a (S,K)-channel machine. Due to lack of space, the detailed proofs of the lemmas and theorems presented in this section can be founded in [15].

5.1 Marked Violation of (k-Bounded) TSO-to-SC Linearizability

Recall that call and return actions cannot be handled directly by the reduction method in [2]. We introduce a fresh new process to captures the call and return actions, which occur along the traces (or k-traces) of $[\![\mathcal{L}, n]\!]_{tso}$ by the specific *cas* actions. In this way, the behaviors of a concurrent system $[\![\mathcal{L}, n]\!]_{tso}$ can be characterized exactly by the extended concurrent system $[\![Clt(\mathcal{L}), n+1]\!]_{tso}$ (defined below), with the benefit that the call and return actions need not be involved for reduction later.

Let $markedVal(\mathcal{M}, \mathcal{D}_\mathcal{L}, n) = \{call(i,m,a), return(i,m,a) | 1 \leq i \leq n, m \in \mathcal{M}, a \in \mathcal{D}_\mathcal{L}\}$ denote the set of values that are used by the specific *cas* actions to mark the call and return actions in $[\![\mathcal{L}, n]\!]_{tso}$. Then, the concurrent system $Clt(\mathcal{L})$ consists of $n+1$ processes P_i $(1 \leq i \leq n+1)$. For each $1 \leq i \leq n$, process P_i runs the most general client program $(\{x_{wit}\}, \mathcal{M}, \mathcal{D}_\mathcal{L}, \{q_c\}, \rightarrow_{mgc})$. The process P_{n+1} runs the client program $C_{marked} = (\{x_{wit}\}, \mathcal{M}, markedVal(\mathcal{M}, \mathcal{D}_\mathcal{L}, n), \{q_{wit}\}, \rightarrow_{wit})$, where $x_{wit} \notin \mathcal{X}_\mathcal{L}$ is the memory location used by the specific *cas* actions; $\rightarrow_{wit} = \{(q_{wit}, cas\_suc(x_{wit}, \_, a), q_{wit}) | a \in markedVal(\mathcal{M}, \mathcal{D}_\mathcal{L}, n)\}$ is the transition relation of C_{marked}.

A marked violation is a trace of $[\![Clt(\mathcal{L}), n+1]\!]_{tso}$ that can witness the violation of TSO-to-SC linearizability. It correctly captures the corresponding call and return actions, stops immediately when a non-linearizable action takes place and flushes all the stored items so far. Formally, a trace $t \in trace([\![Clt(\mathcal{L}), n+1]\!]_{tso})$ is a *marked violation* of TSO-to-SC linearizability between libraries $\mathcal{L}$ and $\mathcal{L}'$ for n processes, if

– The specific *cas* actions mark correctly the call and return actions, i.e., for each $1 \leq i \leq |t| - 1$, $m \in \mathcal{M}$ and $a \in \mathcal{D}_\mathcal{L}$, the following conditions hold:
  1. $t(i) = cas(n+1, x_{wit}, call(i, m, a))$ if and only if $t(i+1) = call(i, m, a)$.
  2. $t(i) = cas(n+1, x_{wit}, return(i, m, a))$ if and only if $t(i+1) = return(i, m, a)$.
– $history(t) \notin history(\llbracket \mathcal{L}', n \rrbracket_{sc})$, and for each prefix t' of t such that $history(t) \neq history(t')$, $history(t') \in history(\llbracket \mathcal{L}', n \rrbracket_{sc})$.
– $t = t_1 \cdot t_2$ such that t_1 ends with a call or return action, and t_2 is a sequence of flush actions. Moreover, all the write actions in t have been flushed.

Furthermore, the trace t is a *marked violation* of k-bounded TSO-to-SC linearizability between libraries $\mathcal{L}$ and $\mathcal{L}'$ for n processes, if t is a k-trace. For two libraries $\mathcal{L}$, $\mathcal{L}'$, and $n, k \geq 0$, a (k-bounded) TSO-to-SC marked violation problem is to determine whether there is a marked violation of (k-bounded) TSO-to-SC linearizability between libraries $\mathcal{L}$ and $\mathcal{L}'$ for n processes. The following lemma relates a (k-bounded) TSO-to-SC marked violation problem with the complement problem of (k-bounded) TSO-to-SC linearizability.

Lemma 3. *$\mathcal{L}'$ does not (k-bounded) TSO-to-SC linearizes $\mathcal{L}$ for n processes, if and only if there is a marked violation of (k-bounded) TSO-to-SC linearizability between libraries $\mathcal{L}$ and $\mathcal{L}'$ for n processes.*

The specific *cas* actions are launched nondeterministically in $\llbracket Clt(\mathcal{L}), n+1 \rrbracket_{tso}$ and hence may result in many incorrectly guessed traces that do not occur in $\llbracket \mathcal{L}, n \rrbracket_{tso}$. However, the channel machines M_i^k we constructed can guarantee that the incorrectly guessed traces will be safely excluded during the verification procedure.

5.2   Simulating $\llbracket Clt(\mathcal{L}), n+1 \rrbracket_{tso}$ with A Channel Machine

In the rest of this section, we show that for libraries $\mathcal{L}$ and $\mathcal{L}'$, how the k-bounded behaviors of the concurrent system $\llbracket Clt(\mathcal{L}), n+1 \rrbracket_{tso}$ can be further characterized by a (S, K)-channel machine. As in [2], this amounts to construct a channel machines M_i^k corresponding to each process P_i in $\llbracket Clt(\mathcal{L}), n+1 \rrbracket_{tso}$.

Each M_i^k ($1 \leq i \leq n+1$) launches actions of process P_i according to the control state of this process, and nondeterministically guesses the write, call or return actions of the other processes. It contains only one channel c_i that is used to store the pending written items according to the total store orders in $\llbracket Clt(\mathcal{L}), n+1 \rrbracket_{tso}$. Each item sent to each channel c_i contains the current valuation of all the memory locations, i.e., the current snapshot of the memory.

We first use the example shown in Fig. 1 to illustrate the main idea of our construction method. Figure 1(a) presents a k-trace t of a concurrent system $\llbracket Clt(\mathcal{L}), 3 \rrbracket_{tso}$ with $k = 4$, while Fig. 1(b), (c) and (d) present the corresponding traces of M_1^k, M_2^k, M_3^k, respectively. Each pair of a call and its accompanying return action is associated with a (dashed) line interval. In Fig. 1, $r(x)0$ is an action that reads 0 from x; $w(x)1$ is an action that writes 1 to x; $f(x)1$ is a flush action that changes the value of x to 1; $c(y)1$ is a *cas* action that changes the value of y to 1 successfully; $c_1, \ldots, c_4$ are the specific *cas* actions for marking the corresponding call and return actions; $g(x)1$ and $f(x)1$ are the guessed write action and its accompanying flush action for $w(x)1$; $g(y)1$ and $f(y)1$ are the guessed write action and its accompanying flush actions for $c(y)1$; g_i and f_i are the guessed write action and its accompanying flush actions for the action c_i $(1 \leq i \leq 4)$; Noted that the actions in Fig. 1(a) contain only values, while the actions in Fig. 1(b), (c) and (d) contain the snapshots of the memory.

In this example, M_1^k first guesses a marked write action g_1, performs the accompanying flush action f_1 and the call action of process P_1 and then reads 0 from x. Before M_1^k performs the $w(x)1$ action, it need to guess the write and *cas* actions of processes P_2 and P_3. These actions need to occur later than $w(x)1$ but their accompanying flush actions need to occur earlier than $f(x)1$ in t. Therefore, it guesses g_2, g_3 and $g(y)1$ accordingly. Then, M_1^k flushes g_2 (with f_2), guesses the call action of process P_2, flushes g_3 (with f_3), performs the return action of process P_1, and flushes $g(y)1$ (with $f(y)1$). At last, M_1^k flushes $w(x)1$ (with $f(x)1$), guesses the marked write action g_4, performs the accompanying flush action f_4 and guesses the return action of process P_2.

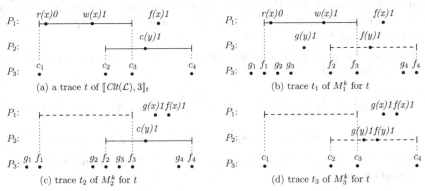

Fig. 1. Traces of M_1^k, M_2^k and M_3^k for a trace t of $[\![Clt(\mathcal{L}), 3]\!]_{tso}$

We now present the formal definition of the channel machine M_1^k. The channel machines M_2^k and M_3^k can be defined in a similar way. Note that $history([\![\mathcal{L}', 2]\!]_{sc})$ is a regular language, because the LTS $[\![\mathcal{L}', 2]\!]_{sc}$ contains a finite number of states. Let $\mathcal{A}_{Spec} = (Q_s, \Sigma_s, \rightarrow_s, q_{is})$ be a deterministic finite state automaton that accepts $history([\![\mathcal{L}', 2]\!]_{sc})$, where Q_s is a set of states, Σ_s is a set of transition labels, $\rightarrow_s \subseteq Q_s \times \Sigma_s \times Q_s$ is a transition relation and q_{is} is the initial state. It can be seen that each state in Q_s can be assumed as a final state because $history([\![\mathcal{L}', 2]\!]_{sc})$ is prefix-closed. Let $q_{error} \notin Q_s$ be a trap state. A new transition relation $\rightarrow_{s'}$ can be derived from $\rightarrow_s$ such that $q_1 \xrightarrow{\alpha}_{s'} q_2$ if either $q_1 \xrightarrow{\alpha}_s q_2$, or $q_1 \in Q_s$, $q_2 = q_{error}$ and there is no outgoing transitions from q_1 in $\xrightarrow{\alpha}_s$.

Let *Val* be the set of valuation functions that map a memory location in $\mathcal{X}_{\mathcal{L}}$ to a value in $\mathcal{D}_{\mathcal{L}}$ and x_r to a value in $markedVal(\mathcal{M}, \mathcal{D}_{\mathcal{L}}, 2)$. Then, the channel machine M_1^k is defined as a tuple $(Q_1, \{c_1\}, \Sigma, \Lambda, \Delta_1)$, where

- $Q_1 = (\{q_c\} \cup (Q_{\mathcal{L}} \times \{q_c\})) \times Val \times Val \times (Q_s \cup \{q_{error}\}) \times (marked Val(\mathcal{M}, \mathcal{D}_{\mathcal{L}}, 2) \cup \{\epsilon\}) \times \{0, \ldots, k\}$ is the state set. A configuration $(q, d_c, d_g, q_s, mak, cnt) \in Q_1$ consists of a control state q, a valuation d_c of the memory, a valuation d_g of the memory with all the stored items in c_1 applied, a state q_s for monitoring the violation of the linearizability condition, a marker mak indicating that each marked cas action is immediately followed by a corresponding call or return action, and the number cnt of the call and return actions already occurred in the whole trace.

- $\Sigma = \Sigma_1 \cup \Sigma_2 \cup \Sigma_3$ is the alphabet of channel contents with $\Sigma_1 = \{(3, x_{wit}, d) | d \in Val\}$, $\Sigma_2 = \{((i, x, d), \sharp) | 1 \leq i \leq 2, x \in \mathcal{X}_{\mathcal{L}}, d \in Val\}$ and $\Sigma_3 = \{a | (a, \sharp) \in \Sigma_2\}$. Σ_1 contains the items appended by guessing the marked cas actions. Σ_2 are the newest item in c_1 or the newest one for a memory location. Σ_3 is similar to Σ_2 except the symbols $\sharp$ are removed. In case that M_1^k is interpreted with a lossy channel, Σ_1 and Σ_2 are the sets of strong symbols of M_1^k.

- Λ is the set of transition labels. It contains write, cas, flush, call and return actions, but not read or τ actions (seen as ϵ transition).

- $\Delta_1 \subseteq Q_1 \times (\Lambda \cup \{\epsilon\}) \times Guard(\{c_1\}) \times Op(\{c_1\}) \times Q_1$ is the minimum transition relation. we report below the read and call transition rules in Δ_1, and other transition rules in Δ_1 can be found in [15]. For any $q_1, q_2 \in Q_{\mathcal{L}}$, $d_c, d_g \in Val$, $q_s \in Q_s$ and $cnt < k$,

  - For the $read(x, a)$ action of process P_1: if $q_1 \xrightarrow{read(x,a)}_{\mathcal{L}} q_2$, then for each $d \in Val$ with $d(x) = a$,

$$((q_1, q_c), d_c, d_g, q_s, \epsilon, cnt) \xrightarrow{\epsilon, c_1 \in \Sigma^* \cdot (\beta, \sharp) \cdot \Sigma^*, nop}_{\Delta_1} ((q_2, q_c), d_c, d_g, q_s, \epsilon, cnt)$$

$$((q_1, q_c), d, d_g, q_s, \epsilon, cnt) \xrightarrow{\epsilon, c_1 \in \Theta^*, nop}_{\Delta_1} ((q_2, q_c), d, d_g, q_s, \epsilon, cnt)$$

where $\beta = (1, x, d)$, $\Theta = \Sigma \backslash \{((1, x, d'), \sharp) | d' \in Val\}$ and nop means no operation on c_1.

  - For the $call(m, a)$ action of process P_1, if $q_s \xrightarrow{call(1,m,a)}_{s'} q_s'$, then

$$(q_c, d_c, d_g, q_s, call(1,m,a), cnt) \xrightarrow{call(1,m,a), c_1 \in \Sigma^*, nop}_{\Delta_1} ((is_{(m,a)}, q_c), d_c, d_g, q_s', \epsilon, cnt+1)$$

  - For the guessed $call(m, a)$ of process P_2, if $q_s \xrightarrow{call(2,m,a)}_{s'} q_s'$, then

$$(q, d_c, d_g, q_s, call(2,m,a), cnt) \xrightarrow{call(2,m,a), c_1 \in \Sigma^*, nop}_{\Delta_1} (q, d_c, d_g, q_s', \epsilon, cnt+1)$$

5.3   Reducing to a Control State Reachability Problem

Let $M_i^{k\text{-}w}$ be a channel machine that is resulted from M_i^k by replacing its all but write and cas transitions with internal transitions. The remaining cas actions can be regarded as write actions. Since a k-trace contains at most k call and return actions, and the first marked item can be guessed and flushed as in t_1 of Fig. 1(b) without influence subsequent executions, the number of marked items in a k-trace can be always less than k at any time. Let $S = \langle \Sigma_1, \Sigma_2 \rangle$, $K_1 = \langle k\text{-}1, |\mathcal{X}_{\mathcal{L}}|\text{+}1 \rangle$, the following lemma states that a control state reachability problem of a (S, K_1)-channel machine is enough to capture the complement problem of k-bounded TSO-to-SC linearizability.

Lemma 4. *There exists a marked violation t of k-bounded TSO-to-SC lineariz-ability between libraries $\mathcal{L}$ and $\mathcal{L}'$ for n processes from $(p_{init}, d_{init}, \epsilon^n)$ to (p_w, d_w, ϵ^n) in $[\![Clt(\mathcal{L}), n\text{+}1]\!]_{tso}$, if and only if $\bigcap_{i=1}^{n+1} T_{(q_i, q_i')}^{(S, K_1)} M_i^{k\text{-}w} \neq \emptyset$, where for each $1 \leq i \leq n\text{+}1$, $q_i = (p_{init}(i), d_{init}, d_{init}, q_{is}, \epsilon, 0)$, $q_i' = (p_w(i), d_w, d_w, q_{error}, \epsilon, |t \uparrow_{(\Sigma_{cal} \cup \Sigma_{ret})}|)$.*

Since the number of the specific configurations of $[\![Clt(\mathcal{L}), n\text{+}1]\!]_{tso}$ is finite, the complement problem of k-bounded TSO-to-SC linearizability can be further reduced to a finite number of control state reachability problems of the same (S, K_1)-channel machine but interpreted with lossy channels. Then, by Lemma 3 and the fact that a control state reachability problem of a (S, K)-lossy channel machine is decidable, we obtain the decidability result of k-bounded TSO-to-SC linearizability.

Theorem 1. *The decision problem of k-bounded TSO-to-SC linearizability is decidable.*

We reduce the reachability problem of a single-channel channel machine, which is known to have non-primitive recursive complexity [14], to a 3-bounded TSO-to-SC linearizability problem. Therefore, the k-bounded TSO-to-SC lin-earizability problem also has non-primitive recursive complexity.

Let $K_2 = \langle \infty, |\mathcal{X}_{\mathcal{L}}| + 1 \rangle$. Similar to Lemma 4, the complement problem of TSO-to-SC linearizability can be reduced to the control state reachability prob-lems of a channel machine where the amount of marked items in a channel is unbounded, or specifically, a (S, K_2)-channel machine that is the product of $M_1^{ts\text{-}w}, \ldots, M_{n+1}^{ts\text{-}w}$. Each $M_i^{ts\text{-}w}$ $(1 \leq i \leq n\text{+}1)$ is the same as $M_i^{k\text{-}w}$ except discard-ing counting the call and return actions.

Theorem 2. *The decision problem of TSO-to-SC linearizability can be reduced to a control state reachability problem of a (S, K_2)-channel machine.*

6 Conclusion and Future Work

We have shown in this paper that the decision problem of k-bounded TSO-to-SC linearizability is decidable for a concurrent system with $n \geq 1$ processes. The proof method is essentially by a reduction to a control state reachability problem of a lossy channel machine, which is already known to be decidable. To facilitate the reduction, a new process is introduced to use the specific *cas* actions to cap-ture the call and return actions of the original concurrent system. In this way, the complement problem of TSO-to-SC linearizability on the n processes can be transformed to a marked violation problem on the $n+1$ processes. Then, a channel machine M_i^k ($1 \leq i \leq n+1$) is constructed to simulate the k-bounded behaviors of the extended concurrent system from the perspective of each process P_i. We then demonstrate that the product of $M_1^{k\text{-}w}, \ldots, M_{n+1}^{k\text{-}w}$, when interpreted with lossy channels, can characterize the TSO behaviors of the original concurrent system. Furthermore, we show that the k-bounded TSO-to-SC linearizability problem has non-primitive recursive complexity.

Since the notion of k-bounded TSO-to-SC linearizability does not require the size of a store buffer or the length of a trace of a concurrent system to be bounded, it still allows infinite-state behaviors. Hence, our decidability result is non-trivial. It sheds light on developing algorithms for automatically verifying concurrent libraries on relaxed memory models.

We have successfully reduced the decision problem of TSO-to-SC linearizability to a control state reachability problem of a lossy-channel machine with unbounded number of strong symbols. However, the decidability of this problem still remains open. As future work, we would like to pursue this problem further with other possible heuristics. Also we would like to continue investigating the decidability of other correctness conditions of concurrent libraries and programs.

References

1. Alur, R., McMillan, K., Peled, D.: Model-checking of correctness conditions for concurrent objects. In: LICS 1996, pp. 219–228. IEEE Computer Society (1996)
2. Atig, M.F., Bouajjani, A., Burckhardt, S., Musuvathi, M.: On the verification problem for weak memory models. In: Hermenegildo, M., et al. (eds.) POPL 2010, pp. 7–18. ACM (2010)
3. Batty, M., Dodds, M., Gotsman, A.: Library abstraction for C/C++ concurrency. In: Giacobazzi, R., Cousot, R. (eds.) POPL 2013, pp. 235–248. ACM (2013)
4. Bouajjani, A., Emmi, M., Enea, C., Hamza, J.: Verifying concurrent programs against sequential specifications. In: Felleisen, M., Gardner, P. (eds.) ESOP 2013. LNCS, vol. 7792, pp. 290–309. Springer, Heidelberg (2013)
5. Bouajjani, A., Emmi, M., Enea, C., Hamza, J.: On reducing linearizability to state reachability. In: Halldórsson, M.M., Iwama, K., Kobayashi, N., Speckmann, B. (eds.) ICALP 2015. LNCS, vol. 9135, pp. 95–107. Springer, Heidelberg (2015)
6. Bouajjani, A., Emmi, M., Enea, C., Hamza, J.: Tractable refinement checking for concurrent objects. In: Rajamani, S.K., et al. (eds.) POPL 2015, pp. 651–662. ACM (2015)
7. Burckhardt, S., Gotsman, A., Musuvathi, M., Yang, H.: Concurrent library correctness on the TSO memory model. In: Seidl, H. (ed.) Programming Languages and Systems. LNCS, vol. 7211, pp. 87–107. Springer, Heidelberg (2012)
8. Gotsman, A., Musuvathi, M., Yang, H.: Show no weakness: sequentially consistent specifications of TSO libraries. In: Aguilera, M.K. (ed.) DISC 2012. LNCS, vol. 7611, pp. 31–45. Springer, Heidelberg (2012)
9. Herlihy, M.P., Wing, J.M.: Linearizability: a correctness condition for concurrent objects. ACM Trans. Program. Lang. Syst. **12**(3), 463–492 (1990)
10. Lamport, L.: How to make a multiprocessor computer that correctly executes multiprocess program. IEEE Trans. Comput. **28**(9), 690–691 (1979)

11. Manson, J., Pugh, W., Adve, S.V.: The Java memory model. In: Palsberg, J., Abadi, M. (eds.) POPL 2005, pp. 378–391. ACM (2005)

12. Owens, S., Sarkar, S., Sewell, P.: A better x86 memory model: x86-TSO. In: Berghofer, S., Nipkow, T., Urban, C., Wenzel, M. (eds.) TPHOLs 2009. LNCS, vol. 5674, pp. 391–407. Springer, Heidelberg (2009)

13. Sarkar, S., Sewell, P., Alglave, J., Maranget, L., Williams, D.: Understanding POWER multiprocessors. In: Hall, M.W., Padua, D.A. (eds.) PLDI 2011, pp. 175–186. ACM (2011)

14. Schnoebelen, P.: Verifying lossy channel systems has nonprimitive recursive complexity. Inf. Process. Lett. **83**(5), 251–261 (2002)

15. Wang, C., Lv, Y., Wu, P.: Bounded TSO-to-SC linearizability is decidable. Technical report ISCAS-SKLCS-15-11, State Key Laboratory of Computer Science, ISCAS, CAS (2015). http://lcs.ios.ac.cn/~lvyi/files/ISCAS-SKLCS-15-11.pdf

16. Wang, C., Lv, Y., Wu, P.: TSO-to-TSO Linearizability Is Undecidable. In: Finkbeiner, B., Pu, G., Zhang, L. (eds.) ATVA 2015. LNCS, vol. 9364, pp. 309–325. Springer, Heidelberg (2015)

Probabilistic Autoreductions

Liyu Zhang[1]$^{(\boxtimes)}$, Chen Yuan[3], and Haibin Kan[2]

[1] Department of Computer Science,
University of Texas Rio Grande Valley, Brownsville, TX 78520, USA
`liyu.zhang@utrgv.edu`
[2] School of Physical Mathematical Sciences,
Nanyang Technological University, Singapore, Singapore
`hbkan@fudan.edu.cn`
[3] Shanghai Key Lab of Intelligent Information Processing,
School of Computer Science, Fudan University, Shanghai 200433, China
`yuan0064@e.ntu.edu.sg`

Abstract. We consider autoreducibility of complete sets for the two common types of *probabilistic* polynomial-time reductions: RP reductions containing one-sided errors on positive input instances only, and BPP reductions containing two-sided errors. Specifically, we focus on the probabilistic counterparts of the deterministic many-one and truth-table autoreductions. We prove that non-trivial complete sets of NP are autoreducible for the *RP many-one reduction*. This extends the result by Glaßer et al. [9] that complete sets of NP are autoreducible for the deterministic many-one reduction. We also prove that complete sets of classes in the *truth-table Polynomial Hierarchy*, which is the polynomial hierarchy defined using the truth-table reduction instead of the general Turing reduction, are autoreducible with respect to the *BPP truth-table reductions*. This generalizes the result by Buhrman et al. [3] that truth-table-complete sets for NP are probabilistically truth-table autoreducible to multiple classes of higher complexity although for a weaker reduction.

Keywords: Computational complexity · Complete sets · Probabilistic polynomial-time autoreductions · Probabilistic many-one and truth-table reductions

1   Introduction

Let r be a reduction between two languages as defined in computational complexity such as the common *many-one* and *Turing* reductions. We say that a language L is *r-autoreducible* if L is reducible to itself via the reduction r where the reduction does not query on the same string as the input. In case that r is the many-one reduction, we require that r outputs a string different from the input in order to be an autoreduction. Researchers started investigating on autoreducibility as early as 1970's [18] although much of the work done then was in

L. Zhang—Research supported in part by NSF grant CCF-1218093.

the recursive setting. Ambos-Spies [1] translated the notion of autoreducibility to the polynomial-time setting, and Yao [21] considered autoreducibility in the probabilistic polynomial-time setting, which he called *coherence*.

More recently polynomial-time autoreducibilities, which correspond to polynomial-time reductions, gained attention due to its candidacy as a structural property that can be used in the *"Post's program for complexity theory"* [5] that aims at finding a structural/computational property that complete sets of two complexity classes don't share, hereby separating the two complexity classes. Autoreducibility is believed to be possibly one of such properties that will lead to new separation results in the future [3]. Autoreducibility certainly is also an interesting topic in its own right as knowing whether a language is autoreducible or not helps us better understand its computational complexity/structure. This is especially valuable when the language is a complete set for a complexity class for then autoreducibility of that language informs about the intrinsic computational properties that *all* languages in that class might have since they are all reduced to the complete set [4].

During the past two decades or so many exciting results have been obtained regarding autoreducibility of complete sets of common complexity classes. In particular we know now that many-one/Turing complete sets for most natural complexity classes including P, NP, Polynomial Hierarchy, PSPACE and EXP, are many-one/Turing autoreducible [2,3,9]. We refer the reader to Glaßer et al. [8] for a survey. There were also several more recent papers that investigate (non-)autoreducibility of complete sets for high-complexity classes such as NEXP [16] or more restricted reductions such as log-space reductions [11]. Most of those results, however, concern only autoreducibilities under *deterministic* reductions.

In this paper we consider autoreducibilities under *probabilistic* reductions and attempt to see whether and how the probabilistic counterparts of deterministic autoreductions might exhibit different or similar behaviors. Towards that goal we examined two common types of probabilistic reductions, the RP-type reductions, which have one-sided errors on positive input instances only, and BPP-type reductions, which have two-sided errors. We were able to prove that complete sets for NP are autoreducible for the *RP many-one reduction*, the RP-type probabilistic version of the common many-one reduction. This extends the result by Glaßer et al. [9] that complete sets of NP are autoreducible for the (deterministic) polynomial-time many-one reduction, to its RP-type probabilistic counterpart. We also proved that all complete sets of classes in the *truth-table Polynomial Hierarchy*, which is the polynomial hierarchy defined in terms of the polynomial-time truth-table reductions instead of the general Turing reduction, are autoreducible for the *BPP truth-table reduction*, the BPP-type probabilistic version of the common polynomial-time truth-table reduction. This generalizes the result by Buhrman et. al. [3] that truth-table-complete sets for NP are RP truth-table autoreducible to multiple classes of higher complexity but for a weaker reduction (BPP reduction instead of RP reduction). Wagner [20] introduced Θ^P-*levels* to the standard Polynomial Hierarchy, which coincides with the Δ^P-levels in the truth-table Polynomial Hierarchy. Hence, our result also indicates that all truth-table complete sets in the Θ^P-levels are truth-table autoreducible.

We give necessary definitions and notations in Sect. 2 and present our results in Sects. 3 and 4.

2   Definitions and Notations

We assume familiarity with basic notions in complexity theory and particularly, common complexity classes such as P, RP, NP, PH and BPP, and polynomial-time reductions including many-one ($\leq_m^p$), truth-table ($\leq_{tt}^p$) and Turing reductions ($\leq_T^p$) [13,14]. Without loss of generality, we use the alphabet $\Sigma = \{0,1\}$ and all sets we referred to are languages over Σ. We also use *Turing machines and algorithms interchangeably*. Following Glaßer et al. [10], we define a non-trivial set to be a set L where both L and $\overline{L}$ contain at least two distinct elements. This allows us present our results in a simple and concise way. All reductions used in this paper are *polynomial-time computable* unless otherwise specified. A language L is *complete* for a complexity class $\mathcal{C}$ for a reduction r if every language in $\mathcal{C}$ is reducible to L via r. For any algorithm $\mathcal{A}$, we use $\mathcal{A}(x)$ to denote both the execution and output of $\mathcal{A}$ on input x, i.e., "$\mathcal{A}(x)$ accepts" has the same meaning as "$\mathcal{A}(x) = accept$". We use $\mathcal{A}^B(x)$ for the similar meaning if the algorithm/Turing machine $\mathcal{A}$ has oracle access to a set B.

We consider two types of probabilistic reductions, those with one-sided errors on positive input instances only (RP-type) and those with two-sided errors (BPP-type), that correspond to the common deterministic *many-one* and *truth-table* reductions. The truth-table reduction is often also called *non-adaptive Turing reduction* in the literature.

Definition 1 [19]. *Define a language A to be RP (randomized polynomial-time) many-one reducible ($\leq_m^{rp}$) to a language B, if there exists a probabilistic polynomial-time algorithm $\mathcal{A}$ and a polynomial q such that the following hold for every $x \in \Sigma^*$:*

- *If $x \in A$, then $\Pr[\mathcal{A}(x) \in B] \geq \frac{1}{q(|x|)}$.*
- *If $x \notin A$, then $\Pr[\mathcal{A}(x) \in B] = 0$.*

Note that we cannot use a fixed polynomial or constant in the definition of RP many-one reductions for otherwise the reduction will not be transitive.

Definition 2. *Define a language A to be BPP (bounded-error probabilistic and polynomial-time) truth-table reducible ($\leq_{tt}^{bpp}$) to a language B, if there exists a probabilistic polynomial-time algorithm $\mathcal{A}$ with oracle access to B and a (deterministically) polynomial-time computable function g such that the following hold for every $x \in \Sigma^*$:*

- *On input x, $g(x)$ outputs all queries $\mathcal{A}$ will make to B.*
- *If $x \in A$, then $\Pr[\mathcal{A}^B \ accepts \ x] \geq \frac{2}{3}$.*
- *If $x \notin A$, then $\Pr[\mathcal{A}^B \ accepts \ x] \leq \frac{1}{3}$.*

Note that in the above definition Algorithm $\mathcal{A}$ is a Monte-Carlo algorithm [6,15], i.e., always runs in polynomial time on all inputs regardless of the probabilistic execution. In addition, Algorithm $\mathcal{A}$ on any fixed input makes the same set of queries to B that can be computed from the input deterministically in polynomial time by the function g. For the latter, we also say that algorithm $\mathcal{A}$ has *truth-table oracle access* to the language B.

Using the standard probability amplification technique we immediately obtain the following:

Corollary 1. *A language A is BPP truth-table reducible to a language B if and only if there exists a probabilistic polynomial-time algorithm $\mathcal{A}$ with truth-table oracle access to B such that the following hold for every $x \in \Sigma^*$:*

- *If $x \in A$, then $\Pr[\mathcal{A}^B$ accepts $x] \geq 1 - \frac{1}{2^{\lceil x \rceil}}$.*
- *If $x \notin A$, then $\Pr[\mathcal{A}^B$ accepts $x] < \frac{1}{2^{\lceil x \rceil}}$.*

With the above definition of the BPP truth-table reduction, the notion of *BPP truth-table hard* sets can be defined accordingly, which will be used for the proof of our results in Sect. 4.

Definition 3. *A language A is* BPP truth-table hard *for a complexity class $\mathcal{C}$ if every language $B \in \mathcal{C}$ is BPP truth-table reducible to A.*

Again using the standard probability amplification technique we immediately have the following:

Corollary 2. *A language A is BPP truth-table hard for a complexity class $\mathcal{C}$ if and only if for every language $B \in \mathcal{C}$, there exists a probabilistic polynomial-time algorithm $\mathcal{A}$ with truth-table oracle access to A that decides B with error probability 2^{-n}, where n is the input size.*

Now that we have defined RP many-one reductions and BPP truth-table reductions, the definition of the corresponding autoreductions follow straightforwardly.

3 RP Many-One Autoreductions

In this section we extend the result by Glaßer et al. [9] that many-one complete sets for NP are many-one autoreducible to the RP many-one reduction. Although the overall proof strategy is similar to that of the previous result, the additional part that argues why the adapted autoreduction output a correct value with the desired probability is not trivial at all.

Theorem 1. *If L is a non-trivial $\leq_m^{rp}$-complete set for NP, then L is $\leq_m^{rp}$-autoreducible.*

Proof. Let L be a nontrivial $\leq_m^{rp}$-complete set for NP. Then there exist strings a_1 and a_2 that belong to L, and b_1 and b_2 that belong to $\overline{L}$. Let N be a nondeterministic polynomial-time Turing machine that accepts L in time $p(n)$ for some polynomial p. Without loss of generality, we assume that all computation paths of N are of length $m = p(n)$ on inputs of length n. Define the following "left-set" [17] for L:

$$left(L) = \{\langle x, u \rangle | x \in L, \text{ and there is an accepting path } v \text{ of } N \text{ on } x \text{ where } u \leq v.\}$$

Here $\leq$ represents the common dictionary order among strings. Clearly $left(L) \in$ NP. Hence, there is a RP many-one reduction f from $left(L)$ to L such that a polynomial q exists, where

- for every $\langle x, y \rangle \in left(L)$, $\Pr[f(\langle x, y \rangle) \in L] \geq \frac{1}{q(|x|)}$, and
- for every $\langle x, y \rangle \notin left(L)$, $\Pr[f(\langle x, y \rangle) \in L] = 0$.

Now define a probabilistic polynomial-time computable function f' as follows: On input $\langle x, y \rangle$, where $|x| = n$ and $|y| = p(n)$, run f on $\langle x, y \rangle$ for $2q(n) \ln p(n)$ times and output x if at least one of the $2q(n) \ln p(n)$ runs of f outputs x, and output z if otherwise, where $z \neq x$ is one of the outputs by the $2q(n) \ln p(n)$ runs of f.

Consider the following function g:

```
1        Input x
2        Let m ← p(|x|) and z ← f'(⟨x, 0^m⟩)
3        If z ≠ x then output z
4        If f'(⟨x, 1^m⟩) = x then
5            If N(x) accepts along the path 1^m then
6                Output a string in {a_1, a_2} − {x}
7            Else
8                Output a string in {b_1, b_2} − {x}
9        Determine w of length m such that
             f'(⟨x, w⟩) = x ≠ f'(⟨x, w + 1⟩) = y
10       If N(x) accepts along w then
                output a string in {a_1, a_2} − {x}
11       Else
12           Run f' on ⟨x, w + 1⟩ again and yield output y'
13           Output y' if y' ≠ x, or {b_1, b_2} − {x} otherwise
```

Line 9 of the above function g can be executed in polynomial time since f' is a polynomial-time computable function and a standard binary search can be used to determine w as specified. Hence, the function g is polynomial-time computable. It is also clear that $g(x)$ always outputs a string $s \neq x$. Now it remains to show that $L \leq_m^{rp} L$ via g.

Assume $x \notin L$. Then for every $w \in \Sigma^m$, $\langle x, w \rangle \notin left(L)$ and hence, $f(\langle x, w \rangle) \notin L$. Therefore, the function g on input x outputs $z = f'(\langle x, w \rangle) \notin L$ in line 3, or output a string not in L in either line 8 or line 13.

Now assume $x \in L$. Let us define a computation path u (of length m) to be *good* if $\Pr[f(\langle x, u \rangle) = x] \leq 1/2q(n)$, and *bad* otherwise. We immediately observe the following claim:

Claim 1. For every $u \in \Sigma^m$ and $s = f'(\langle x, u \rangle)$,

i. if u is good, then
  (a) $\Pr[s \neq x] \geq \frac{1}{2p(n)}$, and
  (b) $\langle x, u \rangle \in left(L) \Rightarrow \Pr[s \in L \mid s \neq x] \geq \frac{1}{r(n)}$ for some polynomial $r(n)$.
ii. if u is bad, then $\Pr[s = x] > 1 - \frac{2}{p(n)}$.

Due to space limit, we omit the proof of Claim 1. Now we consider the following cases:

Case 1: 0^m is good. Let $z = f'(\langle x, 0^m \rangle)$. By i(a) of Claim 1, we have $\Pr[z \neq x] \geq \frac{1}{2p(n)}$. Hence, with probability at least $\frac{1}{2p(n)}$ the function g outputs $z = f'(\langle x, 0^m \rangle) \neq x$ in line 3. Furthermore, the probability that $z \in L$ in this case is $\Pr[z \in L \mid z \neq x] \geq 1/r(n)$ for some polynomial r, by i(b) of Claim 1. Therefore, the function g outputs a string $z \in L$ where $z \neq x$ with probability at least $\frac{1}{2p(n)} \cdot \frac{1}{r(n)}$ in case 1.

Case 2: Both 0^m and 1^m are bad. By ii of Claim 1,

$$\Pr[f'(\langle x, 0^m \rangle) = x] > 1 - \frac{2}{p(n)}, \text{ and}$$

$$\Pr[f'(\langle x, 1^m \rangle) = x] > 1 - \frac{2}{p(n)}.$$

Hence, with probability at least $(1 - \frac{2}{p(n)})^2 > 1 - \frac{4}{p(n)}$ the function g reaches line 5. Since $x \in L$, it follows that $\langle x, 1^m \rangle \in left(L)$ and so 1^m is an accepting path of N on x. Consequently g outputs in line 6 a string in $\{a1, a2\} - \{x\}$. Therefore, the function g outputs a string in L that does not equal to x with probability at least $1 - \frac{4}{p(n)}$ in case 2.

Case 3: 0^m is bad and 1^m is good. By Claim 1 again, we have

$$\Pr[f'(\langle x, 0^m \rangle) = x] > 1 - \frac{2}{p(n)}, \text{ and}$$

$$\Pr[f'(\langle x, 1^m \rangle) \neq x] \geq \frac{1}{2p(n)}.$$

Hence, with probability at least $(1 - \frac{2}{p(n)})\frac{1}{2p(n)}$ function g reaches line 9. Note that using a standard binary search to determine a string $w \in \Sigma^m$, where $f'(\langle x, w \rangle) = x$ and $f'(\langle x, w + 1 \rangle) \neq x$, requires computing $f'(\langle x, u \rangle)$ for $m - 1$ strings u in Σ^m in addition to 0^m and 1^m. We denote those strings by $u_1, \ldots, u_{m-1}$ and also let $u_0 = 0^m$ and $u_m = 1^m$. Now consider the following statement S_i for each $i \in \{0, 1, \ldots, m\}$:

$$S_i : u_i \text{ is bad} \Rightarrow f'(\langle x, u_i \rangle) = x.$$

Assume that the execution of g has reached line 9. Then $f'(\langle x, u_0 \rangle) = x$ and $f'(\langle x, u_m \rangle) \neq x$ as computed in line 2 and 4 of function g, respectively. Hence, both S_0 and S_m hold since u_0 is bad and u_m is good. Now for each $i \in \{1, \ldots, m-1\}$, let $z_i = f'(\langle x, u_i \rangle)$. Then Statement S_i fails with the following probability:

$$\Pr[\overline{S_i}] = \Pr[z_i \neq x \text{ and } u_i \text{ is bad}] \leq \Pr[z_i \neq x \mid u_i \text{ is bad}] < \frac{2}{p(n)}.$$

Therefore, all S_i's hold with probability at least $(1 - 2/p(n))^{p(n)-1} \geq 1/(2e^2)$. Now assume that all S_i's hold. Let w be the string determined in line 9 of function g. If w is an accepting path of N on x, then function g output a string in $L - \{x\}$ in line 10. Otherwise, it holds that $\langle x, w \rangle \in left(L)$ since $f'(\langle x, w \rangle) = x \in L$, and hence it must be the case that $\langle x, w + 1 \rangle \in left(L)$. Also, $w+1$ must be good since $f'(\langle x, w+1 \rangle) \neq x$. Thus, for the string y' produced in line 12, $\Pr[y' \in L \mid y' \neq x] \geq \frac{1}{r(n)}$ for some polynomial r by i(b) of Claim 1. It follows in this case that the function g outputs a string in L that does not equal x with probability at least that all the following events occur:

- The function g reaches line 9,
- All statements S_i hold for $i \in \{0, 1, 2, \ldots, m\}$, and
- The function g outputs a correct string in line 10 or line 13 given that the string $w + 1$ determined in line 9 is good.

By our previous discussion, the probability referred to above is at least

$$\left(\left(1 - \frac{2}{p(n)}\right) \frac{1}{2p(n)} \right) \cdot \frac{1}{2e^2} \cdot \left(\left(1 - \frac{1}{2q(n)}\right) \frac{1}{r(n)} \right) \geq \frac{1}{16e^2 p(n) r(n)}.$$

We have shown in all cases that the function g on an input $x \in L$ produces a string $y \neq x$ where $y \in L$ with probability no less than $\frac{1}{q'(n)}$ for some polynomial $q'(n)$. This finishes the proof of Theorem 1.   □

We believe that similar results to Theorem 1 would hold for RP versions of those reductions such as 1-tt and dtt, as well as for RP many-one complete sets of all the complexity classes as listed in Glaßer et al. [9] for which deterministic autoreducibilities of complete sets have been proved. However, we have not considered all the proof details for those sets yet.

4   BPP Truth-Table Autoreductions

In this section we consider *BPP truth-table reductions*. We prove that complete sets of classes in the *truth-table Polynomial Hierarchy* (PH^{tt}), which is defined below, are autoreducible for the BPP truth-table reductions. This generalizes the result by Buhrman et. al. [3] that truth-table complete sets for NP are probabilistically (in fact, RP) truth-table autoreducible.

Given a complexity class $\mathcal{C}$, we use P$^{tt[\mathcal{C}]}$ (NP$^{tt[\mathcal{C}]}$) to denote the class of languages decidable by a (nondeterministic) polynomial-time Turing machine with truth-table oracle access to a language in $\mathcal{C}$.

Definition 4.

$$\Sigma_0^{P,tt} = \Pi_0^{P,tt} = \Delta_0^{P,tt} = P$$

For $k \geq 1$,

$$\Sigma_k^{P,tt} = NP^{tt[\Sigma_{k-1}^{P,tt}]},$$

$$\Pi_k^{P,tt} = co\Sigma_k^{P,tt} = \{L \mid \overline{L} \in \Sigma_k^{P,tt}\}, \text{ and}$$

$$\Delta_k^{P,tt} = P^{tt[\Sigma_{k-1}^{P,tt}]}$$

$$PH^{tt} = \bigcup_{k \geq 0} \Sigma_k^{P,tt} = \bigcup_{k \geq 0} \Pi_k^{P,tt}$$

Clearly, PH^{tt} as defined above is the same as the standard PH except thats truth-table reductions are used instead of the general Turing reductions.

Towards of the goal of proving the main result of this section, we first observe that Valiant and Vazirani's result [19] can be extended to *relativized CNF formulas* in a straightforward manner.

Definition 5 [7]. *For any language A, a CNF formula relative to A, ϕ^A, is a CNF formula with each clause of the following form*

$$x_{i_1} \vee x_{i_2} \cdots \vee x_{i_u} \vee y_{i_1} \vee y_{i_2} \vee \cdots \vee y_{i_v},$$

where x_{i_j}'s are literals, and each y_{i_j} is a predicate of the form $A(w)$ or $\overline{A}(w)$ for some string w consisting of literals, 0's, 1's and other predicates of the form $A(w')$ or $\overline{A}(w')$.

Definition 6. *A relativized formula is a formula truth-table relative to a language A, $\phi^{tt[A]}$, if the following conditions hold:*

1. *$\phi^{tt[A]}$ is of the form $x_{i_1} \wedge x_{i_2} \wedge \cdots \wedge x_{i_k} \wedge TRUE \wedge F$, where $x_{i_1}, x_{i_2}, \ldots, x_{i_k}$ are literals and F is a formula relative to A.*
2. *Every variable appearing inside the predicate A in F does not appear outside predicate A unless it is one of those x_{i_j}'s $(1 \leq j \leq k)$ or its negation as stated in 1.*

If a CNF formula ϕ is one relative to some language A, we also say that ϕ is a *relativized (CNF) formula*. In case ϕ is truth-table relative to to A, ϕ is also called a *truth-table relativized (CNF) formula*, or simply *truth-table formula*.

Note that in order to satisfy a truth-table formula $x_{i_1} \wedge x_{i_2} \wedge \cdots \wedge x_{i_k} \wedge TRUE \wedge F$, all x_{i_j}'s must be true. This induces a polynomial-time algorithm that on a truth-table relativized formula $\phi^{tt[A]}$, outputs all queries to A that are needed for evaluating $\phi^{tt[A]}$ under a satisfying assignment and do not depend on any particular assignment of $\phi^{tt[A]}$.

With the above definitions, we say that a relativized formula ϕ^A is *satisfiable* if $\phi^A(a)$ evaluates to true for some assignment a, where for each occurrence of a predicate $A(w)$, $A(w) = 1$ if and only if $w \in A$ with the value of w determined by a and values of other predicates of the form $A(u)$ that appears in w. In addition, we say that ϕ^A has a unique satisfying assignment a if $\phi^A(a)$ evaluates to true and for every other a' where $\phi^A(a')$ is true, a and a' coincides on variables appearing outside of the predicate A.

Definition 7. *We define the following languages.*

- SAT *is the set of satisfiable CNF formulas.*
- SAT^A *(*$\text{SAT}^{tt[A]}$*) is the set of satisfiable CNF formulas (truth-table) relative to A.*
- USAT^A *(*$\text{USAT}^{tt[A]}$*) is the set of CNF formula (truth-table) relative to A that have unique satisfying assignments.*

Let NP^A ($\text{NP}^{tt[A]}$) denote the class of languages decidable by nondeterministic polynomial-time Turing machines with (truth-table) oracle A. Goldsmith and Joseph [12] proved that for every language A, SAT^A is complete for NP^A via a many-one reduction that does not use any oracle. A straightforward adaption of their proof yields a similar result for $\text{SAT}^{tt[A]}$.

Lemma 1. *For any language A, $\text{SAT}^{tt[A]}$ is complete for $\text{NP}^{tt[A]}$ via a many-one reduction that does not use any oracle.*

Theorem 2 [19]. *SAT is reducible to USAT via a RP many-one reduction r such that $r(\phi) \notin \overline{\text{SAT}}$ with probability 1 if $\phi \notin \text{SAT}$, and $r(\phi) \in \text{USAT}$ with probability at least $1/(4|\phi|)$ otherwise.*

Note that the proof of Theorem 2 is relativizable and hence we observe the following corollary immediately.

Corollary 3. *For any language A, $\text{SAT}^{tt[A]}$ is reducible to $\text{USAT}^{tt[A]}$ via a RP many-one reduction r such that $r(\phi) \in \overline{\text{SAT}^{tt[A]}}$ with probability 1 if $\phi \in \overline{\text{SAT}^{tt[A]}}$, and $r(\phi) \in \text{USAT}^{tt[A]}$ with probability at least $1/(4|\phi|)$ if $\phi \in \text{SAT}^{tt[A]}$. In addition, the reduction r does not use any oracle or change w for each occurrence of $A(w)$ or $\overline{A}(w)$ in ϕ.*

Buhrman et al. [3] proved that truth-table complete sets for NP are autoreducible for the RP truth-table reductions. The key element of the proof is a probabilistic algorithm that utilizes Theorem 2 and decides the satisfiability of a CNF formula with oracle access to a truth-table complete set for NP where a particular query is avoided. With Corollary 3, we are able to prove a result similar to Buhrman et al.'s for all complete sets in PH^{tt}.

Our proof uses the following languages consisting of relativizable propositional formulas.

Definition 8.

- $\text{SAT}^{(1),tt} = \text{SAT}$. *For every $k \geq 2$, $\text{SAT}^{(k),tt} = \text{SAT}^{tt[\text{SAT}^{(k-1),tt}]}$.*
- $\text{USAT}^{(1),tt} = \text{USAT}$. *For every $k \geq 2$, $\text{USAT}^{(k),tt} = \text{USAT}^{tt[\text{SAT}^{(k-1),tt}]}$.*
- $\text{F}^{(1)} = \text{F}^{(1),tt}$ *is the set of CNF propositional formulas. For every $k \geq 2$.*

– $F^{(k)}$ ($F^{(k),tt}$) is the set of CNF propositional formulas (truth-table) relative to $\mathrm{SAT}^{(k-1),tt}$.

Theorem 3. *For every $k \geq 1$ and every BPP truth-table hard set L_k for $\Sigma_k^{\mathrm{P},tt}$, there is a probabilistic algorithm $\mathcal{A}_k$ that on input $\langle \phi, y, 0^n \rangle$ runs in polynomial time in $|\phi|$ and n, where $\phi \in F^{(k),tt}$, and decides the satisfiability of ϕ with error probability at most $2^{-\max(n,|\phi|)}$. In addition, $\mathcal{A}_k$ makes only truth-table queries to L_k and does not query on y.*

Proof. We prove the theorem by induction. Theorem 4.10 in Buhrman et al. [3] essentially established the proof for the base case $k = 1$.

Now we prove the induction step. Let L be a BPP truth-table hard set for $\Sigma_k^{\mathrm{P},tt}$ where $k > 1$. Clearly, L is also BPP truth-table hard for $\Sigma_i^{\mathrm{P},tt}$ for $1 \leq i \leq k - 1$. Let $\mathcal{A}_1, \mathcal{A}_1, \cdots, \mathcal{A}_{k-1}$ be the probabilistic algorithms that satisfy the properties as stated in the lemma for $1 \leq i \leq k - 1$ and use L as the oracle.

Define

$$T = \left\{ \langle \phi, 0^i \rangle \mid \begin{array}{l} \phi \in \mathrm{SAT}^{(k),tt} \text{ has a satisfying assignment} \\ \text{where the i-th variable is true.} \end{array} \right\}$$

Since $T \in \Sigma_k^{\mathrm{P},tt}$, there is a BPP truth-table reduction g from T to L. Now let r be the RP many-one reduction of $\mathrm{SAT}^{(k),tt}$ to $\mathrm{USAT}^{(k),tt}$ as stated in Corollary 3. Hence, if $\phi \in \mathrm{SAT}^{(k),tt}$, then $r(\phi) \in \mathrm{USAT}^{(k),tt}$ with probability at least $1/4|\phi|$. Also, $r(\phi) \in \overline{\mathrm{SAT}^{(k),tt}}$ with probability 1 if $\phi \in \overline{\mathrm{SAT}^{(k),tt}}$.

Consider the following algorithm $\mathcal{B}$:

```
1    Input ⟨φ, y, 0ⁿ⟩
2    If φ ∉ F^(k),tt, REJECT
3    ψ := r(φ)
4    Use g to determine the memberships of ⟨ψ, 0ⁱ⟩ in T for
     1 ≤ i ≤ m, where m is the number of variables in ψ,
     with oracles L − {y} and L ∪ {y}, respectively.
5    Let a₀ and a₁ be the two assignments induced by the two
     sets of memberships of ⟨ψ, 0ⁱ⟩ in T, respectively,
     as determined in Line 4.
6    Evaluating ψ(a₀) and ψ(a₁) by calling 𝒜_{k−1} on
     ⟨q, y, 0^max(|φ|,n)⟩ for each membership query SAT^(k−1),tt(q).
7    If the above ψ(a₀) or ψ(a₁) evaluate to true, ACCEPT.
8    REJECT.
```

We again omit some part of the proof here due to space limit, but just state that one can show that the algorithm $\mathcal{B}$ has the following properties on input $\langle \phi, y, 0^n \rangle$:

– runs in polynomial time in $|\phi|$ and n,
– makes nonadaptive queries only to L, none of which is y,
– accepts ϕ with probability at most ϵ_1 if $\phi \notin \mathrm{SAT}^{(k),tt}$, and

- accepts ϕ with probability at least ϵ_2 if $\phi \in \mathrm{SAT}^{(k),tt}$,
- where $\epsilon_1 = o(1/\max(|\phi|, n))$ and $\epsilon_2 = \Omega(1/\max(|\phi|, n))$ for sufficiently large ϕ and n.

Then we use the standard amplification technique to obtain an algorithm $\mathcal{A}_k$ that consists of multiple runs of $\mathcal{B}'$ and has the error probability as stated in the lemma. $\qquad\square$

Corollary 4. *Every BPP truth-table-complete set of every class in PH^{tt} is BPP truth-table autoreducible.*

Proof. The corollary is trivially true for $\Sigma_0^{\mathrm{P},tt} = \Pi_0^{\mathrm{P},tt} = \Delta_0^{\mathrm{P},tt} = \mathrm{P}$ since every language in P can be decided by a deterministic polynomial-time Turing machine without any oracle access.

Now let L_a be a BPP truth-table-complete set for $\Sigma_k^{\mathrm{P},tt}$ for $k \geq 1$. Then L_a reduces to $\mathrm{SAT}^{(k),tt}$ via a many-one reduction f that does not use any oracle. Now let $\mathcal{A}_k$ be the probabilistic polynomial-time algorithm as stated in Theorem 3 that makes nonadaptive queries to L_a. Consider the following algorithm $\mathcal{A}_a$: On input x, compute $\phi = f(x)$. Then run $\mathcal{A}_k$ on $\langle \phi, x, 0^{|x|} \rangle$ and accept if and only if $\mathcal{A}_k$ accepts. It is not hard to show that $\mathcal{A}_a$ is a BPP truth-table autoreduction for L_a. It also follows immediately that a BPP truth-table-complete set L_b for $\Pi_k^{\mathrm{P},tt}$ is BPP truth-table autoreducible since $\overline{L_b}$ is a BPP truth-table-complete set for $\Sigma_k^{\mathrm{P},tt}$ and a set is autoreducible if and only if the complement of the set is autoreducible for the same reduction.

Now consider a BPP Truth-table-complete set L_c for $\Delta_k^{\mathrm{P},tt}$ for $k \geq 1$. Then there exists a deterministic polynomial-time Turing machine M that decides L_c with truth-table oracle access to $\mathrm{SAT}^{(k-1),tt}$. Let $\mathcal{A}_{k-1}$ be the probabilistic polynomial-time algorithm as stated in Theorem 3 that makes nonadaptive queries to L_c. Now consider the following algorithm $\mathcal{A}_c$: On input x, run M on x and resolve each query on q by running $\mathcal{A}_{k-1}$ on $\langle q, x, 0^{|x|} \rangle$; accept x if and only if M accepts x. Using the properties that $\mathcal{A}_{k-1}$ has according to Theorem 3 we can show that $\mathcal{A}_c$ is a BPP truth-table autoreduction for L_c. $\qquad\square$

Wagner [20] defined the Θ^{P}-levels by $\Theta_0^{\mathrm{P}} = \mathrm{P}$ and $\Theta_{k+1}^{\mathrm{P}} = L^{\Sigma_k^{\mathrm{P}}}$, where L denotes a log-space oracle Turing machine, and recommended including Θ^{P}-levels in the standard PH. He also proved for every $k \geq 1$ that $\Theta_k^{\mathrm{P}} = \mathrm{P}^{tt[\Sigma_{k-1}^{\mathrm{P}}]} = \Delta_k^{\mathrm{P},tt}$. Hence, we immediately have the following corollary.

Corollary 5. *For every $k \geq 0$, every truth-table complete set for each class in Θ_k^P is probabilistically truth-table autoreducible.*

References

1. Ambos-Spies, K.: On the structure of the polynomial time degrees of recursive sets. Habilitationsschrift, Zur Erlangung der Venia Legendi Für das Fach Informatik an der Abteilung Informatik der Universität Dortmund, September 1984

2. Beigel, R., Feigenbaum, J.: On being incoherent without being hard. Comput. Complex. **2**(1), 1–17 (1992)
3. Buhrman, H., Fortnow, L., van Melkebeek, D., Torenvliet, L.: Using autoreducibility to separate complexity classes. SIAM J. Comput. **29**(5), 1497–1520 (2000)
4. Buhrman, H., Torenvliet, L.: On the structure of complete sets. In: Proceedings 9th Structure in Complexity Theory, pp. 118–133 (1994)
5. Buhrman, H., Torenvliet, L.: A post's program for complexity theory. Bull. EATCS **85**, 41–51 (2005)
6. Cormen, T., Leiserson, C., Rivest, R., Stein, C.: Introduction to Algorithms, 3rd edn. The MIT Press, Cambridge (2009)
7. Fortnow, L.: The role of relativization in complexity theory. Bull. Eur. Assoc. Theor. Comput. Sci. **52**, 52–229 (1994)
8. Glaßer, C., Ogihara, M., Pavan, A., Selman, A., Zhang, L.: Autoreducibility and mitoticity. ACM SIGACT News **40**(3), 60–76 (2009)
9. Glaßer, C., Ogihara, M., Pavan, A., Selman, A.L., Zhang, L.: Autoreducibility, mitoticity, and immunity. J. Comput. Syst. Sci. **73**, 735–754 (2007)
10. Glaßer, C., Pavan, A., Selman, A., Zhang, L.: Splitting NP-complete sets. SIAM J. Comput. **37**(5), 1517–1535 (2008)
11. Glaßer, C., Witek, M.: Autoreducibility and mitoticity of logspace-complete sets for NP and other classes. In: Proceedings 39th International Symposium on Mathematical Foundations of Computer Science, part II, pp. 311–323. Budapest, Hungary, August 2014
12. Goldsmith, J., Joseph, D.: Three results on the polynomial isomorphism of complete sets. In: Proceedings of the 27th IEEE Symposium on Foundations of Computer Science, pp. 390–397. IEEE, New York (1986)
13. Hemaspaandra, L., Ogihara, M.: The Complexity Theory Companion. Springer, Berlin (2002)
14. Homer, S., Selman, A.: Computability and Complexity Theory. Texts in Computer Science, 2nd edn. Springer, New York (2011)
15. Motwani, R., Raghavan, P.: Randomized Algorithms. Cambridge University Press, Cambridge (1995)
16. Nguyen, D., Selman, A.: Non-autoreducible sets for NEXP. In: Proceedings 31st Symposium on Theoretical Aspects of Computer Science, pp. 590–601. LIPICS, Lyon, France, March 2014
17. Ogiwara, M., Watanabe, O.: On polynomial-time bounded truth-table reducibility of NP sets to sparse sets. SIAM J. Comput. **20**(3), 471–483 (1991)
18. Trakhtenbrot, B.: On autoreducibility. Dokl. Akad. Nauk SSSR **192**(6), 1224–1227 (1970). (translation in Soviet Math. Dokl. **11**(3), 814C817 (1790))
19. Valiant, L., Vazirani, V.: NP is as easy as detecting unique solutions. Theoret. Comput. Sci. **47**, 85–93 (1986)
20. Wagner, K.: Bounded query classes. SIAM J. Comput. **19**(5), 833–846 (1990)
21. Yao, A.: Coherent functions and program checkers. In: Proceedings of the 22nd Annual Symposium on Theory of Computing, pp. 89–94 (1990)

Software Engineering: Methods, Tools, Applications (Regular Papers)

ABS: A High-Level Modeling Language for Cloud-Aware Programming

Nikolaos Bezirgiannis[✉] and Frank de Boer

Centrum Wiskunde & Informatica (CWI), Amsterdam, The Netherlands
{n.bezirgiannis,f.s.de.boer}@cwi.nl

Abstract. Cloud technology has become an invaluable tool to the IT business, because of its attractive economic model. Yet, from the programmers' perspective, the development of cloud applications remains a major challenge. In this paper we introduce a programming language that allows Cloud applications to monitor and control their own deployment. Our language originates from the Abstract Behavioral Specification (ABS) language: a high-level object-oriented language for modeling concurrent systems. We extend the ABS language with Deployment Components which abstract over Virtual Machines of the Cloud and which enable any ABS application to distribute itself among multiple Cloud-machines. ABS models are executed by transforming them to distributed-object Haskell code. As a result, we obtain a Cloud-aware programming language which supports a full development cycle including modeling, resource analysis and code generation.

1 Introduction

The IT industry, always looking for cutting operational costs, has been increasingly relying on virtualized resources offered by the "Cloud". Besides being more economically attractive, the Cloud can allow certain software to benefit in security and execution speed. For these reasons, software applications are steadily being migrated to run on virtualized hardware, essentially turning cloud computing into a hot topic among the software community.

Recent research has led to numerous methodologies, tools, and technologies being proposed to help the migration and execution of software in the cloud, ranging from (static) configuration management tools to (live) orchestration middleware, and from simple resource monitoring services to the dynamic (elastic) provisioning of resources. Unfortunately, the (so-called) DevOps engineers are now burdened with developing and maintaining an extra logic for such cloud tools, besides the usual application logic. These cloud tools may be best described as semi-automatic and it is often the case that an engineer has to manually intervene to apply the desired configuration & deployment of a cloud application.

These cloud applications are migrated unchanged: monolithic boxes of code which are transferred from a non-cloud setting to the new cloud environment

Partly funded by the EU project FP7-610582 ENVISAGE. This work was carried out on the Dutch national e-infrastructure with the support of SURF Foundation.

© Springer-Verlag Berlin Heidelberg 2016
R.M. Freivalds et al. (Eds.): SOFSEM 2016, LNCS 9587, pp. 433–444, 2016.
DOI: 10.1007/978-3-662-49192-8_35

by the DevOps engineers. Such separation of the application from its execution is traditionally believed to be an advantage, long before Cloud came to existence. However, one would expect that with the introduction of the virtualized (dynamic) hardware of the Cloud, and since software logic is inherently dynamic, an application could "become aware" and leverage its own execution for managing its cloud resources & deployment in an optimal way, and without the constant administering of an engineer.

In this paper, we aim to address the challenges of engineering cloud applications by introducing a "cloud-aware" programming language that provides certain high-level abstractions for unifying the application logic together with its deployment logic in a single integrated environment, while in the same-time, hiding any lower-level hardware and cloud-provider considerations. The language is intended for DevOps engineers and (potentially) computational scientists who are responsible for both the development and execution of software residing in the Cloud but would rather focus more on the application's logic than manage continuously its deployment. Applications written in the proposed language are christened "cloud-aware" in the sense that they can *actively* monitor and control their own deployment.

The proposed language is based on the *Abstract Behavioral Specification language* (ABS), a formally-specified, object-oriented modeling language that has been used for both analyzing [1], verifying ([8]), and simulating [5] software programs, as well as running them in production through the various backends developed (currently targeting Java, Erlang, and Haskell). We extend ABS with *Deployment Components* that serve as a suitable abstraction over Cloud Virtual Machines and which allow the application to distribute itself among multiple (provider-agnostic) computing systems. The ABS developer writes code that can dynamically create, monitor and shutdown such Deployment Components (Virtual Machines) and most importantly bring up new objects inside them. To this end, an ABS cloud-application forms a cloud-aware *distributed-object* system, which consists of a number of inter-VM objects that communicate asynchronously, while recording any failures that may happen in the cloud.

An implementation of this extension must be efficient and safe so that it can be put in production code. For this, the Haskell backend of ABS is chosen for translating ABS code to Haskell intermediate code, which is again typechecked and transformed to an executable by an external Haskell compiler. We augment this backend with support for *Cloud-Haskell*, a framework for type-safe, fault-tolerant distributed programming in the Haskell ecosystem. The implementation, although in its infancy, is already being tested in a real cloud environment, exhibiting promising results which are also presented.

2   ABS Language and its Cloud Extension

The ABS (for "Abstract Behavioral Specification language") [5] is a statically-typed, executable modeling language with formal operational semantics. The language consists of a purely-functional programming core and an imperative, object-oriented layer. The syntax and behaviour resembles that of Java with two clear differences: side-effectful code cannot be mixed with pure expressions, and class inheritance is abolished in favour of code reuse via delta models [3]. ABS adds, next to the Java-like (passive) objects, builtin support for active (concurrent) objects coupled with cooperative scheduling.

The *functional core* provides a declarative way to describe computation which abstracts from possible imperative implementations of data structures. The primitive types (Int and Rational) can be extended with (possibly recursive) algebraic data types (ADTs) (e.g. `data Bool = True | False`) that can exhibit parametric polymorphism (`List < A >`) and Hindley-Milner type inference. Pure expressions are formed by successive λ-let abstractions and applications over values of the defined datatypes (`let x = 3 in x > 2 || True`). Function definitions associate a name to a pure expression which is evaluated in the scope where the the expression's free variables are bound to the function's arguments. The functional core supports pattern matching with a **case**-expression which matches a given expression against a list of branches.

The *imperative layer* specifies the interlaced control flow of the concurrent objects in terms of communication, synchronization, and internal computation. This layer extends the functional core (datatype and function definitions) with interface definitions, class definitions, and a main block. Interfaces declare a set of method names to their type-signatures. An interface **extends** other interfaces, in this case inheriting the methods of of its super-interfaces. A class definition declares its (private-only) attributes and a set of interfaces it **implements**. Method implementation bodies are comprised of statements of standard sequential composition $s; s$, assignment $x = rhs$, conditionals, while-loops, and return. Statements can mutate private attributes of the current class, locally-defined variables, and the method's formal parameters. The read-only variable **this** evaluates to the object in which computation occurs. A program's main block is a special method body with no *this* associated object. Classes are *not* types and used only to create object instances that instead are typed-by-interface. Note that interfaces support subtype polymorphism while ensuring strong encapsulation of implementation details.

Methods calls are either synchronous (`v = obj.method(args);`) where the statement is blocked until the method has finished with result v, or asynchronous (`f = obj!method(args);`) where the statement returns immediately with a *future* f (with type `Fut < A >`), without waiting for the method's completion. Each asynchronous method call creates a new *process* which will eventually store the result of the method call into the future reference. The caller can use this future reference to retrieve the result by calling the blocking statement `v = f.get;`. Objects may form a so-called *Concurrent Object Group* (COG), where objects (and their processes) share the same thread of control: at each point in time, only one process of the COG is executing. This process may decide to willfully pass control to another same-group process, by waiting until a future is ready (`await f?;`) or a boolean expression is met (`await exp;`). ABS does not specify any concrete policy for this cooperative scheduling of processes; it is left to the particular implementation (backend) to decide.

2.1 Extending to the Cloud

We extend the ABS language with syntactic and library support for Deployment Components. A *Deployment Component* (DC), first described in [7], is "an abstraction from the number and speed of the physical processors available to the underlying ABS program by a notion of concurrent resource". Simply put, a DC corresponds to a single (properly-quantified) Virtual Machine which executes ABS code. We restrict the definition of DC to correspond only to a Platform Virtual Machine (VM) residing inside the boundaries of a Cloud infrastructure. Multiple inter-communicating VMs effectively form an ABS cloud application.

To be able to programmatically (at will) create and delete VMs in any language, would require modeling them as first-class citizens of that language. As such, we introduce DCs as first-class citizens to the already-existing language of ABS in the least-intrusive way: by modeling them as objects. All created DC objects are typed by the interface DC. Minimal implementation for the methods of the DC interface are **shutdown** for shutting down and releasing the cloud resources of a virtual machine, and **load** for probing its average *system load*, i.e. a metric for how busy the underlying computing-power stays in a period of time. We use the Unix-style convention of returning 3 average values of 1, 5 and 15 min. The DC interface resides in the augmented standard library:

```
module StandardLibrary.CloudAPI;
interface DC {
    Unit shutdown();
    Triple<Rat,Rat,Rat> load();
}
```

By having a common DC interface the different cloud backends can agree on a basic service, while still being able to provide additional functionality through sub-interfaces and distinct DC-interfaced classes. Each DC-interfaced class implements the connection to a distinct cloud provider (e.g. Amazon, OpenStack). A code skeleton of such a class follows, where the DC (VM) is parameterized by the number of CPU cores and main RAM memory:

```
module StandardLibrary.SomeProvider;

data CpuSpec = Micro | Small | Large;
data MemSpec = GB(Int) | MB(Int);

class SomeProvider(CpuSpec c, MemSpec m) implements DC {
    Unit shutdown() { /*omitted*/ }
    Triple<Rat,Rat,Rat> load() { /*omitted*/ }
}
```

The implementor can expose other properties to DCs, such as, network, number of IO operations, VM region location. A concrete implementation, which is omitted for brevity, usually involves some high-level ABS logic coupled with low-level code written in a foreign language (in our case Haskell). The average ABS user will not have to provide such connections to the cloud, since we (the implementors) intend to provide class implementations for most major cloud providers/technologies, in an accompanying ABS library. With this approach, we lift the low-level API of the cloud provider (usually XML-RPC) to a *typed* high-level API (e.g. CpuSpec and MemSpec datatypes).

Moving on, we create an object of the SomeProvider class by passing the number of cores and memory measured in GBs as class' formal parameters. The call to "new SomeProvider" contacts the specific cloud provider in the background for bringing up a new VM instance. The provider responds with a unique identifier (commonly the public IP address of the created VM) which is stored in the DC object. Finally, the machine is released by calling shutdown(), making the DC object point to null.

```
DC dc1 = new SomeProvider(Large, GB(8));
_ future_l1 = dc1 ! load();// underscore infers the type
_ l1 = future_l1.get;
dc1 ! shutdown();
```

The creation of a DC object reference is usually fast, since it involves a single network communication between the current ABS node and the cloud provider. Still, the underlying VM requires considerably more time to boot up and be responsive, depending on factors such as provider's availability, congestion and hardware. To address this, we allow the creation of new dc objects to continue, but we require the program to potentially block when executing the first operation of the newly-created DC, as shown in the example:

```
DC mail_server = new Amazon(..);
DC web_server = new Azure(..);
DC db_server = new Rackspace(..);
mail_server!load(); // will block if DC is not up yet
```

Similar to this identifier, a method context contains the **thisDC** read-only variable (with type DC) that points to the VM host of the current executing object. A running ABS node can thus control itself (or any other nodes), by getting its system load or shutting down its own machine. However, after its creation, a running ABS node will remain idle until some objects are created/assigned to it. The **spawns** keyword is added to create objects that "live" and execute in a remote DC:

```
Interf1 o1 = dc1 spawns Cls1(args..);
o1 ! method1(args..);
this.method2(o1);
```

The spawns behaves similar to the new keyword: it creates a new object (inside a new COG), initializes it, and optionally calls its run method. Indeed, the expression new Cls1(params) is equivalent to thisDC spawns Cls1(params). The keyword spawns returns a *remote object reference*, (also called a proxy object; o1 in the above example) that can be called asynchronously for its methods and passed around as a parameter. Every remote object reference is a single "address" *uniquely identified* across the whole network of nodes. Calls to spawns will also (besides shutdown, load) block a until the VM is up and running. From a theoretical standpoint, a remotely-spawned object must point to the same code (attributes and methods) as in a local object; a remark that is reinforced in the Subsect. 3.1.

Whereas the development of ABS code is by-definition provider-dependent — the user has to explicitly specify the class of the cloud provider —, the communication and interaction between the spawned remote objects is (in principle) *provider-agnostic*. To this extent, an ABS user could write an ABS cloud application that spans over multiple cloud providers and, most importantly, different cloud technologies.

Cloud Failures. In cloud computing, and in any distributed system in general, failures are more frequent, mostly because of unreliable networks. Based on this fact, we further extend ABS with proper support for extensible, asynchronous exceptions. At the language level, exceptions are pure expressions modeled as single-constructor values of the ADT `Exception`. To define new exceptions the user writes a declaration similar to an ADT declaration, e.g. `exception MyException(Int, List < String >);`. Our cloud extension predefines certain common "local" exceptions (e.g. NullPointerException, DivisionByZeroException) and cloud-related exceptions (e.g. NetworkErrorException, DCAllocationException, DecodingException).

Exception values are either implicitly raised by a primitive operation (e.g. DivisionByZeroException) or explicitly signaled using the `throw` keyword. To recover from exceptions the user writes a `try/catch/finally` block as in Java, the only difference being that the user can pattern-match on each catch-clause for the exception-constructor arguments. Normally, if an exception reaches the outermost caller without being handled, its process will stop. We introduce a special built-in keyword named `die` that changes this behaviour and causes an object to be nullified and *all* of its processes to stop. With this in hand, a distributed application can easily model objects that can be remotely killed:

```
interface Killable { Unit kill(); }
class K implements Killable { Unit kill() { die; } }
Killable obj = dc1 spawns K();
obj ! kill();
```

Exceptions originating from asynchronous method calls are recorded in the future values and propagated to their callers. When a user calls "future.get;", an exception matching the exception of the callee-process will be raised. If on the other hand, the user does not call "future.get;", the exception will *not* be raised to the caller node. This design choice was a pragmatic one, to allow for fire-and-forget method calls versus method calls requiring confirmation. In our extension, we name this behaviour "lazy remote exceptions", analogous to lazy evaluation strategy.

3   Implementation

For the implementation, we rely on our abs2haskell backend/transcompiler. Haskell is a statically-typed, purely-functional language and, as such, it becomes straightforward to translate the ABS' functional core to Haskell. In the imperative layer, we model interfaces as Haskell's typeclasses, objects as references to mutable data (IORef in the Haskell world), and futures as synchronizing variables (MVar in Haskell). Nominal subtyping is achieved through an upcasting typeclass. An alternative would be to encode OO using extensible records [6], although this method widens the spectrum to structural subtyping.

At runtime, each COG becomes a Haskell lightweight thread (with SMP parallelism). The COG-thread holds a process-enabled *queue*, a process-disabled *table*, and a local *mailbox*. Upon an asynchronous method call, a new process is created and put in the end of the process-enabled queue; note that processes are not threads, they are coroutines (first-class continuations) and thus can be stored as data. The COG resumes the next process from the queue until it reaches an `await` (on a future or a condition), where the process is suspended and moved to the process-disabled table. Later, another process informs the COG (by writing to its mailbox) that the await-condition is met; the COG will move back the process to the enabled queue. This strategy avoids *busy-wait* polling the await conditions of processes.

Moving on to distributed programming, we extend our backend with layered support for Cloud-Haskell [4], a framework for Haskell that replicates Erlang's concurrency & distribution model (message passing) but in a type-safe manner. We reuse the network transports and serialization protocols defined in Cloud Haskell for the ABS transmitted data between Virtual Machines. Each COG-thread is accompanied with a separate Cloud-Haskell thread (also lightweight) that listens for messages in *public* mailbox and *forwards* them to the local mailbox of its associate COG-thread. This approach was chosen to firstly, avoid needless network-serialization between local communication and secondly, treat our distributed extension as optional to our (previously SMP-only) haskell backend.

Serialization. ABS data have to be serialized to a standard format before transmitting them between DCs. The serialization of values of primitives and algebraic datatypes are automatically done by Haskell. We serialize object/future references to *proxy references* by serializing their Cloud-Haskell thread ID (network-unique) together with a COG-unique ID, and leaving out their actual attributes/future results. Each asynchronous method call is serialized to a *static closure*, i.e. a static code-pointer to the method (known at compile-time and platform-independent) and a serialized environment of its free variables (method arguments and local variables). No kind of code (source-, byte- or machine-code) corresponding to the method body is transferred. All described-above serializations are type-safe and version-safe, in the sense that we include next to the payload of an ABS datum, its serialized type signature and the library-versions of any types involved; thus, we avoid decoding bugs because of type and library-version mismatches.

Garbage Collection. In a local-only setting, all ABS-based values, i.e. ADTs, futures, objects are automatically garbage-collected by the underlying Haskell GC. However, in our distributed setting some object/future references may have to be transmitted outside as proxy references, which results to the local ABS system garbage-collecting "too-early". An obvious solution would be to abolish automatic GC altogether, but that would hinder the development of software applications, especially those supposed to be long-running (as is the norm in cloud applications). On the other hand, introducing *distributed garbage collection* to ABS would allow both local and remote objects to be automatically GC'ed. The downside is that the user can no longer reason about the GC-incurred performance penalty which may be considerable. We chose a middleground where objects are by default GC-enabled and only become disabled when they are remotely communicated over (to another DC). The implementation has been straight-forward: a process appends the local object reference(s) that are transmitted remotely to a locally-held list of GC-disabled objects. This global list is held during the lifetime of the node, effectively surpassing the Haskell's garbage collector underneath. Our design choice was based on best practice; we believe that a distributed cloud ABS application of many DCs would contain a combination of a lot of local ephemeral objects, yet a few long-lived remote objects.

DCs, being special objects, are treated differently: when falling out of context they are automatically GC'ed. That does not mean that the attached VM is shut down. The user that wants to shutdown a DC but holds no reference to it anymore, has to contact a remote object residing there to return a reference to the DC (with `thisDC`), or to shut it down on user's behalf. If the executing program holds (now and in the future) no reference to a DC and its objects, we consider its VM unreachable and fallen out of scope of the ABS application.

Futures are garbage-collected in a publish-subscribe pattern: the caller of an asynchronous method is a subscriber, while the callee is the publisher. When the callee has finished computing the future, it "pushes" the resulted value to its caller (the direct subscriber) and may now locally garbage-collect that value. A subscriber that "passes over" a remote future reference to other nodes becomes an intermediate broker with the responsibility to later also "push" that future value to all others *before* it is allowed to locally garbage-collect it. This forwarding strategy avoids unnecessary tracking and network communication between the initial node and all (directly and indirectly) subscribed nodes.

Cloud Architecture. When creating a new DC, a cloud provider is on the background contacted (usually via an XML-RPC API) and asked to bring up a new VM with the given characteristics. After the machine has booted, the caller replicates itself (the current ABS application) by transmitting its machine code to the newly-created machine. In case the cloud provider offers heterogeneous platforms (different OS or CPU architecture), we instead transmit the ABS source code and compile it in-place with our compiler toolset (that prior reside in the VM's image). The new machine runs the transmitted ABS application and sends an acknowledgment signal to its creator, which can now start computations to the new DC by spawning new objects in it.

When it comes to network communication between machines, Cloud-Haskell does not enforce any particular network transport; even different transports can be composed together. Some existing implementations are TCP, AMQP, CCI, in-memory, etc. In ABS, the particular transport used depends on the implementation of the DC-interfaced class: we currently have DC-class implementations for OpenNebula (TCP), Azure (TCP) and Local (in-memory).

4 Experimental Results

We tested two instances of a real-world load-balancer: one with a static deployment of workers, and an adaptive (dynamic) load-balancer with worker VMs created on-demand based on how "well" the workers can keep up with incoming requests. Clients were submitting job requests (of approximately of equal size) to the balancer in a steady rate; workers were distinct Cloud VMs that were continuously computing the results for their incoming job requests.

The *static* load-balancer case is a fairly straight-forward cloud ABS application, consisting of 3 classes of LoadBalancer, Worker, and Client, exchanging asynchronous method calls of job requests/results. The LoadBalancer runs the main block and initially creates N number of Worker DCs (VMs) before starting accepting requests and forwarding to workers in round-robin. We ran this static deployment against varying size (N=1..16) of worker VMs. The results of the runs are shown in Fig. 1(a) stripped from the initial boot time of VMs. What we can draw from these results is that the completed jobs (per minute) nearly doubles when we double the number of worker VMs until we reach 5 workers. After that, we still increase the completed jobs but with a slower pace. This observation can be attributed to the fact that a point is reached where there is not a significant benefit from adding more worker VMs; the rate of job requests is always steady, thus worker VMs are "slacking".

We modified the static load-balancer to an *adaptive* version, that takes full advantage of the expressivity of the cloud extension. The LoadBalancer creates now only 1 initial VM. We accommodate the LoadBalancer with a HeartBeater object which periodically retrieves the load of each worker in the VM "farm". The HeartBeater computes the average `load` of all VMs and if this average exceeds 80 %, it creates a new DC (VM), adds it to the current farm, and remotely spawns a Worker in the new DC. We illustrate a particular run of this configuration in Fig. 1(b) (NB: VM boot times are not subtracted from the result). Each asterisk ∗ in (b) is a point where the HeartBeater decides to create a new DC. This run stabilizes on 6 workers, which is a good approximation of maximum speed (according to Fig. 1(a)), and possibly a good choice if we took into account any VM costs. As an extra, the HeartBeater could potentially `shutdown` machines if their load remained small (under a threshold) for a certain time.

The tests were conducted on the SURF cloud-provider with OpenNebula IaaS, modern 8-cores, each with 8GB RAM and 20Gbps Ethernet. Interesting to mention is that each worker can benefit from ABS multicore (SMP) parallelism. A snippet of the HeartBeater follows with the full ABS code at our repository[1]:

---

[1] Upstream abs2haskell repository at http://github.com/bezirg/abs2haskell.

```
class HeartBeater(List<Worker> farm, Balancer b) {
  Unit beat() {
    Rat avg = this.
    if (avg > 80/100) {
      DC dc = new NebulaDC(8,8192); // 8-core, 8GB RAM
      Worker w = dc spawns Worker();
      farm = Cons(w,farm);
      b ! updateFarm(farm); } } }
```

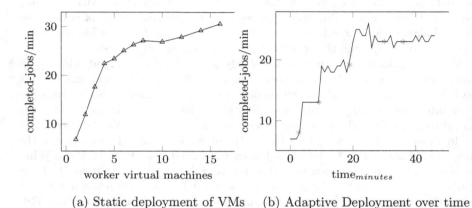

(a) Static deployment of VMs   (b) Adaptive Deployment over time

Fig. 1. (a) Static deployment of VMs. (b) Adaptive Deployment over time

5   Related Work

With the introduction of the Cloud, a plethora of cloud technologies & tools have appeared in the software community. We distinguish two categories of technologies related to our work: distributed-prog. languages and cloud middleware.

Distributed languages. Erlang is one of the first distributed-oriented languages that next to the canonical message-passing communication, offers distinct features, such as hot-code loading and serialization of arbitrary closures. This comes with a cost in safety since the serialized Erlang data are untyped and usually unversioned. Erlang's builtin processes are lightweight threads whereas ABS processes are coroutines (even more lightweight). The Akka framework brings (typed) actors to the Scala language. Although Akka provides a rich library and toolkit, it currently lacks a cloud-aware API. At runtime Akka is constrained by a threadpool (since JVM threads are expensive) and actors are not able to use cooperative scheduling and instead resort to a form of message routing. The Java RMI (Remote Method Invocation) is a library bundled in the Java platform for communication between remote objects. The product pioneered in areas such as bytecode downloading and distributed-GC. The method invocation is strictly synchronous (the caller has to wait for the remote method to finish) and thread-unsafe. JADE [2] is an active distributed multiagent system also built in Java; agents are more expressive than actors at the expense of program complexity and, possibly, performance.

Cloud middleware. Ubuntu JuJu is a tool primarily for scaling and orchestrating a system's deployment on the cloud. Juju also comes with a GUI for modeling and visualizing a cloud deployment and saving it to a "recipe" for later reuse. It is usually accompanied by a configuration-management tool (such as Puppet) for the provisioning of cloud machines. CoreOS is a container-based OS that provides service and configuration discovery. It can be thought as a low-level infrastructure, primarily targeted to system administrators, for managing system services across a cluster of cloud machines, The Aeolus research project has built various tools that can derive an optimized deployment from the constraint-based model of a desired deployment, and automatically deploy that derivation. Finally, general SaaS supported by cloud providers eases the migration of existing software to the cloud and its automatic scaling of deployment. Albeit dynamic, a SaaS deployment can only vary on the CPU consumption, whereas our proposal would allow a much more expressive deployment that can depend on arbitrary application logic.

6   Conclusion and Future Work

We presented an extension to the ABS language that permits the management of an application's own cloud-deployment inside the language itself. We discussed the realization of such extension (by a Haskell transcompiler) and the execution of an ABS cloud application (based on Cloud-Haskell). Results showed that ABS can benefit from the extra performance that the Cloud offers. Moreover, the extension gives to ABS the expression power it needs to fuse the application logic with the application's own (dynamic) deployment logic. A positive side-effect of the proposed extension is that, ABS being primarily a modeling language, could now be used to model also an application's deployment. Indeed, such cloud-aware software models could be simulated against different and dynamically-varying cloud deployment scenarios.

For future work we are considering additions both at the language and run-time level. At the language level, it would be beneficial to include, besides the system load, other metrics such as memory, disk usage, object count, process count, exceptions raised. In this way, an ABS application would enhance its monitor and cloud-control logic. In a different direction, we plan to work on adding a basic *service discovery* mechanism to the standard library of ABS. This can be simply realized by extending the DC interface with two extra methods: an `acquire(Interface obj)` method that returns a reference to a remote object implementing the provided Interface; an `expose(Interface obj)` that subscribes the passed object together with its current interface-view to the service registry of the DC.

At the system level, we are first interested in expanding our library support for other common cloud providers (such as Amazon EC2, OpenStack). Besides the current open (peer-to-peer) topology of DCs we want to add support for other cloud topologies, such as provider-specific, slave-master, or supervision topologies – a crude solution to topologies would be to introduce to the DC interface a method `List < DC > neighbours()` that lists all ABS nodes residing in the same private cloud network. A second consideration is to extend our virtualization technology support. With the introduction of micro-kernels (see the Xen hypervisor and unikernels), the cloud user no longer needs an OS underneath the application/service. By packaging the application into the kernel itself, the startup time of the VM is greatly improved, as is its management & distribution. The Haskell Lightweight Virtual Machine (HaLVM) is a promising such technology that allows the user to: "run Haskell programs without a host operating system". Likewise, *containers* (e.g. Docker), with its OS-level virtualization, would allow us to offer a more fine-grained control of deployment.

We believe that the cloud extension of ABS leads to new opportunities for furthering the application of formal methods to cloud computing, for example: specifying, verifying, and monitoring Service Level Agreements (SLA) of software systems — with that being the overall goal of ENVISAGE, our current research project. Indeed, we like to envisage software that is aware of its deployment and thus can control it, while its users merely monitor its behaviour via SLAs signed between the interested parties.

References

1. Albert, E., Arenas, P., Genaim, S., Gómez-Zamalloa, M., Puebla, G.: Costabs: a cost and termination analyzer for ABS. In: PEPM, pp. 151–154 (2012)
2. Bellifemine, F., Poggi, A., Rimassa, G.: Jade-a fipa-compliant agent framework. In: Proceedings of PAAM, vol. 99, pp. 33, London (1999)
3. Clarke, D., Helvensteijn, M., Schaefer, I.: Abstract delta modeling. ACM Sigplan Not. **46**(2), 13–22 (2011)
4. Epstein, J., Black, A.P., Peyton-Jones, S.: Towards haskell in the cloud. In: ACM SIGPLAN Notices, vol. 46, ACM (2011)
5. Johnsen, E.B., Hähnle, R., Schäfer, J., Schlatte, R., Steffen, M.: ABS: a core language for abstract behavioral specification. In: Aichernig, B.K., de Boer, F.S., Bonsangue, M.M. (eds.) Formal Methods for Components and Objects. LNCS, vol. 6957, pp. 142–164. Springer, Heidelberg (2011)
6. Kiselyov, O., Lmmel, R., Schupke, K.: Strongly typed heterogeneous collections. In: Proceedings of the 2004 ACM SIGPLAN workshop on Haskell, pp. 96–107 (2004)
7. Schäfer, J., Poetzsch-Heffter, A.: JCoBox: generalizing active objects to concurrent components. In: D'Hondt, T. (ed.) ECOOP 2010. LNCS, vol. 6183, pp. 275–299. Springer, Heidelberg (2010)
8. Wong, P.Y.H., Albert, E., Muschevici, R., Proença, J., Schäfer, J., Schlatte, R.: The abs tool suite: modelling, executing and analysing distributed adaptable objectoriented systems. STTT **14**(5), 567–588 (2012)

Aspect, Rich, and Anemic Domain Models in Enterprise Information Systems

Karel Cemus[1]([✉]), Tomas Cerny[1], Lubos Matl[1], and Michael J. Donahoo[2]

[1] Department of Computer Science, Czech Technical University,
Technická 2, 166 27 Praha, Czech Republic
{cemuskar,tomas.cerny,matllubo}@fel.cvut.cz
[2] Department of Computer Science, Baylor University,
One Bear Place #97356, Waco, TX, USA
jeff_donahoo@baylor.edu

Abstract. The research shows that maintenance of enterprise information systems consumes about 65–75 % of the software development time and about 40–60 % of maintenance efforts are devoted to software understanding. This paper compares the Anemic Domain Model used by the three-layered architecture followed by Java EE and .NET platforms and the Rich Domain Model often deployed into many conventional MVC-like web frameworks to a novel Aspect Domain Model followed by the Aspect-driven design. While all these models strive to avoid information restatement, they greatly differ in the underlying idea and resulting efficiency. This research compares considered models based on development efficacy, maintainability and their impact on the rest of the system. We evaluate qualities such as information cohesion, coupling and restatement, and discuss related maintenance efforts of the novel approach in the context of existing approaches.

Keywords: Model-view-controller · Three-layered architecture · Enterprise information systems · Design approach comparison · Aspect-oriented programming · Rich domain model · Anemic domain model

1 Introduction

Enterprise Information Systems (EISs) reflect various information to fulfill rapidly growing requirements, e.g., a domain model, selected business rules, validation rules, and presentation widgets. A design approach significantly impacts development and maintenance, because the resulting architecture influences information cohesion, coupling, and encapsulation [13]. Specifically, it determines readability, system's learning curve and affects the difficulty in making changes. In an early phase of a project, architects make critical design decisions to determine system architecture. This decision impacts project's future success as consequence of development and maintenance efforts.

Despite the existence of many approaches suggesting division of high-level responsibilities into layers or dedicated components, *cross-cutting concerns* are

© Springer-Verlag Berlin Heidelberg 2016
R.M. Freivalds et al. (Eds.): SOFSEM 2016, LNCS 9587, pp. 445–456, 2016.
DOI: 10.1007/978-3-662-49192-8_36

usually hard to separate from others. They tend to negatively impact component maintenance and reuse because standard object-oriented approaches fail to selectively address them [12]. In consequence, this results in emerging information restatement, low reuse, high duplication [7,10], and manual concern distribution throughout the whole system. Such design leads to difficult, tedious, and error-prone maintenance [2]. Therefore, an efficient design approach must provide mechanisms to separate all sorts of concerns, avoid cross-cuts, and provide the ability to centralize concerns to a single location.

Contemporary systems usually follow either the *three-layered architecture* with a *Anemic Domain Model* (ADM) or the *MVC-like architecture* with a *Rich Domain Model* (RDM) [3] with their *primary* system design. While the former is well-known[1], by the other we mean any *primary* system architecture derived from the *Model-View-Controller (MVC)* architectural pattern with RDM. For illustration, many conventional web frameworks belong to this category, such as Nette for PHP, Django for Python, Rails for Ruby, and Play for Java/Scala[2].

Unfortunately, the three-layered architecture with ADM often fails to address cross-cutting concerns. However, there exists its aspect-oriented extension, i.e., *Aspect-driven design approach* [6] with *Aspect Domain Model* (AsDM), tailored to deal with cross-cutting concerns. In this paper, we evaluate qualities and the impact of AsDM by its comparison to ADM and RDM used by common approaches. In Sect. 2, we discuss challenges in EIS design; while in Sect. 3, we elaborate the models and their underlying concepts. For better illustration of the differences, Sect. 4 shows a small case study highlighting qualities of all models and discusses the results. Section 5 provides an overview of related work, and we conclude the paper in Sect. 6.

2   Cross-Cutting Concerns

An EIS optimizes business processes and maintains large amounts of data with a respect to a given business domain [3,13]. The data management in an EIS often involves a User Interface (UI) and/or a web service component [7,10]. To satisfy all given requirements, a system covers various types of information, such as a domain model, presentation widgets, page layouts, business (domain) rules, and text localization. Unfortunately, most of these concerns apply to multiple locations throughout the both horizontal and vertical dimensions of the system [7].

Typical representatives of cross-cutting concerns in terms of Aspect-Oriented Programming (AOP) [12] are business rules. For example, we consider them in all layers of the three-layered architecture:

- input validation in the presentation layer
- business operations in the application layer
- constraint verification in the persistence layer

---

[1] Java EE and Microsoft .NET platforms build on it.
[2] http://nette.org,   http://djangoproject.com,   http://rubyonrails.org,   http://playframework.com.

Furthermore, the three-layered architecture supports component fragmentation inside of each layer, i.e., a system can have three presentation layer components: a UI, a Web Service and a Console. The business rules tangle throughout all components as they determine validation and access restrictions.

Besides business rules, there are other cross-cutting concerns [10]. The challenge lies in their addressing and reuse as they are considered in various places throughout the system. Furthermore, when we consider different technologies and possibly programming languages used for the implementation of different parts of a system [7], concern reuse becomes even more difficult. There are some attempts to overcome this gap [1, 7], although there are no architectures, design approaches, or frameworks providing a generic mechanism. Consequently, in most cases, developers are unable to capture a concern in a single focal point and then reuse it anywhere it is needed [10]; thus it usually results in high information restatement and code duplication [7, 12]. Unaddressed, tangled concerns are responsible for low cohesion of components [10], which deteriorates readability. The maintenance of such code is highly error-prone and inefficient [2].

3   Design Approaches and Domain Models

Despite the absence of a standard solution, there are approaches addressing the challenge and minimizing information restatement. This section summarizes their fundamental ideas and discusses their benefits and limitations in the context of cross-cutting concern encapsulation, cohesion, coupling, and reuse.

3.1   Anemic Domain Model in the Three-Layered Architecture

The three-layered architecture [3] splits up the application into three different layers: persistence, application and presentation. System functionality is thereby distributed with each layer owning a subset of the responsibilities. The key concepts of the three-layered architecture involve ADM [4, 9] and the *Transaction Script* [3] design patterns. They both determine the structure and qualities of a system as they directly define responsibilities and information distribution. ADM captures only data in a domain model with neither additional functionality nor dependencies. It is pushed into upper layers such as application and presentation, including business rules and logic.

Migration of business rules from ADM to the upper layers causes inconsistencies in the rules because they must be restated in all operations (transactions) performed over the model [8]. Furthermore, there are other cross-cutting concerns, which apply to more than one location, e.g., security policy, presentation widgets, or localization, but with this model, there is no single location to capture them. The approach strictly limits capabilities of the model as it does not capture anything but data. In consequence, all these additions have to be mixed in upper layers, which tangles them through code at the cost of low information encapsulation and high restatement [10].

3.2   Aspect Domain Model

The motivation for use of the transaction script pattern within the three-layered architecture lies in business operations (transactions), which reflect user's intentions. Each operation defines its own assumptions (preconditions) about the application state and a user's context. In this paper, we refer these preconditions as *business rules* and the operation-specific set of preconditions as a *business context*. For each business context, we are able to put down assumptions and attach them to an operation (e.g., in a use case scenario); however there is no easy way to reuse a context among multiple operations [8]. Similar issues apply also to UI development and maintenance as it also strongly depends on business rules [11].

The Aspect-oriented extension [6] of the three-layered architecture resolves its inability to deal with the cross-cutting concerns. This approach decomposes cross-cutting concerns into their isolated descriptions (aspects) expressed in convenient, e.g., domain-specific, languages (DSL) and aggregates them in a single place, the Aspect Domain Model (AsDM). It provides a single focal point and a single place to update. For example, it decomposes business rules into independent business contexts. Then it integrates (weaves) those descriptions into the system at runtime. Depending on transformation rules, a weaver produces multiple, distinct output components, such as a modified persistent layer or a modified application layer. Figure 1 illustrates an example of a decomposed system.

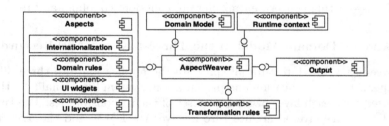

Fig. 1. Aspect-driven decomposition of a system

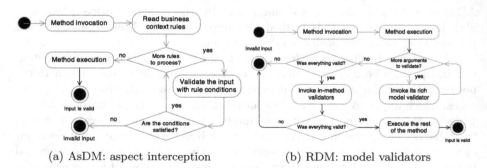

(a) AsDM: aspect interception   (b) RDM: model validators

Fig. 2. Input validation execution in different models

In consequence, we design and develop a system without any cross-cutting concerns involvement. For example, imagine input validation in the application layer. The conventional three-layered architecture tangles business rules into operations, but the Aspect-driven approach designs operations themselves and business rules captures in the AsDM. Then, it addresses the considered rules from operations through a business context. Figure 2a shows an example of such a method invocation with an aspect interception.

3.3 Rich Domain Model

RDM [4,9], contrary to ADM, suggests capturing all concerns in a model or in its dependencies through highly-decorated classes and fields; e.g., domain logic, database access, or field presentation widgets. Basically, each class carries information on how to persist, validate, and render itself.

Many fields and classes share a lot of configuration and mechanisms. The model uses inheritance and complex field data types to avoid code duplication and information restatement. This means that for class validation, persistence, or presentation, there usually exist super-classes already doing that; for fields, there are predefined complex data types. For example, there is an EmailAttribute, a StringAttribute or a NumberAttribute. Each class carries its own default validation rules, database constraints, presentation renderer, etc.

In consequence, RDM deals with cross-cutting concerns through rich data types. It captures them already in model fields, validators, and renderers, and tries to avoid their entangling into the rest of the application. However, there are some context-specific concerns and concerns cross-cutting multiple operations at once. As this model does not provide any mechanism to efficiently express and reuse them, it falls back to their repetition and object-oriented improvements described in [8]. For example, consider business operation-specific preconditions, which apply to multiple but not all operations of the class. As such, including those preconditions into the class validator is not an option and we are forced to duplicate rules in the operations themselves. Invocation and validation of such a business operation is shown in Fig. 2b. It shows regular invocation of standard object-specific validators at first followed by the invocation of operation-specific rules defined in the body of the operation.

4 Case Study

In order to evaluate the models, we conduct a case study demonstrating the efficacy of all three types. There are significant differences between them in design so the study focuses on system modeling rather than implementation as implementation differences are consequences of design. We use platform-specific models [13] because there are differences the most visible.

4.1   Assumptions

We model an issue tracker with multiple projects composed of issues. Each issue has a set of work logs and comments. The system maintains a catalog of users, where each user might have a different role in each project. The behavior is as usual, system implements following use cases: Users browse, report, edit, and resolve issues, make comments and work logs depending on their project role. Administrators also maintain catalogs of users and projects. There are no other roles or use cases. In overall, we identify 92 business rules, i.e., 63 model constraints, and 29 operation preconditions. Considering nonfunctional requirements, the application has three different layouts (desktops, tablets, cell phones), supports two languages (English and French), and provides two different sets of UI widgets; one for touch devices and one for the rest.

We consider the Java EE 7 platform with JDK 7 for future implementation of the plain three-layered architecture with ADM and AsDM. The model of the latter assumes JBoss Drools Expert 5.5[3] for business rules definition, and the Aspect Faces 1.4[4] framework implementing the concept proposed in [7] for the UI implementation. We will use our own implementation of the aspect weaving core based on AspectJ[5] as there is no other public implementation. Figure 3a shows an ADM of the application. The schema is identical for both these designs as their overall system architectures follow the same concept, because all the enhancements (business rules, localization, layouts, widgets) are described separately as standalone aspects, and they are weaved in at runtime.

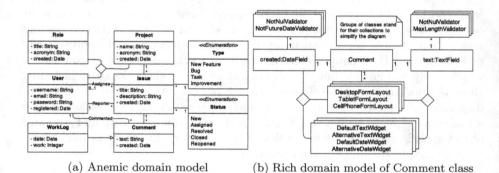

(a) Anemic domain model   (b) Rich domain model of Comment class

Fig. 3. Application domain model

MVC-like web framework Django 1.8 built on the Python language deploys RDM and we use it as a target platform in our case study. While the basic domain model structure remains same, i.e., classes are identical, there are differences

---

[3] http://www.drools.org/.

[4] http://www.aspectfaces.com/.

[5] http://www.eclipse.org/aspectj/.

reflecting the RDM character of the design. The detailed model diagram is very complex because fields of classes are complex data types and each of them has a relation to its own widgets and validators. We publish there only a small part in Fig. 3b. It shows the UML object diagram of the simplest class, the Comment. The rest of the diagram follows the structure from Fig. 3a with objects from Fig. 3b.

4.2 Model Efficiency

Our case study consists of 8 model classes restricted by 63 constraints, and 29 operation preconditions. We design this application three times, always for a different model. Table 1 shows ADM has no support for concerns separation and reuse, so we end up with manual repetition of the 29 constraints. The total 63 constraints located in 96 places throughout the application layer transactions and the presentation layer views. *JSR 303: Bean Validation* [1] enables partial business rules reuse, which improves the results. Without this the total number of constraints locations would be much greater. We repeat business rules as well as UI widgets and layouts to deliver all combinations of required UI. While we want to provide 3 layouts and 2 different widget sets for 10 views, we have to implement 3 times 2 times 10 views, i.e., 60 views in total. As we see, this architecture is unable to efficiently address cross-cutting concerns and usually ends up with highly tangled code with plenty duplications.

Table 1. The overall design efficiency

Criterion/approach	Anemic DM	Aspect DM	Rich DM
Restated rules	29	0	20[A,B]
Rules locations	96	1	38[A]
Business services	5	5	Unsupported
Validators	0	0	11
UI widgets	Unsupported	12	12
UI layouts	Unsupported	3	3
UI views	60	10	10
Client-side valid	Limited	Supported	Supported[C]

[A] Only in the domain model.
[B] Restated only references to validators.
[C] Requires mapping of validators to client-side technology.

The results in Table 1 confirm that AsDM describes all business rules in a single place and ensures their automatic runtime transformation as there are no restated rules and their are located in a single focal point. Our implementation of the aspect weaver combines rules, UI widgets and layout templates together with the annotated behavior classes to produce the application.

Finally, RDM uses complex data types and reusable field and class-specific validators, which is captured in the results as 20 restated validator references in a model. We model business operations as methods over the model, which allows us to reuse model validators and combine them with operation-specific conditions. Customized renderers produce the UI as we discuss earlier.

Table 1 summarizes overall efficiency of considered models. We see ADM delivers the worst result, as there is major business rules and UI views repetition. Maintenance of such a system is error-prone and requires many efforts. It results from its inability to reuse cross-cutting concerns. The table also shows significant improvement of the design by deploying AsDM. While it preserves the system's model structure and architecture, it removes all sorts of repetitions and establishes maximal concerns reuse. For example, contemporary implementations of the approach allow business rules reuse in all layers of the system including the client-side UI. The overall results in Table 1 suggest, the approach delivers even better design than the other contemporary approach. Although RDM delivers much better results than ADM, there is still some repetition. The most significant disadvantage is absence of the single focal point, which is shown in the results as 38 rules places.

4.3   Concerns Representation

Coupling, cohesion and encapsulation are crucial qualities [13] strongly influencing development efficiency. While high encapsulation is needed when adding new features, low coupling and high cohesion strongly simplify system modification. Otherwise these operations are highly tedious and error-prone.

Unfortunately, ADM does not provide any mechanism to represent and reuse cross-cuts such as business rules, widgets, and layouts. The results show, it results in high information restatement (96 locations, 29 restated rules) and concerns tangling. Although there is a standardized technique JSR 303 to reuse some of the model constraints, it does not cover operation preconditions. This concern tangling preserves information encapsulation inside transactions but all concerns are highly coupled, which significantly reduces code cohesion.

This limitation is resolved by AsDM, which deploys DLSs to efficiently describe these concerns. It aggregates all concerns and provides a single focal point, single place to maintain, which results confirm. This model significantly increases cohesion and reduces coupling. However, it deploys multiple DLSs and distributes concerns into multiple aspects, which significantly reduces their encapsulation. In consequence, although there is a single place to update, it might be a bit difficult to track relations among model and all aspects. Especially when the weaving is context-aware and performed at runtime.

RDM keeps everything tangled in the model as is suggested in notes to the measured repetitions in Table 1. This approach ensures very good encapsulation in the object-oriented manner, but as every class maintains all sorts of concerns and does not have direct support of the runtime context, the concerns are tangled. Coupling is high and cohesion low. The benefit is, that these worsen qualities apply only inside of a class. Inter-class coupling is low. Also the strong encapsulation significantly simplifies views and controllers. Unfortunately, concerns representation in the domain model includes use of the programming language for their expression, which makes them difficult to transform. In consequence, this design leads to server-centric system, where all logic and rules are in the model in the server we are unable to automatically transform it into the other technology or to propagate it in the client's side.

These qualities apply also in UI. While neither ADM nor standard technologies support automated UI generation (60 resulting views), AsDM enables us to generate the UI from collected aspects. It weaves them together on every request to deliver the context-aware UI. RDM uses widget and layout renderers bound to a class and its fields. The build process triggers UI generation methods invoking all particular renderers composing the resulting UI in the class interface. This method enables partially self-maintainable UI, but as the model is unable to automatically transform other concerns, the resulting UI lacks the functionality. In consequence, we must manually customize renderers to mix it in.

4.4 Usage Efforts

The three-layered architecture with ADM is very straightforward to deploy. It uses only one programming language and usually one mark-up for the UI description. On the other hand, in order to use AsDM, it is necessary to apply multiple supportive tools and frameworks. As the key idea relies on multiple DLSs, it needs compilers/interprets, and complex aspect weaver to describe and integrate all captured concerns together. Furthermore, there are various transformation rules producing the whole application. Such overhead and technological dependency indicate a significant barrier impeding the smooth concept adaptation. Another issue is the steep learning curve for development. The more languages in use, the more complicated the development becomes. The strong advantage of DSLs lies in the possibility to delegate the work to domain experts with limited programming knowledge [14].

Contrary, RDM is the central source of information for most of systems concerns. Such a model does not require complex tools or frameworks since it uses a single (programming) language. Also the learning curve is much less. However, every change in the system must be done in the domain model, which must be done only by developers. Then it requires recompilation and new deployment of the application.

4.5 Threats to Validity

We identify several internal and external threats possibly affecting the results. Among internal threats we consider validity of compared designs following the considered model types. A Java expert proposes ADM and AsDM designs as we design for Java EE platform. A Python expert with long professional experience proposes RDM design. All designs are peer reviewed. Two peers conduct manual measurement of the results independently, the results are double-checked.

We recognize validity of the overall case study as an external threat. Our case study is a small representative of a real enterprise system. All use cases, scenarios, and model classes create a core of production-size issue trackers, we just reduce the scope of the system. In consequence, the measured results are scalable to the production-size system including system out of the issue tracking domain as we do not use anything specific to this domain.

5   Related Work

We inspect the models from their ability to encapsulate the information and preserve high cohesion and low coupling to deliver the easiest possible maintenance. These challenges are well known, and authors discuss them e.g., in [15] where state that about 65–75 % of total project lifetime is consumed by maintenance.

The difficulty of business logic description and maintenance is discussed in [8]. It states that efficient isolation and application of business logic is very complicated using pure object-oriented techniques, even when we restrict our focus only to the application layer. It proposes a new concept to transform business logic into a presentation layer in [7]. The method relies on a domain model decorated by additional information, such as simple business rules as is introduced in [1]. It transforms a decorated model according to a given dynamic user's context into a user interface at runtime. It shows significant source code reduction in the UI (up to 32 %) and easy information/template maintenance. Later, the concept has been generalized into one of the compared approaches and introduced in [6].

The focus on maintainable, object-oriented software belongs among best practices, which are covered by architectural and design patterns explained in [3,5]. The premise is correct for many kinds of use cases, but there are cases where it fails. Furthermore, design patterns are meant for lower level of abstraction as they represent a component rather than system architecture.

Cross-cutting concerns impact design and code because they represent functionality and features, which affect multiple classes, components, and layers at once. Common object-oriented techniques fail here because they are unable to clearly assign the responsibility to a single object [12]. *Aspect-Oriented Programming* is an alternative concept, whose the fundamental idea lies in isolated description of all cross-cutting concerns and then their automated integration into the rest of an application. This technique lies in the core of the Aspect-driven approach to deal with cross-cutting concerns. The value of AOP in the three-layered architecture is also shown by the Java EE Spring framework[6], which enhances the Java EE platform to support requirements as mentioned above.

Seemingly, the *Model-driven development (MDD)* [16] is another major approach to consider. This concept uses application modeling in an independent, possibly graphical, language and semi-automated transformations into source code in a target platform. The approach name indicates that this concept is on

---

[6] http://projects.spring.io/spring-framework/.

a different layer of abstraction. While this paper focuses on types of domain models, the MDD is aimed to simplify the development process with respect to design approach and target architecture. Thus it can be used with all considered models as long as we are able to transform the abstract model into the chosen platform.

6   Conclusion

EISs development faces the pressure of fast-growing complexity and scope. Selection of an approach significantly impacts system development and maintenance efforts. In this paper, we evaluate three types of domain models and discuss their efficiency to deal with the cross-cutting concerns.

Our case study shows that the Anemic Domain Model usually used within the three-layered architecture fails to address cross-cutting concerns such as business rules. It has no mechanism to represent and reuse them, which leads to high information repetition, low cohesion and high coupling. Maintenance of such a system is very expensive and error-prone. Bright side of this model is its very straightforward use. This standardized design relies on a single programming language and is vastly supported by many tools and frameworks.

The Rich Domain Model used by many MVC-like web frameworks shows to be much more efficient with cross-cuts. It relies on rich classes and fields containing all possible information. Every field carries its own validators, renderers, labels, etc., which is the way to represent cross-cutting concerns. Finally, although the RDM has some boundaries and tends to high coupling and low cohesion, the supportive tools and frameworks aims to remain as simple as possible which flattens learning curve and makes its use very efficient in most cases.

Finally, the Aspect Domain Model implemented by Aspect-driven design approach extends the Anemic Domain Model within the three-layered architecture. It decomposes cross-cuts in the system and describes them individually in multiple DSLs in the model itself. Such design delivers a single focal point, a single place to update. Concerns are automatically weaved into the rest of the system at runtime. Furthermore, it transforms them into various components and technologies. Such isolation efficiently avoid information repetition and coupling. Easy maintenance, low development efforts, low coupling and high cohesion are direct consequences of the model. Unfortunately, there are significant initial costs; steep learning curve and complex tools and frameworks are required. Furthermore, contemporary implementations limit us in concerns transformation.

The results presented in this paper are solid input for more extensive, industry-related, case study. In future work, we conduct a case study measuring efficiency of AsDM in the real production-size application.

Acknowledgements. We would like to thank the Baylor University in Waco, Texas for the support during the research. This research was supported by the Grant Agency of the Czech Technical University in Prague, grant No. SGS14/198/OHK3/3T/13.

456   K. Cemus et al.

References

1. Bernard, E.: JSR 303: Bean validation. http://jcp.org/en/jsr/detail?id=303, November 2009
2. Fowler, M., Beck, K.: Refactoring: Improving the Design of Existing Code. Object Technology Series. Addison-Wesley, Boston (1999)
3. Fowler, M.: Patterns of Enterprise Application Architecture. Addison-Wesley Longman Publishing Co., Inc., Boston (2002)
4. Fowler, M.: Anemic domain model. http://martinfowler.com/bliki/AnemicDomainModel.html, November 2003
5. Gamma, E., Helm, R., Johnson, R., Vlissides, J.: Design Patterns: Elements of Reusable Object-Oriented Software. Addison-Wesley Longman Publishing Co., Inc., Boston (1995)
6. Cemus, K., Cerny, T.: Aspect-driven design of information systems. In: Geffert, V., Preneel, B., Rovan, B., Štuller, J., Tjoa, A.M. (eds.) SOFSEM 2014. LNCS, vol. 8327, pp. 174–186. Springer, Heidelberg (2014)
7. Cerny, T., Cemus, K., Donahoo, M.J., Song, E.: Aspect-driven, Data-reflective and Context-aware User Interfaces Design. Appl. Comput. Rev. 13(4), 53–66 (2013)
8. Cerny, T., Donahoo, M.J.: How to reduce costs of business logic maintenance. In: 2011 IEEE International Conference on Computer Science and Automation Engineering (CSAE), June 2011
9. Iglesias, C.A., Fernández-Villamor, J.I., Del Pozo, D., Garulli, L., García, B.: Combining Domain-Driven Design and Mashups for Service Development. Springer, Vienna (2011)
10. Kennard, R., Edmonds, E., Leaney, J.: Separation anxiety: stresses of developing a modern day separable user interface. In: 2nd Conference on Human System Interactions, HSI 2009, pp. 228–235. IEEE (2009)
11. Kennard, R., Edmonds, E., Leaney, J.: Separation anxiety: stresses of developing a modern day separable user interface. In: Proceedings of the 2nd Conference on Human System Interactions, HSI 2009, pp. 225–232. IEEE Press, Piscataway, NJ, USA (2009). http://portal.acm.org/citation.cfm?id=1689359.1689399
12. Kiczales, G., Irwin, J., Lamping, J., Loingtier, J.M., Lopes, C.V., Maeda, C., Mendhekar, A.: Aspect-oriented programming. In: Akşit, M., Matsuoka, S. (eds.) ECOOP 1997. Lecture Notes in Computer Science, vol. 1241, pp. 220–242. Springer, Berlin (1997)
13. Larman, C.: Applying UML and Patterns: An Introduction to Object-Oriented Analysis and Design and the Unified Process, 2nd edn. Prentice Hall PTR, Upper Saddle River (2001)
14. Mernik, M., Heering, J., Sloane, A.M.: When and how to develop domain-specific languages. ACM Comput. Surv. 37(4), 316–344 (2005). http://doi.acm.org/10.1145/1118890.1118892
15. Muthanna, S., Ponnambalam, K., Kontogiannis, K., Stacey, B.: A maintainability model for industrial software systems using design level metrics. In: Proceedings of the Seventh Working Conference on Reverse Engineering (WCRE 2000), pp. 248-256. IEEE Computer Society, Washington, DC, USA (2000). http://dl.acm.org/citation.cfm?id=832307.837117
16. Sendall, S., Kozaczynski, W.: Model transformation: the heart and soul of model-driven software development. IEEE Softw. 20(5), 42–45 (2003)

Finding Optimal Compatible Set of Software Components Using Integer Linear Programming

Jakub Danek$^{(\boxtimes)}$ and Premek Brada

NTIS – New Technologies for the Information Society, Faculty of Applied Sciences,
University of West Bohemia, Univerzitní 8, 306 14 Plzeň, Czech Republic
{danekja,brada}@kiv.zcu.cz

Abstract. Reusable components and libraries reduce costs in software development but also bring challenges like ensuring that application's components form a consistent and working set. While dependency management and build tools provide assistance in creating the set, they can't guarantee its correctness in terms of interoperability. On the other hand, the methods which detect component interoperability issues do not provide guidance in finding the proper set of components to fix any uncovered inconsistencies. In this work we present a method for finding such set of components which provides the required functionality, is free from type-level inconsistencies, and at the same time is optimal according to a given criterion. The method is based on pre-computed compatibility data and integer linear programming and allows to optimize the found solution set with respect to an arbitrary cost function.

Keywords: Component · Library · Compatibility · Composition · Integer linear programming · Verification

1 Introduction

Most of the currently developed software systems are a combination of the application code and (third-party) libraries or components. While these parts – collectively called "components" in this paper – considerably reduce the effort required for development of the desired functionality thanks to their re-usability, they also bring several new challenges.

As applications may consist of tens or hundreds of components, it is important to ensure they form a consistent set. However, finding a compatible set of components is not a trivial task. The term "compatibility" is rather wide and its meaning depends on the particular level of contract. For basic interoperability, the compatibility on syntactic level is required. To meet user's requirements fully, the components also need to behave as expected and have a desired quality level. Both integration testing and static verification techniques are used to check interoperability at various contract levels.

This publication was supported by the project LO1506 of the Czech Ministry of Education, Youth and Sports.

R.M. Freivalds et al. (Eds.): SOFSEM 2016, LNCS 9587, pp. 457–468, 2016.
DOI: 10.1007/978-3-662-49192-8_37

Easier management of components and their dependencies is supported by tools available for many common environments, e.g. Maven or Gradle for the JVM ecosystem, RubyGems for Ruby packages, etc. These tools employ declarative meta-data to ease the retrieval of the components and their transitive dependencies used in a project, but they don't ensure the resulting set is consistent. The issues involved are, among other ones, that

1. Correctness of the declared dependencies heavily depends on the component developers and thoroughness of the (library) integration tests.
2. The transitive dependencies may result in multiple (potentially incompatible) versions of the same component being fetched into the application. This may cause unexpected runtime errors in environments where only a single version may be linked at a given time (such as Java).

In our previous work [9] we have designed a method (and its implementation for Java bytecode) for the detection of interface-level incompatibilities between libraries used as components in an application as well as a means of storing related results of matching in a meta-data repository called CRCE [5]. However, solving the detected incompatibility issues requires the developers to make decisions for which they do not have enough information. In particular, trying to manually find the proper combination of component versions results in a series of trial and error attempts, which is slow, costly and doesn't ensure positive outcome.

In this work we present a method for finding a set of components which are mutually compatible and fit user requirements, while optimizing a selected metric value (e.g. the number of components used or total size of their binaries). The optimization in respect to a metric is useful e.g. when the resulting application is meant for devices with limited resources (operating memory size, disc size).

Importantly, we aim for creating an approach which can be used by developers without requiring them to modify their existing code, models or making major changes to their development process.

The text is structured as follows: after a motivation example and an overview of related work, Sect. 3.1 specifies the terms and relationships used by the proposed method. The following two sections provide a logical model of the problem (Sect. 3.2) and its implementation as an integer linear programming model (Sect. 3.3). Section 4 provides a validation of the method's performance requirements using simulation. In the Conclusion we summarise the paper and discuss future work.

1.1   Motivation

As mentioned, we address the situation when an unresolved dependency is detected in an application – i.e. it is known what its components require but the providers of the functionality are missing – and a suitable set of providing components needs to be supplied.

Our method should help achieve the following goals, in an automated way:

1. Finding an set of additional or upgraded application components such that the compatibility with the rest of the system is maintained.
2. Removing incompatible component duplicates, leaving only a single (version of) providing component in cases where the system contains dependencies on its multiple versions neither of which can satisfy all the requirements.
3. Optimizing the set of used components in respect to a given metric depending on the particular use-case.

Example 1. The following code demonstrates problem of missing dependencies in Java. All the examples throughout this paper are based on the presented piece of code.

```
//Library-A version 1              //client
class Foo {                        class Main {
    public Collection foo()            public static void
        { ... }                        main(String[] args) {
}                                          Foo f = new Foo();
class Bar {                                Bar b = new Bar();
    public void bar() { ... }              f.foo();  b.bar();
}                                  } }
```

In this example we want to upgrade *Library-A* to a newer version due to a e.g. better performance of the updated version. However, the developers of the library have decided to split it into two units:

```
//Library-A version 2              //Library-B version 1
class Foo {                        class Bar {  // moved here
    public Collection foo()            public void bar()
        { ... }                            { ... }
}                                  }
```

As a result, simple replacement of the *Library-A* old version with the new one would lead to runtime errors (`MethodNotFound` exception in the Java system).

While the *problem* can be detected by re-compilation (which is a possibility for own code, but not the 3rd-party components) and by static analysis for binary components [9], finding a *solution* is not a trivial task. Unless the change is well documented, developers don't have a simple way to reliably determine the proper set of components which can replace the *Library-A version 1* in the system even if they have access to a large component repository.

2   Related Work

The areas of research most relevant to the presented work are software build and integration order methods, component and module compatibility or interoperability checking approaches, and use of repositories and component meta-data in these approaches.

Concerning build and integration issues, Jezek [10] analyses transitive dependencies of libraries and highlights problems occurring during dependency resolution, noting that selecting the set of mutually compatible libraries is a difficult task on which both automated tools and human developers fail. The proposed solution is to use static analysis to verify compatibility within the dependency graph obtained by a given dependency resolution mechanism; however, no remedy is proposed when an incompatibility is found. Steindl et al. [13] describe a method to optimize the integration order of components so that some criteria (e.g. development effort) is minimized, using simulated annealing.

The survey of search-based techniques by Harman et al. [8] notes that many tasks in software engineering are essentially optimization problems. They list several cases of search-based optimisation techniques used in software model checking (for state space reduction), modularization (finding clusters of software elements) and reuse (selecting components best matching given requirements).

Within the last group, which is interesting from our point of view, Desnos et al. [6] propose a model-based approach to (incrementally) build a complete consistent assembly of components from a universe available in a repository, using a branch-and-bound algorithm with constraints representing functional and compatibility objectives. They optimize the solution only in respect to the number of dependencies and connections among them within the resulting assembly, not in respect to an arbitrary cost function (which is our goal).

In an interesting study, Olaechea et al. [11] evaluated exact and approximate algorithms for multi-objective optimization in the context of product line configurations, finding that it is possible to quickly find a Pareto-optimal solution for systems with low tens of features. However, when the size of the problem grows in the number of features and/or objectives, both types of algorithms lead to execution times of (tens of) hours to produce good accuracy results. This would make them unusable for our case of a part of normal development process.

In the area of component compatibility, Flores [7] addresses the replacement substituability verification by semi-automated creation and use of blackbox tests. Matching of interface syntactical structure is used as a compatibility sentry to the testing process, allowing for non-strict matches to be accepted (and resolved manually, e.g. by writing adaptors). The method uses ranking of the interface matching results and heuristics for adaptor generation to reduce the test suite set.

In our previous work [5] we described an approach for storing compatibility data in an efficient manner using an enhanced component repository, so as to enable their fast evaluation. The presented method is meant to directly utilise this approach for finding the optimal consistent set of interoperable components using static analysis.

3   The Search-Optimization Approach

This section introduces our method used to find an optimal set of mutually compatible components required for the application assembly. While we primarily aim for the syntax/binary-level compatibility, the method itself is usable at other contract levels as long as we are able to use the data model provided further to describe the dependencies between components and our requirements.

3.1 Component Representation

For this work, we use the same library and component representation model as presented by Jezek in [9]. It is based on notation by Brada [4] where component type $C = (P, R)$ with P and R representing the sets of provided and required parts of the component's interface. A software system can be described as $S = \{C_i = (P_i, R_i)\}_{i \in I}$. Dependencies between the components are described as pairs of appropriate provided-required parts not belonging to the same component: $D = \{(p, r) | p \in P_i \wedge r \in R_j \wedge i \neq j \wedge p \approx r)\}$ where $p \approx r$ denotes that the pair connects matching provided-required parts.

There is no restriction on what the model represents – the C can be a plain library, a component (independently of the component model) or a web service endpoint; the actual (p, r) pairs then represent library API usage, system specific means such as OSGi Import/Export package wiring [1], or general service invocation [2].

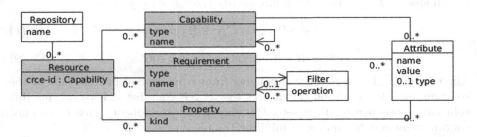

Fig. 1. CRCE meta-data model

For practical purposes, this formal model has been transformed into a meta-data scheme (Fig. 1) which describes a component as a *resource* with sets of *capabilities* i.e. the provided features (public classes of a library, etc.), *requirements* (e.g.imported package names or desired extra-functional properties of a matching provider) and other *properties* (e.g. values of selected metrics). Capabilities and requirements form a hierarchical structure, which allows to represent several levels of detail about the given feature – for example, a provided package (a top-level capability of an OSGi component), its contained public classes, and their non-private methods and attributes.

Using the described model we can represent the provided parts of existing components and prepare search queries based on the required parts.

Obtaining the Component Representation. In this work, we do not discuss the means of obtaining component interface representation. However, the relevant techniques mostly employ analysis of component implementation (source or binary form) [4,12] which is resource and time consuming and therefore unsuitable for potentially infinite searchable spaces like large software repositories.

We therefore persist the obtained representations as meta-data in CRCE, a Component Repository for Compatibility Evaluation [5], using the above scheme. Since the repository can perform arbitrary functions on the meta-data, it can be used to support advanced search methods like the one presented in this paper.

3.2 Search Model

The component interface representation model allows us to represent the query when searching for a solution (compatible components) from the set (repository) as described in Sect. 1.1. As already mentioned, the space of all available components Γ is theoretically infinite. From this space, candidate components must be selected that can contribute to the solution.

Given a set of requirements R, any component which provides at least a single item p which satisfies a requirement $r \in R$ is a candidate. The actual search space $SP \subseteq \Gamma$ must therefore satisfy the following conditions:

$$\forall C \in SP : \exists p \in C, r \in R : p \approx r \tag{1}$$

$$\forall r \in R : \exists C \in SP : p \approx r \wedge p \in C \tag{2}$$

The condition (1) selects all components from Γ which can provide part of the solution, and the condition (2) ensures the set SP contains at least one feasible solution. Subsequently, if the set SP is empty, the component space Γ does not contain components capable of fulfilling the requirements R.

The set SP is a superset of components satisfying the requirements R. Our goal is to find the *optimal* subset Res in respect to a chosen quality function. Given the definition of the quality function for a single component

$$cost(C) \rightarrow \mathbb{R} \tag{3}$$

the subset Res is defined using the following conditions:

$$Res \subseteq SP \tag{4}$$

$$\forall r \in R : \exists C \in Res : p \in C \wedge p \approx r \tag{5}$$

$$c_{Res} \text{ is } minimal/maximal, \text{ where } c_{Res} = \sum_{C_i \in Res} cost(C_i) \tag{6}$$

The condition (5) has the same purpose as the condition (2) for the set SP. It ensures the resulting subset still contains a feasible solution.

The condition (6) ensures the given quality value of the set is optimal depending on the nature of the *cost* function. Its definition remains abstract here, as it is bound to the specific use case. For example, for devices with limited resources the search might focus on minimal binary size or memory footprint. The *cost* function can also be used to minimize the number of selected components (similarly to [6]), in which case its value is the same for all components.

Depending on the use case, it may also be desirable to ensure a given requirement is satisfied only by a single component in the set *Res*, hence:

$$\forall(C_i, C_j) \in Res : p_k \in C_i \wedge p_l \in C_j \wedge p_k \approx r \wedge p_l \approx r \Rightarrow i = j \qquad (7)$$

This restriction is optional, since it need not be necessary at all times. For example while duplicate providers on Java classpath may result in linkage errors, duplicate web service endpoints do not interfere with each other (therefore the restriction is not necessary).

Example 2. To create the model corresponding to Example 1, the set Γ can in practice be represented by a repository (Maven repository or a meta-data providing repository such as CRCE). The remaining sets then are defined as:

```
R = {Foo, Bar}    SP = {Library-A-v1, Library-A-v2, Library-B-v1}

P[Library-A-v1] = {Foo, Bar}    P[Library-B-v1] = {Bar}
P[Library-A-v2] = {Foo}

cost(c) = {100, 300, 10}    //values are in the same order as in SP
optimization function: max //cost marks performance score

// Library-A-v1 satisfies all requirements, but combination of the
// remaining two has better score (300 + 10 > 100)
Res = {Library-A-v2, Library-B-v1}
```

3.3   Building an Integer Linear Programming Model

The presented task of finding the optimal set of components in the search space can be described using integer linear programming (ILP) model [3]. Linear programs consist of *variables* (also called *decisions*), *constraints* and an *objective function*. The *variables* take numeric values. *Constraints* define a feasible region for the values. The *objective function* specifies which of the feasible solutions is the optimal one. In integer programming, there is an additional constraint that some or all of the variables must take integer values.

The model can be written in its canonical form as

$$\min c^T x \text{ or } \max c^T x$$
$$Ax \geq b \qquad (8)$$
$$x \in Z^n$$

where A is a matrix and c, x, b are vectors.

In our case, the *variables* vector x represents components from the SP set and the following applies:

$$|x| = |SP|$$
$$\forall i : x_i \in \{0, 1\}$$
$$x_i = 1 \Leftrightarrow C_i \in Res$$

(9)

When the solution is found, those components whose value is 1 in vector x constitute the Res set.

Constraints $Ax \geq b$ of the model are derived from the set of requirements R. The matrix A describes which requirements are satisfied by which components of the set SP. It has dimensions $|R| \times |SP|$ and is defined as

- $a_{ij} = 1$ when the requirement r_i is satisfied by the component C_j.
- $a_{ij} = 0$ otherwise

The vector b ensures that every requirement is satisfied in the resulting component set. For that to hold, every requirement must be provided by the components in the resulting set at least once, therefore $b = \{1\}^{|R|}$.

If the condition (7) is in place, the constraints change from $Ax \geq b$ to $Ax = b$.

The *objective function* equals to the *cost* function described earlier. Elements of the vector c represent the *cost* value for each component C in the set SP (vector x):

$$|c| = |x|$$
$$\forall i : c_i = cost(C_i)$$

(10)

Table 1. Example ILP model

	Library-A-v1	Library-A-v2	Library-B-v1
Foo	1	1	0
Bar	1	0	1
Cost	100	300	10

Example 3. Using the definitions above, we can construct the integer linear programming search model for the case previously used in Examples 1 and 2. The matrix A and the cost vector c are shown in Table 1. The final solution vector x is to be computed. For this example it is $x = [0, 1, 1]$ which corresponds to the components *Library-A-v2* and *Library-B-v1*.

The resulting model representation as an ILP problem allows the use of existing solvers such as Gurobi[1] or CPLEX[2] to retrieve set of components which satisfies given requirements and is optimal in respect to given quality function.

---

[1] Gurobi - http://www.gurobi.com/.
[2] IBM   CPLEX   -   http://www-01.ibm.com/software/commerce/optimization/cplex-optimizer/.

4 Evaluation of the Approach

While the basic task of obtaining the set of components that can satisfy the missing dependencies is achievable by current technology, the practical uses of the full method as described in the preceding section depend on the ability to provide the optimised result set "fast enough". To examine the general time frame needed to find optimal solutions, we have performed a simulation using an artificially created data set representing an application and its dependencies.

To provide a solid foundation for the data set, its parameters were based on the data gained during experiments performed in our previous research [9] with the Qualitas Corpus set of applications [14]. Its evaluation had shown that the number of relationships between the corpus components vary from units to tens of connections and that for each component there are up to tens of versions available.

The data set used for the simulation consists of the search engine library *lucene* and its dependency on libraries *ant, ant-junit* and *junit*. Table 2 shows the exact count of references to these components from *lucene* as well as the number of different library versions used during the simulation. Library size in kB was chosen as the *cost* function, with values publicly available from the Maven Central repository[3], and the goal was to find the solution with minimal size.

Table 2. Test data parameters

	References from lucene	Versions in repository
Ant	5	21
Ant-junit	3	15
Junit	26	20

4.1 Simulation Scenarios

Since the goal of the simulation was to provide data on the performance of the method, real interoperability of concrete library versions was not tested – in a full implementation of the method, this information would be available from the CRCE repository using a single query. The speed of the query against the repository will depend on repository contents and particular implementation of the search algorithm, used indexing, etc., and is out of scope of this paper.

Two variants of the simulation were performed to cover the typical cases: (1) There are many possible solutions in the searched space – this was reduced to the case that all the versions of a particular library were considered as feasible (interoperable) solutions; (2) There is only a small number of possible solutions – a single version was considered feasible in the simulation.

---

[3] Maven Central - http://mvnrepository.com/.

The simulation was run for each subset of the dependency list – i.e. the goal was to find a solution for each library on its own, for each pair and for all three at once. This design was used to show the difference between scenarios in which a single library can provide all the required functionality and those in which a set of libraries must be used to satisfy all requirements (similarly to the many-to-one substitution in [6]).

To test method behaviour when working with larger data sets, the simulation was run multiple times with 1, 10, 100 and 1000 magnifier arguments applied to both the number of available library versions and their reference count from *lucene*. This represented (with some loss of faithfulness) progressively more complex applications and larger search spaces.

4.2   Results and Discussion

The simulation was run 100-times for each set-up, and mean value and standard deviation of computation time were calculated. The simulation machine had the following specifications: CPU Intel Core i7 3612QM, 4 cores / 8 threads, 2.1 GHz; RAM 8 GB; OS Gentoo Linux, 64bit, kernel 4.0.5; Solver Gurobi 6.0.4 (used via *Python 2.7* interface).

The Tables 3 and 4 display the values obtained for the *junit* library, *junit–ant-junit* pair and the whole triple[4]. The results show that for up to hundreds of candidate components and required constraints the computation of the model takes less than 1 second. Even for thousands of components (which is presumably an unlikely scenario) the computation takes several seconds, but less than 10.

Table 3. Simulation results for multiple data-size magnifiers - multiple feasible solutions. Values in [s]

Magnifier	1	10	100	1000
Junit	0.008 ± 0.0001	0.053 ± 0.0006	0.513 ± 0.0227	5.402 ± 0.2421
Junit, ant-junit	0.011 ± 0.0015	0.063 ± 0.0033	0.587 ± 0.0132	6.192 ± 0.2482
Junit, ant-junit, ant	0.019 ± 0.0040	0.077 ± 0.0061	0.730 ± 0.0416	7.664 ± 0.3757

Table 3 displays results for the scenario in which all candidate components were part of a feasible solution. Table 4 displays results for the scenario in which only a single instance of the candidate components was part of a feasible solution. It can be seen that the second scenario was solved more quickly for larger amounts of data. This is caused by lower optimization needs of the second scenario due to a limited number of feasible solutions.

The results of the simulation are promising in respect to practical use of the method: they show that the optimization does not introduce a performance

---

[4] Full results are available at http://relisa-dev.kiv.zcu.cz/data/experiments/ optimal-set-ilp-2015-07/.

Table 4. Simulation results for multiple data-size magnifiers - single feasible solution. Values in [s].

Magnifier	1	10	100	1000
Junit	0.008 ± 0.0001	0.050 ± 0.0078	0.444 ± 0.0037	4.630 ± 0.1173
Junit, ant-junit	0.011 ± 0.0003	0.055 ± 0.0033	0.536 ± 0.0614	5.319 ± 0.1485
Junit, ant-junit, ant	0.014 ± 0.0014	0.071 ± 0.0091	0.640 ± 0.0476	6.639 ± 0.2775

bottleneck in the approach. The method should be usable in developer assistance tools where it is important to provide results in time the users are willing to wait.

On the other hand, the current design of the method solves the initial problem only for the direct dependencies of the client code (represented by the R set in the search model) – the SP set does not include the dependencies of the candidate components themselves. To obtain a complete, consistent assembly the current method may need to be applied multiple times.

5   Conclusion

This paper has discussed the issues created by using libraries and components to assemble applications, with focus on choosing a compatible set of components optimized in respect to a chosen metrics. Even if developers are provided with enough information about the components, choosing an optimal and compatible set tends to be a difficult task.

As a solution, we have proposed an integer linear programming model which can be used to select a set of components providing required functionality while optimizing the set against an arbitrary function. Although the method has been demonstrated on Java examples, it is generic under the condition the component representation and interoperability (compatibility) data can be provided in the described data model; this condition can be satisfied using an advanced meta-data repository such as the CRCE tool we have designed previously.

Our simulation has shown that solving the optimisation task itself – and therefore the problem faced by the developers in case of repairing broken component or library dependencies – should be possible in a tolerable time frame. In particular, even in the (rather unrealistic) case of thousands of candidate components and broken component dependencies the time to find an optimal solution set was less than ten seconds; solutions for normal situations can be computed in less than a second. However, because the simulation was done using artificially created data, as a next step we need to perform more thorough evaluation based on real-world data to gain more detailed view on the method's performance. At the same time we would like to compare with the methods presented by Olaechea et al. [11] to validate the usefulness of our approach.

In the future, we would like to combine the method with our static analysis tools for linkage error detection and use it to find suitable solutions to the detected issues. Also, the current method does not take transitive dependencies into account and therefore may require multiple walk-throughs to result in a complete consistent set. We would like to address this problem in our future research by finding a suitable extension to the model so that the dependencies of the components are considered during the selection step as well.

References

1. OSGi Service Platform - Core Specification. Release 4, version 4.3, The OSGi Alliance, June 2011
2. Belguidoum, M., Dagnat, F.: Dependency management in software component deployment. In: Formal Aspects of Component Software (FACS 2006), Prague, Czech Republic, September 2006
3. Bosch, R., Trick, M.: Integer programming. In: Burke, E., Kendall, G. (eds.) Search Methodologies, pp. 69–95. Springer, New York (2005)
4. Brada, P.: Enhanced type-based component compatibility using deployment context information. Electron. Notes Theoret. Comput. Sci. **279**(2), 17–31 (2011)
5. Brada, P., Jezek, K.: Repository and meta-data design for efficient component consistency verification. Sci. Comput. Program. **97**, Part 3, 349–365 (2015)
6. Desnos, N., Huchard, M., Tremblay, G., Urtado, C., Vauttier, S.: Search-based many-to-one component substitution. J. Softw. Maintenance Evol. Res. Pract. **20**(5), 321–344 (2008)
7. Flores, A., Polo, M.: Testing-based process for component substitutability. Softw. Test. Verification Reliab. **22**(8), 529–561 (2012)
8. Harman, M., Mansouri, S.A., Zhang, Y.: Search-based software engineering. ACM Comput. Surv. **45**(1), 1–61 (2012)
9. Jezek, K., Ambroz, J.: Detecting incompatibilities concealed in duplicated software libraries. In: Proceedings of 41st Euromicro SEAA Conference (August 2015, to appear)
10. Jezek, K., Dietrich, J.: On the use of static analysis to safeguard recursive dependency resolution. In: 40th Euromicro SEAA Conference, pp. 166–173. IEEE, August 2014
11. Olaechea, R., Rayside, D., Guo, J., Czarnecki, K.: Comparison of exact and approximate multi-objective optimization for software product lines. In: Proceedings of the 18th International Software Product Line Conference, vol. 1, pp. 92–101. ACM (2014)
12. Parsons, T., Mos, A., Trofin, M., Gschwind, T., Murphy, J.: Extracting interactions in component-based systems. IEEE Trans. Software Eng. **34**(6), 783–799 (2008)
13. Steindl, M., Niemetz, M., Mottok, J., Racek, S.: Optimizing software integration in component-based embedded systems by using simulated annealing. In: Eurocon 2013, pp. 446–451. IEEE, July 2013
14. Tempero, E., Anslow, C., Dietrich, J., Han, T., Li, J., Lumpe, M., Melton, H., Noble, J.: The qualitas corpus: a curated collection of Java code for empirical studies. In: 2010 Asia Pacific Software Engineering Conference, pp. 336–345. IEEE, November 2010

Effective Parallel Multicore-Optimized K-mers Counting Algorithm

Tomáš Farkaš, Peter Kubán[(✉)], and Mária Lucká

Faculty of Informatics and Information Technologies,
Slovak University of Technology in Bratislava, Ilkovičova 2,
842 16 Bratislava, Slovakia
{xfarkast,peter_kuban,maria.lucka}@stuba.sk

Abstract. For many bioinformatics applications it is crucial to know frequencies of all subsequences of length k (k-mers) constructed from reads (short-reads) that are obtained in process of DNA sequencing. We present an effective parallel algorithm for k-mers counting that is based on nested bucket sort algorithm, whereby sizes of partitions and number of buckets per partition are precomputed. The proposed algorithm is designed for multicore architecture and properly combines MPI framework (OpenMPI) with POSIX threads achieving very good performance. According to our experiments it overcomes existing solutions in running time when compared on the genome of Drosophila melanogaster (SRX040485).

Keywords: Sorting · Bucket sort · K-mers counting · Parallel computation · Genome assembly

1 Introduction

Modern sequencing technologies are able to decode a genome (DNA) of an organism as a large number of small fragments, called *reads*. These reads are represented as strings over the alphabet {A, C, G, T} and their count can easily reach millions. Counting frequencies of *k-mers* (substrings of length k) in a long string, or in a set of strings (reads), is an ubiquitous task and it is important step in many applications. There are many bioinformatics tasks that are based on the knowledge of *k-mers* frequencies, such as data preprocessing for the *de novo* genome assembly, error correction of reads, repeat sequences detection, finding mutations in sequencing data, multiple sequence alignment. Determining the *k-mers* count is widely used with the currently most popular methods for large genome assembly using de Bruijn graphs [4] or Overlap Layout Consensus method [14] as well.

The paper is organized as follows: in Sect. 2 we formulate the problem and present the simplest method of *k-mers* counting and related work. In Sect. 3 we describe the proposed method and in Sects. 4 and 5 we present the achieved results and outline the future work.

© Springer-Verlag Berlin Heidelberg 2016
R.M. Freivalds et al. (Eds.): SOFSEM 2016, LNCS 9587, pp. 469–477, 2016.
DOI: 10.1007/978-3-662-49192-8_38

2   Problem Formulation and Related Work

In bioinformatics a read (short-read) is a finite string over the alphabet $\sum =$ {A, C, G, T}. A *k-mer* is a short string over the alphabet $\sum$ taken from a read whose length is k. A short-read s of the length n can be broken into $(n - k + 1)$ *k-mers*.

The concept of counting *k-mers* is straightforward and it is solvable for example with brute force algorithms using a hash table, with *k-mers* as keys and counts as values. However, brute force algorithms are often slow and unusable for large amounts of data. The most of *k-mer* counting algorithms are very similar to each other. In the beginning of the process of *k-mer* counting they process each read and extract all possible *k-mers* one by one. They differ mostly in the following steps: preprocessing and the *k-mers* counting itself. Several methods for *k-mer* counting rely on hashing or/and on suffix arrays (e.g. Jellyfish [11]) or/and on using Bloom filters [2] (e.g. BFCounter [13], Khmer [18]). Some methods use sorting (e.g. KAnalyze [1], Turtle [16]) and other methods are disk based with focus on low memory usage (e.g. KMC [6], KMC 2 [7], DSK [15]).

The Whole Genome Sequencing [17] and the Next-Generation Sequencing [17] technologies are currently more available, much cheaper and faster than old technologies and they produce enormous amount of sequencing reads on daily basis. For speeding up the processing of data used in bioinformatics, researchers are looking for new effective algorithms using parallelization techniques. They search for efficient and scalable parallel algorithms that take advantages of huge computational power offered by current high performance computers. They are focusing on massively parallel approaches and methods including multicore computers, clusters with multiple nodes or other types of high performance computing architectures.

The goal of this paper is to propose and design an efficient and scalable *k-mers* counting algorithm with focus on massively parallel computing and processing.

3   Proposal of the Algorithm

We propose a new parallel multithreaded algorithm for *k-mers* counting that is executed in three steps: (1) preprocessing (k-mers generation, partitions identification and *k-mers* distribution), (2) sorting, (3) counting. This algorithm is suitable for multicore architecture and it is based on and relies on fast bucket sort algorithm [5].

Bucket sort algorithm stands on basic idea of sorting the incoming data into mutual exclusive groups according to their values and then ordering each group separately.[1] Therefore we can identify two stages - data distribution phase and group processing phase.

---

[1] In this case by *sorting* we mean classification by some criteria, *ordering* means arranging the data into non-increasing or non-decreasing order.

3.1   Data Distribution Phase

As noted above, the bucket sort algorithm requires splitting the data into smaller groups. For achieving efficiency it is also important to uniformly distribute data over processes. Therefore, it is important (I): To split input data (reads) into multiple partitions of approximately same size and (II): To perform bucket sort splitting phase simultaneously. Ideally, for P partitions and datasets of size S each partition should have size S/P. After all k-mers are generated they are distributed into P different partitions in such a way that each partition gets k-mers that are unique for that partition. It means that k-mers in all parts are mutually exclusive and can be processed independently.

To achieve the goal, all processes in the communicator's rank (MPI_comm_rank) are divided into 2 groups of the same size - the group of masters and the group of workers. At the beginning input file(s) is read and processed into k-mers by the masters. After all k-mers are generated they are transformed into 64-bit numbers. Therefore, the size of k-mers is limited to 32 (one base/letter of DNA is encoded using 2 bits). As the workers are now inactive, every master reads and processes input file using two CPU cores. For to determine the partition where each k-mer must be sent, we have to look to the highest N bits (e.g. N=8) of each generated k-mer. Depending on the value of these bits (i.e. values from 0 to 255 for N=8 bits) the k-mers are distributed mutually exclusive to procesive nodes satisfying (II). However, the distribution based only on the highest N-bits of a k-mer is proven to provide non-uniform sizes of data groups, varying about 20 %. For instance, on a random 17 GB large fraction of human genome reads, sequences starting with 'CGA' occur about 6-times less frequent than those starting with 'TTT'. To solve this, we have used a load-balanced version of the algorithm. In this version, the data are still split according to the highest N bits, however N is set to produce approximately 16 times more groups than is the worker (processing node) count and those are then distributed to workers not in a fixed way according to the prefix, but according to the cumulative in-group count. For instance, using 16 CPUs, we do not set $N = 4$ as $2^4 = 16$, but $N = 8$. In this case it is possible for some workers to process for example 20 smaller groups in comparison with the others that process 10 larger ones and the amount of work is equal.

3.2   Group Processing Phase

Parallel sorting of each partition starts individually after all k-mers are distributed to the desired partitions. The process of sorting starts after all workers have got their data and every master process has finished. After then we have separate mutually exclusive groups of k-mers - represented as numbers, that can be processed independently. During this phase the masters are idle so every worker can perform ordering of the groups using two cores.

To sort the groups, every sorting algorithm can be used, however some of them perform better and in our case bucket sort is proved to be the most powerful one.

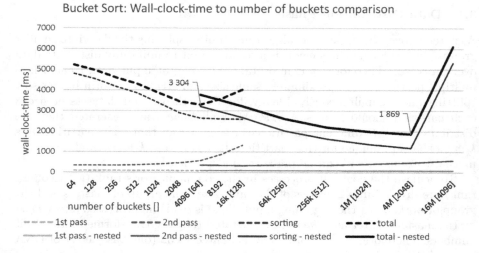

Fig. 1. Bucket sort: wall-clock-time to number of buckets comparison. 5 runs average data. Number in parentheses on axis x shows number of nested buckets. Graph shows improvements in wall-clock-time for higher number of buckets.

Nested Bucket Sort. The graph (Fig. 1) shows running times required for performing all three steps. *First pass (1st pass)* is the time required for counting the elements in groups to be created. Cumulative counts form group starting indices. *Second pass (2nd pass)* represents time needed for actual shifting the elements into groups, *sorting* indicates the time required for sorting the newly created groups. It is apparent that the time required for the first pass completion is constant, the time needed for sorting the buckets is linearly decreasing, what is correct due to the logarithmic axis x scale.[2] However, the time required for the second pass tends to grow rapidly to the end of measured interval and when reaching the border of 8192 buckets, its growth cancels the benefit of decreasing logarithmic factor of complexity and overall time starts to rise.

The number and corresponding size of buckets cannot be set to be optimal. It is apparent that *Sorting* performs well in millions of buckets, but *Second pass* limits it to a few thousands. However, there is a possibility to use the bucket sort algorithm once again and use it for sorting buckets themselves if needed. This version we will call *nested bucket sort* and it does not only allow to use much more buckets effectively, but it also reduces the total memory overhead coupled with using huge amount of buckets in the non-nested version. The basic idea

---

[2] Decreasing time effect is produced by comparative $O(n.log(n))$ final sorting algorithm that uses less time to sort k groups of l elements than 1 group of $k*l$ elements. This implies overall time complexity to be $O(n + n.log(\frac{n}{b}))$, b to be the number of buckets. For reasonable high b values the complexity tends to be $O(n)$. Please keep in mind that time complexity is much less accurate than actual real performance measuring.

of this algorithm, is a mixture of bucket, radix [10] and american flag sort [12] algorithms. The algorithm may look similar to the MSD (Most Significant Digit) radix sort algorithm, however the idea is slightly different. Our algorithm does not rejoin the split partitions before the next distribution step and uses indices to produce approximately $b*b$ buckets, where b is the original number of buckets. It is 40 % faster than both - the original bucket sort and the radix sort algorithms [9].

3.3 Frequencies Counting

For our algorithm it is critical to sort the k-mers before we start to count their abundances. The fact that they are sorted makes the counting very easy. In the following text we will assume that we have an sorted array of k-mers.

The simplest and the most straightforward algorithm could work like this: to start at the first element of the sorted array, to remember the k-mer at this element, to set the counter to one and to move to the next element. If the next element is same as previous one, the counter of this k-mer is increased, otherwise this new k-mer is remembered and the process is repeated until the last element is reached.

Another possibility is to use a step-based algorithm, which proceeds similarly as the binary search. It starts with pointing to first element, marking it as starting index, and remembering its first (current) value. Then it proceeds similarly as in binary search, except that the pointer does not move to the half of the array, but to the selected position in the array. The current value is compared with the value selected by the moved pointer. If these values are the same, the previous steps are repeated and the pointer is moved further, otherwise it returns back, dividing the step by two and moving the pointer again by this changed value and repeating the previous steps. However, the pointer is firstly checked, if the value of the next element (element at the position index +1) is different from current value. If it is true, the difference between the starting and current indeces is calculated, providing so the count of k-mers.

The second approach is not suitable for small datasets, where the number of unique k-mers is too large in comparison with the total number of k-mers. Efficiency of this algorithm depends on the ratio of the number of unique k-mers to the total number of k-mers. The lower this ratio is, the better. But in general, this algorithm does not bring great improvements. Considerably better improvements are expected on massive datasets, therefore the best algorithm for our experiments was the first straightforward algorithm.

4   Experiments and Results

Testing and time measuring was performed on the IBM iDataPlex cluster consisting of 52 computing nodes connected on the high-speed network: 2×10 Gb/s Ethernet (RoCE) with CPU 2x 6 cores Intel Xeon X5670 2.93 GHz and main memory of 48 GB (24 GB per processor/socket, NUMA architecture) per node.

The disks are 1x 2TB 7200 rpm SATA per node. Operation system is Scientific Linux 6.4 (kernel 2.6.32-358.el6) and the OpenMPI [8] version 1.6.5 was used.

The main limitation of our solution is that the current implementation runs only on datasets that do fit into available main memory. As a testing case we have selected the genome of Drosophila melanogaster (SRX040485), where the size of input dataset is suitable. Sequencing reads were taken from the European Nucleotide Archive (ENA, https://www.ebi.ac.uk/ena). Reads of length 76 were produced by Illumina Genome Analyzer II. The total size of processed reads was ≈9.8 GB (48 432 878 reads), from the genome containing 139.5 millions base pairs. Sequences were split into subsets of 10 millions, 20 millions, 30 millions, 40 millions and over 48 millions reads. For each tool we ran experiments several times and took average values as results.

Table 1. Benchmark Drosophila melanogaster datasets used for counting k-mers

Dataset	size of fastq file (GB)	number of reads	number of unique k-mers	number of distinct k-mers	total number of k-mers
subset 1	2.1	10 000 000	140 948 976	199 840 105	458 926 804
subset 2	4.1	20 000 000	187 804 272	272 601 565	906 605 778
subset 3	6.1	30 000 000	223 636 699	321 865 881	1 364 337 258
subset 4	8.1	40 000 000	259 588 501	365 877 941	1 821 845 026
whole dataset	9,8	48 432 878	289 202 942	394 953 130	2 207 175 063

We have measured the time required for calculation of the *k-mers* frequencies in five read data subsets using Jellyfish 2.2.0 [11], KMC 2 [7], BFCounter [13] and the proposed algorithm. We have chosen to benchmark frequencies, because this functionality is common to the most of tools, and it is a common analysis approach for determining assembly parameters [3]. We selected $k = 31$, as it is commonly used value. Then we tried to tweak parameters for each tested program to get the best results of them and applied each one to progressively increasing subsets of a ≈50 million reads genome dataset (Table 1).

In our experiments we have restricted the computational resources to one node with 12 cores, because we wanted to compare our solution with other existing software packages that do not run on multiple nodes. As Table 2 shows, the time usage of our approach is comparable to KMC 2 and Jellyfish. BFCounter is the slowest of the four tools compared in this tests, while KMC 2, Jellyfish and our solution are the fastest. Jellyfish is a winner in preprocessing time over all other tested programs. However, it is not possible to do more precise comparisons because the tested programs vary in used algorithms and methods. Our solution

Table 2. K-mers counting comparison of Drosophila melanogaster data subsets, running on one node with 12 cores

Number of reads	Time [s]			
-	Jellyfish	KMC 2	BFCounter	Our algorithm
10 000 000	17.1	7.4	257	8.1
20 000 000	32.4	14.3	443	12.4
30 000 000	48.9	21.4	688	17.2
40 000 000	54.5	28.9	887	22.3
48 432 878	61.2	36.8	980	26.5

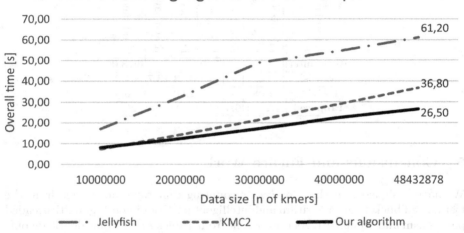

Fig. 2. K-mers counting comparison of Drosophila melanogaster data subsets (excluding BFCounter) running on one node with 12 cores.

outperformed all other tested algorithms in running time for the selected dataset on used hardware (Fig. 2).

Table 3 shows time required for the separate parts of algorithms. The measurements were done on the whole genome sequencing data of Drosophila melanogaster[3]. We could not measure all parts of every tested software, only those that had time measurements available directly in them as well as overall running time. As we have mentioned before, doing comparisons of tools based on different approaches, algorithms and methods is difficult and the results might be hardware dependent.

---

[3] Drosophila   melanogaster   (SRX040485)   http://www.ebi.ac.uk/ena/data/view/
SRX040485.

Table 3. Comparison of running wall-clock-times of k-mer frequency counting algorithms, running on one node with 12 cores

Algorithm	Initialization and data loading time dt [s]	Preprocessing time [s]	Data distribution time [s]	Data sorting time [s]	Frequency counting time [s]	Result saving time [s]	Total time [s]
BFCounter	978.0						978.0
Jellyfish	0.2	-	-		47.3	13.7	61.2
KMC 2	14.8	-		22.0			36.8
Proposed	4.2	1.1	6.4	13.5	1.0	0.3	26.5

5   Conclusions and Future Work

We have designed a method for *k-mer* counting that takes advantage from the fast nested bucket sort algorithm and intelligent partitioning using multithreaded parallelism and OpenMPI. As is apparent from Tables 2 and 3 our algorithm outperforms tested software (in running time) mainly thanks to fast nested bucket sort algorithm.

In the future we plan to improve our method to address its drawbacks. For example, to be able to use our method with lower memory requirements, so it can be used also for larger datasets that do not fit into available main memory. Therefore, we plan to extend our method for bigger genomes (e.g. human genome) and to run and test it on multiple nodes in parallel.

Acknowledgements. This work was partially supported by the Institute of Informatics and Software Engineering, FIIT STU, Intelligent analysis of big data by semantic-oriented and bio-inspired methods in parallel environment, the scientific Grant Agency of the Slovak Republic, grant No. VG 1/0752/14, project DNApuzzleDNA, FIIT STU that allowed us to use high performance computing on cluster of STU (project number 26230120002) and by the Research and Development Operational Programme as part of the project "International Centre of Excellence for Research of Intelligent and Secure Information-Communication Technologies and Systems", ITMS 26240120039.

References

1. Audano, P., Vannberg, F.: Kanalyze: a fast versatile pipelined k-mer toolkit. Bioinformatics (2014). doi:10.1093/bioinformatics/btu152. Accessed 18 March 2014
2. Bloom, B.H.: Space/time trade-offs in hash coding with allowable errors. Commun. ACM **13**(7), 422–426 (1970). doi:10.1145/362686.362692
3. Chikhi, R., Medvedev, P.: Informed and automated k-mer size selection for genome assembly. Bioinformatics **30**(1), 31–37 (2014)
4. Compeau, P.E., Pevzner, P.A., Tesler, G.: How to apply de Bruijn graphs to genome assembly. Nat. Biotechnol. **29**(11), 987–991 (2011). doi:10.1038/nbt.2023
5. Cormen, T.H., Leiserson, C.E., Rivest, R.L., Stein, C.: Introduction to Algorithms, 2nd edn., pp. 174–177. MIT Press and McGraw-Hill, Cambridge, New York (2001). ISBN: 0-262-03293-7. Section 8.4: Bucket sort
6. Deorowicz, S., Debudaj-Grabysz, A., Grabowski, S.: Disk-based k-mer counting on a PC. BMC Bioinf. **14**, 160 (2013)
7. Deorowicz, S., Kokot, M., Grabowski, S., Debudaj, A.: KMC 2: fast and resource-frugal k-mer counting. abs/1407.1507 (2014)
8. Edgar, G., Fagg, G.E., Bosilca, G.: Open MPI: goals, concept, and design of a next generation mpi implementation. In: Proceedings: 11th European PVM/MPI Users' Group Meeting, Budapest, Hungary (2004)
9. Farkaš, T.: Parallel Bucket sort algorithm for ordering short DNA sequences. In: IIT.SRC 2015: Student Research Conference, Bratislava, pp. 77–82 (2015). ISBN: 978-80-227-4342-6
10. Hollerith, H.: US. pat. Nr. 395781, 395782, 395783
11. Marais, G., Kingsford, C.: A fast, lock-free approach for efficient parallel counting of occurrences of k-mers. Bioinformatics **27**(6), 764–770 (2011)
12. McIlroy, P.M., et al.: Engineering radix sort. Comput. Syst. **6**(1), 5–27 (1993)
13. Melsted, P., Pritchard, J.K.: Efficient counting of k-mers in DNA sequences using a bloom filter. BMC Bioinform. **12**, 333 (2011)
14. Pevzner, P.A., Tang, H., Waterman, M.S.: An eulerian path approach to DNA fragment assembly. Proc. Nat. Acad. Sci. U.S.A. **98**(17), 9748–9753 (2001)
15. Rizk, G., Lavenier, D., Chikhi, R.: DSK: k-mer counting with very low memory usage. Bioinformatics **29**(5), 652–653 (2013)
16. Roy, R.S., Bhattacharya, D., Schliep, A.: Turtle: identifying frequent k-mers with cache-efficient algorithms. Bioinformatics (2014). doi:10.1093/bioinformatics/btu132
17. Shendure, J., Ji, H.: Next-generation DNS sequencing. Nat. Biotechnol. **26**(10), 1135–1145 (2008)
18. Zhang, Q., Pell, J., Canino-Koning, R., Howe, A.C., Brown, C.T.: These are not the k-mers you are looking for: efficient online k-mer counting using a probabilistic data structure. PLoS ONE **9**(7), e101271 (2014). doi:10.1371/journal.pone.0101271

Meta-Evolution Style for Software Architecture Evolution

Adel Hassan[✉] and Mourad Oussalah

Faculty of Science, Nantes University, LINA-FRE CNRS, 2729, Nantes, France
{adel.hassan,Mourad.Oussalah}@univ-nantes.fr

Abstract. Changes over time are commonplace and inevitable for any software system if it is to remain effective. Since the system changes fairly frequently, it is essential that its architecture is restructured to keep abreast of these changes. Recently the term 'evolution style' has emerged in some studies as a technique for modeling potential architecture evolution scenarios in a particular domain that can provide reusable knowledge that encapsulates the best practices in this domain. Analysis and comparison of these alternatives assists architects in planning and thinking about architecture evolution. Our approach endeavors to unify the solutions and standardize the modeling concepts in order to develop evolution styles library that exploits the best methods and elements in the existing approaches. To this end, the main contribution of this paper is a Meta-Evolution Style (MES) for software architecture evolution, which promotes mapping and comparing of evolution styles, as well as it will help in approaching issues like reuse and interchange elements among evolution styles.

Keywords: Software architecture · Architecture evolution · Evolution styles

1 Introduction

Software evolution is a process whereby a software product requires continual updating, maintenance, and improvement over time in order for it to remain a viable product (law of software evolution) [12]. This phenomenon has led to an increase in the importance of and research into software evolution. A number of studies have been carried out to improve in the means, processes, activities, methods, and tools whereby evolution is implemented.

Since software systems change fairly frequently, it is essential that their architecture is restructured to keep abreast of these changes. Software architecture evolution has emerged as an important precursor phase in the evolution cycles, because it can permit planning and system restructuring at a high level of modeling where business requirements and quality goals can be ensured and alternative solutions can be explored. Therefore, equipping the architect with the necessary techniques, methods and tools that will assist him to carry out this process has become an important focus of research.

© Springer-Verlag Berlin Heidelberg 2016
R.M. Freivalds et al. (Eds.): SOFSEM 2016, LNCS 9587, pp. 478–489, 2016.
DOI: 10.1007/978-3-662-49192-8_39

Recently, some researcher have introduced the term "evolution style" [1–4]: an evolution style tries to capture the main characteristics of the set of activities performed in evolving software architecture, and to provide the vocabulary of concepts necessary to model the potential scenarios in evolving a domain-specific software architecture. These modeled scenarios can be grouped together in an evolution style as a reusable body of knowledge. The analysis and the trade-off between these alternatives will assist architects in planning and reasoning about the future evolution of the software architecture.

The ultimate goal of the evolution style approaches, as we mentioned, is to equip the architect with a library comprising a variety of evolution styles in different domains that model the best practices and knowledge in software architecture evolution related to these domains. To get the optimal practices and methods to build this library, we need to exploit, transfer and share the best results that have been achieved by different approaches; therefore, we should be working towards unifying the concepts of this process to allow us to efficiently exploit these practices and knowledge.

To this end, this paper presents an attempt to explore a different but complementary approach: rather than focus on a specific architecture evolution style, it considers the more general problem of bringing together all the work in software architecture evolution modeling. Thus, we will exploit the metamodeling technique to devise a meta-evolution style (MES) that works as a unified solution for software architecture evolution that will permit the standardization of concepts this will broaden our horizons and allow us to exploit the best practices and knowledge in the software architecture evolution.

The contribution of this paper is twofold. Firstly, it aims to devise a meta-evolution style which can define the core conceptual elements for software architecture evolution regardless of the approaches to implementation. Secondly, it endeavors to explain how MES could promote in enacting the mapping rules that manage the transformations among different evolution styles, as well as between the disciplines of software architecture evolution and the object modeling.

This paper describes our approach to devising a meta-evolution style for software architecture evolution. It is organized as follows. Section 2 reviews some related work. Section 3 describes the basic concepts underlying our approach. Section 4 describes mapping and comarison strategies and how it can be applied among evolution styles. Finally Sect. 5 presents the conclusion.

2 Related Work

In this paper, we are concerned with software architecture evolution, and particularly with evolution styles. In recent years, a number of software architecture researchers have turned their attention to evolution, and some evolution styles have been developed, in this section we are going to review some of the approaches according to the classification by Pahl et al. [5].

2.1   The Garlan et al. Evolution Style [2]

This approach provides a model for software architecture evolution activities to support architects in planning and reasoning about domain-specific architecture evolution. The main idea of the evolution style is to model potential scenarios of software architecture evolution from the current style to the target style, possibly through sequence of transitional architectural states (evolutionary steps) known as evolution path (perhaps several ones), along with evolution operators that characterize the transitions among these states; constraints are imposed on evolution paths which allow these paths to be in the evolution style. Finally, an evaluation function is developed to allow comparisons among these paths.

2.2   The Cuesta et al. Evolution Style [3]

Unlike the previous approach, this combines the novel concept of evolution style with the basis provided by Architecture Knowledge (AK) to simplify the reuse of practices; by capturing the information and decisions related to architectural design. Therefore, topological information and architecture knowledge are both constraints and triggers for evolution in software architecture. Moreover, this approach emphasizes that architecture knowledge is an equally critical driver in the decision-making process in architecture evolution. The evolution style is applied as part of a semi-automatic process, in which an evolution style is a set of conceptually related evolution patterns. An evolution pattern describes the answer to a specific evolution condition, which triggers an evolution decision process where a particular configuration is chosen among several alternatives. Once the configuration has been selected which means that evolution decision is taken. Thus, the system executes an evolutionary step, modifying the AK structure in the process. Thereafter, if the evolution condition has been satisfied, the process finishes; otherwise, the evolution may continue, until the termination condition is met.

2.3   The Oussalah et al. Evolution Style [1]

The main idea of this approach is to model software architecture evolution tasks to provide a classification scheme to describe evolution styles libraries. And enable the libraries to acquire, update and retrieve evolution knowledge matching domain-specific evolution. Their approach considers that architecture evolution consists of these operations; addition, remove, and modification as first class entities in the context of component-based architectures in which component, connector, interface, and configuration are first class elements of the design. These elements provide the main concepts for describing domain-specific architecture evolution. They define a meta-model represented in a context of object-oriented, in which a set of architecture element types and constraints provide a domain-specific design vocabulary, and a particular type of configuration represents an evolution style.

3   Software Architecture Evolution: Modeling and Style

Evolution style is a modeling approach for software architecture evolution. In more detail, we can say that an evolution style is a modeling language for software architecture evolution, and each style is defined in a way that reflects the perspectives of the modeler with his own conceptual vocabulary. In essence, each style possesses specific mechanism, vocabulary and set of interrelations for modeling software architecture evolution. In the modeling methodology we can consider evolution style as a metamodel that specifies a modeling language to define a particular model of evolution. Therefore, we have to find a way to approach these concepts, and one of the best adopted techniques which provides many benefits in different modeling engineering fields is a metamodeling.

3.1   Meta-Evolution Style for Software Architecture Evolution

Software architecture evolution process modelling aims to capture the main characteristics of the set of activities that are performed to evolve software architecture. For better process modeling which expands the reuse of experience and knowledge, a variety of styles have been created. To reason about and unify the modeling concepts that formulate these styles and represent the modeler's perspectives, a meta-modeling language (meta-evolution style) comprise the necessary vocabularies is needed.

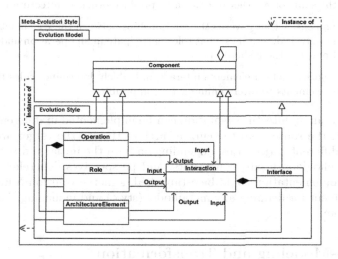

Fig. 1. Meta-evolution style MES

Metamodeling as we mentioned, is ascending techniques that are used to define a higher level where is a minimum concepts in order to define and reason about models in the level below. Meta-evolution style (meta-metamodel) is an intended layer which has the essential concepts to specify a meta-evolution modeling language. To remain compatible with the four modeling levels of the OMG (Object Management Group), each evolution model is an instance of the model in the level above (its meta), including the meta-evolution style, which is an instance of itself. The meta-evolution style introduces the essential elements and their interrelationships that represent the concepts required for modeling architecture evolution process. These concepts are: operations, roles, architecture elements, interfaces, and interactions. These essential concepts are comprised in the meta-evolution style called MES, illustrated in Fig. 1. MES is a component-oriented concept for modeling the software architecture evolution process. Indeed, it is a reflective concept whereby the concept of the component reflects the modeling of the process that evolves it. In the sense that everything is a component, like Class in object oriented modeling where everything is subclass of the abstract class "modelElement". Thus, the component is the basic evolution entity of the MES model.

Role: the stakeholders, tools and techniques that participate in performing the architecture evolution operation.

Operation: one of the units of process that consumes or produces an architecture element. The operation are associated with roles and architecture element through the interaction element.

Architecture Element: the inputs and the outputs of an architecture evolution operation. An architecture element produced by an operation can be used later as raw material for the same or another operation to produce other architecture elements.

Interaction: an entity that governs the relationships between these elements and their behavior. Generally, it determines what roles participate in an operation and what kind of inputs and outputs there should be.

Interface: a place in which elements interact and which determines rules that should be followed by elements to intercommunicate compatibly.

Normally, an architecture evolution can be implemented in different ways. In this context, the route from the current architecture to the target architecture might take different trajectories, depending on both the initial and target styles, and on stakeholders' needs, constraints, and perspectives. Therefore, different models of implementation could be established; a meta-evolution MES provides the core element for defining any metamodel (style) and managing the mapping between them.

4   Meta-Modeling and Transformation

Metamodeling, as we mentioned, is a powerful technique which enables us to understand and reason about the mapping between models. One of the important reasons for defining a meta-evolution style is to manage the transformations among evolution styles; therefore, our approach attempts to provides solid principles for evolution style transformation with two methods of mappings. One is mapping among software architecture evolution styles [1–4], while the other transfers these styles to MDA's environment (e.g. Software & Systems Process Engineering Metamodel SPEM, Eclipse Process Framework Composer EPFC), to improve the efficiency of the process, and to exploit the best practices and knowledge in the domain of object modeling.

4.1   Vertical Mapping

The basic assumption in using modeling levels is the possibility of serializing the view of the model at different levels and utilizing the reflection of elements at a higher level to determine their counterparts in the succeeding level. The vertical mapping among MES and evolution styles provides two benefits: firstly, it can be used to evaluate the ability of MES to define an evolution style; secondly it can facilitate the matching of corresponding elements between two evolution styles, where every counterpart should have the same meta-element in MES. Moreover, it can also be exploited in mapping an evolution style with one MOF-compatible models (SPEM, Eclipse process framework).

Mapping Between the MES and Garlan et al. Evolution Style [2]. One of the vital reasons for defining MES is to set up the rules to govern the transfer of evolution styles; thus, to investigate the efficiency of MES in defining and transforming an evolution style model, vertical mapping between MES and one evolution style (Garlan) is reported in this section.

Garlan et al. defined an evolution style as explained above (Sect. 2.1). An architectural concept for Garlan et al. style has been derived to facilitate the mapping, as it is shown in Fig. 2.

Mapping Between the MES and Cuesta et al. Evolution Style [3]. Cuesta et al. have defined an evolution style, as mentioned before (Sect. 2.2). MES can be mapped to Cuesta's evolution style as following: the Evolution Pattern and Evolution Step can be defined by the Operation element in MES, the Evolution Decision, Evolution Condition and AK can be defined by the Role element in MES, and the Architectural Element can be defined by the Architecture Element in MES. Figure 3 illustrate this mapping.

Mapping Between the MES and SAEM Metamodel [1]. The SAEM (Style-based Architectural Evolution Model) was developed by Oussalah et al. and has also been explained above (Sect. 2.3). They defined a metamodel for their evolution style. Conforming MES by the SAEM evolution style can be done in this way: the Competence and Header can be defined by Operation, Constraint can be defined by the Role, and Architectural Element Type can be defined by the Architecture Element in the MES. Figure 4 illustrates this mapping.

4.2   Horizontal Mapping

Horizontal mapping is mapping among models at the same modeling level and which might be defined by the same metamodel or by different metamodel; this mapping could be at any level.

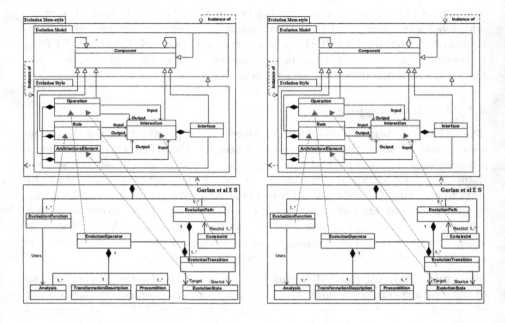

Fig. 2. Mapping the Garlan style with MES

Fig. 3. Mapping the Cuesta style with MES

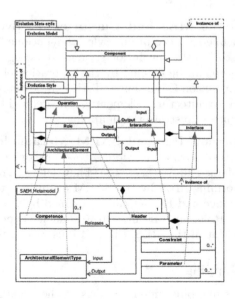

Fig. 4. Mapping SAEM with MAS

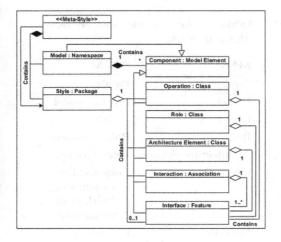

Table 1. Mapping MES with MOF

MES	MOF
Meta-style	Stereotype package
Style	Package
Model	Namespace
Component	Model element
Operation	Class
Role	Class
Architecture element	Class
Interaction	Association
Interface	Feature

Fig. 5. Mapping MES with MOF

In general, mapping is a complex task. To be precise, mapping is a particular directional or bidirectional translation solution defined for two specific models, so it cannot be reused for other models. Indeed, metamodeling provide a many opportunities for facilitating this process, whereas in the upper level there are always fewer conceptual elements to be translated.

Mapping MES Concepts to MOF Concepts. MOF (MetaObject Facility) is the foundation of OMG's industry-standard environment; and it is a meta-language for many modeling languages (UML, SPEM, CWM, etc.), and supports translation from one to another (for example, the set of rules used to transform a MOF model into a UML model has been derived from the OMG UML Profile for MOF specification) [9].

It can be stated that mapping rules between MDA/MOF and software Architecture Evolution/MES have more benefits at meta-metamodel level than in the levels below.

- At this level there are fewer elements to be translated.
- These mapping rules can be reused by every metamodel (style) in the domain of software architecture evolution that is compatible with MES.
- The translations from MOF to its metamodels (e.g. UML, SPEM) already exist and are well defined (Fig. 5).

Table 1 represents the corresponding elements in MES and MOF, which are useful for translating a model of evolution style to model-driven development MDD area to be implemented by one of its modeling environments.

Mapping the Garlan et al. Evolution Style to the SPEM Metamodel. SPEM is a standard metamodel based on MOF and developed by Object Man-

Table 2. MOF instantiation by SPEM

MOF	SPEM
Class	Activity, Work product, Role, Guidance
Association	Work product relationship, Work sequence, Process responsibility assignment
Feature	Process parameter
Package	Package

Table 3. Matching elements between MES and SPEM

MES	SPEM
Style	Package
Model	Process model
Operation	Activity
Role	Role, Guidance
Architecture element	Work product
Interaction	Responsibility assignment, Work product relationship
Interface	Parameter

agement Group. SPEM is a process engineering metamodel as well as a conceptual framework, which can provide the necessary concepts for modeling, documenting, presenting, managing, interchanging, and enacting development methods and processes [10].

Our aim is to bring the evolution style concept closer to the MOF-compatible model, in order to exploit that their achievements, methods and tools. To this end, we are going to carry out mapping between the Garlan evolution style and SPEM.

The first mapping step must be independently defined for each evolution style, while the other steps are defined once and can then be used by any style compatible with MES. The second step has already been defined as the mapping between MES and MOF. The third step is the instantiation of the MOF by SPEM; this operation is specified by the OMG [OMG Document Number: formal/2008-04-01]. Furthermore, SPEM is defined as a UML profile, and each element (stereotype) of a SPEM is defined by a corresponding UML element. Table 3 shows the possible SPEM instances for each element of MOF that we used in the mapping with MES.

The Class is the fundamental element of MOF, and is the meta-element for Activity, Work Product, Role, and Guidance in SPEM. In the mapping between MES and MOF: the Class is also the equivalent element that fits the four core elements in MES. Consequently, from Tables 1 and 2 we notice that:

Operation, Role, Architecture Element, and Interface → Class → Activity, Work Product, Role, and Guidance

We have mapped between MES and MOF those elements belonging to the same modeling level (meta-metamodel), to minimize the elements and reason the matching. Therefore, to make the transformation more specific we could match the MES and SPEM, considering both the mapping of super-classes (meta-element) and the ontology of the elements in the two approaches. Thus, Table 3 illustrates the matching between their elements that could provide a more accurate translation.

Table 4. Garlan style and SPEM: corresponding elements

Garlan et al. style	SPEM metamodel
Operator, Transition	Activity
Architect	Role
Evaluation function	Guidance
Constraint	Responsibility, Relationship, Guidance
Evolution state	Work product

Table 5. Evolution styles quantitative comparison

MES concepts	Garlan et al.	Cuesta et al.	Oussalah et al.
Operation	Operator, Transition	Evolution step	Header, Competence
Role	Architect, Evaluation function	Architect, Decision, Architecture knowledge	Architect
Architecture element	Evolution state	Architectural element	Architecture element type
Interaction	Constraint	Condition	Constraint
Interface	Does not have explicit element	Does not have explicit element	Parameter

Finally, we reach the last step in this mapping with the selection of the final concepts. At the end of the matching of MES elements with SPEM elements that were instantiated by MOF, we reach a reduced number of choices for each conceptual element in the evolution style. This step is specific for each evolution style, while all evolution styles that are compatible with MES reuse the three previous steps in this mapping strategy. To this end, the end user can define and apply some criteria that will assist and guide him in his choices (the last step), to transfer his style to a target environment according to these rules.

For instance, the final step could be transforming the Garlan style to SPEM: selecting corresponding concepts for the Garlan style that are the most representative and semantically closest, with consideration of the rules that were enacted in previous steps. Table 4 illustrates this step.

4.3 MES and Evolution Styles Comparison

The purpose of this section is to present our methodology for evolution style comparison. Indeed our approach attempts to support the qualitative and quantitative comparison between styles. These comparisons can offer enough information for architect to explore their strengths and weaknesses (Table 5).

The Quantitative Comparison: Our approach presumes that MES has all the meta-conceopts needed to model an evolution style. Therefore, quantitatively compare an evolution style with MES allow us to know whether an evolution style covers all these necessary concepts to model this process. Quantitative comparison also identifies the similarity and the difference in the concepts among evolution styles and it helps in matching concepts for model transformation between them.

The Qualitative Comparison: The qualitative comparison requires us to firstly determine attributes of the quality and clarify how they could be measured. The quality can often be difficult to measure due to ambiguity and overlap many of these attributes in its calculation. However, the quality measurement is the foundation of software development (and evaluation) and gives a better insight into the factors that influence software quality by defining a consistent terminology for software quality and by clarifying the mechanism to measure it. To this end, a variety of general quality models have been built, such as McCall's model, Dromey's model, Boehm's model, and the ISO 9126 model. However, in order to use one of these quality models in a specific domain, it has to be extended to include the particularities of that domain or project.

Our future work will include developing a quality model to assess and measure evolution style basis on the ISO/ICE 25010 model, and its characteristics and sub-characteristics will be customized to fit the need of evolution style. We assume that instead of using the quality model to assess the final product (evolution style model), we will integrate the quality model with the three main aspects of evolution styles: specification, modeling and application.

5   Conclusions

Software architecture is considered as a blueprint to predict and guide the process of software production and evolution. Recently, a number of evolution styles have been developed to model software architecture evolution in a specific domains in order to support architects in planning and reasoning software architecture evolution. Notwithstanding all these considerable achievements, some important concepts related to standardization, comparison, and mapping among evolution styles are still absent, and need more exploration.

Actually, metamodeling techniques are widely used in engineering design, and are concerned with documenting, reasoning, comparing, reusing, and unifying models. Accordingly, this study has exploited the metamodeling techniques in the field of evolution style, ascending to the meta-metamodel level, to devise a meta-evolution style called MES which works as a unified solution for software architecture evolution modeling.

This study also aimed to explore the ability of MES to specify the metamodeling language that has the necessary vocabulary to define an evolution style, in other words, that has the meta-concepts that could instantiate any conceptual style element. Therefore, mappings and comparison between MES and an evolution style were carried out. Furthermore, our future work will be devoted to develop a framework based on MES that could model a variety of evolution styles.

References

1. LeGoaer, O., Tamzalit, D., Oussalah, M.C., Seriai, A.-D.: Evolution styles to the rescue of architectural evolution knowledge. In: Proceedings of the 3rd International Workshop on Sharing and Reusing Architecture Knowledge (SHARK 2008), pp. 31–36 (2008)
2. Barnes, J.M., Garlan, D., Schmerl, B.: Evolution styles: foundations and models for software architecture evolution. Softw. Syst. Model. **13**(2), 649–678 (2013)
3. Cuesta, C.E., Navarro, E., Perry, D.E., Roda, C.: Evolution styles: using architectural knowledge as an evolution driver. J. Softw.: Evol. Process **25**(9), 957–980 (2013)
4. Oussalah, M., Sadou, N., Tamzalit, D.: SAEV: a model to face evolution problem in software architecture. In: Proceedings of ERCIM Workshop on Software Evolution, Lille, France, pp. 137–146 (2006)
5. Aakash, A., Jamshidi, P., Pahl, C.: Classification and comparison of architecture evolution reuse knowledge - a systematic review. J. Softw.: Evol. Process **26**(7), 654–691 (2014)
6. Barais, O., et al.: Software architecture evolution. In: Mens, T., Demeyer, S. (eds.) Software Evolution, pp. 233–262. Springer, Heidelberg (2008)
7. Nguyen, T.N.: Managing software architectural evolution at multiple levels of abstraction. J. Softw. **3**(3), 60–70 (2008)
8. Oussalah, M. (ed.): Software Architecture. Wiley, Hoboken (2014)
9. META-OBJECT FACILITY-MOF, version 1.4., Object Management Group, Document Formal/2002-04-03 (2002)
10. OMG: software and systems process engineering meta-model specification, version 2.0. Object Management Group (OMG), Document formal/2008-04-01 (2008). http://www.omg.org/spec/SPEM/2.0/PDF/
11. Jeusfeld, M.A., Jarke, M., Mylopoulos, J.: Metamodeling for Method Engineering. The MIT Press, Cambridge (2009)
12. Lehman, M.M.: Laws of software evolution revisited. In: Montangero, C. (ed.) EWSPT 1996. LNCS, vol. 1149, pp. 108–124. Springer, Heidelberg (1996)

The Simulation Relation for Formal E-Contracts

Luis Llana[1]($\boxtimes$), María-Emilia Cambronero[2], and Gregorio Díaz[2]

[1] Computer Science Department, Complutensis University of Madrid, Madrid, Spain
llana@dfi.uclm.es
[2] Computer Science Department, University of Castilla-La Mancha,
Ciudad Real, Spain
{memilia.cambronero,gregorio.diaz}@uclm.es

Abstract. Relationships between entities in today's increasingly inter-connected context have grown in complexity and evolved from simple communication processes to more complicated distributed systems. Electronics contracts (*e-contracts*) are of general purpose and aimed to specify relationships in a wide variety of scenario, including web and cloud services, inter and intra organization, electronic banking, etc. It is in this context that we aim to develop a consistent definition for these relationships together with a set of techniques to check their proper use. In this paper we present a process algebra to describe these contract relationships and a set of formal machinery to determine whether an implementation follows the rules established by these contracts. The main formal technique used is a simulation relation where an implementation is checked step by step against a given contract. Several toy examples are provided to facilitate understanding of the formal definitions.

1 Introduction

A well-known formalism to specify contracts, deontic logic [20], is concerned (among other things) with the formalization of contracts, specifically with moral and legal obligations, permissions, and prohibitions, their interrelation and properties, as well as events/consequences resulting from violations of obligations and prohibitions. In this context, it is important to take the time to analyze whether these contracts, in the form of permissions, obligations and prohibitions, are fulfilled by the different parties involved in the contract. In this paper, we present a formalism to analyze *e-contracts* based on timed process algebras [10,18]. In our case, the underlying semantics is that of timed simulation [19]. Simulation is an important in many areas of computer science (e.g., model checking, concurrency theory, and formal verification). It is both a theoretical (e.g. [1,13]) and practical (e.g. [17]) area of research. The use of practical implementation applications for e-contracts is particularly interesting in model checking minimization [3,11]

Research partially supported by the Spanish MEC projects ESTuDIo (TIN2012-36812-C02-01, TIN2012-36812-C02-02), DArDOS (TIN2015-65845-C3-01, TIN2015-65845-C3-02), the Comunidad de Madrid project SICOMORo-CM (S2013/ICE-3006) and the UCM-Santander program to fund research groups (group 910606).

R.M. Freivalds et al. (Eds.): SOFSEM 2016, LNCS 9587, pp. 490–502, 2016.
DOI: 10.1007/978-3-662-49192-8_40

as a technique to overcome the state explosion problem. Then, one of the main motivations of this paper is to use the simulation relation to model and analyze *e-contracts* taking advantage of this model checking minimization.

This paper defines a process algebra to describe *e-contracts* relationships and a set of formal techniques to determine whether an implementation follows the rules established by these *e-contracts*. In particular we present a formal semantics for *e-contracts* in order to capture the deontic normative concepts, as well as the penalties in cases of certain violations. In our formal framework each party involved in a contract is represented by a contract agent. We have defined a simulation preorder between agents. This preorder can be used as an implementation preorder: if the contract agent P simulates the contract agent Q, then Q can be replaced by P in any contract.

Deontic logic was first proposed in [20], since then several works have proposed its use for reasoning about contracts. Prisacariu and Schneider in [15] present a formal language for *e-contract*, which is based on deontic notions.

But their contract language lacks the possibility to express time constraints and they propose adding real-time as an extension to express and reason about contracts with deadlines. Governatory et al. [7] present a formalism for the representation of contracts also using deontic logic, by including the representation and reasoning about violations of obligations in contracts. In [12] C. Prisacariu et al. shows how to obtain a run-time monitor for contracts written in Contract Language (CL [16]), which is an action-based formal language tailored for writing e-contracts and which allows to write conditional obligations, permissions, and prohibitions to be written over actions.

A generic construction for obtaining a contract framework based on assume-guarantee (AG) pairs from a component-based specification theory is presented by Bauer et al. [2].

None of the previous works consider real-time restrictions and their focus is on verification or monitoring. Our approach is based on the simulation relation, since previous works [8,9] show that simulation can be a useful technique to reduce the number of states of a model. They utilize a variant of simulation that can be used as a conformance relation, with the aim of extracting tests from a specification. We propose to translate these results to our framework, but considering timed restrictions in the system.

In this paper we have chosen a visual model for the design of contracts [5] since it implements many of the desirable properties of a good formal language for normative texts, as presented in [14]. This model is composed of Contract-Oriented Diagrams (*C-O Diagrams*), which allow the representation of complex clauses describing the obligations, permissions, and prohibitions of different signatories, as well as reparations. Also, C-O Diagrams permit users to define real-time constraints. We use these diagrams only for representation issues, since a graphical representation helps to more easily understand complex contracts, which can be composed of many clauses.

The paper is structured as follows: Sect. 2 presents an overview of the visual model for e-contracts. Next, in Sect. 3, we present the formal language: its syntax and its operational semantics. Section 4 develops the formal semantics of language. Section 5 presents a case study of a *Coffee Machine*. Finally, the conclusions and future work are described in Sect. 6. Due to lack of space, we have removed the proofs of the results in this paper, they can be found in the following link: http://antares.sip.ucm.es/~luis/sofsem16.pdf

2 The Visual Model for E-Contracts: C-O Diagrams

C-O diagrams presented in [5] are a simplified visual model to represent e-contracts. We will use this model to facilitate the task of specifying contracts. The diagram type used here differs slightly from the one presented in [5] to adapt it to the specific goals of this work. Figure 1 depicts a typical example of a coffee machine, which consists of two agents (the machine and a customer). This example is explained in detail in the case study section, but first we will present the elements these diagrams consists of, so readers can start to familiarize themselves with them. The main difference is the use of variables, which are not considered in this first stage of this proposal.

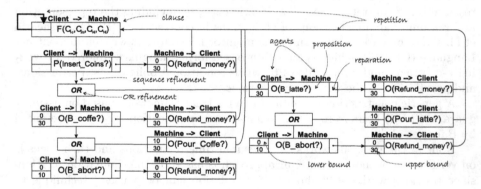

Fig. 1. C-O diagram for coffee machine specification

The diagram depicts a set of boxes and arrows to interconnect boxes. A box, known as "**clause**", can be divided in four different cells. The central cell, the "**proposition**", specifies the obligations, prohibitions, permissions, actions and refinements. The left-hand cells define the upper (top) and lower (bottom) bounds defining the interval in which the clause must be enacted. In some cases, it might not be necessary to declare bounds and $[0, \infty]$ is taken as default interval. The last cell on the right-hand side of the clause is the recuperation cell, which is only declared for obligations and prohibitions[1], which is activated whenever the behaviour declared in the central cell, the proposition, is violated. An extra element is added to the clauses at the top when actions are used at the proposition. This element specifies two agents, from left to right: the performer

---

[1] For these two types of propositions when the main proposition fails and no reparation is defined the contract is violated.

and the receiver of the action. An Arrow connecting clauses specifies a sequence, but if it targets a predecessor clause it defines a recursion refinement. If an arrow targets a clause with either OR or AND, then this targeted clause must connect with two other clauses behaving as the disjunctive or conjunctive refinements.

As in process algebra, any communication involves two agents: one that emits the communication and another one that receives. This is similar in our case. For instance in the case of an obligation there is one or more agents that have the obligation to perform an action and another agent receives. The agent that is supposed to receive an action can detect that the contract is not valid if the action is not received. For instance let us suppose that we have a simple contract where agent P has the obligation to perform an action received by agent Q within the time interval 2 and 8. This can be specified in our formalism by:

Example 1. $\boxed{\begin{array}{c} \mathbf{P} \dashrightarrow \mathbf{Q} \\ \begin{array}{c|c} 2 \\ 8 \end{array} O(a?) \end{array}}$ $P = a![2,8], \quad Q = O_b(a?)[2,8]$

Permissions allow agents to receive actions from other agents in a specific time interval. For instance, Example 2 shows an agent P allowing other agent Q to perform the action a in the interval $[3,5]$. An implementation of Q_1 where the action is performed during this implementation will be valid. On the contrary, if an implementation specifies a longer or shorter interval for instance $[4,7]$ or $[1,4]$, respectively, then it might lead to the contract violation and, therefore to an invalid implementation. In this case, $P_r(a?)[3,5]$ in P grants the other process the permission of execute the input $a!$.

Example 2. $\boxed{\begin{array}{c} \mathbf{P} \dashrightarrow \mathbf{Q} \\ \begin{array}{c|c} 3 \\ 5 \end{array} P(a?) \end{array}}$ $P = P_r(a?)[3,5] \qquad Q = a![3,5]$

Prohibitions are the most complex deontic operator since they combine the idea of not allowing a certain set of actions during a given interval of time. Example 3 specifies a e-contract where action a is forbidden during the interval $[3,5]$ in the behavior of an AND refinement where the left clause permits action a during $[1,5]$ while on the right part action b must be performed in $[1,4]$. Action a is only permitted in the interval $[1,2]$ by agent Q as a result of the interaction between the prohibition and the permission as it can be observed in the specification of process Q. An implementation performing a outside this interval leads to a violation of this contract.

Example 3.

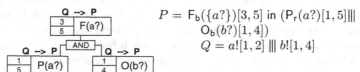

$P = F_b(\{a?\})[3,5] \text{ in } (P_r(a?)[1,5] \|\| O_b(b?)[1,4])$
$Q = a![1,2] \|\| b![1,4]$

3   The Language

In this section we first establish the *e-contract* model, which captures the main deontic aspects of *e-contracts* that have been briefly described in the previous section, after which we define an operational semantics for it. For our purposes, a specification is described by Definition 1. In this definition we are considering a set of inputs I $(a?, b?, \ldots$ will denote individual inputs), a set of outputs O $(a!, b! \ldots$ will denote outputs). Each input and output has a counterpart in the other set: $a? \in I$ iff $a! \in O$. For any $a \in I \cup O$, we define its counter part as $\bar{a}$. We need two special actions vc, ic $\notin I \cup O$ to signal the correct finalization of a contract and an invalid contract, the set of actions $\mathsf{Act} = I \cup O \cup \{\mathsf{ic}, \mathsf{vc}\}$ $(a, b, \ldots$ denote individual actions), a silent action $\tau \notin \mathsf{Act}$, the set $\mathsf{Act}_\tau = \mathsf{Act} \cup \{\tau\}$, and a set of agent variables Var $(x, y, z$ will denote individual agent variables). We will also need a discrete time domain T $(t, t_1, t_2 \ldots$ will denote time units), we will assume that there is a minimum time unit δ and all the time units are multiple of that minimum time unit, we need a variation of the subtraction operation: $t_1 \dot{-} t_2 = \max(0, t_1 - t_2)$. We will need the infinity, $\infty \notin T$, to represent a behavior that is always available; $\infty \dot{-} t = \infty$.

Definition 1. The set of contract agent terms is defined by the following BNF:

$$P :: = \mathsf{P_r}(a?)[t_1, t_2] \mid \mathsf{O_b}(a?)[t_1, t_2] \rhd R \mid \mathsf{F_b}(A)[t_1, t_2] \text{ in } P \rhd R \mid a![t_1, t_2] \mid$$
$$\mathsf{ic} \mid \mathsf{vc} \mid \mathsf{rec}\, x.P \mid x \mid P_1; P_2 \mid P_1 \square P_2 \mid P_1 \sqcap P_2 \mid P_1 \;|||\; P_2$$

Where $\mathsf{P_r}$, $\mathsf{O_b}$ and $\mathsf{F_b}$ are the deontic operators for the permissions, obligations and prohibitions, respectively. $a? \in I$, $a! \in O$, $t_1 \in T$, $t_2 \in T \cup \{\infty\}$, $A \subseteq I$, P, R, P_1, P_2 are agent terms, and $x \in \mathsf{Var}$. We will consider the set of contract agents (or simply agents), written $\mathcal{AG}$, as the set of closed contract agents terms.

We define the language of $P \in \mathcal{AG}$, written $L(P)$, as the set of inputs and outputs appearing in P.   □

Therefore, $\mathsf{P_r}(a?)[t_1, t_2]$ models the permission to perform action a in a time interval $[t_1, t_2]$. $\mathsf{O_b}(a?)[t_1, t_2] \rhd R$ represents the obligation to perform action a in a time interval $[t_1, t_2]$; if the obligation is not fulfilled it is necessary to follow the behavior specified by the reparation R. $\mathsf{F_b}(A)[t_1, t_2]$ in $P \rhd R$ depicts the prohibition of performing an input action subset (A) in a time interval $[t_1, t_2]$, in case this prohibition is violated a reparation specified by R must be performed. $a![t_1, t_2]$ can execute the output action at any time in the time interval $[t_1, t_2]$; this operator represents an agent that can perform the input at any time in the interval, so the choice of the time is an internal choice of the agent. vc defines a valid contract, while ic defines an invalid contract. $x \in \mathsf{Var}$, where Var defines a set of agent variables, is used to define the recursion operator $\mathsf{rec}\, x.P$. This operator models the recursion of P, that is, the contract repetition. $P_1; P_2$ represents the concatenation in sequence. $P_1 \square P_2$ defines a deterministic choice (or external choice), where the environment has the possibility to make the choice between P_1 and P_2. $P_1 \sqcap P_2$ defines a nondeterministic choice (also called internal choice), where the system is taking the decision and the environment has no control over it. $P_1 \; ||| \; P_2$ models the parallel execution of two agents, P_1 and P_2.

In our examples in the paper we will use some shortcuts: If the time interval is omitted (for instance $P_r(a?)$), the time interval is $[0, \infty]$ ($P_r(a?)[0, \infty]$). If the reparation is omitted (for instance $O_b(a?)$), we assume that is ic ($O_b(a?) \rhd$ ic).

3.1 Operational Semantics

Figure 2 shows the operational semantics. There are two kinds of transitions: timed transitions and action transitions. Timed transitions expresses the time delay while action transitions expresses the execution of an action. The base operator vc can delay time (rule vc2) and signals the correct termination of a contract (rule vc1). Similarly, the operator ic can signal an incorrect contract (rule ic1) and delay time (rule ic2). Rule rec1 indicates the unfolding of the recursion operator. This unfolding is urgent, so must be done immediately (rule rec2).

The input action, the permission operator and the obligation operator can only delay time until the lower bound of the interval is reached (rules act1, perm1, obl1). When an output action is enabled, the operator chooses internally the time t' when the action will be executed (rule act2), then, due to rule act1, the time passes until t'. When this time is reached (the time interval of the operator is $[0,0]$) and the action is executed the computation proceeds normally (rule act3), otherwise if the action cannot be executed (because there are no other agents willing to synchronized) the invalid contract signal is raised (rule act4).

The permission and obligation operator behaves in the same way until the upper bound of the interval is reached: while the input action is enabled, both operators can execute the action (rules perm3 and obl3) and both operators can delay time until the upper bound of the interval (rules perm2 and obl2). The main difference between perm2 and rule obl2 comes from their behaviors when time overtakes the upper bound of the interval. In the case of the permission the computation evolves normally, while in the case of the obligation the part of the recuperation is enabled.

The forbidden operator behaves normally as long as the time interval is not enabled (rules forb1 and forb4). When time reaches the upper bound of the interval, the operator is disabled (rule forb2). While the time interval is enabled only actions not belonging to the set of forbidden actions can be performed (rule forb5) and if a forbidden action is performed, the part of the recuperation is enabled (rule forb3).

Next we have the rules dealing with the compound operators. They are the typical rules of a timed process algebra [18]. The internal choice operator can only make the internal decision to behave as one of its components (rules icho1 and icho2), rule icho3 is necessary for Proposition 2. The external choice can let time pass if both components can (rule cho1); when one of the components of the choice is able to execute an action then the other component is disabled

$$\text{rec1} \frac{}{\text{rec } x.P \xrightarrow{\tau} P[\text{rec } x.P/x]} \qquad \text{rec2} \frac{}{\text{rec } x.P \overset{0}{\rightsquigarrow} \text{rec } x.P}$$

$$\text{vc1} \frac{}{\text{vc} \xrightarrow{\text{vc}} \text{vc}} \qquad \text{vc2} \frac{}{\text{vc} \overset{t}{\rightsquigarrow} \text{vc}} \qquad \text{ic1} \frac{}{\text{ic} \xrightarrow{\text{ic}} \text{ic}} \qquad \text{ic2} \frac{}{\text{ic} \overset{t}{\rightsquigarrow} \text{ic}}$$

$$\text{act1} \frac{0 \leq t \leq t_1}{a![t_1, t_2] \overset{t}{\rightsquigarrow} a![t_1 - t, t_2 - t]} \qquad \text{act2} \frac{t' \leq t, \ t >= 0}{a![0, t] \xrightarrow{\tau} a![t', t']}$$

$$\text{act3} \frac{}{a![0, 0] \xrightarrow{a!} \text{vc}} \qquad \text{act4} \frac{t > 0}{a![0, 0] \overset{t}{\rightsquigarrow} \text{ic}}$$

$$\text{perm1} \frac{0 \leq t \leq t_2}{\mathsf{P}_r(a?)[t_1, t_2] \overset{t}{\rightsquigarrow} \mathsf{P}_r(a?)[t_1 \doteq t, t_2 \doteq t]} \qquad \text{perm2} \frac{t' > t}{\mathsf{P}_r(a?)[0, t] \overset{t+t'}{\rightsquigarrow} \text{vc}}$$

$$\text{perm3} \frac{}{\mathsf{P}_r(a?)[0, t] \xrightarrow{a?} \text{vc}}$$

$$\text{obl1} \frac{0 \leq t \leq t_2}{\mathsf{O}_b(a?)[t_1, t_2] \triangleright R \overset{t}{\rightsquigarrow} \mathsf{O}_b(a?)[t_1 \doteq t, t_2 \doteq t] \triangleright R}$$

$$\text{obl2} \frac{R \overset{t_1}{\rightsquigarrow} R' , \ t_1 > 0}{\mathsf{O}_b(a?)[0, t_2] \triangleright R \overset{t_1 + t_2}{\rightsquigarrow} R'} \qquad \text{obl3} \frac{}{\mathsf{O}_b(a?)[0, t] \triangleright R \xrightarrow{a?} \text{vc}}$$

$$\text{forb1} \frac{0 \leq t \leq t_2 , \ P_1 \overset{t}{\rightsquigarrow} P_1'}{\mathsf{F}_b(A)[t_1, t_2] \text{ in } P_1 \triangleright P_2 \overset{t}{\rightsquigarrow} \mathsf{F}_b(a?)[t_1 \doteq t, t_2 \doteq t] \text{ in } P_1' \triangleright P_2}$$

$$\text{forb2} \frac{t > t_2 , \ P_1 \overset{t}{\rightsquigarrow} P_1'}{\mathsf{F}_b(A)[t_1, t_2] \text{ in } P_1 \triangleright P_2 \overset{t}{\rightsquigarrow} P_1'} \qquad \text{forb3} \frac{a \in A , \ P_1 \xrightarrow{a} P_1'}{\mathsf{F}_b(A)[0, t_2] \text{ in } P_1 \triangleright P_2 \xrightarrow{a} P_2}$$

$$\text{forb4} \frac{t_1 > 0 , \ P_1 \xrightarrow{a} P_1'}{\mathsf{F}_b(A)[t_1, t_2] \text{ in } P_1 \triangleright P_2 \xrightarrow{a} \mathsf{F}_b(A)[t_1, t_2] \text{ in } P_1' \triangleright P_2}$$

$$\text{forb5} \frac{b \notin A , \ P_1 \xrightarrow{b} P_1'}{\mathsf{F}_b(A)[0, t_2] \text{ in } P_1 \triangleright P_2 \xrightarrow{b} \mathsf{F}_b(A)[0, t_2] \text{ in } P_1' \triangleright P_2}$$

$$\text{icho1} \frac{}{P \sqcap Q \xrightarrow{\tau} P} \qquad \text{icho2} \frac{}{Q \sqcap P \xrightarrow{\tau} P} \qquad \text{icho3} \frac{}{Q \sqcap P \overset{0}{\rightsquigarrow} P \sqcap Q}$$

$$\text{cho1} \frac{P \overset{t}{\rightsquigarrow} P' \quad Q \overset{t}{\rightsquigarrow} Q'}{P \square Q \overset{t}{\rightsquigarrow} P' \square Q'} \qquad \text{cho2} \frac{P \xrightarrow{a} P', a \in \text{Act}}{P \square Q \xrightarrow{a} P'} \qquad \text{cho3} \frac{Q \xrightarrow{a} Q', a \in \text{Act}}{P \square Q \xrightarrow{a} Q'}$$

$$\text{cho4} \frac{P \xrightarrow{\tau} P'}{P \square Q \xrightarrow{\tau} P' \square Q} \qquad \text{cho5} \frac{P \xrightarrow{\tau} P'}{Q \square P \xrightarrow{\tau} Q \square P'}$$

$$\text{par1} \frac{\begin{array}{c} P_1 \overset{t}{\rightsquigarrow} P_1' \quad P_2 \overset{t}{\rightsquigarrow} P_2' \\ \forall t' < t : P_1 \overset{t'}{\rightsquigarrow} P_1'', P_2 \overset{t'}{\rightsquigarrow} P_2'' \Longrightarrow (P_1'' \parallel\!\parallel P_2'' \xrightarrow{\tau} \!\!\!\!/ \ \wedge P_1'' \parallel\!\parallel P_2'' \xrightarrow{\text{vc}} \!\!\!\!/ \ ) \end{array}}{P_1 \parallel\!\parallel P_2 \overset{t}{\rightsquigarrow} P_1' \parallel\!\parallel P_2'}$$

$$\text{par2} \frac{P_1 \xrightarrow{a} P_1', a \in I \cup O \cup \{\text{ic}\}}{P_1 \parallel\!\parallel P_2 \xrightarrow{a} P_1' \parallel\!\parallel P_2} \qquad \text{par3} \frac{P_2 \xrightarrow{a} P_2', a \in I \cup O \cup \{\text{ic}\}}{P_1 \parallel\!\parallel P_2 \xrightarrow{a} P_1 \parallel\!\parallel P_2'}$$

$$\text{par4} \frac{P_1 \xrightarrow{a} P_1' \quad P_2 \xrightarrow{\bar{a}} P_2'}{P_1 \parallel\!\parallel P_2 \xrightarrow{\tau} P_1' \parallel\!\parallel P_2'} \qquad \text{par5} \frac{P_1 \xrightarrow{\text{vc}} P_1' \quad P_2 \xrightarrow{\text{vc}} P_2'}{P_1 \parallel\!\parallel P_2 \xrightarrow{\text{vc}} \text{vc}}$$

$$\text{seq1} \frac{P_1 \xrightarrow{a} P_1', a \in \text{Act}}{P_1 ; P_2 \xrightarrow{a} P_1' ; P_2} \qquad \text{seq2} \frac{P_1 \xrightarrow{\text{vc}} P_1'}{P_1 ; P_2 \xrightarrow{\tau} P_2}$$

$$\text{seq3} \frac{P_1 \overset{t}{\rightsquigarrow} P_1' \quad \forall t' < t : P_1 \overset{t'}{\rightsquigarrow} P_1 \Longrightarrow P_1 \xrightarrow{\text{vc}} \!\!\!\!/}{P_1 ; P_2 \overset{t}{\rightsquigarrow} P_1' ; P_2}$$

Fig. 2. Operational semantics

(rules cho2 and cho3), while the execution of internal actions does not disable the choice (rules cho4 and cho5).

The timed rule of the parallel operator is the most complex of the timed rules (rule par1). There are two obvious conditions for the parallel composition being able to let time pass: both components can let time pass. But if at some intermediate point a synchronization or termination is available, it must be executed before letting time pass. The synchronization of two components is translated by a silent move (rule par4). In order to terminate, both components of the operator must terminate (rule par5). Finally any of the components can evolve autonomously (rules par2 and par3).

The sequence operator behaves as follows: First it executes all actions that the first component can execute (rule seq1). When the first component finishes successfully, the control passes to the second component (rule seq3). The operator can let time pass as so the first component can. But termination is urgent, so the operator cannot let time pass if the first component finishes successfully.

First we will prove some basic properties of the operational semantics. It is important to note that some rules have negative premises that could lead to an inconsistent semantic (for instance $P \xrightarrow{a} Q$ iff $P \xrightarrow{a}\!\!\!\!\!/\ Q$).

Proposition 1. The operational semantics is consistent. $\qquad\square$

Other propositions that the semantic of the timed transitions verifies are the following basic properties:

Proposition 2. Let $P, P', P'' \in \mathcal{AG}$, $t, t' \in \mathcal{T}$, then the following properties:

- $P \stackrel{0}{\rightsquigarrow} P$.
- If $P \stackrel{t}{\rightsquigarrow} P'$ and $P \stackrel{t}{\rightsquigarrow} P''$, then $P' = P''$.
- If $P \stackrel{t}{\rightsquigarrow} P'$. Then for any $0 < t' < t$ there exists P'' such that $P \stackrel{t'}{\rightsquigarrow} P''$.
- If $P \stackrel{t}{\rightsquigarrow} P' \stackrel{t'}{\rightsquigarrow} P''$, then $P \stackrel{t+t'}{\rightsquigarrow} P''$. $\qquad\square$

The associative and commutative properties of the binary operators help to avoid writing excessive parenthesis in the terms.

Proposition 3. The operators $\square$, $\sqcap$, and $|||$ are commutative and associative. $\square$

Next we can define the notion of contract, valid contract and valid implementation. First we have the notion of contract, which is the parallel composition of several agents. Since the parallel operator is associative and commutative, the order among the agents is not relevant and the parenthesis are not necessary.

Definition 2. Let $n \in \mathbb{N}$ be a natural number and let P_i be a contract agent for $1 \leq i \leq n$. A contract is specified by the parallel composition of the agents: $C = P_1 \,|||\, P_2 \,|||\, \cdots \,|||\, P_n$. $\qquad\square$

In order to define the notion of valid contract we need the following notation to simplify the definition.

Definition 3. Let P, P' be contract agents, $a \in \mathsf{Act}_\tau$ and $t \in \mathcal{T}$, we write $P \xrightarrow{t\,a} P'$ iff there is a contract agent P'' such that $P \rightsquigarrow^t P'' \xrightarrow{a} P'$.   □

So a valid contract is one that when executed never yields the signal ic of invalid contract.

Definition 4. Let C be a contract, an *interaction of the contract* is a sequence of contracts $C_0 = C, C_1 \ldots C_n$ such that there are transitions

$$C_0 \xrightarrow{t_0\,\tau} C_1 \xrightarrow{t_1\,\tau} \ldots \xrightarrow{t_{n-2}\,\tau} C_{n-1} \xrightarrow{t_n} \rightsquigarrow C_n$$

A contract is *invalid* if there is a interaction of the contract $C_0 = C, C_1 \ldots C_n$ such that $C_n \xrightarrow{\text{ic}}$. A contract is *valid* if it is not invalid.   □

Definition 5. Let $C = A_1 ||| A_2 ||| \cdots ||| A_k ||| \cdots ||| A_n$, we say that an implementation I *is correct for agent* A_k iff the contract C' obtained by substituting the agent A_k by the implementation I, ($C' = A_1 ||| A_2 ||| \cdots ||| I ||| \cdots ||| A_n$), is valid.   □

4   Simulation Semantics

In this section we are going to provide and discuss the simulation relation for the agents of a contract. The semantics defined in the previous section (Definition 5) is difficult to check. We believe that the simulation relation is a good alternative. It can be computed efficiently [6] and can be used to reduce the number of states as a previous step in other formal techniques such as model checking.

Before the definition of the simulation relation we need some notation.

Definition 6. Let $P \in \mathcal{AG}$, $t \in \mathcal{T}$ and $a \in I \cup O$, we define the transitions:

- $P \xRightarrow{t\,\epsilon} P'$ iff there are $P_i \in \mathcal{AG}$ ($0 \leq i \leq n+1$) and $t_i \in \mathcal{T}$ ($0 \leq i \leq n$) such that: $P = P_0 \xrightarrow{t_0\,\tau} P_1 \cdots \xrightarrow{t_{n-1}\,\tau} P_n \xrightarrow{t_n} \rightsquigarrow P_{n+1} = P'$ where $t = \sum_{i=0}^{n} t_i$.
- $P \xRightarrow{t\,a} P'$ iff there are $P_i \in \mathcal{AG}$ ($0 \leq i \leq n+1$) and $t_i \in \mathcal{T}$ ($0 \leq i \leq n$) such that: $P = P_0 \xrightarrow{t_0\,\tau} P_1 \cdots \xrightarrow{t_{n-2}\,\tau} P_{n-1} \xrightarrow{t_{n-1}\,a} P_n \xrightarrow{t_n} \rightsquigarrow P_{n+1} = P'$ where $t = \sum_{i=0}^{n} t_i$.
- $\mathsf{Act}_\epsilon(P) = \{(a, t) \mid \exists P' \in \mathcal{AG}, a \in I \cup O \cup \{\epsilon\} : P \xRightarrow{t\,a} P'\}$

     □

Next we define our concept of simulation. It is a timed simulation semantics and follows the classical co-inductive schema. The only particularity is that we want to restrict the simulation to the set of actions specified in the contract; since the implementation could perform actions not specified in the contract[2]. Let us recall that $L(C)$ is the set of inputs and outputs appearing in C (Definition 1).

---

[2] There are not restrictions for the execution of those actions.

Definition 7. Let S be a relation of agents ($S \subseteq \mathcal{AG} \times \mathcal{AG}$). S is a *simulation relation for a contract* C iff whenever $(P, Q) \in S$ the following conditions hold:

- $\mathsf{Act}(Q) \cap L(C) \times \mathcal{T} = \mathsf{Act}(P) \cap L(C) \times \mathcal{T}$.
- For any $(a, t) \in \mathsf{Act}_\epsilon(Q)$, and $t \in \mathcal{T}$, if $P \overset{t\,a}{\Longrightarrow} P'$, then there exists Q' such that $Q \overset{t\,a}{\Longrightarrow} Q'$ and $(P', Q') \in S$.

$\square$

Definition 8. Let $i, s \in \mathcal{AG}$ two agents for a contract C, we say that P *simulates* Q, written $P \precsim Q$, iff there exists a simulation S for the contract C such that $(P, Q) \in S$.

$\square$

The simulation we have defined is a preorder so it can be used as an implementation relation.

Proposition 4. The relation $\precsim$ for a contract C is a reflexive and transitive relation.

$\square$

Next we present the main result in this section. The simulation relation can be used to obtain a valid implementation of an agent.

Theorem 1. Let $C = AG_1 \mathbin{|||} AG_2 \mathbin{|||} \cdots \mathbin{|||} AG_k \mathbin{|||} \cdots \mathbin{|||} AG_n$ a valid contract. If $P \precsim AG_K$, then P is a valid implementation of AG_k.

$\square$

Example 4. The reciprocal of Theorem 1 is not true: there are valid implementations that not simulate the corresponding agent. To prove it is enough to consider a valid contract where the agents never yields the invalid contract signal, for instance let us consider the contract agents $P_1 = \mathsf{P_r}(a?)[0, 10]$, $P_2 = a![0, 10]$, and $P_3 = P(b?[0, 10])$. Then, $i = \mathsf{vc}$ is a valid implementation of P_3. It is also clear that i does not simulate P_3.

5   Complete Example

The example presented in Fig. 1 is inspired by the one described in [4,5]. It consists of *"A Coffee Machine"* involving the interaction between two different agents: *a customer* and *a coffee machine*. The coffee machine system starts when a customer orders a drink by inserting money and selecting a beverage. Coffee can be chosen either with or without milk. The machine proceeds to pour the selected drink, provided the money paid covers its exact price and the correct coins are used. If not, the machine refunds the inserted coins. After payment customers have 30 s to choose between either a *coffee* or a *latte*. The money is refunded if no option is selected in this interval. The order can be cancelled and the customer refunded in an interval of 10 s or the selected drink is poured. Note that a machine only accepts 10, 20 or 50 cent coins. Following the syntax given above, a specification of this example is given next:

Example 5. Coffee machine contract specification

$Machine := \mathbf{rec}\ x.\mathsf{F_b}\{C_{1c}?, C_{2c}?, C_{1\epsilon}?, C_{2\epsilon}?\})$ in
$In\_Coins_{50};$
$\Big(\ (\mathsf{P_r}(B\_abort?)[0,10]; refund![0,30]; x)\square$
   $(\mathsf{O_b}(B\_coffee?)[0,30] \rhd (refund![0,30]; x); (P\_coffee![10,30]) \sqcap refund![30_\delta, 60]))\square$
   $(\ \mathsf{O_b}(B\_latte?)[0,30] \rhd (refund![0,30]; x); (P\_latte![10,30]) \sqcap refund![30_\delta, 60]))\Big); x$
$\rhd (refund![0,30]; x)$

where the number 30_δ represents the number $30 + \delta$ and the In_Coins_{50} agent
is defined as follows:

$$In\_Coins_{50} := P(C_{50c}?)\ \square\ (P(C_{20c}?); In\_Coins_{30})\ \square\ (P(C_{10c}?); In\_Coins_{40})\ \square$$
$$(P(C_{5c}?); In\_Coins_{45})\ \square\ (P(C_{2c}?); In\_Coins_{48})\ \square\ (P(C_{1c}?); In\_Coins_{49})$$

where the other agents In_Coins_x are defined similarly. The customer is specified
as follows:

$Client := \mathbf{rec}\ x.\ Out\_Coins_{50}; \Big($
   $(B\_abort![0,10]; \mathsf{O_b}(refund?[0,30]))\sqcap$
   $(B\_coffe![0,30]; \mathsf{O_b}(P\_coffe?[10,30]) \rhd \mathsf{O_b}(refund?[0,30]))\sqcap$
   $(B\_latte![0,30]; \mathsf{O_b}(P\_latte?[10,30]) \rhd \mathsf{O_b}(refund?[0,30]))\Big); x$

The agent Out_Coins_{50} is similar to In_Coins_x:

$$Out\_Coins_{50} := C_{50c}!\ \square\ (C_{20c}!; Out\_Coins_{30})\ \square\ (C_{10c}!; Out\_Coins_{40})\ \square$$
$$(C_{5c}!; Out\_Coins_{45})\ \square\ (C_{2c}!; Out\_Coins_{48})\ \square\ (C_{1c}!; Out\_Coins_{49})$$

where the other agents Out_Coins_x are defined similarly.

This Example 5 is a simple transcription from the C-O diagram. This e-contract model consists of the two agents described in the diagram the coffee machine and the customer. The next step in analyzing this contract is to compare a set of implementations. We check whether these implementations can play the role of one of the specified agents.

Example 6 consists of three implementations for the customer agent. In the first implementation the customer inserts a 50 cent coin, pushes the coffee button in the interval $[10, 20]$ and waits for the coffee in the interval $[10, 30]$. This specification coincides with the one given for the customer, however the interval in which the customer pushes the coffee button is smaller than the one given by the specification. An implementation is correct either when outputs are produced in an interval that fully coincides with the specification or when this interval is included into the contract specified interval. The second implementation also varies with the customer specification in the interval given for the obligation of pouring coffee $[10, 20]$ instead of $[10, 30]$. Here as in the first implementation a smaller interval is defined. However now the contract behaviour is not guaranteed since, as specified in the contract, the machine might produce the coffee during the last ten time units. In the third implementation, it is clear that the contract is not fulfilled by this implementation since the latte button might be pushed after the upper bound specified in the contract.

Example 6. Client implementations.

$$Imp^1_{Client} = \text{rec } x.C_{50c}!; B\_coffee![10, 20]; O_b(P\_coffee?)[10, 30] \rhd refund[0, 30]; x$$

$$Imp^2_{Client} = \text{rec } x.C_{50c}!; B\_coffee![0, 30]; O_b(P\_coffee?)[10, 20] \rhd refund[0, 30]; x$$

$$Imp^3_{Client} = \text{rec } x.C_{50c}!; B\_latte![20, 40]; O_b(P\_latte?)[10, 30] \rhd refund[0, 30]; x$$

6 Conclusion and Future Work

In this paper we have presented a formalism based on process algebra to express e-contracts. We have defined an operational semantic, a notion of valid contract and the notion of implementation. Furthermore, we have defined a notion of simulation that is correct with regard to the notion of implementation. However, the simulation relation is not complete: not all valid implementation of an agent simulates the agent. One of our future work lines is to find a relation that reduces the gap between the notion of correct implementation and the notion of simulation. The implementation issues have not been discussed in detail in this proposal. We expect to implement the features defined in this paper by using the mCRL2 tool set. The mCRL2 tool set is based in a process algebra. It has the possibility to check the simulation relation.

References

1. Aceto, L., de Frutos Escrig, D., Gregorio-Rodríguez, C., Ingolfsdottir, A.: Axiomatizing weak ready simulation semantics over BCCSP. In: Cerone, A., Pihlajasaari, P. (eds.) ICTAC 2011. LNCS, vol. 6916, pp. 7–24. Springer, Heidelberg (2011)
2. Bauer, S.S., David, A., Hennicker, R., Guldstrand Larsen, K., Legay, A., Nyman, U., Wąsowski, A.: Moving from specifications to contracts in component-based design. In: de Lara, J., Zisman, A. (eds.) Fundamental Approaches to Software Engineering. LNCS, vol. 7212, pp. 43–58. Springer, Heidelberg (2012)
3. Bustan, D., Grumberg, O.: Simulation-based minimization. ACM Trans. Comput. Logic 4(2), 181–206 (2003). ACM, New York. http://doi.acm.org/10.1145/635499.635502
4. Camilleri, J.J., Paganelli, G., Schneider, G.: A CNL for contract-oriented diagrams. In: Davis, B., Kaljurand, K., Kuhn, T. (eds.) CNL 2014. LNCS, vol. 8625, pp. 135–146. Springer, Heidelberg (2014)
5. Díaz, G., Cambronero, M.E., Martínez, E., Schneider, G.: Specification and verification of normativetexts using C-O diagrams. IEEE Trans. Softw. Eng. 40(8), 795–817 (2014). http://doi.ieeecomputersociety.org/10.1109/TSE.2013.54
6. Gentilini, R., Piazza, C., Policriti, A.: From bisimulation to simulation: coarsest partition problems. J. Autom. Reasoning 31(1), 73–103 (2003)
7. Governatori, G., Milosevic, Z.: A formal analysis of a business contract language. Int. J. Coop. Inf. Syst. 15(4), 659–685 (2006). http://dx.doi.org/10.1142/S0218843006001529
8. Gregorio-Rodríguez, C., Llana, L., Martínez-Torres, R.: Extending mCRL2 with ready simulation and iocos input-output conformance simulation. In: The 30th ACM/SIGAPP Symposium on Applied Computing, April 2015, to appear

9. Gregorio-Rodríguez, C., Llana, L., Martínez-Torres, R.: Effectiveness for inputoutput conformance simulation iocos. In: Ábrahám, E., Palamidessi, C. (eds.) FORTE 2014. LNCS, vol. 8461, pp. 100–116. Springer, Heidelberg (2014). http://dx.doi.org/10.1007/978-3-662-43613-4_7

10. Hennessy, M., Regan, T.: A process algebra for timed systems. Inf. Comput. **117**(2), 221–239 (1995). http://dx.doi.org/10.1006/inco.1995.1041

11. Katoen, J.-P., Kemna, T., Zapreev, I., Jansen, D.N.: Bisimulation minimisation mostly speeds up probabilistic model checking. In: Grumberg, O., Huth, M. (eds.) TACAS 2007. LNCS, vol. 4424, pp. 87–101. Springer, Heidelberg (2007). http://dx.doi.org/10.1007/978-3-540-71209-1_9

12. Kyas, M., Prisacariu, C., Schneider, G.: Run-time monitoring of electronic contracts. In: Cha, S.S., Choi, J.-Y., Kim, M., Lee, I., Viswanathan, M. (eds.) ATVA 2008. LNCS, vol. 5311, pp. 397–407. Springer, Heidelberg (2008)

13. Lüttgen, G., Vogler, W.: Ready simulation for concurrency: It's logical!. Inf. Comput. **208**(7), 845–867 (2010)

14. Pace, G.J., Schneider, G.: Challenges in the specification of full contracts. In: Leuschel, M., Wehrheim, H. (eds.) IFM 2009. LNCS, vol. 5423, pp. 292–306. Springer, Heidelberg (2009)

15. Prisacariu, C., Schneider, G.: A formal language for electronic contracts. In: Bonsangue, M.M., Johnsen, E.B. (eds.) FMOODS 2007. LNCS, vol. 4468, pp. 174–189. Springer, Heidelberg (2007)

16. Prisacariu, C., Schneider, G.: CL: A Logic for Reasoning about Legal Contracts Semantics. Technical report, University of Oslo (2008)

17. Ranzato, F.: A more efficient simulation algorithm on kripke structures. In: Chatterjee, K., Sgall, J. (eds.) MFCS 2013. LNCS, vol. 8087, pp. 753–764. Springer, Heidelberg (2013)

18. Schneider, S.: An operational semantics for timed CSP. Inf. Comput. **116**(2), 193–213 (1995). http://dx.doi.org/10.1006/inco.1995.1014

19. Taşiran, S., Alur, R., Kurshan, R.P., Brayton, R.K.: Verifying abstractions of timed systems. In: Sassone, V., Montanari, U. (eds.) CONCUR 1996. LNCS, vol. 1119, pp. 546–562. Springer, Heidelberg (1996). http://dx.doi.org/10.1007/3-540-61604-7_75

20. von Wright, G.H.: Deontic logic. Mind **60**, 1–15 (1951)

Data, Information, and Knowledge Engineering (Regular Papers)

Solving the Problem of Selecting Suitable Objective Measures by Clustering Association Rules Through the Measures Themselves

Veronica Oliveira de Carvalho[1]([✉]), Renan de Padua[2],
and Solange Oliveira Rezende[2]

[1] Instituto de Geociências e Ciências Exatas, UNESP - Univ Estadual Paulista,
Rio Claro, Brazil
veronica@rc.unesp.br
[2] Instituto de Ciências Matemáticas e de Computação,
USP - Universidade de São Paulo, São Carlos, Brazil
{padua,solange}@icmc.usp.br

Abstract. Many objective measures (OMs) were proposed since they are frequently used to discover interesting association rules. Therefore, an important challenge is to decide which OM to use. For that, one can: (a) reduce the number of OMs to be chosen; (b) aggregate OMs' values in only one importance value as a mean of not selecting a suitable OM. The problem with (a) is that many OMs can remain. Regarding (b), the problem is that the obtained values cannot be well understandable. This work proposes a process to solve the problem related to the identification of a suitable OM to direct the users towards the interesting patterns. The goal is to find the same interesting patterns, as if the most suitable OM had been used, also trying to reduce the exploration space to minimize the user's effort.

Keywords: Association rules · Post-processing · Objective evaluation measures · Clustering

1 Introduction

Among the data mining tasks association rules mining is one of the most widely used due to its easy comprehensibility even for non-data miners. An association rule expresses a relation between items that occur in a given data set. The relations are of the type $A \Rightarrow B$, A representing the antecedent and B the consequent of the rule. However, due to the nature of the task, a major problem related with it is the amount of association rules that are usually generated. Even with a small data set, many rules can be extracted. Therefore, many works have been done in the area of post-processing. The idea of these works is to aid the user to discover, among all the extracted patterns, the ones that are interesting for him. To do so, many approaches exist. One of the most used technique relies on objective evaluation measures.

© Springer-Verlag Berlin Heidelberg 2016
R.M. Freivalds et al. (Eds.): SOFSEM 2016, LNCS 9587, pp. 505–517, 2016.
DOI: 10.1007/978-3-662-49192-8_41

Objective evaluation measures, or simply objective measures (OMs), compute the importance of a rule $A \Rightarrow B$ based on the information available in the data set. Based on this importance value, a ranking is built, ordering the rules from the most to the least important ones, and the n top rules of the ranking are considered to be the most relevant to the user. Therefore, the user is directed towards the interesting patterns. Due to their frequent use, many OMs have been proposed (above 50), like support, defined as $P(AB)$, and confidence, defined as $P(B|A)$ (many of these OMs are defined and described in [1]). Therefore, an important challenge in a post-processing process is to decide which OM to use. To overcome this problem, some solutions have been proposed: (a) some of them aim at filtering the OMs to reduce the number of measures to be chosen, as in [1–4]; (b) some others aim at aggregating many OMs' values in only one importance value as a mean of not selecting a suitable OM to rank the rules, as in [5,6]. One way of solving (a) is to cluster the OMs to split them in groups. The problem is that even splitting the measures, many of them can remain (at least one of each group). The problem related with (b) is that the obtained values cannot be well understandable. However, some works, as discussed in [7], state that better results can be obtained through the combined use of OMs compared to their individual use. Finally, it is important to mention that the selection of a good association rule through a suitable OM is an important issue in many areas, as in recommender systems [8].

Based on what stated above, this work proposes a process to solve the problem related to the identification of a suitable OM to direct the users towards the interesting patterns. The goal is to find the same interesting patterns, as if the most suitable OM had been used, also trying to reduce the exploration space to minimize the user's effort. Therefore, with this process (i) it is not necessary to select a set of suitable OMs, regarding (a), nor explicit aggregate many OMs, regarding (b), to rank the rules to find the interesting ones; (ii) the exploration space might be reduced since we assume that there is a subset of groups that contains all the interesting rules, so that a small number of groups have to be explored. The proposed solution is based on a clustering of association rules considering the existing similarity among the rankings obtained by a set of OMs. To the best of our knowledge, this is the first work that discusses the current problem as presented here. It is important to mention that many works that clusters association rules exist, as in [9,10]. However, the aim, in these cases, is to organize the rules in order to provide the user with a better understanding of the domain, helping him to find which are, in fact, the interesting patterns.

The paper is structured as follows: Sect. 2 presents some related works; Sect. 3 shows the proposed process; Sect. 4 analyzes the configurations used in experiments to apply the proposed process; Sect. 5 presents the results and discussion; Sect. 6 draws the conclusions and signals future works.

2   Related Works

Since many OMs exist, some solutions have been proposed to aid the user to decide which OM has to be used. These solutions basically aim at: (a) filtering the OMs to reduce the number of measures to be chosen; (b) aggregating many OMs' values in only one importance value as a mean of not selecting a suitable OM to rank the rules. A review discussing these ideas is presented in [7]. It is important to mention that other approaches can be found in [7], regarding the proposed problem, and for this reason and due to the lack of space, only some recent works are discussed here.

As stated before, one way of solving (a) is to cluster the OMs to split them in groups. In this case, each cluster represents a group of OMs that present similar behavior, i.e., that chooses the same rules as the interesting ones (it can be viewed as a form to reduce the redundancy that can occur among the OMs). Therefore, after selecting the OMs, from the clusters, the rules are ranked. Based on these ideas, [4] present a tool, named ARQAT, that implements an approach that clusters the OMs, from a Rl x M matrix ($Rules$ by $Measures$), to aid the user to select a set of suitable OMs and, consequently, the most interesting rules within a specific rule set (the approach must be executed for each rule set). Since the user can select one OM from each cluster, some operators are available, as union and intersection, to support the identification of the most interesting rules. To do the clustering, the Pearson's correlation coefficient is used to compute the similarity among the OMs, considering that each cell $[r][m]$ of the matrix stores an OM's value. On the other hand, [1] aim to group similar OMs in one cluster, focusing on "excluding" the redundant ones. In other words, the work reduces the selection space, i.e., it tries to present fewer OMs to the user. However, as in the previous work, the user can select one OM from each cluster and, so, many OMs can remain. To do the clustering, a Rl x M matrix is used, in which each cell $[r][m]$ stores the ranking of the rule associated with the OM, and the Spearman's correlation coefficient computes the similarity among the OMs. Similar to the work of [1,2] present the same ideas, but the difference is that the clusters are not crisp, i.e., one OM can belong to more than one cluster. They state that an OM may belong to a cluster because it shares some kinds of properties and to another because it shares some other properties. To do the clustering, boolean factor analysis is used based on a M x P matrix ($Measures$ by $Properties$), in which each cell $[m][p]$ stores 0 or 1 (the OM presents (1) or not (0) the property). Another work that uses a M x P matrix, maintaining the same goal, is [3], using as similarity the Euclidean distance.

The solutions regarding (b) can be seen as the proposal of a "new" OM that generates a unique value for each rule. In this case, the problem is converted to an optimization one, and, generally, an optimal equation is obtained to combine many values in a single value. Based on this single value the ranking is constructed. Considering the exposed, [5] aim to find the most interesting association rules considering, at the same time, many OMs. For that, the skyline operator, used to resolve mathematical and economics problems, is used. Based on this operator, a set of undominated rules, considered as the interesting ones by all the OMs in use, is found. In the end, all of these undominated rules are ranked according to the existing similarity among them and a fictitious rule, the most interesting one considering all the OMs. On the other hand, [6] use

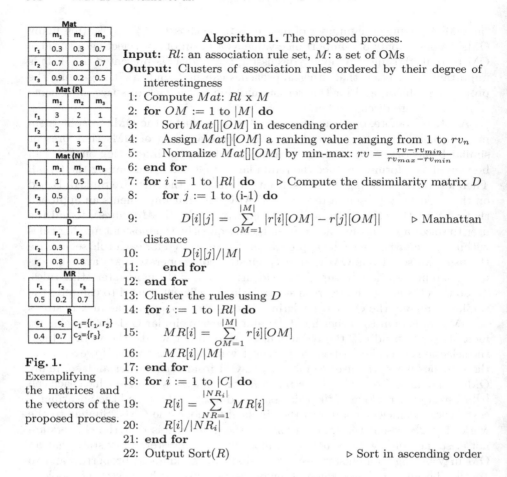

Mat

	m_1	m_2	m_3
r_1	0.3	0.3	0.7
r_2	0.7	0.8	0.7
r_3	0.9	0.2	0.5

Mat (R)

	m_1	m_2	m_3
r_1	3	2	1
r_2	2	1	1
r_3	1	3	2

Mat (N)

	m_1	m_2	m_3
r_1	1	0.5	0
r_2	0.5	0	0
r_3	0	1	1

D

	r_1	r_2
r_2	0.3	
r_3	0.8	0.8

MR

r_1	r_2	r_3
0.5	0.2	0.7

R

c_1	c_2	
0.4	0.7	$c_1=\{r_1, r_2\}$
		$c_2=\{r_3\}$

Fig. 1.
Exemplifying the matrices and the vectors of the proposed process.

Algorithm 1. The proposed process.

Input: Rl: an association rule set, M: a set of OMs
Output: Clusters of association rules ordered by their degree of interestingness

1: Compute Mat: $Rl \times M$
2: **for** $OM := 1$ to $|M|$ **do**
3:     Sort $Mat[][OM]$ in descending order
4:     Assign $Mat[][OM]$ a ranking value ranging from 1 to rv_n
5:     Normalize $Mat[][OM]$ by min-max: $rv = \frac{rv - rv_{min}}{rv_{max} - rv_{min}}$
6: **end for**
7: **for** $i := 1$ to $|Rl|$ **do**   ▷ Compute the dissimilarity matrix D
8:     **for** $j := 1$ to (i-1) **do**
9:         $D[i][j] = \sum_{OM=1}^{|M|} |r[i][OM] - r[j][OM]|$ ▷ Manhattan distance
10:         $D[i][j]/|M|$
11:     **end for**
12: **end for**
13: Cluster the rules using D
14: **for** $i := 1$ to $|Rl|$ **do**
15:     $MR[i] = \sum_{OM=1}^{|M|} r[i][OM]$
16:     $MR[i]/|M|$
17: **end for**
18: **for** $i := 1$ to $|C|$ **do**
19:     $R[i] = \sum_{NR=1}^{|NR_i|} MR[i]$
20:     $R[i]/|NR_i|$
21: **end for**
22: Output Sort(R)   ▷ Sort in ascending order

Choquet integral to combine the values of different OMs into a single interestingness value to rank the rules.

3   The Proposed Process

As it can be seen through the previous sections, the problem related with the works regarding (a), i.e., filtering the OMs to reduce the number of measures to be chosen, is that even clustering the measures, many of them can remain (at least one of each group); with the works of (b), i.e., aggregating the OMs values in only one importance value, that the obtained values cannot be well understandable. Based on what stated above, this work proposes a process to solve the problem related to the identification of a suitable OM to direct the users towards the interesting patterns. The goal is to find the same interesting patterns, as if the most suitable OM had been used, also trying to reduce the exploration space to minimize the user's effort. Therefore, (i) it is not necessary to select a set of suitable OMs, regarding (a), nor explicit aggregate many OMs, regarding (b). The solution is based on a clustering of association rules considering the existing similarity among the rankings obtained by a set of OMs.

The proposed process, seen in Algorithm 1, receives as input an association rule set Rl and a set M of OMs. The rules in Rl are obtained by applying an association rule extraction algorithm. The OMs in M are a group of OMs the user has interest to use. Not to bias the process regarding any OM, it is important to extract the rules only using the support measure (see Sect. 4). The output of the process is a group of clusters ordered by their degree of interestingness, in a manner that only a small number of groups have to be explored, trying to reduce the exploration space – it is expected that the first clusters be the ones that will contain the interesting patterns. For better understanding the clustering process, described below, Fig. 1 presents the output of some matrices and vectors that are stored in Algorithm 1. Besides, the following notation is used: $|M|$ stands for the number of OMs in M; $|Rl|$ stands for the number of rules in Rl; $|C|$ stands for the number of clusters obtained after the clustering process; NR_i stands for the number of rules contained in a cluster i.

First of all, as seen in Algorithm 1, it is computed, in line 1, the matrix $Mat: Rl$ x M, in which each line represents a rule $r \in Rl$ and each column an $OM \in M$. The value of each cell, $[r][OM]$, stores the value of OM in r. An example of such matrix is presented in Fig. 1 as Mat. After that, from lines 2 to 6, a normalization is done, since not all OMs have a pre-defined range, i.e., some of them range, for example, from $[0..1]$, as support, and others from $[0..\infty]$, as collective strength. Thus, if the ranges were transformed to $[0..1]$, for example, a value of 0.5 in one measure could be much more important than a value of 0.5 in another. For that, in lines 3 and 4, each column's values are replaced with their ranking's values as in [1]. As stated by them [1], we also agree that "...what matters most to the practitioner is how an interesting measure ranks rules so she can get the most interesting rules out of her data.". Therefore, each r receives a value that is proportional to its importance. It is important to remember, from OMs context, that the higher the value the more interesting the rule (as occurs in support). That way, in Algorithm 1, rv_n represents the highest ranking in a given OM. Finally, in line 5, the rankings are normalized, in range $[0..1]$, in order to provide equal weighting for the OMs, since each one can have different values of rv_n. It is important to mention that this block of code (lines 2 to 6) processes the data, i.e., Mat, as presented in [11] regarding how to handle ordinal variables. An example of the results obtained from lines 3 to 4 and line 5 are presented in Fig. 1, respectively, as $Mat(R)$ and $Mat(N)$.

To cluster the rules, the dissimilarity matrix D is computed, from lines 7 to 12, based on the existing similarity among the rules' rankings. For that, the Manhattan distance is used (line 9). Note that since the matrix is symmetric, only part of it is stored in memory. Other measures of group cohesiveness, as Kendall's Coefficient of Concordance W [12], could be used. However, we preferred to use the absolute differences, among all the rankings, than the mean square differences, since we think this is more intuitive to the presented problem. To keep the range between [0..1], each distance's value is divided by the number of OMs used in the computation (line 10). The idea is to find the existing agreement among the rules. Therefore, it is possible to obtain groups of rules that agree with their classifications considering different semantics (one of each OM). An example of such matrix is presented in Fig. 1 as D. In line 13 the rules are clustered based on D.

To finish the process, the clusters are ranked (lines 14 to 21). This work considers that there is a subset of groups that contains all the interesting rules, so that a small number of them have to be explored. That way, trying to minimize the user's effort by reducing the exploration space, without knowing the most suitable OM to be used, the user can discover the relevant patterns. Therefore, after clustering the rules, these groups must be ordered, since the aim is to explore only the first ones. The ranking is built as follows: for each rule r, all of its rankings, in all of the OMs, are summed and the mean taken (lines 15 and 16). These values are stored in the vector MR. Based on these means, the interestingness of each cluster is computed by taking the mean of the values obtained for each rule (lines 19 and 20). These values are stored in the vector R. Both MR and R are presented in Fig. 1 as MR and R. At the end, in line 22, the clusters are sorted and output to the user. The idea is to compute the average ranking of each cluster, in which the lower the value the better the cluster, since all the rules inside the cluster present similar agreements regarding different OMs.

4   Experiments

Some experiments were carried out to demonstrate the feasibility of the proposed process. Algorithm 1 shows that, first of all, a rule set Rl and an OM set M are needed. Thus, to evaluate the process, a real data set, provided by the Civil Defense of Rio Claro city, São Paulo state, was used. The data set contains the occurrences attended by them from 2008 to 2012. The kinds of attendance are classified in "investigation", "no criminal", "infraction act", "criminal", "traffic accident" and "others". The aim was to discover interesting patterns regarding the behavior of population in the attendances "others". The data set contained 2323 transactions and 16 features. After pre-processing, with the domain expert aid, 1043 transactions remained along with 9 features. Due to the privacy of the data, details about the features and other particularities will not be provided. The association rules were extracted with an Apriori implementation[1] from the pre-processed data set, using 5 % of minimum support, 0 % of minimum confidence (not to bias the results when using this OM to cluster the rules), rules with a minimum of 2 and a maximum of 5 items considering only one item in the consequent. Before extracting the rules, the data set was converted to a transactional format, where each transaction was composed by pairs of the form "attribute=value". A total of 2215 rules were obtained.

---

[1] Developed by Christian Borgelt: http://www.borgelt.net/apriori.html.

In order to construct a "gold standard" rule set (named here as G set), to enable the analysis of the experiments, the expert analyzed all of the 2215 extracted rules to identify, among them, the ones considered as interesting, coming up with a total of 181 ($|G|=181$). However, to analyze if the size of the rules affects the results, the original rule set was used to generate 4 different rule sets: the first one, R_2, containing only the extracted rules with 2 items; the second one, R_3, containing the extracted rules with a minimum of 2 and a maximum of 3 items; the third one, R_4, containing the extracted rules with a minimum of 2 and a maximum of 4 items; the last one, R_5, containing all the extracted rules. That way, each rule set R_x has its own G_x set: G_2 containing the gold rules with only 2 items, G_3 containing the gold rules with a minimum of 2 and a maximum of 3 items, G_4 containing the gold rules with a minimum of 2 and a maximum of 4 items and G_5 containing all the gold rules. The number of rules contained in each set is presented in Table 1.

Regarding the OM set M, 20 OMs were used, the ones presented in Table 1. All of these OMs are described in [1]. In their work, the authors analyzed 61 OMs and clustered them in 20 groups, demonstrating that they are a good representative for the OMs. Based on that analysis, we selected one OM from each group. The choice was made considering the computational cost to compute the OM; therefore, the simplest ones were selected.

To do the clustering, we used the complete linkage algorithm and cut the dendrograms in 0.25 and 0.50[2], as presented in Table 1. It is important to remember that the result of a hierarchical clustering can be expressed by a tree named dendrogram. This representation allows cutting the tree in a given height to obtain a set of clusters as in the partitional case. The height represents the distance between clusters. Therefore, we decided to use a hierarchical algorithm since we do not have to specify the number of k, finding this number through the cut. The values of 0.25 and 0.50 were chosen since they indicate a similarity inside each group above 0.75 (1–0.25) and 0.50 (1–0.50), respectively. The complete linkage was chosen, as in [1], among the others of the family, as it has a tendency to create more compact clusters. To execute the proposed process, we used, among others, the *daisy* function in the *cluster* package (metric ="gower") and the *hclust* function in the *stats* package, both available in R[3].

Based on what stated above, the proposed process was executed for each rule set R_x. To evaluate the results, four values were computed (see Table 1): *Recall in Clustering* (*Rc.Cl*) (Eq. 1), *Reduction in Clustering* (*Rd.Cl*) (Eq. 2), *Recall in OM* (*Rc.OM*) (Eq. 3) and *Reduction in OM* (*Rd.OM*) (Eq. 4). In all of them, $|x|$ stands for the number of elements in x. *Rc.Cl* stores, for a given rule set R_x, the number of G_x rules that were found, from the total that should be found, for all of its clusters. The order the clusters are processed is related to their degree

---

[2] In fact, the cuts 0.25, 0.30, 0.35, 0.40, 0.45 and 0.50 were tested to see the impact of them in the results. As all of them behaved similarly, being the analysis described in Sect. 5 basically the same to all of them, we decided to present here only the results obtained in the first and in the last cut.

[3] http://www.r-project.org/.

Table 1. Configurations used in the experiments.

Rule sets	$|R_2|{=}408$, $|R_3|{=}1317$, $|R_4|{=}1985$, $|R_5|{=}2215$
Gold sets	$|G_2|{=}45$, $|G_3|{=}116$, $|G_4|{=}160$, $|G_5|{=}181$
OMs	Support, Prevalence, K-measure, Least Contradiction, Confidence, TIC, EII1, Leverage, Directed Information Ratio, Loevinger, Odds Ratio, Dilated Chi-square, Added Value, Cosine, Lift, J-measure, Recall, Specificity, Conditional Entropy, Coverage
Algorithm	Complete linkage (cuts of 0.25, 0.50)
Evaluation measures	$Rc.Cl$, $Rd.Cl$, $Rc.OM$, $Rd.OM$

of interestingness (see Sect. 3). Therefore, t, in Eq. 1, is related to the order of the clusters, being $t = 1$ related to the best classified cluster. $Rd.Cl$ stores, for all the clusters of a given rule set R_x, the exploration space reduction, i.e., the number of rules the user does not have to explore to find the knowledge he considers as interesting. As in $Rc.Cl$, the order the clusters are processed is related to their interestingness. A $Rc.Cl$ of 10 %, for a given cluster C_t of R_x, indicates that from all the G_x rules related with R_x only 10 % of them were found. On the other hand, a $Rd.Cl$ of 70 % indicates that the user only needed to explore (explicit evaluate) 30 % of the rules (100 %–70 %) to find the relevant patterns. Thus, in both cases, the higher the values the better the results. Note, in Eqs. 1 and 2, that the values are cumulative as the clusters are explored. For better understanding, consider that three clusters were obtained and ranked in the following order: c_3 with 50 rules, being 10 of them interesting; c_1 with 20 rules, being 20 of them interesting; c_2 with 30 rules, being 20 of them interesting (total of 100 rules, being 50 of them interesting). Therefore, the following values would be computed: $Rc.Cl = \{20\,\%, 60\,\%, 100\,\%\}$ $(20\,\%{=}(10/50)$, $60\,\%{=}((20{+}10)/50)$, $100\,\%{=}((20{+}10{+}20)/50))$ and $Rd.Cl = \{50\,\%, 30\,\%, 0\,\%\}$ $(50\,\%{=}(1{-}(50/100))$, $30\,\%{=}(1{-}((20{+}50)/100))$, $0\,\%{=}(1{-}((30{+}50{+}20)/100)))$.

$$Rc.Cl(R_x) = \bigcup_{t=1}^{|C|} Rc.Cl_{C_t}(R_x),$$

$$Rc.Cl_{C_t}(R_x) = \frac{|Gold\ Rules\ in\ C_t| + \sum_{i=1}^{t-1} |Gold\ Rules\ in\ C_i|}{|G_x|} \tag{1}$$

$$Rd.Cl(R_x) = \bigcup_{t=1}^{|C|} Rd.Cl_{C_t}(R_x),$$

$$Rd.Cl_{C_t}(R_x) = 1 - \frac{|Rules\ in\ C_t| + \sum_{i=1}^{t-1} |Rules\ in\ C_i|}{|R_x|} \tag{2}$$

$$Rc.OM(R_x) = Rc.OM(R_x)_{OM} = Rc.Cl(R_x) \tag{3}$$

$$Rd.OM(R_x) = Rd.OM(R_x)_{OM} = Rd.Cl(R_x) \tag{4}$$

To compare the results of the proposed process with the results that would be obtained by the traditional OM post-processing approach, $Rc.OM$ (Eq. 3) and $Rd.OM$ (Eq. 4) were computed. Both are based on the previous equations and have, therefore, the same meaning. The difference here is that, as many OMs exist, it is necessary to compute the values for each one of the 20 OMs used to cluster the rules. The aim is to discover the best OM, i.e., the one that reaches a $Rc.OM$ of 100 % with less effort, i.e., with a high $Rd.OM$. The results related with the best OM are stored to be used as the baseline to compare the results. For that, to compute the values, for a given OM, the rules are ranked, and each position of the ranking is considered as a cluster containing one rule. For better understanding the procedure, consider the three rules and their rankings presented in Fig. 1. Besides, that r_3 is the only interesting rule in the set ($|G|=1$). In this case, three rankings would be obtained: r_1, r_2, r_3 for m_1; r_3, r_1, r_2 for m_2; r_1, r_2, r_3 for m_3. Therefore, the following values would be computed for each measure: m_1: $Rc.OM = \{0\,\%, 0\,\%, 100\,\%\}$ and $Rd.OM = \{67\,\%, 33\,\%, 0\,\%\}$; m_2: $Rc.OM = \{\underline{100\,\%}, 100\,\%, 100\,\%\}$ and $Rd.OM = \{\underline{67\,\%}, 33\,\%, 0\,\%\}$; m_3: $Rc.OM = \{0\,\%, 0\,\%, 100\,\%\}$ and $Rd.OM = \{67\,\%, 33\,\%, 0\,\%\}$. Therefore, in the end, the results of m_2 would be stored, since it is the measure in which the user reaches a $Rc.OM$ of 100 % with less effort ($Rd.OM = 67\,\%$).

As it can be observed, what is being analyzed is the user's effort to achieve a good recall. Therefore, it is possible to compare the results of the proposed process considering the best OM results, i.e., when the user knows which is the most suitable OM that would recover the relevant patterns. However, it is important to remember that this information is not known during real applications. Thus, it is expected that the proposed process behaves in a similar or better way than the traditional one.

5   Results and Discussion

The results obtained from the experiments are presented in Fig. 2. Figure 2(a), for example, presents the results of R_2 considering a cut of 0.25 (R_2:0.25). All graphics present the four information described before: $Rd.Cl$, $Rd.OM$ and $R = Rc.Cl = Rc.OM$. In fact, the values of $Rc.Cl$ and $Rc.OM$ are represented by R, since it is presented, for the same recalls' values, the exploration space reductions in the proposed process and in the traditional OM post-processing approach. Therefore, it is possible to evaluate, in each graphic, if the user can, with the proposed process, recover the same interesting patterns as if the best OM had been used, also considering if it would be possible to minimize the user's effort by reducing the exploration space. Remember that the information regarding the best OM to be used is not known during real applications and, so, it is expected that the proposed process behaves in a similar or better way

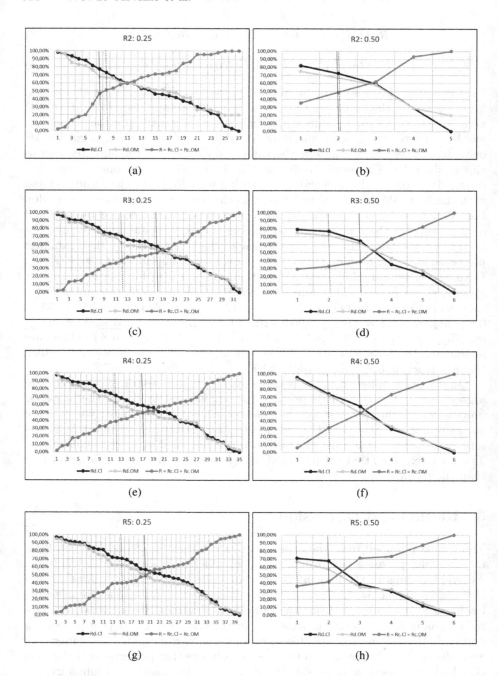

Fig. 2. (a), (c), (e) and (g): results, for each rule set R_x, considering a cut of 0.25 in the dendrogram; (b), (d), (f) and (h): results, for each rule set R_x, considering a cut of 0.50 in the dendrogram.

than the traditional one. In all graphics, R ranges from 0 % to 100 %, expressing the increasing of the value along the clusters. On the other hand, in all of them, $Rd.Cl$ and $Rd.OM$ range from 100 % to 0 %, expressing the decreasing of the values along the clusters. Therefore, the R line is always in the inverse direction of $Rd.Cl$ and $Rd.OM$. The y-axis expresses the percentage of each presented information. The x-axis represents the obtained clusters, ordered by interestingness (from best to worst, 1 meaning the results of the best cluster), and, so, it is possible to see how many clusters were obtained in each case (in Fig. 2(a), for example, 27). Considering Fig. 2(b), for example, it can be observed that exploring only the first interesting cluster, a $Rd.Cl$ value of 82.11 % is obtained with a R ($Rc.Cl$) of 35.56 %. If the best OM measure had been used, considering the same R ($Rc.OM$), a $Rd.OM$ value of 75 % would be obtained. In other words, to achieve the same R more effort would be necessary. Observe that when all the clusters have been explored, a $Rd.Cl$ value of 0 % is obtained with a R of 100 %. Evaluating the results, it can be noticed that:

- the results of the proposed process substantially overlap with the ones that would be obtained with the best OM. This means that the user's effort to achieve a good recall is almost the same in both the procedures, although the proposed process can achieve a good recall with less effort, i.e., with a better exploration space reduction value (see next discussion). In other words, the user can find the same interesting patterns, as if the most suitable OM had been used, also minimizing the user's effort. Therefore, good results were obtained, since, in real applications, the user does not know which OM to use to achieve reasonable results.
- in real applications, the user may not reach a R of 100 %, since only the first n rules are explored to obtain a good exploration space reduction. Therefore, considering the recalls' values close to 50 % (solid line), the following exploration space reductions are obtained:
  $R_2 : 0.25$: $Rd.Cl \approx 77\% > Rd.OM \approx 68\%$; $R_2 : 0.50$: $Rd.Cl \approx 73\% > Rd.OM \approx 67\%$;
  $R_3 : 0.25$: $Rd.Cl \approx 57\% > Rd.OM \approx 54\%$; $R_3 : 0.50$: $Rd.Cl \approx 65\% > Rd.OM \approx 62\%$;
  $R_4 : 0.25$: $Rd.Cl \approx 59\% > Rd.OM \approx 51\%$; $R_4 : 0.50$: $Rd.Cl \approx 59\% > Rd.OM \approx 50\%$;
  $R_5 : 0.25$: $Rd.Cl \approx 57\% > Rd.OM \approx 51\%$; $R_5 : 0.50$: $Rd.Cl \approx 68\% > Rd.OM \approx 58\%$.

Thus, in all of these cases, the proposed process can provide the user with the same relevant information (i.e., the same R) with little less effort, since the exploration space reductions are higher than the ones obtained through the best OM. In fact, for recalls below 50 %, the proposed process generally presents better exploration space reductions in relation to the ones obtained by the best OM. Besides, to achieve recalls close to 50 %, in general, only half or less of all the clusters have to be explored: 25.93 % in R_2:0.25 (7/27), 40 % in R_2:0.50 (2/5), 56.25 % in R_3:0.25 (18/32), 50 % in R_2:0.50 (3/6), 48.57 % in R_4:0.25 (17/35), 50 % in R_4:0.50 (3/6), 51.28 % in R_5:0.25 (20/39) and 33.33 %

in R_5:0.50 (2/6). Finally, for recalls above 50 %, the results of the best OM are, in general, a little better than the ones obtained by the proposed process. However, for higher recalls, none of them are interesting, since only the first n rules are explored, hardly achieving a R of 100 %.

– regarding the rule size effect on the results, it can be seen that the above discussions are true for all of the R_x sets. Therefore, the rule size does not seem to have a high effect on results, as well as the cut on the dendrogram. However, in general, the lower the rule size the higher the recall for the same exploration space reduction. Looking at reductions close to 70 % in the proposed process (dotted line), for example, it can be seen that $R_{R_2}(\approx 51\%; \approx 49\%) \geq R_{R_3}(\approx 40\%; \approx 33\%) \geq R_{R_4}(\approx 39\%; \approx 31\%)$ (exception to $R_{R_5}(\approx 40\%; \approx 36\%)$).

As it can be seen, the proposed process contribution, in automating the exploration process, is highlighted by the reduction that is achieved in relation to the user's effort and by the achieved results that are closer or better than the ones that could be obtained regarding the best OM. That way, without knowing the best OM to apply, in order to reduce the user's effort in finding the relevant patterns, the proposed process provides a reasonable way to achieve this goal. Thus, through experimentation and real case study, the importance of the work can be visualized.

6   Conclusion

This work proposed a process to solve the problem related to the identification of a suitable OM to direct the users towards the interesting patterns. The goal is to find the same interesting patterns, as if the most suitable OM had been used, also trying to reduce the exploration space to minimize the user's effort. That way, (i) it is not necessary to select a set of suitable OMs nor explicitly aggregate many OMs; (ii) the exploration space might be reduced.

Some future works can be done to improve the described process: (a) check if other clustering algorithms can lead to better results; (b) explore ways to rank the rules inside the clusters to try to obtain better exploration space reductions; (c) check if the use of redundant OMs (the ones that lead to the same rankings) affects the results of the proposed process, since only non-redundant OMs were used in the experiments (only the set extracted from the clustering process proposed by [1]). Besides, ways to improve the efficiency of the process, considering the space complexity aspect (due to the matrices and vectors that need to be computed), will be explored.

Acknowledgments. We wish to thank FAPESP and CAPES for the financial support.

References

1. Tew, C., Giraud-Carrier, C., Tanner, K., Burton, S.: Behavior-based clustering and analysis of interestingness measures for association rule mining. Data Min. Knowl. Disc. **28**(4), 1004–1045 (2014)

2. Belohlavek, R., Grissa, D., Guillaume, S., Nguifo, E.M., Outrata, J.: Boolean factors as a means of clustering of interestingness measures of association rules. ann. math. artif. intell. **70**(1–2), 151–184 (2014)
3. Guillaume, S., Grissa, D., Mephu Nguifo, E.: Categorization of Interestingness Measures for Knowledge Extraction. CoRR, vol. ArXiv e-prints (2012). abs/1206.6741
4. Huynh, X.-H., Guillet, F., Blanchard, J., Kuntz, P., Briand, H., Gras, R.: A graph-based clustering approach to evaluate interestingness measures: a tool and a comparative study. Qual. Measures Data Min. Stud. Comput. Intell. **43**, 25–50 (2007)
5. Bouker, S., Saidi, R., Yahia, S.B., Nguifo, E.M.: Mining undominated association rules through interestingness measures. Int. J. Artif. Intell. Tools **23**(4), 22 (2014)
6. Nguyen Le, T.T., Huynh, H.X., Guillet, F.: Finding the most interesting association rules by aggregating objective interestingness measures. In: Richards, D., Kang, B.-H. (eds.) PKAW 2008. LNCS, vol. 5465, pp. 40–49. Springer, Heidelberg (2009)
7. Bong, K.K., Joest, M., Quix, C., Anwar, T., Manickam, S.: Selection and aggregation of interestingnes measures: a review. J. Theor. Appl. Inf. technol. **59**(1), 146–166 (2014)
8. Bong, K.K., Joest, M., Quix, C., Anwar, T.: Automated interestingness measure selection for exhibition recommender systems. In: Nguyen, N.T., Attachoo, B., Trawiński, B., Somboonviwat, K. (eds.) ACIIDS 2014, Part I. LNCS, vol. 8397, pp. 221–231. Springer, Heidelberg (2014)
9. Djenouri, Y., Drias, H., Habbas, Z., Chemchem, A.: Organizing association rules with meta-rules using knowledge clustering. In: Proceedings of the 11th International Symposium on Programming and Systems, pp. 109–115 (2013)
10. de Carvalho, V.O., dos Santos, F.F., Rezende, S.O., de Padua, R.: PAR-COM: a new methodology for post-processing association rules. In: Zhang, R., Zhang, J., Zhang, Z., Filipe, J., Cordeiro, J. (eds.) ICEIS 2011. LNBIP, vol. 102, pp. 66–80. Springer, Heidelberg (2012)
11. Kaufman, L., Rousseeuw, P.J.: Finding Groups in Data: an Introduction to Cluster Analysis. John Wiley and Sons, Hoboken (2005)
12. Siegel, S., Castellan Jr, N.J.: Nonparametric Statistics for the Behavioral Sciences. McGraw-Hill, New York (1988)

Survey on Concern Separation in Service Integration

Tomas Cerny[1]([✉]) and Michael J. Donahoo[2]

[1] Department of Computer Science, Czech Technical University,
Charles square 13, Prague, Czech Republic
tomas.cerny@fel.cvut.cz
[2] Department of Computer Science, Baylor University, Waco, TX, USA
jeff_donahoo@baylor.edu

Abstract. Ever-changing business processes in large software systems, integration of heterogeneous data sources as well as the desire for legacy service integration drive software design towards reusable, platform-independent, web-accessible microservices. Such independently deployable services provide an interface for retrieval and data manipulation in machine-readable formats. While this approach brings many advantages from the perspective of service integration aiming to separate data manipulation from business processing, the standard approaches provide only limited structural semantics and constraints provided through the interface. This leads to considerable information restatement and repeated decisions in integrating components, which considerably impacts development and maintenance efforts. Integration component operability becomes highly sensitive to interaction with underlying services, which are possibly composed of other services. The sensitivity is especially apparent in the structural semantics of produced and consumed information that must correlate on both sides of the interaction. This paper surveys service integration from the perspective of separation of concerns. In order to reduce the coupling and information restatement on the integration component side, it suggests introducing multiple communication channels with additional information that apply in the service interaction, extending the integration component's ability to derive service expected information structural semantics, constraints or business rules. Finally, we consider the impact of this new approach from the development and maintenance perspectives.

Keywords: Web services · Service integration · Aspect-Oriented programming

1 Introduction

Software design of large systems, which integrate functionality from different heterogeneous data sources and provide decentralize governance, utilize reusable, independently-replaceable, scalable and deployable microservices [13]. Such

© Springer-Verlag Berlin Heidelberg 2016
R.M. Freivalds et al. (Eds.): SOFSEM 2016, LNCS 9587, pp. 518–531, 2016.
DOI: 10.1007/978-3-662-49192-8_42

services provide composable functionality addressing the disadvantages of monolithic design. [13]. They provide platform independence, support interoperable [1] interaction among different components using standard machine-readable formats. These services emphasize well-defined interfaces and availability on network, with location transparency.

The emphasis on service interface enables easy exchange of service providers and thus reduces coupling between the integration components (or generally peers) and services. An independent self-deployable component that integrates services is in the text referred as an Integration Component (IC). The IC-service interaction can be further mediated to multiple providers to support service availability, scalability, and performance. Services can recursively integrate other services, making such service composition transparent to ICs. Market changes, innovation, and/or evolution in business requirements make it easy for the ICs to use novel services or apply a new business processes on top of an existing service infrastructure. At the same time, such design, especially in the early development stage, may demand higher investments than code-centric monolithic design [1], while opening the services for broader future reuse. Thus from the long-term perspective, the overall costs are expected to reduce. ICs are fragile with respect to failure, flaws, or performance bottlenecks in any of its underlying services. Furthermore, the standard format of communication brings additional performance demands related to serialization and deserialization of information between the machine-readable format and the internal platform-specific format.

Services built on existing technologies provide information formatted in a specific structure, which the ICs must strictly follow. Current standards-based approaches provide data values and some semantics of the internal structure, although it is insufficient to automatically derive the internal structure at the IC side. For example, property data types are limited and their constraints or validation rules are missing. Furthermore, it is not possible to automatically determine the expected structural semantics of information a service consumes. From the design perspective, ICs must implement appropriate internal structural representation for service-provided information. These are usually Data Transfer Objects (DTO) [10], or map structures. These components define structure for the native platform, although this is a restatement since the structure exists and is co-defined at the web service side. This gives a commitment to the IC. Any time service information defining structure changes, its ICs must reflect the change, creating a difficulty in maintenance, as there is no mechanism preventing such inconsistencies.

The situation becomes worse when considering multiple communicating ICs or middlewares that all process the same information representation, considering the same constraints, validation rules, etc. The maintenance becomes complex and correlation fragility grows. Different individuals might manage particular services or ICs at heterogeneous locations and follow different evolution and changes. Service is usually unaware of its ICs, and thus its internal change may lead to catastrophic consequences. Due to such difficulties, it is now common practice[1] to leave the existing service as it is when structural changes in data take place. Instead of extending the particular service, its copy with changes is made. Consequently, the two services run simultaneously. Such an approach does not naturally scale[2] since multiple such services must be monitored for operation and backups, driving up operation costs.

This paper considers service integration from the perspective of separation of concerns. The motivation is to decrease maintenance effort and mitigate the impact on ICs related to changes in service structural representation, constraints, and validation or business rules. The ICs become adaptable to changes in the above service concerns.

The conventional approaches use a single channel for communicating the information to ICs. As mentioned earlier, only a limited amount of structural information can be derived. For instance, consider a web service producing information in XML or JSON. These formats carry data values and partially describe the data structure with property names, leaving a gap for types, constraints, etc. Thus internal IC representation must exist to provide the missing pieces of information, introducing restatements.

A concern-separating approach suggests a different form of communication. It provides novel meta information that can be used to derive the internal IC structural representations at runtime. This meta information relate to different concerns. Besides the complete data structure information, the channels may consider input validation rules, user context, and business rules. Providing these concerns in a single communication channel, would lead to inefficiencies, as some concerns tend to change more often than others. Instead, we propose the use of multiple communication channels to avoid repetition and support separation of concerns at the communication-level. This extends concern reuse at the IC side and improves caching capabilities [6]. All the novel concerns are provided in a machine-readable format. The ICs become capable to derive internal structural representation, constraints, validation and business rules at runtime based on the service provided information. Later service changes are adopted by ICs and all their communication successors, which help to avoid consistency errors.

The rest of the paper is organized as follows. Section 2 provides background. Details on concern separation analysis in service integration are provided in Sect. 3. Section 4 details the concern separating design. Related work is mentioned throughout Sects. 2 till 4. Section 5 concludes the paper.

2   Background

Web services produce and consume data. A web service is the only component with access to the data source, and the only component that can persist or provide data. Usually, such a service persists data to a relational database, although

---

[1] Based on experience, while technically consulting with software architects of Czech banks.

[2] The highest number of simultaneously running service copied versions was reported[3] 22.

its design most likely uses object-oriented programming (OOP). Even legacy services designed in non-conventional style can be extended to provide a web service interface [1]. Such services communicate in machine-readable formats, such as XML or JSON.

Contemporary trends in OOP design are apparent from the Java Enterprise Edition (Java EE) platform [9]. The platform has a standard for dealing with Object-Relational Mapping (ORM) for persistence called Java Persistence API (JPA), input validation (Bean Validation), and even for serialization/deserialization of data represented by objects to JSON or XML formats[3] and backwards.

A service in the Java EE platform represents its data model with classes called *entities* that are associated with each other and extended with JPA descriptors for ORM as well as with Bean Validation descriptors to enforce input validation. A service can enforce business rules on the top of the data model by referencing particular entities and their properties. As mentioned in [3,10], a standardization or generally accepted approach for defining business rules is missing. One possibility is to define such rules using OOP [2]; unfortunately this leads to significant restatements of rules across various system modules or layers [3], extending maintenance efforts. Alternative approaches suggest using frameworks describing the rules in Domain-Specific Languages (DSLs), such as Drools [16] or MPS [17]. These approaches isolate rule definitions and enforce their application throughout the system.

A web-service hides its internals from other ICs, even though the information structure it provides or consumes is influenced by its entities and their data structure. Sometimes services aggregate entities or filter their properties using DTOs indirection [10]. The entities or DTOs then determine the desired format (XML/JSON). When considering the produced format, it consists of information relevant to a particular data instance, as well as to the data structure since each data value is provided together with its property. The product thus contains limited structural information; however, it does not provide the expected property data type, constraints, etc.

The IC must follow the service-expected data structure. The IC internal data representation, similarly to the service, uses either DTOs designed and compiled for this purpose, or map data structure, derived from the service provided information. The map structure may seem more flexible, but it provides no type safety or assurance on correct data structure, property types, constraints, etc. when submitting data to the service. From the perspective of service data consumption the map structure may seem impractical as the internal structure accepts any input, but the service may reject the information due to typological errors.

Both DTO and map structure properties must correlate with the runtime service representation to avoid inconsistency errors and rejected service submissions. The issue is that service and IC are independent components with different evolution time spans. Thus any time the service-side data representation changes,

---

[3] Java Architecture for XML Binding, Java API for {RESTful/XML} Web Services.

the IC representation must change accordingly. The ICs' DTO property restatement suffers from tight coupling and the inability to adapt to service changes, its structure is determined at compile time. A mechanism is missing to indicate or prevent inconsistency errors due to changed service internal structure.

Restatement problems are not limited to data representation only; a similar issue arises with input validation. For performance and usability reasons, we apply input validation on the IC before incurring the cost of communication and service-side processing, although to do so, we need to manually apply the input validation at the IC. This negatively impacts development and maintenance efforts since the same validation rules are restated on all tiers. Furthermore, when the IC integrates multiple services together, it might be intended to apply service business rules already at the IC level. This again leads to their replication across tiers.

The situation is exacerbated by context-awareness [4]. For instance, consider various user roles that are authorized to access different data properties. From the service autonomy perspective [1], this should apply at the service, but it cannot be omitted at the IC side, due to usability perspective [14]. The context-awareness is more complex [4] than just security. User may come from different geo locations, at different times, with various devices that all may impact the provided data, data representation, its structure, validation rules or business rules. A considerable number of decisions might be repeated at different tiers, tangling through other application concerns [6].

Current web service system design allows service reuse, composition, distribution, replication, and cross-platform compatibility. On the other hand, it does not effectively handle integration component development or service evolution. As we have shown, many concerns are considered at different tiers, although a mechanism that shares concerns across tiers is missing. This paper proposes a concern separation approach applied to the communication among ICs. The advantage is that concerns considered at the service level can be reused by other ICs, which simplifies service evolution and brings ICs' better adaptability to service changes. The context-awareness causes concern tangling in conventional approaches, deteriorating the complexity, development and maintenance efforts. The proposed approach provides concern distribution through multiple channels and handles context-awareness more effectively than a single channel communication. The multi-channel concern distribution extends reuse [6] and supports caching abilities.

3   Analysis and Discussion on Concern Separation in Services

We identify several problems with conventional service integration design. No matter the internal service design, the interaction with other ICs only provides limited information about service concerns. It focuses primarily on data value interaction. Data representation must structurally correlate among ICs and services. The validation rules and constraints must be replicated on the IC side.

When a service does not expose its source code, IC design can only consider service documentation to apply service business rules in its design to improve usability. Even when code is provided, rule derivation might be very difficult, since one service can capture business rules tangled in the OOP design [3], another may use Drools and another MPS. All later changes must again correlate with ICs, which makes global business rule maintenance hard. Runtime context, such as security, time, IC location, etc., may influence the produced or consumed data. From the concern perspective, we consider the following elements:

① Data values   ② Data structure representation
③ Data input validation and constraint   ④ Business rules
⑤ Context-awareness (geo location, time, IP address, security, etc.)

Each IC that uses a service and wants to process its data values ①, restates the data structure representation ②and most likely its validation rules ③. The IC may need to restate business rules ④ or even integrate context ⑤. Besides the data values ①, this presents significant responsibility and burden for development, service evolution, and maintenance. Even a small change to the above service elements ②–⑤ may cause inconsistency in multiple ICs that integrate the service, thus requiring manual change propagation. The issue with change propagation is that a service rarely knows its consumers or has only limited capability to control them.

Naturally, the question is whether there exists a way to loosen the coupling between ICs and service with respect to data structure representation ② or even other concerns ③–⑤. In order to do so, the data structure representation ② at the IC side cannot be determined statically at compile/deploy time. Instead the structural representation should be provided in a separate channel of communication and determined at runtime.

Let us assume a form of communication where an IC requests the data structure representation ② and then maps the data values ① to it. What is the consequence? *First* the IC must compose the representation at runtime, which either requires metaprogramming [7] or the use of map data structure. Since the structure is not determined at compile time, field references may loose type safety [4] within the scope of IC. This may negatively impact the programming style. On the other hand, since a service may change at any time after the IC deploys, the type safety only helps the initial IC design. *Second*, the runtime, on-demand representation derivation enforces consistency with the service. Thus the IC reflects later changes to the service structures, which improves maintenance and evolution. *Third*, there might be performance degradation due to the communication overhead and metaprogramming. On the other hand, the data structure representation ② request can be issued concurrently with the data value request ① [6]. At the same time, the data structure representation derivation could consider caching scheme similar to HTTP [6], where IC requests a particular representation version and reuses the derived structure until it changes on the service side.

With the proposed design, the data structure representation ② is provided by service in a separate channel of communication. The same approach can apply to data input validation ③. The properties of this concern has although lot of similarities with the previous ② and when we consider the Beans Validation standard from Java EE, it is even part of the data model (its extension) [6]. This suggests the possibility to integrate the concern within the channel for the data structure representation ②.

The service business rules ④ might be unknown to the ICs throughout their execution. Such rules can be documented, although ICs cannot use the business rules separately from the service, unless restating the rules. If the rules were known or there was a way to provide them to IC in machine-readable format, the IC could take advantage of such knowledge for usability or performance improvements. For instance, consider a situation restricting an airplane selection for particular flight based on the current passenger occupancy. If IC has the knowledge of such restriction, it may avoid additional requests to the service or rejected submission attempts. Alternatively, consider a service business rule being interpreted at the client-side, resulting in a web browser JavaScript execution that verifies constraints before the submission takes place. In order to provide ICs the business rules in a special channel of communication, the service must capture them in a format that allows not only their evaluation but also their transformation onto format suitable for transmission. For instance, a DSL solution that exposes separated parser, internal representation and execution would fit such purpose. From the ICs' perspective, a business rule definition usually references data structure representations ② and their attributes by name. This introduces a coupling and limits the versatility of IC adaptation to service changes. The proposed IC runtime derivation of data structure representations may hinder the definitions of business rules at the IC side.

Context-awareness might be the next evolutionary step in software system abilities [5]. Nowadays production systems only rarely deal with context-aware features [4] and if so, then only in limited scope [6], such as interactive consoles, due to the increased costs of development and maintenance efforts [4]. For instance, Human-Computer Interaction shows existing context-aware prototypes [14] and sleek features; however these prototypes are missing production experience either due to performance requirements [5] or large development efforts [4]. Survey in [4] suggests that the complexity behind context-awareness is related to poor separation of concerns. Concerns that cannot be cleanly decomposed from the rest of the system are called cross-cutting concerns [12]. Conventional programming languages cannot effectively address cross-cutting concerns, and cause code tangling. The state of the art suggests addressing these concerns through Generative Programming (GP) [8] or Aspect-Oriented Programming [12]. Unfortunately, the program structure and the over all design must change.

GP suggests designing applications from conventional components and integrate models, DSLs descriptions, or alternative problem description formats. It takes all the above as input and then, based on a configuration script and templates, produces various combinations from the inputs. The result may produce a large amount of combinations that are later compiled. The difficulty comes when certain inputs present exponential dependency on its composition [4]. In such case, the produced result becomes impractical. Another deficiency is that the approach targets compile time product derivation. Furthermore, the execution uses generated code, which complicates debugging.

AOP proposes to design application from two building blocks. The base functionality is captured through conventional components and their extensions that are separable concerns, or even cross-cutting concerns that are captured by another building block called aspect. Aspects can use DSL or the same programming language. The aspect brings a mechanism to separate a particular concern from the base program. The way components and aspects connect together is the main AOP instrument. The base component program is transformed onto a join point representation [12]. A join point might be a name of a method, method call, location in the program or a method extended with annotation. It indicates a location in the program where an aspect may extend the program execution. Such a join point representation is a simplified skeleton of the program or a particular subsystem. An aspect has a condition formed from join points that indicates when and under what context it becomes active. The condition may use any logical or arithmetical operators to generalize the condition. The aspect integration can be compile time or runtime [4], and thus only aspects activated by given context are applied to the program execution. The component and aspect integration performs an aspect weaver, an instrument similar to a compiler [12] or renderer [4]. Since it is possible to apply the approach at runtime, the produced result is not affected by exponential concern dependency, since only context-selected concerns apply for given request, although the complexity related to debugging remains.

Thus extending the service with context-awareness while aiming for efficient design, the service should consider separating out the basic functionality from the contextual extension through aspects. Although different from other service concerns ①–④, the context might not be something we aim to provide to ICs as a separate channel of communication. Instead we might expect that context is something related to or provided by the IC requesting the service (e.g., request parameters, location, access rights) or something derived at runtime at the service side (resource usage, time). Thus context may influence the provided result of ①–④.

To demonstrate, consider that the service is requested by a IC with a low level of authorization. The service should only consume or produce a subset of data values ① or to expose the IC limited scope of data representation ②. Alternatively, consider ICs requesting personal information sensitive to the geolocation. One IC may receive information including *country*, *state* and custom date formatting, while another only receives *country* and general date formatting.

4   Design and References to Concern-Separating Approaches

In order to separate concerns mentioned in previous section and stream them in separate communication channels, we must be able to interpret concern description at the service side. However, we should avoid reinventing existing solutions and not expect the industry to make big changes in conventional development or programmer attitudes. For these reasons, we may tend to avoid Model-Driven Development (MDD) [4]. Next option is to design custom DSL for the service description on higher level of abstraction [15], but this approach is not much different from MDD since developer must learn a new language and change the design abstraction. Instead, a minimal impact on developer should be expected for easy adoption and transition to production development.

One possibility is to use a code-inspection mechanism that uses metaprogramming. These approaches allow reusing existing code for the purpose of transformation, which is in our case the ② data structure representation. For example, [11] uses this approach in MetaWidget framework that derives User Interfaces (UIs) from the data model. Similarly [4] uses metaprogramming to derive join point representation [12] for later use in AOP-based transformation. Applying code-inspection does not considerably affect the service development perspective, as its use is transparent.

Similarly, when considering the existing validation or ORM standards, code-inspection can derive the validation rules and constraints ③ and thus further extend the join point representation. Additional data structure representation extensions can be considered in the same way (access, presentation extension [6], etc.).

At this point, the development impact does not involve significant changes, even though the service provides communication channels for the concerns ①–③. The AspectFaces framework [4] provides an example code-inspection tool. [6] shows its use for the separated concern delivery for UI derivation. In order to apply it at the service level, the frameworks' aspect weaver is pointed to application data model from which it derives the join point representation that can be bidirectionally transformed to XML/JSON formats. An IC derives the service internal, platform-specific data structure representation from the received join point model and feeds it with the provided data values ①. [5,6] show the usage for UI derivation for mobile, standalone and web clients (Google Web Toolkit, AngularJS, HTML5) and demonstrate platform-independence for the concern delivery. Changes to the service data structure representation are adopted by all the derived UIs across various platforms. Furthermore, [6] gives details on caching options and performance, which can be applied. For instance, as mentioned earlier, the requests to various channels (data values and join point representation) can be done concurrently. The join point representation can be cached and reused with invalidation mechanism using versioning similar to HTTP.

Usage at an IC impacts the development perspective. The IC usually aims to integrate multiple services and works with multiple ② data structure representations. The internal IC structure representation is a proxy with defined name. Its properties, such as fields are received at runtime, which deteriorates the type-safety at the development time. This is the trade-off for the ability to adapt to service structural changes. The benefit is that validation and constraint enforcement ③ on data values is part of the proxy and can be performed at the IC side. When the IC applies business rules and processes, explicit references bind to the proxy by property names. This limits IC adaptivity to changes in data structures, since service changes to property name do not update the ICs' named binding. Usage of DSLs for business rule definition and enforcement at the IC side is not impacted by the approach, since such DSL has already limited type safety. The IC may apply business processing and forward the proxy to another ICs (e.g., client providing presentation and UI). A proxy propagation is not different from the above description. The data consumption is equivalent to the conventional approach at the service side, with the difference that IC knows what the service expects, what properties it has, which types, constraints and validations are considered, etc.

Common use cases [4] for service maintenance consider changes in structure naming, property naming, property constraint/validation modification, property removal and mostly addition of a novel property. How does the service using ICs react to them?

The conventional approach using DTOs is impacted by any service structural change that is promoted to the machine-readable format and thus causes inconsistency at the IC side. The constraint/validation change is not known and thus may occur at production environment as an inconsistency.

The proposed concern-separating approach cannot deal with changed naming of the given structure since the IC's proxy is determined by the name, although a key-based indirection would address adaptability. The proxy reflects all service changes of property names, constraints and validation rules. It further reflects property additions or removals. Although it has the ability to deal with the changes, the IC application may explicitly reference given properties to apply business processing or enforce business rules, which limits the adaptivity. [5,6] show that this is rarely the case for UIs, even though local coupling may exist. In UIs, it usually uses generic approach to access all provided fields rather than to make explicit references. In the UI, the structural representation can be seen as a logical unit.

When assuming that local reference to given structure exists, then the adaptivity degrades. The IC application no longer adapts to change of property names or property removals. Although the proxy still adapts, the reference to changed property may fail, similarly to conventional design. On the other hand, the adaptivity to changes in constraints and validation rules promote all changes to ICs. Thus when the service maintenance follows the policy that only allows increments in properties and allows constraint/validation changes, then the integrity is preserved and reflected by all ICs. This avoids consistency errors and preserves functionality.

From the above it is apparent that, for the maintainability purposes, the most suitable approach would embed business rules to services rather than to ICs, which [13] suggests for microservices. Then the IC adaptivity promotes to most of the structural changes. However, not all situations allow promoting business rules to services. Furthermore, it might be the target design to use business process and rule indirection to promote flexibility in business changes and evolution. Naturally, structural changes must promote to referencing business rules no matter the origin of the business rule, and thus next we consider the ability to reuse single business rule definition across multiple tiers.

The service ability to provide its business rules to other ICs requires design changes in the way the rules are captured. [2,3] show possible approach that captures the rules in DSL and binds rule to data through annotations. The approach brings the ability to perform business rules inspection and transformation. This can use transformation to machine-readable format and thus be used by a separate distribution channel. An IC can interpret the provided rules locally and avoid rejected submission or improve usability. In service composition, this brings the advantage of combining business rules from various services as well as a centralized view on variety of rules from heterogeneous service environment. The ability of sharing business rules across services can be utilized in business process modeling and execution, although this is left for future work.

The most challenging perspective is the service context-awareness. [4] suggests that many contemporary systems follow the "one for all" approach in the UI design due to its development and maintenance demands. Context can influence the production as well as the consumption of data. We may hardly imagine significant changes in data values ① or structure representations ②, beyond property access restriction or conditional rendering. The ③ input validation and constraints or business rules ④ may differ more significantly basing on the context. For instance, company vendor agent may submit orders with a given delivery date only until a certain time before the distribution stage starts. The accepted order time may differ based on the geo-location of the order destination as the shipment time from a central warehouse differs in the delivery time. As a bonus for customers with high turnover, the agent is able to submit the order after the deadline. When agent updates customers profile he/she must provide all personal information and follow the validation rules on provided formats. An administrator can update customers with a subset of information, while skipping all the validation rules.

Context can more or less impact any of the other considered concerns ①–④, and this impacts the service design. [4] suggests an AOP-based approach, which is not significantly different from conventional design from the development perspective. The data model together with its extensions referencing validation, business rules, etc. is transformed to the join point representation. This representation is produced (once) and utilized by the aspect weaver that mediates service requests. On each request, the weaver clones the join point representation and considers separately-defined aspects that may modify the join point representation. Aspects trigger based on supplied context and join points found in the particular processed representation section. As a consequence, this may modify given constraints, hide properties, etc. The result of the weaving, as shown in [5,6,14], is transformed to a machine-readable format. The corresponding data values ① follow the same cycle in order to determine which data properties are authorized for the delivery.

Context-awareness is mostly notable in UIs where the UI adjusts to particular user, browsing device abilities, location, etc. Earlier we mentioned that [5,6] show multiple prototypes for mobiles, standalone and web clients that can process the communication channels for the concerns ①–③. When service considers context-awareness, these prototypes adjust to the provided output and become context-aware. The impact is on caching abilities, which applies more strict invalidation [6]. The usage across different platforms demonstrates the approach versatility. Furthermore, [6] shows that, for web delivery, the approach untangles UI concerns and supports their reuse at the IC side, which positively impacts service performance. [6] provides evaluation that show the impact in production environment where concern separating approach outperforms the conventional single channel delivery regarding to UI responsiveness as well as reduces service side resources.

5   Conclusion

This survey discusses service integration from the perspective of separation of concerns. Conventional approaches for web service design bring many advantages over the code-centric, monolithic approaches. Unfortunately, service integration posses multiple deficiencies. Data structures considered by services must be understood and followed by all ICs that become tightly coupled to the data structures. This disallows service evolution and usually results in new, slightly modified service introduction to avoid correlation errors with legacy. Moreover, ICs are unaware of service internal constraints, validation or even business rules. Service knowledge distribution would improve performance, consistency, and usability as well as provide a centralized view on combined services.

Separation of concerns and concern distribution addresses the above deficiencies. Services can provide additional information to ICs in multiple distribution channels that are utilized, providing data structure semantics to derive the service data structure representation at runtime. This overwhelms the tight coupling regarding of data properties. This has two sides. Runtime derivation allows the IC to adapt service-side structural changes and thus open the ability for the service to evolve. On the contrary, the IC uses structure proxies unaware of its properties at compile time, unless referencing the service. IC reference to particular properties introduce coupling and limit its adaptivity to structural services changes, although as [5,6] show generic referencing can be applied as demonstrates the usage in UIs. The benefit is that the proxy representation comes with all constraints and validation rules that can be applied at the IC side, avoiding restatement. Using a suitable form of business rule definition at the service side brings the ability to inspect rule definitions and provide them in a separate distribution channel in machine-readable format for the IC. Such rules can be applied earlier in the request processing or even combined with other service rules. Novel business rule definitions at the ICs level might be affected by the weak type safety introduced by the approach. Another solution is to promote business rules to a particular service supporting their reuse, although deteriorating the flexibility of their modification and evolution.

Context-awareness pushes the service design towards the direction of AOP, since it fosters efficient design that avoids tangled code and replication. The context influences both production and consumption of service data values, as well as impacts the resulting data structure representation, validation rules, and business rules.

The main advantages of our approach are its ability to adapt ICs to service changes in data structure, although the extent of the adaptation is influenced by IC references to data structure properties. The approach opens meta information and thus shares its constraints, validation rules, and business rules with other ICs, which provides them with extended abilities that impact performance, composability, and usability. Context-awareness security involvement enforces authorization to all distribution channels.

Future work will address the business rule involvement in business processes definitions in distributed environment. The limited type safety in IC development could use verification mechanism with respect to the service provided meta information. For instance, the IC could use a DSL language, such as MPS [17], as a verification instrument. It could use metaprogramming and suitable constructs to resolve valid references. Alternatively, in case of Java, we can even avoid the impact on the IC design and apply bytecode manipulation framework such as Apache BCEL or ASM [18] to enforce the property correlation at the development/compile time. Our research will also consider the AOP approach to the service backwards compatibility. Each time a service changes, novel aspect is introduced, responsible for backwards compatibility transformation. ICs may use given services and indicate compatibility version. The version triggers a chain of aspects that apply transformation rules to the latest service version and mediate the communication with the IC acting as the older service version.

Acknowledgments. This work was supported by the Grant Agency of the Czech Technical University in Prague, grant No. SGS14/198/OHK3/3T/13.

References

1. Buelow, H., Deb, M., Kasi, J., LHer, D., Palvankar, P.: Getting Started with Oracle SOA Suite 11G R1 a Hands-On Tutorial. Packt Publishing, Birmingham (2009)
2. Cemus, K., Cerny, T.: Aspect-driven design of information systems. In: Geffert, V., Preneel, B., Rovan, B., Štuller, J., Tjoa, A.M. (eds.) SOFSEM 2014. LNCS, vol. 8327, pp. 174–186. Springer, heidelberg (2014)

3. Cemus, K., Cerny, T., Donahoo, M.J.: Automated business rules transformation into a persistence layer. Procedia Comput. Sci. J. **62**, 312–318. Elsevier (2015)

4. Cerny, T., Cemus, K., Donahoo, M.J., Song, E.: Aspect-driven, data-reflective and context-aware user interfaces design. In: Applied Computing Review, vol. 13, no. 4, pp. 53–65. ACM (2013)

5. Cerny, T., Donahoo, M.J.: On separation of platform-independent particles in user interfaces. Cluster Comput. **18**(3), 1215–1228. Springer, USA (2015). http://dx. doi.org/10.1007/s10586-015-0471-7

6. Cerny, T., Macik, M., Donahoo, J., Janousek, J.: On distributed concern delivery in user interface design. Comput. Sci. Inf. Syst. **12**(2), 655–681. ComSIS Consortium (2015)

7. Chiba, S.: Proceedings of the ACM OOPSLA 1998 workshop on reflective programming in C++ and java. In: Javassist - A Reflection-Based Programming Wizard for Java (1998). http://www.csg.is.titech.ac.jp/~chiba/oopsla98/proc/chiba.pdf

8. Czarnecki, K., Eisenecker, U.W.: Generative Programming: Methods, Tools, and Applications. ACM Press/Addison-Wesley Publishing Co., New York (2000)

9. DeMichiel, L., Shannon, B.: JSR 342: Java^{TM} Platform, Enterprise Edn. 7 Spec (2013). https://jcp.org/en/jsr/detail?id=342

10. Fowler, M.: Patterns of Enterprise Application Architecture. Addison-Wesley Longman Publishing Co. Inc., Boston (2002)

11. Kennard, R., Leaney, J.: Towards a general purpose architecture for UI generation. J. Syst. Softw. **83**(10), 1896–1906 (2010). http://www.sciencedirect.com/science/article/pii/S0164121210001597

12. Kiczales, G., Irwin, J., Lamping, J., Loingtier, J.-M., Lopes, C.V., Maeda, C., Mendhekar, A.: Aspect-oriented programming. In: Akşit, M., Matsuoka, S. (eds.) ECOOP 1997. LNCS, vol. 1241. Springer, Heidelberg (1997)

13. Lewis, J., Fowler, M.: Microservices (2014). http://martinfowler.com/articles/microservices.html

14. Macik, M., Cerny, T., Slavik, P.: Context-sensitive, cross-platform user interface generation. J. Multimodal User Interfaces, **8**(2), 217–229. Springer, Heidelberg (2014). http://dx.doi.org/10.1007/s12193-013-0141-0

15. Mernik, M., Heering, J., Sloane, A.M.: When and how to develop domain-specific languages. ACM Comput. Surv. **37**(4), 316–344. ACM, New York (2005). http://doi.acm.org/10.1145/1118890.1118892

16. Proctor, M.: Drools: a rule engine for complex event processing. In: Schürr, A., Varró, D., Varró, G. (eds.) AGTIVE 2011. LNCS, vol. 7233, pp. 2–2. Springer, Heidelberg (2012). http://dx.doi.org/10.1007/978-3-642-34176-2_2

17. Voelter, M., Kolb, B., Warmer, J.: Projecting a modular future. IEEE Softw. **99**, 1. IEEE Computer Society, Los Alamitos, CA, USA (2014)

18. Wu, J., Huang, L., Wang, D.: ASM-based model of dynamic service update in OSGi. SIGSOFT Softw. Eng. Notes **33**(2), 8:1–8:8. ACM, New York (2008). http://doi.acm.org/10.1145/1350802.1350815

Utilizing Vector Models for Automatic Text Lemmatization

Ladislav Gallay and Marián Šimko[✉]

Faculty of Informatics and Information Technologies, Slovak University of Technology in Bratislava, Ilkovičova 2, 842 16 Bratislava, Slovakia
ladislav.gallay@lentil.sk, marian.simko@stuba.sk

Abstract. In this paper we tackle the problem of lemmatization of inflectional languages. We introduce a new algorithm which utilizes vector models of words. Current approaches in this area are limited to knowing either full grammar rules or the translation matrix between the word and its basic form. However, this information is encoded in natural text. Our solution uses text corpora to build vector models of words and a small amount of user input to infer lemmas. We have evaluated our approach on the Slovak language and present interesting findings on its feasibility for real-world utilization.

Keywords: Lemmatization · Vector space model · word2vec · Natural language processing · Slovak language · Word embedding

1 Introduction

Lemmatization, similarly to stemming, is a crucial step in text preprocessing for a variety of core NLP and text processing tasks. Keyword extraction, ontology construction, word sense disambiguation, machine translation, user modeling, web search – all these tasks are connected with lemmatization.

The complexity of lemmatization differs among languages. The task is particularly difficult for inflectional languages such as Slovak. They can be even harder to understand by the computer. The same piece of information can have various forms. The words can have different order and the format of each word is dependent on the position and grammar connection with surrounding words. For inflectional languages that usually include inflectional rules with many exceptions (such as Slovak), the lemmatization rules are not always viable. There is often no other option than using a dictionary approach – exploiting a dictionary of word form-lemma mappings.

The problem is that such a dictionary cannot be easily created automatically. Hence dictionaries are created manually or at best semi-automatically by utilizing large corpora annotations [2]. Since a human author (or supervisor) is required, the task of dictionary creation and maintenance is very demanding. Moreover, due to the volume of the dictionary, it is very difficult to involve special types of words such as neologisms or domain specific terms. This makes

© Springer-Verlag Berlin Heidelberg 2016
R.M. Freivalds et al. (Eds.): SOFSEM 2016, LNCS 9587, pp. 532–543, 2016.
DOI: 10.1007/978-3-662-49192-8_43

creation and maintenance of a universally usable lemmatization dictionary even harder. For many small yet rich languages, efforts to create such a dictionary are due to the various reasons not feasible.

We seek after alternatives to a dictionary approach that will address the aforementioned issues. Recently, there has been an emergence of vector space word models (also known as word embeddings) induced i.a. by advances in computational performance of neural networks [3,6]. It has already been shown that vector space word representations are useful in syntactic and semantic analysis [1,4] and other tasks related with language processing [11]. In this paper, we explore the possibilities of the vector space model for automatic word lemmatization. We propose a method for automatic word lemmatization by utilizing vector space models trained on Slovak national corpus [5]. We evaluate our method by conducting a series of experiments and show that with a well trained model, it is possible to attain 80 % correctness.

2 Related Work

There are two basic approaches to finding a basic form of words in text processing applications: lemmatization and stemming. The Porter algorithm is the most common algorithm for stemming in English and has been repeatedly shown empirically to be very effective [7]. It is based on the several rules that fit the English language. However, inflectional languages such as the Slovak language are often complex enough to make this algorithm unusable because of the variety of different suffixes and their mapping to the root word.

For the Slovak language a morphological dictionary has been created [2] which covers more than 98,000 lemmas and all of their word forms. Despite the input costs for creation, this is the best solution for the Slovak so far. The disadvantage is that the dictionary needs to be built manually and it cannot work with the word context. There are some words that can have the same form but different lemmas because of the context they are used in. This cannot be determined with the dictionary only and requires a more advanced algorithm.

Other techniques include some heuristic functions and guessing the similarity between two words based on the letter frequency or affix similarity [10]. If these words pass the condition they are considered to have the same meaning. This technique is accurate only for specific domains.

High precision stemmer was built for other Slavic languages including Czech or Polish [8]. It shows reasonable results but it is still only a stemmer which can unify some non-related words with common prefix.

The problem is that current approaches are very hard to employ due to the language variety or the massive human input which is required. Yet they are not 100 % accurate – the manually created dictionaries cannot cover all words. Our aim is to create a lemmatizer that will not demand such a human power. This would be accomplished by utilizing a vector space model created automatically from large corpora of texts, while keeping a reasonably high accuracy.

3   Utilizing Vector Model for Lemmatization

It has been shown that the vector word model is powerful enough to hold syntactic and semantic information [4]. They can be automatically trained on large texts without a supervisor and allow us to perform easy mathematical operations. As an example of the model created by Mikolov with the word2vec tool [3] the following math operation is presented:

$$vector(king) - vector(man) + vector(woman) = vector(queen) \qquad (1)$$

We can interpret this equation as: The word *man* is related to *woman* like *king* is related to *queen*, which encodes a relationship between male and female nouns.

Our approach is based on an assumption that vector space word representations encode not only syntactic and semantic regularities, but also morphological ones. In other words, we expect that the words *happiness* and *happy* have the same connection as the words *illness* and *ill*.

The main idea behind our algorithm is depicted in Fig. 1. It is based on finding the correct vector shift of word to lemma reference in order to get the correct lemma for the input word. The vector shift can be obtained from word pairs that have similar relationships (encoded in vector representation). For example, let us have an input word *rybníkom* (instrumental case, meaning "(with) pond"). Potentially similar word *vodníkom* (instrumental case, meaning "(with) vodyanoy") matches the form of the input word *rybníkom*. If the shift from *vodníkom* to *vodník* (lemma) is known, we can apply it to the word *rybníkom* to get *rybník* (lemma). The two words from which the vector shift is calculated, we refer to as a *reference pair*.

Our method consists of the following steps (considering an input word w):

1. Relevant reference pairs selection.
2. Lemma candidates retrieval based on reference pairs.
3. Lemma candidate weight computation.
4. The most suitable lemma selection.

First, we choose reference word-lemma pairs from reference lexicon suitable for lemmatizing input word w. We experiment with various methods for automatic reference pair selection based on morphological, grammatical and other properties.

Secondly, we apply reference pair vector shift to the input word and we retrieve lemma candidates for each reference pair. The vector model may contain mistakes (as a result of large real-world corpora processing) or it still may not be precise enough for words occurring infrequently in the corpus. As a result, vector shift would not necessarily lead into the one correct lemma word. Hence we investigate the neighborhood (surrounding words) of resulting vector shift (dotted circle in Fig. 1), obtaining more lemma candidates, which potentially include the correct lemma.

Thirdly, we need to select the correct lemma from all candidates. Each candidate is given a weight based on its similarity with the input word. The idea behind this is that there should be some connection between the input word and correct lemma. We propose several different methods for calculating this weight.

Finally, the weights for the same candidate are summed together and the candidate with the highest total weight is selected to be the correct lemma.

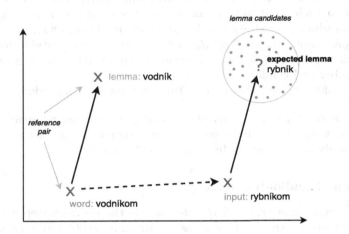

Fig. 1. Relationship between words *vodníkom* (vodyanoy, instrumental case) and *vodník* (vodyanoy, nominative case) is similiar to words *rybníkom* (fish pond, instrumental case) and *rybník* (fish pond, nominative case). However the expected lemma may be somewhere around the centroid vector, therefore we need to explore surrounding words which we call lemma candidates.

3.1   Relevant Reference Pairs Selection

Our algorithm expects reference lexicon R containing (r, r_L) pairs, where r is reference word and r_L is reference word's lemma. The reference lexicon should cover all morphological variations utilizable for lemmatization. For example, for Slovak nouns it would be all combinations of grammatical genders and cases. More than one reference pair for each combination is expected as the algorithm matches the input word to the same grammatical category. We experiment with various methods for choosing the suitable pairs $R_w \subseteq R$ with regards to the input word w.

– **R1: Suffix length**. We select pairs from R based on length of common suffix. In many fusional languages such as Slovak, the grammatical category is determined by the suffix. The words with different lemma but the same suffix may belong to the same category and have the same vector shift. The word is selected based on the longest common suffix. We normalized the length of longest suffix by the words' average length.
Examples of (r, r_L) for $w = autom$: (mestom, mesto), (vlakom, vlak).

- **R2: Cosine similarity**. We have empirically discovered that the closest words in the model usually match the grammatical categories of w. In this variant, we select pairs from R based on the cosine distance from w. The closer the word is, the stronger latent relationship is formed.

  Examples of (r, r_L) for $w = autom$: (autobusom, autobus), (svetlom, svetlo).
- **R3: Grammatical categories**. The reference pairs are divided into groups by the grammatical categories such as the grammatical case. We expect in this variant that each input word w comes with the same information so that we know from which group the pairs should be selected from. This variant requires quite more manual input (big R) and therefore is less suitable for strictly automatic lemmatization. We consider as grammatical categories grammatical case, number and gender.

  Examples of (r, r_L) for $w = autom$: (mestom, mesto), (hniezdom, hniezdo).

The result of this step is a set of automatically selected reference pairs R_w for the input word w. The reference pairs are ordered. To order reference pairs in R3, we apply naive ordering by common suffix length.

3.2 Lemma Candidates Retrieval

For each reference pair $(r, r_L) \in R_w$ obtained in the previous step we compute lemma candidate centroid vector c_c as vector shift: $vector(w) + vector(r_L) - vector(r)$. Then we retrieve lemma candidates – the closest words according to cosine distance from lemma candidate centroid in vector space. The cosine distance is obtained from the word2vec tool. The result of this step is set C containing lemma candidates and their distance from lemma candidate centroid $(c, d_c) \in C$, where c is lemma candidate and $d_c = cossim(c, c_c)$ is cosine distance between c and c_c.

3.3 Lemma Candidate Weight Computation

We compute lemma candidate weights by combining cosine distance from lemma candidate centroid vector and considering morphological attributes of the input word w. We compute weight of lemma candidate c as:

$$d_f(c, c_c, w) = d_c * d_m(c, c_c, w) \tag{2}$$

where d_f is the final weight of lemma candidate c, d_c is the cosine distance between c and c_c from the previous step and d_m is the morphological weight of lemma candidate c with regards to w.

We experiment with various alternatives on how to compute morphological weight of lemma candidate c:

- **DM0: Ignored.** As a baseline, we ignore morphological attributes of lemma candidate and consider only cosine distance, i.e., $d_m = 1$.

- **DM1: Levenshtein distance.** The normalized Levenshtein distance is calculated for each c. The number of common letters is divided by the maximum length of c or w. The problem is with anagrams as they are evaluated as identical.
- **DM2: Jaro-Winkler distance.** The candidates are given the weight based on the Jaro-Winkler distance. It is similar to Levenshtein distance with some improvements. The theory behind this is that the root word w_L and w have many letters in common. Jaro-Winkler algorithm prefers the words that have a common prefix which is similar to the DM3 method.
- **DM3: Relative prefix length.** Various morphological forms are created by changing the suffix while the prefix remains the same, e.g. *auto, autami*. The relative prefix is calculated as the prefix divided by the average length of input word w and lemma candidate c.

$$d_m(c, c_c, w) = \frac{lcp(w, c)}{(strlen(w) + strlen(c))/2} \tag{3}$$

where $lcp(w, c)$ is the longest common prefix of w and c.

The result of this step is a set of weighted lemma candidates for each reference pair (r, r_L).

3.4 Lemma Selection

In this step we examine lemma candidate weights and select the most suitable lemma. Each reference pair is mapped with a unique list of lemma candidates. The correct lemma candidate is expected to appear in connection with different reference pairs. Therefore the weights of the same lemma candidates are summed up. The lemma candidates' final weights d_f are also normalized by the number of reference pairs in R_w.

The word with the highest normalized final weight is selected to be a correct lemma w_L.

4 Evaluation

To evaluate the proposed method, we have performed several experiments. First we describe data used, method configuration and measures employed, then we describe experiments and discuss the results.

4.1 Experimental Setup

Data. We trained vector model on prim-6.0-public-all corpus from Slovak National Corpus [5]. The corpus contains 829,771,945 tokens and 655,572,511 words. The standard vector size used in word2vec is around 100. However, we selected the size of 300 to get more precise results. We utilized continuous bag-of-words as a learning algorithm to train the model, because it shows to be more

Table 1. Description of the default test configuration

Reference lexicon	Extended Slovak grammatical paradigms
Reference pairs selection variant	R1: Suffix length
Lemma candidate weighting	DM3: Relative prefix length
Reference lexicon size	112 word pairs
Number of selected reference pairs ($|R_w|$)	3
Number of candidates for each pair ($|C|$)	40

accurate for less frequent words. Currently we do not provide any method to add new words to the model, however, the original training text can be extended and a new vector model can be trained and used in our algorithm.

The input words to be used in experiments were extracted from the annotated lexicon by Ľudovít Štúr Institute of Linguistics, Slovak Academy of Sciences. The lexicon contains 1.1 million tokens followed by their root form and 98,782 unique lemmas. As these tokens were annotated with grammatical categories, we were able to perform further evaluation based on these categories. We selected only non-nominative words, as our algorithm is not able to determine whether the input is already a lemma. It always tries to find the lemma which is expected to have a different form from the input word. We have focused on nouns only.

Default Configuration. We conducted experiments by using the default method configuration (Table 1). We gradually experimented with changing one parameter at a time, while preserving all others.

Evaluation Measures. We measure the method's performance as correctness of lemma selection (i.e., the ratio of correctly returned lemmas). Our algorithm is supposed to return the correct lemma on the first position – with the heighest weight d_f. We measure correctness considering the position of the correct lemma – if it appeared within top 1, top 5 or top 10 (referred to as *correctness@k*, where k is the number of positions to check) results – to get a better insight into method's performance. In addition, we used the mean reciprocal rank (MRR) to measure the average rank of returned lemmas [9].

4.2   Assessing Relevant Reference Pair Selection Variant

In the first experiment, we evaluated method's performance with regards to the relevant reference pair selection method.

We applied our method as described in default configuration and compared the reference pairs selection variants. The results are shown in Fig. 2. We were able to automatically lemmatize 59.9 % of the input words (R2) where the correct lemma had the highest weight (correctness@1) and 84.3 % where the lemma was among 5 results. Having a better selection method could lead to the correctness

of 92.7 % (correctness@5) as it was in the R3-all. The results show that there is
a significant difference between variants R1/R2, which are fully automated, and
variant R3, which need some additional information. The method employing
variant R3 with all 3 input word attributes (case, number, gender) has the
highest correctness, as expected. In this case, the most relevant words were
selected from the reference lexicon in comparison with R1/R2.

We observe that the correct lemma was in many cases among the candidates but not always in the first position (compare correctness@1 vs. correct-
ness@5). This means that the correct lemma did not have the highest weight.
This can be improved by changing the weighting function or training the model
on larger/better corpus. However, there is little difference between correctness@5
and correctness@10. As seen in Fig. 2, MRR measure correlates with the correctness and supports aforementioned observations.

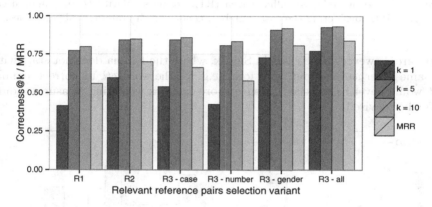

Fig. 2. Comparing various automatic selection methods of reference pairs for most
frequent words in corpus, based on: relative suffix length (R1), cosine similarity (R2),
grammatical categories (R3) considering the case, number, gender or all.

As a part of this experiment, we also evaluated our method by using 1 000
random input words to assess the method for real-world input. The results (see
Fig. 3) are worse and suggest that overall performance in a real-world setting is
lower than expected.

4.3   Assessing Weight Computation Algorithms

In the next experiment, we compared performance of different weight computation algorithms. In Fig. 4 the weighting variants DM0-DM3 are compared. The
results confirm that the simple cosine distance (DM0) performs the worst and
that more complex weighting methods are relevant here. The best performance
was yielded by the relative prefix length variant (D3), confirming that morphological attributes should be considered when weighting lemma candidates.

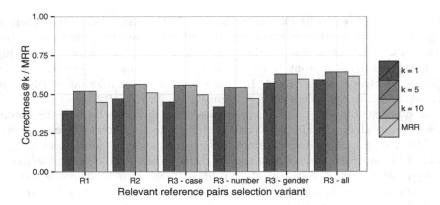

Fig. 3. Comparing various variants of automatic reference pairs selection for random input set, based on: relative suffix length (R1), cosine similarity (R2), grammatical categories (R3) considering the case, number, gender or all grammatical categories.

There are, however, some cases in Slovak, where the lemma does not start with the same letter as the input word form (e.g. for the word *cti*, its correct lemma, *čest*, would never be matched). Hence, more complex weighting measures should be further devised.

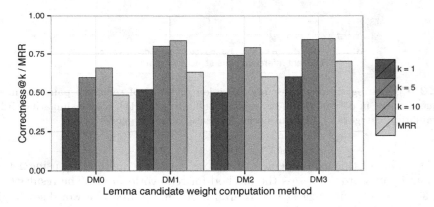

Fig. 4. Comparing various automatic weighting methods: ignored (DM0), Levenshtein distance (DM1), Jaro-Winkler distance (DM2), and relative prefix length (DM3).

4.4 Assessing Number of Reference Pairs Involved

In the next experiment, we examined the number of reference pairs to be included in lemmatization. As our algorithm selects the reference pairs from the reference lexicon, we can change the number of pairs being selected that match the input word, i.e., the size of the set R_w. Having the ideal vector model, we would be able to lemmatize with only one reference pair. The size of R_w depends on the examples for each grammatical category that we want to cover. If we have the word in Genitive case, plural and male gender, we should have at least three reference pairs in the reference lexicon, having R_w size 3. If less reference pairs are available the wrong pair could be selected, which will produce incorrect vector shift. Having more reference pairs (even with the same grammatical categories), we should more likely get the correct lemma. We experimented with the number of reference pairs to be selected with an aim to examine more lemma candidates obtained by performing more vector shifts. Figure 5 shows that bigger R_w improves the correctness up to a certain point and then the correctness decreases.

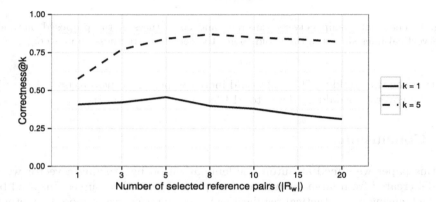

Fig. 5. Correctness of lemmatization with respect to the number of reference pairs selected for lemmatization. The more reference pairs are selected, the better the correctness is – this is valid only to a certain point.

4.5   Assessing Lemmatization Candidate Weight Impact

In the previous experiments, we evaluated all the results as relevant and expected to always find the correct lemma for each input. However, we can set a threshold during the weighting and consider only highly-weighted lemma candidates to be correct lemmas. This would drop the coverage as some input words would not be lemmatized, but the correctness for those we are able to lemmatize would increase.

We have ordered the results based on the weight d_f. We can then set the final weight threshold $d_f = D$ so that everything above the D will be treated as *correct* and everything under D will be treated as *not found*. Then, we can calculate the correctness only for the *correct* outputs and we can see the ratio of *correct* to all outputs which we call coverage.

Figure 6 illustrates how correctness of lemmatized words and coverage of lemmatizable words are related. The vertical lines show some interesting threshold points D. We can see that the total correctness for 100 % coverage is 52.3 %.

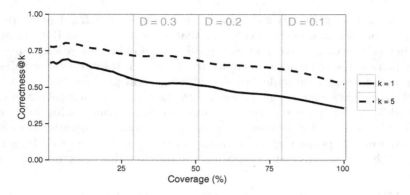

Fig. 6. The relationship between coverage and correctness of the proposed method. The vertical lines show the threshold to get the certain correctness and coverage.

Setting the threshold to $D = 0.2$ would increase the correctness to 68.4 % (k=5), but the coverage would go down to 51.0 %.

5   Conclusions

In this paper we aimed at automatic lemmatization by utilizing a vector word model created from large unannotated corpora of textual resources. Inspired by works focusing on syntactical regularities in vector space, we proposed a method which utilizes small reference lexicon (e.g., declension paradigms) to traverse vector space when seeking for correct lemmas of input words. We evaluated the method on the Slovak language, however, the algorithm is applicable to various languages (from language families where the inflectional word forms are created by changing the suffix while the prefix remains the same). We examined various parameters showing reasonable accuracy for several configurations.

The main advantage of the proposed method resides in the small human input that is required to build the reference lexicon. In contrast with algorithmic lemmatization, dictionary-based lemmatizers need each word to be manually annotated and verified. When compared to dictionary-based lemmatization, automated lemmatization returns potentially correct lemma for each input, including neologisms or domain specific terms. The experiments showed that the bigger the initial reference lexicon, the better results. Though this means more human input is needed, it is a much better option than a static dictionary created and maintained manually.

There is still much space to further explore and improve automatic lemmatization based on vector word models. We tested the algorithm only on nouns as the lemmatization is a huge topic and this is only the first stage of our research. Additional variants for reference pair selection and lemma candidates weighting should be devised to better utilize morphological or language-specific regularities in word to lemma relationships. The perspective research direction may be the combination of dictionary-based and automatic methods to create universal automatic lemmatizer handling language exceptions and irregularities.

The proposed method is not language-dependent and may be beneficial for languages where only little work on automatic lemmatization has been done and only limited (unannotated) language resources are available.

Acknowledgments. This work was partially supported by the Scientific Grant Agency of Slovak Republic, grant No. VG 1/0646/15 and the Cultural and Educational Grant Agency of the Slovak Republic, grant No. KEGA 009STU-4/2014.

References

1. Bansal, M., Gimpel, K., Livescu, K.: Tailoring continuous word representations for dependency parsing. In: Proceedings of the Annual Meeting of the Association for Computational Linguistics (2014)
2. Garabík, R.: Slovak morphology analyzer based on Levenshtein edit operations. In: Proceedings of 1st Workshop on Intelligent and Knowledge-Oriented Technologies, pp. 2–5 (2006)
3. Mikolov, T., Chen, K., Corrado, G., Dean, J.: Efficient estimation of word representations in vector space (2013). arXiv preprint arXiv:1301.3781
4. Mikolov, T., Yih, W.t., Zweig, G.: Linguistic regularities in continuous space word representations. In: Proceedings of Conference of the North American Chapter of the ACL: Human Language Technologies, HLT-NAACL 2013, pp. 746–751 (2013)
5. JÚĽŠ: Slovak national corpus - prim-6.0-public-all. Bratislava: Ľ. Štúr Institute of Linguistics SAS (2013). http://korpus.juls.savba.sk
6. Cortes, C., Vapnik, V.: Support-vector networks. In: Machine learning, p. 99 (1995)
7. Porter, M.F.: An algorithm for suffix stripping. Program **14**(3), 130–137 (1980)
8. Brychcín, T., Konopík, M.: Hps: high precision stemmer. Inf. Process. Manage. **51**(1), 68–91 (2015)
9. Chapelle, O., Metlzer, D., Zhang, Y., Grinspan, P.: Expected reciprocal rank for graded relevance. In: Proceedings of the 18th ACM Conference on Information and Knowledge Management, pp. 621–630. ACM (2009)
10. Krajči, S., Novotný, R.: Hľadanie základného tvaru slovenského slova na základe spoločného konca slov (In Slovak). In: 1st Workshop on Intelligent and Knowledge Oriented Technologies, pp. 99–101 (2006)
11. Šajgalík, M., Barla, M., Bieliková, M.: Exploring multidimensional continuous feature space to extract relevant words. In: Besacier, L., Dediu, A.-H., Martín-Vide, C. (eds.) SLSP 2014. LNCS, vol. 8791, pp. 159–170. Springer, Heidelberg (2014)

Improving Keyword Extraction from Movie Subtitles by Utilizing Temporal Properties

Matúš Košút and Marián Šimko(⊠)

Faculty of Informatics and Information Technologies, Slovak University of Technology in Bratislava, Ilkovičova 2, 842 16 Bratislava, Slovakia
matuskosut@gmail.com, marian.simko@stuba.sk

Abstract. In our work we aim at keyword extraction from movie subtitles. Keywords and key phrases although missing the context can be found very helpful in finding, understanding, organizing and recommending the media content. Generally, they are used by search engines to help find the relevant information. Movies and video content are becoming massively available and widespread. The ability to automatically describe and classify videos has a vast domain of application. In our work we select movie subtitles as a source of information to process. We proposed a method for keyword extraction from movie subtitles by analysing their temporal properties and detecting conversations. We evaluated our method by conducting two experiments (a priori synthetic experiment and a posteriori user experiment) involving 200 movies and show that conversation analysis can improve traditional approaches based on automatic term extraction algorithms.

Keywords: Keyword extraction · Subtitles analysis · Movie metadata · ATR algorithm

1 Introduction

The importance of having video data annotated and ordered grows exponentially with rising amounts of data created every day. In order to enable efficient access to media content (filtering, search, recommendation), it has to be annotated with at least basic semantic descriptions – keywords. It became almost impossible to assign necessary keywords to the entire movie data manually, as it is a demanding and time-consuming task. Although a research about video and audio analysis advanced recently [12], yet it is computationally intensive to process such data and the results are not always as expected [13]. It may be beneficial to consider alternative sources of information such as subtitles. Original subtitles are usually provided with movies on the media (e.g., DVD, Blu-ray, etc.) and also a lot of educational video content available online is available with subtitles provided either by authors or communities.

To the best of our knowledge, there still is lack of research focused on utilizing alternative sources of information for keyword extraction in movies. Several

© Springer-Verlag Berlin Heidelberg 2016
R.M. Freivalds et al. (Eds.): SOFSEM 2016, LNCS 9587, pp. 544–555, 2016.
DOI: 10.1007/978-3-662-49192-8_44

works use keyword extraction algorithms designed to work with pure text, which they obtain from subtitles treated as ordinary documents [1,2,11]. We are not aware of works that consider other than textual information from movie subtitles; no processing of subtitle-specific attributes is reported in the field of information extraction.

The contribution of our work is a method for keyword extraction from movie subtitles by utilizing temporal properties encoded in subtitles: we perform conversation analysis to identify words with potentially higher importance for the movie and boost results of traditional keyword extraction algorithms. Our aim was to show to what extent the processing of subtitle-specific characteristics can improve traditional text-based approaches to keyword extraction from movie subtitles. We performed several experiments that confirm our expectation and show that our method can improve traditional automatic term extraction algorithms by far more than tenths or hundredths of percent.

Note that approaches utilising audio and video analysis is beyond the scope of this paper. Our primary aim is to evaluate the potential of subtitles as "cheap", community-created form of information to enrich movie descriptive metadata which can be further utilized for tasks such as recommendation of movies, TV shows or other video content with subtitles.

The rest of the paper is structured as follows. In Sect. 2, we discuss related work. We describe our method for movie keywords extraction in Sect. 3. We evaluate our method by conducting two experiments: a priori and a posteriori, which we describe in Sect. 4. In Sect. 5, we discuss the implications of our research and conclude the paper.

2   Related Work

Only a few works deal with subtitle analysis to extract semantic information. Langlois et al. utilize subtitles to support video information retrieval [1]. Within the VIRUS project (Video Information Retrieval Using Subtitles), users may look up for movie scenes featuring specific characteristics (e.g., love scenes, violent scenes, funny situations). The authors of the paper perform simultaneous analysis of video, audio and subtitles to allow indexing and retrieval of such video scenes.

Katsiouli et al. tackle semantic video classification based on subtitles [2]. The authors introduce a service that assigns a category label to a movie. The proposed method involves keywords extraction from subtitles as one of the method's steps. It uses TextRank algorithm with its default settings to extract keywords [3]. After that, the method assigns a category to keywords extracted in the previous step by using Word-Net domains and mappings of video categories to WordNet domains.

Similarly, Demirtas et al. proposed automatic categorization and summarization of documentaries using subtitles of videos [11]. Both categorization and summarization are based on text processing. The first one utilizes the existing video categorization algorithm [2]. But in addition to the source method, the authors also use the title of the documentary to get important clues about the video. The second method is using learning module to perform categorization. The authors use TextRank to extract a summary of video subtitles and then they make a video summary by matching it with the corresponding parts of video.

Besides external subtitle processing, several works deal with in-video text extraction and processing. For example, Mai and Hoang present a system for automatic caption text- and keyframe-based retrieval as a solution to acquire a caption from video content [9].

When acquiring semantics from vast text document corpora, traditional automatic term recognition (ATR) algorithms such as TF-IDF, HighScore, RAKE [6], TextRank [3] and ExpandRank [10] and others [5] are very popular. They are being constantly improved and adopted for specific domains [4,8,14]. However, we are not aware of research works exploring performance of ATR algorithms on movie subtitles, nor examination of subtitles specifics such as timings and closed captioning to improve the original keyword extraction methods or their enhancements. In comparison with general text documents, subtitle text is not as fluent or smooth – speech (encoded in subtitles) is supplemented with video and audio "language" and these are missing in traditional subtitle processing. Hence, keyword extraction from subtitles should take advantage of other information present in subtitles. Timings are one of them.

With growing amounts of movie and TV show subtitles produced by communities (in different languages), the potential of subtitle databases such as OpenSubtitles, Addic7ed, SubtitleSeeker and SubScene[1] for information extraction increases. In our work, we aim to examine the potential of subtitles for movie keywords extraction. We propose a method utilizing temporal properties contained within movie subtitles to improve performance of the state-of-the-art automatic term extraction algorithms.

In our method we analyze and process specifics of subtitles to split them into conversations, to rate them and to evaluate their impact on the keyword extraction.

3   Method for Keywords Extraction from Movie Subtitles

Subtitles consist of pure text apportioned to individual movie scenes with the timings of when and how long to display. Our idea behind keywords extraction from subtitles is to recognise individual segments of movie. We refer them to as conversations. We suppose that we could divide dialogues in subtitles into conversations and get the scenes separately, approaching the natural distribution of scenes as it is seen by viewers. In the sense that when there are different characters talking in dialogues or scene is changed, we assume a new conversation occurred.

---

[1] available at www.opensubtitles.org, www.addic7ed.com, www.subtitleseeker.com, subscene.com, respectively.

We explore possibilities for rating conversations according to the relevance for the movie. The higher the relevance is, the higher the ratings of keywords extracted from the conversations become.

These conversations could also possibly help us by joining subtitles created by different authors in case we need to get more information about the scene. That would result into more keywords, assuming different authors use different sentences and on condition that they are not an exact transcript.

To split the subtitles into conversations, we use timings included in subtitles. We detect the gaps between individual subtitles, supposing that if there is a gap bigger than the gaps in the surrounding titles, the conversation has changed. We also experiment with the speech rate (words per minute) in individual titles and conversations as way to differentiate between individual conversations.

Our method to keyword extraction from movie subtitles consists of the following steps (see Fig. 1):

1. Subtitles pre-processing
2. Named entity recognition
3. Conversations identification
4. Keyword extraction
5. Keyword weight adjustment

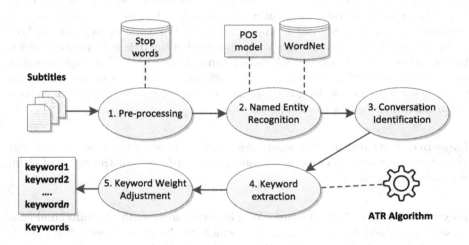

Fig. 1. Scheme of our method for keywords extraction

Subtitles Pre-processing. Pre-processing consists of processing information about timings of individual titles, computing the speech rate and counting gaps between them. Irrelevant symbols and subtitle-specific content, such as music symbols and colour tags, are filtered out. Text inside hearing impaired tags and HTML tags is labelled. Then all the text in titles is segmented, tokenized and POS tagged. We use Rake Text Smart List [6] of stop words to filter out unnecessary and irrelevant words from the text of subtitles.

Named Entity Recognition. Some named entities such as places and locations from movies often appear as keywords we want to extract. We identify named entities in the pre-processed content. First, we select named entity candidates - all unique words in which the first letter is uppercase. We count for every word the frequency of the word's appearance in lowercase. We use POS tagger and WordNet to get more information about these words. Out of these words, we select only proper nouns that never appeared in lowercase to be named entities. For our needs all of these entities were added to banned words list for keywords except those that belong to a certain lexical domain (location, object, group, communication, etc.). These we consider acceptable as keyword candidates. On the other hand, the banned words are mostly person names or unknown entities that are not usually used as keywords. Acceptable candidates stay within the text and they are treated by ATR algorithms in the same way as the rest of the text.

Conversations Identification. We experiment with subtitles splitting into conversation using gaps between individual titles and conversation rate. We suppose that the gaps between the titles of different conversations are wider than those of the same conversation. We also suppose that speech rate of titles differ between conversations according to characters, intensity and tense of conversation. We experiment on ratings using the location, length and speech rate of conversation and filtering out unimportant ones. We count speech rate as how many words (at least 2 characters long) appeared in the title per minute (defined by the difference of starting and ending time of the title). The importance of conversation is rated according to how much time of the conversation showing the text spends and how long text appears on the screen. Text appearance time is normalised according to the speech rate.

Keyword Extraction. We apply standard ATR algorithms to extract and weight relevant words. These words are usually ordered and the top n of them is chosen as keywords.

Keyword Weight Adjustment. We also experiment with the individual conversations utilised for word extraction, which results in more sets of keywords for one subtitles to be processed. Ratings of conversations are used to calculate the final weights of words. We experiment with combination of these ratings and joining the conversation results from the previous steps.

4   Evaluation

The major aim in evaluation was to assess whether or not our method improves accuracy of keyword extraction. We have implemented web services and a research environment for experimentation with our keyword extraction methods, which we have made public[2]. It was implemented using web framework Sinatra based on Ruby language. Our service implements REST API to allow users upload their own subtitles. For the purpose of extraction we utilised existing implementations of ATR algorithms (as gems) according to the description of our method. To evaluate our method, we performed several experiments.

First, we compared the results of our method with the gold standard (a priori evaluation). Secondly, we performed a small user experiment to examine real users' perception of keywords for given movies (a posteriori evaluation). In the both experiments, we compared the performance of original ATR algorithms (using only text to extract keywords) with the results of the method we proposed (i.e., improved ATR algorithms).

4.1 Comparison with the Gold Standard

In the first experiment, we have randomly chosen 200 movies from the IMDb popular movies list[3]. For each movie, we downloaded subtitles from OpenSubtitles.

The gold standard in this experiment consists of keywords from the Moviecus movie search engine[4] (see Table 1 for examples of keywords). Since movie keywords were created prior to applying our method, this experiment is a priori evaluation of our method. Since the gold standard was created independently on our method, from the methodological point of view, this is a more significant form of evaluation.

Table 1. Examples of keywords in the gold standard

Movie	Keywords
Whiplash (2014)	School, Band, Whiplash, Jazz, Student, Drama, Drummer, Instructor, New Jersey, Abuse, Drum, Class, Relationship, New York, Caravan, Talent, Buddy, Rich, Car, Ambition, Competition, Lawyer, Music, High-school, Humiliation, Parents
Jane Eyre (2011)	Romantic drama, Romantic, British, Drama, Love, Aunt, Running, Running away, Alone, Master, Uncle, Sister, Marriage, Horse, River, Sisters, School, Home, Rain, Strange, Dark, Doctor, Flashback, Fire, Secret, Happiness
The Matrix (1999)	Hacker, Computer, Programming, Australia, Action, Reality, Simulation, Machine, Science-Fiction, Truth, Rebel, Police, Security, Future, Pills, Agent, Ship, Dystopia, Animated, Ambush, War, Energy, Computer hacker, Zion, Attack, Rescue

---

[2] http://text.fiit.stuba.sk/subkex.
[3] http://www.imdb.com/search/title?at=0\&sort=moviemeter.
[4] http://moviecus.com.

To compare the extracted keywords with gold standard keywords, we measured standard information retrieval metrics Precision (P), Recall (R) and F-score (F). In addition, we measured 'extractable' Recall (R') to consider only keywords present in subtitles to be relevant (by omitting the keywords in the gold standards that are not extractable from subtitles; to reveal the actual limits of the method). 'Extractable' F-score (denoted F') is weighted average of P and R'.

We employed three traditional ATR algorithms in our experiments: well-known TF-IDF, TextRank [3] and HighScore[5]. For every used method we considered three setups:

- Basic (ATR) – application of the basic ATR algorithm as proposed by original authors (steps 1 and 4 of our method).
- Advanced (ATR+NER) – application of the basic algorithm enriched with named entities recognition (steps 1, 2 and 4 of our method).
- Full (ATR+NER+C) – application of the full method as described in previous section, i.e., basic ATR algorithm + named entity recognition + conversation analysis (i.e., all steps of our method).

ATR in the list represents one of the aforementioned ATR algorithms. The reason for considering the advanced setup of the method (ATR+NER only) was to evaluate the improved performance of ATR algorithm approaching to the form as it would be applied in a real world setting for the task of keyword extraction.

We applied our method with each ATR algorithm to extract keywords. We have selected the number of extracted keywords to evaluate the method's performance to be 10. The results showing P, R, F, R' and F' metrics are presented in Table 2.

We can see that both named entity recognition and conversation rating improve performance of each ATR algorithm in a basic setup. This result is very promising.

Table 2. The results of comparison with the gold standard (%). Each evaluation measure is computed as *Measure*@10.

Method setup	P	R	F	R'	F'
TF-IDF	9.75	3.78	5.45	6.05	7.41
TF-IDF+NER	20.65	8.05	11.58	14.02	16.57
TF-IDF+NER+C	24.92	9.70	13.96	18.23	20.34
TextRank	19.10	7.41	10.68	12.65	14.80
TextRank+NER	24.47	9.53	13.72	17.20	19.74
TextRank+NER+C	**30.30**	**11.80**	**16.99**	21.82	24.60
HighScore	18.04	7.02	10.11	12.99	14.68
HighScore+NER	21.46	8.36	12.03	15.71	17.68
HighScore+NER+C	28.64	11.14	16.04	**23.24**	**24.87**

---

[5] https://github.com/domnikl/highscore.

The best results in this experiment were achieved using the TextRank together with named entity recognition and conversations rating. We achieved the increase of precision by 11.20 % and F-score by 6.31 %. Although the results of TF-IDF algorithm are lowest in comparison to other algorithms, TF-IDF achieved the best increase of precision by 15.17 % and F-score increase by 8.51 % in combination with named entity recognition and conversations rating. Our method utilizing HighScore algorithm achieved best performance when considering R' and F' metrics. The results also confirm that named entities are very important to describe movie content as keywords since they improved each basic ATR algorithm on their own.

Overall, the results confirm that conversation processing improves ATR algorithms (in both basic and advanced setups). Splitting all movies into conversation, selecting the most relevant keywords for conversations and adjusting overall keywords weights turned out to be very useful.

However, the overall results are lower than expected. F-score and F'-score did not exceed 17 % and 25 %, respectively. The results reveal that the baseline performance of ATR algorithms is much worse than reported in other domains. This confirms our expectation that traditional ATR algorithms may not perform well on not "fully-fledged" texts such as subtitles. The obtained performance shows that conversation analysis offers a reasonable option how to improve performance of ATR algorithms. The results support our assumptions that named entities are particularly important as movie descriptors and that timing of subtitles affects the relevance of words contained within subtitles.

The results also indicate that keyword extraction from subtitles may be not sufficient when using subtitles as the only source of information. Nevertheless, it can be very useful to combine our method with different forms of movie data (video, audio) processing. The potential also lies in processing multiple subtitles for a movie created by different authors, either of the same or other languages. We expect that this would lead to improved recall.

It is also important to realize that final performance of both the basic ATR algorithms and our method might have been affected by the choice of the gold standard keywords. Although the Moviecus keywords have been used for advanced search in the movie database since 2011, they still may be – though relevant for movie description assessment (with most quality data so far) – not perfect baseline for other information processing tasks such as movie exploration or recommendation (the keywords may be too general or too specific). In order to assess the method's performance from user's perspective, we conducted a live user experiment.

4.2   User Experiment

The results of the first experiment were strongly dependent on keywords provided in the gold standard hence those might not really match keywords we extracted, supposing keywords extracted by our method may be more precise.

We conducted an experiment involving real users. We were interested how people perceive extracted words and how those words describe the movie according to them. 17 respondents participated in the experiment during the 17th Spring Workshop on Personalized Web [15]. They were asked to assess how relevant to the movie they find the presented keywords on condition that they have seen the movie. Since the participants assessed the results obtained by our method, this experiment constitutes an a posteriori evaluation.

To get the most proper results we have chosen the newest movies from the list of the most favourite movies in the IMDb database[6] (as of March 2015). For every movie we extracted the set of the keywords using 1) basic ATR algorithm and 2) our method (full). There were 20 movies selected in the experiment. For each movie we provided approximately 20 keywords to cover all the ATR algorithms. We created them by merging the extracted keyword sets (union) from all ATR algorithms – top 5 keywords per method – together, hiding information about their origin from the participants. Keywords always appeared in random order to the participants.

The participants were presented movies sequentially. For each movie, they were presented the movie title and short description extracted from the IMDb. They were asked to answer the questions about how they had liked the movie and when they had seen it so we could weigh their answers. Then they were asked to assess the extracted keywords. For every provided keyword, they answered how it describes the movie using the four point Likert scale.

Precision of the keywords was calculated as the weighted average of the participants' ratings obtained for every movie by normalizing the scale to [0;1] interval. Keywords were mapped back to the methods (ATR algorithms) and precision of every method was calculated as the average of all the precisions of keywords.

Since it is not possible to employ recall measure in such setup, we employed P@n measure (Precision at n first results) to observe how the performance of our method changes with the increasing number of extracted keywords. The experiment's results are shown in Table 3.

The best results were achieved by employing our full method utilizing Text-Rank algorithm. The average precision (computed from P@n) achieved was 71.40 %. Although the TextRank results were the best, the highest increase was attained by our full method utilizing TF-IDF algorithm.

We can conclude that our method has improved the performance of the method with regard to precision of extracted keywords. The improvement was achieved for all the ATR algorithms when including the conversation detection as account of subtitles specifics processing. We can see the most significant improvement for small n. This suggests that our method succeeds in "boosting" the relevance of true keywords.

Overall, the results reveal that in the case of user-based assessment, precision of keywords yielded by our method is much higher. This observation supports the potential of our method for extraction of relevant keywords describing the

---

[6] http://www.imdb.com/search/title?title_type=feature.

Table 3. Results of user experiment showing precision of keyword extraction P@n (%)

Number of keywords (n)	TF-IDF		TextRank		HighScore	
	Basic	Full	Basic	Full	Basic	Full
1	6.90	69.06	20.69	78.00	10.34	75.06
2	6.90	67.31	17.24	79.00	17.24	68.08
3	6.20	66.13	21.84	76.00	22.99	64.19
4	9.48	64.76	21.55	73.00	20.69	63.21
5	10.34	63.33	19.31	72.00	17.93	62.99
6	13.79	62.22	20.69	69.00	20.11	65.12
7	13.79	61.40	22.17	68.00	20.69	62.52
8	13.36	60.53	21.55	66.00	21.12	61.63
9	13.41	60.22	20.69	67.00	21.46	59.78
10	14.83	58.63	21.38	66.00	21.03	57.99

movie from movie subtitles. In this type of evaluation, we were however not able to assess recall, i.e., we were not able to evaluate what the method did not produce.

5   Conclusions

In this paper we tackled the issues of keyword extraction from movie subtitles. Keywords as a form of metadata are useful for providing intelligent services such as advanced search or recommendation of movies. However, not much research is done in this field; we are aware of only a limited number of works exploring the potential of subtitles for movie-related information processing.

The contribution of this paper is our subtitles-based keyword extraction method that enriches standard ATR algorithms with processing of temporal properties in movie subtitles. We proposed the detection and rating of so called conversations to adjust keyword weights produced by original ATR algorithms.

The evaluation showed that traditional ATR algorithms designed to text documents yield poor results and has to be further adopted for subtitle processing. Both named entity recognition (particularly important for movie description) and conversation analysis succeeded in improving performance of the original ATR algorithms.

We suppose that our method could be used with advantage in combination with methods for text/keyword extraction based on audio and video analysis, which were proposed and described in numerous related works. Results of audio and video analysis keyword extraction could be combined with keywords extracted by our method by employing a variety of weighting algorithms to get even more accurate results.

The presented work is a part of a research project aimed to provide missing movie and TV show metadata for purposes of recommendation. The information available in open linked data repositories is often not sufficient. Enrichment of movie and TV show descriptive metadata is important for efficient employment of traditional content-based recommendation methods [7] or their combination with other types of recommendation [16].

More research has to be carried out to further improve results of keyword extraction, e.g., by examining impact of processing multiple subtitles per movie or combining subtitles processing with video and audio processing. Our future work will cover improvement of baseline ATR algorithms and further improvement of subtitle specifics utilisation. It will be also important to assess the extracted keywords with regards to a particular information-processing task they are used for: movie/TV show recommendation (in vivo evaluation).

Acknowledgments. This work was partially supported by the Scientific Grant Agency of Slovak Republic, grant No. VG 1/0646/15 and the Cultural and Educational Grant Agency of the Slovak Republic, grant No. KEGA 009STU-4/2014.

References

1. Langlois, T., Chambel, T., Oliveira, E., Carvalho, P. et al.: VIRUS: video information retrieval using subtitles. In: Proceedings of the 14th International Academic Mind Trek Conference: Envisioning Future Media Environments, pp. 197–200, Tampere, Finland. ACM (2010)
2. Katsiouli, P., Tsetsos, V., Hadjifethymiades, S.: Semantic video classification based on subtitles and domain terminologies. In: Proceedings of the KAMC. http://ceur-ws.org/Vol-253/paper05.pdf (2007)
3. Mihalcea, R., Tarau, P.: TextRank: bringing order into texts. In: Proceedings of EMNLP, Association for Computational Linguistics, pp. 404–411, Barcelona, Spain. ACL (2004)
4. Harinek, J., Šimko, M.: Improving term extraction by utilizing user annotations. In: Proceedings of 13th ACM Symposium on Document Engineering, pp. 185–188. ACM (2013)
5. Hasan, K.S., Ng, V.: Conundrums in unsupervised keyphrase extraction: making sense of the state-of-the-art. In: Proceedings of the 23rd International Conference on Computational Linguistics, 365–373, Beijing, China (2010)
6. Dagan, I., Church, K.: Termight: identifying and translating technical terminology. In: Proceedings of the Fourth Conference on Applied Natural Language Processing, pp. 34–40, Stuttgart, Germany (1994)
7. Kompan, M., Bieliková, M.: Content-based news recommendation. In: Buccafurri, F., Semeraro, G. (eds.) EC-Web 2010. LNBIP, vol. 61, pp. 61–72. Springer, Berlin (2010)
8. Lučanský, M., Šimko, M.: Improving relevance of keyword extraction from the web utilizing visual style information. In: SOFSEM Emde Boas, P., Groen, F.C.A., Italiano, G.F., Nawrocki, J., Sack, H. (eds.) SOFSEM 2013. LNCS, vol. 7741, pp. 445–456. Springer, Heidelberg (2013)
9. Mai, D., Hoang, K.: Caption text and keyframe based video retrieval system. In: Proceedings of the 4th International Conference, ICCCI, pp. 244–252, Ho Chi Minh City, Vietnam (2012)

10. Wan, X., Xiao, J.: Exploiting neighbourhood knowledge for single document summarization and keyphrase extraction. ACM Trans. Inf. Syst. **28**(2), 244–252 (2010)
11. Demirtas, K., Cicekli, N.K., Cicekli, I.: Automatic categorization and summarization of documentaries. J. Inf. Sci. **36**(6), 671–689 (2010)
12. Hu, W., Xie, N., Li, L., Zeng, X., Maybank, S.: A survey on visual content-based video indexing and retrieval. IEEE Trans. Syst. Man Cybern. Part C: Appl. Rev. **41**(6), 797–819 (2011)
13. Nguyen, P.X., Wang, K., Belongie, S.: Video text detection and recognition: dataset and benchmark. In: Proceedings of 2014 IEEE Winter Conference on Applications of Computer Vision (WACV), pp. 776–778 (2014)
14. Uherčík, T., Šimko, M., Bieliková, M.: Utilizing microblogs for web page relevant term acquisition. In: SOFSEM Emde Boas, P., Groen, F.C.A., Italiano, G.F., Nawrocki, J., Sack, H. (eds.) SOFSEM 2013. LNCS, vol. 7741, pp. 457–468. Springer, Heidelberg (2013)
15. Bieliková, M.: Proceedings in Informatics and Information Technologies 17th Spring 2015 PeWe Workshop Gabčíkovo, Nakladateľstvo STU, Vazovova 5. Bratislava, Slovakia, 11 April 2015. ISBN 978-80-227-4340-2
16. Kompan, M., Bieliková, M.: Group recommendation: survey and perspectives. Comput. Inform. **33**(2), 446–476 (2014)

Identification of Navigation Lead Candidates Using Citation and Co-Citation Analysis

Robert Moro(✉), Mate Vangel, and Maria Bielikova

Faculty of Informatics and Information Technologies,
Slovak University of Technology in Bratislava,
Ilkovičova 2, 842 16 Bratislava, Slovakia
{robert.moro,xvangel,maria.bielikova}@stuba.sk

Abstract. Query refinement is an integral part of search, especially for the exploratory search scenarios, which assume that the users start with ill-defined information needs that change over time. In order to support exploratory search and navigation, we have proposed an approach of exploratory navigation in digital libraries using navigation leads. In this paper, we focus specifically on the identification of the navigation lead candidates using keyword extraction. For this purpose, we utilize the citation sentences as well as the co-citations. We hypothesize that they can improve the quality of the extracted keywords in terms of finding new keywords (that would not be otherwise discovered) as well as promoting the important keywords by increasing their relevance. We have quantitatively evaluated our method in the domain of digital libraries using experts' judgement on the relevance of the extracted keywords. Based on our results, we can conclude that using the citations and the co-citations improves the results of extraction of the most relevant terms over the TF-IDF baseline.

Keywords: Navigation leads · Keyword extraction · Domain modeling · Citation analysis · Co-citations · Digital libraries

1 Introduction

There are various needs that motivate users to search vast information spaces, such as the Web or digital libraries. A classical taxonomy of Web search by Broder [2] differentiates three types of needs, namely (*i*) *navigational*, the goal of which is to locate a specific web page whose existence is known to the user, (*ii*) *informational*, the goal of which is to acquire certain information the whereabouts of which are unknown to the user, and (iii) *transactional*, the goal of which is to locate a web page where further transaction will occur, e.g., online shopping. Out of these, the informational need is the most general, but we always assume that users know exactly what kind of information they need (e.g., the title of an article written by an author).

However, the information need of users is often ill-defined at the beginning and it tends to change in the light of new information that they gather during

© Springer-Verlag Berlin Heidelberg 2016
R.M. Freivalds et al. (Eds.): SOFSEM 2016, LNCS 9587, pp. 556–568, 2016.
DOI: 10.1007/978-3-662-49192-8_45

the search. Thus, their search tasks tend to be open-ended and more exploratory in their essence; the term *exploratory search* was coined by Marchionini [10] for this type of searches.

An integral part of exploratory search is *sense-making* [16], i.e., making sense of a problem at hand or a new domain, learning the basic concepts and relationships between them, etc. A typical example of this behavior is *researching a new domain*, a task that researcher novices (e.g., doctoral students) often have to face. Their goal is not to find the specific facts, but to *learn* about the given domain and *investigate* the topics, the existing approaches as well as the gaps in the current state of knowledge.

In order to support exploratory search and navigation, we have proposed a method of exploratory navigation using the *navigation leads*, i.e., the automatically extracted keywords, which help the users to filter the information space [11]. The conceptual overview of the method can be seen in Fig. 1. It is a modification of the classical model of web information retrieval (IR) as defined in [2]. The users do not have to refine their query manually, but we augment the search results with the navigation leads. When the users choose a specific lead visualized in a summary of a search result or underneath it, their query gets modified with the lead so that only documents containing the selected lead are retrieved. The idea is similar to the probabilistic (or blind) relevance feedback [15], but in contrast our approach does not expand the query automatically, but lets the users to decide which terms to use. Also, lacking the relevance judgements, we rely on the topical relevance of the extracted terms [11].

The process of augmentation of the search results with the navigation leads consists of three main steps (see Fig. 1):

1. *Identification of the navigation lead candidates* – it includes automatic keyword extraction from the documents as well as assessment of their document relevance; it results into a set of the navigation lead candidates.
2. *Selection of the navigation leads* – in this step, the document relevance of the individual keywords is combined with their navigational value, i.e., how relevant the lead candidates are for the whole information (sub)space. The result of this step is a set of the selected navigation leads.
3. *Visualization of the leads with the search results* – the selected navigation leads need to be placed into the search results list, preferably within the summaries (abstracts) of the search results or underneath them. The result of this step is a list of search results augmented with the navigation leads.

While in [11] we have introduced the concept of navigation leads and provided the preliminary results on the step two, i.e., selection of leads, in this paper, we focus mainly on the first step, i.e., the identification of the lead candidates. For this purpose, the document metadata can be utilized alongside the actual contents of the documents. We are interested in the researcher novice scenario in the digital libraries (or more precisely digital library systems that provide access to scientific publications). Therefore, we consider also the metadata that are specific for this domain (see next section). The most prominent of these

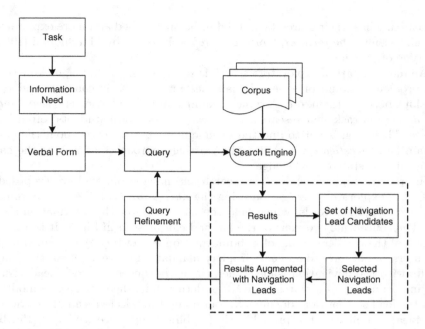

Fig. 1. A model of web IR augmented with the navigation leads to support exploratory search.

are citations, i.e., *citing sentences* that provide a unique source of information; they highlight different aspects of the articles (documents) that were deemed important or interesting by other researchers.

We propose a method of keyword extraction using citation analysis that serves as a means of navigation lead candidates identification. Besides the direct citations that have been to some extent examined in the related works, e.g., [1,4,9], we consider also the *co-citations* (two articles are co-cited, if there is a third article that cites them both) that to the best of our knowledge have not yet been utilized for this purpose.

In this paper, we examine the following research questions:

1. Does the use of citations and co-citations in the process of keyword extraction for the purpose of navigation lead identification helps to improve the overall quality of the extracted keyword set? Are they capable of finding new words, or promoting the important words by boosting their relevance in comparison with the content-based (TF-IDF) baseline?
2. What are the limitations of using the citation and co-citation analysis with respect to the number of citations of an individual article?

We provide a quantitative evaluation of our method in the domain of digital libraries using experts' judgement on the relevance of the extracted keywords.

2   Identification of Keywords in Digital Libraries

Although our proposed approach of exploratory navigation using navigation leads can be in general applied to an information space in any domain, the knowledge of domain specifics allows us to tailor the method to them and thus, to improve the overall navigation process. Our focus is on the domain of digital libraries, or more specifically (as we have already mentioned) on the digital library systems of journal (or in general research) articles.

2.1   Specifics and Similarities Between Digital Libraries and the Web

When designing a method of keyword extraction, we need to consider several specifics of the domain of digital libraries, which make it distinct from the wild Web:

- *Size and structure of information space* – the size of information space of digital libraries is much smaller in comparison with the whole Web, also the rate with which there is a new content is lower. Because the content is in most cases protected by copyrights, we can observe a separation of metadata, which are publicly available and easily processed, from the actual content of the documents.
- *Structure of documents* – in contrast to the wild Web, the documents in digital libraries tend to follow a predefined structure. This structure can differentiate among different publishers and journals, but it is always possible to identify the basic building blocks of the documents, such as title, authors, abstract, etc. There are approaches to automatically extract a table of contents of an article, as well as the actual contents of its sections [7], which can be utilized to reweigh the extracted keywords based on the section of the article in which they occur.
- *Unique set of metadata* – the articles in digital libraries have various metadata associated, which differ from the other sources on the Web in general, such as authors, publishers, where it was published and more interestingly, keywords that are identified by the authors themselves. This all can help to identify the most important aspects of the articles by taking into consideration not only the actual content of an article, but, e.g., also other similar articles from the same authors, etc.

There are also several aspects of this domain that are analogous to those of the Web:

- *User-added tags* – they represent a special type of metadata, because they are added by the users that use them in order to organize the articles for their later retrieval. It is a unique source of metadata, because users tend to use their own vocabulary and can highlight different aspects of the articles than their authors.

- *Links between documents* – the articles are linked by the use of citations. In contrast to the hyperlinks, their mining often requires advanced text mining techniques, because the reference can occur in the text after it has been explained.

The focus of this work is on the latter, i.e., on the citations, or more specifically, on the citation sentences and on their use for extraction of keywords from research articles for the purpose of the navigation lead candidates identification. They provide a unique view of the article content from a point of view of other researchers; citations cover different aspects of an article, but the amount of unique information converges as the number of citations increases [4]. There is an overlap between the topics (and the keywords) that we can extract from the abstract of an article (which has a special place in the domain of digital libraries, because the abstracts are usually freely accessible, unlike the article contents), but the topics of the abstract tend to be more general than those present in the citation sentences [9].

In our work, we examine also the co-citations assuming that there is a stronger relationships between frequently co-cited documents; however, there are other aspects in play, such as proximity of the co-cited articles in the document [8].

2.2 Domain Modeling in Annota

The specifics of digital libraries discussed in the previous section are reflected in the domain model of a bookmarking service Annota (http://annota.fiit.stuba. sk), which we have developed. The users can bookmark and annotate research articles in the digital libraries, such as ACM DL, IEEE Xplore, Springer Links, etc. The metadata of these articles are automatically extracted and processed into our domain model.

The model is two-layered; the domain model at the first layer is overlaid by the user model at the second [6]. It is a graph: the vertices consist of the normalized extracted terms and of the research articles; the weighed edges model the associations between the terms and the articles as well as between the terms themselves. Figure 2 shows a conceptual model of the domain representation used in Annota.

The main relation is between an *Article* and a *Term* entity; it is modelled as a combination of partial relations coming from various sources, e.g. the user-added tags, the folder names into which the users organize their articles, author-added keywords, etc. Additionally, we automatically extract keywords from articles (*Keyword* entity) that are used as the navigation lead candidates. The model allows various keyword extraction services to be used and combined into new types of *KeywordRelation*. It is also possible to configure which partial relations (and with which weights) should be used in a combination of *ArticleHasTermRelation*. This makes the model flexible and enables us to test various settings of the model as well as to quantitatively compare various keyword extraction services by using Annota as an A/B testing platform.

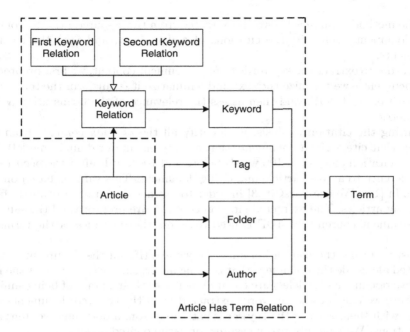

Fig. 2. A conceptual model of the domain representation in Annota. A relation between an article and a term consists of a combination of various relations that can be configured.

3 Method of Keyword Extraction Using (Co-)Citation Analysis

We have proposed a method of keyword extraction using citation and co-citation analysis that we employ for the purpose of identification of a set of potential navigation lead candidates as an extension of the existing domain model. A conceptual model of this extension can be seen in Fig. 3.

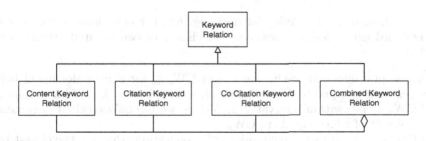

Fig. 3. An extension of the *KeywordRelation* that combines keywords extracted from a document content, its citations, and the co-cited documents.

The method combines keywords extracted from three sources: (*i*) the content of a document itself, (*ii*) the citations of a document, and (*iii*) the co-cited documents.

In order to extract the keywords from a document content, we first preprocess the document—we tokenize the text and lemmatize it (transform the terms into their dictionary form)—and then assess the relevance of the terms using a TF-IDF metric.

During the citation analysis, we identify all the *citation contexts* from the articles that cite a given document (that are present in our domain model). We define a citation context as 100 words before and 100 words after the occurrence of a reference to a given article in a citing document. This value is based on the results in [14]. We use ParsCit [3] in order to extract the citation contexts from the citing articles. The extracted citation contexts are preprocessed the same as the document content; TF-IDF is used to assess the relevance of the terms as well.

Lastly, during the co-citation analysis, we identify all the documents which are cited alongside the document that is being analyzed. We assume that the fact that two documents (articles) are co-cited increases their chance of being similar; therefore, we can use the keywords extracted from the co-cited documents and extend with them a set of the keywords extracted from a document content and its citations. We compute two measures for each co-cited article:

- *Co-citation weight (CW)* – it represents a frequency with which two articles are co-cited. If, e.g., the articles A and B are both cited by the articles C, D, and E, then the value of CW of B w.r.t to A (and vice versa) equals to 3. It is a global (or aggregated) measure of the co-citation relevance.
- *Co-citation proximity index (CPI)* – it is based on [5]; if the articles A and B are referenced in the same sentence of the article C, than CPI equals to 1, if in the same paragraph, it equals to $1/2$, if in the same section, it equals to $1/4$ and if they occur in the same article, CPI equals to $1/8$, which is the smallest possible value for any co-citation. In other words, the closer the references to the articles occur in the text, the stronger the relationship we assume to be. As it characterizes the individual co-cited articles, it is a local measure of the co-citation relevance.

We have defined a set of rules for deciding, which co-citations to consider for the keyword extraction, to maximize the chance of two co-cited articles being similar:

1. In case of a maximal co-citation weight CW_{max} for a given document being larger than or equal to N, we ignore all co-citations with CW lower than N.
2. If CW_{max} lies within interval < 2, N-1 >, we use only co-cited documents, the weight of which equals to CW_{max}.
3. If CW_{max} equals to 1, we consider only co-citations that are the closest to a given document for each citing document separately based on their CPI. In other words, if there were, e.g., five citing documents, we would consider for each document only co-citations with the maximal CPI.

The specific values of a threshold value N can differ based on the domain and dataset used. We have empirically chosen N to be 5 based on the standard dataset that we used during the evaluation (see next section). It should be fine-tuned if a distribution of citations and co-citations differs significantly from the one in our dataset.

As the last step, we combine the keywords extracted from the content of the document with those extracted from the citation contexts as well as from the selected subset of co-cited articles (applying the same treatment during the text processing). We use a linear combination of weights which we normalize using the min-max normalization prior to the combination itself:

$$w = \alpha w_D + \beta w_C + \gamma w_{CC} \tag{1}$$

where w_D is a weight of a keyword extracted from the document content, w_C is a weight of a keyword extracted from the citation contexts and w_{CC} is a weight of a keyword extracted from the co-cited documents. The weight is for all three sources determined using a TF-IDF metric. The coefficients α, β, and γ are real numbers from the interval $< 0, 1 >$; we used a value of 1 in all our experiments.

The result is a set of extracted keywords, which serve as navigation lead candidates, but can be used also for other purposes, as they are a part of the domain model.

4   Evaluation

We have evaluated our proposed method of keyword extraction using the citation and the co-citation analysis on a standard dataset of articles in ACL Anthology Network [13] that we imported into Annota. Overall, the dataset consists of 18,290 articles with 84,237 citations. We have conducted a quantitative evaluation using the experts' judgements; therefore, we have limited a number of articles from the original dataset to those which have already been known by the experts, i.e., which they had bookmarked in their Annota personal libraries. This way, we could use a subset of dataset consisting of 250 articles.

The goal of the experiment was to assess the relevance of the extracted keywords based on the experts' judgements. We have extracted three keyword sets for each document:

- *M1*: Keywords extracted from a content of a document using TF-IDF; this served us as a baseline.
- *M2*: Keywords from M1 enriched with the keywords extracted from the citation contexts of a given article.
- *M3*: Keywords from M2 enriched with the keywords extracted from the co-cited articles; this represents our proposed method discussed in the previous section.

We have hypothesized that adding the keywords extracted from the citation contexts and from the co-cited articles will improve the overall precision P, i.e., formally:

$$P(M1) < P(M2) \leq P(M3) \tag{2}$$

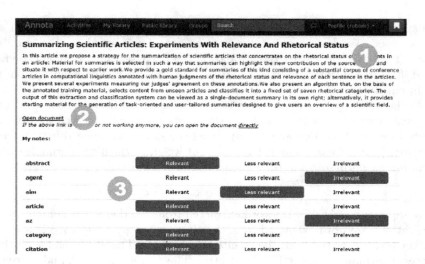

Fig. 4. An evaluation interface in Annota. The experts could have read the title and the abstract of an article (1), or navigate to its fulltext (2). They assessed the relevance of the presented keywords by choosing one of the possible values – *relevant, less relevant,* or *irrelevant* (3).

We have prepared an evaluation interface in Annota for the experts to be able to easily assess the relevance of the presented keywords (see Fig. 4). The experts could have chosen one of the possible assessments of a keyword relevance: *relevant, less relevant* (meaning *somewhat relevant*), or *irrelevant*.

Under each article, we have presented top ten scoring keywords extracted by each compared variant (M1, M2, M3), i.e., together up to 30 keywords for each document merged into a single list. All the keywords were presented only once in the list even if there was an overlap between the sets of extracted keywords. The keywords in the list were sorted alphabetically so that the experiment participants could not have found out which keywords were extracted by which method (and with which relevance).

There were 8 domain experts who participated in the experiment. Together, they assessed 844 unique extracted keywords from 45 different articles; 7 articles were assessed by more than one expert. We have evaluated each variant (M1, M2, M3) when considering only relevant or also less relevant keywords using a standard information retrieval metric P@N (precision at N); results are shown in Table 1. The metric computes a ratio of relevant keywords among the top N scoring ones. We can see that the variant M2 and M3 outperformed the TF-IDF baseline (M1) in all P@N measures for N = 1, 2, 5, and 10, thus confirming our hypothesis. As to the comparison of M2 and M3 variant, considering just citations provided in general a slightly better precision, except of P@5, where co-citations significantly improved the results (there was an improvement also for P@10, although marginal). A closer analysis of the extracted keywords revealed that using M3 variant in some cases promoted names of the authors or acronyms of the method names which were deemed irrelevant by the judges, as they were too specific and not really describing the content of the articles.

Table 1. The results of precision for each variant when considering only the keywords assessed by the experts as *relevant* (R), or when considering also those assessed as *less relevant* $(R + L)$.

	P@1		P@2		P@5		P@10	
Variant	R	R+L	R	R+L	R	R+L	R	R+L
M1	78.85	88.46	69.23	79.81	60.08	73.26	51.35	68.73
M2	88.46	96.15	81.73	95.19	59.46	74.90	53.49	69.77
M3	84.62	94.23	79.81	93.27	63.85	80.77	52.13	70.35

We have analyzed also the limitations of our proposed method concerning the second research question. For this purpose, we have compared values of the P@10 measure (when considering the relevant as well as the less relevant keywords) for each variant (M1, M2, M3) with respect to the number of citations of an article (see Fig. 5). Because the number of the evaluated keywords in the selected citation groups was not the same, we have reported also the confidence intervals.

We can see that though the overall scores of the P@10 measure are almost identical for all the variants (see Table 1), there are differences based on the number of citations. The TF-IDF baseline (M1) slightly outperforms the other two (M2, M3) if the number of citations is lower than 20, although the measured differences are not significant. The situation changes with the increasing number of citations in favor of variants M2 and M3 with M3 (employing co-citations) being better. Again, the differences are in the most cases not significant (with the exception of articles with at least 20 and less than 100 citations), but the trend is clear. This is in agreement with the previous finding in [4] that 20 citations can in general cover all the important aspects of an article. Novelty lies in the comparison with the co-citations that seem even better suited for the keyword extraction in this case.

Lastly, we have found out that out of 844 extracted keywords that have been assessed by the experts, 119 were identified by the M3 variant, meaning that they did not occur at all among the keywords extracted by M1 and M2 or they were deemed irrelevant be them. Out of these 119 keywords, 54.6% were assessed as relevant or less relevant by the experts. This suggests that considering the co-citations is capable of finding new important words, although there is still some noise which could be reduced by making the rules for deciding, which co-cited articles to consider, more strict.

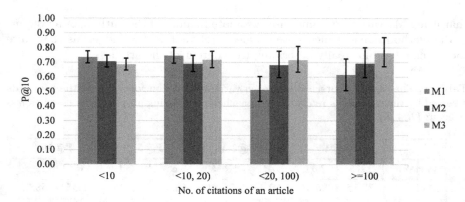

Fig. 5. A comparison of P@10 for each variant w.r.t. the number of citations of an article. The error bars around the mean values represent the confidence intervals.

5   Related Work

As we have already established throughout this article, citations reflect value, impact, and importance of research works, which makes them interesting in areas such as scientometrics. They also represent a judgement of other researchers on the actual content of a research work, which is a reason, why they are researched also in the field of natural language processing (e.g., for keyword extraction and summarization).

It has been found out that as much as 20 citations is enough to cover all the important aspects of an article [4]. The topics extracted from the citations differ significantly from those that can be extracted from the abstracts [9] and using at least one citation sentence for keyword extraction renders better results in comparison with the keywords extracted solely from the document content, while the optimal citation context is 100 words before and after a reference in a citing article [14].

The keywords (or keyphrases) extracted from the citation contexts are also capable of producing better summaries of researcher articles as shown in [12] on 25 manually annotated articles from ACL Anthology Network.

There is still much work to be done on the categorization and characterization of citations and their role within an article; Bertin and Atanassova [1] analyzed verbs used in the citations in different sections of articles and found out that there are significant differences which can be attributed to their different roles. This remains a challenge, because the most of the works do not differentiate between these roles when using citations for keyword extraction or other natural language processing related tasks. Also, the exact role and potential contribution of the co-citations for these tasks remains an open problem, which we have tried to tackle also in this work.

6   Discussion and Conclusions

In this work, we have presented a general model of Web search that we have augmented for exploratory search and navigation in an information space using navigation leads. In order to select navigation leads, we first identify a set of potential lead candidates using automatic keyword extraction.

For this purpose, we have proposed a method of keyword extraction in digital libraries domain employing citation and co-citation analysis and tried to answer two research questions, namely whether the citations and the co-citations improve keyword extraction and what are their limitations. Our main contributions are as follows:

1. We have extended a domain model of a digital library system with keywords extracted from the co-cited articles and proposed a set of rules for deciding which co-citations to consider, and which not.
2. We have evaluated usefulness of citation and co-citation analysis on a standard dataset and examined their limitations. Based on the results of our quantitative experiment, we can conclude that using citations and co-citations significantly outperforms the TF-IDF baseline; in addition, the co-citations are capable of finding new keywords that would not have been otherwise extracted. As to the limitations of our proposed method, its precision depends on the number of citations of an article with the ideal number being above 20.

Although co-citation analysis is capable of finding new keywords, there is still some noise in the form of irrelevant keywords that should be addressed in the future (e.g. by filtering out the names of authors or method acronyms by setting a higher threshold for minimal number of occurrences of a term in the documents when using TF-IDF). Open question remains, how to automatically adapt the rules of co-citation selection so that they would take into consideration citation specifics of different domains.

In addition, our method depends on a number of citations of an article, i.e., it performs well when there is enough information in the form of citations and co-citations and is outperformed by the TF-IDF baseline, if there is not. This could be reduced by modifying the method to automatically adapt the weights based on the number of citations, thus promoting content keywords for articles with only a few citations and gradually increasing the weight of the citation and the co-citation keywords with the increasing number of citations. Other promising direction is to analyze the citation intention and its role within an article.

Acknowledgement. This work was partially supported by the Cultural and Educational Grant Agency of the Slovak Republic, grant No. KEGA 009STU-4/2014, the Scientific Grant Agency of the Slovak Republic, grant No. VG 1/0646/15, and by the Slovak Research and Development Agency under the contract No. APVV-0208-10.

References

1. Bertin, M., Atanassova, I.: A Study of Lexical Distribution in Citation Contexts through the IMRaD Standarda. In: Proceedings of the 1st Workshop on Bibliometric-Enhanced Information Retrieval Co-located with 36th European Conference on Information Retrieval (ECIR 2014), pp. 5–12. CEUR-WS (2014)
2. Broder, A.: A taxonomy of web search. ACM SIGIR Forum. **36**, 3–10 (2002)
3. Councill, I.G., Giles, C.L., Kan, M.: ParsCit: an open-source CRF reference string parsing package. In: LREC 2008: Proceedings of the 6th International Conference on Language Resources and Evaluation, pp. 661–667. ELRA (2008)
4. Elkiss, A., Shen, S., Fader, A., Erkan, G., States, D., Radev, D.R.: Blind men and elephants: what do citation summaries tell us about a research article? J. Am. Soc. Inf. Sci. Technol. **59**, 51–62 (2008)
5. Gipp, B., Beel, J.: Citation proximity analysis (CPA) - a new approach for identifying related work based on co-citation analysis. In: ISSI 2009: Proceedings of the 12th International Conference on Scientometrics and Informetrics, pp. 571–575. ISSI (2009)
6. Holub, M., Moro, R., Sevcech, J., Liptak, M., Bielikova, M.: Annota: towards enriching scientific publications with semantics and user annotations. D-Lib Mag. **20** (2014). http://www.dlib.org/dlib/november14/holub/11holub.html
7. Klampfl, S., Kern, R.: An unsupervised machine learning approach to body text and table of contents extraction from digital scientific articles. In: Aalberg, T., Papatheodorou, C., Dobreva, M., Tsakonas, G., Farrugia, C.J. (eds.) TPDL 2013. LNCS, vol. 8092, pp. 144–155. Springer, Heidelberg (2013)
8. Liu, S., Chen, C.: The effects of co-citation proximity on co-citation analysis. In: ISSI 2011: Proceedings of the 13th International Conference of the International Society for Scientometrics and Informetrics, pp. 474–484 (2011)
9. Liu, S., Chen, C.: The differences between latent topics in abstracts and citation contexts of citing papers. J. Am. Soc. Inf. Sci. Technol. **64**, 627–639 (2013)
10. Marchionini, G.: Exploratory search: from finding to understanding. Commun. ACM. **49**, 41–46 (2006)
11. Moro, R., Bielikova, M.: Navigation leads selection considering navigational value of keywords. In: WWW 2015 Companion: Proceedings of the 24th International Conference on World Wide Web Companion, pp. 79–80. IW3C2, Geneva (2015)
12. Qazvinian, V., Radev, D.R., Özgür, A.: Citation summarization through keyphrase extraction. In: COLING 2010: Proceedings of the 23rd International Conference on Computational Linguistics, pp. 895–903. Association for Computational Linguistics (2010)
13. Radev, D.R., Muthukrishnan, P., Qazvinian, V., Abu-Jbara, A.: The ACL anthology network corpus. Lang. Resour. Eval. **47**, 919–944 (2013)
14. Ritchie, A., Robertson, S., Teufel, S.: Comparing citation contexts for information retrieval. In: CIKM 2008: Proceedings of the 17th ACM Conference on Information and Knowledge Mining, pp. 213–222. ACM Press, New York (2008)
15. Robertson, S., Zaragoza, H.: The probabilistic relevance framework: BM25 and beyond. Found. Trendsin Inf. Retrieval **3**(4), 333–389 (2009)
16. White, R.W., Roth, R.A.: Exploratory Search: Beyond the Query-Response Paradigm. Morgan & Claypool, San Rafael (2009)

Summarizing Online User Reviews Using Bicliques

Azam Sheikh Muhammad[1], Peter Damaschke[2]([⊠]), and Olof Mogren[2]

[1] College of Computing and Informatics, Saudi Electronic University (SEU),
Riyadh 13316, Kingdom of Saudi Arabia
m.sheikh@seu.edu.sa
[2] Department of Computer Science and Engineering,
Chalmers University, 41296 Göteborg, Sweden
{ptr,mogren}@chalmers.se

Abstract. With vast amounts of text being available in electronic format, such as news and social media, automatic multi-document summarization can help extract the most important information. We present and evaluate a novel method for automatic extractive multi-document summarization. The method is purely combinatorial, based on bicliques in the bipartite word-sentence occurrence graph. It is particularly suited for collections of very short, independently written texts (often single sentences) with many repeated phrases, such as customer reviews of products. The method can run in subquadratic time in the number of documents, which is relevant for the application to large collections of documents.

1 Introduction

Extractive summarization, i.e., selection of a representative subset of sentences from input documents, is an important component of modern information retrieval systems. Existing methods usually first derive some intermediate representation, then score the input sentences according to some formula and finally select sentences for the summary [13,16].

The field of automatic summarization was founded in 1958 by Luhn [14], who related the importance of words to their frequency of occurring in the text. This importance scoring was then used to extract sentences containing important words. With the advent of the World Wide Web, large amounts of public text became available and research on multi-document summarization took off. Luhn's idea of a frequency threshold measure for selecting topic words in a document has lived on. It was later superseded by $tf \times idf$, which measures the specificity of a word to a document and has been used extensively in document summarization efforts.

Radev et al. pioneered the use of cluster centroids in summarization in their work [18], with an algorithm called MEAD that generates a number of clusters of similar sentences. To measure the similarity between a pair of sentences, the authors use the cosine similarity measure where sentences are represented as

© Springer-Verlag Berlin Heidelberg 2016
R.M. Freivalds et al. (Eds.): SOFSEM 2016, LNCS 9587, pp. 569–579, 2016.
DOI: 10.1007/978-3-662-49192-8_46

a weighted vector of $tf \times idf$ terms. Once sentences are clustered, a subset from each cluster is selected. MEAD is among the state-of-the-art extractive summarization techniques and frequently used as baseline method for comparing new algorithms.

Summarization is abstractive when new content is generated while summarizing the input text. Ganesan et al. [8], considered online user reviews, which are typically short (one or two sentences) and opinionated. They presented an abstractive approach that used parts of the input sentences to generate the output. For evaluations they provided the Opinosis dataset, consisting of user reviews on 51 different topics. Their system performed well evaluating generated summaries against those written by human experts. They also compare their results with the aforementioned MEAD [18] method.

Bonzanini et al. [4] introduced an iterative sentence removal procedure that proved good in summarizing the same short online user reviews from the Opinosis dataset. Usually, an extractive summarization method is focused on deciding which sentences are important in a document and considered for inclusion in the summary. The sentence removal algorithm by Bonzanini et al. [4] would instead iteratively choose sentences that are unimportant and remove them, starting with the set of all sentences in the input. The process continues until the required summary length is reached.

SumView [19] is also specialized on collections of short texts and uses feature extraction and a matrix factorization approach to decide on the most informative sentences. Besides the aforementioned work we refer to an extensive survey [16] discussing different approaches to sentence representation, scoring, and summary selection, and their effects on the performance of a summarization system.

Our Contribution: We propose a novel, purely combinatorial approach aimed at extractive summarization of collections of sentences. Since we will consider online user reviews as input, that typically are single sentences from independent authors, we speak of *sentences* rather than *documents* from now on. The idea is simply to find the key combinations of words appearing in several sentences. We work with a bipartite graph where the two vertex sets correspond to sentences and words, respectively, and edges exist between words and the sentences where they appear. Then, finding sets of sentences that share the same combination of words is equivalent to finding the bicliques in this graph. (Formal definitions follow in Sect. 2.) Finally we select bicliques for the summary according to further criteria. Leveraging recent advances in fast algorithms for determining the most similar sets, the approach is also computationally fast. Altogether this enables us to present a method for extractive summarization of short independent texts that should be attractive due to its conceptual simplicity and direct interpretability of the output. We show that it performs well on the benchmark Opinosis dataset [8]. It also outperforms state-of-the-art systems achieving the best precision, F_1 and F_2 measures. (See definitions and results in Sect. 6.2.)

As a delimitation, due to its very idea the method cannot be expected to extract good summaries of single complex texts, but this type of application is beyond the scope of this work.

2   Preliminaries

The following notation will be used throughout this paper. We consider any sentence as a bag of words, that is, as the set of distinct words, disregarding order and multiplicity of words. The ith sentence is denoted s_i. and $|s_i|$ is the number of distinct words, after removal of multiple occurrences and stopwords. The given set of sentences is T, and a summary is denoted $S \subset T$. Note that an extractive summary is merely a selection of the most informative sentences; no new text is generated. The length of a summary in terms of the number of sentences is denoted by ℓ.

Symbol $M_{s_i;s_j}$ stands for a sentence similarity measure. Several measures for sentence similarity have been explored for extractive summarization. The simplest ones are called the surface matching sentence similarity measures [15], because they do not require any sophisticated linguistic processing, and the similarity value is merely based on the number of words that occur in both of the sentences. In the present paper we are considering two of them:

- the *match* measure $M_{s_i;s_j} = |s_i \cap s_j|$,
- the *cosine* measure $M_{s_i;s_j} = |s_i \cap s_j| / \sqrt{|s_i| \times |s_j|}$.

Note that the cosine measure can be viewed as the match measure normalized by the sentence size.

Given a collection of sentences, we consider the *word-sentence occurrence graph*, that is, the bipartite graph $B = (T, W; E)$ with T and W being the set of sentences and words, respectively, where $sw \in E$ is an edge if and only if word w occurs in sentence s. Here the number of occurrences is disregarded, that is, edges are not weighted.

We use the standard notation $N(v)$ for the set of neighbored vertices of a vertex v in a graph. It naturally extends to vertex sets V, by defining $N(V) := \bigcup_{v \in V} N(v)$. A bipartite clique, *biclique* for short, is a pair (X, Y) of sets $X \subseteq T$ and $Y \subseteq W$ that induces a complete bipartite subgraph of B, that is, $N(X) \supseteq Y$ and $N(Y) \supseteq X$. A maximal biclique (not to confuse with a biclique of maximum size) is a biclique not contained in other bicliques. We call a subset Y of W of the form $Y = N(s_i) \cap N(s_j)$, with $s_i, s_j \in T$, a *2-intersection*. It corresponds to a biclique (X, Y) with $|X| = 2$.

3   Overall Idea

Customer reviews and similar text collections consist of sentences written independently by many authors. Intuitively, combinations of words appearing in several sentences should be important for a summary. Obviously they are bicliques in the word-sentence occurrence graph.

The problem of enumerating the maximal bicliques (and hence, implicitly, all bicliques) in a graph is well investigated, in particular, several output-sensitive algorithms with different running times are known [1,6,9,11]. However, in general a graph has very many bicliques, and we need to select those which appear most relevant for a small extractive summary. We found that the large 2-intersections (see definitions above) are good candidates, by the following reasoning.

The word set Y of a biclique (X, Y) with $|X| = 2$ has been used by at least two independent authors, hence it has not only occurred by chance, in one sporadic statement. Moreover, Y is a maximal word set with this property. More repetitions of less specific word sets, that is, bicliques with $|X| > 2$ but smaller Y, do not seem to add much to a summary. The restriction to $|X| = 2$ also allows the use of standard pairwise similarity measures in the heuristic rules that afterwards select the sentences to be put in the summary.

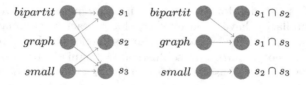

Fig. 1. Biclique Summarization. Example sentences s_1: "This is a bipartite graph", s_2: "Look how small it is", s_3: "This bipartite graph is very small". After stemming and stopwords removal: s_1: "bipartit graph", s_2: "small", s_3: "bipartit graph small"

An example of the concept is visualized in Fig. 1. The bipartite graph is shown on the left while all possible 2-intersections are on the right. Since s_3 is present in both of the nonempty 2-intersections, it may be regarded as representative of the other two and thus selected for the summary.

4   Implementation Details

The given set of sentences T undergoes a preprocessing before passing it to the main algorithm. This involves stopwords removal and stemming. Words such as "am", "are", "by", "is", are called stopwords. They are common to many sentences and usually do not carry meaningful information when comparing texts in the context of summarization. Stemming reduces a word to its stem, base or root form. Stemming prevents mismatch between two words which apparently differ but are actually grammatical variations of the same word, such as singular and plural. In many cases stemming is achieved by chopping off the ends of words. Details about the stopwords list and stemming technique used in our implementation (which, however, do not affect the core ideas of the method) are described in Sect. 6.2.

In Algorithm 1 we give a pseudocode description of the method. As said in Sect. 3, in a biclique (X, Y), the Y part is the word set and X with the restriction

$|X| = 2$ (because we need 2-intersections only), is the set containing i, j corresponding to a pair of sentences (s_i, s_j), $i \neq j$, in T. Our implementation using the subroutine $FindBiclique(s_i, s_j, sim)$ collects bicliques (X, Y) with $|Y| \geq 2$, that is, there are at least two words common in the corresponding pair of sentences (s_i, s_j). For every such pair we also compute the value of the similarity measure specified by the parameter sim which is either equal to match or cosine; see Sect. 2. We find all such bicliques and this concludes the first part of the algorithm. There are two more major parts in our algorithm.

Algorithm 1. Biclique Summarization

Input: sim, top, ℓ, T
Output: S
 1: **Part-1: Find Bicliques:**
 2: $nBicliques \leftarrow 0$
 3: **repeat**
 4:     $Choose\ a\ new\ pair\ (s_i, s_j), i \neq j,\ in\ T$
 5:     $[X, Y, simValue] \leftarrow FindBiclique(s_i, s_j, sim)$
 6:     $Bicliques.add(X, Y, simValue)$
 7:     $nBicliques \leftarrow nBicliques + 1$
 8: **until** $NoMoreBicliques$
 9: **Part-2: Filter Important Bicliques:**
10: $Bicliques \leftarrow SortBySimValue(Bicliques)$
11: $max \leftarrow \lceil \frac{nBicliques \times top}{100} \rceil$
12: $impBicliques \leftarrow Select(Bicliques[1...max])$
13: **Part-3: Summary Selection:**
14: $SentenceIDs \leftarrow UniqueSortedByFreq(impBicliques.X)$
15: $S \leftarrow GetSentences(T, SentenceIDs[1...\ell])$
16: **return** S

In the second part we filter out important bicliques, making use of the parameter top which defines percentage of the bicliques that should be kept. Typical values used are 10, 30 and 50.

First, the bicliques are sorted with respect to their decreasing similarity value and then we select best top% of the entire bicliques collection. In this way, the most informative bicliques are kept while the rest are discarded. For example when top=10, we select top 10 percent of the bicliques from the sorted list.

Finally, to select a summary, we need a ranking of the sentences appearing in the filtered important bicliques. In our implementation we simply count the occurrences of sentences in the filtered bicliques and take the ℓ most frequent sentences.

5 Processing Time

The time complexity is dominated by the time needed to determine the sentence pairs (2-intersections) with top similarity scores. This subproblem is known as *top-k set similarity joins* (more precisely: self-joins), where our k is the number of bicliques to keep. A basic implementation as displayed in Algorithm 1 loops through all pairs of sentences and therefore costs quadratic time in the number of sentences. However, the top-k set similarity joins problem is well studied for different standard similarity measures, and fairly simple heuristic algorithms as proposed in [2,7,20] and related work have experimental running times far below the naive quadratic time bound. They can replace Part 1 and 2 of Algorithm 1. These heuristics rely on the very different word frequencies in texts, which essentially follow power laws [10]. Some further theoretical analysis of time bounds is given in [5].

For the sake of completeness we briefly outline these techniques. The first main idea is to process the words by increasing frequencies f and collect the pairs of sentences where any common words have been detected. A word appearing in f sentences trivially appears in fewer than $f^2/2$ of the 2-intersections. Since most words have low frequencies, the total number of such sentence pairs does not grow much in the beginning. The second main idea is called prefix filtering, which is actually a branch-and-bound heuristic. It bounds for every considered sentence pair the maximum possible value of the similarity measure, in terms of the number of (frequent) words not yet considered. This allows early exclusion of many pairs of sentences that cannot be among the top k pairs any more. Building on these principles, the "segment bounding algorithm" for the overlap measure [3] divides the set of words into segments according to their frequencies. Then the largest 2-intersections within the segments are computed by subset hashing, and finally the partial results are combined following some simple priority rules, until the top k pairs are established for a prescribed k. The authors of [3] report that their algorithm runs faster than those from [20] especially on large datasets. Finally, for the largest 2-intersections found, one can also filter those where other similarity measures are maximized.

Thanks to these techniques, our method can be implemented to run in sub-quadratic time. By way of contrast, this is apparently not possible for other summarization approaches like sentence removal [4] which is intrinsically more than quadratic.

6   Experimental Results

In this section, we present an empirical evaluation of the proposed method.

6.1   Dataset

Considering large amounts of highly redundant short text opinions expressed on the web, our experimental study focuses on assessing the performance of the proposed method on such a dataset which includes users stating their single line opinions about products or services, or commenting on some hot topics or issues on certain discussion forums and social media sites.

The Opinosis dataset [8] is particularly relevant to our purpose because it contains short user reviews in 51 different topics. Each of these topics consists of between 50 and 575 one-sentence user reviews made by different authors about a certain characteristic of a hotel, a car or a product (e.g. *"Location of Holiday Inn, London"* and *"Fonts, Amazon Kindle"*). The dataset includes 4 to 5 gold-standard summaries created by human authors for each topic. The length of the gold-standard summaries is around 2 sentences.

6.2   Evaluation Method and Baseline Selection

Following standard procedure, we use ROUGE [12] for evaluation. ROUGE compares the system-generated summary with the corresponding gold-standard summaries for a topic and reports the assessment in terms of quantitative measures: recall, precision and F-measure. Recall is the number of words in the intersection of system-generated and the gold-standard summaries divided by the number of words in the gold-standard summary. Precision is the number of words in the intersection divided by the number of words in the system-generated summary.

F-measure, also called F_1 score, is a composite measure defined as the harmonic average of precision and recall. Sometimes, in order to emphasize the importance of recall over precision, another F-measure called F_2 score is also computed. On our benchmark dataset, Opinosis, F_2 scores are also reported in state-of-the-art results.

The general definition of F-measure for positive real β is:

$$F_\beta = (1 + \beta^2) \cdot \frac{\text{precision} \cdot \text{recall}}{(\beta^2 \cdot \text{precision}) + \text{recall}}$$

ROUGE works by counting n-gram overlaps between generated summaries and the gold standard. Our results show ROUGE-1, ROUGE-2 and ROUGE-SU4 scores, representing matches in unigrams, bigrams and skip-bigrams respectively. The skip-bigrams allow four words in between.

The experiments are aligned (in terms of ROUGE settings etc.,) with those of Bonzanini et al. [4], which provide state-of-the-art results on extractive summarization on the Opinosis dataset. As mentioned in Sect. 1, they use a sentence removal (SR) algorithm. They used ROUGE to evaluate their methods, and MEAD [18], an extractive multi-document summarizer (see Sect. 1, too), as a baseline.

There are two different versions of the SR algorithm in [4], one based on similarity (SR_{SIM}) and the other one based on diversity (SR_{DIV}). We compare our method to both of them.

Summary length ℓ was fixed at 2 sentences. This matches the supplied gold-standard summaries and is also necessary to align our results to [4].

In addition to MEAD [18], they [4] use a brute-force method as a baseline which, for any given topic, enumerates all combinations of 2 sentences and chooses the pair that optimizes on the same scoring functions as used in their sentence removal algorithm.

Our implementation maximizes ROUGE scores: We consider all possible pairs of sentences within each topic, compute ROUGE-1, ROUGE-2 and ROUGE-SU4 scores (of recall, precision, F_1, and F_2) and choose the sentence pair with highest value. Choosing such pairs for all topics in the dataset gives us the maximum scores, denoted with OPTIMAL, in our evaluations. These are the ideal scores attainable by an extractive summarization algorithm on this dataset.

We evaluate an implementation of the Biclique algorithm with sentence similarity measures *match* and *cosine*. For the *cosine* measure we let top vary between 10 and 30. For *match* we evaluate with top fixed at 50, which gave us the highest scores. Accordingly, the methods are abbreviated $Biclique_{Cosine1}$, $Biclique_{Cosine3}$, and $Biclique_{Match5}$, respectively.

The systems are evaluated with ROUGE version 1.5.5 using the following options: -a - m -s -x -n 2 -2 4 -u. For F_2, alongside the previous option, we also add: -p 0.2.

For preprocessing we make use of a Porter stemmer [17]. Stopwords were removed using the stopword list distributed as part of the ROUGE evaluation system [12].

6.3   Results

In Table 1 we show results for the experiments. R-1, R-2 and R-SU4 represent scores of ROUGE-1, ROUGE-2 and ROUGE-SU4 respectively. The best results (with the exception of the brute-force optimal scores) are shown in bold. The best scores among biclique alternatives are shown in italic. The brute-force optimal scores with respect to the evaluated measure in Table 1, marked in gray in the first row in each sub-table, are the maximum attainable scores. Baselines are shown at the bottom of each sub-table.

Recall: With respect to recall, $Biclique_{Match5}$ has attained a ROUGE-1 value of 43.54 and ROUGE-SU4 of 16.36. This is better than the corresponding values (37.46 and 13.80) of the baseline method SR_{SIM} making an improvement of 16.23 % and 18.55 %, respectively. Similarly, with respect to ROUGE-2 value, $Biclique_{Match5}$ has attained 8.91 compared to the value 8.67 of the baseline method SR_{DIV}. Comparing ROUGE-1 and ROUGE-SU4, SR_{DIV} has attained better scores, and $MEAD$ has overall higher recall scores. However, it should be kept in mind that both of the latter are biased towards recall and do not perform well on precision compared to our method $Biclique_{Match5}$ in all three ROUGE settings.

Precision: $Biclique_{Cosine1}$ is the best system for precision in all three settings of ROUGE, increasing the best scores among the baseline methods by over 85 % for the ROUGE-1, over 90 % for ROUGE-2, and over 223 % for ROUGE-SU4.

F_1: The performance of $Biclique_{Cosine1}$ is consistent when we consider the measures F_1 and F_2, showing that the high precision is also combined with a high recall. For example, using $Biclique_{Cosine1}$, the best F_1-scores of the best performing baseline methods are improved by at least 34 %, 35 %, and 120 % for ROUGE-1, ROUGE-2, and ROUGE-SU4, respectively.

F_2: Here the best results are achieved by our method $Biclique_{Cosine3}$. We consistently improve on the best scoring method among the baselines by at least 6 %, 8 % and 64 % for ROUGE-1, ROUGE-2, and ROUGE-SU4, respectively.

Table 1. ROUGE scores for $Biclique_{Dice}$, $Biclique_{Cosine}$ - Biclique (with Match and Cosine sentence similarity, respectively) obtains the highest Precision, as well as F_1 and F_2 scores. OPTIMAL scores (gray) contain the corresponding score for an optimal summary for each cell. SR_{SIM}, SR_{DIV} - Bonzanini et al. (2013)

Recall	R-1	R-2	R-SU4	Precision	R-1	R-2	R-SU4
OPTIMAL	57.86	22.96	29.73	OPTIMAL	57.35	22.07	36.07
$Biclique_{Cosine1}$	30.48	7.62	12.40	$Biclique_{Cosine1}$	**36.85**	**9.88**	**17.58**
$Biclique_{Cosine3}$	31.94	8.13	13.40	$Biclique_{Cosine3}$	33.29	9.07	15.03
$Biclique_{Match5}$	*43.54*	*8.91*	*16.36*	$Biclique_{Match5}$	11.70	2.26	3.36
MEAD	**49.32**	**10.58**	**23.16**	MEAD	9.16	1.84	1.02
SR_{DIV}	46.05	8.67	20.10	SR_{DIV}	9.64	1.77	1.10
SR_{SIM}	37.46	9.29	13.80	SR_{SIM}	19.87	5.18	5.44
F_1	R-1	R-2	R-SU4	F_2	R-1	R-2	R-SU4
OPTIMAL	46.57	19.49	23.76	OPTIMAL	48.41	20.45	24.72
$Biclique_{Cosine1}$	**32.67**	**8.41**	**13.93**	$Biclique_{Cosine1}$	31.16	7.88	12.86
$Biclique_{Cosine3}$	31.85	8.35	13.46	$Biclique_{Cosine3}$	**31.73**	**8.17**	**13.28**
$Biclique_{Match5}$	17.93	3.50	5.40	$Biclique_{Match5}$	26.91	5.36	8.66
MEAD	15.15	3.08	1.89	MEAD	26.27	5.43	4.34
SR_{DIV}	15.64	2.88	2.03	SR_{DIV}	25.39	4.70	4.16
SR_{SIM}	24.38	6.23	6.31	SR_{SIM}	29.92	7.54	8.08

6.4   Discussion

Empirical evaluation of the method proposed in this paper suggests that we have a clear improvement over state-of-the-art extractive summarization results on the Opinosis dataset. Our method has shown substantial improvement compared to the existing results, especially for precision, F_1, and F_2 on all ROUGE settings. The only result we cannot beat is the recall scores of the baseline methods MEAD and SR_{DIV}, which achieve the high recall at the expense of a sharp drop in precision (explained by the fact that these methods tend to choose larger sentences which results in high recall only).

Generally our method provides a balance between the two metrics, recall and precision, which is clear from the F_1-scores (Table 1). Still the biclique method has the flexibility of optimizing a certain metric, e.g., $Biclique_{Match5}$ is obtained using a parameter setting favoring the recall.

To conclude, supported by the best scores for all ROUGE settings for precision, F_1, and F_2 on the Opinosis dataset, our biclique method should be a good addition to the existing multi-document summarization systems, and it is particularly well suited to summarizing short independent texts like user reviews. With all its strengths, the method should also be appealing because of its simplicity and good time complexity. However, it is not expected to perform equally well on datasets of more complex texts which are not in the focus of our study.

7   Conclusions

We have proposed a novel method for extractive multi-document summarization, and showed with empirical results that it outperforms state-of-the-art summarization systems. Our method is based on the detection of bicliques in a bipartite graph of sentences and their words. To keep it simple and to highlight the strength of the main idea, we have evaluated only a basic version of the method which is already better than existing top-performing systems. The technique is also flexible as it can be easily adapted for a higher recall, a higher precision, or a balance between the two metrics. Considering the time efficiency, our proposed biclique algorithm offers, for standard similarity measures, the possibility of subquadratic running time, as opposed to the at least quadratic running time of the baseline sentence removal method.

We believe that more elaborate versions making use of deep similarity measures and combining with ideas from other methods, such as MEAD, can further enhance the performance. A natural extension of the preprocessing would be to cluster semantically related words (synonyms, etc.) and to replace the words from each cluster with one representative. As mentioned in Sect. 6.2 we use stopwords from ROUGE that also include negations. As the method does not rely on the meaning of words this is not an issue, still one could study the effect of different stopword lists.

Acknowledgments. This work has been supported by Grant IIS11-0089 from the Swedish Foundation for Strategic Research (SSF), for the project "Data-driven secure business intelligence". We thank our former master's students Emma Bogren and Johan Toft for drawing our attention to similarity joins, and the members of our Algorithms group and collaborators at the companies Recorded Future and Findwise for many discussions.

References

1. Alexe, G., Alexe, S., Crama, Y., Foldes, S., Hammer, P.L., Simeone, B.: Consensus algorithms for the generation of all maximal bicliques. Discr. Appl. Math. **145**, 11–21 (2004)
2. Arasu, A., Ganti, V., Kaushik, R.: Efficient exact set-similarity joins. In: Dayal, U. et al. (eds.) VLDB 2006, pp. 918–929, ACM (2006)

3. Bogren, E., Toft, J.: Finding top-k similar document pairs - speeding up a multi-document summarization approach. Master's thesis, Department of Computer Science and Engineering, Chalmers, Göteborg (2014)
4. Bonzanini, M., Martinez-Alvarez, M., Roelleke, T.: Extractive summarisation via sentence removal: condensing relevant sentences into a short summary. In: Jones, G.J.F. et al. (eds.) SIGIR 2013, pp. 893–896, ACM (2013)
5. Damaschke, P.: Finding and enumerating large intersections. Theor. Comp. Sci. **580**, 75–82 (2015)
6. Dias, V.M.F., de Figueiredo, C.M.H., Szwarcfiter, J.L.: On the generation of bicliques of a graph. Discr. Appl. Math. **155**, 1826–1832 (2007)
7. Elsayed, T., Lin, J., Oard, D.W.: Pairwise document similarity in large collections with MapReduce. In: ACL 2008: HLT, Short Papers (Companion Volume), pp. 265–268, Association for Computational Linguistics (2008)
8. Ganesan, K., Zhai, C., Han, J.: Opinosis: a graph based approach to abstractive summarization of highly redundant opinions. In: Huang, C.R., Jurafsky, D. (eds.) COLING 2010, pp. 340–348, Tsinghua University Press (2010)
9. Gely, A., Nourine, L., Sadi, B.: Enumeration aspects of maximal cliques and bicliques. Discr. Appl. Math. **157**, 1447–1459 (2009)
10. Li, W.: Random texts exhibit Zipf's-law-like word frequency distribution. IEEE Trans. Inf. Theor. **38**, 1842–1845 (1992)
11. Li, J., Liu, G., Li, H., Wong, L.: Maximal biclique subgraphs and closed pattern pairs of the adjacency matrix: a one-to-one correspondence and mining algorithms. IEEE Trans. Knowl. Data Eng. **19**, 1625–1637 (2007)
12. Lin, C.Y.: ROUGE: a package for automatic evaluation of summaries. In: Moens, M.F., Szpakowicz (eds.) ACL Workshop "Text Summarization Branches Out", pp. 74–81 (2004)
13. Lin, H., Bilmes, J.A.: A class of submodular functions for document summarization. In: Lin, D., Matsumoto, Y., Mihalcea, R. (eds.) ACL 2011, pp. 510–520, Association for Computational Linguistics (2011)
14. Luhn, H.P.: The automatic creation of literature abstracts. IBM J. **2**, 159–165 (1958)
15. Manning, C.D., Schütze, H.: Foundations of Statistical Natural Language Processing. MIT Press, Cambridge (1999)
16. Nenkova, A., McKeown, K.: A survey of text summarization techniques. In: Aggarwal, C.C., Zhai, C. (eds.) Mining Text Data, pp. 43–76. Springer, Berlin (2012)
17. Porter, M.F.: An algorithm for suffix stripping. Program **14**, 130–137 (1980)
18. Radev, D.R., Allison, T., Blair-Goldensohn, S., Blitzer, J., Celebi, A., Dimitrov, S., Drábek, E., Hakim, A., Lam, W., Liu, D., Otterbacher, J., Qi, H., Saggion, H., Teufel, S., Topper, M., Winkel, A., Zhang, Z.: MEAD - a platform for multidocument multilingual text summarization. In: LREC(2004)
19. Wang, D., Zhu, S., Li, T.: SumView: a web-based engine for summarizing product reviews and customer opinions. Expert Syst. Appl. **40**, 27–33 (2013)
20. Xiao, C., Wang, W., Lin, X. Haichuan Shang, H.: Top-k set similarity joins. In: Ioannidis, Y.E., Lee, D.L., Ng, R.T. (eds.) ICDE 2009, pp. 916–927, IEEE (2009)

Post-processing Association Rules: A Network Based Label Propagation Approach

Renan de Padua[1]([⊠]), Veronica Oliveira de Carvalho[2],
and Solange Oliveira Rezende[1]

[1] Instituto de Ciências Matemáticas e de Computação,
USP - Universidade de São Paulo, São Carlos, Brazil
{padua,solange}@icmc.usp.br
[2] Instituto de Geociências e Ciências Exatas,
UNESP - Univerdidade Estadual Paulista, Rio Claro, Brazil
veronica@rc.unesp.br

Abstract. Association rules are widely used to find relations among items in a given database. However, the amount of generated rules is too large to be manually explored. Traditionally, this task is done by post-processing approaches that explore and direct the user to the interesting rules. Recently, the user's knowledge has been considered to post-process the rules, directing the exploration to the knowledge he considers interesting. However, sometimes the user wants to explore the rule set without adding his prior knowledge BIAS, exploring the rule set according to its features. Aiming to solve this problem, this paper presents an approach, named PAR_{LP} (**P**ost-processing **A**ssociation **R**ules using **L**abel **P**ropagation), that explores the entire rule set, suggesting rules to be classified by the user as "Interesting" or "Non-Interesting". In this way, the user is directed to analyze the rules that have some importance on the rule set, so the user does not need to explore the entire rule set. Moreover, the user's classification is propagated to all the rules using label propagation approaches, so the most similar rules will likely be on the same class. The results show that the PAR_{LP} succeeds to direct the exploration to a set of rules considered interesting, reducing the amount of association rules to be explored.

1 Introduction

Association is widely used in data mining due its simplicity and comprehensibility. This task aims to extract the correlations among items in a given database [1]. However, a large number of rules can be generated even on small data sets. Therefore, the manual exploration of the rules to find interesting patterns is unfeasible. Generally, the number of interesting rules is very small compared to the total number of patterns and, in most of the times, the user must search many rules to find the rules that are considered interesting to him/her.

Some research has been done to direct the users on the exploration and help them finding the interesting rules. Some authors propose the use of networks to

© Springer-Verlag Berlin Heidelberg 2016
R.M. Freivalds et al. (Eds.): SOFSEM 2016, LNCS 9587, pp. 580–591, 2016.
DOI: 10.1007/978-3-662-49192-8_47

direct the user's exploration, as [6, 10]. Networks are well known on the literature by its capability to model data, preserving the connection among the items on the data set. The networks are used as a mean to facilitate the rules' exploration, modeling it and pruning the rules that are not interesting to the user. The problem of these approaches is that the user must inform, beforehand, what he considers interesting, forcing him to have the knowledge apriori.

Based on the exposed, this paper presents a subjective post-processing approach, named PAR_{LP}, that interacts with the user to extract his knowledge according to the importance of the association rules on the rule set. This interaction is iterative made during the exploration, suggesting rules that are considered most relevant according to some network measures. The approach can be divided into 4 steps: (1) the most important rules, according to a network measure, are discovered; (2) user assigns labels to the extracted rules; (3) label propagation is performed to obtain the interesting rules according to the user classification; (4) the user decides if the approach needs to execute again or finish the processing. The approach is based on the idea of using classification algorithms to post-processes association rules, as proposed by [2, 7], learning with the prior interactions.

On the second iteration, the user interaction is made considering only the rules that are classified as "Interesting" by the approach. Networks are well known to model relationships among data set objects, extract properties about the importance of the objects and patterns of the data set [5]. This peculiarity allows extracting important rules for labeling and also allows propagating the labels to classify other rules. After the user interaction classifying rules, the current labels are propagated to all the other rules using a label propagation algorithm. The classification algorithm, used in this work, takes as input a data set containing a huge amount of unlabeled data and a small amount of labeled data and propagates the labels to all the data, classifying the entire data set. The PAR_{LP} uses these characteristics to propagate the user's knowledge to the entire rule set.

In summary, the main contribution of this paper is a new post-processing approach that helps the user to find the most relevant rules according to his knowledge and the rule's importance in the rule set. The user interaction is made in a way that he does not need to have a prior knowledge on the data set, suggesting a few rules to be classified by him based on the rule's relevance in the network. By doing so, the approach excludes the necessity to have all the knowledge beforehand, making the classification process easier based on only a few rules, not considering the entire rule set. Besides, this paper also shows that post-processing association rules using label propagation is effective to find the interesting rules according to user's knowledge. This paper presents an experimental evaluation to demonstrate that PAR_{LP} is capable of finding the rules that the user considers as interesting. The experiments also show that the proposed approach can reduce the amount of rules to be explored by the user to find those interesting rules.

This paper is organized as follows. Section 2 describes related research and basic concepts. Section 3 presents the PAR_{LP} and its motivation. Section 4 describes the experiments that were carried out to analyze the approach. Section 5 discusses the results obtained in the experiments. Finally, conclusion and future works are given in Sect. 6.

2   Background and Related Works

The transductive post-processing approach, proposed in [7] and extended in this paper, comes up with a different way to explore the rules. The approach, named PAR_{LP}, models the association rules into networks. This modeling allows the approach to explore the rules according to their importance on the network, selecting the rules to be classified by the user, directing the user's effort. The approach uses label propagation to find the interesting rules based on the users' classification. The label propagation was chosen considering that if a rule is interesting, the rules similar to it are also interesting.

A network can be characterized by a set of elements and the relations among them. Formally, a network can be represented by $N = (V, E, W)$ where V is the set of vertices (elements), E the set of links between the vertices and W the weight of the links [5]. When all the vertices represent the same kind of object, only web sites, for example, the network is called homogeneous network. On the other hand, when the representation of different kinds of objects is considered, like persons and communities in a social network, the network is called heterogeneous network. Besides, there is a variation of the conventional network called bipartite network. The bipartite networks have two different kinds of vertices: groupers (G) and items (H). The groupers can only connect to items vertices and the items can only connect to groupers vertices. The groupers vertices allow the propagation of information among the items vertices. Those different representations allow a large variety of exploration. One example of a bipartite network is the connection among documents (groupers) and terms (items).

One way to classify all vertices of a network considering user's knowledge is through label propagation. Label propagation algorithms classify a data set based on few classified examples (the training set is composed of labeled and unlabeled instances). The classification is made based on a similarity measure. The elements that are considered similar are classified on the same class whereas the elements that are not considered similar are classified on different classes [13].

In this paper, we use two network types to apply label propagation: homogeneous and heterogeneous. The homogeneous networks are building using similarity measures to connect the rules. The bipartite network considers an objective measure to connect the items to the groupers vertices. To propagate the labels two algorithms were used per network type: Learning with Local and Global Consistency (LLGC) and Gaussian Fields and Harmonic Function (GFHF) for the homogeneous networks and Label Propagation using Bipartite Heterogeneous Network (LPBHN) and GNetMine for the bipartite networks. The propagation is made in a way to minimize a regularization function.

The LLGC and GFHF algorithms used to classify the homogeneous networks, were selected because of their great results obtained on the literature. The GFHF algorithm [12] considers the labeled elements as unchanging truth and propagates the class without the need of a parameter. The LLGC algorithm [11] considers that a labeled element can change its labels based on its neighbors. The LLGC algorithm needs the definition of a learning rate parameter α.

The GNetMine and LPBHN used to classify the bipartite heterogeneous networks, were also selected of their results obtained on the literature. The LPBHN algorithm [9] works in the same way the GFHF does. It considers the element's labels as unchanging truth and propagates these labels without the need of a parameter. The GNetMine [4] works in the same the LLGC does. It can change the original label based on its neighbors and need the definition of a learning rate parameter $alpha$. The algorithms' functions can be seen on their respective papers.

[6] proposed an approach that uses networks as a mean to post-process association rules. In their work, the authors force the user to select an item (let's call it objective item) to be the main objective of the exploration. After the selection, a directed hypergraph is constructed considering the objective item as the consequent of the rules and the antecedents are modeled, constructing the level 1. The level 2 considers the antecedents of all the rules modeled on the level 1 as consequent and repeat the process. The hypergraph is constructed until all the items are modeled or the maximum level is reached (informed by the user). Using that hypergraph, all the rules modeled according to an objective item and the user can see how the items interact with the selected item. This approach forces the user to explore only one item per time and, yet, the user must know beforehand which item he/she wants to study.

3 PAR_{LP}: Post-processing Association Rules Using Label Propagation

The main concept behind the PAR_{LP} is that if the user considers a rule R_x as interesting, then the rules similar to it will also be interesting to be explored. The same goes to the case that the user finds a rule R_y not interesting, considering the similar rules also as not interesting. Figure 1 shows how the exploration is made. Consider "1" and "2" as the rules to be found. On (a) user selects "2" as interesting and "5" as not interesting. This classification is made aided by the proposed approach. On (b) the label propagation is done, considering the squares as not interesting and circles as interesting. Note that, after applying the classification algorithm, the interesting rules ("1" and "2", plus "3") are classified as interesting and will be presented to the user. In this case, the reduction of the exploration space is 50 % (the user only needs to explore 50 % of the existing rules).

The PAR_{LP} works as shown in Fig. 2. The input of the approach is a rule set, containing the rules to be explored. The approach starts on step 1, modeling the association rules in a network. To do so, it's necessary to define the

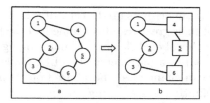

Fig. 1. The PAR_{LP} concept

network type [NT] and the similarity measure [S] to be used. The network type defines the network structure, i.e., homogeneous or heterogeneous networks. In case of homogeneous networks, the user also needs to define a way to link the rules considering their similarity, such as connecting a rule with their K-Nearest Neighbors or connect all rules through an RBF kernel [3]. The similarity measure [S], like hoJaccard (Eq. 2) and hoConfidence (Eq. 3), is used to calculate the similarity between two rules, setting the weight of their connections. In the beginning, the modeled association rules have no defined class, containing only the rules and the connections among them.

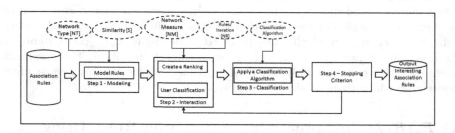

Fig. 2. The PAR_{LP} approach

In step 2, the PAR_{LP} interacts with the user, aiming to classify a few rules that will be used on the label propagation process to discover other interesting and non-interesting rules. This interaction aims to capture the user's knowledge and direct the exploration to the rules he considers "Interesting". To do so, a ranking is created to select the best rules according to a centrality measure [NM]. On the bipartite networks, a projection on a homogeneous network was created to apply the [NM]. This project was based on the bibliographic coupling concept [5]. The approach selects the N best rules and the N worst rules to be classified by the user; that way, the selected rules are not in the same dense point in the network. The number of selected rules is defined by the parameter rules/iteration [NR], where $NR = N + N$ (N-best rules and N worst rules). In this experiment, the [NR] was set to 10 to all the configurations. This number was selected so the user can evaluate just a small amount of rules and the classifiers still have a satisfactory amount of classified elements. The selected rules are shown to the user to be classified in 2 different classes: "Interesting" and "Non-Interesting". The "Interesting" class contains the rules considered as interesting and will be explored by the user when the approach finishes. On the first iteration, the entire rule set will be explored to make the ranking. From the second iteration, only the rules classified as "Interesting" are considered to make the ranking. Since the ranking is created to suggest rules to be classified by the user, the rules already classified by the user are not considered on the ranking.

In step 3, the unlabeled association rules are classified using a network based label propagation algorithm. It is important to distinguish the classifications made by the user from the ones made by the classifier: the user's classifications can not be changed, i.e., over the iterations these classifications will be maintained by the approach and used in the training set over all the iterations; on the other hand, the classifier's classifications can change over the iterations. The change on the rules classified by the label propagation algorithm is made so the approach can refine the results, adapting to the new knowledge informed by the user.

On the last step, a stopping criterion is checked to decide if the approach will finish or will execute again. If it's decided to continue, the rules classified by the label propagation algorithm will be explored again from Step 2, considering only the ones classified as "Interesting" to create the ranking. However, all the rules are considered on the label propagation phase. The process continues until the stopping criterion is met. In the end, the rules considered as "Interesting", either by the classification algorithm and by the user, are outputted to the user.

4   Experiments

The experiments were carried out with 2 different objectives: validate the approach and find the best configurations. The evaluation measure is the exploration space reduction, i.e., the percentage of rules that doesnt need to be explored in order to find all the interesting rules. High values of exploration space reduction means that the user needs to explore fewer rules to find the ones he is looking for; if the reduction space is 60 % for example, the user needs to explore only 40 % of the entire rule set. Both the approach validation and the analysis of the best configurations were performed on the exploration space reduction. The analysis is made considering a small set of rules, called "objective set". This objective set contains the rules to be found on the exploration, simulating the user's interests.

The approach was analyzed on 8 data sets (balance-scale, breast-cancer, car, ecoli, habermann, iris, tic-tac-toe and zoo), all of them available on UCI[1]. These data sets were processed and converted to transactions. The transactions with missing values were removed. The support and confidence values were empirically defined to generate an amount of rules between 1.000 and 2.000. This number of rules was selected to obtain a good trade-off between exploration space (the higher the better) and the number of rules on the objective set (explained below).

---

[1] http://archive.ics.uci.edu/ml/.

Two different kinds of networks were selected to carry out the experiments: homogeneous and bipartite heterogeneous. The homogeneous network allows an exploration based on the similarity among the rules. The bipartite network allows an exploration based on the items shared by the rules and based on the rules shared by the items. Therefore, the homogeneous network have three different network types [NT]: kNN, in which a rule is linked to its K most similar rules; Gaussian, that changes the weight among the connections by applying the weight to a gaussian function; Conventional, that makes no change on the generated network, maintaining all the original connections and weights. No modifiers were used on the bipartite heterogeneous network.

The number of rules per iteration [NR] on both networks were set to 10 (5 best + 5 worst rules per iteration) aiming to reduce the number of rules to be classified by the user without losing performance on the label propagation algorithms. The centralities measures [NM], used to create rankings on both homogeneous and bipartite heterogeneous, and were: degree, which analyses the rule's importance based on its connections to its neighbor; PageRank [8] that measures the rule's importance based on the entire network connections.

To generate the homogeneous networks, the similarities [S] were used: hoJaccard (Eq. 2, that uses as base Eq. 1) that calculates the similarity between two rules based on the items they share, considering the rule's antecedent and consequent separately; hoConfidence (Eq. 3) that calculates the similarity between two rules based on the transactions they share. The confidence was adapted here to work as a similarity measure because of it can calculate how the occurrence of a rule items can contribute to the occurrence of the items on the other rules. In Eqs. 1, 2 and 3, RL_x represents the x-est rule on the rule set, $LHS(RL_x)$ and $RHS(RL_x)$ represent the RL_x antecedent and consequent, respectively, $\#(T(RL_x))$ represents the number of transaction that contains the rule RL_x. To classify the homogeneous networks the GFHF and LLGC classification algorithms were used. These algorithms were selected due to their great results obtained on the literature.

$$Jacc(RL_1, RL_2) = \frac{RL_1 \cap RL_2}{RL_1 \cup RL_2} \tag{1}$$

$$hoJaccard(RL_1, RL_2) = \frac{Jacc(LHS(RL_1), LHS(RL_2)) + Jacc(RHS(RL_1), RHS(RL_2))}{2} \tag{2}$$

$$hoConfidence(RL_1, RL_2) = \frac{\#(T(RL_1) \cup T(RL_2))}{\#T(RL_1)} \tag{3}$$

The bipartite heterogeneous network was modeled so the weight among the connections between items and their respective rules was an objective measure. Two similarity measures [S] were selected: bipartiteJaccard (Eq. 4) that calculates the amount of transactions shared by the rule's LHS and RHS and the traditional confidence measure. The bipartiteJaccard uses the transactions, instead of the items, so it can get the similarity among the items that are on the LHS and the item on the RHS of the rule. The connection among two different rules was made by the items they share, where the item is a connection point that connects two rules with two different weights. In equation, $LHS(RL_1)$ returns the rule's left-hand side, $RHS(RL_1)$ return the rule's right-hand side and $\#T(RL_X)$ returns the number of transactions that contains the items on RL_X. To classify the bipartite heterogeneous networks the LPBHN and GNetMine algorithms were used. The GNetMine have the same classification bias as LLGC. Also, the LPBHN have the same bias as GFHF. This allows us to make a fair comparison and analysis to which network type provides a better space reduction.

$$bipartiteJaccard(RL_1) = \frac{\#T(LHS(RL_1)) \cap \#T(RHS(RL_1))}{\#T(LHS(RL_1)) \cup \#T(RHS(RL_1))} \quad (4)$$

In this paper, the user's classification, Step 2 in Fig. 2, was simulated using a set of rules as an objective set. These objective sets are the set of rules to be found on the rule set, simulating the user's interests. Two different groups of objective sets were generated to simulate different types of users. The first group is the random objective set. This set is generated by randomly selecting a total of 1 % of the rule set size. The other group consisted on randomly selecting one rule in the rule set and creating a similarity ranking among the selected rule and the entire rule set. The ranking is used to select 1 % of the most similar rules to be added into the objective set. The similarity used to calculate the objective set was Jaccard (Eq. 1), that calculates the number of items the rules share divided by the number of distinct items. It is important to emphasize that this measure was not directly used to generate the networks. Because the objective sets were randomly generated, 30 sets were created using each approach. So, for each rule set and each PAR_{LP} configuration, the experiments were executed considering the two groups of user's interests simulation. The random objective set (first group) simulates the users that have a more widely interest on the rule set, with the interesting rules spread across the network. The similarity objective set (second group) simulates the users that have a more specific interest on the rule set.

Based on the objective sets, the user's classification is simulated considering a threshold to be calculated on the first iteration. This threshold is the mean similarity among the rules to be classified by the user and the rules on the objective set, based on the closest similarity of each rule, i.e., for each rule to be classified by the user, the highest similarity is considered to all rules on the objective set. The threshold can be seen in Eq. 5, where $ruleObj$ is the set of rules on the objective set, $ruleCl$ is the set of rules to be classified by the user on the current iteration and nCl is the number of rules to be classified by the user (equals $N + N$ as previously explained). Therefore, in each iteration, the mean similarity among the rules to be classified and the objective set is calculated and compared to the threshold, labeling the rule as "Interesting" if the mean similarity is greater than or equal to the threshold or "Non-Interesting" if the similarity is smaller.

$$Threshold = \frac{1}{nCl} \sum_{i=1}^{nCl} max(similarity(ruleCl_i, ruleObj)) \qquad (5)$$

Finally, regarding the stopping criteria, the approach was executed until all the rules on an objective set were classified as "Interesting" either by the user or by the classifier. The experiments were carried out and an analysis was made, aiming to find all the rules on each objective set and looking for the reduction on the exploration space, remembering that the greater the reduction the better the result. In the end, a mean reduction was calculated to each PAR_{LP} configuration (Table 1) and each objective set configuration (random and similarity). Therefore, the results shown in the next section are the mean of 30 executions considering each kind of objective set individually, for each PAR_{LP} configuration.

Table 1. PAR_{LP} configurations

Network	Network type [NT]	Network measure [NM]	Similarity [S]	Classifier
Homogeneous	Knn (K = 10, 20, 30, 40, 50); Gaussian (σ = 0.25, 0.50, 0.75); conventional	Output degree; PageRank	hoJaccard; hoConfidence	LLGC (α = 0.1, 0.3, 0.5, 0.7, 0.9); GFHF
Bipartite heterogeneous	Conventional	Output degree; PageRank	bipartideJaccard; confidence	GNetMine (α = 0.1, 0.3, 0.5, 0.7, 0.9); LPBHN

5   Results and Discussion

The experiments were carried out on each data set using all the described PAR_{LP} configurations (Table 1). For each configuration, the PAR_{LP} approach was applied over all the 60 objective sets to test two different strategies of simulating the user's interests: one containing scattered rules, for users that want to explore the rule set without looking for a specific "theme", and one containing more similar rules, for users that have a specific area of interest. The results were analyzed considering the mean of the 30 executions on each PAR_{LP} configuration and each group of objective sets, which means that each PAR_{LP} configuration has 2 different results: one for the randomly generated objective sets and one for the objective sets generated based on the similarity.

Table 2 shows the best and the worst results obtained on each data set. The first column presents the data set used to generate the rules. The second column shows the best result obtained using the random objective sets. The third column shows the worst result obtained using the random objective sets. The fourth and fifth columns show, respectively, the best and worst results obtained using the similarity objective sets. Remember that the experiments were analyzed based on the exploration space reduction, which means that the values on these columns represent the percentage of rules that the user doesn't need to explore. For example, a 60 % exploration space reduction means that the user needs to explore only 40 % of the rules to find all the interesting ones.

The results show that the objective sets generated by the similarity strategy obtained better results in comparison to the random objective sets. These results can show that an exploration guided by some "theme" or by some related topics will result in a higher reduction than an exploration where the user explores by selecting dissimilar rules as "Interesting". These better results occurred because of the label propagation algorithms, used to classify the rules, since these algorithms put similar knowledge on the same class and dissimilar knowledge on different classes. By simulating the user's classification and selecting dissimilar rules as "Interesting" the label propagation algorithms classify different parts of the network as "Interesting", increasing the number of rules that are considered "Interesting" in comparison to the objective sets that classify only one point on the network. Also, the best reduction was 66.49 %, which means that the user will only have to explore about $\frac{1}{3}$ of the entire rule set. The table also shows a variation among the data sets, which means that the domain can contribute to the exploration. The best data set (iris) reduced almost 50 % more in comparison to the worst data set (breast-cancer).

As seen, Table 2 doesn't link the results to the PAR_{LP} configurations, since the analysis performed were based on the overall quality of the results. Therefore, a second analysis was performed aiming to find the best PAR_{LP} configurations on the rule sets. Initially, all the results (considering all the 8 data sets, PAR_{LP} configurations and both objectives sets strategies) were divided into 20 groups considering network, network type [NT] and classifier. From these 20 groups, the 2 best results of each group were selected through a statistical test, coming up with 40 PAR_{LP} configurations. After that, another statistical test was done, based on these 40 PAR_{LP} configurations, through Friedman N × N with Nemenyi as post-test, selecting the 20 best configurations. Figure 3 shows the results of the final statistical analysis. The horizontal lines show that the PAR_{LP} configurations have no statistical difference. The configurations are described using the following pattern: network type [NT] - network parameter

Table 2. Best and worst exploration space reductions obtained on each data set

Data set	# Gen. Rules	Best Rd	Worst Rd	Best Sim	Worst Sim
balance-scale	1746	40.66%	4.81%	63.69%	29.50%
breast-cancer	1602	19.98%	5.37%	42.45%	4.56%
car	1326	15.91%	4.68%	52.64%	22.17%
ecoli	1685	28.66%	4.87%	51.57%	21.01%
habermann	1006	46.12%	9.15%	58.45%	29.72%
iris	967	51.71%	10.13%	66.49%	39.50%
tic-tac-toe	1317	37.05%	4.02%	61.88%	16.02%
zoo	1658	30.88%	4.40%	46.38%	17.13%

(when applied) - classifier - classifier parameter (when applied) - similarity measure [S] - rank measure [NM]. The first configuration, for example, is kNN, with $k = 50$, using the GFHF classifier, hoJaccard as a similarity measure and PageRank as the ranking measure. It is possible to see that the kNN network, together with the GFHF classifier, obtained the overall best results, being on 9 out of 10 best results.

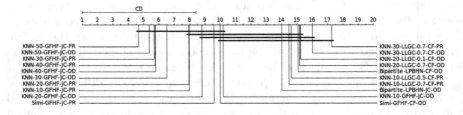

Fig. 3. The statistical analysis results

6   Conclusion

This paper proposed the PAR_{LP}, an iterative and interactive post-processing association rules approach that uses networks and label propagation algorithms to extract interesting and non-interesting rules according to user's knowledge. The paper presented the PAR_{LP} structure and carried out some experiments using a set of configurations and a user classification simulation. The obtained results were shown and discussed according to the total reduction on the exploration space, i.e., the number of rules not to be exploited by the user.

The results show that the PAR_{LP} is capable of finding a set of rules according to the interactions made during the process, meaning that the PAR_{LP} can be used successfully to direct the exploration of the rules according to the interactions made with the algorithm, reducing the amount of rules to be explored and directing the exploration to the rules chosen by the user interaction. This reduction can be increased when the rules to be found by the user are more similar, or share the same "theme". Even on the random objective sets, that contain more widespread rules, the obtained reductions were satisfactory. Also, the analysis on the 20 best configurations shows that the kNN network, together with GFHF classifier, are the most promising configuration.

As future works a further analysis on the way the rules sets are modeled will be done aiming to find the characteristics that result on greater exploration space reduction. This analysis will be carried out aiming to discover beforehand which similarity measure to be used according to the rule set features. Also, the algorithms complexity will be reduced. The actual configuration is expensive e can not be used on larger rule sets. To solve the "theme" guided exploration, the proposed approach will be extended to consider different user's interests.

Acknowledgments. We would like to thank CAPES (PROEX-8434242/D) and FAPESP: Grant 2014/08996-0, São Paulo Research Foundation (FAPESP) for the financial aid.

References

1. Agrawal, R., Imielinski, T., Swami, A.: Mining association rules between sets of items in large databases. In: ACM SIGMOD (1993)
2. de Carvalho, V.O., de Padua, R., Rezende, S.O.: Semi-supervised learning to support the exploration of association rules. In: Bellatreche, L., Mohania, M.K. (eds.) DaWaK 2014. LNCS, vol. 8646, pp. 452–464. Springer, Heidelberg (2014)
3. de Sousa, C.A.R., Rezende, S.O., Batista, G.E.A.P.A.: Influence of graph construction on semi-supervised learning. In: Blockeel, H., Kersting, K., Nijssen, S., Železný, F. (eds.) ECML PKDD 2013, Part III. LNCS, vol. 8190, pp. 160–175. Springer, Heidelberg (2013)
4. Ji, M., Sun, Y., Danilevsky, M., Han, J., Gao, J.: Graph regularized transductive classification on heterogeneous information networks. In: Balcázar, J.L., Bonchi, F., Gionis, A., Sebag, M. (eds.) ECML PKDD 2010, Part I. LNCS, vol. 6321, pp. 570–586. Springer, Heidelberg (2010)
5. Newman, M.E.J.: Networks: An Introduction. Oxford University Press, Oxford (2010)
6. Pandey, G., Chawla, S., Poon, S., Arunasalam, B., Davis, J.G.: Association rules network: definition and applications. Stat. Anal. Data Min. **1**, 260–279 (2009)
7. Padua, R., Carvalho, V.O., Rezende, S.O.: Post-processing association rules using networks and transductive learning. In: 13th ICMLA, pp. 318–323 (2014)
8. Page, L., Brin, S., Motwani, R., Winograd, T.: The PageRank citation ranking: bringing order to the web. Techical report, Stanford InfoLab (1999)
9. Rossi, R.G., Lopes, A.A., Rezende, S.O.: A prameter-free label propagation algorithm using bipartite heterogeneous networks for text classification. In: ACM SAC, pp. 79–84 (2009)
10. Yang, G., Shimada, K., Maby, S., Hirasawa, K., Hu, J.: A genetic network programming based method to mine generalized association rules with ontology. JACIII **12**, 63–76 (2007)
11. Zhou, D., Bousquet, O., Navin Lal, T., Weston, J., Schölkopf, B.: Learning with local and global consistency. NIPS **16**, 321–328 (2004)
12. Zhu, X., Ghahramani, Z., Lafferty, J.: Semi-supervised learning using Gaussian fields and harmonic functions. In: ICML, pp. 912–919 (2003)
13. Zhu, X., Goldberg, A.B.: Introduction to Semi-Supervised Learning, vol. 6. Morgan & Claypool Publishers, San Rafael (2009)

Application of Multiple Sound Representations in Multipitch Estimation Using Shift-Invariant Probabilistic Latent Component Analysis

Krzysztof Rychlicki-Kicior$^{(\boxtimes)}$, Bartłomiej Stasiak, and Mykhaylo Yatsymirskyy

Institute of Information Technology, Technical University of Łódź,
Ul. Wólczańska 215, 90-924 Łódź, Poland
krzysztof.rychlicki-kicior@dokt.p.lodz.pl,
{bartlomiej.stasiak,mykhaylo.yatsymirskyy}@p.lodz.pl
http://it.p.lodz.pl

Abstract. Probabilistic analysis has become one of the most important directions for development of new methods in Music Information Retrieval (MIR) field. Its ability to correctly find necessary information in the music audio recordings is especially useful in multipitch estimation, a vital task belonging to the MIR field. Since the multipitch estimation is still far from being resolved, it is important to enhance the existing state-of-the-art methods. Usually, a spectrogram, generated from the Constant-Q transform (CQT) is used as a basis for the SI-PLCA method. The new approach involves application of more than one method (cepstrum and CQT) in association of the shift-invariant probabilistic latent component analysis approach and additional processing of all the sound representations, in order to achieve better results.

Keywords: MIR · Fundamental frequency estimation · Multi F_0 · Multipitch · Polyphony · PLCA

1 Introduction

Multiple fundamental frequency (*multi-F_0*) estimation is a low-level task defined within the Music Information Retrieval (MIR) field. It forms a foundation for more complex and high-level problems, such as Audio Chord Estimation, Audio Melody Extraction or Real-time Audio to Score Alignment [1]. This task should not be confused with simpler problem of recognizing only one fundamental frequency, which also has numerous practical applications, i.a. in pitch tracking for query-by-humming search interface [2] or in speech emotion recognition [4]. More similar, yet distinct task is melody extraction from polyphonic music signal. Although many different pitches can be detected there, mostly the main, the strongest pitches constituting a melody are carefully analysed [5].

It is very important to note the distinction between a fundamental frequency and a pitch. The F_0 of a signal is its physical property, independent on the observer. If we treat a signal as a sum of sinusoidals (and this is valid for most of

© Springer-Verlag Berlin Heidelberg 2016
R.M. Freivalds et al. (Eds.): SOFSEM 2016, LNCS 9587, pp. 592–601, 2016.
DOI: 10.1007/978-3-662-49192-8_48

musical sounds), then F_0 is the lowest frequency that appears in this sum. The pitch of a sound, on the other hand, is a quality perceived by the observer. It is often described as a frequency of a sinusoidal that a group of human listeners would match with a given sound.

The main goal of the multipitch estimation is to detect correct pitches in a signal generated by several independent, concurrent sound sources. The number of the sources can be known (the algorithm usually assumes that it should find a constant number of sources) or not. The latter problem is more complex and involves an additional step called polyphony inference. This process is not considered in this work, because we assume the constant number of the sources [6].

One of the most important sources of information for multipitch estimation algorithms is the frequency spectrum of a sound. Although the first idea could be that the multipitch estimation should be simplified in order to find only the biggest peaks in the spectrum, unfortunately, this is mostly untrue. This stems from the fact that spectra of sounds generated by musical instruments are usually very complex.

Complexity of such spectra is associated with the existence of harmonics – the partials of a sound, which have frequencies defined by the following formula:

$$f_i = (i + 1)f_0 \tag{1}$$

where f_i represents the consecutive harmonics, i is the i-th partial number and f_0 is the fundamental frequency.
In this work we assume that the first harmonic is the fundamental frequency (f_0), the second harmonic is f_1, and so on.

2 Known Approaches

Many approaches have been proposed to address the multipitch estimation problem. Basically, two aspects of solution should be considered: how should the sound be represented in order to get the most valuable information and how to analyse this information to achieve the best results.

The most popular representation of sound is its spectrum. Discrete Fourier Transform (DFT) has been used since the beginning of the research on the problem of the multipitch estimation [8], however, recently other spectral forms are getting more and more popular, such as Multiresolution FFT (MRFFT) [9], wavelets [1,3] and Constant-Q Transform, which is used in this work.

A very interesting extension to the concept of the spectrum is the *salience*, which describes the power of each frequency component in the sound spectrum:

$$s(\tau) = \sum_{m=1}^{M} g(\tau, m)|Y(f_{\tau,m})| \tag{2}$$

where Y is a sound spectrum, $f_{\tau,m}$ represents a certain frequency corresponding to the given τ. $g(\tau, m)$ is a weight function that decreases meaning of further partials. The exact form of this function depends on parameter values, which are a subject of optimization. M defines number of partials to be summed and τ in the above equation represents *lag* and is directly related to frequency component:

$$\tau = \frac{f_s}{f} \tag{3}$$

where f_s represents a sample rate of the input signal and f is a given frequency.

Salience is much better representation of frequencies power, because it represents weighted sum of powers of all partials of given frequency. Unfortunately, this approach often yields poor results, especially when one sound is louder than others. It results in yielding not only the fundamental frequency of the louder sound, but also its partial, whereas fundamental frequencies of other sounds are often omitted [6].

Application of a certain sound transformation, in order to achieve appropriate representation of the sound, is just the first part of the process. What happens with the representation is even more important, because no matter what sound representation is chosen, there are certain problems that might arise during their analysis.

2.1 Methods of Sound Representation Analysis

Having chosen a sound representation(s), the analysis method must be chosen in order to transform this representation into a set of fundamental frequencies. Initially, two major algorithms for this task have been used: iterative cancellation and joint estimation [6].

Iterative cancellation is a method, in which the strongest component in the spectrum is found. After finding the strongest component, it is removed from the spectrum, along with the components representing its harmonics. The procedure is repeated until the assumed number of sound sources is found or until the residue of the spectrum is smaller than a certain threshold. This is quite fast approach; unfortunately, overlapping of the partials worsens results achieved with this method. Overlapping happens when two independent sounds have one or more common harmonics (for example 200 Hz sound and 300 Hz have common harmonic 600 Hz). In the iterative approach, overlapping partials are usually assigned only to one of the sounds and that makes it difficult to find others.

Joint estimation approach resolves this problem, since all possible combinations of frequency candidates are analysed jointly – not sequentially. Each set of candidates is removed from the original spectrum, and the winner set is the one having the smallest residue after performing the removal. The computional complexity is higher, especially when more complicated chords are analysed, but the overall precision of this approach is better than the iterative cancellation [10]. This approach is analysed more in the following sections.

In the last few years, however, different methods have been chosen for the sound analysis. Since the spectrogram (showing the changes of a spectrum in time) contains only non-negative values, it can be analysed using non-negative spectogram factorisation techniques, such as NMF (Non-negative Matrix Factorization) [1] or SI-PLCA (Shift-invariant Probabilistic Latent Component

Analysis) [11]. The latter method has been used often lately, since its ability to decompose spectrogram, treated as a bivariate probability distribution, to a number of marginal distributions. Each marginal distribution may be treated as a separate sound; moreover, the SI-PLCA method is able to show the presence of the particular sound within the whole time, without additional operations to be done. The SI-PLCA method is described more carefully in the following sections.

3 Research Database

The database used in this research has been constructed using state-of-the-art RWC Music Database: Musical Instrument Sound [7]. The RWC database contains high quality recordings of many instruments. In this work, these recordings have been preprocessed, in order to obtain single sounds and to associate frequencies to them on the basis of the metadata attached to the RWC database.

Next, these sounds have been mixed with each other, in order to obtain intervals. In this work, the database of around 1750 intervals and more complicated chords has been established. Various instruments have been chosen, such as piano, electric piano, flute, alto sax, viola, violin, trumpet. Large number of intervals made it possible to check almost all possible combinations of instruments.

The frequency range of the database is from 100 Hz to 1500 Hz. Therefore, intervals from almost four octaves have been covered. It is very important to note that often databases of simpler structure are proposed, e.g. where only intervals of the same instruments are checked or where a smaller range of frequencies is checked. Vast frequency range makes it difficult to recognize sounds, because one of the biggest problem of multipitch estimation – octave errors – can happen a lot more often. The octave error is a situation, in which a sound is recognized as having a pitch n times bigger than it really has, that is, is mistaken for its harmonic (partial). The difference between a sound and its following harmonic (twice bigger frequency) is called an octave in music, hence the name.

4 Proposed Solution

The approach presented in this work consists of a few fundamental steps. At first, the signal is divided into separate frames using the Hanning window function. Next, the constant-Q transform and cepstrum of each frame are calculated, creating a spectrogram and a cepstrogram. Furthermore, both representations are analysed using the shift-invariant probabilistic latent component analysis. Each representations yields a set of frequency candidates, from which the special algorithm, called the judge, selects the final result of the algorithm.

4.1 Constant-Q Transform

After a windowed frame of sound signal has been retrieved, it is transformed to the frequency domain. Instead of the popular DFT, the Constant-Q Transform

(CQT) has been applied. The CQT differs from the regular DFT, in that it results in the spectrum in the logarithmic scale, i.e. frequency bins, which are distributed linearly within the DFT, become distributed logarithmically within the CQT. The frequency of the k-th CQT frequency bin is defined as:

$$F_k = F_{min} 2^{\frac{k}{n}} \qquad (4)$$

where F_{min} is the lower bound of the frequency range, F_k is the frequency centre of the k-th bin and n is the number of bins per octave. For any given frequency, n is the number of bins that will cover the range in a spectrum between frequencies f and $2f$. This range (interval) in the musical terminology is called an octave.

The Q from the CQT name is the Quality factor. It describes how accurately the spectrum is described using the particular instance of the transform. It is defined as a ratio of a center frequency of any bin to its width:

$$Q = \frac{F_k}{\delta F_k} \qquad (5)$$

Of course the higher the Q is, the better the quality of the transform becomes.

In order to calculate the Constant-Q transform, certain parameters must be defined. The frequency range has been set to (50, 4000) Hz, in order to be able to find at least two harmonics of high frequency pitches. Classic piano keyboard contains 12 keys within a one octave and often this is the default number of bins per octave. However, in this work the number of bins per octave has been set to 48. This lengthens calculations, but gives much bigger precision. The sampling frequency has been set to 44,100 Hz, because this was the sampling frequency of the database sound files.

The importance of the CQT transform stems from the fact that, when compared to the DFT, it gives much more information about the lower frequency band of the analysed frequency range. This is associated with the bins in the lower band being distributed much more tightly than in the upper band. Better low-frequency resolution gives a possibility to detect spectral peaks more precisely, which leads to obtaining better results of the multipitch estimation. Therefore, this method is known to yields much better results than the regular Discrete Fourier Transform (DFT).

4.2   Cepstrum

Cepstrum has been used for a long time in sound processing, however, it has been mostly used for speech processing. For instance, Mel-Frequency Cepstral Coefficients have become the state-of-the-art tool used for human voice pitch recognition [4]. They are also useful in speaker identification.

Application of cepstrum in multipitch estimation may seem controversial at first, however, as results show, it helps to achieve much better accuracy, than in case of using just the Constant-Q Transform. This stems from the fact that the regular Constant-Q Transform (and other purely spectral representation, such as regular Discrete Fourier Transform or Multi-Resolution Fast Fourier Transform) might be problematic when the frequency, which is a true pitch, is not the strongest one. Cepstrum, which intuitively shows the rate of changes in the spectrum (and is therefore called *a spectrum of a spectrum*) concentrates on the existence of components with harmonics and relationship between them rather than on certain, particular components.

Cepstrum usually yields little worse results than CQT in case of high-frequency harmonics, due to less accuracy in those areas, therefore using both methods and taking their advantages is important aspect of this research.

4.3 SI-PLCA

Each sound representation – a spectrogram and a cepstrogram – is processed using the shift-invariant probabilistic latent component analysis method. It has become one of the most important approaches in the Music Information Retrieval, especially in multipitch estimation [11,12]. It might be used both with preexisting knowledge (in this particular situation this means previously obtained spectral templates of single musical notes) and without it. The general idea, however, stays very similar, and in this research one of the unsupervised approaches have been used [13].

The general idea behind the Shift-invariant Probabilistic Latent Component Analysis is to decompose a distribution $P(x)$ (x being the N-dimensional random variable) to a certain number of N-dimensional latent distributions, defined by their marginal distributions [12]:

$$P(x) = \sum_z P(z) \prod_{j=1}^N P(x_j \mid z) \tag{6}$$

Marginals in this model are calculated using the Expectation-Maximization algorithm. It is a simple model, however, it does not react well with shifted patterns, therefore usually its modified version is used, which includes the *shift invariance* assumption:

$$P(x) = \sum_z \left(P(z) \int P(w, \tau \mid z) P(h - \tau \mid z) d\tau \right) \tag{7}$$

This way the *kernel* distributions ($P(w, \tau \mid z)$) and *impulse* distributions ($P(h - \tau \mid z)$) are obtained, where w and h are mutually exclusive subsets of components and τ is a random variable [12]. In this research, kernel distributions represent spectral templates of single notes (i.e. one kernel distribution should contain all harmonics of a single sound), whereas impulse distribusions present where these templates show up in the original spectrogram (or cepstrogram).

After obtaining marginal distributions from a single sound representation, maximum values are retrieved and the frequency of these maxima are calculated. Each marginal distribution contributes two candidate frequencies and since the original representation is decomposed to two marginal distributions, each sound representations contributes four candidate frequencies.

Table 1. Percentage of correctly detected intervals by the type of interval (intervals higher than an octave have been reduced to their equivalents within an octave)

Interval (semitones)	Accuracy (%)
0	88.83
1	87.82
2	85.77
3	87.62
4	89.11
5	88.24
6	84.62
7	83.00
8	85.11
9	89.67
10	90.85
11	83.75

For each candidate frequency, its normalized power is calculated as a ratio of candidate frequency (quefrency in case of cepstrum) magnitude to the maximum magnitude in the marginal distribution where this particular candidate frequency appeared.

4.4 The Judge

Having the eight frequencies with their calculated power, a decision must be made which frequencies should be chosen as the final result of the algorithm. The part of the whole approach responsible for merging all the input received from the SI-PLCA method performed for both cepstrum and CQT is called the *judge*.

First of all, all similar candidates (i.e. those which differ by less than 6 % from each other, 6 % being a distance of one semitone) are grouped as one candidate. Their power is summed and their count is included (by default each candidate frequency has count equal to one). Then, candidate frequencies are sorted using the following criteria: multiplication of the count of candidates and their summed power. The first candidates are chosen as a result of the algorithm and are compared to the ground truth.

5   Results

Results have been obtained for polyphony from two up to four simultaneous voices. Intervals (two sounds played at the same time) have been checked the most carefully (with 1245 different intervals checked), since it is possible to check different interesting dependencies, showing accuracy for different kinds of intervals (Table 1) and for particular pairs of instruments (Table 2).

Table 2. Percentages of correctly detected intervals by pairs of instruments (examples)

Instrument #1	Instrument #2	Accuracy (%)
Violin	Classic guitar	100.0
Horn	Trumpet	80.56
Alto sax	Horn	95.0
Trumpet	Alto sax	95.45
Oboe	Alto sax	83.33
Classic guitar	Alto sax	96.15
Cello	Horn	83.33
Oboe	Oboe	54.54
Violin	Electric piano	90.00
Electric piano	Alto sax	100.00
Classic guitar	Cello	96.67
Violin	Violin	68.42
Horn	Oboe	70.00
Cello	Flute	100.00
Alto sax	Flute	97.36
Flute	Trumpet	88.64
Trumpet	Cello	88.89
Violin	Horn	92.10
Violin	Piano	96.88
Horn	Flute	71.88
Oboe	Horn	88.89
Alto Sax	Electric piano	93.75
Electric piano	Trumpet	100.0
Flute	Flute	86.84
Trumpet	Horn	87.50
Oboe	Cello	76.47
Flute	Oboe	85.00
Electric piano	Cello	100.00
Classic guitar	Oboe	82.14
Trumpet	Classic guitar	94.74
Flute	Electric piano	92.86
Alto sax	Oboe	88.10
Horn	Cello	86.36
Alto sax	Trumpet	92.86
Violin	Alto sax	86.36
Electric piano	Oboe	87.50
Classic guitar	Trumpet	96.15
Violin	Cello	83.33
Cello	Trumpet	88.89
Trumpet	Violin	83.33
Cello	Oboe	100.00
Trumpet	Piano	80.09
Oboe	Piano	89.29
Flute	Piano	100.00
Flute	Violin	83.33
Classic guitar	Electric piano	64.29
Oboe	Classic guitar	86.36
Violin	Oboe	79.55

The accuracy is defined as the ratio of correctly recognized frequencies to the total number of frequencies in all analysed sound samples. Since the algorithm always returns chosen number of frequency candidates and there is always assumed ground truth number of frequencies (two for intervals, three, and so on), such as simple aforementioned ratio is enough to determine the quality of the algorithm.

The accuracy for intervals has achieved 87 %, for three-sound chords – 81,5 % (251 chords) and for four-sound chords – 75,2 % (255 chords). As the number of sound sources rises, the multipitch estimation becomes more and more difficult, therefore the decrease in the accuracy is absolutely expected. It is noticeable that the difference in the accuracy in both cases achieves about 6–7 %.

It should be also noted that when all frequency candidates were checked with the ground truth (i.e. all possible frequency candidates before the work of the judge), the accuracy in case of intervals achieved 95,2 %. A significant increase was noticed also in case of higher polyphonies (93,6 % in three-sound polyphony and 88,9 % in four-sound polyphony). Therefore, it clearly show that there is a lot of possibility to further improve presented approach in the future.

Both tables show interesting information about the accuracy of this approach. There are noticeable differences between particular kinds of intervals. The fifth (a seven-semitone interval) achieved the poorest accuracy, mostly due to the fact that fifths tend to have more common harmonics. Relatively good result of the 0 interval (which represents all the octaves in the database) might be explained due to the existence of intervals with the distance of two or more octaves. Therefore, the number of common harmonics might be lower in this case and the obtained results are better.

It is certainly interesting to notice the relationship betweens pairs of instruments and the results in the Table 2. The results are usually better, when the instruments from different groups are checked (e.g. Eletric Piano and Cello, Cello and Oboe, Violin and Horn), whereas instruments from the same group tend to yield a little worse results (e.g. Horn and Flute, Violin and Cello, etc.). It might be a result of the fact that instruments from the same group have similar spectral properties. Therefore, it might be a little more difficult to recognize them, compared to instruments from different groups.

6 Conclusions

As shown in the previous section, application of two sound representations, even with the same analysis method, allowed to achieve high accuracy of the whole approach, even in case of higher polyphonies. Usually, only one sound representation is used, therefore it is advisable to check different possible multipitch estimation solutions in terms of application of multiple sound representations. It must be noted that, although there have been conducted experiments with polyphonies with even higher number of simultaneous sounds used [6,10], the vast range of frequencies and a high number of sounds chosen with frequencies from both ends of this range greatly increased the difficulty of performed task and prove that the proposed method is worth to be analysed and developed in the future. This method will be also tested more with specific databases, such as instrument-specific databases (e.g. only for piano), because such databases are also used in various researches [1].

References

1. Benetos, E., Dixon, S., Giannoulis, D., Kirchoff, H., Klapuri, A.: Automatic music transcription: challenges and future directions. J. Intell. Inf. Syst. **41**(3), 407–434 (2013)
2. Stasiak, B.: Follow that tune - dynamic time warping refinement for query by Humming. In: Proceeding of Joint conference NTAV/SPA 2012. New Trends in Audio and Video Signal Processing: Algorithms, Architectures, Arrangements, and Applications, pp. 109–114 (2012)
3. Stolarek, J., Lipiński, P.: Improving digital watermarking fidelity using fast neural network for adaptive wavelet synthesis. J. Appl. Comput. Sci. **18**(1), 61–74 (2010)
4. Stasiak, B., Rychlicki-Kicior, K.: Fundamental frequency extraction in speech emotion recognition. In: Dziech, A., Czyżewski, A. (eds.) MCSS 2012. CCIS, vol. 287, pp. 292–303. Springer, Heidelberg (2012)
5. Salomon, J., Gomez, E., Ellis, D.P.W., Richard, G.: Melody extraction from polyphonic music signals. IEEE Signal Process. Mag. **31**(2), 118–134 (2014)
6. Davy, M., Klapuri, A.: Signal Processing Methods for Music Transcription. Springer-Verlag, Heidelberg (2006)
7. Goto, M., Hashiguchi, H., Nishimura, T., Oka, R.: RWC Music Database: music genre database and musical instrument sound database. In: Proceedings of the 4th International Conference on Music Information Retrieval (ISMIR 2003), pp. 229–230 (2003)
8. Argenti, F., Nesi, P., Pantaleo, G.: Automatic music transcription: from monophonic to polyphonic. In: Solis, J., Ng, K. (eds.) Musical Robots and Interactive Multimodal Systems. STAR, vol. 74, pp. 27–46. Springer, Heidelberg (2011)
9. Dressler, K.: Multiple fundamental frequency extraction for MIREX 2012. In: The 13th International Conference on Music Information Retrieval (2012)
10. Klapuri, A.: Multiple fundamental frequency estimation by summing harmonic amplitudes. In: Proceedings of 7th International Conference on Music Information Retrieval, pp. 216–221 (2006)
11. Benetos, E., Dixon, S.: Multiple-F0 estimation and note tracking for MIREX 2012 using a shift-invariant latent variable model. In: Music Information Retrieval Evaluation Exchange (2015). http://www.music-ir.org/mirex/abstracts/2012/BD1.pdf
12. Smaragdis, P., Bhiksha, R.: Shift-invariant probabilistic latent component analysis. Technical report (2007)
13. Weiss, R.J., Bello., J.P.: Identifying repeated patterns in music using sparse convolutive non-negative matrix factorization. In: Proceedings of International Conference on Music Information Retrieval (ISMIR) (2010)

Projection for Nested Word Automata Speeds up XPath Evaluation on XML Streams

Tom Sebastian[1]([✉]) and Joachim Niehren[2]

[1] Innovimax and Links Team of Inria Lille and Cristal Lab, Lille, France
tom.sebastian@inria.fr
[2] Inria and Links Team of Inria Lille and Cristal Lab, Lille, France

Abstract. We present an evaluator for navigational XPath on XML streams with projection. The idea is to project away those parts of an XML stream that are irrelevant for evaluating a given XPath query. This task is relevant for processing XML streams in general since all XML standard languages are based on XPath. The best existing streaming algorithm for navigational XPath queries runs nested word automata. Therefore, we develop a projection algorithm for nested word automata, for the first time to the best of our knowledge. It turns out that projection can speed up the evaluation of navigational XPath queries on XML streams by a factor of 4 in average on the usual XPath benchmarks.

1 Introduction

Projection is most relevant for efficient XML processing algorithms, as shown for in-memory evaluators for XQUERY in [11] and for a fragment of in XPATH [10]. The projection algorithm for XQUERY runs in Saxon [7], today's most used XML processing tool. Projection algorithms for the in-memory evaluation of XSLT are missing though.

The objective of the present paper is to initiate the development of projection algorithms for processing XML streams. Given that a single program written in one of the XML standards XQUERY, XSLT, or XPROC contains a collection of XPATH queries, we are interested in the evaluation of a collection of XPATH queries on a single input stream. The parsing time can be shared between many XPATH queries, and thus be should counted seperately. Therefore, we are mainly interested in the parsing-free time for query evaluation. Note however, that the parsing-free time for a single query is often dominated by the parsing time.

We will restrict ourselves to projection for navigational XPATH queries, since these are fundamental to all others. For instance, in order to check whether the root of a tree has at least 5 a-children, all other children of the root can be projected. The computation of the projection still requires to read the entire input tree, but the time for this can be shared similarly to the parsing time.

The most efficient evaluation algorithm for navigational XPATH queries on XML streams so far was presented in [2]. Similarly to many recent evaluation algorithms for XPATH on XML streams [6,9,12], it is based on the compilation

© Springer-Verlag Berlin Heidelberg 2016
R.M. Freivalds et al. (Eds.): SOFSEM 2016, LNCS 9587, pp. 602–614, 2016.
DOI: 10.1007/978-3-662-49192-8_49

of navigational XPATH queries to nested word automata (NWAs) [1]. Given that XML streams are nested words, NWAs provide a canonical formalism for defining algorithms on XML streams. This leads to highly efficient algorithms based on first principles as argued in [2]: In particular, one can rely on the nondeterminism of NWAs to express XPATH queries with recursive axes, such as "descendant" or "following", and then use on-the-fly determinization for their evaluation. The evaluation of an XPATH query can then be reduced to running an NWA on all possible answer candidates. Furthermore, the runs of multiple answer candidates in the same state can be shared.

Projection for finite automata is well known [5,10]. It amounts to project away all letters of the input word that do not change the state. Projection for NWAs is more tedious, since such automata have a stack by which they can pass information from opening tags to corresponding closing tags. Therefore, one cannot simply project an opening tag away without taking care of the corresponding closing tag. Our idea is that a projected nested word should contain jump symbols $.^i.$ for projected factors, where the integer i stands for the excess of the factor, i.e., the difference between the number of opening and closing tags. We present *projection nested word automata (PNWAs)*, a kind of mixed pushdown and counting automata, that input projected nested words which beside others contain integers as letters. These integers allow the automaton to compute the depth of the current node of the tree at any time, and also the excess of the last jump. Conversely, a projection of a nested word with respect to a given NWA can be computed by any corresponding PNWA. It may be surprising, but it turns out there may exist different PNWAs with maximal projection for the same NWA. Therefore, our projection algorithm has to make its choices.

We then lift NWA projection to the evaluation of navigational XPATH queries on XML streams. It turns out that the parsing-free time for query answering is reduced by a factor of 4 on average on the usual XPathMark benchmark compared to the previously existing algorithm [2].

Outline. In Sect. 2, we recall NWAs and their usage for XPATH evaluation on XML streams. In Sect. 3, we introduce PNWAs. In Sect. 4, we introduce notions of irrelevant labels and prefixes of nested words for states of NWAs. In Sect. 5, we use them to project NWAs to PNWAs. In Sect. 6, we present our experimental results for XPATH evaluation on XML streams. The appendix of the present paper at hal.inria.fr/hal-01182529 contains additional examples of PNWAs, an extension of NWA projection for node selection, and the collection of queries used in our experiments.

2   Nested Word Automata

We recall the definition of NWAs, while pointing out the close relationship between nested words and XML streams.

Let Σ be a finite alphabet. Let P_Σ be the set of parenthesis with labels in Σ, that is the set of opening tags $\langle a \rangle$ and closing tags $\langle /a \rangle$ where $a \in \Sigma$. A nested word over Σ is a word over P_Σ which is well-balanced, so that any opening tag

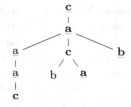

Fig. 1. An unranked tree

$$\langle c \rangle \ \langle a \rangle \ \ \langle a \rangle \ \langle a \rangle \ \ \langle c \rangle \ \ \langle /c \rangle \ \langle /a \rangle \ \langle /a \rangle$$
$$\langle c \rangle \ \langle b \rangle \ \langle /b \rangle$$
$$\langle a \rangle \ \langle /a \rangle \ \langle /c \rangle$$
$$\langle b \rangle \ \langle /b \rangle \ \langle /a \rangle \ \langle /c \rangle$$

Fig. 2. The corresponding nested word is an XML stream

$\langle a \rangle$ can be assigned to a unique corresponding closing tag $\langle /a \rangle$, and vice versa, and such that the initial opening tag of the word corresponds to the closing tag at its end. Our nested words are more restricted than in the general case [1], in that internal symbols are omitted, corresponding opening and closing tags must have the same label, and initial opening tags cannot be closed before the end.

An XML stream is a nested word that is obtained by linearizing an unranked tree in document order. This is a strong simplification of the XML data model, in that we ignore data values (internal symbols) and the different types of nodes (text, element, attribute, etc.). An example of an unranked tree is given in Fig. 1. The XML stream obtained by linearizing this unranked tree into a nested word is given in Fig. 2. It should be mentioned that we cannot assume any a priori knowledge on the set of tags of an XML document in practice (where no XML schemas are available). Instead, the finite alphabet Σ is determined by the tags appearing in the XPATH query of interest [2].

An NWA is a pushdown automaton on nested words [1], whose stack is "visible" in the sense that only a single symbol is pushed at opening events and popped at closing events. Here we assume that NWAs are early [2], so that whenever a final state is reached, any continuation completing the nested word will be accepted. More formally, an (early) NWA is a tuple $A = (\Sigma, Q, I, F, \Gamma, R)$ where Σ is a finite alphabet, Q a finite set of states with subsets $I, F \subseteq Q$ of initial and final states, Γ a finite set of stack symbols, and R is a set of transition rules of the following two types, where $q, q' \in Q, a \in \Sigma$, and $\gamma \in \Gamma$:

Open: $q \xrightarrow{\langle a \rangle \downarrow \gamma} q'$. When processing an opening tag $\langle a \rangle$, γ is pushed onto the stack, and the state is changed from q to q'.

Close: $q \xrightarrow{\langle /a \rangle \uparrow \gamma} q'$. When processing a closing tag $\langle /a \rangle$, γ is popped from the stack and the state is changed from q to q'.

A configuration of an NWA is a word in $Q\Gamma^*$ consisting of a state $q \in Q$ and a stack $S \in \Gamma^*$. A run of an NWA on a nested word $w \in P_\Sigma^*$ is a function r that maps prefixes w' of w to configurations. The initial configuration must contain an initial state and the empty stack, i.e. $r(\epsilon) \in I$. The NWA then rewrites this configuration: for any prefix $w'p$ of w, $r(w'p)$ is produced from $r(w')$ by applying some rules consuming tag $p \in P_\Sigma$. A run on w is successful if $r(w) \in F$, i.e. if it reaches at the end a final state and the empty stack. Since we assume early

NWAs, any run reaching a configuration with a final state on some prefix of a nested word can always be continued into a successful run. The *language* $\mathcal{L}(A)$ of an NWA A is the set of all nested words that permit a successful run by A.

An NWA is called *deterministic* if it is deterministic as a pushdown automaton. Note that NWAs can always be determinized [1] in contrast to more general pushdown automata. Our streaming algorithms will determinize NWAs constructed from XPATH expressions on the fly (as explained in [2]), so that we will only have to project deterministic NWAs but this while creating them on the fly.

An example for a deterministic NWAs is given in Fig. 3. It defines the XPATH filter [//a/b] which accepts all XML trees that contain some a-descendant with a b-child. Rules containing label sets represent sets of rules, one for each label. Node selection XPATH queries can be compiled to deterministic NWAs in a similar manner [2] by adding variables to the alphabet. This requires some minor extensions for NWA projection which are out of the scope of the present paper.

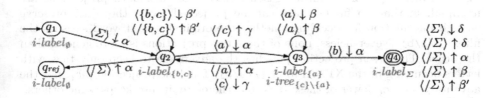

Fig. 3. A deterministic NWA over $\Sigma = \{a, b, c\}$ for XPATH filter [//a/b]

3   Projection NWAs

We next introduce projected nested words. Let $\mathbb{N}$ be the set of natural numbers, $\mathbb{N}_0 = \mathbb{N} \uplus \{0\}$, and $\mathbb{Z}$ the set of integers. For any unranked tree, we are interested in the binary node relations child ch, descendant ch^+, n-th grand parents ch^{-n} where $n \in \mathbb{N}$, descendants of n-th grand parents ch^{-n}/ch^+, children of n-th grand parents ch^{-n}/ch, and stay at *self*. So let:

$$Rels = \{ch, ch^+, ch^{-n}, ch^{-n}/ch^+, ch^{-n}/ch, self \mid n \in \mathbb{N}\}.$$

A *projected nested word* is a word whose letters are jump symbols $.^i.$ where $i \in \mathbb{Z}$ and jump targets $p@r$ where $p \in P_\Sigma$ and $r \in Rels$. We write $P_{\Sigma}^{..}$ for the set of all these letters. We assume that any jump target is proceeded by a jump symbol that indicates the excess of the jump, that is the depth difference in the tree or equivalently, the difference of the numbers of opening and closing tags in the nested word. We also assume that projected nested words are well-nested up to jumping.

Two examples for projected nested words are given in Fig. 4. Both are valid descriptions of the nested word in Fig. 2: pw_1 projects to the letters drawn in blue, while pw_2 projects to the letters drawn in green. As we will see, both

pw_1: for all a-nodes without an a-parent and all non-a-children of a-nodes keep the opening and closing tags, until the opening tag of the first match of //a/b:

$$\underline{\langle c \rangle} \; .\overset{0}{.} \; \underline{\langle a \rangle}@ch^+ \; .\overset{2}{.} \; \underline{\langle c \rangle}@ch^+ \; .\overset{0}{.} \; \underline{\langle /c \rangle}@self \; \overset{-2}{.} \; \underline{\langle c \rangle}@ch^{-3}/ch^+$$
$$.\overset{0}{.} \; \underline{\langle a \rangle}@ch^+ \; .\overset{0}{.} \; \underline{\langle /a \rangle}@self \; .\overset{0}{.} \; \underline{\langle /c \rangle}@ch^{-1} \; .\overset{0}{.} \; \underline{\langle b \rangle}@ch^{-1}/ch^+$$

pw_2: for all a-nodes and all b-children of a-nodes keep the opening and closing tags, until the opening tag of the first match of //a/b:

$$\underline{\langle c \rangle} \; .\overset{0}{.} \; \underline{\langle a \rangle}@ch^+ \; .\overset{0}{.} \; \underline{\langle a \rangle}@ch^+ \; .\overset{0}{.} \; \underline{\langle a \rangle}@ch^+ \; .\overset{0}{.} \; \underline{\langle /a \rangle}@self$$
$$.\overset{0}{.} \; \underline{\langle /a \rangle}@ch^{-1} \; .\overset{1}{.} \; \underline{\langle a \rangle}@ch^{-1}/ch^+ \; .\overset{0}{.} \; \underline{\langle /a \rangle}@self \; \overset{-1}{.} \; \underline{\langle b \rangle}@ch^{-2}/ch$$

Fig. 4. Two projected nested words describing the nested word in Fig. 2

projections can be obtained from the NWA in Fig. 3. Note that the initial opening tag is always kept for technical reasons. Except of this, both projections are maximal, in that no further tags can be projected away: they just preserve enough information for deciding whether the original nested word satisfies the filter [//a/b]. Nevertheless, none of these two projections is more general than the other. The green projection pw_2 has the advantage to keep only tags with letters occurring in the XPATH filter [//a/b]. The blue projection pw_1, has the advantage to keep fewer of these tags, but therefore, it also keeps some others.

The blue projection pw_1 starts with $\langle c \rangle$, meaning that any matching nested word must start with $\langle c \rangle$. The next factor $.\overset{0}{.}\; \langle a \rangle@ch^+$ describes a nested word with excess 0 that is followed by $\langle a \rangle$ in descendant position, i.e., by the opening tag of an a-child of the root. The next factor $.\overset{2}{.}\; \langle c \rangle@ch^+$ describes a nested word with excess 2 followed by $\langle c \rangle$ opening a descendant. Then $.\overset{0}{.}\; \langle /c \rangle@self$ requires to jump with excess 0 to the closing tag $\langle /c \rangle$ of the same node. Next, $\overset{-2}{.}\; \langle c \rangle@ch^{-3}/ch^+$ asks to jump with excess -2 to an opening tag $\langle c \rangle$ of a descendant of a grand-grand-grand-parent, etc.

We next introduce PNWAS as a mixture of a pushdown and a counting automaton, that receive projected nested words as input. The counting serves for updating the depths of nodes when jumping, so that the depth of the current node can always be deduced from the current stack. Whenever jumping over a projected factor, the excess of this factor is pushed. This is an integer that is popped when trying to close the jump.

Definition 1. *A* PNWA *is a tuple* $A = (\Sigma, Q, I, F, \Gamma, R)$ *like an* NWA *but with different kinds of transition rules: given* $a \in \Sigma$, $\gamma \in \Gamma$, *and* $q, q' \in Q$, *there are rules of the following types in* R, *for changing the state from* q *to* q':

Open: $q \xrightarrow{\langle a \rangle \downarrow \gamma} q'$ *Like for* NWAs.

Close: $q \xrightarrow{\langle /a \rangle \uparrow \gamma} q'$. *Like for* NWAs.

Jump to a child or a descendant: $q \xrightarrow[\forall z \geq 0]{\overset{z}{.}\langle a \rangle@r \downarrow z \downarrow \gamma} q'$, *where* $r \in \{ch, ch^+\}$.

When $r = ch$ *then* z *must be* 0, *and we jump to the opening tag of an* a-child *and push first 0 and then γ onto the stack. When $r = ch^+$ then we jump over z descendants to the opening tag of an a-descendant, and push first z and then γ onto the stack. For short we denote this transition as $q \xrightarrow{ju(\langle a \rangle, r, \gamma)} q'$.*

Rejump to another child or descendant: $q \xrightarrow[\forall z, z'. \ z' \geq 0, z + z' \geq 0]{.\overset{z}{.}\langle a \rangle @ ch^{-(z'+1)}/r \uparrow z' \downarrow z + z' \downarrow \gamma} q'$,

where $r \in \{ch, ch^+\}$.

While trying to close a jump from some grand parent to some node one can rejump to another opening a-tag of a child or a descendant of the same grand parent. The excess of the jump to the first node z' on the stack is updated to the excess of the second node $z + z'$. Furthermore, γ is pushed. For short, we write this transition as $q \xrightarrow{reju(\langle a \rangle, r, \gamma)} q'$.

Jump to the closing tag of the self node: $q \xrightarrow{.\overset{0}{.}\langle /a \rangle @ self \uparrow \gamma} q'$. *Jump to the closing tag of the self a-node. In this case, γ is popped from the stack.*

Jump back to the jump's origin: $q \xrightarrow[\forall z \geq 0]{.\overset{z}{.}\langle /a \rangle @ ch^{-(z+1)} \uparrow z \uparrow \gamma} q'$. *When trying to close a jump, one may jump back to the closing tag of the a-node where the current jump started. The excess of $-z$ is popped from the stack together with the symbol γ which was pushed for the non-jumped a-node. For short we write $q \xrightarrow{ju\text{-}back(\langle /a \rangle, \gamma)} q'$.*

Close last jump step: $q \xrightarrow[\forall z > 0]{\langle /a \rangle \uparrow z \downarrow z - 1} q'$. *When trying to close a jump, one may read a closing a-tag for which the corresponding opening a-tag was jumped, so that no stack symbol was pushed. In this case the excess of the jump on the stack must be updated from z to $z - 1$.*

A configuration of a PNWA is a word in $Q(\Gamma \uplus \mathbb{N}_0)^*$ consisting of a state in Q and a stack in $(\Gamma \uplus \mathbb{N}_0)^*$. A run r of a PNWA A on a projected nested word over Σ is a function that maps any prefix of the projected nested word to a configuration. The run must start in some configuration with some initial state and the empty stack, i.e., $r(\epsilon) \in I$. Furthermore, for any prefix wl where $l \in P_{\ddot{\Sigma}}$, the configuration $r(w)$ must be transformed into $r(wl)$ by applying some rule consuming letter l. A run on a projected nested word w is called successful if it eventually reaches a configuration with a final state, i.e., if $r(w') \in F(\Gamma \uplus \mathbb{N}_0)^*$ for some prefix w' of w. The language $\mathcal{L}(A)$ of a PNWA A is the set of all projected nested words that permit a successful run on A (Fig. 6).

In Fig. 5 we present PNWA $\mathbf{A}_1$ that is a projection of the NWA in Fig. 3 for the XPATH filter [//a/b]. This automaton accepts the blue projection pw_1 in Fig. 4 of the nested word in Fig. 2. Automaton $\mathbf{A}_1$ visits the opening and closing tags of all a-nodes with no a-parent, and of all non-a-children of these a-nodes, and jumps over all other nodes. Automaton $\mathbf{A}_1$ accepts when the first match of [//a/b] arrives. In Fig. 10, we illustrate a successful run of $\mathbf{A}_1$ on pw_1. The states of configurations are placed below the tags, while the stack consists of the labels on the subedges above the state. Edges between tags indicate their correspondence. Furthermore there are edges for jumps to children and descendants, where the excess is pushed, while jumps to the jump origin close the jump, and

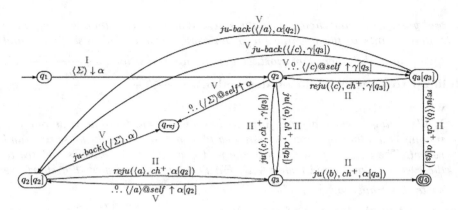

Fig. 5. PNWA $\mathbf{A}_1$ for the XPATH filter [//a/b]

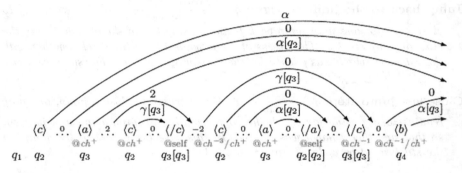

Fig. 6. A successful run of the PNWA $\mathbf{A}_1$ of Fig. 5 on pw_1

rejumps update the excess on the stack. The only exceptions are jumps to the closing tag of self nodes, where no excess is pushed. In general PNWA $\mathbf{A}_1$ works as follows. It starts in the initial state q_1, it opens the root and goes into q_2, where either the root can be closed to q_{rej} or where it can jump over b and c nodes to the opening tag of an a-descendant and go to q_3. There are 3 possibilities depending on what happens first: (1) close the a-node, and go to $q_2[q_2]$, (2) jump down over a sequence of a-nodes to the opening tag of a c-descendant and go to q_2, or (3) jump down over a sequence of a-nodes to the opening tag of a b-descendant and accept in q_4. In state q_2 a c-node with a sequence of a-grand parents can be closed to $q_3[q_3]$. The sequence of a-grand parents consists of a sequence of jumped a-nodes and one not jumped a-node π at the top. Continuing depending on what comes first, the following can happen in $q_3[q_3]$: (1) jump back to the closing tag of the a-grand parent π and go to $q_2[q_2]$, (2) rejump over a sequence of a-nodes, while staying below π, to the opening tag of a c-descendant and go to q_2, or (3) rejump over a sequence of a-nodes, while staying below π, to the opening tag of a b-descendant, and accept in state q_4. In state $q_2[q_2]$ there

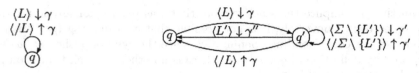

Fig. 7. $q \in i\text{-}label_L$ **Fig. 8.** $q \in i\text{-}tree_{L \backslash L'}$

are 3 possibilities depending on what happens first: (1) rejump over a sequence of b and c nodes to the opening tag of an a-descendant and goto q_3, while staying below a not-jumped c-grand parent with a sequence of a-grand parents if exists, (2) jump back to the closing tag of a not-jumped c-grand parent with the a-grand parents sequence if exists, or else (3) close the root to q_{rej}.

Next we are interested to evaluate a collection of PNwas obtained from deterministic Nwas on a single nested word. For this, we need to project the nested word with respect to the PNwas, and run the PNwas on the respective projected nested word. Therefore, we have to define how to project a nested word w with respect to a deterministic PNwa. More generally, we define a projection $\pi_q(w)$ for any suffix w of some nested word in P_Σ^* and state q of a PNwa A:

$$\pi_q(w) \; = \; .^i. \; p@r \, \pi_{q'}(w'')$$

such that $w = w'pw''$ for some $p \in P_\Sigma$ and $w', w'' \in P_\Sigma^*$, where w' is the shortest prefix, so that there exists a rule of A from q to q' consuming letter $p@r$ for some $r \in Rels$, and i is the excess of w'.

4   Irrelevant Labels and Prefixes of Nested Words

In this section, we define properties of Nwa states which allow to skip parenthesis with irrelevant labels and irrelevant prefixes of nested words, that is prefixes of linearizations of subtrees.

Definition 2. *An Nwa E can jump over parenthesis with labels in L and incoming state q – in formulas $q \in i\text{-}label_L$ – if there exists a stack symbol γ, such that E has all transitions shown in Fig. 7, no other opening transition pushing γ, no other a-opening transition in q, and no other a-closing transition with γ.*

If $q \in i\text{-}label_L$ then any sequence of letters in P_L is irrelevant in state q, so that it can be removed from the nested word and replaced by a jump symbol. Consider a run of E on a nested word w and assume $q \in i\text{-}label_L$. We next argue, that we can replace all letters in P_L of w with ingoing state q by jump symbols, while "repairing" the run. The first point is that the state is not changed when reading such letters, so that their removal keeps the states correct. But we must also take care of the stack. If an opening tag $\langle a \rangle$ is removed but not the corresponding closing tag, then we have to repair the run, in order to be able to reproduce the missing stack symbol when needed. The idea is to memoize the state before jumping. Since this state does not change while jumping, one can then recompute the stack symbol that was pushed for any letter that was jumped over. Conversely, it is not possible that a closing tag $\langle /a \rangle$ was removed but not the corresponding opening tag, since the symbol pushed at $\langle a \rangle$ must be γ, and by definition of $q \in i\text{-}label_L$ there is no other opening transition pushing γ than that started in q.

Definition 3. *An* NWA *E in state q can jump over prefixes of nested words (subtrees) that start in $\langle L \rangle$, do not contain letters in $P_{L'}$, and either end with the closing tag of the subtree's root or with a letter in L', if there exist three different stack symbols $\gamma, \gamma', \gamma''$ and a state q' such that the transitions shown in Fig. 8 exist, but no further opening transitions with γ, no further transitions with γ', and no further opening transitions in q' for L', and no further closing transition in q for L popping γ. In this case, we write $q \in i\text{-}tree_{L \setminus L'}$ and call q a state of irrelevant subtrees.*

In the easiest case where $q \in i\text{-}tree_{L \setminus \emptyset}$ one can jump over nested words linearizing subtrees, with incoming state q and labels in L only. When opening the root of the subtree, the state changes to q' and stays there until closing the root and going back to q. So the removal of the subtree does not change the state globally. In this case, the full nested word of the subtree is read, so the stack difference is zero. In the case where $L' \neq \emptyset$ it is more tricky to repair the run, in order to deal with missing stack symbols. But it remains possible, since the state used within the subtree does not change, so that it can be memoized and so that missing stack symbols can be recomputed at closing time.

For illustration, we have annotated the state of the NWA in Fig. 3 with the properties that they satisfy. It turns out that state q_3 satisfies both properties $i\text{-}label_{\{a\}}$ and $i\text{-}tree_{\{c\} \setminus \{a\}}$, but that we cannot perform the two corresponding projections at the same time. When choosing projection with $i\text{-}label_{\{a\}}$ then we obtain the PNWA $\mathbf{A}_1$ from Fig. 5.

5   Projection from Nwas to PNwas

We show how to project deterministic NWAs E to a PNWA A. For any state q of E, we chose a projection property $choice(p)$, which is either $i\text{-}label_L$ or $i\text{-}tree_{L \setminus L'}$ for some sets $L, L' \subseteq \Sigma$. Note that $i\text{-}label_\emptyset$ can always be assigned, so that this assumption can always be satisfied, but not always in a unique manner.

Any state of A is either a state of q of E or a pair of states of E that we write as $q[q']$. Such a pair means that one is in state q and that on the top of the stack is a jump symbol i that was pushed from a jump over i descendants that started in state q'. Any stack symbol of A is either a stack symbol γ of E or a pair written as $\gamma[q]$ of a stack symbol and a state of E. γ serves as the stack symbol that was pushed before at opening tags, while q is the state where a previous jump started. Whenever such a pair $\gamma[q]$ is on the stack then the symbol below is always a jump symbol i that was pushed by a jump over i descendants that started in state q. The sets of initial and final states remain unchanged.

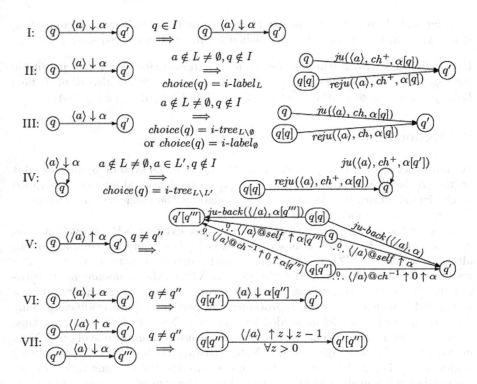

Fig. 9. Rewriting system for rules of a deterministic NWA to rules of the PNWA

Every transition rule of E gives rise to a possible empty set of transition rules of A, according to rules I–VII in Fig. 9. In PNWA $\mathbf{A}_1$ from Fig. 5 we annotated transitions accordingly. Transitions from an initial state are translated to non-jumping transitions that open the root. If $choice(q) = i\text{-}label_L$, then all looping transitions required by $i\text{-}label_L$ are removed. The other opening transitions starting from q are translated to jumping and rejumping transitions to descendants and descendants of grand parents. If $choice(q) = i\text{-}tree_{L\setminus\emptyset}$ then the opening and closing L transitions, and looping transitions required by $i\text{-}tree_{L\setminus\emptyset}$ are removed. The other opening transitions starting from q are translated to jumping and rejumping rules to children and children of grand parents. If $choice(q) = i\text{-}tree_{L\to L'}$ then the opening and closing L transitions, and looping transitions required by $i\text{-}tree_{L\to L'}$ are removed. All other transitions with opening tag $a \in L'$ departing q are translated to jumping and rejumping rules for descendants. Closing transitions are translated to six rules: Two rules to close self nodes, two rules to jump back to jump's origins, and two last rules that close parents in a state $q[q'']$ for $q \neq q''$. Those states do not allow to rejump, since the previous jump started in a different state q'' than the current state q, and therefore they also do not allow to jump back to the jump's origin. For these states opening and closing transitions are translated as indicated, while recomputing stack symbols, that have not been pushed for jumped grand parents.

Proposition 1 (Soundness). *Let E be a deterministic NWA E with initial state q_0 and A be a PNWA obtained from E by our projection algorithm. It then holds for any nested word w that $w \in \mathcal{L}(E)$ if and only if $\pi_{q_0}(w) \in \mathcal{L}(A)$.*

6   Experiments

We implemented NWA projection within the QUIXPATH system [3] and tested it on the (revised) XPATHMARK query set [4] for navigational queries. As argued in the introduction, it is most natural to measure the efficiency in *parsing-free time* which can be measured as described in [2].

In a first experiment, we start from the best existing XPATH evaluator on XML streams so far which is based on NWAs [2] (see there for comparisons to alternative tools by [7,12] and others), and enhance it with projection. The results are presented in Fig. 11 for a 559 MB XPATHMARK document. It turns out that projection reduces the parsing-free running time for this query set by a factor of 4.3, which is a major improvement, in particular when evaluating many XPATH queries in parallel as needed for streaming XSLT or XQUERY programs.

In our second experiment, we compare the overall running time of our PNWA evaluator of XPATH queries on XML streams with SAXON's in-memory evaluator [7]. For each of our queries, we compare the full running times including parsing, when evaluating the query n-times. The results are given in Fig. 11. It turns out that QUIXPATH with projection for NWAs can answer on average a query up to 12 times in parallel, in no more time than needed by SAXON for the same task.

One observes that running less than 12 queries in parallel with PNWAS is a lot quicker than running them with SAXON, mostly due to the expensive in-memory tree creation. But when running more than 12 queries on small documents, the advantage of in-memory evaluation takes over. Indeed, without the time for parsing and in-memory tree construction, SAXON in-memory evaluation is still faster by a factor 20 in average than streaming with PNWAS. With the improvements of the present paper, it now seems possible that stream processing can become more efficient than in-memory evaluation in practice in the future.

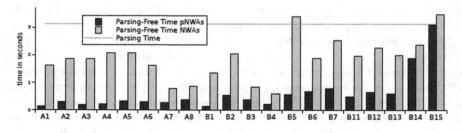

Fig. 10. A successful run of the PNWA $\mathbf{A}_1$ of Fig. 5 on pw_1

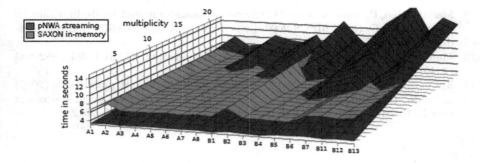

Fig. 11. Improvement by NWA projection of XPATH query evaluation

Conclusion and Future Work

We have developed a projection algorithm for evaluation navigational XPATH queries on XML streams. The next step will be to lift this algorithm to all of XPATH 3.0. We believe that this can be done by decomposing general XPATH queries into a network of navigational XPATH queries. Such a decomposition underlies the implementation of XPATH 3.0 in our QUIXPATH tool [3], which is unpublished so far. Once this is done, one can hope to lift our XPATH projection to XSLT and XQUERY, by using X-FUN as an intermediate language [8].

References

1. Alu, R., Madhusudan, P.: Adding nesting structure to words. J. ACM **56**(3), 1–43 (2009)
2. Debarbieux, D., Gauwin, O., Niehren, J., Sebastian, T., Zergaoui, M.: Early nested word automata for XPath query answering on XML streams. TCS **578**, 100–125 (2015)
3. Debarbieux, D., Sebastian, T., Zergaoui, M., Niehren, J.: Quix-tool suite (2014). https://project.inria.fr/quix-tool-suite/
4. Franceschet, M.: XPathMark: an XPath benchmark for the XMark generated data. In: Bressan, S., Ceri, S., Hunt, E., Ives, Z.G., Bellahsène, Z., Rys, M., Unland, R. (eds.) XSym 2005. LNCS, vol. 3671, pp. 129–143. Springer, Heidelberg (2005)
5. Frisch, A.: Regular tree language recognition with static information. In: Levy, J.-J., Mayr, E.W., Mitchell, J.C. (eds.) TCS2004. IFIP, vol. 155, pp. 661–674. Springer, Heidelberg (2004)
6. Gauwin, O., Niehren, J.: Streamable fragments of forward XPath. In: Bouchou-Markhoff, B., Caron, P., Champarnaud, J.-M., Maurel, D. (eds.) CIAA 2011. LNCS, vol. 6807, pp. 3–15. Springer, Heidelberg (2011)
7. Kay, M.: The saxon XSLT and XQuery processor. https://www.saxonica.com
8. Labath, P., Niehren, J.: A uniform programmning language for implementing XML standards. In: Italiano, G.F., Margaria-Steffen, T., Pokorný, J., Quisquater, J.-J., Wattenhofer, R. (eds.) SOFSEM 2015-Testing. LNCS, vol. 8939, pp. 543–554. Springer, Heidelberg (2015)

9. Madhusudan, P., Viswanathan, M.: Query automata for nested words. In: Královič, R., Niwiński, D. (eds.) MFCS 2009. LNCS, vol. 5734, pp. 561–573. Springer, Heidelberg (2009)

10. Maneth, S., Nguyen, K.: XPath whole query optimization. VLPB J. **3**(1), 882–893 (2010)

11. Marian, A., Simeon, J.: Projecting XML documents. In: VLDB, pp. 213–224 (2003)

12. Mozafari, B., Zeng, K., Zaniolo, C.: High-performance complex event processing over XML streams. In: SIGMOD Conference, pp. 253–264. ACM (2012)

Evaluation of Static/Dynamic Cache
for Similarity Search Engines

R. Solar[1], V. Gil-Costa[2]([✉]), and M. Marín[3,4]

[1] CITIAPS, Universidad de Santiago de Chile, Santiago, Chile
roberto.solar@usach.cl
[2] Yahoo! Research Latin America, UNSL-CONICET, San Luis, Argentina
gvcosta@unsl.edu.ar
[3] DIINF, University of Santiago, Santiago, Chile
[4] Center for Biotechnology and Bioengineering, University of Chile, Santiago, Chile
mauricio.marin@usach.cl

Abstract. In large scale search systems, where it is important to achieve a high query throughput, cache strategies are a feasible tool to achieve this goal. A number of efficient cache strategies devised for exact query search in different application domains have been proposed so far. In similarity query search on metric spaces it is necessary to consider additional design requirements devised to produce good quality approximate results from the cache content. In this paper, we propose a Static/Dynamic cache strategy for metric spaces which takes advantage of results of static cache miss operations and their associated distance evaluations for increasing the overall performance of the cache. We present an experimental evaluation of the performance obtained with our strategy for different query selection/replacement strategies.

Keywords: Approximate similarity search · Metric cache

1 Introduction

Nowadays, large scale similarity search engines are deployed into clusters of distributed memory processors connected by a high-speed communication infrastructure. One critical factor involved in the development of large-scale metric-space similarity search engines is how to handle with sudden peaks in query traffic. We have to take into account that the users behavior is highly unpredictable, complex, dynamic and often influenced by other users. Therefore, we have to be able to incorporate robust resource management mechanisms in order to prevent from system saturation and maintain the quality of service within certain acceptable ranges.

In this context, caching is a mechanism devised to improve the performance of search systems which have to access large-scale metric-space indices (typically stored in disk) to retrieve the query results. The main idea of caching is to keep frequent data in a small storage device that can be accessed faster than retrieving data from the index. There are different works [7,11,13] which consider using

© Springer-Verlag Berlin Heidelberg 2016
R.M. Freivalds et al. (Eds.): SOFSEM 2016, LNCS 9587, pp. 615–627, 2016.
DOI: 10.1007/978-3-662-49192-8_50

result cache or a distance cache. However, none of these works take advantage of the cooperative interaction between different hierarchical levels of cache.

In particular, in this paper we work with a metric-space index called *List of Clusters (LC)* [4] as the index data structure. The LC has been shown to be more efficient than other alternatives in high dimensional spaces. Moreover, we focus on approximate search by applying the algorithm of [9] originally presented for distributed metric spaces search engines.

The contribution of this paper is a *Static/Dynamic* Cache (*SDCache*) approach for approximate similarity search engines. In the static cache, we store queries that remain popular over time. Dynamic cache keeps queries that are popular for a short period of time. In both cases, we analyze the effect of combining the popularity priority of queries with their covering area on the metric space. Our SDCache approach takes advantage of the hierarchical level of caches and its contribution are two-fold: (1) re-use of results of static cache miss operations (*direct cooperation*), and (2) re-use of distance evaluations involved in a static cache miss operation (*indirect cooperation*). Direct cooperation helps to increase the number of dynamic cache hits and their accuracy. On the other hand, indirect cooperation helps to reduce the computation costs when traversing the dynamic cache. Furthermore, we propose a simple metric to determine whether a new query reports a hit cache and we evaluate the performance of our proposal with different query selection/replacement strategies.

The remaining of this paper is organized as follows. Section 2 reviews related work on approximate similarity search and cache strategies. In Sect. 3 we present the SDCache approach. Section 4 presents the results and Sect. 5 summarizes the main conclusions from our work.

2   Preliminaries and Related Work

A *metric space* $(\mathcal{U}, d)$ is composed of a universe of objects $\mathcal{U}$ and a *distance function* $d : \mathcal{U} \times \mathcal{U} \rightarrow \mathbf{R}^+$ which fulfills the following properties: strictly positiveness ($d(x, y) > 0$ and if $d(x, y) = 0$ then $x = y$), symmetry ($d(x, y) = d(y, x)$), and the triangle inequality ($d(x, z) \leq d(x, y) + d(y, z)$). The distance function determines the similarity between two given objects. There are two main types of queries. Given a database $\mathcal{X} \in \mathcal{U}$ a *range search query* ($R_{\mathcal{X}}(q, r)$) retrieves all the objects $x \in \mathcal{X}$ within a radius r of the query q; and a *k-nearest neighbors query* retrieves the k most similar objects to q.

Many metric index structures have been proposed and studied. This work is based on the List of Clusters (LC) [4], which partitions the set of objects into a set of disjoint clusters as follows. We first choose a cluster center $c \in \mathcal{X}$ and a radius r_c. The cluster ball (c, r_c) contains the subset of elements of $\mathcal{X}$ at distance at most r_c from c. From the remaining set of objects we select the next center as the one that maximize the sum of distance to the previous centers. This process is recursively repeated until all objects in the database are indexed.

During the processing of a range query $R(q, r_q)$, q is sequentially compared with the cluster centers of the LC. If $d(c, q) - r_q \leq r_c$ (the query ball intersects the cluster) we compare the query with the objects inside the cluster c and the search process continues. If $d(q,c) + r_q \leq r_c$ (the query is completely contained by the cluster) we compare the query with the objects inside the cluster c and the search stops.

2.1 Metric Cache

In [5] is introduced the *similarity cache problem* as a generalization of the classical cache approach. Authors focused on buffer management for approximate nearest-neighbor (ANN) applications, such as multimedia systems and contextual advertising. In the past years, many research work have been presented to address approximate and cache-based algorithms [1,7,8,10,12]. However, They do not combine the benefits of keeping queries along its results which are persistent over time, and queries that are relevant just for a brief period of time.

A metric space cache $\mathcal{C}$ consists of a collection of past queries along their results. Given a metric space $(\mathcal{U}, d)$ and a database $\mathcal{X} \in \mathcal{U}$, a cached query $q_i \in \mathcal{C}$ with its results $kNN_{\mathcal{X}}(q_i, k)$ and a query q, the *safe radius* of the query q with respect to q_i is defined as $s_q(q_i) = r_{q_i} - d(q, q_i)$, where r_{q_i} is the radius of the query q_i and $d(q, q_i)$ is the distance between q and q_i. If $s_q(q_i)$ is a positive value, then all objects within the sphere $\mathcal{R}_{\mathcal{X}}(q, s_q(q_i))$ can be used to solve q. Also, the $k' \leq k$ objects within the sphere $\mathcal{R}_{\mathcal{X}}(q, s_q(q_i))$ are the nearest neighbors of q [7].

The authors in [7,8] evaluated two algorithms for a cache system. Results Cache (*RCache*) and Query Cache (*QCache*). Both algorithms use a hash table $\mathcal{H}$ for storing query objects and their respective results. The *RCache* builds a metric index $\mathcal{M}$ with the kNN object results of the queries stored in $\mathcal{H}$. If a query q is not found in $\mathcal{H}$, an approximate similarity search $k\text{ANN}_{\mathcal{C}}(q, k)$ over the metric index $\mathcal{M}$ is performed to find the k closest objects in $\mathcal{M}$. The *QCache* algorithm builds $\mathcal{M}$ by indexing query objects. The *QCache* algorithm solves a kNN query q in an approximate way by merging the results of the closest queries to q found in $\mathcal{M}$. In [11] different similarity cache policies for *contextual advertising* systems are evaluated. The cache is implemented on the basis of a locality sensitive hashing (LSH) using a LRU replacement policy. The authors in [13] presented the *D-cache* (distance cache), which stores precomputed distances between objects.

The work in [3] re-uses distance evaluations involved in cache miss operations in order to produce good approximate results. If a *cache miss* occurs, the distance evaluations performed by the cache lookup procedure are used for traversing the index as fast as possible.

2.2 Approximate Nearest Neighbors Algorithms

An approximate answer consists of all those objects that are close to the current query, but not all of them are the k closest objects. The approach presented in [1] use the inverted files under the hypothesis that if two objects o_1 and o_2 are very similar then their view of the surrounding space is also similar. To this end, authors select a set of reference objects called *permutants* (representations of the surrounding space). Distances between objects are computed in an approximate way by comparing the order of each component of the permutants vectors.

The work in [10] proposed indexing and searching algorithms using suffix array (MSA) and a *permutation-based* index. In [6] a permutation-based index called PP-Index is presented. The PP-index uses permutation prefixes to quickly select a small set of candidate objects that can be close to the query.

In [12] the LSH approach is extended to general metric spaces. An object x is inserted into $\mathcal{L}$ hash tables $\mathcal{H}_i$. Each hash table $\mathcal{H}_i$ is accessed by means of a hash function $h_{c_i}(x)$, such that x is stored in the position $\mathcal{H}_i[h_{c_i}(x)]$ of each hash table $\mathcal{H}_i$. Each hash function h is associated with a Voronoi seed (c_i). A query search for q consists of applying $\mathcal{L}$ hash functions h_{c_i} to q. Each object from $\mathcal{H}_i[h_{c_i}(q)]$ is stored as a candidate. Finally, only the k-NN objects from the candidate set are returned as result.

Te work in [9] uses the LC index and it is devised for distributed search engines. The approximate algorithm works as follows: (1) It determines the main cluster c_i, i.e. the cluster in which the query q lies, (2) It computes the safe radius sf $(r_{c_i} - d(q, c_i))$, and search for similar objects within the query ball (q, sf) by traversing the clusters of the LC that intersect (q, sf), and (3) The sf is iteratively incremented by means of a parameter α given by the system engineer. The algorithm stops when the M closest objects to q are founded. The remaining $k - M$ objects are approximate results. The algorithm is compared against the QCache and the RCache. Results show that the proposal outperforms both state of the art algorithms by 60 %.

3   Proposed SDCache Approach

The SDCache approach consists of two lists, one for static queries and another for dynamic queries. Both lists are stored within a single LC index. Figure 1(b) shows the general scheme, where red queries belong to the static cache and green queries belong to the dynamic cache. Notice that the static cache can be built and updated off-line when most frequent queries changes, without affecting the performance of the current index.

Both static and dynamic cache lists cooperate directly and indirectly to resolve a query. Direct cooperation happens when the object results of a cache miss on the static cache are re-used for complementing the results obtained with the dynamic cache (see Fig. 1(a)). Results obtained with the static and the dynamic caches are merged to improve the quality of results. An indirect cooperation occurs when previously processed distance evaluations involved in a hit miss on the static cache are re-used during the search process on the DCache.

To this end a LC index is kept in secondary memory. This index is built with objects from the metric space database $\mathcal{X}$. In main memory, a second smaller LC index called $\mathcal{M}$ is used as a cache. The cache keeps the centers of the LC and their covering radius (c, c_r). Static and dynamic cached queries are inserted into $\mathcal{M}$ and marked as *read-only* and *read-write* respectively (there are not duplicated cached queries/objects in $\mathcal{M}$). The approximate algorithm presented in [9] is used to process incoming queries.

Algorithm 1 shows the main steps executed at running time. An auxiliary data structure $\mathcal{A}$ is used to share the information (distance evaluations, candidate objects, etc.) between the static and the dynamic cache. In line 3, the SCache is used to search for the query q. All possible results are stored in $\mathcal{R}$. Then, the top_k results are build using the information stored in $\mathcal{A}$ and $\mathcal{R}$. If the quality of the results is good enough, the algorithm reports a HIT cache. Otherwise, the search continues in the DCache re-using the information stored in $\mathcal{A}$. Finally, if the query is not found in the SDCache (the algorithm returns a MISS cache), it is processed in the LC index. Afterwards, the query along its top-k results are inserted into the DCache as follows. A query object is inserted into the first cluster containing it, namely the first cluster which covering radius covers the query object ($d(q, c) < r_c$). In Fig. 1(b), q_1 is inserted into (c_0, r_{c_0}) no matter its results are contained by others clusters of the LC. Queries in the dynamic cache are removed/promoted depending on the replacement policy. Results objects are efficiently managed with a hash table $\mathcal{H}_r$. Empty clusters of the cache are not explored during the search.

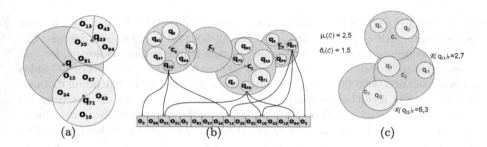

Fig. 1. Proposed SDCache lists scheme

In contrast to classical cache approaches, in a metric cache approach defining a criteria to determine whether a cache hit occurs is not a trivial task. In this work we propose a metric to measure the quality of a set of object results $\mathcal{R}(q)$, based on the information of past queries that are currently stored in cache. The proposed metric is $\mathcal{R}(q)_r \leq \tau$. The threshold τ is defined as $\tau = \mu_r(\mathcal{C}) + \sigma_r(\mathcal{C})$, where $\mu_r(\mathcal{C})$ and $\sigma_r(\mathcal{C})$ are the mean and the standard deviation of the radius of the k-nearest neighbors to queries $q \in \mathcal{C}$, that is $kNN_\mathcal{C}(q, k)$. Then, if the covering radius of the k results selected for an incoming query q, namely $\mathcal{R}(q)_r \leq \tau$ we report a hit for the query q. Notice that this metric prevents to retrieval results which covers a large area of the metric space. On the contrary, it tends to report cache hits for compact queries results.

Figure 1(c) shows an example of the proposed metric. The covering radius of the incoming query q_{I1} is $\mathcal{R}(q_{I1})_r = 2,7$ lower than $\tau = 4,0$ thus reporting a cache hit. However the incoming query q_{I2} reports a cache miss because $\mathcal{R}(q_{I2})_r > \tau$. In a static cache the value of τ is computed off-line. In a dynamic cache the value of τ is updated at running time.

Algorithm 1. SDCache Algorithm $(q, K, \mathcal{M})$

$\mathcal{A} \leftarrow$ Initialize(q, K)	$\triangleright \mathcal{A}$ has its own distance table
$top_k \leftarrow \emptyset$	$\triangleright$ create an empty TopKResults
$\mathcal{R} \leftarrow$ Static.Index.Search$(\mathcal{A}, \mathcal{M})$	$\triangleright$ static index search
$top_k \leftarrow$ SDCache.BuildTopK$(\mathcal{A}, \mathcal{R}, top_k)$	
if SDCache.SufficientQuality(top_k) **then**	
SDCache.Return(top_k, HIT)	$\triangleright$ Static Cache Hit
else	
$\mathcal{R} \leftarrow$ Dynamic.Index.Search$(\mathcal{A}, \mathcal{M})$	$\triangleright$ dynamic index search (reusing $\mathcal{A}$)
$top_k \leftarrow$ SDCache.BuildTopK$(q, \mathcal{R}, top_k)$	$\triangleright$ reusing top_k
if SDCache.SufficientQuality(top_k) **then**	
SDCache.Return(top_k, HIT)	$\triangleright$ Static Dynamic Cache Hit
end if	
end if	
SDCache.Return(top_k, MISS)	$\triangleright$ Static Dynamic Cache Miss

4   Experimental Results

We performed experiments with cluster sizes {2500, 5000, 7500}. We also run experiments with query logs $\mathcal{Q}_{log}$ of 250.000, 500.000 and 750.000 queries. The algorithms presented the same tendency in each configuration. However, for lack of space we show results for cluster size of 5000 and 750.000 queries. The LC index is composed of 5.000.000 images obtained from [2]. We use the Euclidian distance because it has intuitive meaning and the computation scales. It has reported low performance only a few times with an underlying (cartesian) coordinate system [14] which is not the case of our dataset. We used the following metrics: the relative error on the maximal distance $REM = \frac{max_{x \in \mathcal{R}_{hit}} d(q,x)}{r_q} - 1.0$, the relative error on the sum of distances $RES = \frac{\sum_{x \in \mathcal{R}_{hit}} d(q,x)}{\sum_{y \in kNN} d(q,y)} - 1.0$, where $\mathcal{R}_{hit}$ is the approximate result obtained from a cache hit, r_q is the query radius and $k - NN$ are the k nearest neighbor objects retrieve from the index. We also evaluated the traditional F-measure or balanced F-score (F_1 score) computed as $F_1 = 2 \times \frac{precision \times recall}{precision + recall}$. It can be interpreted as a weighted average of the precision and recall, where an F_1 score reaches its best value at 1 and worst score at 0. To better illustrate the results we show normalized results (we divide each value by the maximum reported in the experiment). First we evaluate the performance of the SCache and the DCache individually, then we present results for the hierarchical cooperation.

Query log analysis is a crucial factor for improving the effectiveness of static cache. In this work, we propose several strategies used to select a suitable set of queries to be stored in the static cache, which take advantage of past queries typically saved along its k nearest objects results in a query log $\mathcal{Q}_{log}$:

simulated cache (simcache): an index $\mathcal{M}$ for the SDCache is built with a sub-set of queries in $\mathcal{Q}_{log}$. Next, a search is performed in $\mathcal{M}$ for all q in $\mathcal{Q}_{log}$ (we explicitly avoid retrieving q as exact result). If a *cache hit* occurs (namely $\mathcal{R}(q)_r \leq \tau$), all queries involved in a cache result $\mathcal{R}(q)$ are stored in a priority queue $\mathcal{PQ}$. The most frequent queries in $\mathcal{PQ}$ are used to build the static cache. This approach tends to keep popular queries in the SCache.

k-means: we use the k-means algorithm in a recursive fashion by partitioning $\mathcal{Q}_{log}$ into two clusters until obtaining N_q queries as candidates for caching. Query candidates are stored in a priority queue which will be used to build the static LC cache.

lc-centers: this approach consists of using the center selection heuristic of the LC, which maximize the sum of distances to previously selected center, to determine which queries must be cached. The center selection heuristic is recursively applied by partitioning $\mathcal{Q}_{log}$ into *equal-sized* clusters until obtaining N_q queries for caching.

wse query log (wseqlog): each object of the query log $\mathcal{Q}_{log}$ is matched with a text query of a real web search engine query log $\mathcal{W}_{log}$. Most frequent queries are selected to be part of the static LC cache. The main idea of this approach is to imitate the behavior of real users.

Replacement policies are used for determining which query along its k results must be evicted from the dynamic cache. In this work, we evaluate the following replacement policies for dynamic cache: (a) maxsf: promotes the query $q_c \in \mathcal{M}$ which maximizes the safe radius to the new query q; and (b) promall: promotes all queries $q_c \in \mathcal{M}$ which k nearest results are part of the result for the new query being processed.

4.1   Evaluation of the Static Cache

In this section we present experimental results for the static cache (SCache) using the query selection strategies described above. In addition, we compare our proposal with a LSH implementation of a static cache.

Table 1. Number of LC clusters with cached queries.

Strategy	Clusters		$\mu_q(LC)$		$\sigma_q(LC)$	
	1 GB	2 GB	1 GB	2 GB	1 GB	2 GB
simcache	796	1109	13.47	27.56	14.28	31.06
wseqlog	1594	1660	3.23	6.85	1.66	2.55
lc-centers	1461	1651	3.96	7.57	3.55	5.45
k-means	1491	1637	4.31	9.23	3.85	7.94

Figure 2(a) shows the normalized number of distance evaluations reported with the SCache. We observe that strategies based on k-means, wseqlog and lc-centers perform similar number of distance evaluations. These strategies tend to cover a larger area of the metric space. Each cluster of the LC allocates few

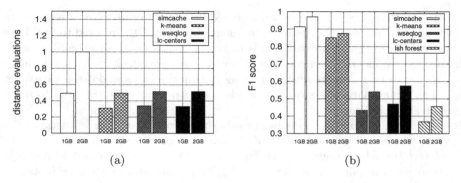

Fig. 2. (a) Distance evaluations and (b) F_1 score

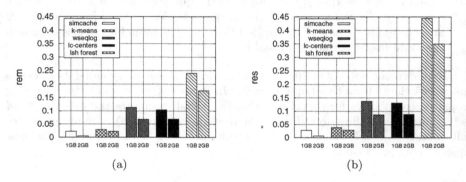

Fig. 3. Relative errors reported by the SCache: (a) REM and (b) RES

queries as shown in Table 1. In other words, cached queries are distributed among a large number of clusters. Thus a more aggressive prune is performed, due to incoming queries are compared with a few number of cached queries when the cluster satisfying $d(q, c) < r_c$ is visited.

The simcache-based strategy selects popular queries to be cached, and those queries tend to be allocated in a few number of LC clusters. Thus, on the contrary to other strategies, incoming queries are compared with a large number of cached queries when the cluster satisfying $d(q, c) < r_c$ is visited, according to the approximate search algorithm presented in [9].

Figure 2(b) shows the F_1 score reported by the SCache. As expected, the simcache selection strategy presents high F_1 score and very low REM and RES values (in Fig. 3) because queries stored in the cache are very popular, which allows more accurate cache hits. The k-means strategy also reports a good F_1 score and low REM and RES errors. That is because the metric space is partitioned with hyperplanes recursively, thus cached queries tend to be evenly spread in the space. Table 1 shows that the k-means strategy is the second strategy with larger $\mu_q(LC)$ values. Regarding the wseqlog-based and the lc-centers-based strategies,

REM values are close to 10 % with a cache of 1 GB and 6 % with a cache of 2 GB. RES values are close to 14 % and 9 % respectively.

On the other hand, the state of the art LSH approximate search strategy reports a low F_1 score and the highest REM and RES errors. Notice that the LSH does not report distance evaluations, because a hash function is computed for each incoming query to determine the bucket with objects reporting the same value for the hash function. According to the hash function, that bucket contains objects that can be similar to the query.

4.2   Evaluation of the Dynamic Cache

In this section we present experimental results obtained for the DCache. We compare the results achieved by our proposal with the results reported by the LSH strategy implemented for a dynamic cache.

Figure 4(a) shows that both replacement strategies reports similar number of distance evaluations. Figure 4(b) shows that the promall strategy slightly improves the maxsf by 3 %. The LSH reports the lowest F_1 score values around 0.06.

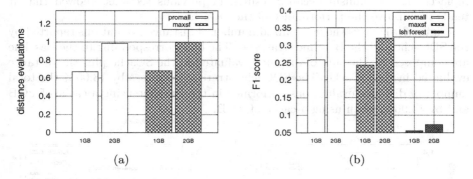

Fig. 4. (a) Normalized number of distance evaluations and (b) F_1 score

Figure 5 shows the values obtained for REM and RES errors. There are no major differences between results achieved by the promall and the maxsf replacement strategies. However, the LSH strategy has very high errors as it depends on a hash function to gather queries that are going to be selected as similar for a given query. The other strategies select similar results objects for queries based on distance evaluations. Thus, there is a trade-off between LSH and strategies based on distance evaluations. Reducing the computation cost, with the LSH, has the disadvantage of reducing the quality of results.

4.3   Evaluation of the Hierarchical SDCache

In this section we present experimental results of the cooperation achieved by our SDCache strategy and we compare them with results obtained from isolated

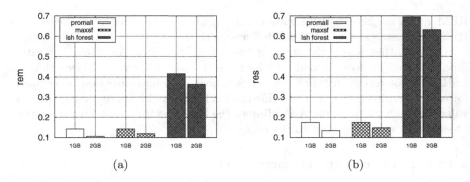

Fig. 5. Relative errors reported by the DCache: (a) REM and (b) RES

executions of the SCache and DCache strategies (Baseline). In both cases, the SCache was implemented with the k-means selection strategy, because it reports a low number of distance evaluations, presents a good F_1 score (85 %), and low REM and RES values. The DCache was implemented with the promall replacement mechanism because results, in previous sections, showed that it tends to improve the performance of the DCache.

Figure 6(a) shows the normalized number of distance evaluations reported by the SDCache and the Baseline strategies. The bars corresponding to the Baseline are composed by the sum of distance evaluated in the SCache plus the distance evaluated in the DCache. The SDCache strategy decreases by 13.05 % the total number of distance evaluations performed in the Baseline using a cache of 1 GB and by 14.96 % when using a cache of 2 GB.

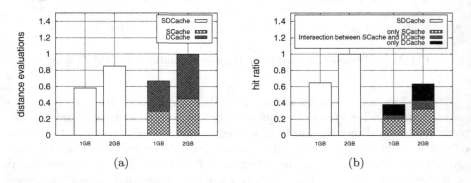

Fig. 6. (a) Distance evaluations and (b) cache hits ratio reported by the SDCache and the baseline strategies

Figure 6(b) shows cache hits rates. The bars corresponding to the Baseline are composed by three sections: query hits obtained with the SCache (bottom); intersection of the query hits reported by the SCache and the DCache (middle);

query hits reported by the DCache. This experiment shows the effect of re-using the distance evaluations and the results of static cache miss operations. In other words, the number of SDCache hits is greater than the sum of hits reported by the SCache and the DCache working independently. The benefit of the SDCache is about 41.27 % with a cache of size 1 GB and 36.79 % with a cache size of 2 GB.

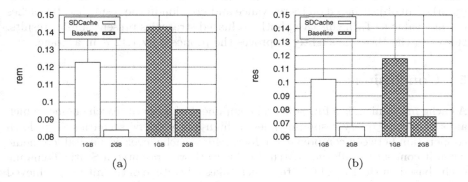

Fig. 7. Relative errors reported by the SDCache and the baseline strategies (a) REM and (b) RES

Figure 7 shows the REM and RES errors reported by the SDCache and the Baseline strategies. For the Baseline, errors are computed as the average of the error reported by the SCache and the DCache treated independently. As shown in Fig. 7(a), the SDCache reduces the value of REM by 18 % in average. On the other hand, Fig. 7(b) shows that the SDCache reduces by 10 % in average the value of RES. From Fig. 7 we conclude that SDCache results are good enough in terms of the query results quality.

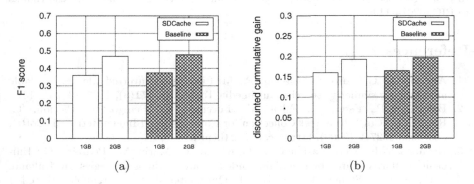

Fig. 8. (a) F_1 score and (b) DCG reported by the SDCache and the baseline strategies

Figure 8(a) shows the F_1 score achieved by both metric cache schemes. The Baseline F_1 score is computed as the average of the F_1 scores reported by the

SCache and the DCache. Results are very similar for both schemes. There is a slightly disadvantage for the SDCache of 1.25 % with a cache of 1 GB and of 0.97 % with a cache of 2 GB. Nevertheless, this small loss in terms of F_1 score is negligible in comparison with the benefits obtained in terms of cache hits (see Fig. 6(b)) and number of distance evaluations (see Fig. 6(a)). Finally, Fig. 8(b) shows the Discounted cumulative gain (DCG) which measures the usefulness (gain) of an object based on its relevance and position in the result list. It reaches its best value at 1. In this case, both evaluated strategies present similar results. However, the Baseline slightly improves the proposal by 1.12 % in average.

5   Conclusions

A cache mechanism is a fundamental component of modern search engines which aims to reduce query response times in high-query traffic scenarios. Top-down cooperation between components of hierarchical cache levels is a suitable scheme when it comes to similarity search. In this work we presented a Static/Dynamic cache based on the List of Cluster which takes advantage of the hit miss achieved by queries in the higher level of the cache scheme. We also proposed a simple but effective hit metric based on the mean value of the radius of cached queries.

Our proposal was evaluated with different strategies for selecting queries to be allocated in the static and dynamic caches. Results show that our cooperative hierarchical scheme can be easily adapted to different selection/replacement strategies reporting a higher effectiveness than the state of the art LSH. Furthermore, our scheme drastically reduces the number of distance evaluation performed by cache structures working independently.

Acknowledgment. Powered@NLHPC: This research was partially supported by the supercomputing infrastructure of the NLHPC (ECM-02). The authors would also like to thank to Basal funds FB0001, Conicyt, Chile; and Veronica Gil-Costa also thanks to PICT-2014-1146.

References

1. Amato, G., Gennaro, C., Savino, P.: MI-File: using inverted files for scalable approximate similarity search. Multimedia Tools Appl. **71**(3), 1333–1362 (2014)
2. Bolettieri, P., Esuli, A., Falchi, F., Lucchese, C., Perego, R., Piccioli, T., Rabitti, F.: CoPhIR: a Test Collection for Content-Based Image Retrieval. CoRR, abs/0905.4627v2 (2009). http://cophir.isti.cnr.it
3. Brisaboa, N.R., Cerdeira-Pena, A., Gil-Costa, V., Marin, M., Pedreira, O.: Efficient similarity search by combining indexing and caching strategies. In: Italiano, G.F., Margaria-Steffen, T., Pokorný, J., Quisquater, J.-J., Wattenhofer, R. (eds.) SOFSEM 2015. LNCS, vol. 8939, pp. 486–497. Springer, Heidelberg (2015)
4. Chavez, E., Navarro, G.: An effective clustering algorithm to index high dimensional metric spaces. In: SPIRE, p. 75 (2000)
5. Chierichetti, F., Kumar, R., Vassilvitskii, S.: similarity caching. In: PODS, pp. 127–136 (2009)

6. Esuli, A.: Use of permutation prefixes for efficient and scalable approximate similarity search. IPM J. **48**, 889–902 (2012)

7. Falchi, F., Lucchese, C., Orlando, S., Perego, R., Rabitti, F.: A metric cache for similarity search. In: LSDS-IR, pp. 43–50 (2008)

8. Falchi, F., Lucchese, C., Orlando, S., Perego, R., Rabitti, F.: Similarity caching in large-scale image retrieval. IPM J. **48**, 803–818 (2012)

9. Gil-Costa, V., Marin, M.: Approximate distributed metric-space search. In: LSDS-IR, pp. 15–20 (2011)

10. Mohamed, H., Marchand-Maillet, S.: Permutation-based pruning for approximate K-NN search. In: Decker, H., Lhotská, L., Link, S., Basl, J., Tjoa, A.M. (eds.) DEXA 2013, Part I. LNCS, vol. 8055, pp. 40–47. Springer, Heidelberg (2013)

11. Pandey, S., Broder, A., Chierichetti, F., Josifovski, V., Kumar, R., Vassilvitskii, S.: Nearest-neighbor caching for content-match applications. In: WWW, pp. 441–450 (2009)

12. Silva, E., Teixeira, T., Teodoro, G., Valle, E.: Large-scale distributed locality-sensitive hashing for general metric data. In: Traina, A.J.M., Traina, Jr., C., Cordeiro, R.L.F. (eds.) SISAP 2014. LNCS, vol. 8821, pp. 82–93. Springer, Heidelberg (2014)

13. Skopal, T., Lokoc, J., Bustos, B.: D-Cache: universal distance cache for metric access methods. TKDE **24**, 868–881 (2012)

14. Walters-Williams, J., Li, Y.: Comparative study of distance functions for nearest neighbors. In: Advanced Techniques in Computing Sciences and Software Engineering, pp. 79–84 (2010)

Author Index

Printed in the United States
By Bookmasters